Springer Series in Solid and Structural Mechanics

Volume 11

Series Editors

Michel Frémond, Rome, Italy
Franco Maceri, Department of Civil Engineering and Computer Science,
University of Rome "Tor Vergata", Rome, Italy

The Springer Series in Solid and Structural Mechanics (SSSSM) publishes new developments and advances dealing with any aspect of mechanics of materials and structures, with a high quality. It features original works dealing with mechanical, mathematical, numerical and experimental analysis of structures and structural materials, both taken in the broadest sense. The series covers multi-scale, multi-field and multiple-media problems, including static and dynamic interaction. It also illustrates advanced and innovative applications to structural problems from science and engineering, including aerospace, civil, materials, mechanical engineering and living materials and structures. Within the scope of the series are monographs, lectures notes, references, textbooks and selected contributions from specialized conferences and workshops.

More information about this series at http://www.springer.com/series/10616

Vincenzo Vullo

Gears

Volume 2: Analysis of Load Carrying Capacity and Strength Design

 Springer

Vincenzo Vullo
University of Rome "Tor Vergata"
Rome, Italy

ISSN 2195-3511   ISSN 2195-352X (electronic)
Springer Series in Solid and Structural Mechanics
ISBN 978-3-030-38634-4   ISBN 978-3-030-38632-0 (eBook)
https://doi.org/10.1007/978-3-030-38632-0

This Springer imprint is published by the registered company Springer Nature Switzerland AG
The registered company address is: Gewerbestrasse 11, 6330 Cham, Switzerland

Aphorism

L'analisi teorica è il più nobile parto della mente di uno scienziato. Essa però è zoppa quando venga usata da sola nelle scienze ingegneristiche, che sono eminentemente scienze applicate. Per sorreggersi in queste scienze, essa ha bisogno del bastone dell'indagine sperimentale, senza la quale corre il rischio di diventare speculazione filosofica.

(Theoretical analysis is the noblest birth of a scientist's mind. However, it is lame when used alone in engineering sciences, which are eminently applied sciences. To support itself in these sciences, it needs the stick of experimental investigation, without which it runs the risk of becoming philosophical speculation).

Vincenzo Vullo

To my wife Maria Giovanna,
my sons Luca and Alberto,
my nephew Nicolò
and my students

Preface

Gears and gear power transmissions used in most mechanical applications and, especially, in the automotive and aerospace industry, constitute one of the classical topics of Machine Design Theory and Methodology. In the opinion of many scholars, including the author of this monothematic textbook, the gears are, by far, one of the most complicated, but fascinating subjects of the mechanical engineering, because it is extremely polyhedral, as it calls into question a remarkable variety of disciplines, of which the main ones are: plane and space geometry; various branches of theoretical and applied mechanics, such as kinematics, dynamics, elastohydrodynamics, vibrations, noise, etc.; continuum mechanics and machine design theory and methodology; static and dynamic material strength; structural optimization; surface contact and lubrication theories, and tribology; materials science and metallurgy; cutting processes and, more generally, technological processes of manufacture; maintenance; etc.

Despite the fact that, from a historical perspective (see *Gears—Vol. 3: A Concise History*), the gears constitute, after the potter's wheel, the oldest mechanism invented by Homo Sapiens Sapiens, we can say that a unified scientific theory of the gears, able to consider all the aforementioned disciplines, amalgamating them in a *"unicum"*, does not yet exist at present. Currently, according to the most widespread thinking that is shared by historians of science and scientists, a science, to be such, must have at least the following essential features:

- All that a science claims should not involve real objects, but specific theoretical entities.
- The theory on which a science is based must have a strictly deductive structure, i.e. it must be characterized by a few fundamental statements (called axioms or postulates or principles), and by a unified method, universally accepted, to deduce from them a unlimited number of consequential properties.
- Its applications to real objects must be based on rules of correspondence between the theoretical entities and real objects.

Well, a unified and comprehensive science of the gears, which meets all these requirements with reference to multiplicity of the disciplines mentioned above, does not exist and, perhaps, it will not exist for a long time yet. The way to achieve this important goal is still long and fraught with difficulties, since it is necessary to agree the different scientific principles that form the bases of the various disciplines involved. Based on the current state of knowledge, we can only say that there are different scientific theories of the gears, as many as the disciplines that contribute to defining a gear design.

Gear design scholars and experts are well aware that, with a design based on the fundamental statements related to only one of the disciplines involved, there is a reasonable certainty of not meeting the postulates related to other disciplines. The difficulties to make a theoretically perfect design of a gear power transmission, i.e. respectful of the countless design requirements related to the various disciplines involved, are actually insurmountable, so the designer must fall back to a compromise design (the so-called design optimized, but often very far from the theoretical one), able to reconcile as best as possible the different and almost always conflicting requirements, dictated by these disciplines.

The scholars of vaunted credit, who claim to know everything and to be able to speak or write about any subject, and who boast or attribute themselves knowledge in any field, consider the gears as a synonym of obsolescence, a symbol of the past or, when they are benevolent, a nineteenth century old stuff. This way of thinking of those we have benevolently referred to as scholars of vaunted credit (no one can therefore accuse this author of not being kindly gentlemen) implies at least the ignorance of the historical evidence that the theory of the gears, considered as mechanisms, is much older, having it characterized the birth of science, in the first Hellenism (see *Gears*—Vol. 3, Chap. 3). But the unjustified conviction of these so-called scholars hides a far more serious ignorance. In fact, they prove not to know that the most significant contributions for the calculation of the load carrying capacity of the gears were brought gradually in the entire 20th century (with the exception of Lewis, 1892—see references in Chap. 3) and that in this beginning of the third millennium new and equally significant contributions of high scientific value appeared and continue to appear at the horizon (see, for example, Chaps. 10 and 11).

On the contrary, the true scholars and experts of gear power transmission systems are well aware that the gears were yesterday, continue to be today and for a long time yet will continue to be an ongoing scientific and technological challenge. Even today, surely it is worth investing significant financial resources, in terms of manpower, tools and means, in R&S activities on this important area, as all the technologically advanced countries continue to do. This depends on the fact that the gears are a very complex multidisciplinary field, as few in mechanical engineering, and knowledge still to be acquired are numerous. We can affirm, without fear of being denied, that gears are the result of ancient knowledge in continuous updating.

Knowledge still to be acquired concern not only those specific to each of the disciplines involved (geometry and kinematics; static and dynamic loads, including those due to impact; friction and efficiency; dynamic response and noise emission;

static and fatigue tooth root strength; contact stresses and surface fatigue durability; nucleation of fractures of any kind and their propagation until breakage; full film, mixed and boundary lubrication; scuffing and wear; materials and heat treatments; new materials; cutting processes and other manufacturing processes; etc.), but also those arising from their mutual interactions. Other important challenges are those related to the new fields of application (see, for example, those related to the helicopter and aerospace industry and wind power generators), which require the introduction and design of new types of gear drives that are not reflected in the current technological landscape.

Researchers working in this field are well aware of being in front of an important goal: to develop a comprehensive and unified scientific theory of the gears. Really, at present, a scientific theory of gears having a general horizon, capable of considering and balancing all the aspects involved in their design (i.e. geometric and kinematic aspects, mechanical strength aspects and technological and production aspects) does not yet exist. Only attempts of partial scientific theories of gearing exist, which are limited only to the first of the three aforementioned aspects. At the moment, partial scientific theories concerning the other two aspects mentioned above cannot unfortunately be reported, as non-existent.

Among the partial scientific theories concerning the geometric and kinematic aspects, it is however worth mentioning the laudable attempt by Radzevich (2013, 2018: see references in *Gears*—Vol. 1, Chaps. 1 and 11) who, for the first time, tried to develop a scientific theory of gearing. In reality, Radzevich is heavily indebted to Litvin (1994: see references in *Gears*—Vol. 1, Chap. 12) and his followers, inasmuch as he treasures the substantial contributions on the subject in terms of differential geometry brought by them, presenting already known knowledge concerning the above aspects in the systematic form of a scientific theory. However, Radzevich has the great merit of having set up a general scientific theory of gearing that, although limited to geometric and kinematic aspects, is capable of treating all types of gears, including hypoid gears that include, as specially cases, all other types of gears.

Notoriously, gears constitute a classic multidisciplinary machine design topic. In this multidisciplinary framework, numerous and different areas can be identified, which however are between them interdependent, since the variation of a parameter or quantity relative to a given area, inexorably affects a parameter or quantity of other areas involved. Usually three areas (it is to be noted that each of these areas is composed of two or more disciplines) are introduced, which identify three characteristic aspects of the gears. They are as follows:

- Geometric-kinematic area, which considers the gears as ideal theoretical entities, and that, in these ideal conditions, tries to optimize their geometric and kinematic quantities.
- Stress analysis area, which aims to stretch as much as possible the lifetime of the gear, considered as an actual structural machine element, working under real operating conditions.

- Technological-productive area that, in relation to the material used and its chemical and metallurgical properties, including heat treatment, or in relation to the development of new and more efficient materials, has as main objective to make the final product in accordance with the requirements imposed by the designer.

Notoriously, a good design of gears or gear power transmission systems should consider, in a balanced manner, all aspects and specific issues of the three afore-mentioned areas. In other words, it should try to balance optimally the pros and cons of different design choices to be made. Usually, textbooks on gears address one or two of these areas, with some mention to the third area. As it is known to the writer author, a textbook which in an equal manner addresses the subjects of all three of these areas does not exist today. Only gear handbooks (at not all) discuss all the issues concerning these three areas as well as those of the various disciplines that characterize each area. However, in most cases, the different problems are discussed with a sectorial vision, without enlarging the horizons for evaluating the consequences of the decisions pertaining to them on issues relating to the other areas.

This textbook also favors the first two of the aforementioned areas, but, in the discussion of the typical topics of the disciplines of each of these two areas, special attention is paid to issues of third area, namely that concerning the technological-productive problems of the gears, with a special reference to those for cutting of the gears. Issues concerning this third area would deserve at least the same attention and the same space as those reserved to the first two areas, but the limitations of a usual textbook generally do not allow a discussion of such a broad horizon. To try to satisfy at least partially this need, this author has considered it appropriate to discuss the most salient aspects of gear cutting in the framework of related topics concerning the other two areas; this is done whenever these aspects can significantly influence the gear design, so they cannot be overlooked by the designer.

This textbook fits into the long trail of gear monographs. It starts with the textbook entitled *Théorie Géométrique Des Engrenages Destinés a Transmetter Le Mouvement De Rotation Entre Deux Axes Situés Ou Non Situés Dans En Même Plan* by Théodore Olivier (1842: see references in *Gears*—Vol. 3, Chap. 5), which is likely to be considered the first monograph ever written and published on the gears. Other textbooks and monographs on this subject have gradually written and published that, however, are not specifically mentioned here, as the main ones are referred to several times in the references of this monothematic texbook. All these textbooks and monographs collect and synthesize, sometimes in remarkable fash-ion, the theoretical bases of the gears, that is, the geometric and kinematic concepts as well as practical experiences and technologies to manufacture them. However, with the exception of the aforementioned textbooks of Radzevich and Litvin, none of these monographs has for its subject a real *"Theory of Gearing"*, i.e. a scientific theory of the gears. In fact, none of the other textbooks and monographs is really in line with the deepest and fullest meaning of scientific theory that, as we said before, is based on a set of postulates from which we can deduce the entire body of knowledge of a given area concerning the gears.

Even this monograph on the gears is substantially aligned with those published previously. However, unlike the latter, this monograph favors the basic concepts of the calculation of the gear strength, inclusive of the rating of their load carrying capacity in terms of tooth root fatigue strength, surface durability (pitting), micropitting, tooth flank fatigue fracture, tooth interior fatigue fracture, scuffing, wear, etc. However, these calculations cannot be performed accurately without in-depth knowledge of the geometric-kinematic aspects of the mating tooth flank surfaces as well as those concerning the tooth cutting processes. In this perspective, a large part of this textbook (the first volume) is dedicated to the fundamental geometric-kinematic aspects, while the second volume of the same textbook is reserved to the basic concepts of strength and durability design, which constitute the basis of international standards on the topic. The technological aspects of the tooth cutting is not however discussed in a specific part of this monograph, comparable by extension and deepening with the parties reserved to the two aforementioned aspects; they are just called up and described here and there, in the context of the first two aspects, just enough because the designer has a picture as complete and comprehensive as possible to be fulfilled for a good gear design.

This textbook mainly collects the lectures held by the author, in the four Universities where his academic career took place, namely in chronological order: Polytechnic of Turin, first as an Assistant Professor and later as Associate Professor, from 1971 to 1980; University of Naples "Federico II", as Full Professor, from 1980 to 1986; University of Rome, "*Tor Vergata*", as Full Professor from 1986 to 2013, and as Full Professor retired from 2014 to 2017; Cusano Telematic University of Rome, as Contract Professor from 2018 to date.

However, it is noteworthy that this textbook takes into account the lectures held in these four Universities, in the framework of many courses of Bachelor's degree, Master's degree and Ph.D./D.Phil. terminal degree in Mechanical Engineering Science, but also lectures for the training of researchers and research technicians, specialized in gear design, which the author held for Costamasnaga S.p.A. The textbook also collects and builds on many years' experience (over 15 years) gained from the author at this company, as a consultant, especially for industrial research programs and projects of technological innovation regarding special and custom-made gear systems.

This textbook has the two-fold aim of meeting the needs of university education and those of the engineering profession. In the first area, the goal is to provide a link between the matters covered in the most basic textbooks on gears intended for students in three-year first-cycle degree programs, and those addressed in the more advanced monographs on gear theory to be used in second-cycle and third-cycle or doctoral programs. In the second area, the purpose is to provide practitioners and professional engineers working in research and industry with an advanced knowledge on the subject that can serve as a basis for designing gear systems efficient and technologically sophisticated, or for developing new and innovative applications of the gears.

Anyone who has ever worked with the analysis and design of gears is well aware that their actual engineering applications, with no exception, are characterized by the fact that a gear or a gear train is part of a complex mechanical system, which must be analyzed in its entirety, that is considering all the possible interactions with other mechanical parts or mechanical units that make up the entire system. These interactions between the components or groups of the system enhance the resulting stress and strain states, including those that arise under dynamic operating conditions, such as impact loading, vibration, high- and low-cycle fatigue, and so forth. The deviations between theoretical and actual operating conditions are the greater the more the transmission error is high.

The analysis of such complex systems requires the use of sophisticated and refined mathematical models. To save time and costs for the calculations to be made, these mathematical models can use simultaneously numerical models, based both on the discretization of the continuum, such as FEM (Finite Element Method) and BEM (Boundary Element Method), and on the discretization of the governing differential equations, such as FD (Finite Differences), and analytical models, such as step-by-step integration methods of the differential equations governing the dynamic behavior of the system to be examined and modal analysis methods for the study of the system's response to the dynamic loads applied to it.

All the problems presented in this textbook on the gears are approached and solved preferably using, whenever possible, theoretical methods, and attention is mainly focused on analytical and methodological aspects. The analytical definition of the tooth flank surface geometry, including that of the tip and fillet with all or part of the rim of the gear members to be analyzed, has another great advantage compared to traditional methods that do not use the differential analytical geometry. In fact, the equations that analytically define the aforementioned surfaces including their intentional modifications allow to automatically generate refined FE-models or BE-models, with which to perform the contact analysis as well as the stress analysis of the gear pair or gear drive under consideration, thus avoiding the loss of accuracy due to the development of solid models by CAD (Computer Aided Design) computer programs.

Some topic is discussed using different methods, especially when each of the considered methods allows to examine it from different points of view that allow to better evaluate significant design aspects. In this framework, also some method that may appear a little dated is described, especially when it allows us to better understand the physical phenomenon underlying the topic under discussion. Moreover, this choice has also the advantage of keeping alive the historical memory of the pioneers who have traced new paths that have facilitated the task of developing the most sophisticated current calculation methods.

When necessary for the understanding of the phenomena of interest, hints are made to methods of experimental analysis, and to numerical methods. However, generally, these are not extended to the whole mechanical system, but only limited to the single gear pair. This is because the attention wants to be called on the understanding of the specific phenomena, which are proper of the same gear pair, without the risk that they are more or less substantially altered or overshadowed by the above-mentioned interaction effects with other components and groups of the mechanical system to be examined.

The analytical and numerical solutions proposed here are formulated so as to be of interest not only to academics, but also to the designer who deals with actual engineering problems concerning the gears. This is because such solutions, though sophisticated and complex (see for example the solution of the matrix equations that define the contact surfaces between the mating tooth flanks), have become immediately usable by practitioners in the gear field thanks to today's computers, which can readily solve demanding equations. For the reader, moreover, these solutions provide the grounding needed to achieve a full understanding of the requirements laid out in the standards applying in this area. In most cases, the analytical relationships developed for use in the design analysis and/or in the response analysis (the latter analysis, also called verification analysis, is used mainly by the standards) are also presented in graphical form. This gives the reader an immediate grasp of the underlying physical phenomena that these formulae explain, and clarifies exactly which major quantities must be borne in mind by the designer.

Each topic is addressed from a theoretical standpoint, but in such a way as not to lose sight of the physical phenomena that characterize the various types of gears gradually examined, up to hypoid gears, which are those with a more complex geometry. The study of the gears proceeds in steps, starting from the geometric-kinematic aspects, which are related to the cutting conditions of the teeth, and gradually coming to the analysis of the loads and, subsequently, to the analysis of stress and strain states, which allow to evaluate the tooth bending strength, surface durability in terms of macropitting (pitting) and micropitting, scuffing load capacity, tooth flank fracture load capacity, tooth interior fatigue fracture load capacity, wear strength, and service lifetime even under variable loading. The material is thus organized so that the knowledge gained in the beginning chapters provides the grounding needed to understand the topics covered in the chapters that follow.

This monothematic textbook, entitled *Gears*, is also intended for the students in the course on Machine Design Theory and Methodology (Progettazione Meccanica e Costruzione di Macchine) at the Università degli Studi di Roma "Tor Vergata". It consists of the following three volumes:

- *Gears—Vol. 1: Geometric and Kinematic Design*. This volume is organized in 14 Chapters, which tackle the geometric and kinematic design of the various types of gears most commonly used in practical applications, also considering the problems concerning their cutting processes.
- *Gears—Vol. 2: Analysis of Load Carrying Capacity and Strength Design*. This volume is organized in 11 Chapters and an Annex. The eleven chapters address the main problems related to the strength and load carrying capacity of almost all the gears (with the only exception of the worm gears), providing the theoretical bases for a better understanding of the calculation relationships formulated by the current ISO standards. A brief outline of the same problems related to worm gears is made in the Annex.

- *Gears—Vol. 3: A Concise History*. This volume is organized in five short chapters, which summarize the main stages of the development of the gears, and the gradual acquisition of knowledge inherent to them, since the down of the history of *Homo Sapiens Sapiens* to this day.

In this Volume 2, the aspects concerning the calculation of the load carrying capacity of the gears are discussed, first describing the analytical bases, which are routed in the various theories involved (elasticity theory; when necessary, also plasticity theory; elastohydrodynamic theory; tribology; theories of damage; strength theory of materials; etc.). The attention is particularly focused on the surface, subsurface and core three-dimensional stress states that are generated under load, on the multiaxial fatigue criteria and on the multiaxial strength criteria, without which it would become problematic to understand the complex mechanism that determine fatigue damages and breakages in gear teeth. The main calculation methods that use these theoretical bases are then described. Furthermore, the available experimental evidences that support these calculation methods are referred to from time to time. In the final sections of the chapters dealing with these topics, or in specific chapters when the theoretical bases have already been explained in other chapters, the procedures to calculate the load carrying capacity of the gears in accordance with the ISO standards are finally described, taking care to highlight how the formulae of standardized calculations have their roots in the previously discussed theoretical and experimental bases.

The problems of load carrying capacity and strength design of almost all the gears currently used in practical applications are discussed in detail. Only the corresponding problems of worm gears are not discussed in the same detail as they, already very complex for the gears considered here, become much more complex for these types of gears. In fact, for these gears, it would be necessary not only to discuss aspects common to other types of gears (safety against pitting, tooth root fatigue breakage, scuffing, etc.), specializing them in the particular case of interest, but also very specific aspects, such those concerning safety against heating, shear tooth breakage, wear of worm threads and worm wheel teeth, bending and torsional deflections of the worm shaft, etc. Well, for almost all these particular aspects concerning these gears, satisfactory theoretical calculation bases have not yet developed and the methods proposed so far, however elaborated, have not emerged from a framework that is still to be considered substantially empirical. For these reasons, in this volume only a brief mention on the load carrying capacity and strength design assessment of these types of gears is done, referring the reader to specialized publications and to the reference technical standards.

Regarding the aspects discussed in this second volume, it should be pointed out that it is not possible here to present and discuss examples concerning a complete and exhaustive calculation procedure of the load carrying capacity and strength design of the gears considered. This impossibility follows from the fact that any possible example of this kind that could be presented and discussed would occupy a space comparable with that devoted to the theoretical aspects of the topic of interest. It is clear that such calculation examples would conflict with the editorial requirements of the present textbook. On the other hand, calculation examples of this kind can be solved by diligently following the indications of the reference technical standards, which do not fail to add to the general calculation formulae related to the long and complex procedure to be used equally extensive application examples. The few exercises introduced here concern very limited problems of load carrying capacity and strength design of the gears considered, while for the aforementioned complete calculation examples we refer the reader to the afore-mentioned reference technical standards. Even here we do not consider it necessary to present and discuss exercises in cases where the problems could be solved immediately with the relationships presented in each chapter, as it was felt that they would add nothing to an understanding of the text.

About the formulae, diagrams and figures taken or derived from the ISO standards, it should be pointed out that, in the form in which they are discussed and presented here, they do not guarantee the accuracy of the results obtained. They are to be considered as important points of reference for the calculations concerning the determination of the geometric and kinematic parameters and the assessment of the load carrying capacity of the gears as well as a clear demonstration of the usefulness of the theoretical bases previously discussed. The guarantee of reliability of the results with formulae, diagrams and figures drawn from ISO standards is in any case given by the use of the original ISO standards, which moreover almost always add to the standard specification, technical specification or technical report concerning a certain topic an equally standard specification, technical specification or technical report regarding interesting and clarifying examples of calculation for the same topic, to which the user can directly draw.

This textbook is born from the continuous and incessant stimulation of the numerous students of the author in the three public Universities where he taught. They thirst for knowledge led them to treasure the following famous saying of a Aristotle, quoted by Diogenes Laërtius "τῆς παιδείας τὰς μὲν ῥίζας εἶναι πικράς, τὸν δὲ καρπὸν γλυκύν", i.e. *"The roots of education are bitter but the fruit is sweet"*. It has stimulated the author's treatment of unusual and even little difficult subjects, which the students have always shown to appreciable for their growth in view of their inclusion in the world of profession or research. The author's greatest satisfaction is to feel their gratitude when, for some occasional reason, they come back to visit him. They have never let the author miss their attachment and affection, and they still do not miss him. The deepest and heartfelt thanks of the author go to them, who represents the most authentic, sincere and vital component of the University.

Last but not least, the author wishes to express his affectionate, warm and heartfelt thanks to Dr. Ing. Alberto Maria Vullo, his son, who in addition to showing enthusiasm and meticulous care in the drafting of the drawings and graphs of this textbook, has validly collaborated in the preparation of its format. The author then wishes to express is equally affectionate thanks to his wife, for her great willingness to transfer in print format the textbook manuscript and formulae as well as for her uncommon patience shown in having the author taken away from the family needs, given the commitment made in almost four-year work. Finally, the author's thanks are also due to his first-born son, Dr. Ing. Luca Vullo, who even from a distance has been able to give a small but valuable contribution. These heartfelt thanks are a small gesture of gratitude from the author to ask for an ocean of excuses for having stolen time and affection from the whole family, to which, however, he has the pleasure of expressing the deepest movement of his soul.

Rome, Italy   Vincenzo Vullo

Contents

Symbols, Notations and Units

a'	Dimensional empirical material constant (mm)
a_2	Specific coefficient of lubricant oils
a_{BS}	Auxiliary factor (–)
$a_{CP,core}$	Fatigue sensitivity to normal stress at core
$a_{CP,surface}$	Fatigue sensitivity to normal stress at surface
$a_{CP}(z)$	Fatigue sensitivity to normal stress
$a_{i(i=0,1)}$	Oil sump temperature coefficients
a_{rel}	Relative hypoid offset (–)
a_v	Center distance of virtual cylindrical gear (mm)
a_{vn}	Center distance of virtual cylindrical gear in normal section (mm)
a_{vn}	Relative center distance (–)
a	Center distance (mm)
a	Dimensional Neuber constant or Neuber equivalent grain half-length (mm)
a	Hypoid offset (mm)
A^*	Dimensional constant (mm μm/N)
A^*	Related area for calculating the load sharing factor, Z_{LS} (mm)
$A_{FF,CP}(z)$	Local material exposure at contact point, CP, and material depth, z (–)
$A_{FF,max}$	Maximum material exposure (–)
A_{FF}	Local material exposure (–)
A_{att}	Area of active tooth flank surfaces (mm^2)

A_{sne}	Outer tooth thickness allowance (mm)
A	Amplitude ratio (–)
A	Area of contact (mm^2)
A	Auxiliary factor for calculating the dynamic factor, K_{v-C} (–)
A	Constant (–)
A	Instantaneous line of contact (–)
A	Lower end point of the path of contact (–)
A	Tolerance class according to ISO standard (–)
$b_{(b)}$	Reduced face width (mm)
b^*	Tooth width coordinate of contact point, CP (mm)
$b_{1,2}$	Face width of pinion or wheel (mm)
$b_{2,eff}$	Length of contact pattern along the direction of wheel face width (mm)
b_{2H}	Effective worm wheel face width (mm)
b_B	Face width of one helix on a double helical gear (mm)
b_{BS}	Auxiliary factor (–)
$b_{H,CP}$	Half-width of Hertzian contact band (or width) at contact point, CP (mm)
b_H	Half of the Hertzian contact width or semi-width of Hertzian contact band (mm)
$b_{I(II)}$	Length of end relief (mm)
b_a	Developed length of a curved tooth, assumed as face width (mm)
b_b	Relative base face width (–)
b_{bb}	Length part of face width that bears load (mm)
b_{c0}	Length of tooth bearing pattern with unloaded gear (mm)
b_{cal}	Calculated face width (mm)
b_{ce}	Calculated effective face width (mm)
b_{eB}	Effective face width for scuffing (mm)
b_{eH}	Effective face width (mm)
b_{eff}	Effective face width (mm)
$b_{k1,2}$	Mean face width of pinion or wheel (mm)

b_{red}	Reduced face width or face width without end reliefs (mm)
b_s	Web thickness (mm)
$b_{v,eff}$	Effective face width of virtual cylindrical gear (mm)
b_v	Relative face width, active face width or face width of virtual cylindrical gear (mm)
b_{vir}	Virtual base width (mm)
b	Face width (mm)
b	Tooth width coordinate for straight bevel gear, starting from the back cone (mm)
B^*	Constant (–)
B_M	Thermal contact coefficient, $[\mathrm{N}/(\mathrm{mm}^{1/2}\mathrm{m}^{1/2}\mathrm{s}^{1/2}\mathrm{K})]$ or $(\mathrm{N}/\mathrm{ms}^{1/2}\mathrm{K})$
$B_{M1,2}$	Thermal contact coefficient of pinion or wheel $[\mathrm{N}/(\mathrm{mm}^{1/2}\mathrm{m}^{1/2}\mathrm{s}^{1/2}\mathrm{K})]$ or $(\mathrm{N}/\mathrm{m}^{1/2}\mathrm{K})$
$B_{i(i=p,f,k)}$	Dimensionless parameters (–)
$B_{i(i=1,2)}$	Constants (–)
B	Accuracy grade according to ISO 17485
B	Constant (–)
B	Lower point of single pair tooth contact (–)
c^*	Node where load point and field point for BEM coincide
c'_0	Single stiffness for average conditions $[\mathrm{N}/(\mathrm{mm}\ \mu\mathrm{m})]$
c_1	Material exposure calibration factor (–)
c_{1T}	Linear wear coefficient of the test gear (mm/rev)
c_{BS}	Auxiliary factor (–)
$c_{M1,2}$	Specific heat capacity or specific heat per unit mass of pinion or wheel, $[\mathrm{J}/(\mathrm{kgK})]$
$c_{i(i=0,1,2,3)}$	Constants (–)
$c_m(t)$	Gear mesh viscous damping (Ns/mm)
$c_m(t)$	Time-varying viscous damping element (–)
c_{oil}	Specific heat capacity of oil (Ws/kgk)

$c'_{st/st}$	Single stiffness of steel teeth (N/mm μm)
c_t	Damping coefficient of a torsional damping element (kgm^2/s)
$c'_{th,n}$	Normal theoretical single stiffness (N/mm μm)
c'_{th} or $c'_{th,t}$	Theoretical single stiffness or transverse theoretical single stiffness [N/(mm μm)]
c_v	Specific heat capacity per unit volume [N/(mm^2 K)]
$c_{vi(i=1\text{ to }7)}$	Empirical parameter to determine the dynamic factor (–)
$c_{x,y}$	Damping coefficient of dashpot along x or y direction (Ns/m)
c_γ	Mean value of mesh stiffness per unit face width [N/(mm μm)]
$c_{\gamma 0}$	Mesh stiffness for average conditions [N/mm μm]
$c_{\gamma\alpha}$	Mean value of mesh stiffness per unit width for calculations of K_v, $K_{H\alpha}$, $K_{F\alpha}$ [N/mm μm]
$c_{\gamma\beta}$	Mean value of mesh stiffness per unit width for calculations of $K_{H\beta}$, $K_{F\beta}$ [N/mm μm]
c	Constant (–)
c	Node of a boundary element
c	Specified height or cutting height (μm)
c'	Maximum tooth stiffness per unit face width or single stiffness of a teeth pair [N/(mm μm)]
C_1, C_2, C_{2H}	Weighting factors (–)
C_B	Basic rack factor (–)
$C_{B1,2}$	Basic rack factor for pinion or wheel (–)
C_F	Correction factor of mesh stiffness for non-average conditions (–)
$C_{I(II)}$	End relief (μm)
C_M	Correction factor (–)
C_R	Gear blank factor (–)
C_{ZL}, C_{ZR}, C_{Zv}	Factors for determining lubricant film factors (–)
C_a	Tip relief (μm)
$C_{a1,2}$	Tip relief of pinion or wheel (μm)

C_{ay}	Tip relief by running-in (μm)
C_{eff}	Effective tip relief (μm)
$C_{eq1,2}$	Equivalent tip relief of pinion or wheel (μm)
C_f	Root relief (μm)
$C_{f1,2}$	Root relief of pinion or wheel (μm)
$C_{i(i=1\ to\ 9)}$	Coefficients of series expansion expressing q' (–)
$C_{ij}(P)$	Free-term coefficient in boundary integral equation (–)
C_{lb}	Correction factor for the length of contact lines (–)
C_s	Gradient of the scuffing temperature (°C/μs or K/μs)
C_s	Surface factor (–)
$C_{vi(i=1\ to\ 7)}$	Dimensionless factors (–)
C_β	Crowning height (μm)
C	Auxiliary constant or numerical constant (–)
C	Pitch point (–)
CP	Local contact point (–)
$d_{1,2}$	Pitch or reference diameter of pinion or wheel (mm)
$d_{CP1,2}$	Diameter of pinion or wheel at the contact point CP (mm)
d_{Na}	Effective tip diameter (mm)
d_{Nf}	Root form diameter (mm)
d_T	Tolerance diameter according to ISO 17485 (mm)
$d_{Y1,2}$	Y-Circle diameter of pinion or wheel (mm)
d_a	Tip diameter (mm)
$d_{a1,2}$	Addendum, tip or outside diameter of pinion or wheel (mm)
d_{an}	Tip diameter in normal section (mm)
$d_{b1,2}$	Base diameter of pinion or wheel (mm)
d_{bn}	Base diameter in normal section (mm)
d_e	Diameter of circle through outer point of single pair tooth contact or outer pitch diameter (mm)
$d_{e1,2}$	Outer pitch diameter of pinion or wheel (mm)

d_{en}	Diameter of circle through outer point of single pair tooth contact in normal plane or outer pitch diameter in normal section (mm)
d_f	Root diameter (mm)
d_{f2}	Root diameter of internal gear (mm)
d_m	Mean pitch diameter or diameter at mid-face width (mm)
$d_{m1,2}$	Average value of tip and root diameters or mean pitch diameter of pinion or wheel (mm)
d_{m1}	Worm reference diameter (mm)
d_n	Reference or pitch diameter in normal section (mm)
d_s	Diameter of reference circle of virtual crossed axes helical gear (mm)
$d_{sh,1}$	Nominal external diameter of pinion solid shaft (mm)
d_{sh}	External diameter of shaft (mm)
d_{shi}	Internal diameter of a hallow shaft (mm)
d_{soi}	Diameter at start of involute (mm)
d_v	Reference diameter of virtual cylindrical gear (mm)
d_{va}	Tip diameter of virtual cylindrical gear (mm)
d_{van}	Tip diameter of virtual cylindrical gear in normal section (mm)
d_{vb}	Base diameter of virtual cylindrical gear (mm)
d_{vbn}	Base diameter of virtual cylindrical gear in normal section (mm)
d_{vf}	Root diameter of virtual cylindrical gear (mm)
d_{vn}	Reference diameter of virtual cylindrical gear in normal section (mm)
$d_{w1,2}$	Pitch diameter of pinion or wheel (mm)
d	Distance between a load point and a field point (mm)
d	Reference or pitch diameter of a gear wheel (mm)
D_{be}	Bearing bore diameter of plain bearings (mm)

D_{sh}	Journal diameter of plain bearings (mm)
D	Decay function
D	Diagonal of actual rectangle of contact (–)
D	Upper (or outer) point of single pair tooth contact (–)
e	Auxiliary quantity (–)
e	Exponent for peak load distribution along the path of contact (–)
e	User-defined error threshold (–)
E'	Effective modulus of elasticity (GPa)
$E_{1,2}$	Modulus of elasticity of pinion material or wheel material (N/mm^2)
E_r	Reduced modulus of elasticity or equivalent modulus of elasticity (N/mm^2)
EAP	End of active profile, i.e. Contact points E or A respectively for driving pinion or driving wheel (–)
Eh	Material designation for case-hardened wrought steel (–)
Eht	Case depth according to ISO 6336-5
E	Auxiliary quantity for tooth form factor, with Method $B1$ (mm)
E	Modulus of elasticity, Young's modulus (N/mm^2)
E	Upper end point of the path of contact (–)
f'	Bending deflection of wheel (mm)
f^*	Relative distance (–)
f_1	Teeth contact frequency or frequency of main excitation (Hz)
$f_{\Sigma\beta}$	Alignment-of-axes tolerance (μm)
$f_{ACF}(t_x, t_y)$	Autocorrelation function
f_{E1}	Natural frequency of single gear pair (Hz)
f_F	Load correction factor (–)
$f_{H\beta}$	Helix slope deviation (μm)
$f_{H\beta1,2}$	Helix slope deviation tolerance of pinion or wheel (μm)
$f_{H\beta6,5}$	Tolerance on helix slope deviation for ISO accuracy grade 6 or 5 (μm)

f_{be}	Component of equivalent misalignment due to bearing deformation (μm)
f_{ca}	Component of equivalent misalignment due to case deformation (μm)
$f_{f\alpha}$	Profile form deviation (μm)
$f_{f\alpha eff}$	Effective profile form deviation (μm)
$f_{i(i=t,m,r)}$	Distance of tip, middle and root contact lines from the center point of the zone of action (mm)
f_{ma}	Mesh misalignment due to manufacturing deviations (μm)
f_{max}	Maximum distance to middle contact line (mm)
f_{max0}	Maximum distance to middle contact line at left side of contact pattern (mm)
f_{maxB}	Maximum distance to middle contact line at right side of contact pattern (mm)
f_n	Normal bending deflection (μm)
$f_{p,eff}$	Effective pitch deviation (μm)
f_{pb}	Transverse base pitch deviation (μm)
f_{pbeff}	Effective base pitch deviation (μm)
f_{pt}	Transverse single pitch deviation (μm)
f_{ptT}	Absolute single pitch tolerance (μm)
$f_{p\alpha r}$	Deviation from parallelism between pinion and wheel axes (μm)
$f_{sh,b}$	Component of equivalent misalignment due to bending or bending deflection (μm)
$f_{sh,t}$	Component of equivalent misalignment due to twisting or torsional deflection (μm)
f_{sh}	Component of equivalent misalignment due to deformations of pinion and wheel shafts (μm)
f_{sh0}	Shaft deformation under specific load (μm mm/N)
f_{shT}	Component of equivalent misalignment due to shaft and pinion deformation measured at a partial load (μm)

f_t	Transverse bending deflection (μm)
f	Bending deflection of a rectangle plate or bending deflection of pinion (μm)
f	Distance from the center of the zone of contact to a contact line (mm)
f	Tooth deformation or deviation (μm)
$f(x,t)$	Instantaneous local elastic deformation (mm)
$F_{1,2}$	Auxiliary factors for mid-zone factor (–)
$F_{1,2}(t)$	Instantaneous surface friction force on pinion or wheel tooth profile (N)
F_μ	Friction force (N)
F_b	Nominal tangential load at base cylinder (N)
F_{ber}	Radial force on bearing (N)
F_{bn}	Nominal load, normal to the line of contact (N)
F_{bt}	Nominal transverse load in plane of action (N)
F_{ex}	External axial force (N)
F_{ex}	External axial force acting on double helical gears (N)
$F_{m,dyn}(t)$	Dynamic gear mesh force (N)
$F_{m,stat}(t)$	Quasi-static gear mesh force or actual static mesh force transmitted (N)
$F_m = F_t K_A K_v$	Mean transverse tangential load at the reference circle relevant to mesh calculations (N)
F_{mT}	Mean transverse tangential part load at reference circle (N)
F_{max}	Maximum tangential tooth load for the mesh calculated (N)
F_{mt}	Nominal transverse tangential force at mid-face width of the reference cone (N)
F_{mtH}	Determinant tangential force at mid-face width of the reference cone (N)
$F_{n,dyn}(t)$	Individual dynamic normal tooth force (N)
$F_{n,stat}(t)$	Quasi-static normal tooth force (N)

$F_{n,u}$	Specific normal tooth load or unit normal load (N/mm)
F_n	Normal tooth load, nominal normal force or tooth force or normal load in wear test (N)
$F'_n(t)$	Instantaneous normal tooth load per unit face width (N/mm)
F_r	Radial component of full load when tooth axis passes through pitch point (N)
$F_r(t)$	Rolling friction force (N)
F'_r	Radial component of full load considered by Lewis (N)
$F_s(t)$	Sliding friction force (N)
F_t	Nominal transverse tangential force at reference (or pitch) cylinder per mesh or transmitted load (N)
F_t	Tangential component of full load when tooth axis passes through pitch point (N)
F'_t	Tangential component of full load considered by Lewis (N)
$F_{tH} = F_t K_A K_v K_{H\beta}$	Determinant tangential load in a transverse plane for $K_{H\alpha}$ and $K_{F\alpha}$ (N)
$F_{ti(i=1,2,\ldots)}$	Partial load on the ith pair of teeth in simultaneous meshing (N)
F_{ti}	Transverse tangential load per bin, i (N)
F_{tm2}	Tangential force applied to the worm wheel (N)
F_{vmt}	Nominal tangential force of virtual cylindrical gear (N)
F_w	Nominal tangential load at pitch cylinder (N)
F_α	Total profile deviation (μm)
F_β	Total helix deviation (μm)
$F_{\beta 6}$	Tolerance on total helix deviation for ISO accuracy grade 6 (μm)
$F_{\beta x}$	Initial equivalent misalignment, before running-in (μm)
$F_{\beta xT}$	Equivalent misalignment measured under a partial load (μm)
$F_{\beta xcv}$	Initial equivalent misalignment for the crowning height estimation (μm)

$F_{\beta y}$	Effective equivalent misalignment after running-in (μm)
$F(p,q)$	Fourier transform
F	Composite and cumulative deviations (μm)
F	Force or load or instantaneous force or load (N)
g^*	Sliding factor (–)
$g_0(x,t)$	Lubricant film thickness component related to parabolic function of gap (mm)
g_1	Intermediate factor (–)
g_{CP}	Distance of local contact point CP from point A, assumed as parameter on path of contact (mm)
g_J	Relative length of contact to point of load application for Method $B2$ (–)
g'_J	Modified relative length of contact to point of load application for Method $B2$ (–)
g_P	Algebraic value of distance between the point of contact and the instantaneous center of rotation (mm)
g_Y	Distance coordinate of point Y from point A on path of contact (mm)
$g_{an1,2}$	Recess path of contact of pinion or wheel (mm)
g_c	Length of contact line for Method $B2$ (mm)
$g_{fn1,2}$	Approach path of contact of pinion or wheel (mm)
g_{rb}	Relative distance from blade edge to centerline (–)
$g_{v\alpha}$	Length of path of contact of virtual cylindrical gear in transverse section (mm)
$g_{v\alpha}$	Relative length of path of contact of virtual cylindrical gear in transverse section (–)
$g_{v\alpha n}$	Length of path of contact of virtual cylindrical gear in normal section (mm)
$g_{v\alpha n}$	Relative length of path of contact of virtual cylindrical gear in normal section (–)

$g_{v\alpha na}$	Relative length of path of contact of virtual cylindrical gear in normal section, from pinion tip to pitch circle (–)
$g_{v\alpha nr}$	Relative length of path of contact of virtual cylindrical gear in normal section, from wheel tip to pitch circle (–)
g_α or g	Length of path of contact (mm)
g_η	Relative length of contact within the contact ellipse (–)
$g_{\eta\Delta}$	Change of relative position along the path of contact (–)
$g_{\eta I}$	Relative length of contact at critical point within the contact ellipse (–)
$g_{\eta I\Sigma}$	Relative length of path of contact (–)
$g(z/\bar{z})$	Depth function
G_M	Material parameter (–)
GG	Material designation for grey cast iron (–)
GGG	Material designation for nodular cast iron, with perlitic, bainitic or ferritic structure (–)
GTS	Material designation for black malleable cast iron, with perlitic structure (–)
G	Auxiliary quantity for tooth form factor for Method $B1$ (–)
G	Modulus of rigidity, shear modulus (N/mm^2)
G	Non-dimensional number for materials (–)
$h_0(t)$	Reference lubricant film thickness (mm)
h_{Fa}'	Bending moment arm for tooth root stress with load application at tooth tip (mm)
h_{Fe}	Bending moment arm for tooth root stress with load application at the outer point of single pair tooth contact (mm)
h_Y	Local lubricant film thickness (μm)
h_a	Addendum (mm)
h_{a0}	Tool addendum or cutter active addendum (mm)
h_{aP}	Addendum of basic rack of cylindrical gears (mm)

h_{am}	Mean addendum or addendum at mid-face width of hypoid gear (mm)
$h_{am1,2}$	Tip height in mean cone of pinion or wheel (mm)
h_c	Minimum thickness of lubricant film at pitch point (mm)
h_f	Dedendum (mm)
h_{f2}	Dedendum of internal gear tooth (mm)
h_{fP}	Dedendum of basic rack of cylindrical gears (mm)
h_{fm}	Mean dedendum or dedendum at mid-face width of hypoid gear (mm)
h_{iv}	Lubricant film thickness under iso-viscous conditions (mm)
h_m	Mean whole depth for determination of the bevel spiral angle factor (mm)
h_{min}	Minimum lubricant film thickness (mm)
h_t	Tooth height (mm)
h_{vfm}	Relative mean virtual dedendum (–)
h_N	Load height from critical section for Method $B2$ (mm)
h	Average tooth depth (mm)
h	Lubricant film thickness (mm)
h	Number of hours of total lifetime of a gear (h)
h	Tooth depth from root circle to tip circle or from root cone to tip cone (mm)
$H_{i(i=1,2)}$	Auxiliary dimensions (mm)
HV_{core}	Core hardness (–)
$HV_{surface}$	Surface hardness (–)
H_v	Load losses factor (–)
HB	Brinell hardness (–)
HR 30N	Rockwell hardness, with 30N-scale (–)
HRC	Rockwell hardness, with C-scale (–)
HV	Vickers hardness (–)
HV1	Vickers hardness at load $F = 9.81$ N (–)
HV10	Vickers hardness at load $F = 98.1$ N (–)
H	Auxiliary quantity for tooth form factor with Method $B1$ (–)
H	Hardness (MPa)

H	Hydrodynamic line (–)
H	Mid-point of the line of action (–)
H	Non-dimensional number for oil film thickness (–)
$HV(z)$	Hardness at the material depth, z (HV)
i	Bin (–)
i	Transmission ratio (–)
$I_{i(i=1,2)}$	First and second tensor invariants for principal axes
$I_{ij,c}^T$	Integral function of function $N_c ds$ by kernel T_{ij}
$I_{ij,c}^U$	Integral function of function $N_c ds$ by kernel U_{ij}
IF	Material designation for flame or induction hardened wrought special steel (–)
I	Area moment of inertia, moment of inertia of cross section or second area moment of cross section (mm^4)
I	Grid elements (–)
I	Stress integral
j_n	Normal backlash (mm)
j_t	Transverse circular backlash (mm)
J^*	Moment of inertia per unit face width (kgmm^2/mm)
$J_{1,2}$	Mass moment of inertia per unit face width of pinion or wheel (kgmm^2/mm)
J_2	Second deviator invariant for principal axes
J	Jacobian of coordinate transformation
J	Jominy hardenability (–)
J	Polar mass moment of inertia (kgmm^2)
k'	Contact shift factor (–)
$k_{i(i=1,2)}$	Parameter related to material properties (–)
$k_{i(i=1,2,3,4)}$	Coefficients (–)
$k_m(t)$	Dashpot or time-varying mesh spring element (–)
$k_{i(i=x,y)}$	Stiffness of time-varying mesh spring element along x or y direction (N/m)
k	Constant (–)

k	Dimensional constant (mm μm/N)
k	Dimensional quantity in expression of minimum thickness of lubricant film at pitch point $\left[s^{1.4} / \left(mm^{1.66} N^{0.13} \right) \right]$
k	Dimensional Reye's wear constant (mm^2/N)
k	Findley's parameter (–)
k	Iteration index (–)
k	Positive integer (–)
k	Weibull's exponent (–)
K'	Constant of the pinion offset (–)
K_{350}	Influence factor of gear accuracy grade at specific reference load of 350 N/mm (–)
K_A	Application factor (–)
$K_{B\alpha} = K_{H\alpha}$	Transverse load factor for scuffing (–)
$K_{B\beta} = K_{H\beta}$	Face load factor for scuffing (–)
$K_{B\beta be}$	Bearing factor (–)
$K_{B\gamma}$	Helical load factor for scuffing (–)
K_F	Numbers of BE (Boundary Elements) away from the load point (–)
K_F^*	Numbers of BE (Boundary Elements) containing the load point (–)
K_{F0}	Lengthwise curvature factor for bending stress (–)
$K_{F\alpha}$	Transverse load factor for root bending stress (–)
$K_{F\beta}$	Face load factor for root bending stress (–)
$K_{H\alpha}$	Transverse load factor for contact stress (–)
$K_{H\alpha}^*$	Preliminary transverse load factor for contact stress for non-hypoid gears (–)
$K_{H\beta}$	Face load factor for contact stress (–)
$K_{H\beta-be}$	Mounting factor (–)
K_f	Fatigue stress concentration factor, fatigue strength reduction factor or fatigue notch factor (–)
$K_{i(i=1,2)}$	Adjustment factors (–)

$K_{i(i=1,2,3)}$	Dimensionless factors (–)
$K_{material}$	Material factor (–)
K_{mp}	Multiple path factor (–)
K_t	Theoretical (or geometric) stress concentration factor (–)
K_v	Dynamic factor (–)
K_v^*	Preliminary dynamic factor for non-hypoid gears (–)
$K_{v\alpha}$	Dynamic factor for spur gears (–)
$K_{v\beta}$	Dynamic factor for helical gears (–)
K_γ	Mesh load factor for multiple transmission paths (–)
$K_{\tau,per}$	Hardness conversion factor (–)
$K_{\tau,per}$	Proportional factor (–)
$K_{\tau per}$	Hardness conversion factor (–)
K	Constant that converts square millimeters into square meters (–)
K	Constant, tooth load factors, constant factor for determination of dynamic factor K_{v-B} (–)
K	Wear coefficient (–)
$K(x)$	Coefficient of influence
$K(\xi)$	Function expressing the ideal helix angle deviation
l_1	Spacing between worm shaft bearings (mm)
l_{11} and l_{12}	Portions of spacing between worm shaft bearings (mm)
l_F	Length of finite region SF (mm)
l_a	Effective length of roller of a roller bearing (mm)
l_b	Length of contact line for Method $B1$ (mm)
l_{b0}	Theoretical length of contact line (mm)
l_{bm}	Theoretical length of middle contact line (mm)
l_m	Mean value of length of the line of contact (mm)
l_t	Total length of line of contact (mm)
l	Bearing span (mm)
l	Distance between feeler gauges (mm)
l	Length of flank line or tooth trace (mm)

L_{hw}	Lifetime of a gear subjected to abrasive wear (h)
$L_a = s_{Fn}/h_{Fa}$	Dimensionless parameter (–)
L	Contact parameter (–)
L	Sampling length for roughness measurements (mm)
$L = s_{Fn}/h_{Fe}$	Dimensionless parameter (–)
Ln	Evaluation length (mm)
Lr	Sampling length (mm)
$m_{1,2}$	Individual mass per unit face width of pinion or wheel referenced to the line of action (kg/mm)
$m_{1z,2x}$	Individual gear mass per unit face width reduced to the line of action, for pinion or wheel of dynamically equivalent cylindrical gear (kg/mm)
m_T	Module of the test gear (mm)
m_{et}	Outer transverse module (mm)
m_{mn}	Mean normal module or normal module of hypoid gear at mid-face width (mm)
m_{mt}	Mean transverse module or transverse module in the mean cone (mm)
m_n	Normal module (mm)
m_{red}	Mass per unit face width reduced to line of action of dynamically equivalent cylindrical gear (kg/mm)
m_{sn}	Normal module of virtual crossed axes helical gear (mm)
m_t	Transverse module (mm)
m_{vt}	Transverse module of virtual cylindrical gear (mm)
m_{x1}	Worm axial module (mm)
m	Geometric parameter (–)
m	Mass (kg)
m	Module or transverse module (mm)
m	Number of teeth pairs in simultaneous meshing (–)
mr	Material ratio (%)
M_b	Bending moment (Nm)
ME, MQ, ML	Symbols identifying material and heat-treatment requirements (–)
M	Mean stress ratio factor (–)
M	Moment of a force or bending moment (Nm)

$n_{1,2}$	Rotational speed of pinion or wheel (s^{-1} or min^{-1})
n_1	Rotational speed of worm shaft (min^{-1})
n_E	Resonance speed (min^{-1})
n_{E1}	Resonance speed of pinion (min^{-1})
n_c	Number of mesh contacts per revolution (–)
n_i	Number of cycles for bin i (–)
n_p	Number of mesh contacts or number of meshing gears (–)
n	Generic tooth pair in meshing (–)
n	Geometric parameter (–)
n	Number of load cycles (–)
n	Rotational speed (s^{-1} or min^{-1})
$\boldsymbol{n}$	Unit outward normal vector
N_{E1}	Rotational speed corresponding to natural frequency (min^{-1})
N_F	Dimensionless factor depending on the ratio (b/h) or exponent (–)
N_L	Number of load cycles, cumulative number of applied cycles or service life (–)
$N_{L1,2}$	Number of load cycles, cumulative number of applied cycles or service life of pinion or wheel (–)
N_f	Exponent (–)
$N_{i(i=1,2,3)}$	Quadratic shape functions
N_i	Number of cycles to failure per bin, i (–)
N_s	Resonance ratio in the main resonance range (–)
NT	Material designation for nitrided wrought steel, nitriding steel (–)
NV	Material designation for through-hardened wrought steel, nitrided, nitrocarburized (–)
N	Number of tooth pairs in simultaneous meshing (–)
N	Number or frequency ratio (–)
N	Reference speed related to resonance speed n_{E1} (–)
p^*	Relative peak load for calculating the load sharing factor for Method $B1$ (–)

p_0	Reference local load per unit length (N/mm)
$p_0 = \sigma_H$	Maximum normal pressure (N/mm²)
$p_{H,CP}$	Local nominal Hertzian contact stress (N/mm²)
$p_{H,Y}$	Local nominal Hertzian contact stress (N/mm²)
p_H	Nominal Hertzian contact stress (N/mm²)
p_{bmn}	Relative mean normal base pitch (–)
p_{bn}	Normal base pitch (mm)
p_{bt}	Transverse base pitch (mm)
$p_{dyn,CP}$	Local Hertzian contact stress at contact point (N/mm²)
$p_{dyn,Y}$	Local Hertzian contact stress including the load factor, K (N/mm²)
p_{dyn}	Hertzian contact stress including load factor, K (N/mm²)
p_{en}	Normal base pitch (mm)
p_{et}	Transverse base pitch on the path of contact or transverse base pitch for Method $B2$ (mm)
p_m^*	Parameter for mean Hertzian stress (–)
p_{max}	Maximum peak load (N/mm)
p_{mn}	Mean normal pitch (mm)
p_{mn}	Relative mean normal pitch (–)
p_{vet}	Transverse base pitch of virtual cylindrical gear for Method $B1$ (mm)
p or p_t	Pitch, circular pitch, transverse circular pitch or transverse pitch (mm)
p	Local load per unit length (N/mm)
p	Peak load (N/mm)
p	Pressure (MPa)
p	Pressure in the lubricant film (Pa)
p	Slope of the Wöhler-damage line (–)
pr	Protuberance of the tooth (mm)
$Pé_{1,2}$	Péclet numbers of pinion or wheel (–)
P_K	Cooling capacity of oil spray lubrication (W)
P_V	Total power loss of a worm gear unit (W)
P_{VD}	Sealing power loss (W)
P_{VLP}	Bearing power loss (W)

P_m	Flow pressure of a worm surface (N/mm^2)
P_s	Probability of survival of a mechanical component (%)
P_{v0}	Idle running power loss (W)
P_{vz}	Mesh power loss (W)
P_z	Power loss at contact between the mating teeth (kW)
P	Hertzian stress (GPa)
P	Load point for BEM
P	Transmitted power or nominal power (kW)
q'	Minimum value for the flexibility of a pair of meshing teeth [(mm µm)/N]
q_0	Maximum tangential surface force or maximum friction traction force (N/mm^2)
q_{pr}	Tool protuberance (mm)
$q_s = s_{Fn}/2\rho_F$	Notch parameter (–)
$q_{sT} = 2,5$	Notch parameter of the standard reference test gear (–)
q_{sk}	Notch parameter of the notched test piece (–)
q_α	Auxiliary factor (–)
q_α	Dimensionless quantity (–)
q	Auxiliary factor (–)
q	Exponent in the formula for lengthwise curvature factor (–)
q	Flexibility of pair of meshing teeth [(mm µm)/N]
q	Machining stock (mm)
q	Material allowance for finish machining (mm)
q	Notch sensitivity (–)
q	Tangential friction surface load or friction traction load (N/mm^2)
Q_{oil}	Oil spray quantity (m^3/s)
Q	Field point for BEM
Q	Quality grade (–)
$r_{1,2}$	Pitch radius of pinion or wheel (mm)
r_{CP}	Local contact radius (mm)
r_b	Base radius (mm)
r_{c0}	Cutter radius (mm)
r_e	Radial distance of intersection point between pitch and back cone generatrices from the bevel gear axis (mm)

r_i	Radial distance of intersection point between pitch and inner cone generatrices from the bevel gear axis (mm)
$r_{m1,2}$	Pitch or reference radius in mean cone of pinion or wheel (mm)
r_{mf}	Tooth fillet radius at the root diameter in mean section (mm)
r_{mpt}	Mean transverse pitch radius (mm)
r_{my0}	Mean transverse radius to point of load application, for Method $B2$ (mm)
r_{va}	Relative mean virtual tip radius (–)
r_{vbn}	Relative mean virtual base radius (–)
r_{vn}	Relative mean virtual pitch radius (–)
r	Notch radius (mm)
r	Radial distance of intersection point between pitch and intermediate cone generatrices from the bevel gear axis (mm)
$R_{D,CL1}$	Relative radius from tool center to critical fillet point of pinion drive side/coast side (–)
$R_{Y1,2}$	Distance of current point of contact Y on pinion or wheel tooth profile from center (mm)
$Ra_{1,2}$ (or $R_{a1,2}$)	Arithmetic average roughness of pinion or wheel (μm)
R_e	Outer cone distance (mm)
R_m	Mean cone distance (mm)
R_m	Tensile strength of gear material (N/mm^2)
$R_{m1,2}$	Mean cone distance of pinion or wheel (mm)
R_{mpt}	Relative mean back cone distance (–)
$Rq_{1,2}$	Root-mean-square roughness of pinion or wheel (μm)
$Rz_{1,2}$	Mean peak-to-valley roughness of pinion or wheel (μm)
Rz_{10} (or R_{z10})	Relative mean peak-to-valley roughness for gear pairs with relative curvature radius $\rho_{red} = 10$ mm (μm)
Rz_H	Equivalent roughness (μm)
$Rz_T (Rz_T = 10)$	Mean peak-to-valley roughness of the standard reference test piece (μm)

Rz_k	Mean peak-to-valley roughness of the notched rough test piece (μm)
R	Distance from origin of a point of ovoidal boundary obtained by means of a threshold autocorrelation (μm)
R	Stress ratio (–)
Ra (or R_a) $= CLA = AA = Rz/6$ (or $R_z/6$)	Arithmetic average roughness or arithmetic mean roughness value (μm)
Ra	Effective arithmetic mean roughness (μm)
Rq	Root-mean-square roughness (μm)
Rz (or R_z)	Mean peak-to-valley roughness (μm)
s_h	Thickness of hardened surface layer (mm)
s_{Fn}	Tooth root chordal thickness at the critical or calculation sections (mm)
s_N	One-half tooth thickness at critical section for Method $B2$ (mm)
s_R	Rim thickness (mm)
s_a	Initial tooth tip thickness (mm)
s_c	Film thickness of marking compound used in contact pattern determination (μm)
s_c	Local tooth chordal thickness (mm)
s_{mn}	Mean normal circular thickness (mm)
s_{pr}	Residual fillet undercut or tool protuberance (mm)
$s_{t,B-D}$	Chordal tooth thickness in transverse section at diameter corresponding to the middle point between points B and D on the path of contact (mm)
s_{vmn}	Relative mean normal circular thickness or relative mean virtual dedendum (–)
s_w	Material thickness worn out away from a differential area of contact in an infinitesimal time (mm/s)
s	Distance between mid-plane of pinion and the middle of the bearing span (mm)
s	Transverse tooth thickness or local tooth root thickness of bevel gears (mm)

$S_{v,min}$	Minimum safety factor for bending stress (–)
S_B	Safety factor for scuffing (–)
S_F	Finite region of infinite half-space containing the contact zone
S_F	Safety factor against bending stress breakage (–)
S_F	Shear tooth breakage safety factor (–)
S_{FF}	Safety factor against tooth flank fatigue fracture (–)
S_{FZG}	Load stage in FZG test (–)
S_{Flimit}	Minimum safety factor against abrasive wear (–)
S_{Fmin}	Minimum shear tooth breakage safety factor (–)
$S_{GF,Y}$	Local sliding parameter (–)
$S_{H,min}$	Minimum safety factor for contact stress (–)
S_H	Safety factor for contact stress against pitting or pitting safety factor (–)
S_{Hmin}	Minimum pitting safety factor (–)
S_I	Infinite region of infinite half-space, unloaded
$S_{N,b}$	Uniaxial fully reversed bending fatigue strength for finite fatigue life cycle number (N/mm^2)
$S_{N,t}$	Uniaxial fully reversed torsion fatigue strength for finite fatigue life cycle number (N/mm^2)
S_S	Infinite region half cylindrical in shape, surrounding the half-space at infinity
S_{Sl}	Load safety factor against scuffing (–)
S_{Smin}	Minimum required scuffing safety factor (–)
S_T	Working temperature safety factor (–)
S_{Tmin}	Minimum working temperature safety factor (–)
S_W	Wear safety factor (–)
S_{Wmin}	Minimum wear safety factor (–)

S_a	Arithmetic mean height of scale-limited surface (μm)
$Sa_{1,2}$	Arithmetic mean height of the scale-limited surface of pinion or wheel (mm)
S_{al}	Autocorrelation length (μm)
S_{dc}	Surface section difference (μm)
S_{dq}	Root mean square gradient of the scale-limited surface (μm/μm)
S_{dr}	Developed interfacial area ratio of the scale-limited surface (%)
S_e	Boundary of a circular region of computational domain centered at load point
S_{intS}	Scuffing safety factor (–)
S_k	Core height (μm)
S_{ku}	Kurtosis of scale-limited surface (μm)
$S_{mc(mr)}$	Inverse areal material ratio (μm)
S_{mq}	Material ratio (%)
$S_{mr(c)}$	Areal material ratio (%)
$S_{mr1,2}$	Material ratios (–)
S_p	Maximum peak height (μm)
S_{pk}	Reduced peak height (μm)
S_{pq}	Plateau root mean square deviation (μm)
S_q	Root mean square height of scale-limited surface (μm)
S_{sh}	Skewness of scale-limited surface (μm)
S_{td}	Texture direction (°)
S_{tr}	Texture aspect ratio (–)
S_v	Maximum pit height (μm)
S_{vk}	Reduced dale height (μm)
S_{vq}	Dale root mean square deviation (μm)
S_{xp}	Peak extreme height (μm)
S_z	Maximum height (μm)
S_δ	Safety factor against worm shaft deflection (–)
$S_{\delta min}$	Minimum value of safety factor against worm shaft deflection (–)
$S_{\lambda,min}$	Minimum required safety factor against micropitting (–)
S_λ	Safety factor against micropitting (–)

St	Material designation for normalized base steel (–)
S	Area of resistant cross section (mm²)
S	Safety factor (–)
S	Spherical surface area (mm²)
S	Total rubbing distance (mm)
SAP	Start of active profile, i.e. Contact points A or E respectively for driving pinion or driving wheel (–)
SIH	Shear stress intensity hypothesis
Sbi	Surface bearing index (–)
Sci	Surface core fluid retention index (–)
Svi	Surface valley fluid retention index (–)
$t_{1,2}$	Contact exposure time of pinion or wheel (μs)
t_K	Contact exposure time at knee of exposure time curve (μs)
t_c	Contact exposure time (μs)
t_{cmax}	Longest contact exposure time (μs)
t_g	Maximum depth of grinding notch (mm)
t_j	Tractions or resultant stresses (N/mm²)
t_k	Contact exposure time at knee of a given distribution curve of scuffing temperature (μs)
$t_{z1,2}$	Pinion or wheel pitch apex beyond crossing point (mm)
T'	Total torque transmitted by a bevel gear wheel (Nm)
$T_{1,2}$	Nominal torque at the pinion or wheel or input and output torque (Nm)
$T_{1,2}$	Points of interference (–)
T_{1T}	Scuffing torque of test pinion (Nm)
T_2	Output torque from worm wheel (Nm)
T_{eq}	Equivalent torque (Nm)
T_i	Torque for bin i (Nm)
$T_{ij}(P, Q)$	Integral kernel function for t_j
T_{max}	Maximum permissible torque (Nm)
T_n	Nominal torque (Nm)
T	Maximum distance along line of action of the relevant point of contact from pitch point (mm)

T	Numerical factor (–)
T	Pitch plane of hypoid gear pair (–)
T	Thermal line (–)
T	Tolerance (μm)
T	Torque, operating torque or total transmitted torque (Nm)
u_j	Displacements (mm)
u_v	Gear ratio of virtual cylindrical gear or virtual gear ratio (–)
u	Gear ratio (–)
U_Y	Local velocity parameter (–)
U_i	Individual damage quotient (–)
$U_{ij}(P, Q)$	Integral kernel function for u_j
U	Elastic strain energy due to bending load (Nm)
U	Miner sum or sum of individual damage quotients (–)
U	Non-dimensional number for speed (–)
$v_{1,2}(t)$	Instantaneous tangential velocity at a current point of contact of pinion or wheel tooth profile (m/s)
$v_{\Sigma,Y}$	Sum of tangential velocities at point, Y (m/s)
$v_{\Sigma,vert}$	Cumulative normal velocity or sum of the sliding velocity components normal to the contact line (m/s)
v_Σ	Cumulative velocity or sum of velocities in the mean point P (m/s)
$v_{\Sigma h}$	Cumulative velocity along the direction of the tooth profile (m/s)
$v_{\Sigma C}$	Mean cumulative velocity or sum of tangential velocities at pitch point (m/s)
$v_{\Sigma l}$	Sum of velocities in lengthwise direction (m/s)
$v_{\Sigma s}$	Cumulative velocity along the lengthwise direction (m/s)
$v_{et,max}$	Maximum pitch line velocity at operating pitch diameter (m/s)
v_{et}	Tangential speed at outer end (heel) of the reference cone (m/s)
$v_{g,Y}$	Local sliding velocity (m/s)
$v_{g,par}$	Sliding velocity component parallel to the contact line (m/s)

$v_{g,vert}$	Sliding velocity component normal to the contact line (m/s)
v_g	Sliding velocity or sliding velocity in the mean point P (m/s)
$v_{g1,2}$	Sliding velocity of pinion or wheel (m/s)
v_{gh}	Profile velocity component of sliding velocity vector (m/s)
v_{gs}	Lengthwise velocity component of sliding velocity vector (m/s)
v_{gs}	Sliding velocity at pitch point (m/s)
$v_{g\alpha1}$	Profile sliding velocity (m/s)
$v_{g\beta1}$	Helical sliding velocity (m/s)
$v_{g\gamma1}$	Total sliding velocity or maximum sliding velocity at tip of pinion (m/s)
v_{mt}	Nominal tangential speed at mid-face width of the reference cone (m/s)
v_r	Relative or sliding velocity (m/s)
$v_{r1,2,Y}$	Local tangential velocity on pinion or wheel (m/s)
v_{rt}	Average rolling velocity or cumulative semi-velocity under dynamic conditions (m/s)
v_{st}	Sliding velocity under dynamic conditions (m/s)
v_t	Pitch line velocity (m/s)
$v_{t1,2}$	Tangential velocity of pinion or wheel of hypoid gear (m/s)
v_w	Cumulative semi-velocity or average rolling velocity (m/s)
v_w	Wear rate or wear velocity (mm/s)
v	Reference line velocity, tangential velocity at reference circle or at pitch circle (m/s)
$V_m(p)$	Material volume ($\mu m^3/mm^2$ or m^3/m^2)
V_{mc}	Core material volume ($\mu m^3/mm^2$ or m^3/m^2)
V_{mp}	Peak material volume ($\mu m^3/mm^2$ or m^3/m^2)
$V_v(p)$	Void volume ($\mu m^3/mm^2$ or m^3/m^2)
V_{vc}	Core void volume ($\mu m^3/mm^2$ or m^3/m^2)
V_{vv}	Dale void volume ($\mu m^3/mm^2$ or m^3/m^2)

V	Material designation for through-hardened wrought special steel, alloy or carbon, with $\sigma_B \geq 800$ N/mm^2 (–)
V	Ratio between semi-elliptic load distribution (–)
V	Sliding velocity (m/s)
V	Volume of a machine component (mm^3)
V	Volume of material removed due to wear (mm^3)
V	Volume of material removed due to wear per unit of time (mm^3/s)
V	Volume of material removed due to wear per unit sliding distance (mm^3/mm)
VI	Viscosity index (–)
w_{ht}	Transverse tangential load per unit face width (N/mm)
w_{Bn}	Normal unit load (N/mm)
w_{Bt}	Transverse unit load or specific tooth load for scuffing (N/mm)
w_{Bteff}	Effective transverse specific load for scuffing (N/mm)
w_{Btmax}	Maximum transverse specific load for scuffing (N/mm)
w_m	Mean specific load per unit face width or per unit middle contact line (N/mm)
w_{max}	Maximum load per unit face width (N/mm)
w_{max}	Maximum specific load per unit middle contact line (N/mm)
w_t	Tangential load per unit tooth width, including overload factors (N/mm)
w	Angle of contact line relative to the root cone (°)
w	Specific load per unit face width (N/mm)
W_1	Wear rate (mm/h)
W_{1P}	Allowable wear thickness (mm)
W_W	Material factor (–)
W_Y	Local load parameter (–)
W_m	Mass of worn particles per unit of time (mg/s)
W_{m2}	Wheel mean slot width (mm)

W_{mP}	Maximum permissible mass of worn particles (mg)
W	Non-dimensional number for load (–)
W	Passive work done by frictional forces per unit of time (J/s)
x_0	Profile shift coefficient of pinion-type cutter (–)
x_{00}	Distance from mean section to point of load application (mm)
$x_{001,2}$	Contact shift or distance from mean section to point of load application for pinion or wheel (mm)
$x_{1,2}$	Profile shift coefficient of pinion or wheel (–)
$x_{1,2}(t)$	Displacement of the instantaneous point of contact on pinion or wheel tooth profile along the OLOA-direction (mm)
x_1	Relative horizontal distance from pitch circle to fillet point (–)
x_{hm}	Profile shift coefficient (–)
x_E	Generation profile shift coefficient (–)
x_N	Tooth strength factor for Method $B2$ (mm)
$x_{i(i=1,2)}$	Quantities in expression of theoretical length of contact line (mm)
x_{sm}	Thickness modification coefficient (–)
x	Profile shift coefficient (–)
x	Quantity in the Lewis form factor (mm)
X_{Γ}	Load sharing factor (–)
X_{ε}	Contact ratio factor (–)
X_{Θ}	Gradient of the scuffing temperature (–)
X_{BE}	Geometry factor at pinion tooth tip (–)
X_{CP}	Local load sharing factor (–)
X_{Ca}	Tip relief factor (–)
X_E	Run-in factor (–)
X_G	Geometry factor or geometry factor of hypoid gears (–)
X_J	Approach factor (–)
X_L	Lubricant factor (–)

X_M Thermo-elastic factor and thermal flash factor $\left(\text{KN}^{-3/4}\text{s}^{-1/2}\text{m}^{-1/2}\text{mm}\right)$

X_Q Approach factor (–)

X_R Roughness factor (–)

X_S Lubrication factor or lubrication system factor (–)

X_W Structural factor or welding factor of actual gear material (–)

X_{WT} Welding factor of test gear (–)

X_{WrelT} Relative welding factor of test gear (–)

X_X Local load sharing factor (–)

$X_{but,A}$ Buttressing value at point A of path of contact (–)

$X_{but,CP}$ Local buttressing factor (–)

$X_{but,E}$ Buttressing value at point E of path of contact (–)

$X_{but,\Gamma}$ Buttressing factor (–)

$X_{but,CP}$ Local buttressing factor (–)

$X_{but,Y}$ Local buttressing factor (–)

X_{mp} Multiple mating pinion factor and contact factor (–)

$X_{\alpha\beta}$ Pressure angle factor (–)

X Factor depending on the accuracy grade (–)

X Intermediate factor (–)

X Parameter of irregularity of transmission (–)

$y_{1,2}(t)$ Displacement of the instantaneous point of contact on pinion or wheel tooth profile along the LOA-direction (mm)

y_3 Location of point of load application on path of contact for maximum root stress (mm)

y_I Distance of critical point where maximum contact stress occurs on tooth surface from midpoint of path of contact (mm)

y_f Running-in allowance for profile form deviation related to the polished test piece (μm)

$y_{i(i=1,2)}$ Quantities in expression of theoretical length of contact line (mm)

y_j	Location of point of load application for maximum bending stress on path of contact, for Method $B2$ (mm)
y_p	Running-in allowance for pitch deviation related to the polished test piece (μm)
y_α	Running-in allowance for pitch error of a gear pair (μm)
y_β	Running-in allowance, equivalent misalignment (μm)
y	Lewis form factor in terms of module (–)
y	Relative vertical distance from pitch circle to fillet point (–)
y	Running-in allowance (μm)
$Y_{1,2}$	Tooth form factor of pinion or wheel, for Method $B2$ (–)
Y_A	Root stress adjustment factor for Method $B2$ (–)
Y_B	Rim thickness factor (–)
Y_{BS}	Bevel spiral angle factor (–)
Y_{DT}	Deep tooth factor (–)
Y_F	Form factor (–)
Y_F	Tooth form factor influencing the nominal tooth root stress, with load applied at the outer point of single pair tooth contact (–)
Y_{FA}	Tooth form factor for load application at the tooth tip, for Method $B1$ (–)
Y_{FS}	Combined tooth form factor for generated gears (–)
Y_J	Bending strength geometry factor for Method $B2$ (–)
Y_K	Rim thickness factor (–)
Y_{LS}	Load sharing factor for bending, for Method $B1$ (–)
Y_M	Mean stress influence factor (–)
Y_N	Life factor for tooth root stress (–)
Y_{NL}	Life factor for shear tooth breakage (–)
Y_{NT}	Life factor for tooth root bending stress for reference test conditions (–)

Y_{Nk}	Life factor for tooth root stress, relevant to the notched test piece (–)
Y_{Np}	Life factor for tooth root stress, relevant to the plain polished test piece (–)
Y_P	Combined geometry factor, for Method B2 (–)
Y_R	Tooth root surface factor of the actual gear (–)
Y_{RT}	Tooth root surface factor, relevant to the plain polished test piece (–)
Y_{RrelT}	Relative surface condition factor (–)
Y_{Rrelk}	Relative roughness factor (–)
Y_S	Stress correction factor (–)
Y_{ST}	Stress correction factor for dimensions of the reference standard test gears (–)
Y_{Sa}	Stress correction factor for load application at the tooth tip, for Method $B1$ (–)
Y_{Sg}	Stress correction factor for teeth with grinding notches (–)
Y_{Sk}	Stress correction factor relevant to the notched test piece (–)
Y_X	Size factor for tooth root bending stress (–)
Y_f	Stress concentration and stress correction factor for Method $B2$ (–)
Y_i	Inertia factor for bending (–)
Y_β	Helix angle factor for tooth root bending stress (–)
Y_γ	Lead factor (–)
Y_δ	Notch sensitivity factor of the actual gear, relative to a polished test piece (–)
$Y_{\delta T}$	Sensitivity factor of standard reference test piece, relative to the smooth polished test piece (–)
$Y_{\delta k}$	Sensitivity factor of a notched test piece, relative to a smooth polished test piece (–)
$Y_{\delta relT}$	Relative notch sensitivity factor (–)
$Y_{\delta relk}$	Test relative notch sensitivity factor (–)

Y_ε	Contact factor or contact ratio for bending, for Method $B1$ (–)
Y	Current point of contact on path of contact (–)
Y	Factor related to tooth root stress (–)
Y	Lewis form factor in terms of circular pitch (–)
$\bar{z}$	Total case depth (mm)
z_0	Number of blade groups of the cutter (–)
z_0	Number of teeth of pinion-type cutter (–)
$z_{1,2}$	Number of teeth of pinion or wheel (–)
z_{CHD}	Case hardening depth at 550 HV
$z_{HV,max}$	Material depth with maximum hardness (mm)
z_{core}	Material depth at z-coordinate where $HV(z) = HV_{core}$ (mm)
z_n	Virtual number of teeth of a helical gear (–)
$z_{n1,2}$	Virtual number of teeth of helical pinion or wheel (–)
z_v	Number of teeth of virtual cylindrical gear or virtual number of teeth (–)
z_{vn}	Number of teeth of virtual cylindrical gear in normal section (–)
$z_{vn1,2D,C}$	Number of teeth of virtual cylindrical pinion or wheel in normal section, for drive or coast side (–)
z	Coordinate along the tooth axis or along the normal to tooth profile (mm)
z	Height of surface above the datum line (μm)
z	Material depth (mm)
z	Number of teeth (–)
Z_h	Life factor (–)
Z_A	Contact stress adjustment factor for Method $B2$ (–)
Z_B, Z_D	Single pair tooth contact factors for the pinion and for the wheel (–)
Z_E	Elasticity factor $(\text{N/mm}^2)^{1/2}$
Z_{FW}	Face width factor (–)

Z_H	Zone factor (–)
Z_{Hyp}	Hypoid factor (–)
Z_I	Pitting resistance geometry factor for Method $B2$ (–)
Z_K	Bevel gear factor for Method $B1$ (–)
Z_L	Lubricant factor (–)
Z_{LS}	Load sharing factor for pitting, according to Method $B1$ (–)
Z_{M-B}	Mid-zone factor (–)
Z_N	Life factor for contact stress (–)
Z_{NT}	Life factor for contact stress for reference test conditions (–)
Z_R	Roughness factor affecting surface durability (–)
Z_S	Bevel slip factor (–)
Z_S	Size factor (–)
Z_W	Work hardening factor (–)
Z_X	Size factor for pitting (–)
Z_i	Inertia factor (–)
Z_{oil}	Lubricant factor (–)
Z_u	Gear ratio factor (–)
Z_v	Velocity factor (–)
Z_β	Helix angle factor (–)
Z_ε	Contact ratio factor or contact ratio factor for pitting (–)
Z	Factor related to contact stress (–)
Z	Probability of removal of material for atomic encounter (%)
Z	Section modulus (mm^3)
α_0	Pressure angle of the cutter (°)
$\alpha_{1,2}$	Transverse tip pressure angle of pinion or wheel (°)
α_{38}	Pressure-viscosity coefficient at 38 ° C (m^2/N)
$\alpha_{D,Cnf}$	Drive or coast flank pressure angle (°)
α_{Fan}	Load application angle at tooth tip of virtual cylindrical gear for Method $B1$ (°)
α_{Fe}	Load application angle at tooth tip of cylindrical gear for Method $B1$ (°)
α_{Fen}	Load direction angle, relevant to direction of application of load at the outer point of single pair tooth contact of virtual cylindrical spur gears (°)

α_L	Normal pressure angle at point of load application for Method $B2$ (°)
α_{LN2}	Generated pressure angle of wheel at fillet point (°)
α_P	Pressure angle of basic rack for cylindrical gears (°)
α_{Pn}	Normal pressure angle of the basic rack for cylindrical gears (°)
α_T	Coefficient of linear thermal expansion of material (K^{-1})
α_a	Adjusted pressure angle for Method $B2$ (°)
α_a	Angle between the full load vector applied to tooth tip, positioned at upper end point of path of contact, and normal to tooth axis through their point of intersection (°)
$\alpha_{a1,2}$	Transverse tip pressure angle of pinion or wheel (°)
α_{an}	Normal pressure angle at tooth tip (°)
α_e	Effective pressure angle (°)
$\alpha_{eD,C}$	Effective pressure angle for drive side/coast side (°)
α_{en}	Form-factor pressure angle, pressure angle at the outer point of single pair tooth contact of virtual spur gears (°)
α_{et}	Effective pressure angle in transverse section (°)
α_f	Limit pressure angle in wheel root coordinates for Method $B2$ (°)
α_{lim}	Limit pressure angle (°)
α_{mn}	Normal pressure angle at mid-face width of hypoid gear pair (°)
α_n	Normal pressure angle (°)
$\alpha_{nD,C}$	Generated normal pressure angle for drive side/coast side (°)
α_{sn}	Normal pressure angle of virtual crossed axes helical gear pair (°)
α_{st}	Transverse pressure angle of virtual crossed axes helical gears (°)
α_t	Nominal transverse pressure angle or pressure angle at the basic rack profile (°)

α_t' (or α_{wt})	Transverse pressure angle at the pitch cylinder or transverse working pressure angle (°)
$\alpha_{vD,Cnf}$	Drive or coast flank virtual pressure angle (°)
α_{vet}	Transverse pressure angle of virtual cylindrical gears for active flank (°)
α_{vt}	Transverse pressure angle of virtual cylindrical gear (°)
α_w	Working pressure angle (°)
α_{wn}	Normal working pressure angle (°)
α_{wt}	Transverse working pressure angle (°)
α_y	Arbitrary angle (°)
α_{y1}	Pinion pressure angle at arbitrary point (°)
$\alpha_{\vartheta B,Y}$	Pressure-viscosity coefficient at local contact temperature (m^2/N)
$\alpha_{\vartheta M}$	Pressure-viscosity coefficient at bulk temperature (m^2/N)
$\alpha(x, y)$	Angle of steepest gradient (°)
α	Pressure angle at reference cylinder (°)
α	Pressure-viscosity coefficient (mm^2/N or Pa^{-1})
β_B	Inclination angle of contact lines (°)
$\beta_{D,C}$	Intermediate angles (°)
β_a	Intermediate angle (°)
β_b	Base helix angle, helix angle at base circle (°)
β_b	Mean spiral angle, helix angle at reference cone at mid-face width of hypoid gear (°)
β_{bm}	Mean base spiral angle or base helix angle in mid-cone (°)
β_{bv}	Base virtual helix angle or helix angle at base circle of virtual cylindrical gear (°)
β_e	Form-factor helix angle, helix angle at the outer point of single tooth contact (°)
$\beta_{m1,2}$	Mean spiral angle or helix angle in mid-cone of pinion or wheel (°)
β_s	Helix angle of virtual crossed axes helical gear (°)

β_v — Helix angle of virtual gear for Method $B1$, virtual spiral angle for Method $B2$ (°)

β_w — Working helix angle (°)

$\beta(x, y)$ — Direction angle of steepest gradient (°)

β — Helix angle at reference or pitch cylinder (°)

γ' — Projected auxiliary angle for length of contact line (°)

$\gamma_{1,2}$ — Angle between tangential velocity vector of pinion or wheel and major axis of actual Hertzian contact area (°)

$\gamma_{1,2}$ — Angle of direction of tangential velocity of pinion or wheel (−)

γ_a — Auxiliary angle for tooth form and tooth correction factor (°)

γ_e — Tooth thickness half angle at outer point of single pair tooth contact (°)

γ_{m1} — Reference lead angle of worm (°)

γ — Auxiliary angle, auxiliary angle for length of contact line calculation with Method $B1$ (°)

Γ_A — Parameter on the line of action at point A (−)

Γ_{AA} — Parameter on the line of action at point AA (−)

Γ_{AB} — Parameter on the line of action at point AB (−)

Γ_{AU} — Parameter on the line of action at point AU (−)

Γ_B — Parameter on the line of action at point B (−)

Γ_{BB} — Parameter on the line of action at point BB (−)

Γ_D — Parameter on the line of action at point D (−)

Γ_{DD} — Parameter on the line of action at point DD (−)

Γ_{DE} — Parameter on the line of action at point DE (−)

Γ_E — Parameter on the line of action at point E (−)

Γ_{EE}	Parameter on the line of action at point EE (–)
Γ_{EU}	Parameter on the line of action at point EU (–)
Γ_M	Parameter on the line of action at point M (–)
Γ_y	Parameter on the line of action at arbitrary point (–)
Γ	Parameter on the line of action (–)
$\delta_{1,2}$	Deflection of bearing 1 or 2 in direction of load (μm or mm)
$\delta_{1,2}$	Pitch cone angle of pinion or wheel (°)
δ_S	Elongation on fracture (%)
δ_{Wlimn}	Limiting value of flank loss in normal section (mm)
δ_{Wn}	Flank loss in normal section (mm)
δ_a	Face angle (°)
$\delta_{a1,2}$	Face angle of pinion or wheel (°)
δ_{bth}	Combined deflection of mating teeth under even load distribution over the face width (μm)
δ_f	Root angle (°)
$\delta_{f1,2}$	Root angle of pinion or wheel (°)
δ_g	Difference in feeler gauge thickness measurement of mesh misalignment f_{ma} (μm)
δ_{ij}	Kronecker delta
δ_{lim}	Limiting value of worm shaft deflection (mm)
δ_m	Effective worm shaft deflection (mm)
δ	Deflection (μm)
δ	Pitch angle of bevel gear, reference cone angle (°)
$\delta(t)$	Displacement function (m)
$\Delta r_{y01,2}$	Relative distance from pitch circle through the point of load application of pinion and the wheel tooth centerline (–)
$\Delta\alpha_1$	Average pressure angle unbalance (°)
$\Delta\tau_{eff,L,RS,CP}(z)$	Influence of residual stresses on the local equivalent stress (N/mm^2)

$\Delta\tau_{eff,L,RS}(z)$	Residual stresses influencing the local equivalent shear stress (N/mm^2)
$\Delta\vartheta_2$	Wheel angle between fillet points of hypoid wheel (rad)
$\Delta\vartheta_{D,C}$	Wheel angles between centerline and fillet point on drive side/coast side (rad)
$\Delta\vartheta_{oil} = (\vartheta_{out} - \vartheta_{in})$	Difference between oil exit temperature and oil entrance temperature (°)
$\Delta\vartheta_1$	Pinion angle unbalance between fillet points (rad)
$\Delta\Sigma$	Shaft angle departure from 90° (°)
$\varepsilon_{1,2}$	Addendum contact ratio of pinion or wheel (–)
ε_1	Transformation strain at surface (–)
ε_2	Maximum transformation strain (–)
ε_N	Load sharing ratio for bending, for Method $B2$ (–)
ε_{NI}	Load sharing ratio for pitting, for Method $B2$ (–)
ε_a	Recess contact ratio (–)
ε_b	Lengthwise load sharing factor (–)
$\varepsilon_d(t)$	Dynamic transmission error or relative dynamic gear mesh displacement (mm)
ε_f	Approach contact ratio (–)
ε_f	Profile load sharing factor (–)
ε_{max}	Maximum addendum contact factor (–)
ε_{max}	Maximum value of addendum contact ratios of pinion and wheel (–)
ε_n	Contact ratio in normal section of virtual crossed axes helical gear (–)
$\varepsilon_s(t)$	Time-varying external displacement excitation (mm)
$\varepsilon_t(z)$	Instantaneous local transformation strain (–)
$\varepsilon_{v1,2}$	Tip contact ratio of virtual cylindrical pinion or wheel (–)
$\varepsilon_{v\alpha}$	Transverse contact ratio of virtual cylindrical gear (–)
$\varepsilon_{v\alpha n}$	Transverse contact ratio of virtual cylindrical gear in normal section (–)

$\varepsilon_{v\beta}$ Face contact ratio of virtual cylindrical gear (–)

$\varepsilon_{v\gamma}$ Virtual contact ratio for Method $B1$ or modified contact ratio for Method $B2$ (–)

ε_{α} Transverse contact ratio (–)

$\varepsilon_{\alpha n}$ Virtual contact ratio, transverse contact ratio of virtual spur gear (–)

ε_{β} Overlap ratio or face contact ratio (–)

ε_{γ} Total contact ratio (–)

ε Contact ratio (–)

ε Radius of a circular region of computational domain centered at load point (mm)

ζ_{A2} Specific sliding of driven wheel at point A of path of contact (–)

ζ_{E1} Specific sliding of driving pinion at point E of path of contact (–)

ζ_{R} Pinion offset angle in root plane (°)

ζ_{aw} Roll angle from working pitch point to tip diameter (°)

ζ_{fw} Roll angle from root diameter to working pitch point (°)

ζ_{m} Pinion offset angle in axial plane (°)

ζ_{mp} Offset angle of pinion and wheel in pitch plane (°)

ζ_{w} Average specific sliding of the actual gear (–)

ζ_{wT} Average specific sliding of the test gear (–)

ζ Roll angle (°)

η' Eyring effective lubricant viscosity (mPas)

η_{0} Absolute viscosity or dynamic viscosity at ambient pressure and temperature (Pas)

η_{1} Second auxiliary angle (°)

η_{38} Dynamic viscosity at 38 °C (Ns/m^2)

$\eta_{D,C}$ Intermediate factors (–)

η_{oil} Absolute viscosity or dynamic viscosity at oil temperature or at oil sump or spray temperature (mPas)

$\eta_{\vartheta B,Y}$ Dynamic viscosity at local contact temperature (Ns/m^2)

$\eta_{\vartheta M}$	Dynamic viscosity at bulk temperature (Ns/m^2)
$\eta_{\vartheta oil}$	Dynamic viscosity at oil inlet or sump temperature (Ns/m^2)
η	Effective dynamic viscosity of the oil wedge at the mean temperature of wedge (mPas)
η	Hertzian auxiliary coefficient (–)
ϑ_0	Ambient temperature (°C)
$\vartheta_{1,2}(t)$	Rotational vibration amplitude about center of the instantaneous point of contact on pinion or wheel tooth profile (°)
$\vartheta_{B,Y}$	Local contact temperature (°C)
$\vartheta_{D,C1}$	Wheel angle from centerline to pinion critical fillet point on drive side/coast side (rad)
$\vartheta_{D,C2}$	Wheel angle from centerline to wheel critical fillet point on drive side/coast side (rad)
$\vartheta_{D,CLS}$	Wheel angle between centerline and critical pinion drive side/coast side fillet point (rad)
ϑ_D	Wheel angle from centerline to tooth surface at pitch point on drive side (rad)
ϑ_{D0}	Wheel angle from centerline to pinion tip on drive side (rad)
ϑ_{D1}	Wheel angle from centerline to pinion tip on drive side (rad)
ϑ_M	Bulk temperature (°C)
ϑ_{M-C}	Test bulk temperature (°C)
ϑ_{MT}	Test bulk temperature (°C)
ϑ_S	Oil sump temperature (°C)
ϑ_{Slim}	Limiting value of oil sump temperature (°C)
$\vartheta_{Y1,2}$	Roll angle at current point of contact Y on pinion or wheel tooth profile (°)
$\vartheta_{f1,2}$	Dedendum angle of pinion or wheel (°)
$\vartheta_{fl,Y}$	Local flash temperature (°C)
ϑ_{mp}	Auxiliary angle for virtual face width, for Method $B1$ (°)
ϑ_{oil}	Oil sump, inlet or spray temperature (°C)

$\vartheta_{v1,2}$	Angular pitch of virtual cylindrical pinion or wheel (rad)
ϑ_{v2}	Angular pitch of virtual cylindrical wheel (radiant)
ϑ	Angle defining the position of characteristic plane with respect to a reference plane (°)
ϑ	Auxiliary quantity for tooth form and tooth correction factors (rad)
ϑ	Hertzian auxiliary angle (°)
ϑ	Numerical factor (–)
ϑ	Temperature (°C)
Θ_B	Contact temperature (°C)
Θ_{Bmax}	Maximum contact temperature (°C)
Θ_M	Bulk temperature or overall bulk temperature (°C)
$\Theta_{M1,2}$	Overall bulk temperature of pinion teeth or wheel teeth (°C)
Θ_{MT}	Test bulk temperature (°C)
Θ_{Mi}	Interfacial bulk temperature (°C)
Θ_S	Scuffing temperature (°C)
Θ_{Sc}	Scuffing temperature at long contact time (°C)
Θ_{fl}	Flash temperature (K)
Θ_{flaE}	Flash temperature at pinion tooth tip when load sharing is neglected (K)
Θ_{flaint}	Mean flash temperature (K)
$\Theta_{flainth}$	Mean flash temperature of hypoid gear (K)
$\Theta_{flaintT}$	Mean flash temperature of the test gear (K)
Θ_{flm}	Average flash temperature (K)
Θ_{flmax}	Maximum flash temperature (K)
Θ_{flmaxT}	Maximum flash temperature at test gear (K)
Θ_{int}	Integral temperature or integral temperature number (K)
Θ_{intP}	Permissible integral temperature (K)
Θ_{intS}	Scuffing integral temperature, allowable integral temperature or scuffing integral temperature number (K)
Θ_{oil}	Oil temperature before reaching the mesh area (°C)
$\lambda_{1,2}$	Dimensional constants (Pa^{-1})

λ_1	Angle between projection of pinion axis and direction of contact in pitch plane (°)
λ_2	Angle between projection of wheel axis and direction of contact in pitch plane (°)
$\lambda_{GF,Y}$	Local specific lubricant film thickness (–)
$\lambda_{GF,min}$	Minimum specific lubricant film thickness in contact area (–)
λ_{GFP}	Permissible specific lubricant film thickness (–)
λ_{GFT}	Limiting specific lubricant film thickness of test gears (–)
λ_M	Heat conductivity [N/(sK)]
$\lambda_{M1,2}$	Specific heat conductivity of pinion or wheel [N/(sK)] or [W/(mK)]
λ_c	Roughness profile cutoff wavelength (µm)
λ_f	Cutoff wavelength for a high-pass filtering for waviness profile (µm)
λ_r	Difference between the pinion mean spiral angle and virtual spiral angle of a hypoid gear (°)
λ_s	Primary profile cutoff wavelength (µm)
λ	First auxiliary angle (°)
λ	Proportionality factor between face width and module (–)
Λ	Specific lubricant oil film thickness, Bodensieck's index or Λ-index (–)
$\mu_{1D.C}$	Relative distance from centerline to tool critical drive side/coast side fillet point (–)
μ_b	Boundary coefficient of friction (–)
$\mu_{i(i=1,2)}$	Exponents (–)
μ_m	Mean coefficient of friction (–)
μ_{mC}	Mean coefficient of friction reduced to pitch point (–)
μ_{zm}	Mean tooth coefficient of friction (–)
μ	Coefficient of friction in pin-and-ring test (–)
μ	Coefficient of sliding friction, coefficient of kinetic friction or coefficient of friction (–)

v'	Effective Poisson's ratio (–)
v_0	Lead angle of cutter (°)
$v_{1,2}$	Poisson's ratio of pinion material or wheel material (–)
v_{100}	Kinematic viscosity of the lubricant oil at 100 °C (mm^2/s)
v_{40}	Kinematic viscosity of the lubricant oil at 40 °C (mm^2/s; cst)
v_{50}	Kinematic viscosity of the lubricant oil at 50 °C (mm^2/s; cst)
v_M	Kinematic viscosity of lubricant oil at ambient pressure and bulk temperature (mm^2/s)
$v_{i(i=1,2)}$	Exponents (–)
v_ϑ	Nominal lubricant kinematic viscosity at lubricant oil temperature (mm^2/s)
$v_{\vartheta B,Y}$	Kinematic viscosity at local contact temperature (mm^2/s)
$v_{\vartheta M}$	Kinematic viscosity at bulk temperature (mm^2/s)
v	Kinematic viscosity of the lubricant oil (mm^2/s; cst)
v	Poisson's ratio (–)
ξ_h	One half of angle subtended by normal circular tooth thickness at point of load application (rad)
ξ	Assumed angle in locating weakest section (rad)
ξ	Dimensional variable (–)
ξ	Hertzian auxiliary coefficient (–)
ξ	Local intrinsic coordinate for BEM (–)
ρ'	Slip layer thickness (mm)
ρ_0	Lubricant density at ambient pressure and temperature (kg/m^3)
ρ_0	Radius of relative profile curvature (mm)
$\rho_{1,2}$	Instantaneous radius of curvature of pinion or wheel with involute profiles (mm)
$\rho_{1,2}$	Relative profile radius of curvature of pinion or wheel (–)

$\rho_{1,2}(t)$	Instantaneous radius of curvature of tooth profile of pinion or wheel at a current point of contact Y (mm)
ρ_{15}	Density of lubricant at 15 °C (kg/m^3)
ρ_{CT}	Equivalent radius of curvature of the text gear at pitch point (mm)
ρ_Δ	Difference of relative radius of curvature between point of load application and mean point (–)
$\rho_{\Delta red}$	Variation of relative radius of curvature (–)
$\rho_{A1,2}$	Radius of curvature of pinion or wheel at point A of path of contact (mm)
$\rho_{B,D}$	Radius of curvature at inner point of single pair tooth contact B of pinion or D of wheel (mm)
ρ_C	Radius of curvature at the pitch point (mm)
ρ_{Cn}	Equivalent radius of curvature at pitch point in normal section (mm)
$\rho_{E1,2}$	Radius of curvature at tip of pinion or wheel (mm)
$\rho_{E1,2}$	Radius of curvature of pinion or wheel at point E of path of contact (mm)
ρ_F	Fillet radius at point of contact of 30° tangent (mm)
ρ_F	Root fillet radius at point of contact of 30° tangent (mm)
ρ_F	Tooth root radius at the critical section (mm)
ρ_{Fn}	Root fillet radius at point of contact of 30° tangent in normal section (mm)
$\rho_{M1,2}$	Material density of pinion or wheel (kg/m^3)
ρ_{a0}	Cutter edge radius (mm)
ρ_{fP}	Root fillet radius of basic rack for cylindrical gears or cutter edge radius (mm)
ρ_{fPn}	Root fillet normal radius of basic rack for cylindrical gears or cutter edge normal radius (mm)

ρ_{fPv}	Root fillet radius of virtual basic rack for cylindrical gears (mm)
ρ_g	Radius of grinding notch (mm)
$\rho_{m\beta}$	Lengthwise tooth mean radius of curvature (mm)
$\rho_{n,C}$	Normal radius of relative curvature at pitch diameter (mm)
$\rho_{n,Y}$	Normal radius of relative curvature at point, Y (mm)
$\rho_{n1,2}$	Radius of curvature at pitch point in normal section of pinion or wheel (mm)
ρ_{oil}	Lubricant density (kg/dm^3)
$\rho_{red,CP}$	Local radius of relative curvature in normal section at contact point (mm)
$\rho_{red,t,CP}$	Local radius of relative curvature in transverse section at contact point CP (mm)
ρ_{red}	Radius of relative curvature (mm)
ρ_{redC}	Relative radius of curvature at pitch point in transverse section (mm)
ρ_{rel}	Radius of relative curvature perpendicular to contact line at virtual cylindrical gears (mm)
ρ_{relC}	Transverse relative radius of curvature at pitch point (mm)
ρ_{rely}	Local relative radius of curvature at arbitrary point y (mm)
$\rho_{t,Y}$	Transverse radius of relative curvature at point, Y (mm)
ρ_t	Relative radius of profile curvature between pinion and wheel, for Method $B2$ (–)
$\rho_{t1,2,CP}$	Local transverse radius of curvature on pinion or wheel at contact point (mm)
$\rho_{t1,2,Y}$	Transverse radius of curvature of pinion or wheel at point, Y (mm)
ρ_{va0}	Relative edge radius of tool (–)
$\rho_{y1,2}$	Local radius of curvature at arbitrary point of pinion or wheel (mm)
$\rho_{\vartheta B,Y}$	Density of lubricant at local contact temperature (kg/m^3)

$\rho_{\vartheta M}$	Density of lubricant at bulk temperature (kg/m^3)
ρ	Density of gear material (kg/mm^3)
ρ	Density of lubricant oil (kg/m^3)
ρ	Equivalent radius of curvature or effective radius of curvature (mm)
ρ	Radius of curvature (mm)
$\sigma_{0.2}$	Proof stress at 0.2% permanent set (N/mm^2)
σ_0	Local static strength (N/mm^2)
σ_B	Tensile strength or tensile stress limit (N/mm^2)
σ_F	Findley stress (N/mm^2)
σ_F	Tooth root stress (N/mm^2)
σ_{F0}	Nominal tooth root stress (N/mm^2)
σ_{FE}	Allowable stress number for bending (N/mm^2)
σ_{FG}	Tooth root stress limit (N/mm^2)
σ_{FP}	Permissible bending stress or permissible tooth root stress (N/mm^2)
σ_{FPref}	Reference permissible bending stress (N/mm^2)
σ_{FPstat}	Static permissible bending stress (N/mm^2)
σ_{Fi}	Tooth root stress for bin i (N/mm^2)
σ_{Flim}	Nominal stress number for bending (N/mm^2)
σ_G	Fatigue stress limit (N/mm^2)
σ_H	Contact stress or Hertzian contact stress (N/mm^2)
σ_{H0}	Nominal contact stress at pitch point (N/mm^2)
σ_{HG}	Limiting value of mean contact stress (N/mm^2)
σ_{HG}	Pitting stress limit (N/mm^2)
σ_{HP}	Permissible contact stress (N/mm^2)
σ_{HT}	Hertzian stress of the test gear (N/mm^2)
σ_{Hi}	Contact stress for bin i (N/mm^2)
σ_{Hlim}	Allowable stress number for contact load (N/mm^2)
σ_{HlimT}	Pitting strength of test gears (N/mm^2)
σ_{Hm}	Mean contact stress (N/mm^2)
$\sigma_{RS,max}$	Maximum residual stress (N/mm^2)

$\sigma_{RS}(z)$	Tangential component of the residual stress at material depth, z (N/mm^2)
σ_S	Yield stress or yield point (N/mm^2)
$\sigma_{a,H}$	Hydrostatic stress amplitude acting on the characteristic plane (N/mm^2)
σ_a	Alternating stress or normal stress amplitude acting on the characteristic plane (N/mm^2)
σ_b	Bending stress (N/mm^2)
$\sigma_{bnom} = \sigma_{b\max}$	Nominal bending stress (N/mm^2)
σ_c	Compressive normal stress (N/mm^2)
σ_{cT}	Maximum value of compressive thermal stress (N/mm^2)
σ_e	Elastic limit (N/mm^2)
σ_e	Equivalent stress or total normal stress (N/mm^2)
$\sigma_{eff,a}$	Equivalent normal stress amplitude (N/mm^2)
$\sigma_{eff,m}$	Equivalent mean normal stress component (N/mm^2)
σ_{eff}	Effective or equivalent stress (N/mm^2)
σ_{falim}	Alternating tensile strength (N/mm^2)
σ_{fplim}	Pulsating tensile strength (N/mm^2)
$\sigma_{i(i=1,2,3)}$	Principal stresses (N/mm^2)
σ_i	Stress for bin i (N/mm^2)
σ_{klim}	Nominal notched-bar stress number for bending (N/mm^2)
σ_m	Mean normal stress (N/mm^2)
$\sigma_{n,\max}$	Maximum normal stress (N/mm^2)
σ_p	Proportional limit (N/mm^2)
σ_{plim}	Nominal plain-bar stress number for bending (N/mm^2)
σ_t	Equivalent stress (N/mm^2)
σ_z	Normal stress along direction of tooth axis (N/mm^2)
$\sigma_{\varphi\psi,\max}$	Maximum normal stress (N/mm^2)
$\sigma_{\varphi\psi a}$	Normal stress amplitude (N/mm^2)
$\sigma_{\varphi\psi e}$	Local equivalent stress in a defect intersection plane (N/mm^2)
$\sigma_{\varphi\psi m}$	Mean normal stress component (N/mm^2)
σ	Normal stress (N/mm^2)

σ	Standard deviation of height of surface from the datum line (μm)
Σ	Shaft angle, crossing angle of virtual crossed axes helical gear ($°$)
τ_0	Lubricant reference shear stress (N/mm^2)
τ_F	Nominal shear stress at tooth root (N/mm^2)
τ_{FG}	Limiting value of shear stress at tooth root (N/mm^2)
τ_{FlimT}	Shear endurance strength of test gears (N/mm^2)
τ_H	Shear stress due to Hertzian contact pressure (N/mm^2)
τ_a	Shear stress amplitude (N/mm^2)
$\tau_{a1,2}$	Amplitudes of shear stress component acting on the characteristic plane (N/mm^2)
τ_b	Boundary shear stress (N/mm^2)
$\tau_{crit,core}$	Critical shear stress at core (N/mm^2)
$\tau_{crit,surface}$	Critical shear stress at surface (N/mm^2)
$\tau_{crit}(z)$	Critical shear stress (N/mm^2)
$\tau_{eff,CP}(z)$	Local equivalent stress (N/mm^2)
$\tau_{eff,EL,RS}(z)$	Local equivalent shear stress due to external loads and residual stresses (N/mm^2)
$\tau_{eff,L,CP}(z)$ or $\tau_{eff,L}(z)$	Local equivalent shear stress without consideration of residual stresses (N/mm^2)
$\tau_{eff,RS}(z)$	Quasi-stationary residual shear stress (N/mm^2)
$\tau_{eff,a}$	Equivalent shear stress amplitude (N/mm^2)
$\tau_{eff,m}$	Equivalent mean shear stress component (N/mm^2)
τ_{eff}	Shear stress intensity (N/mm^2)
$\tau_{eff}(z)$	Local equivalent shear stress (N/mm^2)
τ_{falim}	Alternating torsional strength (N/mm^2)
τ_{fplim}	Pulsating torsional strength (N/mm^2)
τ_m	Average shear stress (N/mm^2)

τ_m	Lubricant mean viscous shear stress (N/mm^2)
τ_{oct}	Octahedral shear stress (N/mm^2)
$\tau_{per,CP}(z)$ or $\tau_{per}(z)$	Local material shear strength (N/mm^2)
$\tau_{residual}(z)$	Local residual shear stress (N/mm^2)
τ_v	Viscous shear stress (N/mm^2)
$\tau_{v1,2}$	Viscous shear stress on pinion or wheel tooth surface (N/mm^2)
$\tau_{\varphi\psi,a}$	Shear stress amplitude (N/mm^2)
$\tau_{\varphi\psi m}$	Mean shear stress component (N/mm^2)
τ	Angle between tangent of root fillet at weakest point and tooth centerline (°)
τ	Angular pitch (rad)
τ	Shear stress (N/mm^2)
φ_E	Run-in grade (–)
φ_R	Auxiliary angle for calculation of ζ_R (°)
φ	Auxiliary angle to determine the position of the pitch point (°)
φ	Elastic deflections (mm)
φ	Shaft angle of virtual crossed axes helical gear (°)
φ	Zenith angle or colatitude (°)
Φ	Quill shaft twist (°)
χ^*	Relative stress gradient in the notch root of a gear wheel (mm^{-1})
χ_P^*	Relative stress gradient in a smooth polished test piece (mm^{-1})
χ_T^*	Relative stress gradient at notch root of reference test gear $\vert$(mm^{-1})
χ_T^X	Relative stress gradient at notch root of standardized test gear $\vert$(mm^{-1})
χ^X	Relative stress gradient at notch root (mm^{-1})
χ_p'	Relative stress gradient in a smooth polished test piece (mm^{-1})
χ_α	Factor characterizing the transverse base pitch deviation after running-in (–)
χ_β	Factor characterizing the equivalent misalignment after running-in (–)

χ	Running-in factor (–)
ψ	Auxiliary angle (°)
ψ	Azimuthal angle (°)
ψ	Quite small length (mm)
$\omega_{1,2}$	Angular velocity of pinion or wheel (rad/s)
ω_Σ	Angle between the sum of velocities vector and tangent to the trace of pitch cone (°)
ω	Angular velocity (rad/s)

Chapter 1
Load Carrying Capacity of Spur and Helical Gears: Influence Factors and Load Analysis

Abstract In this chapter, the main factors (application factor, dynamic factor, face load factors and transverse load factors) that influence the load carrying capacity of spur and helical gears are first defined. The various methods of calculating all these factors are then described, focusing attention on the main quantities they depend on and how to take them into account. In this framework, various problems are addressed, such as: resonance problems, to this end defining the possible operating ranges of a gear pair in relation to its natural frequency; stiffness problems, in this regard defining and quantifying the various stiffnesses that come into play; numerous sources of misalignment, related to the inevitable manufacturing and assembly errors and to torsion and bending elastic deflections under load of the structural members involved; beneficial effects of running-in and intentional profile and flank line modifications; etc. Finally, the focus is on load distribution between teeth pairs in simultaneous meshing and load variability defined by any load spectrum.

1.1 Introduction

As we mentioned in Vol. 1, Sect. 1.1, the *load carrying capacity* of the gears involves the analysis of the static and fatigue strength (tooth bending strength, and surface durability, in terms of pitting and micropitting) as well as the scuffing and wear strength, with the purpose of ensuring the predetermined lifetime under the expected operating conditions. There are some important limiting factors in specifying the load carrying capacity of any gear. The main problems of resistance and related influencing factors can be summarized as follows (for a more complete overview of possible gear damages, see Shipley 1967; JGMA 7001-01:1990; ISO 10825:1995; JIS B0160:1999):

- *Pitting* and *micropitting*, i.e. *fatigue failure* at the tooth flank surfaces.
- *Fatigue breakage*, i.e. *fatigue failure at tooth root*, due to bending load.
- *Tooth flank fracture*, i.e. *fatigue failure at tooth flank* that is differentiated both from the classical fatigue breakage due to bending load and from the classical pitting damage.
- *Overload breakage*, i.e. failure of the teeth, due to static load.

© Springer Nature Switzerland AG 2020
V. Vullo, *Gears*, Springer Series in Solid and Structural Mechanics 11,
https://doi.org/10.1007/978-3-030-38632-0_1

1

- *Scuffing*, i.e. *adhesive wear* of the tooth flank surfaces.
- *Wear*, in the strict sense, i.e. *abrasive wear* of the tooth flank surfaces.
- *Corrosive wear*, i.e. tooth surface deterioration due to chemical action.
- *Fretting corrosion*, which is a combined type of adhesive wear, abrasive wear, and corrosive wear, due to small amplitude vibrations and related relative small motions between the tooth flank mating surfaces.
- *Plastic yielding*, i.e. surface and subsurface plastic deformation of tooth flanks, due to high contact stress and sliding and rolling action during the meshing cycle.
- *Damage from power losses* by friction, and heat generated during operation.
- *Damage from vibrations*, because of heavy loads or high speeds, especially *hammering* between teeth, and consequent noisiness.
- Other types of damage, some of which may be considered as subclasses falling under those just summarized, while others configure specific types of damage, generally considered of minor importance.

All these resistance problems and influence factors are a function of the loads acting on the teeth, and for this reason an accurate determination of these loads becomes very important (see Panetti 1937; Ferrari and Romiti 1966; Scotto Lavina 1990). These loads are almost always dynamic loads, in that they vary during the meshing cycle as well as depending on changing operation conditions during the lifetime. Furthermore, they are repeated with a frequency that depends on the number of revolutions per minute of the gear wheels. The exact evaluation of these loads is still a problem, which has not yet been solved rigorously. The various factors that appear in the equations commonly used for strength calculations of the gears constitute a clear evidence of the fact that the road to the exact solution of this problem is still long and bumpy (see Buckingham 1949; Dudley 1962; Giovannozzi 1965b; Pollone 1970; Merritt 1971; Henriot 1979; Niemann and Winter 1983; Townsend 1991; Maitra 1994; Radzevich 2016).

Generally, the concept design of transmission drives with cylindrical spur and helical gears is based on prior experience, for which the main dimensions are known or can be extrapolated from these experiences. Therefore, the concept design of cylindrical spur and helical gears is set on this basis. Subsequently, the dimensions so obtained are verified, so as to ensure the required load capacity in terms of surface durability, tooth bending strength, scuffing and wear strength, etc. The starting point is constituted by the technical specifications, which summarize the main input data of these types of gear drives, including those of design, manufacturing and operation.

In this framework, by a calculation of first approximation, the initial dimensions, teeth characteristics and schematic structural drawing of kinematic operation, including the configuration of the housing, bearings that support the shafts, seals, type of lubrication, etc., are first determined. The load carrying capacity is therefore verified and, when this last does not give sufficient guarantees in terms of safety, the design is modified until reaching the solutions that satisfy the predetermined input data.

The *tangential load* that a gear can transmit depends on the size and shape of the teeth, tangential velocity of the same gear, lubrication and cooling conditions of the teeth, errors of manufacturing and assembly, stiffnesses of the teeth, shafts and housing, etc. It is not easy to carefully consider all the factors that influence the tangential load. We here follow the directives of ISO 6336-1:2006, concerning the basic principles to determine the appropriate values of the general *influence factors*, which come into play in the calculations of the load carrying capacity of spur and helical gears. It should be borne in mind that these directives may be applied within well-defined areas of practical applications, and with some equally well-defined limitations. On this important subject, we refer directly to the aforementioned ISO standard.

This ISO standard distinguishes the influence factors into two classes. The first class consists of the influence factors determined by gear geometry, or established by convention, and then calculated in accordance with equations specially formulated. The second class consists of the influence factors affected simultaneously by different influences, which are calculated as if these influences were independent of each other, for which, given this mutual influence, their numerical values cannot be uniquely defined. This last class of influence factors includes the factors $K_A, K_v, K_{H\alpha}, K_{H\beta}$ or $K_{F\alpha}$ (see below), and factors influencing the allowable stress. Some of these factors, such as $K_v, K_{H\beta}$ and $K_{H\alpha}$, are affected by the profile and helix modifications, which however must be taken into consideration only when the magnitudes of these modifications are significantly larger than the manufacturing deviations.

For the determination of the influence factors, ISO provides three different methods, called respectively Method A, Method B, and Method C. As regards their accuracy and reliability, Method A is superior to Method B, and Method B is superior to Method C. To avoid misunderstandings, in cases where some doubt may exist, the influence factors determined with these three methods are designated with subscripts A, B or C, and in the case of alternative choices within the same method, with alphanumeric subscripts (e.g., $A1$, $A2$, etc.).

At the primary design stage, the data available are limited. It is so necessary to make use of approximations or empirical values for some factors. In some application fields or for rough calculations, any influence factor can be taken equal to 1 or having a constant value different from 1. Furthermore, in such cases, conservative safety factors must be selected. Of course, a more precise evaluation of the influence factors can be made only when the gear drive has been manufactured and tested, and data obtained by direct measurements are available.

With Method A, the influence factors are derived from results of full-scale load tests, and precise measurements or extensive and comprehensive mathematical analysis of the gear set, on the basis of proven operating experience, or any combination of these methodologies. The accuracy and reliability of this method are obviously out of the question, because it is based on the availability of all data concerning the gear set under consideration and its actual load conditions.

With Method B, the influence factors for most practical applications are derived with sufficient accuracy on the basis of assumptions that must be clearly defined, proving their applicability to operating conditions of the gear drive being planned and designed. Finally, with Method C, the influence factors are derived on the basis of simplified assumptions, which must also be clearly defined, ensuring their applicability to the existing conditions of the gear set under consideration.

Factors K_v, $K_{H\beta}$ or $K_{F\beta}$, and $K_{H\alpha}$ or $K_{F\alpha}$ depend on the *nominal tangential load*, F_t. They are to some extent interdependent, and should be calculated according to the following sequence, which does not allow exceptions: as a first step, we determine K_v, with the load $F_t K_A$; as a second step, we calculate $K_{H\beta}$ or $K_{F\beta}$, with the load $F_t K_A K_v$; as a third and last step, we determine $K_{H\alpha}$ or $K_{F\alpha}$, with the load $F_t K_A K_v K_{H\beta}$ or $F_t K_A K_v K_{F\beta}$. For the determination of $K_{F\alpha}$, the factor $K_{H\beta}$ can be also used, since for tooth bending this factor represents the determinant load due to uneven distribution of F_t over the face width. In the case where a gear wheel drives two or more mating gear wheels, it is necessary to use the product $(K_A K_\gamma)$ instead of K_A.

The determination of the influence factors is based on the nominal tangential load, F_t, also called *transmitted load*, which is defined as the useful component of force that is transferred from one gear wheel to another during action. This tangential force is determined in the transverse plane at the reference cylinder, and is derived by the *nominal torque*, T, or *nominal transmitted power*, P, transmitted by the gear pair. As nominal torque we assume the effective input torque to the driven machine, which is the torque corresponding to the heaviest regular working condition. Alternatively, as nominal torque we can assume the nominal torque of the prime mover, when it corresponds to the torque requirement of the driven machine.

The quantities F_t, T, and P are defined as the nominal tangential load, nominal torque, and nominal transmitted power per mesh, i.e. for the mesh under consideration; their units are respectively N, Nm, and kW. The relationships that correlate these three quantities to the angular velocity, ω (rad/s), rotational speed, n (s^{-1} or min^{-1}), tangential velocity, v (m/s), and diameter of the reference circle, d (mm) are as follows:

$$F_t = \frac{2000 T_{1,2}}{d_{1,2}} = \frac{1000 P}{v} \tag{1.1}$$

$$T_{1,2} = \frac{F_t d_{1,2}}{2000} = \frac{1000 P}{\omega_{1,2}} \tag{1.2}$$

$$P = \frac{F_t v}{1000} = \frac{T_{1,2} \omega_{1,2}}{1000} \tag{1.3}$$

$$v = \frac{d_{1,2} \omega_{1,2}}{2000} = \frac{\pi d_{1,2} n_{1,2}}{60 \times 10^3} \simeq \frac{d_{1,2} n_{1,2}}{19.099 \times 10^3} \tag{1.4}$$

$$\omega_{1,2} = \frac{2000 v}{d_{1,2}} = \frac{2\pi n_{1,2}}{60} \simeq \frac{n_{1,2}}{9.549}, \tag{1.5}$$

where the numerical coefficients homogenize the units of measurement in the SI-System.

In these relationships, as usual, the subscripts 1 and 2 refer respectively to the pinion and wheel. Of course, when the above relationships are used to assess the load carrying capacity of a gear, all loads acting on it must be considered. It is also necessary to consider the type of gear. For example, in the case of double-helical gears, it is assumed conventionally that the total tangential load is divided equally between the two helices, unless this conventional load distribution is in contrast to the externally applied axial loads. In this case, the two halves of the double-helical gear wheels have to be regarded as two separate helical gear wheels arranged in parallel. Moreover, in the case of multi-path transmission drives, to take account of the fact that the total tangential load is not quite evenly divided between the various load paths, it is necessary to introduce a *mesh load factor*, K_γ, to be determined possibly by means of experimental measurements, or to be estimated using data derived from experience or from consolidated literature.

In many practical applications, the transmitted load is not uniform. In this case, the assessing of load carrying capacity of a gear must be carried out based on the *equivalent tangential load*, *equivalent torque*, and *equivalent transmitted power*. For the determination of these equivalent quantities, which characterize the non-uniform transmitted load, it is necessary to consider the *duty cycle* that can be represented by a *load spectrum*. In these cases, the cumulative fatigue effect of the duty cycle is considered. On this subject, we will return in Sect. 1.8.

Finally, to determine the safety from pitting damage and from sudden tooth breakage due to loads corresponding to the static stress limit, it is necessary to consider the *maximum tangential load*, F_{tmax}, *maximum torque*, T_{max}, and *maximum transmitted power*, P_{max}. They represent the maximum values of F_t, T, and P in the variable duty range. To limit these maximum values, it is appropriate to use a suitable safety clutch (Giovannozzi 1965a).

1.2 Application Factor, K_A

The *application factor*, K_A, takes into account the fact that the nominal tangential load, F_t, calculated with Eq. (1.1) undergoes considerable increases, due to external dynamic actions exerted by the driven and driving machines on the gear set to be designed and calculated. These additional dynamic loads, which increase the nominal tangential load, F_t, are largely dependent not only on the characteristics of the driven and driving machines, but also on the masses and stiffnesses of the mechanical system under the operating conditions, including shafts and couplings.

For special applications, such as marine gears and similar gears, which are subject to cyclic peak torque (and thus to torsional vibrations), and are designed for infinite life, the application factor is defined as the quotient of the peak cyclic torque divided by the nominal rated torque, which is determined by the rated power and speed. In cases where the gears are subjected to a limited number of known loads in excess of the amount of the peak cyclic torque, the additional dynamic loads can be evaluated by means of cumulative fatigue or by means of an increased application factor related to the load spectrum (see Palmgren 1924; Miner 1945; Fatemi and Yang 1998).

With Method A, the application factor, K_{A-A}, is determined by means of careful measurements and a comprehensive analysis of the entire mechanical system including the to-be-designed gear drive, or on the basis of reliable operational experiences

Table 1.1 Application factor, $K_A = K_{A-B}$

Working characteristic of driving machine	Working characteristic of driven machine			
	Uniform	Light shocks	Moderate shocks	Heavy shocks
Uniform	1.00	1.25	1.50	1.75
Light shocks	1.10	1.35	1.60	1.85
Moderate shocks	1.25	1.50	1.75	2.00
Heavy shocks	1.50	1.75	2.00	≥ 2.25

in the field of application of interest. This calculation method is described in detail in the ISO 6336-6:2006, to which we refer the reader.

When reliable data on which to base the Method A are not available, as it happens especially when we are in the early stages of design, we can use the Method B. According to this method, the empirical guideline values of the application factor, K_A (in this case we should write K_{A-B}) shown in Table 1.1, taken from the ISO 6336-6:2006, can be used.

These values only apply to gear drives that operate outside the resonance speed under relatively steady-state loading. On the other interesting directives of the ISO 6336-6:2006, it is advisable to refer to the same ISO standard.

1.3 Dynamic Factor, K_v

1.3.1 Generality

The *dynamic factor*, K_v, also called *internal dynamic factor*, takes into account the internal effects, typical of the gear drive under consideration; these effects depend of the accuracy grade of gear teeth and are related to load and speed. This factor has values that are lower the higher the accuracy grade, and vice versa. The internal dynamic load acting on the gear teeth is influenced by both manufacturing quality and design choices, for which not only the accuracy grade of the gear, but also the intentional modifications of teeth profiles and flank lines come into play.

We can define the internal dynamic factor as the quotient of the total mesh torque at operating speed divided by the mesh torque with *perfect gears*, whereas perfect gears we mean gears having zero quasi-static transmission error at the nominal transmitted mesh torque, i.e. the design torque. As an alternative to individual calculation of the application factor, K_A, and internal dynamic factor, K_v, we can introduce their product $(K_A K_v)$, defined as the quotient of the total mesh torque at operating speed divided by nominal transmitted mesh torque (or design torque).

The parameters affecting internal dynamic load and related calculations are numerous and of different nature. They include design and manufacturing parameters, parameters that determine *transmission perturbance*, parameters that influence the dynamic response of the gear system, and parameters related to operating conditions that can trigger resonance.

The design parameters include pitch line velocity, tooth load, inertia and stiffness of the rotating members, tooth stiffness variation, lubricant properties, stiffness of bearing and housing, critical speeds and internal vibration within the gear itself.

The manufacturing parameters include pitch deviations, runout of reference surfaces with respect to the axis of rotation, tooth flank profile and lengthwise deviations, compatibility of the mating toothed members, balance of parts, bearing fit and preload.

The parameters involved in the transmission perturbance are those that determine dynamic tooth loads resulting from vibration phenomena in response to the excitation known as *transmission error*. We define as transmission error any departure from the ideal kinematics of a mating gear pair, which requires a constant ratio between the input and output rotations, i.e. a uniform relative angular motion. This error is influenced by all deviations from the ideal gear tooth shape as well as by pitch and spacing deviations, due to incorrect design choices and manufacturing inaccuracies of the gear, and to operating conditions of the gear itself. These latter include:

- Pitch line velocity, as the frequencies of excitation depend on this quantity as well as on the module.
- Mesh stiffness variations during the meshing cycle of the gear (this source of excitation is especially pronounced in parallel cylindrical spur gears, while the stiffness variations are smaller in parallel cylindrical spur and helical gears having respectively transverse contact ratio $\varepsilon_\alpha > 2.0$ and total contact ratio $\varepsilon_\gamma > 2.0$).
- Transmitted tooth load (since deflections depend on the load, tooth profile modifications can be designed to give uniform velocity ratio only for one value of load, i.e. the design load; therefore, a load different from this value will give increased transmission error).
- Dynamic unbalance of the gears and shafts.
- Application environment (transmission error increases with the excessive wear and plastic deformation of the tooth flank surfaces; in this respect, an appropriate design should be made of the lubrication system, enclosure and seals to maintain a safe operating temperature and a contamination-free environment).
- Shaft alignment, as gear tooth alignment is affected by load and thermal deformations of the gears, shafts, bearings and housing.
- Excitation induced by teeth friction.

The parameters that influence the dynamic tooth loads and, more generally, the dynamic response of the gear drive include mainly:

- Mass of the gears, shafts, and other main internal components.
- Stiffness of the gear teeth, gear blanks, shafts, bearings and housing.
- Damping (the main sources of damping are the seals and shaft bearings, as well as hysteresis of the gear shafts and viscous damping at sliding interfaces and shaft couplings).

Finally, high dynamic tooth loads can be generated when the frequencies of excitation, such as tooth meshing frequency and its harmonics, coincide or nearly coincide with a natural frequency of vibration of the gear system. In these cases, resonance phenomena occur and, if the magnitude of internal dynamic loads at a speed involving resonance becomes large, operation near this resonance speed should be avoided.

We must distinguish between two types of resonance: the gear blank resonance, and system resonance. The first type of resonance may take place in lightweight gear blanks at high speed. The related resonance deflections can cause high dynamic tooth loads. In addition, in some cases, plate or shell vibration modes can be triggered (see Krall 1970b; Warburton 1976; Ventsel and Krauthammer 2001), thus being able to determine the failure of the gear blank. Factors K_v evaluated by Methods B and C described below do not take account of gear blank resonance. The second type of resonance takes account of the fact that a gear transmission is a complex mechanical system composed not only of the gearbox, but also of the driving and driven machines and interconnecting shafts and couplings (see Biezeno and Grammel 1953; Ker Wilson 1956, 1963; Giovannozzi 1965b; Den Hartog 1985). The dynamic response of the gear system depends on its configuration. When resonance conditions can be triggered, the system can become critical. In these cases, a detailed dynamic analysis of the system is necessary using appropriate methods, which furthermore allow evaluating the combined effects of the two factors K_A and K_v.

In descending order of accuracy, the dynamic factor K_v can be determined using Method A, Method B, and Method C, respectively obtaining K_{v-A}, K_{v-B}, and K_{v-C}. A unit value of this factor indicates that the gear is characterized by an optimum profile modification appropriate to the load, large overlap ratio, even load distribution over the face width, high accuracy grade of the teeth, and high *specific tooth load*, defined as $(K_A F_t)/b$.

It is to be noted that, in gear trains including multiple mesh gears (e.g., idler gears, and planet, sun, and annulus gears of planetary gear trains), several natural frequencies exist, which can be higher or lower than the natural frequency of a single gear pair characterized by only one mesh. In these cases, resonance of higher natural frequencies occur, and may lead to a higher value of K_v when such gears run in the supercritical range.

It is also to be noted that transverse vibrations of the shaft-gear systems influence the dynamic load. Due to the transverse compliance of these systems, coupled vibrations can be generated where the pinion and/or the gear wheel combine torsional and lateral vibrations. In these cases, more natural frequencies than the one related to the single gear pair formed by just a pinion and a gear wheel occur, and they may lead to a higher value of K_v.

1.3.2 Method A for Determination of K_{v-A}

With this method, we consider the general transmission system, consisting of gearbox, driving and driven machines, and interconnecting shafts and couplings. For this system, we determine the maximum tooth loads, including the dynamic additional loads internally generated and their uneven distributions, by measurements or by a comprehensive dynamic analysis (see Biezeno and Grammel 1953; Tenot 1953; Ker Wilson 1956, 1963; Buzdugan 1964; Giovannozzi 1965b; Krall 1970a, b; Warburton 1976; Garro and Vullo 1978a, b, 1979; Carmignani 2001). In this way, the internal dynamic factor K_v (as well as $K_{H\alpha}$ and $K_{F\alpha}$) is assumed to have a unit value ($K_v = 1$).

Factor K_v can also be assessed by comparing the results of measurements of the tooth root stresses of the gears when they transmit load at working speed and at a lower speed. It can be also determined by a comprehensive analytical procedure, supported by experience of similar designs, which can be found in the literature. Mathematical models satisfactorily verified by measurements can also provide reliable values of the dynamic factor K_v (see, for example: Houser and Seireg 1970; Seireg and Houser 1970; Cantone et al. 2001; Li and Kahraman 2013).

1.3.3 Method B for Determination of K_{v-B}

This method is based on the following simplifying assumptions:

- A multiple-stage gear drive, however it is configured, is considered to be constituted by single stage gear pairs between them independent, for which the influence on a single stage of the other stages is ignored. This assumption is sufficiently valid if the torsional stiffness of the shafts connecting the gear wheel of one stage with the pinion of the next stage is relatively low.
- The single stage gear pair to be designed is regarded as an elementary vibrating system with one degree of freedom, consisting of a relative mass, representative of the masses of the pinion and wheel, and a spring whose stiffness is the mesh stiffness of the contacting teeth.
- An average value of the gear mesh damping is considered, while other sources of damping such as friction at interfaces between gear members, hysteresis, bearings, couplings, etc., are not taken into consideration (see Lazan 1968; Marguerre and Wölfer 1979; Winter and Kojima 1981; Calderale et al. 1984; Den Hartog 1985; Orban 2011).

In accordance with these assumptions, the dynamic factor K_v assessed by this method does not take into account the loads due to torsional vibrations of shafts and coupled masses, which are then to be included within other externally applied loads (application factor K_A takes into account these loads). Furthermore, given the aforementioned additional sources of damping that are neglected, the actual dynamic loads acting on teeth are usually somewhat smaller compared to those calculated by means of this method (this does not apply in the range of main resonance).

Although less accurate than Method A, this method is suitable for all types of gear drives (spur and helical gears with any basic rack profile and any accuracy grade) and, in principle, for all operating conditions. However, there are restrictions for some fields of application and operation, which we will describe from time to time.

As it is well known, the vibrating behavior of an oscillating system is basically determined by the position of the *excitation frequency* with respect to its *natural frequency* (see Thomson 1965; Den Hartog 1985; Géradin and Rixen 2015). In our case, the frequency of the main excitation or frequency of contact between the teeth, expressed in terms of quantities related to the pinion, is given by the following relationship:

$$f_1 = \frac{\omega_1}{2\pi} z_1 = \frac{n_1}{60} z_1, \tag{1.6}$$

where ω_1, n_1, and z_1 are respectively the angular velocity, rotational speed and number of teeth of the pinion. The natural frequency, f_{E1}, of the single gear pair, considered as a single-degree-of-freedom vibrating system (Niemann and Winter 1983), is given by the relationship:

$$f_{E1} = \frac{1}{2\pi} \sqrt{\frac{c_{\gamma\alpha}}{m_{red}}}, \tag{1.7}$$

where $c_{\gamma\alpha}$ and m_{red} are respectively the mean values of *mesh stiffness per unit face width* and *reduced gear pair mass per unit face width* referenced to the line of action. The natural frequency is expressed in $s^{-1} = Hz$ or min^{-1}, and is often called *resonance running speed* or simply *resonance speed*.

The *frequency ratio*, N, between the two frequencies above, taking into account that the units of measure of $c_{\gamma\alpha}$ and m_{red} are respectively [N/(mm μm)] and (kg/mm), can be expressed as:

$$N = \frac{f_1}{f_{E1}} = \frac{n_1}{n_{E1}} = \frac{\pi n_1 z_1}{30 \times 10^3} \sqrt{\frac{m_{red}}{c_{\gamma\alpha}}}, \tag{1.8}$$

where $n_{E1} = (60 f_{E1}/z_1)$ is the rotational speed corresponding to the natural frequency, f_{E1}.

Since this frequency ratio is an important reference indicator in order to assess and avoid possible resonance conditions, it is also called *resonance ratio*.

The *reduced mass*, m_{red}, is the relative mass of a gear pair per unit face width, referred to its base radius or to the line of action; it is given by:

$$m_{red} = \frac{m_1 m_2}{m_1 + m_2} = \frac{J_1 J_2}{J_1 r_{b2}^2 + J_2 r_{b1}^2}, \tag{1.9}$$

where $J_1 = m_1 r_{b1}^2$ and $J_2 = m_2 r_{b2}^2$, and m_1 and m_2 are respectively the mass moments of inertia per unit face width (in kg mm²/mm), and the individual gear masses per unit face width referenced to the line of action (in kg/mm) of the pinion and wheel (see Timoshenko and Young 1951). This last equation applies to the usual cases, where the diameter of the pinion shaft is less than the diameter of the reference pitch circle. For the evaluation of m_{red}, five special cases are then considered (pinion on large diameter shaft; two neighboring gears rigidly connected together; one big wheel driven by two pinions; simple planetary gears; idler gears), on which we do not think of having to linger. For these cases, we refer the reader directly to the ISO 6336-1:2006. For the calculation of $c_{\gamma\alpha}$, we refer to the Sect. 1.6.

Approximate but sufficiently accurate values of the reduced mass of gear pairs with external teeth can be derived from the following relationship:

$$m_{red} = \frac{\pi}{8}\left(\frac{d_{m1}}{d_{b1}}\right)^2 \frac{d_{m1}^2}{\dfrac{1}{\rho_1\left(1 - q_1^4\right)} + \dfrac{1}{\rho_2 u^2\left(1 - q_2^4\right)}}, \tag{1.10}$$

which applies to external spur gears and external single helical or double helical gears configured as Fig. 1.1 shows. In this relationship, which ignores the masses of web and hub because of their negligible influence on the moment of inertia, d_{b1} is the base diameter of the pinion, $d_{m1} = \left(d_{a1} + d_{f1}\right)/2$, $d_{m2} = \left(d_{a2} + d_{f2}\right)/2$, $q_1 = d_{i1}/d_{m1}$, $q_2 = d_{i2}/d_{m2}$, while u is the gear ratio and ρ_1 and ρ_2 are the densities of materials. For solid pinions and wheels, we have: $\left(1 - q_1^4\right) = 1$, and $\left(1 - q_2^4\right) = 1$. For a gear rim whose rim width differs from the face width, the determination of $\left(1 - q_1^4\right)$ or $\left(1 - q_2^4\right)$ is only valid where the masses of the rim are directly connected to the gear rim. In the same relationship, masses positioned at greater distance on the same shaft are ignored, because the stiffness of the interconnecting shaft is usually of minor significance compared to tooth stiffness.

From a theoretical point of view, the entire running speed range can be divided into three ranges: the subcritical range ($N < 1$), main resonance range ($N = 1$), and supercritical range ($N > 1$). However, in reality, the actual natural frequency of the gear pair can be above or below the one calculated with Eq. (1.7), since the latter, as we have already pointed out above, does not take account of some stiffnesses, which

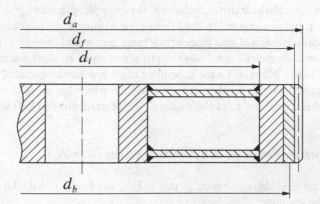

Fig. 1.1 Gear configuration and diameters that appear in Eq. (1.10)

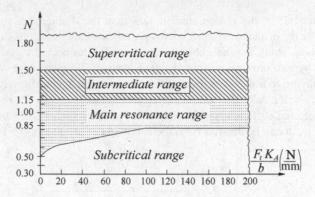

Fig. 1.2   Division of the entire range of the frequency ratio in four sub-ranges

exert a certain influence (e.g., the stiffness of shafts, bearings, housing, etc.), as well as of the damping. For reasons of safety, the frequency ratio, N, which defines the main resonance range, cannot be equal to 1, but it is between a lower limit, N_s, which is variable with the specific load $F_t K_A / b$ (in N/mm), and an upper limit, which is a constant equal to 1.15. For specific loads $(F_t K_A / b) < 100$ N/mm, N_s is given by:

$$N_s = 0.50 + 0.35 \sqrt{\frac{F_t K_A}{100b}}, \tag{1.11}$$

while for specific loads $(F_t K_A / b) \geq 100$ N/mm, N_s is a constant equal to 0.85. For the same reasons, the supercritical range is divided into an intermediate range $1.15 < N < 1.50$, and a real supercritical range $N \geq 1.50$. As Fig. 1.2 shows, the entire running range, represented by the frequency ratio, N, is thus divided into four sub-ranges, to each of which a different procedure of calculation of K_v is related.

What we have said above about the determination of resonance speed refers to common gear designs. However, the designer must often determine the resonance speed for less common gear drives including, for example: planetary gear trains with annulus gear rigidly connected to the housing or with rotating annulus gear; idler gears; two rigidly connected gear wheels; one large gear wheel driven by two pinions; pinion shaft with diameter at mid-tooth depth about equal to shaft diameter; etc. in these special cases, Method A should be used to determine resonance speed. Other less approximate methods can however be used, on which we do not consider here to dwell. In this regard, we refer the reader to the aforementioned ISO standard.

1.3.3.1   Dynamic Factor K_v in Subcritical Range $(N \leq N_s)$

Most industrial gear drives operate in this range, where K_v is calculated with the following relationships:

$$K_v = (NK) + 1 \tag{1.12}$$

$$K = (C_{v1} B_p) + (C_{v2} B_f) + (C_{v3} B_k). \tag{1.13}$$

In these equations, C_{v1}, C_{v2}, and C_{v3} are three dimensionless factors, which depend on the total contact ratio, ε_γ, and take into account, respectively, the pitch deviation effects, tooth profile deviation effects, and cyclic variation effects in mesh stiffness. Their values may be obtained as a function of ε_γ using the corresponding curves shown in Fig. 1.3, or can be read in Table 1.2, or determined by the relationships listed in the same Table 1.2. In Eq. (1.13), B_p, B_f, and B_k are three dimensionless parameters that take into account the effects of tooth deviations and profile modifications on the dynamic load. These parameters are given by the following relationships:

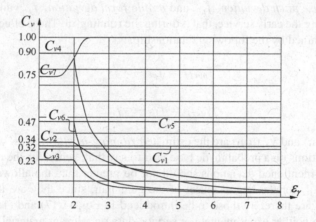

Fig. 1.3 Distribution curves of C_{v1} to C_{v7} for determination of K_{v-B}, as a function of ε_γ

Table 1.2 Values of factors C_{v1} to C_{v7} and relationships for their calculation, for determination of K_{v-B}

Factors C_{vi} (with $i = 1, \ldots, 7$)	$1 < \varepsilon_\gamma \leq 2$	$\varepsilon_\gamma > 2$	
C_{v1}	0.32	0.32	
C_{v2}	0.34	$0.57/(\varepsilon_\gamma - 0.30)$	
C_{v3}	0.23	$0.096/(\varepsilon_\gamma - 1.56)$	
C_{v4}	0.90	$(0.57 - 0.05\varepsilon_\gamma)/(\varepsilon_\gamma - 1.44)$	
C_{v5}	0.47	0.47	
C_{v6}	0.47	$0.12/(\varepsilon_\gamma - 1.74)$	
	$1 < \varepsilon_\gamma \leq 1.5$	$1.5 < \varepsilon_\gamma \leq 2.5$	$\varepsilon_\gamma > 2.5$
C_{v7}	0.75	$0.875 + 0.125 \sin[\pi(\varepsilon_\gamma - 2)]$	1.0

$$B_p = \frac{c' f_{pbeff}}{K_A(F_t/b)} \tag{1.14}$$

$$B_f = \frac{c' f_{faeff}}{K_A(F_t/b)} \tag{1.15}$$

$$B_k = \left| 1 - \frac{c' C_a}{K_A(F_t/b)} \right|, \tag{1.16}$$

where, remaining unchanged the meaning of symbols already introduced, c' [N/(mm μm)] is the *maximum tooth stiffness per unit face width*, or *single stiffness*, of the tooth pair, f_{pbeff} (in μm) is the *effective base pitch deviation*, f_{faeff} (in μm) is the *effective profile form deviation*, and C_a (μm) is the tip relief.

The values of f_{pbeff} and f_{faeff} to be introduced in Eqs. (1.14) and (1.15) are those of the *running-in* of the pinion and wheel, as the initial deviations (i.e., the *transverse base pitch deviation*, f_{pb}, and *profile form deviation*, f_{fa}) are generally changed during the early service, that is during the running-in. These effective values can be determined by the following relationships:

$$f_{pbeff} = f_{pb} - y_p \tag{1.17}$$

$$f_{faeff} = f_{fa} - y_f, \tag{1.18}$$

where y_p (μm) and y_f (μm) are the estimated *running-in allowances*.

Considerations on a probabilistic base lead to say that, generally, the magnitudes of the above-mentioned deviations should not be greater than the allowable values of f_{pb} and f_{fa} for the largest wheel of the gear pair, since they are the highest. These values are therefore those to be introduced in Eqs. (1.17) and (1.18). In the case in which neither experimental nor service data on relevant material running-in characteristics are available, as Method A requires, we can assume $y_p = y_\alpha$, where y_α can be deduced from Fig. 1.9 or Eqs. (1.80) to (1.82). The running-in allowance, y_f, can be determined in the same way as y_α, when the profile deviation, f_{fa}, is used instead of the base pitch deviation, f_{pb}.

The quantity C_a appearing in Eq. (1.16) is the design value of the tip relief at the beginning and end of tooth engagement. This value of the profile modification may only be used in Eq. (1.16) for gears having quality grades in the range 0–5 as specified by ISO 1328-1:2013. For gears with quality grades in the range 6–12, we assume $B_k = 1.0$. In the case of gears without a specified profile modification, the value C_{ay} resulting from running-in is to be substituted for C_a in Eq. (1.16). This value of C_{ay} is given by the following relationship:

$$C_{ay} = \frac{1}{18}\left(\frac{\sigma_{H\,\lim}}{97} - 18.45\right)^2 + 1.50, \tag{1.19}$$

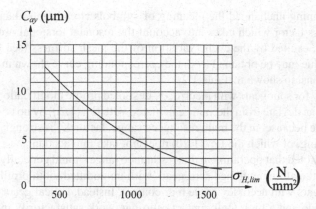

Fig. 1.4 Distribution curve of the average value of tip relief C_{ay} produced by running-in, as a function of $\sigma_{H\,lim}$

where $\sigma_{H\,lim}$ (in N/mm^2) is the *allowable stress number* for contact load. The value of C_{ay} as a function of $\sigma_{H\,lim}$ can be also obtained using the curve shown in Fig. 1.4. It is to be noted that, when the pinion material is different from the wheel material, we must calculate separately C_{ay1} and C_{ay2}, and we have to introduce in Eq. (1.16) their average value $C_{ay} = \left[(C_{ay1} + C_{ay2})/2\right]$. As regards the calculation of c', we refer to the Sect. 1.6.

In conclusion, we must remember that, in the subcritical range, pre-resonances may occur for $N = 1/2$ or $N = 1/3$. In such circumstances, the dynamic factors K_v can exceed the values calculated with Eq. (1.12), but the correlated risk is to be taken into account only for coarse accuracy grades, while it is relatively low for precision spur or helical gears (gears to accuracy grade 5), especially if the latter have suitable profile modifications. The dynamic factor K_v can be just as great as in the main resonance speed range, when the contact ratio of spur gears is small or if the accuracy grade is low. In these cases, the design or operating parameters should be changed.

Pre-resonances at $N = 1/4, 1/5, \ldots$ are most often irrelevant, because the associated vibration amplitudes are usually small. Finally, it is to be noted that, especially for spur and helical gears of coarse accuracy grade running at higher speed, a particular risk of vibration exists when the specific load is low ($F_t K_A/b < 50\,\text{N/mm}$). In these cases, under some circumstances, separation of working tooth flanks can take place.

1.3.3.2 Dynamic Factor K_v in Main Resonance Range ($N_S < N < 1.15$)

Factor K_v in main resonance range is calculated with the following relationship:

$$K_v = (C_{v1} B_p) + (C_{v2} B_f) + (C_{v4} B_k) + 1, \qquad (1.20)$$

where, remaining unchanged the meaning of symbols already introduced, C_{v4} is a dimensionless factor which takes into account the resonant torsional oscillations of the gear pair, excited by the cyclic variation of the mesh stiffness (see Ker Wilson 1963). Its value may be obtained from the corresponding curve shown in Fig. 1.3, or by the relationship shown in Table 1.2.

Especially for spur gears with incorrectly designed profile modification, the actual value of K_v can deviate from the value calculated with Eq. (1.20) by up to 40%. These deviations are because, in the main resonance range, factor K_v is strongly influenced by the damping, of which the Eq. (1.20) does not take into account.

It is to be noted that operation in this resonance range should generally be avoided not only for the aforementioned spur gears with unmodified tooth profiles, but also for helical gears with accuracy grade 6 or coarser. Instead, helical gears with a high accuracy grade and a high total contact ratio can work satisfactorily in this range. Spur gears with accuracy grade 5 or better must have in any case a suitable profile modification.

1.3.3.3   Dynamic Factor K_v in Supercritical Range ($N \geq 1.50$)

Factor K_v in supercritical range is calculated with the following relationship:

$$K_v = (C_{v5} B_p) + (C_{v6} B_f) + C_{v7}, \tag{1.21}$$

where, remaining unchanged the meaning of the symbols already introduced, C_{v5} and C_{v6} are factors corresponding to the factors C_{v1} and C_{v2} in the subcritical range (therefore C_{v5} and C_{v6} have the same function of the latter), while C_{v7} takes into account the component of load deriving from tooth bending deflections during substantially constant speed, due to mesh stiffness variation. Factors C_{v5}, C_{v6}, and C_{v7} may be obtained from the corresponding curves shown in Fig. 1.3, or can be determined by the relationships shown in Table 1.2.

Most high precision gears used in turbine and other high-speed gear transmission systems operate in this supercritical range. For gears operating in this range, the same limitations on gear accuracy grade described for gears operating in the main resonance range apply. Resonance peaks can occur at $N = 2, 3, \ldots$. However, in most cases, the vibration amplitudes are small, because excitation loads with lower frequencies than meshing frequency are generally small. For some gears operating in this range, it is also necessary to consider dynamic loads due to transverse vibrations of the gear and shaft assemblies. The effective value of K_v can exceed the value calculated with Eq. (1.1) by up to 100%, when the critical frequency is near the frequency of rotation. It is necessary to avoid this condition.

1.3.3.4 Dynamic factor K_v in Intermediate Range ($1.15 < N < 1.50$)

Factor K_v in intermediate range is determined by linear interpolation between the values that limit, to the right, the main resonance range ($N = 1.15$), and to the left, the supercritical range ($N = 1, 50$). Therefore, we use the following relationship:

$$K_v = K_{v(N=1.15)} + \frac{K_{v(N=1.15)} - K_{v(N=1.50)}}{0.35} (1.50 - N). \qquad (1.22)$$

1.3.4 Method C for Determination of K_{v-C}

This method is derived from Method B described in the previous section, with the introduction of additional simplifying assumptions, which are summarized as follows:

- subcritical running speed range, i.e. $\left[(vz_1/100)\sqrt{u^2/(1 + u^2)}\right] < 10\text{m/s}$;
- steel gear wheels configured as solid disks;
- transverse pressure angle, $\alpha_t = 20°$, transverse base pitch deviation, $f_{pb} = f_{pt}\cos20°$, and basic rack profile according to ISO 53:1998;
- helix angle $\beta = 20°$, with reference to values of c' and $c_{\gamma\alpha}$ (this is the mean value of mesh stiffness per unit face width, in N/mm μm) for helical gears, and total contact ratio $\varepsilon_\gamma = 2.5$;
- tooth stiffness values, $c' = 14\,\text{N/mm}\,\mu\text{m}$ and $c_{\gamma\alpha} = 20\,\text{N/mm}\,\mu\text{m}$, for spur gears, and $c' = 13.1\,\text{N/mm}\,\mu\text{m}$ and $c_{\gamma\alpha} = 18.7\,\text{N/mm}\,\mu\text{m}$, for helical gears;
- tip relief C_a, and tip relief by running-in C_{ay} both equal to 0;
- effective transverse base pitch deviation f_{pbeff} equal to effective profile form deviation $f_{f\alpha eff}$, i.e. $f_{pbeff} = f_{f\alpha eff}$.

Table 1.3 provides the approximate average values of f_{pbeff} as a function of the accuracy grade as specified by ISO 1328-1:2013. We can take these values in the absence of other more reliable data.

It is to be noted that the considerations described in the conclusion at the end of Sect. 1.3.3.1 are not to be considered when we use this method, which however takes into account the influence of the specific load ($F_t K_A/b$). It is also to be noted that Method C provides average values of K_v, which can be used for industrial gear drives with external and internal spur gears, helical gears with $\beta \leq 30°$, pinions with

Table 1.3 Approximate mean value of f_{pbeff} as a function of accuracy grade, for determination of K_{v-C}

Accuracy grades as specified by ISO 1328-1:2013	3	4	5	6	7	8	9	10	11	12
f_{pbeff}	2.8	5.1	9.8	19.5	35	51	69	100	134	191

relatively low number of teeth ($z_1 < 50$), and solid disk gear wheels or heavy steel gear rim. In this regard, it should be noted that, if the rim is very light or the helical gears have a very large overlap ratio, values of ($K_{350}N$) obtained from Figs. 1.5 and 1.6 are too unfavorable, while calculated values (see Sect. 1.3.4.2) tend to be safe (the same applies when gears are made of cast iron).

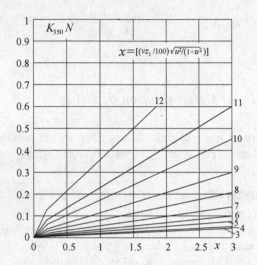

Fig. 1.5 Distribution curves of ($K_{350}N$) as a function of $x = \left[(vz_1/100)\sqrt{u^2/(1+u^2)}\right]$ and gear accuracy grade, for helical gears with $\varepsilon_\beta \geq 1$

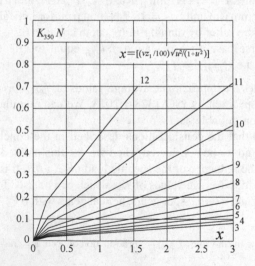

Fig. 1.6 Distribution curves of ($K_{350}N$) as a function of $x = \left[(vz_1/100)\sqrt{u^2/(1+u^2)}\right]$ and gear accuracy grade, for spur gears

This method can also be used for the following fields of application: all types of cylindrical gears, if $\left[(vz_1/100)\sqrt{u^2/(1+u^2)}\right] < 3\,\text{m/s}$; lightweight gear rim, and helical gears with $\beta > 30°$, with the aforementioned restrictions.

Determination of K_{v-C} can be made using graphs or a calculation relationship. Both procedures give similar values of the dynamic factor.

1.3.4.1 Mixed Procedure for Determination of K_{v-C}

This mixed procedure uses, for determination of K_v, the following relationship where two factors appear, the load correction factor, f_F, and $K_{350}N$ (these factors are respectively read out of tables and graphs):

$$K_v = (f_F K_{350} N) + 1. \tag{1.23}$$

In this equation, N is the frequency ratio, while K_{350} takes into account the influence of the gear accuracy grade at the specific reference load of 350 N/mm on the dynamic factor. Instead, the load correction factor f_F takes into account the influence of the actual specific load $(F_t K_A/b)$ on the same dynamic factor. The graphs shown in Figs. 1.5 and 1.6 allow to obtain the product $(K_{350}N)$ as a function of $\left[(vz_1/100)\sqrt{u^2/(1+u^2)}\right]$, for helical gears with $\varepsilon_\beta \geq 1$ and, respectively, for spur gears. In these figures, the curves for gear accuracy grade extend only to the value $\left[(vz_1/100)\sqrt{u^2/(1+u^2)}\right] = 3\,\text{m/s}$, since this value is not generally exceeded for these accuracy grades.

Tables 1.4 and 1.5 provide the values of the load correction factor, f_F, as a function

Table 1.4 Values of the load correction factor f_F for helical gears

Gear accuracy grade	Load correction factor f_F							
	$(F_t K_A/b)$ in N/mm							
	≤ 100	200	350	500	800	1200	1500	2000
3	1.96	1.29	1	0.88	0.78	0.73	0.70	0.68
4	2.21	1.36	1	0.85	0.73	0.66	0.62	0.60
5	2.56	1.47	1	0.81	0.65	0.56	0.52	0.48
6	2.82	1.55	1	0.78	0.59	0.48	0.44	0.39
7	3.03	1.61	1	0.76	0.54	0.42	0.37	0.33
8	3.19	1.66	1	0.74	0.51	0.38	0.33	0.28
9	3.27	1.68	1	0.73	0.49	0.36	0.30	0.25
10	3.35	1.70	1	0.72	0.47	0.33	0.28	0.22
11	3.39	1.72	1	0.71	0.46	0.32	0.27	0.21
12	3.43	1.73	1	0.71	0.45	0.31	0.25	0.20

Table 1.5 Values of the load correction factor f_F for spur gears

Gear accuracy grade	Load correction factor f_F							
	$(F_t K_A / b)$ in N/mm							
	≤100	200	350	500	800	1200	1500	2000
3	1.61	1.18	1	0.93	0.86	0.83	0.81	0.80
4	1.81	1.24	1	0.90	0.82	0.77	0.75	0.73
5	2.15	1.34	1	0.86	0.74	0.67	0.65	0.62
6	2.45	1.43	1	0.83	0.67	0.59	0.55	0.51
7	2.73	1.52	1	0.79	0.61	0.51	0.47	0.43
8	2.95	1.59	1	0.77	0.56	0.45	0.40	0.35
9	3.09	1.63	1	0.75	0.53	0.41	0.36	0.31
10	3.22	1.67	1	0.73	0.50	0.37	0.32	0.27
11	3.30	1.69	1	0.72	0.48	0.35	0.30	0.24
12	3.37	1.71	1	0.72	0.47	0.33	0.27	0.22

of the gear accuracy grade and actual specific load $(F_t K_A / b)$ for helical gears and, respectively, for spur gears. For intermediate values of the actual specific load, the value of f_F can be obtained by linear interpolation between the values that delimit the range within which the actual specific load is included.

For the proper use of aforementioned graphs and tables, the following guidelines are to be considered:

(a) for helical gears with overlap ratio $\varepsilon_\beta \geq 1$, and approximately also for $\varepsilon_\beta > 0.9$, the values of $(K_{350}N)$ and f_F are those obtained respectively from Fig. 1.5 and Table 1.4;

(b) for spur gears, the values of $(K_{350}N)$ and f_F are those obtained respectively from Fig. 1.6 and Table 1.5;

(c) for helical gears with overlap ratio $\varepsilon_\beta < 1$, the value of K_v is determined by linear interpolation between values evaluated in accordance with procedures (a) and (b), using the following relationship:

$$K_v = K_{v\alpha} - \varepsilon_\beta \left(K_{v\alpha} - K_{v\beta} \right), \tag{1.24}$$

where $K_{v\alpha}$ is the dynamic factor for spur gears obtained in accordance with procedure (b), while $K_{v\beta}$ is the dynamic factor for helical gears obtained in accordance with procedure (a).

1.3.4.2 Calculation Relationship for Determination of K_v with Method C

The calculation relationship for determination of K_v with Method C, for spur gears and helical gears with overlap ratio $\varepsilon_\beta \geq 1$, also usable approximately for $\varepsilon_\beta > 0.9$, is as follows:

$$K_v = 1 + \left[\frac{K_1}{(F_t K_A / b)} + K_2 \right] \frac{v z_1}{100} K_3 \sqrt{\frac{u^2}{1 + u^2}}, \qquad (1.25)$$

where K_1 and K_2 are factors whose numerical values must be selected as specified in Table 1.6, while K_3 is a factor whose numerical value must be chosen or calculated in accordance to the two relationships and related inequalities that follow:

$$K_3 = 2 \quad \text{if} \quad \frac{v z_1}{100} \sqrt{\frac{u^2}{1 + u^2}} \leq 0.2 \qquad (1.26)$$

$$K_3 = -0.357 \frac{v z_1}{100} \sqrt{\frac{u^2}{1 + u^2}} + 2.071 \quad \text{if} \quad \frac{v z_1}{100} \sqrt{\frac{u^2}{1 + u^2}} > 0.2. \qquad (1.27)$$

It is to be noted that, in Eq. (1.5), the specific load $(F_t K_A / b)$ must be set equal to 100 N/mm, in the case in which it is lower than 100 N/mm.

For helical gears with overlap ratio $\varepsilon_\beta < 1$, the value of K_v is determined by linear interpolation between values calculated for spur gears, $K_{v\alpha}$, and for helical gears, $K_{v\beta}$, in accordance with Eq. (1.4) used for mixed procedure.

Table 1.6 Values of factors K_1 and K_2 for calculation of K_{v-C}, by Eq. (1.25)

Gear type	K_1										K_2
	Accuracy grade										All accuracy grades
	3	4	5	6	7	8	9	10	11	12	
Spur gears	2.1	3.9	7.5	14.9	26.8	39.1	52.8	76.6	102.6	146.3	0.0193
Helical gears	1.9	3.5	6.7	13.3	23.9	34.8	47.0	68.2	91.4	130.3	0.0087

1.4 Face Load Factors, $K_{H\beta}$ and $K_{F\beta}$

1.4.1 Generality

The face load factors $K_{H\beta}$ and $K_{F\beta}$ take into account the effects of the non-uniform distribution of load along the gear face width on the surface stress, and respectively on the tooth root stress. The first factor, $K_{H\beta}$, is defined as the quotient of the maximum load per unit face width divided by the average load per unit face width, i.e.:

$$K_{H\beta} = \frac{(F_{\max}/b)}{(F_m/b)}. \tag{1.28}$$

In this equation, which is used for an approximate calculation, $F_m/b = (F_t K_A K_v)/b$ is the mean transverse specific load at the reference cylinder relevant to mesh calculations, and $(F_{\max}/b)$ is the corresponding maximum local specific load. The second factor, $K_{F\beta}$, depends on the same variable quantities from which factor $K_{H\beta}$ depends, and also on the ratio (b/h) between the face width and tooth depth.

The non-uniform distribution of load over the gear face width is due to several factors, the main of which are as follows:

- gear tooth manufacturing accuracy, in terms of lead, profile, space width and pitch;
- alignment of the axes of rotation of the mating gear members;
- elastic deflections of members of the gear drive, such as teeth, gear blanks, shafts, bearings, housing and foundation, which support the gear members and gear unit;
- bearing clearances;
- gear geometry;
- Hertzian contact and bending deformations at the tooth surface, including variable tooth stiffness;
- thermal deformations due to operating temperature, which are especially important for gears with large face widths;
- centrifugal deflections due to operating rotational speed;
- helix modifications including tooth crowning and end relief;
- running-in effects;
- total tangential tooth load, including increases due to factors K_A and K_v;
- additional shaft loads as, for example, those due to belt or chain drives.

Uneven load distribution along the face width is due to an *equivalent mesh misalignment* in the plane of action, including the load-induced elastic deformations of gears and housing, displacements of journal bearings, manufacturing deviations and thermal distortions. The combined effects of the gear and housing manufacturing deviations, deflections of the housing and displacements of journal bearings, always result in a linear deviation within the plane of action. Elastic deformations of shafts and gear blanks, as well as the deformations due to thermal distortion produced by uneven temperature distribution along the face width, always result in non-linear deviation within the plane of action. The undulations and the flank shape deviations are then superimposed on the resulting mesh alignment. However, the running-in, which is typical of the material combination, reduces the unevenness of load distribution.

For determination of the face load factors, here also three methods can be used, i.e. Methods A, B, and C. A parameter to be considered is the ratio (b_1/d_1) between the face width and reference pitch diameter of the pinion. In this regard, a careful analysis is recommended when this ratio is greater than 1.5 for through-hardened gears, and greater than 1.2 for case-hardened gears. A nearly uniform load distribution along the face width can be achieved for a given operating condition, if there is a high degree of manufacturing accuracy, and equivalent misalignment due to thermal and mechanical deformations is compensated for by helix modifications, possibly varying along the face width. In this case, the value of the face load factor becomes close to 1. To make uniform the load distribution along the face width, at least within certain limits, the crowning and end relief of gear teeth can be made.

1.4.2 Method A for Determination of $K_{H\beta-A}$ and $K_{F\beta-A}$

With this method, a comprehensive analysis of all influence factors is necessary to determine the load distribution along the face width of gears under load. This distribution can be evaluated by measurements of tooth root strains during operation at working temperature or, with limitations, by a critical examination of the tooth contact pattern obtained by means of a suitable experimental equipment.

The data obtained with both procedures consist of maximum allowable values of the face load factors, or maximum allowable values of the total mesh misalignment under operating load and temperature, from which these factors can be derived using a precise calculation method, which also takes into account all other relevant known influences.

1.4.3 Method B for Determination of $K_{H\beta-B}$ and $K_{F\beta-B}$

This method uses numerical models and computer aided calculations, by means of which the load distribution along the face width is determined. Results obtained by this method depend on the elastic deflections under load, static displacements, and stiffness of the entire elastic system. Since the load distribution during meshing and deformations of the elastic system influence each other, the problem to be solved is a non-linear problem. Consequently, for its solution, iterative methods, influence factors and finite element models are used (see Jaramillo 1950; Roda-Casanova et al. 2013; Radzevich 2016).

The Annex E to the ISO 6336-1:2006 describes a procedure for the analytical determination of the load distribution along the face width, according to this method in incremental terms. The method covers the most important deflections, such as shaft bending and torsional deflections, and tooth deflections; in addition, it gives a guideline on how to take account of other deflections. For details, we refer the reader directly to the ISO 6336-1:2006. However, we want to dwell here, albeit briefly, on the general principles on which this method is based.

The fundamental relationship is the Eq. (1.8) where, however, the maximum load per unit face width, which appears in the numerator, is to be understood in incremental terms, i.e. as the maximum local load intensity per unit length increment of face width, since this last is divided into 10 parts, each corresponding to the same number of successive increments. The basic model of the gear mesh is a spur gear pair having the same number of teeth, transverse module and face width compared to the gear pair being analyzed.

The effective stiffness of the entire elastic system is used for the calculation of the load distribution along the face width. Therefore, this stiffness is the addition of the following contributions: gear mesh stiffness; gear blank stiffnesses; stiffness of shaft/hub connections; stiffness of pinion and gear shaft; stiffnesses of the bearings, housing and foundation.

1.4.4 Method C for Determination of $K_{H\beta-C}$ and $K_{F\beta-C}$

1.4.4.1 Principles, Assumptions, and Basic Equations

This approximate method takes into account the components of equivalent misalignment due to deformations of the pinion and pinion-shaft, as well as those due to manufacturing deviations. For the evaluation of the approximate values of the variable quantities that come into play, this method uses calculations, measurements, and data derived from experience, which may relate to individual variables, or their combinations. It is based on the assumptions that elastic deflections of the gear blanks determine, during the teeth meshing, a linearly increasing separation of the working tooth flanks along the face width, and that equivalent misalignment, inclusive of manufacturing deviations, involves similar separation.

These concepts are summarized in Figs. 1.7 and 1.8, which show the influences on the load distribution of the tooth load, and equivalent misalignment, according to the aforementioned assumption. In particular, Fig. 1.7 shows that the linear load distribution along the face width affects only one part of the length of the latter or its entire length, in the two cases of low load and/or large value of equivalent misalignment (Fig. 1.7b), and high load and/or small value of equivalent misalignment (Fig. 1.7c). Figure 1.8 shows instead how to calculate the maximum load per unit face width, $w_{max} = (F_{max}/b)$, as a function of the transverse specific load corresponding to a linear distribution of the load on the face width (this is given by $w_m = F_m/b = (F_t K_A K_v)/b$), and of the calculated face width, b_{cal}, in the two cases of low load and/or large value of the *effective equivalent misalignment*, $F_{\beta y}$ (Fig. 1.8a), and high load and/or small value of $F_{\beta y}$ (Fig. 1.8b). In these two cases,

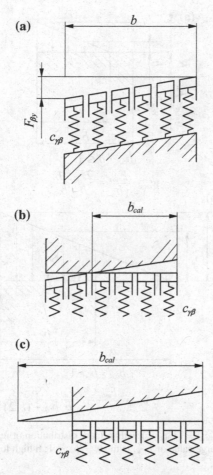

Fig. 1.7 Linear load distribution along face width with linear equivalent misalignment: **a** without load; **b** low load and/or large value of $F_{\beta y}$; **c** high load and/or small value of $F_{\beta y}$

we have respectively $(b_{cal}/b) \leq 1$, and $(b_{cal}/b) > 1$. In Fig. 1.7, $c_{\gamma\beta}$ is the mean value of mesh stiffness for unit face width (see Sect. 1.6).

The face load factor $K_{H\beta-C}$ is calculated as a function of the mean load intensity along the face width, $w_m = (F_m/b)$, mean value of mesh stiffness per unit face width, $c_{\gamma\beta}$, and effective equivalent misalignment, $F_{\beta y}$ (i.e., the *total mesh misalignment after running-in*), using different relationships depending on whether the contact covers only a part of the whole face width (in this case, $(b_{cal}/b) \leq 1$), or the whole face width (in this case, $(b_{cal}/b) > 1$).

Therefore, if $(b_{cal}/b) \leq 1$, corresponding to $\left[F_{\beta y} c_{\gamma\beta}/(2F_m/b) \right] \geq 1$, we have:

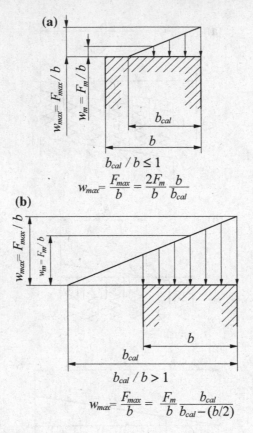

Fig. 1.8 Calculation of $w_{\max} = (F_{\max}/b)$ with a linear distribution of $w_m = (F_m/b)$ on the face width: **a** low load and/or large value of $F_{\beta y}$, with $(b_{cal}/b) \leq 1$; **b** high load and/or small value of $F_{\beta y}$, with $(b_{cal}/b) > 1$

$$K_{H\beta} = \sqrt{\frac{2F_{\beta y}c_{\gamma\beta}}{(F_m/b)}} \geq 2 \tag{1.29}$$

$$\frac{b_{cal}}{b} = \sqrt{\frac{2(F_m/b)}{F_{\beta y}c_{\gamma\beta}}}, \tag{1.30}$$

while, if $(b_{cal}/b) > 1$, corresponding to $\left[F_{\beta y}c_{\gamma\beta}/(2F_m/b)\right] < 1$, we have:

$$K_{H\beta} = 1 + \frac{F_{\beta y}c_{\gamma\beta}}{2(F_m/b)} \tag{1.31}$$

$$\frac{b_{cal}}{b} = 0.5 + \frac{(F_m/b)}{F_{\beta y}c_{\gamma\beta}}. \tag{1.32}$$

1.4.4.2 Effective Equivalent Misalignment, Running-in Allowance, and Running-in Factor

The effective equivalent misalignment after running-in, $F_{\beta y}$, summarizes the combined effects of manufacturing errors of all relevant component members, included in the *mesh misalignment* due to manufacturing deviations, f_{ma}, and elastic deflections of the pinion and pinion-shaft, included in the *component of equivalent misalignment* due to deformations of pinion and pinion-shaft, f_{sh}, reduced by a running-in allowance, y_β. The way in which the two effects are combined depends on the helix modification (crowning, end relief, helix correction, or none) applied to the mesh. The combined effects of f_{ma} and f_{sh} before running-in form the *initial equivalent misalignment*, $F_{\beta x}$, which is the absolute value of the sum of deformations, displacements and manufacturing deviations of pinion and wheel, measured in the plane of action, under load and before running-in. In this framework, $F_{\beta y}$ can be written as follows:

$$F_{\beta y} = F_{\beta x} - y_\beta = F_{\beta x}\chi_\beta, \tag{1.33}$$

where χ_β is the *running-in factor* characterizing the equivalent misalignment after running-in. The use of χ_β in calculations is convenient only as long as y_β is proportional to $F_{\beta x}$. The important influences of *running-in allowance*, y_β, and running-in factor, χ_β, include: pinion and wheel materials; surface hardness; rotational speed at the reference circle; type of lubricant; surface treatment; abrasive particles in the lubricant oil; initial equivalent misalignment, $F_{\beta x}$, due to deformations, displacements and manufacturing deviations. It is to be noted that y_β and χ_β do not take into account the effects of running-in operations due to removal of material obtained by manufacturing processes such as lapping. These effects are included in the value of f_{ma}.

In the absence of reliable data, like those obtained in accordance with Method A, and therefore obtained by experimental measurements or consolidated operational experience, the determination of y_β and χ_β with Method C is carried out with the three pairs of relationships given below, each of which refers to the following groups of ferrous metals:

- 1° group, which includes: normalized low carbon steels and cast steels, such as wrought normalized low carbon steels (St), and cast steels (St(cast)); through-hardened wrought steels, such as carbon steels, and alloy steels (V); through-hardened cast steels, such as carbon steels and alloy steels (V(cast)); cast iron materials, such as nodular cast iron, with perlitic or bainitic structure (GGG(perl., bai.)), and black malleable cast iron with perlitic structure (GTS(perl.)).
- 2° group, which includes: cast iron materials, such as grey cast iron (GG), and nodular cast iron with ferritic structure (GGG(ferr.)).
- 3° group, which includes: case-hardened wrought steels (Eh); flame or induction hardened wrought or cast steels (IF); nitrided wrought steels, nitriding steels,

through-hardening steels nitrided, such as nitriding steels (NT(nitr.)), and through-hardening steels (NV(nitr.)); wrought steels, nitrocarburized, such as through-hardening steels (NV(nitrocar.)).

The pair of relationships related to the first group of ferrous metals is as follows:

$$y_\beta = \frac{320}{\sigma_{H\,\text{lim}}} F_{\beta x} \quad \chi_\beta = 1 - \frac{320}{\sigma_{H\,\text{lim}}}, \tag{1.34}$$

where $\sigma_{H\,\text{lim}}$ is the allowable stress number for contact (in N/mm^2), $y_\beta \leq F_{\beta x}$, $\chi_\beta \geq 0$, and the following conditions are to be met: for $v \leq 5$ m/s, there are no restrictions; for 5 m/s $< v \leq 10$ m/s, the upper limit of y_β is $(25.6 \times 10^3)/\sigma_{H\,\text{lim}}$, corresponding to $F_{\beta x} = 80\,\mu$m; for $v > 10$ m/s, the upper limit of y_β is $(12.8 \times 10^3)/\sigma_{H\,\text{lim}}$, corresponding to $F_{\beta x} = 40\,\mu$m.

The pair of relationships related to the second group of ferrous metals is as follows:

$$y_\beta = 0.55 F_{\beta x} \quad \chi_\beta = 0.45, \tag{1.35}$$

where the following conditions are to be met: for $v \leq 5$ m/s, there are no restrictions; for 5 m/s $< v \leq 10$ m/s, the upper limit of y_β is 45 μm, corresponding to $F_{\beta x} = 80\,\mu$m; for $v > 10$ m/s, the upper limit of y_β is 22 μm, corresponding to $F_{\beta x} = 40\,\mu$m.

Finally, the pair of relationships related to the third group of ferrous metals is as follows:

$$y_\beta = 0.15 F_{\beta x} \quad \chi_\beta = 0.85, \tag{1.36}$$

where the upper limit of y_β is 6 μm for all velocities, and corresponds to $F_{\beta x} = 40\mu$ m.

Relationships (1.34) to (1.36) apply in cases where materials of pinion and wheel are equal. When the two materials are different, first the values $y_{\beta 1}$ and $\chi_{\beta 1}$ of pinion, and values $y_{\beta 2}$ and $\chi_{\beta 2}$ of wheel must be separately determined, and therefore their average values, given by the following pair of relationships, must be used:

$$y_\beta = \frac{1}{2}(y_{\beta 1} + y_{\beta 2}) \quad \chi_\beta = \frac{1}{2}(\chi_{\beta 1} + \chi_{\beta 2}). \tag{1.37}$$

The values of the running-in allowance, y_β, can also be read from distribution curves shown in Fig. 1.9, as a function of the initial equivalent misalignment, $F_{\beta x}$, and allowable stress number, $\sigma_{H\,\text{lim}}$.

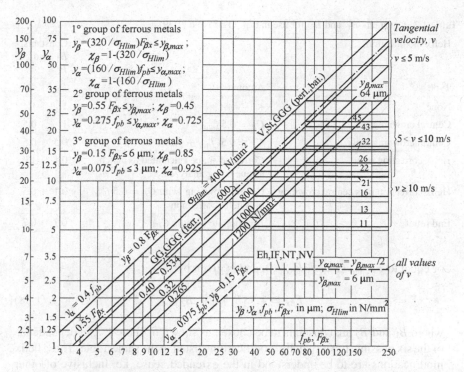

Fig. 1.9 Distribution curves of the running-in allowance for a gear pair, y_α (in μm), and running-in allowance for equivalent misalignment, y_β (in μm), both as a function of the allowance stress number for contact of several types of ferrous metals, $\sigma_{H\,lim}$ (in N/mm^2). Allowance y_α is given as a function of the transverse base pitch deviation, f_{pb} (in μm), while allowance y_β is given as a function of the initial equivalent misalignment before running-in, $F_{\beta x}$ (in μm). The maximum limit values of these quantities, depending on the tangential velocity, v, measured on the reference pitch cylinder, are shown to the right

1.4.4.3 Initial Equivalent Misalignment Before Running-in

Once determined y_β, to calculate $F_{\beta y}$ it is necessary to evaluate $F_{\beta x}$. According to the aforementioned definition of $F_{\beta x}$, only the components of deformation, displacement, and deviation in the plane of action are determinant for the calculation of this quantity. For determination of the initial equivalent misalignment $F_{\beta x}$, the aforementioned ISO standard considers three typical conditions, concerning the distribution and position of the contact pattern between the active flanks of the meshing teeth with respect to the mid-face position, in order of increasing quality, from the most degraded to the ideal one. For each of these three conditions, ISO provides a special calculation relationship. In this regard, we have:

– For gear pairs for which the adequacy of the contact pattern is not proven, in terms of size and positioning with respect to the mid-face position (e.g., edge contact and like), whereby the bearing pattern under load is imperfect, $F_{\beta x}$ is greater than

Table 1.7 Constants B_1 and B_2 for use in Eqs. (1.38) and (1.39)

Helix modification		Constants, B_1 and B_2	
Type	Amount	B_1	B_2
None	–	1	1
Central crowning only	$C_\beta = 0.5 f_{ma}$	1	0.5
Central crowning only	$C_\beta = 0.5(f_{ma} + f_{sh})$	0.5	0.5
Helix correction only	Corrected shape calculated to match torque being analyzed	0.1	1.0
Helix correction plus central crowning	Central crowning with $C_\beta = 0.5 f_{ma}$ plus helix correction	0.1	0.5
End relief	Appropriate amount $C_{\mathrm{I(II)}}$	0.7	0.7

or equal to $F_{\beta x min}$ ($F_{\beta x} \geq F_{\beta x min}$), and is given by:

$$F_{\beta x} = 1.33 B_1 f_{sh} + B_2 f_{ma}, \qquad (1.38)$$

where B_1 and B_2 are constants to be chosen as shown in Table 1.7, in dependence of the six possible types of helix modifications listed in the same Table. These helix modifications are to be understood in the extended sense, i.e. inclusive of other modifications of the tooth axial profile, such as crowning and end relief. The mesh misalignment due to manufacturing deviations, f_{ma}, should take into account the effects of adjustment measures (e.g., lapping, running-in at partial load), crowning or end-relief, or similar, as well as of position of the contact pattern.

– For gear pairs for which a verification of the favorable position of the contact pattern (e.g., by modification of the tooth axial profile or adjustment of bearings) is made, $F_{\beta x}$ is also greater than or equal to $F_{\beta x min}$ ($F_{\beta x} \geq F_{\beta x min}$), and is given by:

$$F_{\beta x} = |1.33 B_1 f_{sh} - f_{H\beta 6}|, \qquad (1.39)$$

where, remaining unchanged the meaning of symbols already introduced, $f_{H\beta 6}$ (in μm) is the tolerance on helix slope deviation for ISO accuracy grade 6, while B_1 is a constant to be chosen in Table 1.7 in the same way as previously indicated. It should be noted that $f_{H\beta}$ is the *helix slop deviation* according to ISO 1328-1:2013, whose value can be alternatively given by the *total helix deviation*, F_β, in the case in which tolerances complying with the same ISO standard are used.

About the choice of the appropriate central crowning between the two types that appear in the Table 1.7 and, more generally, on the subject regarding the helix modification understood in the extended sense, we refer to Sect. 1.4.4.7. About the use of the table, it should be noted that the helix correction only is mainly used for applications characterized by constant load conditions, while the value

$B_1 = 0.1$ regarding the helix correction only and helix correction plus central crowning are valid for very best manufacturing practice, otherwise higher value of B_1 are to be chosen.

About the use of Eqs. (1.38) and (1.39), it should be noted that running clearances in rolling bearings can contribute considerably to initial equivalent misalignment $F_{\beta x}$, for which they must be sufficiently small under working conditions and, in any case, they must be kept under strict control. For double-helical gears, the larger equivalent misalignment of the two helices must be considered. With a favorable position of the contact pattern, the manufacturing deviations and elastic deformations compensate each other. By subtracting in Eq. (1.39) $f_{H\beta 5}$ (i.e., the helix slope deviation tolerance for ISO quality 5), we take into account the compensatory role of the manufacturing deviations and elastic deformations.

However, we can be faced with cases where it is necessary to consider specifically, in addition to pinion blank and pinion shaft deformations, f_{sh1}, also those of the wheel/wheel shaft, f_{sh2}, and the gear case, f_{ca}, as well as the displacements of the bearings, f_{be}. In these cases, f_{ma} is greater than or equal to $f_{H\beta 5}$ $(f_{ma} \geq f_{H\beta 5})$, and instead of Eqs. (1.38) and (1.39), it is necessary to use the following relationship that is more general:

$$F_{\beta x} = 1.33 B_1 f_{sh1} + f_{sh2} + f_{ma} + f_{ca} + f_{be}. \qquad (1.40)$$

In this relationship, the sign of f_{sh2}, f_{ca}, and f_{be} must be carefully evaluated and, if precise information is not available, it is appropriate to choose the positive signs, so as to be on the safety side. In any case, f_{sh2} can only be influenced by the bending deflection of the wheel shaft. Nevertheless, Eqs. (1.38) and (1.39) give satisfactory approximate results in most practical applications.

In order to ensure an even load distribution along the face width, special measures can be taken including gear lapping, bearing setting, or running-in in working conditions. Deformation near mid-face width, and resulting heavy local loads, can be caused by an uneven distribution of the body temperature of large high-speed gears. This deformation can be included in $F_{\beta x}$, or may be offset by a suitable helix modification. Furthermore, since the body temperature of a high-speed helical pinion is usually higher than that of the mating wheel, additional misalignment can occur, which must be taken into account in the calculations. Deformation induced by a large centrifugal force can be treated in the same way. Briefly, as regards the use of Eqs. (1.38) and (1.39), it is to be noted that, when a favorable position of the contact pattern exists, the manufacturing deviations and elastic deformations compensate each other. These effects of compensation are summarized in Fig. 1.10, which shows six different arrangements and, for each of them, provides the appropriate guideline for determining the initial equivalent misalignment before running-in, $F_{\beta x}$. Four of these arrangements refer to the most common mountings with pinion between bearings, while two refer to mountings with overhang pinions. The peak load intensity occurs on the helix near the torqued end. As shown in the next section, we have $B^* = 1$ and $B^* = 1.5$ for spur and single helical gears and, respectively, for double-helical gears.

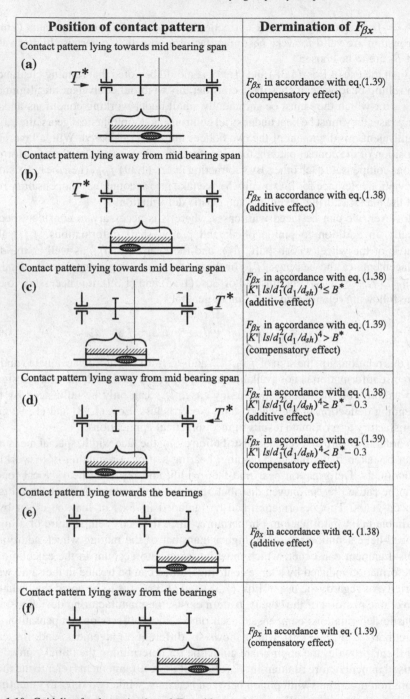

Position of contact pattern	Dermination of $F_{\beta x}$
Contact pattern lying towards mid bearing span **(a)**	$F_{\beta x}$ in accordance with eq.(1.39) (compensatory effect)
Contact pattern lying away from mid bearing span **(b)**	$F_{\beta x}$ in accordance with eq.(1.38) (additive effect)
Contact pattern lying towards mid bearing span **(c)**	$F_{\beta x}$ in accordance with eq. (1.38) $\lvert K'\rvert \, ls/d_1^2 (d_1/d_{sh})^4 \le B^*$ (additive effect) $F_{\beta x}$ in accordance with eq. (1.39) $\lvert K'\rvert \, ls/d_1^2 (d_1/d_{sh})^4 > B^*$ (compensatory effect)
Contact pattern lying away from mid bearing span **(d)**	$F_{\beta x}$ in accordance with eq. (1.38) $\lvert K'\rvert \, ls/d_1^2 (d_1/d_{sh})^4 \ge B^* - 0.3$ (additive effect) $F_{\beta x}$ in accordance with eq. (1.39) $\lvert K'\rvert \, ls/d_1^2 (d_1/d_{sh})^4 < B^* - 0.3$ (compensatory effect)
Contact pattern lying towards the bearings **(e)**	$F_{\beta x}$ in accordance with eq. (1.38) (additive effect)
Contact pattern lying away from the bearings **(f)**	$F_{\beta x}$ in accordance with eq. (1.39) (compensatory effect)

Fig. 1.10 Guideline for determination of $F_{\beta x}$ with regard to contact pattern position

– For gear pairs having ideal contact pattern, due to full helix modification under load, $F_{\beta x}$ is given by:

$$F_{\beta x} = F_{\beta x\min}, \tag{1.41}$$

where $F_{\beta x\min}$ is the greater of the two values $F_{\beta x\min} = 0.5 f_{H\beta}$ or $F_{\beta x\min} = k(F_m/b)$, where k is a dimensional constant given by $k = (5 \times 10^{-3}\,\text{mm}\,\mu\text{m/N})$. As full helix modification, we mean a helix modification intended to compensate the torsion and bending deflections of the pinion and wheel, as well as the deformations and displacements of other components under operating loads and, if known, the tooth alignment deviation of the mating wheel. A gear pair having optimum helix modification under design load should have $F_{\beta y} = 0$, i.e. $K_{H\beta} = 1$. However, for safety reasons, the aforementioned minimum values in Eq. (1.41) are to be used. Similarly, Eq. (1.39) can be used to choose a suitable crowning.

1.4.4.4 Equivalent Misalignment

Preliminarily, it is to be noted that the relationships below, concerning the calculations of elastic deflections of the pinion and pinion-shaft, i.e. the determination of f_{sh}, are simplified and are based on the following assumptions:

– Deflections of the wheel and wheel-shaft, deformations of the housing and bearings (for these deformations, only differences in the deflections are important), and effect of bearing clearances are not included in the basic calculations. Usually, in fact, the wheel, wheel-shaft, housing and bearings are sufficiently stiff, for which their deflections can be ignored. In the case it was required to take account of these deflections, or in the case where the configuration of the gear system is such that the clearances determine a significant shaft tilting, these deflections and shaft tilting must be independently evaluated, and the corresponding values are added to f_{ma} with the proper sign.
– Bending and torsional deflections of the pinion under the actual load distribution does not differ significantly from those determined assuming a uniform load distribution along the face width. This assumption is substantially valid for lower calculated values of $K_{H\beta}$, but becomes increasingly less valid for higher values of $K_{H\beta}$.
– Pinion shaft is solid (or it can be a hollow shaft, but with a ratio between the hole diameter and outside diameter less than 0.5), and pinion position is in accordance with Fig. 1.12. The restriction that the pinion is towards the center of the shaft span ($0 \leq s/l \leq 0.3$) does not apply in the case of suitable helix modification, while factor K' takes into account the stiffing effect of the pinion blank.
– Bearings do not absorb any bending moments, and the additional external loads acting on the pinion shaft, of whatever nature they are (for example, those due to the shaft couplings), have a negligible effect on the bending deflection of the shaft part corresponding to the gear face width.

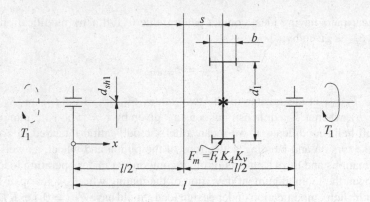

Fig. 1.11 Diagram for calculation of $f_{sh,t}$ and $f_{sh,b}$

– The shaft material is steel.

The equivalent misalignment, f_{sh} (in μm), takes into account the components of equivalent misalignment (the components in the plane of action are determinant) due to bending and twisting of the pinion and pinion shaft. Thus, we have $f_{sh} = \left(f_{sh,b} + f_{sh,t}\right)$, that is f_{sh} is the sum of contributions $f_{sh,b}$ and $f_{sh,t}$ respectively due to bending and torsion.

Assuming a uniform distribution of the mean transverse specific load $F_m/b = (F_t K_A K_v)/b$ on the face width, for a pinion with solid shaft, and with reference to the diagram shown in Fig. 1.11, the component $f_{sh,t}$ is given by:

$$f_{sh,t} = \frac{4}{\pi G} \frac{F_m}{b} \left(\frac{b}{d_1}\right)^2, \tag{1.42}$$

where G is the *shear modulus of elasticity* or *modulus of rigidity* of the material.

Instead, assuming that the load acting on the pinion is concentrated in the mid-face width, and that the pinion solid shaft has nominal external diameter d_{sh1} (in mm), the bending deflection component $f_{sh,b}$ can be approximately determined through the slope of the inflection curve, such as product of b for the deflection $y'(x)$ at the middle point of the shaft. Therefore, with the symbols shown in Fig. 1.11, we can write:

$$f_{sh,b} = by'(x) = \pm b F_m \frac{l^2}{3EI} \left[\frac{x}{l}\left(1 - \frac{x}{l}\right)\left(1 - 2\frac{x}{l}\right)\right], \tag{1.43}$$

where E and I are respectively the Young's modulus of material, and moment of inertia of the cross section.

In the usual variability range $(0 \le s/l \le 0.3)$ of ratio s/l, which characterizes the most interesting practical applications, the quantity within the brackets in Eq. (1.43) can be linearized, and taken as equal to $0.38\,(s/l)$, whereby with $x = [(l/2) + s]$ Eq. (1.43) becomes:

$$f_{sh,b} = \pm \frac{F_m}{b} \frac{64}{3\pi E} \frac{0.38s}{l} \left(\frac{bl}{d_{sh1}^2}\right)^2, \tag{1.44}$$

to be taken with the plus sign, if the input torque, T_1, comes from the right side, and with the minus sign, if it comes from the left side.

In the case where the modulus of elasticity and Poisson's ratio of the material are those of steel, by summing Eqs. (1.42) and (1.44) we obtain:

$$f_{sh} = f_{sh,t} + f_{sh,b} = A^* \frac{F_m}{b} \left[1 \pm 0.8 \frac{sl}{d_1^2} \left(\frac{d_1}{d_{sh1}}\right)^4\right] \left(\frac{b}{d_1}\right)^2, \tag{1.45}$$

where A^* is a dimensional constant equal to 1.55×10^{-2} mm μm/N.

However, since the comparison with the numerical results shows that the effective components $f_{sh,t}$ and $f_{sh,b}$ are not linear, the value of f_{sh} calculated with Eq. (1.45) is multiplied by 1.5 (then the dimensional constant $A^* = 1.55 \times 10^{-2}$ mm μm/N is replaced with the constant $A = 1.5A^* = 0.023$ mm μm/N. A further correction is then introduced, represented by the constant 0.3 appearing in Eqs. (1.46) and (1.47). This constant is related to the fact that the previous two components of the equivalent misalignment cannot be completely eliminated in reality, also because of their non-linear distributions.

In this framework, for both spur and single helical gears, Eq. (1.45) becomes:

$$f_{sh} = 0.023 \frac{F_m}{b} \left[\left|B^* + K' \frac{sl}{d_1^2} \left(\frac{d_1}{d_{sh1}}\right)^4 - 0.3\right| + 0.3\right] \left(\frac{b}{d_1}\right)^2, \tag{1.46}$$

where $B^* = 1$, if the total power is transmitted through a single engagement, and $B^* = [1 + 2(100 - k)/k]$, if there is more than one power path, and only $k\%$ of the input power is transmitted through one gear mesh. Instead, for double helical gears, Eq. (1.45) becomes:

$$f_{sh} = 0.046 \frac{F_m}{b} \left[\left|B^* + K' \frac{sl}{d_1^2} \left(\frac{d_1}{d_{sh1}}\right)^4 - 0.3\right| + 0.3\right] \left(\frac{b_B}{d_1}\right)^2, \tag{1.47}$$

where $B^* = 1.5$, if the total power is transmitted through a single engagement, and $B^* = [0.5 + (200 - k)/k]$, if only $k\%$ of the input power is transmitted through one gear mesh, while the determinant specific load (F_m/b) is calculated with $b = 2b_B$ (see Fig. 8.16a in Vol.1).

In both Eqs. (1.46) and (1.47), the factor K' can be taken from Fig. 1.12, which is valid in the case in which the ratio s/l is less than 0.3. This factor is a constant, which takes into account the position of the pinion with respect to the input or output torqued end, as well as the stiffening effect of the pinion when it is in one piece with its shaft.

Arrangement		Factor K' with | without stiffening

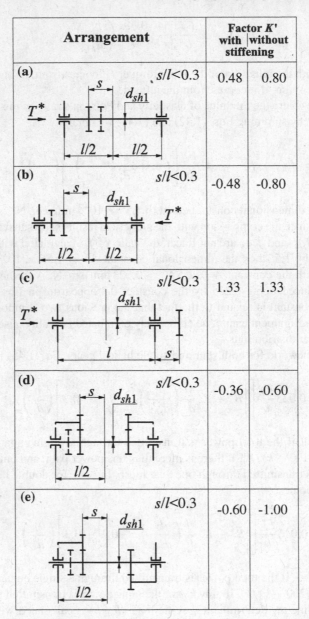

		with	without
(a) $s/l<0.3$		0.48	0.80
(b) $s/l<0.3$		-0.48	-0.80
(c) $s/l<0.3$		1.33	1.33
(d) $s/l<0.3$		-0.36	-0.60
(e) $s/l<0.3$		-0.60	-1.00

Fig. 1.12 Factor K' with and without stiffening effect for five different arrangements of the pinion

Stiffening effects occur depending on whether the ratio (d_1/d_{sh1}) is greater than or equal to 1.15 $(d_1/d_{sh1} \geq 1.15)$, or it is less than 1.15 $(d_1/d_{sh1} < 1.15)$; correspondingly, values of K' with and without stiffening must be chosen. To have the stiffening effect, the pinion needs to show solidarity with its shaft (also a standard shrink fit is effective in this regard; see Vullo 2014), while a poor stiffening, or no stiffening, is to be expected when the pinion slides on the shaft, due to the connection with feather key or a similar fitting. In Fig. 1.12, the dashed lines indicate the less deformed of the two halves of helix of a double-helical gear wheel, which is usually not considered, while T^* is the input or output torqued end, not dependent on direction of rotation.

For other arrangements or in the cases where there are additional shaft loads (e.g., from belt pulleys, or chain wheels) or the values of ratio (s/l) exceeds those specified in Fig. 1.12, a comprehensive analysis is recommended. In Eqs. (1.38) and (1.39), the absolute value of f_{sh} must be introduced.

It is to be noted that crowning and end relief, which are two types of helix modification, influence the face width values to be used in Eqs. (1.46) and (1.17). In fact, crowning is employed to compensate for manufacturing deviations and load-induced deformations, and particularly to relieve the tooth end-loading (generally, gears are crowned symmetrically with respect to the mid-face width). End relief is instead used to protect the tooth ends from the overloading due to equivalent misalignment. Usually, the amount of the end relief is the same at both ends of the teeth. We will deepen this topic in the Sect. 1.4.4.7.

Here we must keep in mind that, if the amount of the crowing or end relief is greater than the value specified in Sect. 1.4.4.7, the reduced face width, $b_{(b)}$, instead of face width, b, must be used in equations for load carrying capacity calculations. The value of this reduced face width, $b_{(b)}$, is determined from values of $C_{\beta(b)}$ or, respectively, $C_{I(II)(b)}$, calculated as shown in Sect. 1.4.4.7. We assume that the teeth ends outside the reduced face width, $b_{(b)}$, are not bearing any load.

For certain gears, the choice of an appropriate value of f_{sh} is made according to the experience with similar gear systems (e.g., $f_{sh} \cong 0 \, \mu m$, for very rigid gear units, for which deformations are neglected, or $f_{sh} \cong 6 \, \mu m$, for some turbine transmissions, for which occasionally the maximum value of the equivalent misalignment is specified). Instead, for other gears, the value of f_{sh} is specified as a percentage of the allowable helix slope deviation, $f_{H\beta}$, that is, in the case of a percentage of 100%, as:

$$f_{sh} = 1.0 f_{H\beta}, \tag{1.48}$$

In both cases, the assumptions must be validated by measurements or computations, and gears are to be designed accordingly.

Finally, it is to be noted that Eqs. (1.46) and (1.47) can be adapted, with slight variations, to address three types of issues for which they would not be strictly valid:

• The first is that in which the gear ratio is equal to unit or close to unit, and the torqued ends of the shaft are at opposite sides of the housing; in this case, the torsional deflections of the pinion and wheel are equal, but of opposite direction, for which they compensate each other, while the bending deflections are additive.

- The second is that of the meshing between the planet gear and annulus gear of a simple planetary gear train. In this case, similarly to what happens in all the idler gears, the planet gear is not subject to torsional deflection, and the main contribution to its bending deflection is that of the pin of the planet carrier, due to loads of the meshing with both the sun and annulus gears.
- The third is that of the meshing between the planet gear and sun gear of a simple planetary gear train. In this case, the sun gear is not subject to bending deflection, but its torsional deflection is the result of the multiple meshing with planet gears (however, deflection of the planet carrier can be significant at this mesh).

1.4.4.5 Mesh Misalignment Due to Manufacturing Deviations

The mesh misalignment due to manufacturing deviations, f_{ma}, is the maximum separation between the meshing teeth flanks, when the teeth are held in contact without significant load, and shaft journal bearings are in their working arrangements. This mesh misalignment is the result of the algebraic sum of all deviations of the individual members in the plane of action, i.e. the way in which these deviations are combined between them. Therefore, it will vary depending on whether the helix slope deviation, $f_{H\beta}$, of each gear member and the alignment deviation of the shafts are added or compensated, or whether the alignment of the shafts is adjustable or not (for example, with adjustable bearings).

For the determination of f_{ma}, machining tolerances come into play. In this regard, we must consider that, rarely, their maximum allowable values are fully used, as a coincidence of extreme and opposite values is very unlikely, and deviations can even compensate. It is therefore necessary that their mean values are introduced. The ISO 6336-1:2006 provides several ways of determination of f_{ma}, but in any case, it is recommended that the values used are verified by checking the contact pattern in the working conditions.

A first way derives f_{ma} from deviations of individual components, and it is to be done after inspection of the gears, bearings and housing, as well as appropriate measurements on them. The maximum and minimum values of the mesh misalignment, related to the most unfavorable and, respectively, to the most favorable combination of individual deviations, are given by the following relationship:

$$f_{ma,\max,\min} = \left(\left| f_{H\beta 1act} + f_{H\beta 2act} + f_{paract} \right| \right)_{\max,\min}, \qquad (1.49)$$

where $f_{H\beta 1act}$ and $f_{H\beta 2act}$ are the measured values of helix slope deviation of pinion and wheel (in accordance with ISO 1328-1:2013), and f_{paract} is the measured value of shaft misalignment, due to in-plane and out-of-plane non-parallelism deviations of both shafts of pinion and wheel. It is to be observed that, in the case of radial runout of one or more journal bearings, f_{paract} can vary with the angle of rotation. For load carrying capacity calculations of gears, the mean value of f_{ma} is to be used; it is given by the relationship:

$$f_{ma} = 0.5(f_{ma,max} + f_{ma,min}).$$ (1.50)

It is to be noted that, in this way, the influence of bearing clearances is neglected.

A second way, sometimes employed for certain gears, is to specify the permissible limits for the total manufacturing deviation, f_{ma}. For example, deviations can be neglected for a high precision manufacturing (in this case, we assume $f_{ma,max} = 0\,\mu m$), while we take $f_{ma,max} = 15\,\mu m$ as a realistic value for certain industrial transmissions. For load carrying capacity calculations of gears, the following mean value of f_{ma} is to be used:

$$f_{ma} = (2/3)f_{ma,max}.$$ (1.51)

A third way is used to determine f_{ma} for assembly of gears of a given accuracy grade, without any modification or adjustment. However, in this case, an inspection after assembly is recommended. According to this way, the most unfavorable combination of deviation for pinion, wheel and housing is given by the following relationship:

$$f_{ma} = f_{H\beta 1} + f_{H\beta 2} + f_{\Sigma\beta}\frac{b}{l},$$ (1.52)

where $f_{H\beta 1}$ and $f_{H\beta 2}$ are the helix slope deviation tolerances of pinion and wheel, for a given gear accuracy grade, and $f_{\Sigma\beta}$ is the alignment-of-axes tolerance. Experience shows that, in many manufacturing environments, the value of f_{ma} determined by means of Eq. (1.52) is the one that occurs with sufficient frequency, such as to justify its use for the calculations. However, statistical studies related to the quality control regime show that in only about 10% of cases the deviations will be combined so as to exceed a total value equal to $1.0 f_{H\beta 2}$, for which the most favorable combination of deviations for pinion, wheel and housing is given by:

$$f_{ma} = 1.0 f_{H\beta 2}.$$ (1.53)

In most cases, the appropriate value of f_{ma} to be used lies between the two extreme values given by Eqs. (1.52) and (1.53), while a useful relationship to calculate f_{ma} in the cases of an average quality control regime is the following:

$$f_{ma} = \sqrt{f_{H\beta 1}^2 + f_{H\beta 2}^2}.$$ (1.54)

The choice of a value of f_{ma} lower than that given by this last equation must be properly justified.

Two other ways of determination of f_{ma} deserve mention. However, both these ways cannot be used in the design stages, as the related application procedures involve experimental measurements on the assembled gear drive.

Fig. 1.13 Lengths of face width b, and contact pattern, b_{c0}

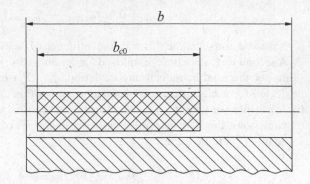

The first of these ways involves determination of f_{ma}, using Eq. (1.50), with gears assembled in the gear-case, without its top cover. Values of $f_{ma,max}$ and $f_{ma,min}$ are determined from measurements made around the circumference, using feeler gauges inserted between the working faces into light contact at both ends of the mesh. With this experimental procedure, f_{ma} is given by the relationship:

$$f_{ma} = \delta_g(b/l), \tag{1.55}$$

where δ_g is the maximum or minimum difference in the feeler gauge indications, b is the face width, and l is the distance between the feeler gauges. Obviously, these measurements include the contributions due to helix modifications.

The second of these ways involves determination of f_{ma} based on no-load contact pattern. Value of f_{ma} is derived by means of the following relationship:

$$f_{ma} = \left(\frac{b}{b_{c0}}\right)s_c, \tag{1.56}$$

where b_{c0} is the length, at low load, of the contact pattern of the assembled gears (Fig. 1.13), b is the face width, and s_c is the coating thickness of marking compound, generally comprised in the range between 2 and 20 μm (usually, a value $s_c = 6$ μm is assumed as a mean value consistent with good working practice).

Depending on whether, in the equation above, b_{c0} is the minimum value (b_{c0min}) or the maximum value (b_{c0max}) of length of the contact pattern, we have, respectively, the maximum value, $f_{ma,max}$, or the minimum value, $f_{ma,min}$, of f_{ma}. We will choose, as an average value suitable for preliminary design calculation, the one given by Eq. (1.51), while as value to recalculate the preliminary rated load carrying capacity, we assume the one given by Eq. (1.50).

1.4.4.6 Components of Mesh Misalignment Due to Housing Deformation and Shaft Displacement

Deformation f_{ca} of the gear case can be ignored in cases where the stiffness of the gear assembly, including the housing, is very high. In all other cases, it is necessary to take account of these deformations, which can be determined by testing and experimental measurements or, approximately, by using numerical models, such as FEM models (see Zienkiewicz 1977; Garro and Vullo 1979; Cook 1981; Cantone et al. 2001).

The components of misalignment in the plane of action, f_{be}, due to bearing deflections and displacements related to clearances in journal bearings, in some cases have effects greater than those of shaft and gear blank deflections. Usually, these components can be neglected when the pinion and wheel of spur and double helical gears are located midway between bearings having equal stiffness and clearance. When this does not happen, or in the case where this happens, but we are faced with single helical gears, or in the case of overhung gears, these components of misalignment can greatly affect the distribution of load along the face width. Moreover, for lubricated sliding journal bearings, misalignment effects can be enhanced due to two well-known phenomena. The first phenomenon consists of the complex actual pressure distributions within the lubricant film thickness, the latter no longer constant, but variable in the lengthwise direction, as the axes of pins and sliding bearings are not actually parallel, due to elastic deformations. The second phenomenon consists of the translation of the reaction radial forces with respect to the theoretical positions, due to the structural deformability of shafts, bearings and housing (Chirone and Vullo 1979).

In such cases, we must pay careful attention to the directions and signs of misalignments of bearing axes, since only the relative misalignments due to bearing deflections and displacements of the common axis of the pinion bearings, f_{be1}, and that of the wheel, f_{be2}, affect the equivalent misalignment. In the simplest cases shown in Fig. 1.14, characterized by a shaft with a single gear wheel and two bearings, f_{be} is given by:

$$f_{be} = f_{be1} \mp f_{be2}, \tag{1.57}$$

where the minus and plus signs are to be used when the effects of the relative misalignment are subtracted (Fig. 1.14a) or added (Fig. 1.14b). In these two cases, we have:

$$f_{be} = \frac{b}{l}(\delta_1 \mp \delta_2), \tag{1.58}$$

where δ_1 and δ_2 are the components of deflections of bearings 1 and 2 in the plane of action. Of course, the effect of tilting moment due to the axial component F_a of the total load acting on the single helical gear wheel must be taken into account.

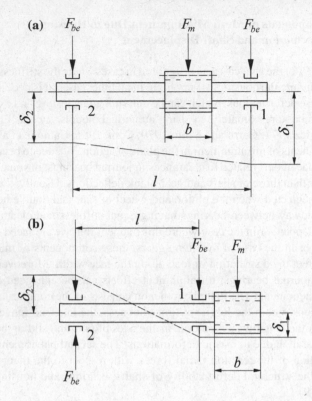

Fig. 1.14 Loads and deflections for: **a** gear wheel mounted between the bearings; **b** overhung mounted gear wheel

1.4.4.7 Intentional Modifications of Flank Line

The effects of manufacturing deviations, elastic deformations, and bearing deflections and displacements on the load carrying capacity and silent operation of the gears may be compensate, at least partially, by intentional modifications of the theoretical tooth traces and flank lines. These deliberate modifications consist of crowning, end relief, and helix modification of teeth, and require an adequate accuracy grade of the gear (at least ISO accuracy grade 7 or better). In the cases of considerable deformations, a helix modification is superposed over crowning or end relief, but a well-designed helix modification is generally preferable.

The topic concerning the intentional modifications of the profile and flank line of gear teeth and their optimization to maximize their surface durability and fatigue bending strength has always constituted a very challenging and fascinating aspect of their design. Scholars' interest in this topic has never ceased, as demonstrated by the numerous scientific papers that continue to be published in this regard (limiting ourselves to mentioning the most recent ones, see for example: Spitas and Spitas 2007; Jia et al. 2018; Sanchez et al. 2019; Zeleny et al. 2019). Here we will focus our

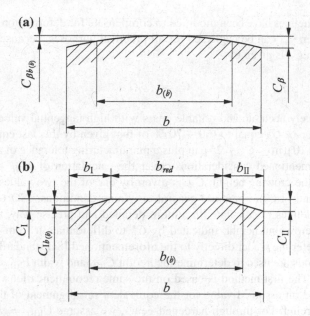

Fig. 1.15 Geometric quantities related to crowning (**a**) and end relief (**b**): height C_β and $C_{\beta b_{(b)}}$, and width $b_{(b)}$ of crowning; height $C_{I(II)}$ and width $b_{(b)}$ of end relief

attention on the results of these works that have been acquired to the ISO standard, not disdaining to do some theoretical deepening, when this is deemed necessary to a better understanding of the advantages related to these intentional modifications in terms of performance to be achieved with the gears to be designed.

As a non-mandatory rule, resulting from experience, the amount of crowning C_β (Fig. 1.15a), necessary to obtain an acceptable load distribution along the face width, can be assumed to be $C_\beta = 0.5F_{\beta xcv}$, where $F_{\beta xcv}$ is the initial equivalent misalignment for the estimation of the crowing height. However, the following limitations exist: (i) C_β must be included in the range $10\,\mu\text{m} \leq C_\beta \leq 40\,\mu\text{m}$ plus a manufacturing tolerance from 5 to $10\,\mu\text{m}$; (ii) the ratio (b_{cal}/b) must be greater than 1, as if the gear was not crowned (Fig. 1.8b). Consequently, also the initial equivalent misalignment $F_{\beta xcv}$ must be calculated as if the gear was not crowned, using a modified expression of Eq. (1.38) in which $1.0f_{sh}$ is substituted for $1.33f_{sh}$, with f_{sh} also determined as if the gear was not crowned.

In order to avoid excessive tooth end loads, the correct value, $f_{mac} = 1.5f_{H\beta}$, of the mesh misalignment due to manufacturing deviations must be used in place of that obtained with the relationships shown in Sect. 1.4.4.5. Thus, the crowning amount is given by:

$$C_\beta = 0.5\left(f_{sh} + 1.5f_{H\beta}\right). \tag{1.59}$$

When the helices have been modified to compensate for deformation at mid-face width, or when f_{sh} can be neglected for all practical purposes, because of the gear high stiffness, C_β is given by:

$$C_\beta = f_{H\beta}. \tag{1.60}$$

For extremely accurate and reliable gears with high tangential velocity, we can assume a value of C_β equal to $(60 \div 70)\%$ of that given by this last equation, with the limitation $10\,\mu m \le C_\beta \le 25\,\mu m$ plus a manufacturing tolerance of about $5\,\mu m$.

The aforementioned considerations cover the calculation of $K_{H\beta}$ for crowned gears where the crowing height, C_β, is given by one of the two values defined by Eqs. (1.59) and (1.60). For more general crowning conditions, ISO 6336-1:2006 provides guidelines to calculate, in a more detailed and precise way, the value of the required crowning height, indicated by C_β^* to differentiate it from C_β. On this subject, we refer the reader directly to the aforementioned ISO standard.

Two methods are used to determine the height $C_{I(II)}$ and width $b_{I(II)}$ of end relief (Fig. 1.15b). The first method is based on the same recommendations for the gear crowning, and on assumed value for the equivalent misalignment of the gear pair without end relief. For through-hardened gears, we assume $C_{I(II)} = F_{\beta xcv}$ plus a manufacturing tolerance of 5 to $10\,\mu m$, for which, by analogy with the value $f_{mac} = 1.5 f_{H\beta}$, we have:

$$C_{I(II)} = f_{sh} + 1.5 f_{H\beta}. \tag{1.61}$$

Instead, for case-hardened and nitrided gears, we assume $C_{I(II)} = 0.5 F_{\beta xcv}$ plus a manufacturing tolerance of 5 to $10\,\mu m$, for which, by analogy with Eq. (1.59), we have:

$$C_{I(II)} = 0.5\big(f_{sh} + 1.5 f_{H\beta}\big). \tag{1.62}$$

When the helices have been modified to compensate for deformation, or when f_{sh} can be neglected for all practical purposes, because of the gear high stiffness, we take $C_{I(II)} = f_{H\beta}$. For very accurate and reliable gears with high tangential velocity, we can take a value of $C_{I(II)}$ equal to $(60 \div 70)\%$ of that given by equations above.

As regards the value of width of end relief, i.e. the length measured along the face width (Fig. 1.5b), we have to distinguish the case of an about constant load and higher tangential velocities, and the case of variable load and low or average tangential velocities. In the first case, we take $b_{I(II)}$ equal to the smaller of the two values $(0.1b)$ and $(1.0m)$, while in the second case, it is appropriate to take $b_{red} = (0.5 \div 0.7)b$, and to determine consequently $b_{I(II)}$, dividing in equal parts the difference $(b - b_{red})$.

The second method, which assumes a uniform distribution of load along the face width, is based on the combined deflection of mating teeth, δ_{bth}, given by:

$$\delta_{bth} = \frac{F_m}{bc_{\gamma\beta}} = \frac{F_t K_A K_v}{bc_{\gamma\beta}}. \tag{1.63}$$

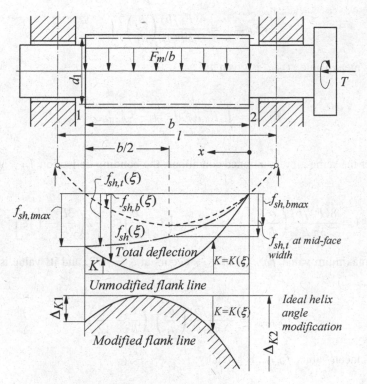

Fig. 1.16 Deflection of pinion shaft and piñion teeth

For highly accurate and reliable gears, with high tangential velocity, we take $C_{I(II)} = (2 \div 3)\delta_{bth}$, and $b_{red} = (0.8 \div 0.9)b$, while for gears with lesser accuracy, we assume $C_{I(II)} = (3 \div 4)\delta_{bth}$, and $b_{red} = (0.7 \div 0.8)b$.

Figure 1.16 shows how to determine, for a simple case, the helix angle modification of the pinion, which is necessary to compensate for deformations due to torsion and bending of pinion shaft, when the specific load (F_m/b) acting on the teeth is distributed uniformly, and bearings are arranged symmetrically with respect to the mid-face width. For cases that are more complex (e.g., bearings arranged asymmetrically, more than one gear wheel mounted on the same shaft, etc.), it is necessary to use more sophisticated numerical models, as well as suitable calculation programs.

In the simple case shown in Fig. 1.16, and under the assumptions described above, the torsion deflection $f_{sh,t}(\xi)$, for Poisson's ratio $\nu = 0.3$, is given by:

$$f_{sh,t}(\xi) = \frac{8(F_m/b)}{0.39\pi E}\left(\frac{b}{d_1}\right)^2 \xi\left(1 - \frac{\xi}{2}\right), \qquad (1.64)$$

where $\xi = (x/b)$ is the dimensionless variable. The maximum value $f_{sh,tmax}$ of $f_{sh,t}$ occurs at $\xi = 1$, and its value is given by:

$$f_{sh,t\max} = \frac{4(F_m/b)}{0.39\pi E}\left(\frac{b}{d_1}\right)^2,$$ · (1.65)

while its mean value, $f_{sh,tm}$ is given by:

$$f_{sh,tm} = \int\limits_0^1 f_{sh,t}(\xi)d\xi = \frac{2}{3}f_{sh,t\max}.$$ (1.66)

Under the same aforementioned conditions, the bending deflection $f_{sh,b}$ is given by:

$$f_{sh,b}(\xi) = \frac{8(F_m/b)}{3\pi E}\left(\frac{b}{d_1}\right)^4\left[\xi^4 - 2\xi^3 + 3\left(1 - \frac{l}{b}\right)\xi^2 + 2\left(\frac{3l}{2b} - 1\right)\xi\right].$$ (1.67)

The maximum value $f_{sh,b\max}$ of $f_{sh,b}$ occurs at $\xi = 1/2$, and its value is given by:

$$f_{sh,b\max} = \frac{2(F_m/b)}{\pi E}\left(\frac{b}{d_1}\right)^4\left(\frac{l}{b} - \frac{7}{12}\right),$$ (1.68)

while its mean value, $f_{sh,bm}$, is given by:

$$f_{sh,bm} = \frac{4(F_m/b)}{3\pi E}\left(\frac{b}{d_1}\right)^4\left(\frac{l}{b} - \frac{3}{5}\right).$$ (1.69)

The following approximate relationship follows:

$$f_{sh,bm} = \frac{2}{3}f_{sh,b\max}.$$ (1.70)

It is noteworthy that $f_{sh,t}(\xi)$ and $f_{sh,b}(\xi)$ are the components of equivalent misalignment in the plane of action, due respectively to torsion and bending deflections of the pinion shaft. Figure 1.16 shows, together with the quantities of interest, the distribution curves of $f_{sh,t}(\xi)$ and $f_{sh,b}(\xi)$ along the face width, and their sum $f_{sh}(\xi) = [f_{sh,t}(\xi) + f_{sh,b}(\xi)]$, which is the total deflection of the pinion shaft. The mirror image of this total deflection curve with respect to the straight-line tangent at point corresponding to its maximum value, and parallel to the shaft axis, represents the modified flank line of the tooth, that is the ideal helix angle modification, expressed by the function $K = K(\xi)$. Of course, the actual helix angle modification will be affected by machining tolerances.

In accordance with the assumption that the elastic deformation f_{sh} is linear along the face width, the mean value of the initial equivalent misalignment before running-in is equal to the sum of the mean values of torsion and bending deflections, i.e.:

$$(1/2)F_{\beta x} = f_{sh,tm} + f_{sh,bm} = \frac{2}{3}\left(f_{sh,tmax} + f_{sh,bmax}\right). \tag{1.71}$$

The effective equivalent misalignment after running-in, $F_{\beta y}$, is obtained multiplying $F_{\beta x}$ by the factor χ_β. Introducing the deflections calculated above into Eq. (1.28), which defines the face load factor $K_{H\beta}$, we get:

$$K_{H\beta} = \frac{c_{\gamma\beta}\left[\dfrac{F_m}{bc_{\gamma\beta}}\left(f_{sh,tm} + f_{sh,bm} - y_\beta\right)1000\right]}{c_{\gamma\beta}\dfrac{(F_m/b)}{c_{\gamma\beta}}} = 1 + \frac{c_{\gamma\beta}\chi_\beta\left(f_{sh,tm} + f_{sh,bm}\right)1000}{(F_m/b)}$$

$$= 1 + \frac{2c_{\gamma\beta}}{3\,(F_m/b)}\,\chi_\beta\left(f_{sh,tm} + f_{sh,bm}\right)1000. \tag{1.72}$$

Finally, it remains to justify the factor 1.33 appearing in Eqs. (1.38)–(1.40). This factor corrects the error arising from the assumption that the elastic deformation f_{sh} is linear. As we have seen (Fig. 1.16), the actualelastic deformation f_{sh} is parabolic. However, using the linear deformation formulation with $1.33 f_{sh}$, we get the same value of $K_{H\beta}$ that would be obtained using the actual parabolic deformation with $1.0 f_{sh}$. With increasing curvature of actual elastic deformation line, the assumption that the elastic deformation is linear may lead to increasing differences between calculated and actual load distributions; therefore, the accuracy of the calculated value of $K_{H\beta-C}$ becomes worse as its magnitude increases. The curvature of actual elastic deformation line increases with heavily loaded gear pairs, or large values of face width to diameter ratio of the pinion, or both.

1.4.4.8 Determination of Face Load Factor for Tooth Root Stress, $K_{F\beta-C}$

As we saw at beginning of Sect. 1.4.1, the face load factor, $K_{F\beta}$, takes into account the effects of the load distribution along the face width on the tooth root stress. Therefore, it is a factor similar to $K_{H\beta}$, which, however, depends not only on the variables that influence $K_{H\beta}$, but also on the ratio (b/h) between the face width and tooth depth.

This factor is determined according to the bending theory of a rectangular plate (in view from above), having variable thickness (and thus variable flexural rigidity), clamped along one edge and loaded by a concentrated load F applied at an arbitrary point of the opposite edge (Fig. 1.17). The usual assumptions are made, according to which any strain in the middle plane of the plate during bending is neglected, and stress component along the normal to this middle plane is also neglected. Consequently, the bending stress state of the plate is a generalized plane stress state (see Jaramillo 1950; Timoshenko and Woinowsky-Krieger 1959; Kagawa 1961; Wellauer and Seireg 1960; Ventsel and Krauthammer 2001; Vullo 2014). The figure shows the

Fig. 1.17 Model of
rectangular plate used for
determination of $K_{F\beta}$

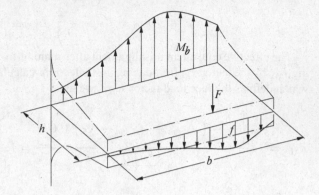

geometry and load condition of this plate, as well as the distribution curves of the
deflection, f, along the loaded edge, and the bending moment, M_b, along the clamped
edge.

Based on the aforementioned theory, we find:

$$K_{F\beta} = \left(K_{H\beta}\right)^{N_F},\tag{1.73}$$

where N_F is a factor depending on the ratio (b/h), given by:

$$N_F = \frac{(b/h)^2}{1 + (b/h) + (b/h)^2} = \frac{1}{1 + (h/b) + (h/b)^2}.\tag{1.74}$$

In this last equation, it is necessary to introduce the smaller of the two values
(b_1/h_1) and (b_2/h_2), and, in the limit condition in which $(b/h) < 3$, we put $(b/h) =
3$. Finally, the face width of one helix, b_B, is to be used instead of b, for double-helical
gears.

1.5   Transverse Load Factors, $K_{H\alpha}$ and $K_{F\alpha}$

1.5.1   Generality

The transverse load factors, $K_{H\alpha}$ and $K_{F\alpha}$, take into account the effects of the non-
uniform distribution of transverse load between the various teeth pairs simultaneously
in meshing on the surface stress, and respectively on the tooth root stress. These
factors are defined as the quotient of the maximum tooth load in the meshing of
a gear pair running at rotational speed close to zero $\left(n \cong 0\,\mathrm{min}^{-1}\right)$ divided by the
corresponding tooth load of an ideal gear pair, i.e. free from inaccuracies, running
under similar conditions.

These two transverse load factors are mainly influenced by the deflections under load, tooth manufacturing accuracy, profile modifications, and running-in effects. For their determination, two methods are provided: Method A and Method B. The values of these factors approach unity in the case of optimum profile modification appropriate for loads, high specific load level, uniform load distribution along the face width, and high manufacturing accuracy.

With Method A, $K_{H\alpha-A}$ and $K_{F\alpha-A}$ are assumed equal to unity when the maximum tooth loads, including the inner dynamic tooth loads and the effect of non-uniform distribution of loads, are determined directly by experimental measurements or by a comprehensive mathematical analysis. In addition, a comprehensive analysis of all influence factors can be used to determine the load distribution in the tangential direction. The sharing of the total tangential load between simultaneously meshing gear pairs can be also obtained by strain gauge measurements, made at the tooth roots of gears transmitting loads at low rotational speeds.

1.5.2 Method B for Determination of $K_{H\alpha-B}$ and $K_{F\alpha-B}$

This method is based on the assumption that the average difference between the base pitches of the pinion and wheel is the main parameter that affects the load distribution between several tooth pairs simultaneously in meshing. Consistently with normal manufacturing practice, the assumption is also made according to which the base pitch deviations appropriated to the specified accuracy grade of the gear are distributed around the circumference of the pinion and wheel. The *transverse base pitch deviation*, f_{pb}, takes into account the total effect of all gear tooth deviations that influence these factors. However, the *profile form deviation*, $f_{f\alpha}$, is to be taken instead of the base pitch deviation in the case in which $f_{f\alpha}$ is greater than f_{pb} ($f_{f\alpha} > f_{pb}$).

With Method B, which is suitable for all types of gears (spur or helical gears with any accuracy, and any basic rack profile), the transverse load factors can be determined by calculation or using special graphic diagrams. Their determination by calculation is made with the following relationships (they do not apply when the gear teeth have some intentional profile deviations):

$$K_{H\alpha} = K_{F\alpha} = \frac{\varepsilon_\gamma}{2}\left[0.9 + 0.4\frac{c_{\gamma\alpha}(f_{pb} - y_\alpha)}{(F_{tH}/b)}\right], \tag{1.75}$$

to be used for gears with total contact ratio $\varepsilon_\gamma \leq 2$, and

$$K_{H\alpha} = K_{F\alpha} = 0.9 + 0.4\frac{c_{\gamma\alpha}(f_{pb} - y_\alpha)}{(F_{tH}/b)}\sqrt{\frac{2(\varepsilon_\gamma - 1)}{\varepsilon_\gamma}}, \tag{1.76}$$

to be used for gears with total contact ratio $\varepsilon_\gamma > 2$.

In these equations, $c_{\gamma\alpha}$ is the mesh stiffness (see next section), y_α is the running-in allowance for a gear pair, $F_{tH} = F_t K_A K_v K_{H\beta}$ is the determinant tangential load in a transverse plane, and f_{pb} is the larger of the transverse base pitch deviations of the pinion or wheel. It is to be noted that $0.5 f_{pb}$ may be used when profile modifications compensate for the teeth deflections at the actual load level. When the two materials differ, first the values $y_{\alpha1}$ and $y_{\alpha2}$ for pinion and wheel materials must be separately determined, and therefore the mean value

$$y_\alpha = \frac{1}{2}(y_{\alpha1} + y_{\alpha2}), \tag{1.77}$$

is to be used for the calculation.

The values $K_{H\alpha}$ and $K_{F\alpha}$ calculated with Eqs. (1.75) and (1.76) are to be used if they fall within the following ranges of variability:

$$1 \le K_{H\alpha} \le \frac{\varepsilon_\gamma}{\varepsilon_\alpha Z_\varepsilon^2} \tag{1.78}$$

$$1 \le K_{F\alpha} \le \frac{\varepsilon_\gamma}{0.25\varepsilon_\alpha + 0.75}, \tag{1.79}$$

where Z_ε is the *contact ratio factor* for pitting (see Sect. 2.9.4). If $K_{H\alpha} > \varepsilon_\gamma/\varepsilon_\alpha Z_\varepsilon^2$ or $K_{H\alpha} < 1$, the values corresponding to the two limit conditions, $K_{H\alpha} = \varepsilon_\gamma/\varepsilon_\alpha Z_\varepsilon^2$ and $K_{H\alpha} = 1$, should be taken. Similarly, if $K_{F\alpha} > \varepsilon_\gamma/(0.25\varepsilon_\alpha + 0.75)$ or $K_{F\alpha} < 1$, the values corresponding to the two limit conditions, $K_{F\alpha} = \varepsilon_\gamma/(0.25\varepsilon_\alpha + 0.75)$ and $K_{F\alpha} = 1$, should be taken.

The first limit condition, $K_{H\alpha} = K_{F\alpha} = 1$, implies a nearly uniform distribution of the total tangential load on the teeth pairs that are simultaneously in meshing. The second limit condition, $K_{H\alpha} = \varepsilon_\gamma/\varepsilon_\alpha Z_\varepsilon^2$ or $K_{F\alpha} = \varepsilon_\gamma/(0.25\varepsilon_\alpha + 0.75)$, implies that the total tangential load is transferred by only one pair of mating teeth. Furthermore, to ensure that the favorable peculiar characteristics of the helical gears are not undermined, but are used to the best, it is recommended that their accuracy be chosen in such a way that $K_{H\alpha}$ and $K_{F\alpha}$ are not greater than ε_α. Consequently, it may be necessary to limit the tolerances of the base pitch deviation of helical gears with coarse accuracy grade.

The transverse load factors $K_{H\alpha}$ and $K_{F\alpha}$ can be also determined as a function of the dimensionless quantity $q_\alpha = [c_{\gamma\alpha}(f_{pb} - y_\alpha)/(F_{tH}/b)]$ using the distribution curves shown in Fig. 1.18, which are consistent with Eqs. (1.75) and (1.76).

1.5.3 Running-in Allowance for a Gear Pair

The value y_α of the running-in allowance for a gear pair is the amount by which the initial transverse base pitch deviation, f_{pb}, is reduced by running-in from the start of operation. It is affected by the same variables that influence the running-in allowance

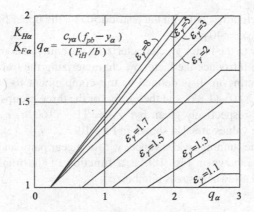

Fig. 1.18 Curves for determination of the transverse load factors, $K_{H\alpha}$ and $K_{F\alpha}$

for equivalent misalignment, y_β (see Sect. 1.4.4.2). As well as y_β, also y_α does not take into account the effects of running-in operations due to removal of material, obtained through particular finishing processes, such as lapping. These effects are taken into consideration with the gear accuracy grade.

Here also, in the absence of reliable data, like those obtained in accordance with Method A, and therefore obtained by experimental measurements or operating experience, the determination of y_α with Method B is carried out with the three relationships given below, each of which refers to the same three groups of ferrous metals described in Sect. 1.4.4.2.

The relationship related to the first group of ferrous metals is as follows:

$$y_\alpha = \frac{160}{\sigma_{H\,\text{lim}}} f_{pb}, \tag{1.80}$$

to be used under the following conditions: for $v \leq 5$ m/s, there are no restrictions; for 5 m/s $< v \leq 10$ m/s, the upper limit of y_α is $12.8 \times 10^3/\sigma_{H\,\text{lim}}$, corresponding to $f_{pb} = 80\,\mu$m; for $v > 10$ m/s, the upper limit of y_α is $6.4 \times 10^3/\sigma_{H\,\text{lim}}$, corresponding to $f_{pb} = 40\,\mu$m.

The relationship related to the second group of ferrous metals is as follows:

$$y_\alpha = 0.275 f_{pb}, \tag{1.81}$$

to be used under the following conditions: for $v \leq 5$ m/s, there are no restrictions; for 5 m/s $< v \leq 10$ m/s, the upper limit of y_α is $22\,\mu$m, corresponding to $f_{pb} = 80\,\mu$m; for $v > 10$ m/s, the upper limit of y_α is $11\,\mu$m corresponding to $f_{pb} = 40\,\mu$m.

The relationship related to the third group of ferrous metals is as follows:

$$y_\alpha = 0.075 f_{pb}, \tag{1.82}$$

to be used for all velocities, but with the restriction that the upper limit of y_α is 3 μm, corresponding to $f_{pb} = 40\,\mu m$.

Again, similarly to the factor χ_β characterizing the equivalent misalignment after running-in, we can introduce the factor χ_α characterizing the transverse base pitch deviation after running-in. It is defined as the complement to unit of the factors appearing in Eqs. (1.80)–(1.82), and therefore, for the three groups of ferrous metals above, its value is respectively given by: $\chi_\alpha = [1 - (160/\sigma_{H\,\lim})]$; $\chi_\alpha = 0.725$; $\chi_\alpha = 0.925$. These values are also shown in Fig. 1.9.

The values of the running-in allowance, y_α, for a gear pair can also be read from distribution curves shown in Fig. 1.9, as a function of the transverse base pitch deviation, f_{pb}, and allowable stress number, $\sigma_{H\,\lim}$.

1.6   Tooth Stiffness Parameters

As *tooth stiffness*, we define the load (in N) acting on 1 mm face width, directed along the line of action, required to produce, in the same direction, the deformation of 1 μ m of one or more pairs of deviation-free teeth in contact. This deformation, in an involute profile toothing, is equal to the length of arc of base circle, corresponding to the load-induced rotation angle of one of the two members of a gear pair when the mating member is held stationary. The load that comes into play is therefore F_{bt}, i.e. the nominal transverse load in plane of action, that is the base tangent plane. Tooth deflection can however be determined approximately using F_t (F_m, F_{tH}, …) instead of F_{bt}, when the differences between these two quantities deriving from known conversion factors can be ignored in comparison to other uncertainties (e.g., tolerances and similar).

The tooth stiffness so defined depends on the position of point of contact along the path of contact. It has a minimum value at points that delimit the path of contact (the marked points A and E in Fig. 2.8 in Vol. 1), corresponding respectively to the beginning and end of the contact, while it reaches its maximum value in the central zone of the path of contact, where the bending moment arms are about equal for pinion and mating wheel. As *single stiffness*, c', we define this maximum value of the tooth stiffness. It represents the maximum stiffness of a single pair of teeth of a spur gear, and is equal approximately to the maximum stiffness of a tooth pair in single pair contact. As approximate maximum value of the single stiffness, c', in the case in which $\varepsilon_\alpha > 1.2$, we can assume the value corresponding to the outer point of single pair tooth contact (i.e., the marked point D in Fig. 1.22).

In the case of helical gears, the tooth stiffness along the path of contact varies even more strongly than is the case in the spur gears. This follows from the fact that, in spur gears with deviation-free teeth in contact, the whole face width of a tooth pair participates in the load carrying capacity of the toothing, and the bending moment arms are constant, while in the helical gears the lengths of the instantaneous lines of contact as well as the bending moment arms are continuously variable. At points A and E' of the rectangle of contact (see Fig. 8.6 in Vol. 1), the length of the instantaneous lines of contact, at least from a theoretical point of view, is equal to zero, so that in these positions of contact the tooth stiffness is very low. It increases with the length of the instantaneous line of contact, and assumes its maximum value when these instantaneous lines cover the entire face width. However, the value c' for helical gears is the maximum stiffness normal to the helix of one tooth pair.

The mean value of stiffness of all the teeth in mesh is the *mesh stiffness*, c_γ. The tooth stiffness parameters, c' and c_γ, are affected by the following main factors: tooth data, in terms of basic rack profile, number of teeth, profile shift, transverse contact ratio, helix angle, etc.; blank design, in terms of rim and web thicknesses; specific load in normal direction to the tooth flank; type of attachment of the hub to the shaft; roughness and waviness of the tooth flank surfaces; mesh misalignment of the gear pair; modulus of elasticity of the materials. For determination of c' and c_γ, two methods are provided: Method A and Method B.

With Method A, c'_A and $c_{\gamma-A}$ are determined by a comprehensive analysis including all influences. This analysis can be performed experimentally, with direct measurements on the gear pair of interest, or by using theoretical models based on the theory of elasticity, or suitable numerical models FEM, or mixed models (see, for example: Vijayakar 1991; Velex and Maatar 1996; Attorri et al. 2001; Cantone et al. 2004).

Method B for determination of c'_B and $c_{\gamma-B}$ is based on study of the elastic behavior of solid disk spur gears. Generally, this method, described in detail below, is sufficient accurate for calculations of the dynamic factor and face load factors as well as for determination of profile and helix modifications for gears in the following conditions:

- external spur and helical gears with $\beta \leq 45°$ and specific load $[(F_t K_A)/b] \geq 100$ N/mm;
- any design of both members of the gear pair, but both made of steel;
- any basic rack profile of the toothing, and type of attachment of the hub to the shaft able to spread uniformly the transmission of torque around the circumference (pinion integral with shaft, interference fitting or splined fitting).

Method B can also be used appropriately or with further auxiliary factors for gears in the following conditions: internal gears; specific load $[(F_t K_A)/b] < 100$ N/mm; combination of materials of the two members of the gear pair other than steel/steel, and type of attachment of the hub to the shaft other than those described above (e.g., with fitted key).

In any case, the virtual numbers of teeth of pinion and wheel of the helical gear pair in the normal section can be evaluated with the approximate relationship (8.47 in Vol. 1), instead of with the exact relationship (8.48 in the same Vol. 1), for which we have $z_{n1} \cong z_1/\cos^3\beta$ and $z_{n2} \cong z_2/\cos^3\beta$.

The first step of the method consists of determining the *theoretical single stiffness*, c'_{th}, defined as the tooth stiffness of a solid disk cylindrical spur gear having teeth with standard basic rack profile defined according to ISO 53:1998 by: $\alpha_P = 20°$, $h_{aP} = m_n$, $h_{fP} = 1.2m_n$ and $\rho_{fP} = 0.2m_n$, where as usual the second subscript, P, indicates the rack profile. For a helical gear, c'_{th} is the theoretical single stiffness of correlated virtual spur gear, having virtual numbers of teeth, z_{n1} and z_{n2}, in the normal section defined above. The theoretical single stiffness c'_{th} is given by:

$$c'_{th} = \frac{1}{q'} \tag{1.83}$$

where q' is the minimum value of the flexibility of a tooth pair, given by the following sample series expansion based on an assumed specific load $F_t/b = 300$ N/mm:

$$q' = C_1 + \frac{C_2}{z_{n1}} + \frac{C_3}{z_{n2}} + C_4 x_1 + \frac{C_5 x_1}{z_{n1}} + C_6 x_2 + \frac{C_7 x_2}{z_{n2}} + C_8 x_1^2 + C_9 x_2^2, \tag{1.84}$$

where x_1 and x_2 are the profile shift coefficients of pinion and wheel, while the nine coefficients C_1 to C_9 can be read in Table 1.8.

Equations (1.83) and (1.84) apply for the range: $x_1 \geq x_2$, and $-0.5 \leq (x_1 + x_2) \leq 2.0$. Deviations of the actual values from those calculated in the range [100 N/mm $\leq (F_{bt}/b) \leq 1600$ N/mm] are between $+5\%$ and -8%.

For the aforementioned external spur and helical gears for which Method B is sufficiently accurate, the single stiffness, c', is obtained using the following relationship:

$$c' = c'_{th} C_M C_R C_B \cos\beta. \tag{1.85}$$

This equation, which provides acceptable average values of c', adapts the theoretical single stiffness, c'_{th}, to the actual conditions of the gear pair of interest, through the introduction of the factors defined below, C_M, C_R, C_B, and $\cos\beta$, each of which takes into account certain influences.

Correction factor, C_M; takes into account the differences between the measured values and the theoretical calculated values for solid disk gears, which are essentially due to the fact that the Hertz theory (Hertz 1882) provides coarsely approximate values of contact deformations, especially when the specific load is low. This correction factor is assumed constant and equal to

$$C_M = 0.8. \tag{1.86}$$

Gear blank factor, C_R, takes into account the flexibility of gear rims and webs. The following two relationships provide average values of C_R, which are suitable for cases in which the blank of the mating wheel has stiffness equal to or greater than that of the gear wheel under consideration. The choice of average values of the gear blank factor is justified by the fact that other uncertainties make their influence felt.

Table 1.8 Coefficients for Eq. (1.84)

C_1	C_2	C_3	C_4	C_5	C_6	C_7	C_8	C_9
0.04723	0.15551	0.25791	−0.00635	−0.11654	−0.00193	−0.24188	0.00529	0.00182

Among these uncertainties, those that determine an uneven tooth stiffness along the face width deserve to be mentioned, as it happens in a gear of webbed design. The two relationships to be used are:

$$C_R = 1,\tag{1.87}$$

for solid disk gear blank, and

$$C_R = 1 + \frac{\ln(b_s/b)}{5e^{(s_R/5m_n)}},\tag{1.88}$$

for gear body with rim and web (Fig. 1.19). This last relationship, where b_s is the web thickness, and s_R is the rim thickness, is to be used with the following limit conditions: take $b_s/b = 0.2$, when $b_s/b < 0.2$; take $b_s/b = 1.2$, when $b_s/b > 1.2$; take $s_R/m_n = 1$, when $s_R/m_n < 1$. Relationship (1.88), where the natural logarithm ln is used, i.e. the logarithm base e, provides results consistent with those that may be obtained using the curves shown in the diagram of Fig. 1.19, with a margin of error in the range -1% to $+7\%$. It should be noted that Eqs. (1.87) and (1.88) provide average values of C_R (and thus of c'_{th}) on the face width. The theoretical single stiffness, however, is not constant, but variable along the face width, since in the

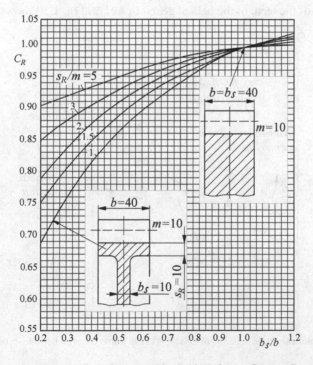

Fig. 1.19 Curves for determination of the wheel basic blank factor, C_R, as a function of the gear rim thickness, s_R, and central web thickness, b_s

area adjacent to the mid-face width, sufficiently distant from the free ends, we have a plane strain state, but, as the distance from the free ends is reduced, a plane stress state is to occur, from which a further fall of the local stiffness follows.

Basic rack factor, C_B, takes into account the deviations of the actual basic rack profile of the gear from the standard basic rack profile that comes into play in the formulation of Eqs. (1.83) and (1.84). This factor is determined by the following relationship:

$$C_B = \left[1 + 0.5\left(1.25 - h_{fP}/m_n\right)\right]\left[1 - 0.02(20° - \alpha_{Pn})\right]. \qquad (1.89)$$

It should be remembered that, when the pinion basic rack dedendum is different from that of the wheel, C_B is given by the following arithmetic mean:

$$C_B = \frac{1}{2}(C_{B1} + C_{B2}), \qquad (1.90)$$

where C_{B1} and C_{B2} are respectively the basic rack factors for a gear pair conjugate to the pinion basic rack, and for a gear pair conjugate to the wheel basic rack.

Finally, the term $\cos\beta$, that is the last factor appearing in Eq. (1.85), transforms the theoretical single stiffness of teeth of virtual spur gear of a helical gear pair from the normal theoretical single stiffness, $c'_{th,n}$, into the transverse theoretical single stiffness, c'_{th}, of the teeth of the helical gears. In fact, as Fig. 1.20 shows, we have:

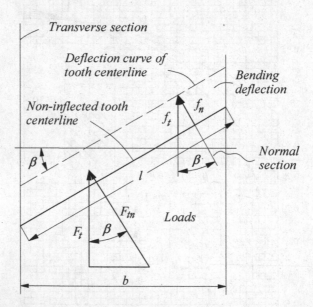

Fig. 1.20 Theoretical single stiffness of helical gears in transverse and normal sections

$$c'_{th,n} = \frac{F_{tn}}{f_n l} \qquad c'_{th,t} = c'_{th} = \frac{F_t}{f_t b} = c'_{th,n} \cos \beta. \tag{1.91}$$

In these equations, F_t and $F_{tn} = (F_t / \cos \beta)$ are the transverse and normal tangential loads at reference cylinder per mesh, f_t and $f_n = f_t \cos \beta$ are the transverse and normal bending deflections, and $l = b / \cos \beta$ is the length of flank line or tooth trace.

The following additional notes are relevant:

- Eqs. (1.83) and (1.84) provide approximate values of the theoretical single stiffness of teeth of internal gears, provided that in these equations we replace z_{n2} with infinity.
- The value of c' for material combinations other than steel/steel can be determined using the following relationship:

$$c' = c'_{St/St}\left(\frac{E}{E_{St}}\right), \tag{1.92}$$

where E_{St} is the Young's modulus of steel and E is the equivalent modulus of elasticity given by:

$$E = \frac{2E_1 E_2}{(E_1 + E_2)}. \tag{1.93}$$

The ratio (E/E_{St}) is equal to 0.74 and 0.59 for material combinations steel/gray cast iron, and respectively gray cast iron/gray cast iron. E_1 and E_2 are respectively the modulus of elasticity of pinion and wheel materials.

- Shaft and gear assembly influences the single stiffness. Under constant load, the single stiffness varies between maximum and minimum values twice per revolution, if the pinion or the wheel or both are connected to the shaft(s) with a fitted key. The minimum value is approximately equal to the single stiffness with shrink or spline fits. The average value of single stiffness is about 5% greater than the minimum one, when only one member of the gear pair is press fitted onto a shaft with a fitted key (the mating member is connected to its shaft by means of a shrink or spline fit). Indeed, it is about 10% greater than the minimum one when both members of the gear pair are press fitted onto their shafts with fitted keys.
- With regard to the specific load, it is to consider the fact that, for $(F_t K_A)/b \geq 100$ N/mm, the single stiffness c' can be assumed to be constant, while it decreases under reduced load, i.e. for $(F_t K_A)/b < 100$ N/mm. For these low specific loads, we can determine an approximate value of c' with the following relationship:

$$c' = c'_{th} C_M C_R C_B \cos\beta \left(\frac{F_t K_A/b}{100}\right)^{0.25}. \tag{1.94}$$

It remains now to calculate the mesh stiffness, c_γ. First of all, we distinguish between $c_{\gamma\alpha}$, which is used to calculate the internal dynamic factor, K_v, and the transverse load factors, $K_{H\alpha}$ and $K_{F\alpha}$, and $c_{\gamma\beta}$, which is used to calculate the face load factors, $K_{H\beta}$ and $K_{F\beta}$.

First, however, to bring the relationships that express $c_{\gamma\alpha}$ and $c_{\gamma\beta}$, it is necessary that we make a very important premise. In the case of spur gears that usually have a transverse contact ratio in the range ($1 < \varepsilon_\alpha < 2$), one or two teeth pairs are simultaneously in contact. Since the total transverse load is constant, the load acting on each teeth pair and the corresponding deflection change throughout the meshing cycle. Consequently, also the total tooth stiffness fluctuates (this total tooth stiffness considers the total effect on the tooth of the instantaneous single stiffness of teeth pairs, and therefore it is equal to the normal total load acting on the tooth profile per mm of face width, and for μm of deflection). Figure 1.21a shows this fluctuation, as well as the distribution curve of c' along the path of contact.

In the case of helical gears, more than two teeth pairs are usually at simultaneous contact, and increasingly with the overlap ratio, ε_β. Therefore, the total tooth stiffness fluctuates less than what happens with spur gears, also due to the nearly sinusoidal distribution curve of the single stiffness. Here also Fig. 1.21b shows this fluctuation as well as the nearly sinusoidal distribution curve of c' along the path of contact. This cyclic variability of total tooth stiffness, higher in spur gears and lower, but also present, in helical gears, involves an irregular transmission of motion, resulting in deviations from the theoretical line of action, as well as the occurrence of secondary dynamic loads. For calculations that interest us here, it is necessary that an average value of the total tooth stiffness during the meshing cycle be introduced: this is the mesh stiffness.

For spur gears with $\varepsilon_\alpha \geq 1.20$ and helical gears with $\beta \leq 30°$, the mesh stiffness, $c_{\gamma\alpha}$, is calculated with the following relationship:

$$c_{\gamma\alpha} = c'(0.75\varepsilon_\alpha + 0.25). \tag{1.95}$$

For spur gears with $\varepsilon_\alpha < 1.20$, the value $c_{\gamma\alpha}$ can be up to 10% less than values calculated with this last equation.

The mesh stiffness $c_{\gamma\beta}$ is instead calculated with the following relationship:

$$c_{\gamma\beta} = 0.85c_{\gamma\alpha}. \tag{1.96}$$

1.7   Load Distribution on Pairs of Teeth in Simultaneous Meshing

Spur gear pairs are always characterized by a transverse contact ratio ε_α greater than 1, usually included in the range ($1 < \varepsilon_\alpha < 2$). Therefore, one or two pairs of teeth

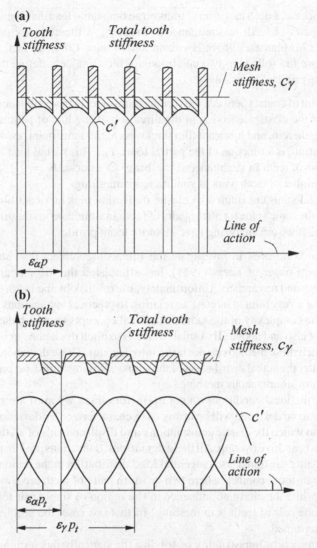

Fig. 1.21 Fluctuations of the total tooth stiffness and mesh stiffness in: **a** spur gears; **b** helical gears

are in simultaneous meshing. For helical gear pairs, which cannot rely only on the transverse contact ratio, ε_α, but also on the overlap ratio, ε_β, resulting in a total contact ratio $\varepsilon_\gamma = \varepsilon_\alpha + \varepsilon_\beta$ that often is greater than 2, we can also have more than two pairs of teeth simultaneously in meshing. In these double or multiple contact conditions, if the toothing was free from deviations compared with those related to the ideal case, the nominal tangential load F_t (see Sect. 1.1) would be shared between the various teeth pairs in simultaneous meshing according to their single stiffness.

In the ideal case, a determination of transverse tangential load distribution between the various pairs of teeth in simultaneous meshing is theoretically possible (see Karas 1941; Giovannozzi 1965b; Niemann and Winter 1983). In fact, to this end, we can impose the following two conditions, which uniquely define the statically indeterminate problem to be solved:

- At each point of contact between the various pairs of teeth in simultaneous meshing, the sum of the elastic deflection in the direction of the line of action, due to the load acting therein, and lubricant film thickness at the same point, evaluated in the same direction, is a function of the partial load, F_{ti}. This partial load is applied to the i-th pair of teeth in simultaneous meshing. Of course, $F_t = \sum_{i=1}^{m} F_{ti}$, where m is the number of teeth pairs in simultaneous meshing.
- All aforesaid sums constituted by elastic deflection plus lubricant film thickness must have the same value for all the pairs of teeth in simultaneous meshing, because the gears that we are analyzing have involute tooth profiles.

Here there is no need to investigate the interesting subject that, starting from the well-known paper of Karas (1941), has stimulated the theoretical interest of many scientists and researchers. Unfortunately, the results obtained with the methods proposed have a very limited interest, in relation to practical applications. This is due not only to the complexity of the same theoretical concepts on which these methods are based, but also and above all to small transverse pitch deviations or other similar errors, of which these methods do not takes into account. These deviations and errors can greatly alter the actual distribution of the transverse tangential load on the various pairs of teeth in simultaneous meshing.

In reality, the ideal conditions do not exist, for which we must always consider that the gears to be designed will be in any case characterized by deviations from the ideal shape, to which the elastic deformations and displacements of all the structural members of a gear drive overlap. All these deviations, deflections, and displacements, significantly alter the transverse tangential load distribution in the regions of double or multiple pair tooth contact, where two or more pairs of teeth are simultaneously in meshing, while they have no influence in the region of single pair tooth contact, where only one pair of teeth is in meshing. In this last case, the problem is in fact statically determined.

Given the practical impossibility of solving the statically indeterminate problem by means of an accurate theoretical method, for calculations of the load carrying capacity of the gears the convention is used according to which the total transverse tangential load, in the double or multiple contact regions, is divided equally between the various pairs of teeth simultaneously in meshing. Figure 1.22 shows this type of conventional load distribution in a spur gear pair with $(1 < \varepsilon_\alpha < 2)$. The same figure shows the inner point B of single pair tooth contact of the pinion, and the inner point D of single pair tooth contact of the wheel, when the pinion is driving. On the path of contact, these points, B and D, are respectively distant from marked points E and A of a length equal to the transverse base pitch, p_{bt}.

The conventional load distribution shown in Fig. 1.22 may be considered sufficiently correct for a toothing without errors. Any departure from the assumed ideal

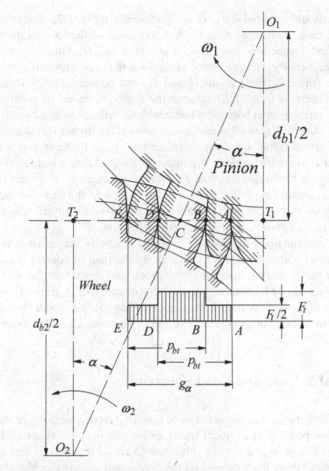

Fig. 1.22 Conventional load distribution in a spur gear pair

conditions of the teeth implies significant changes to this load distribution. The factors that most influence the actual load distribution are mainly the transverse base pitch deviations, but also deviations of profile shape, flank line and base circle, elastic deformations, radial oscillations, and so on, make their influence felt, although to a lesser degree than that caused by the pitch deviation. From time to time, in the following chapters, we will see how this conventional load distribution is modified depending on these main factors.

We come to the same conclusions when we consider the helical gears that, especially if they are characterized by a large value of the overlap ratio, are much more sensitive to deviations and errors than spur gears. In such a case, it often happens that only one pair of teeth is to withstand the partial loads of several pairs of teeth simultaneously in meshing.

Finally, it should be noted that, still with reference to Fig. 1.22 concerning a spur gear pair, all these deviations do not affect the value of the load in the region of single pair tooth contact (the one between points B and D). On the contrary, these deviations significantly affect the value of the load in the two regions of double pair tooth contact (those between points A and B, and between points D and E). In particular, an increase in the load value in the region bounded by points A and B occurs (therefore, the upper horizontal line translates upward, with a consequent load value greater than $F_t/2$), and this increase is greater the higher the aforementioned deviations. Correspondingly, a decrease in the load value in the region bounded by points D and E occurs (therefore, the upper horizontal line translates downward, with a consequent load value less than $F_t/2$), and this decrease is greater the higher the deviations above. Everything finds its justification in the fact that the hatched area in the figure must necessarily remain constant, given the invariance of the total elastic strain energy involved in the problem of interest.

This is a fundamental principle of theory of elasticity that must never be forgotten, whenever a load distribution curve along the path of contact is considered, and whatever the deviations from the conventional one described above. Therefore, these deviations can be homothetic or non-homothetic, that is, translations parallel to themselves of the contour lines in the regions of double pair tooth contact, or true variations of these contour lines (see Love 1944; Timoshenko and Goodier 1951).

1.8   Variable Load and Load Spectrum

Often gears are subjected to *variable loads*, resulting from a working process, starting process or operation at a critical speed, or near to the critical speed. It follows that also the teeth of a gear set are subjected to *variable stresses*. The frequency and magnitude of these loads depend on the driven and driving machines, mass distributions and elastic properties of the entire mechanical system that includes the gear drive under consideration. These loads can be determined by various procedures, such as: experimental measurements under the operating conditions of the gear drive; estimation of the spectrum acquired with a similar gear set, and similarly operating; calculation with models able to properly simulate the external excitation and distribution of elastic masses of the gear system under consideration, preferably followed by experimental testing to validate the calculation.

To assess with a sufficient degree of reliability the effects of load variability on the fatigue damage, the *load spectra* should be available. To obtain these spectra, the range of the measured (or calculated) loads is divided into *bins* or classes of equal or different sizes, each of which contains the number of load occurrences recorded in its load range. Usually, it is preferable to use smaller bin sizes at the upper loads, and larger bin sizes at the lower loads in the range. In this way, the most damaging loads are limited to fewer calculated load cycles, and the resulting gears may have smaller sizes. However, to properly assess the loads acting on the teeth, the torques to be used should include the dynamic effects at different rotational speeds.

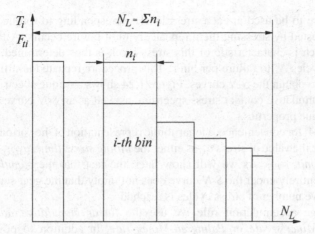

Fig. 1.23 Example of a cumulative torque (or load) spectrum

Here we do not consider appropriate to further dwell on the most suitable procedures to obtain such load spectra. On this subject, we refer the reader to ISO 6336-6:2006. However, here it is necessary to specify that the *torque spectrum* thus obtained is only valid for the measured or evaluated period of time taken into consideration, which should be sufficiently extended to capture extreme load peaks. In any case, this spectrum, measured or calculated, should be extrapolated to represent the required lifetime of the gear set to be designed. Stress spectra concerning pitting and bending can be obtained from the load spectra.

Figure 1.23 shows an example of a *cumulative torque* (or *load*) *spectrum*. On ordinate axis, the torque T_i per bin i or the transverse tangential load F_{ti} per bin i can indifferently be shown. On abscissa axis, n_i is the *number of cycles per bin i*, N_i the *number of cycles to failure per bin i*, and $N_L = \sum n_i$ the *number of load cycles* or *cumulative number of applied cycles*. It is to be noted that n_i is equal to: the number of load cycles of the shaft, in the case of only one contact per revolution; a multiple of the number of load cycles of the shaft, in the case of multiple contacts per revolution, as it happens for the idler gear wheels or the planet gears of a planetary gear train.

The general calculation of service life is based on the theory that every load cycle causes a *damage* of the gear, whose amount depends on the *stress level*. The damage can be considered as zero for lower stress levels. The calculated *fatigue life* of a gear subjected to pitting and bending loads is a measure of its ability to accumulate discrete damage until failure occurs. To calculate the gear fatigue life, the *stress spectrum*, *material fatigue properties*, and a *damage accumulation rule* are required (see Palmgren 1924; Miner 1945; Heywood 1962; Almen and Black 1963; Giovannozzi 1965a; Juvinall 1967; Sors 1971; Stephens et al. 2000; Schijve 2009; Radaj and Vormwald 2013).

Strength values are based on material fatigue properties, and chosen from applicable *S-N* curves. To get a statistical interpretation for a specific probability, many specimens are to be used at each stress level, corresponding to torque T_i at the i-th bin, and tested by stressing them repeatedly until failure occurs. A failure cycle number, which is characteristic of this stress level, is thus determined, that is, the number of cycles N_i to failure per bin i. This procedure repeated at different stress levels leads to obtain the *S-N* curves. Figure 1.24 shows a torque spectrum with the associated cumulative contact stress spectrum, as well as an *S-N* curve for specific material fatigue properties.

In Fig. 1.24, for convenience, a logarithmic representation of the various quantities is used, while the value of stress σ_G is either the *pitting stress limit*, σ_{HG}, or the *tooth root stress limit*, σ_{FG}. As we will show later, the fact that the *cumulative stress spectrum* is entirely under the *S-N* curve does not imply that the gear survives, once the cumulative number of stress cycles is reached.

As damage accumulation rule we use the *linear cumulative damage rule*, also called *Miner's rule* or *Palmgren-Miner rule*, in addition to other rules or modifications. This rule assumes that:

- The damaging effect of each stress cycle at a given stress level is the same, regardless of whether the stress cycle is the first, or last, or an intermediate cycle, that is, irrespective of the temporal order in which it is applied.
- The portion of the useful fatigue life consumed by a number of cycles, n_i, for bin i, is equal to the quotient of n_i divided by the number of cycles to failure, N_i, for bin i; therefore, it is given by the ratio $U_i = (n_i/N_i)$, which is usually called *individual damage quotient*.
- The total fatigue life of a part can be estimated by adding up the percentage of the portions of life consumed by each bin.

Therefore, failure could be expected when the following equality is verified:

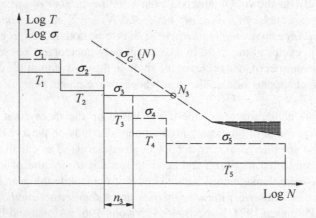

Fig. 1.24 Torque spectrum and associated cumulative contact stress spectrum, and *S–N* curve for specific material fatigue properties

$$\sum_{i=1}^{k} \frac{n_i}{N_i} = \frac{n_1}{N_1} + \frac{n_2}{N_2} + \cdots + \frac{n_k}{N_k} = 1. \qquad (1.97)$$

While many deviations from the Palmgren-Miner rule have been observed, and numerous modifications to this relationship have been proposed, none has been proved better or gained wide acceptance. It is then to be considered that the material fatigue characteristics and endurance data are usually related to a specific and required failure probability, e.g., 1%, 5% or 10%. Depending on whether an endurance limit exists (upper horizontal line beyond the knee in Fig. 1.24), or an endurance limit does not exist (lower line beyond the knee in Fig. 1.24), the calculation is done respectively only for stresses above this endurance limit, or must be done for all stress levels. For each stress level, i, the number of load cycles to failure, N_i, are taken from the corresponding part of the S-N curve.

For the evaluation of the service strength, a widely known method is usually used, based on application of linear cumulative damage calculations according to the Palmgren-Miner rule. For its ease of use, this method is generally preferred to other methods described in the literature, perhaps even more accurate, but more elaborate. It is only valid for recalculation, and implies the preliminary calculation of the stress spectrum from the torque spectrum. To this end, the torques at the upper limit of each torque class, i, and the associated number of cycles, n_i, are considered, and for each level T_i of the torque spectrum, the actual stresses σ_{Hi} and σ_{Fi} for contact and bending loads in accordance with the following relationships are separately calculated:

$$\sigma_{Hi} = Z_H Z_E Z_\varepsilon Z_\beta Z_{BD} \sqrt{\frac{2000 T_i}{d_1^2 b}\left(\frac{1+u}{u}\right) K_{vi} K_{H\beta i} K_{H\alpha i}} \qquad (1.98)$$

$$\sigma_{Fi} = \frac{2000 T_i}{d_1 b m_n} Y_F Y_S Y_\beta K_{vi} K_{F\beta i} K_{F\alpha i}, \qquad (1.99)$$

where all the load-dependent K-factors are calculated for each torque class. In these two equations, the application factor K_A does not appear (it is set equal to 1), since with this method all the application load influences are included in stress levels. The meaning of factors Z_H, Z_E, Z_ε, Z_β and Z_{BD} will be given in Chap. 2, while that of factors Y_F, Y_S and Y_β will be described in Chap. 3.

Once the stress spectra are obtained, using the Palmgren-Miner linear-cumulative-damage rule, the individual damage quotients $U_i = (n_i/N_i)$ are calculated for all values of σ_i (σ_{Hi} or σ_{Fi}). Then, these individual damage quotients are added together, imposing that the following condition is satisfied:

$$U = \sum_i U_i = \sum_i \frac{n_i}{N_i} \leq 1.0, \qquad (1.100)$$

where U is the Miner sum or sum of individual damage quotients, which expresses the *damage condition*.

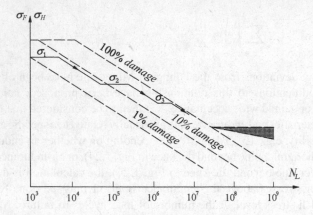

Fig. 1.25 Damage percentage lines, and cumulative damage

The calculated values are compared with the strength values, corresponding to the damage percentage lines of *S-N* curves for pitting and bending strength, determined by experimental tests or with standardized methods. It is to be noted that the calculation of the speed-dependent parameters for each load level, as well as the determination of the *S-N* curve, are based on the average rotational speed. It is also to be noted that, where teeth are loaded in both directions (e.g., idler gear wheels), the values determined for tooth root strength must be reduced (see Chap. 3). Furthermore, the reverse torque influences the contact stress spectrum of the rear flank, for which the damage accumulation has to be considered separately for each flank side.

Figure 1.25 gives a graphical representation of the procedure described above, which is to be applied to each pinion and gear wheel for both contact and bending stress. From diagrams like the one shown in this figure, we can deduce whether the gear wheel examined survive or not to the total number of stress cycles.

The safety factor, *S*, must also be calculated separately for pinion and gear wheel, and for contact and bending stresses. In principle, it cannot directly be inferred from the Palmgren-Miner linear-cumulative-damage rule, but must be determined by iteration methods, on which it is not the place to dwell (see ISO 6336-6:2006). For the highest stress of the design life, the safety factor for static load strength is also to be calculated.

Since the stress values in the range $(0 \div 10^3)$ cycles may exceed those corresponding to the elastic limit for contact or bending load of the tooth, for the calculations related to the accumulation of damage, ISO considers only stress levels by 10^3 cycles up, and then neglects those that fall in that range. The safety factor for static load strength, however, is to be calculated for the highest stress of the design life, which can be either the maximum stress in the load spectrum or an extreme transient load that is not considered in the fatigue analysis. Depending on the load applied and material used, a single stress cycle in the range between $(0 \div 10^3)$ cycles, and having value greater than the limit stress level at 10^3 cycles, could result in plastic yielding of the gear tooth.

1.9 Determination of Application Factor, K_A, from a Given Torque Spectrum

When the geometrical data of the gear set are not fixed yet, as it happens during the gear design stage, a first estimation of value of the application factor, K_A, can be made on the basis of a given torque spectrum, by agreement between purchaser and gear unit manufacturer. In this case, we define the application factor as the quotient of the equivalent torque, T_{eq}, divided by the nominal torque, T_n, that is:

$$K_A = \frac{T_{eq}}{T_n}. \tag{1.101}$$

The equivalent torque, T_{eq}, is given by the following relationship:

$$T_{eq} = \left(\frac{n_1 T_1^p + n_2 T_2^p + \cdots + n_i T_i^p + \cdots}{n_1 + n_2 + \cdots + n_i + \cdots} \right)^{1/p}, \tag{1.102}$$

where n_i and T_i are respectively the number of cycles and torque per ben i (with $i = 1, 2, \ldots$), and p is the slope of the Wöhler-damage line, the values of which for pitting and tooth root strength are given in Table 1.9, as a function of heat treatment of the steel (Wöhler 1870; Fatemi and Yang 1998). It is to be noted that values of p for pitting are given for torque, for which they must be doubled to convert them for stress. Factor K_A must be determined for pinion and gear wheel, each for both pitting and bending, and the highest of these four values has to be used for a gear rating.

The determination of the equivalent torque, T_{eq}, assumes that the torque spectrum, the slopes p of the Wöhler-damage lines, and the number of load cycles at the reference point, N_{Lref}, are known. The Wöhler-damage lines used are the simplified ones, i.e. those obtained by ignoring any damage that may occur at stresses below some limit stress. Moreover, since the position of the endurance limit in terms of stress is not known until the gear design is available, we assume that the position of the same endurance limit in terms of cycles does not change when the gear design changes.

Table 1.9 Exponent p and number of load cycles N_{Lref}

Heat treatment	Surface durability (pitting)		Tooth root strength	
	p	N_{Lref}	p	N_{Lref}
Case-carburized	6.610	5×10^7	8.738	3×10^6
Through-hardened	6.610	5×10^7	6.225	3×10^6
Nitrided	5.709	2×10^6	17.035	3×10^6
Nitro-carburized	15.715	2×10^6	84.003	3×10^6

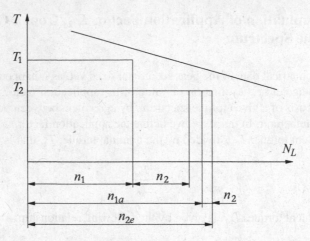

Fig. 1.26 Bin (T_2, n_{2e}) replacing bins (T_1, n_1) and (T_2, n_2)

In similar way to what happens in the rolling-element bearings (Giovannozzi 1965a), we assume that the relationship between n_i and T_i is of the type:

$$T_i^p n_i = T_j^p n_j = const. \tag{1.103}$$

Therefore, a torque T_i in the bin i can be replaced by a torque T_j in a new bin j in such a way that the damage caused by the two torques T_i and T_j is the same. Figure 1.26 shows this equivalence for the two torques T_1 and T_2, and their corresponding numbers of load cycles, n_1 and $(n_1 + n_2)$. On this basis, to determine T_{eq} we can activate the procedure that follows. We denote first of all the torque bins (T_i, n_i), and number them in order of decreasing torque, for which T_1 is the highest torque (n_1 is the corresponding number of cycles). Then, according to Eq. (1.103), a larger number of cycles n_{1a} at lower torque T_2, given by:

$$n_{1a} = n_1 \left(\frac{T_1}{T_2} \right)^p, \tag{1.104}$$

will be equivalent to the number of cycles n_1 at torque T_1. The bins 1 and 2 can be replaced by a single bin (T_2, n_{2e}), if $n_{2e} = n_2 + n_{1a}$, as Fig. 1.26 shows.

Similarly, a larger number of cycles n_{2a} at lower torque T_3, given by

$$n_{2a} = n_{2e} \left(\frac{T_2}{T_3} \right)^p, \tag{1.105}$$

will be equivalent to the number of cycles n_{2e} at torque T_2. Then, bins 1, 2 and 3 can be replaced by a single bin (T_3, n_{3e}), if $n_{3e} = n_3 + n_{2a}$. The procedure is stopped when n_{ie} reaches the endurance limit cycles, N_{Lref}, the values of which are given in Table 1.9.

The equivalent torque, T_{eq}, to be determined will therefore be in the range $(T_i < T_{eq} < T_{i-1})$, for which K_A will be included in the range

$$\frac{T_i}{T_n} < K_A < \frac{T_{i-1}}{T_n}, \qquad (1.106)$$

and its value can be obtained by linear interpolation on a log-log basis.

References

Almen JO, Black PH (1963) Residual stresses and fatigue in metals. McGraw-Hill Book Company Inc, New York

Attorri R, Salvini P, Vivio F, Vullo V (2001) A mixed finite element—numerical solution for mesh stiffnesses evaluation. In: MPT 2001—FUKUOCA, The JSME international conference on motion and power transmissions, Fukuoca, Japan, 15–17 Nov, GDN-10, vol I, pp 51–56

Biezeno CB, Grammel R (1953) Technische Dynamik: Dampfturbinen und Brennkraftmaschinen, Zweiter Band, Zweite Erweiterte Auflage. Springer, Heidelberg

Buckingham E (1949) Analytical mechanics of gears. McGraw-Hill Book Company Inc, New York

Buzdugan G (1964) La Measure des Vibrations Mécaniques. Eyrolles Éditeur, Paris

Calderale PM, Regalzi G, Vullo V (1984) Lo smorzamento interno dei materiali metallici da costruzione, Seminario sui Problemi di Smorzamento delle Vibrazioni nei Materiali, nelle Strutture e nelle Macchine, Atti della Giornata per la Presentazione del Premio Agostino A. Capocaccia, Genova, 30 May

Cantone C, Cantone L, Salvini P, Vullo V (2004) Valutazione della Variabilità della Rigidezza d'Ingranamento lungo la Linea di Contatto negli Ingranaggi Elicoidali, Atti XXXIII Convegno AIAS e XIV Convegno ADM Innovazione nella Progettazione Industriale, Bari, 31 agosto - 2 settembre

Cantone L, Salvini P, Vullo V (2001) A general dynamic modelling procedure for power gear transmission. In: MPT 2001—FUKUOCA, The JSME international conference on motion and power transmissions, Fukuoca, Japan, 15–17 Nov, GDN-19, vol I, pp 102–108

Carmignani C (2001) Dinamica Strutturale. Edizioni ETS, Pisa

Chirone E, Vullo V (1979) Cuscinetti a Strisciamento. Libreria Editrice Universitaria Levrotto&Bella, Torino

Cook RD (1981) Concepts and applications of finite element analysis, 2nd edn. Wiley, New York

Den Hartog JP (1985) Mechanical vibrations. Dover Publications Inc, New York

Dudley DW (1962) Gear handbook. the design, manufacture, and application of gears. McGraw-Hill Book Company, Inc., New York

Fatemi A, Yang L (1998) Cumulative fatigue damage and life prediction theories: a survey of the state of the art for homogeneous materials. Int J Fatigue 20(1):9–34

Ferrari C, Romiti A (1966) Meccanica Applicata alle Macchine. Unione Tipografica–Editrice Torinese (UTET), Torino

Garro A, Vullo V (1979) Acoustic problems of vehicle transmission, Nauka I Motorna Vozila '79, Bled, Slovenija, Jugoslavija, 4–7 June

Garro A, Vullo V (1978a) Alcune considerazioni sul proporzionamento degli ingranaggi, Atti del VI Convegno Nazionale AIAS, Brescia, 22–24 June

Garro A, Vullo V (1978b) Note integrative sulla memoria: Alcune considerazioni sul proporzionamento degli ingranaggi, Atti del VI Convegno Nazionale AIAS, Vol. II, Discussioni, Brescia, 22–24 June

Géradin M, Rixen DJ (2015) Mechanical vibrations: theory and application to structural dynamics, 3rd edn. Wiley, Chichester, UK

Giovannozzi R (1965a) Costruzione di Macchine, vol I, 2nd revised and expanded edn. Casa Editrice Prof. Riccardo Pàtron, Bologna

Giovannozzi R (1965b) Costruzione di Macchine, vol II, 4th edn. Casa Editrice Prof. Riccardo Pàtron, Bologna

Henriot G (1979) Traité théorique et pratique des engrenages, vol 1, 6th edn. Bordas, Paris

Hertz HR (1882) Über die Berührung fester elastischer Körper. Journal für Reine und Angewandte Mathematik (Crelle's J.) 92:156–171

Heywood RB (1962) Designing against fatigue. Chapman & Hall Ltd, London

Houser DR, Seireg A (1970) An experimental investigation of dynamic factors in spur and helical gears. Trans ASME Ser B J Eng. Ind 92:495–503

ISO 1328-1:2013 Cylindrical gears-ISO system of flank tolerance classification-Part 1: definitions and allowable values of deviations relevant to flanks of gear teeth

ISO 53:1998 Cylindrical gears for general and for heavy engineering—standard basic rack tooth profile

ISO 6336-1:2006 Calculation of load capacity of spur and helical gears—Part 1: basic principles, introduction and general influence factors

ISO 6336-6:2006 Calculation of load capacity of spur and helical gears—Part 6: calculation of service life under variable load

ISO 10825:1995 Gears—wear and damage to gear teeth—terminology

Jaramillo TJ (1950) Deflections and moments due to a concentrated load on a cantilever plate of infinite length. Trans ASME Ser E J Appl Mech 17:67–72

JGMA 7001-01 (1990) Terms of gear tooth failure modes

Jia C, Fang ZD, Zhang XJ, Yang XH (2018) Optimum design and analysis of helical gear tooth profile modification. J Huazhong Univ Sci Technol 46(5):66–71 (Natural Science Edition)

JIS B0160 (1999) Gears—wear and damage to gear teeth—terminology

Juvinall RC (1967) Stress, strain, and strength. McGraw-Hill Book Company, New York

Kagawa T (1961) Deflections and moments due to a concentrated edge-load on a cantilever plate of infinite length. In: Proceedings of 11th Jap. Nat. Congr. Appl. Mech., pp 47–52

Karas F (1941) Elastische Formänderung und Lastverteilung beim Doppeleingriff gerader Stirnradzähne, VDI – Forschungheft 406, B, Bd. 12

Ker Wilson W (1956) Practical solution of torsional vibration problems, vol I. Chapman and Hall, London

Ker Wilson W (1963) Practical solution of torsional vibration problems, vol II. Chapman and Hall, London

Krall G (1970a) Meccanica Tecnica delle Vibrazioni: Parte Prima Sistemi Discreti. Eredi Virgilio Veschi, Roma

Krall G (1970b) Meccanica Tecnica delle Vibrazioni: Parte Seconda Sistemi Continui. Eredi Virgilio Veschi, Roma

Lazan BJ (1968) Damping of materials and members in structural mechanics. Pergamon Press, Oxford

Li S, Kahraman A (2013) A tribo-dynamic model for a spur gear pair. J Sound Vib 332:4963–4978

Love AEH (1944) A treatise on the mathematical theory of elasticity, 4th edn. Dover Publications Inc, New York

Maitra GM (1994) Handbook of gear design, 2nd edn. Tata McGraw-Hill Publishing Company Ltd, New Delhi

Marguerre K, Wölfer K (1979) Mechanics of vibration. Sijthoff & Noordhoff, Alphen aan den Rijn, The Netherlands

Merritt HE (1971) Gear Engineering. Sir Isaac Pitman & Sons Ltd, London

Miner MA (1945) Cumulative damage in fatigue. J Appl Mech 12:A159–A164

Niemann G, Winter H (1983) Maschinen-Elemente, Band II: Getriebe allgemein, Zahradgetriebe-Grundlagen, Stirnradgetriebe. Springer, Berlin

Orban F (2011) Damping of materials and members in structures. J Phys Conf Ser 268 (1)

Palmgren A (1924) Die Lebensdauer von Kugellagern. Verfahrenstechinik 68:339–341 (Berlin)

Panetti M (1937) Lezioni di Meccanica Applicata alle Macchine, Parte II, Ruote-Roteggi-Macchine Funicolari-Cingoli. Arti Grafiche Pozzo, Torino

Pollone G (1970) Il Veicolo. Libreria Editrice Universitaria Levrotto & Bella, Torino

Radaj D, Vormwald M (2013) Advanced method of fatigue assessment. Springer International Publishing AG, Heidelberg

Radzevich SP (2016) Dudley's handbook of practical gear design and manufacture, 3rd edn. CRC Press, Taylor&Francis Group, Boca Raton, FL

Roda-Casanova V, Sanchez-Marin FT, Gonzalez-Perez I, Iserte JL, Fuentes A (2013) Determination of the ISO face load factor in spur gear drives by the finite element modelling of gears and shaft. Mech Mach Theor 65:1–13

Sanchez MB, Pleguezuelos M, Pedrero JI (2019) Influence of profile modifications on meshing stiffness, load sharing, and transmission error of involute spur gears. Mech Mach Theor 139:506–525

Schijve J (2009) Fatigue of structures and materials. Springer International Publishing AG, Heidelberg

Scotto Lavina G (1990) Riassunto delle Lezioni di Meccanica Applicata alle Macchine: Cinematica Applicata, Dinamica Applicata - Resistenze Passive - Coppie Inferiori, Coppie Superiori (Ingranaggi – Flessibili – Freni). Edizioni Scientifiche SIDEREA, Roma

Seireg A, Houser DR (1970) Evaluation of dynamic factors for spur and helical gears. Trans ASME Ser B J Eng Ind 92:504–515

Shipley EE (1967) Gear failures: how to recognize them, what causes them, how to avoid them. Machine Design, Dec 7

Sors L (1971) Fatigue design of machine components. Pergamon Press Ltd, Oxford

Spitas V, Spitas C (2007) Optimizing involute gear design for maximum bending strength and equivalent pitting resistance. Proc Inst Mech Eng Part C J Mech Eng Sci 221(4):479–488

Stephens RI, Fatemi A, Stephens RR, Fuchs HO (2000) Metal fatigue in engineering, 2nd edn. Wiley, New York

Tenot A (1953) Measure des Vibrations et Isolation des Assises de Machines. Dunod, Paris

Thomson WT (1965) Vibration theory and applications. Prentice-Hall, Inc., Upper Saddle River, NJ

Timoshenko SP, Goodier JN (1951) Theory of elasticity. McGraw-Hill Book Company Inc, New York

Timoshenko SP, Woinowsky-Krieger S (1959) Theory of plates and shells, 2nd edn. McGraw-Hill International Editions, Singapore

Timoshenko SP, Young DH (1951) Engineering mechanics. McGraw-Hill Book Company, New York

Townsend DP (1991) Dudley's gear handbook. McGraw-Hill Book Company Inc, New York

Velex P, Maatar M (1996) A mathematical model for analyzing the influence on shape deviations and mounting errors on gear dynamic behavior. J Sound Vib 191(5):629–660

Ventsel E, Krauthammer T (2001) Thin plates and shells: theory, analysis, and applications. Marcel Dekker, Inc, New York

Vijayakar S (1991) A combined surface integral and finite element solution for a three-dimensional contact problem. Int J Numer Method Eng 31:525–545

Vullo V (2014) Circular cylinders and pressure vessels: stress analysis and design. Springer International Publishing Switzerland, Heidelberg

Warburton GB (1976) The dynamical behaviour of structures, 2nd edn. Pergamon Press, Oxford

Wellauer EJ, Seireg A (1960) Bending strength of the gear teeth by cantilever-plate theory. ASME J Eng Ind 82(3):213–220

Wöhler A (1870) Über die Festigkeitsversuche mit Eisen und Stahl. Zeitschrift für Bauwesen 20:73–106

Winter H, Kojima M (1981) A study on the dynamics of geared system—estimation of overload on gears in system. In: International symposium on gearing & power transmissions, Tokyo

Zeleny V, Linkeova I, Sykora J, Skalnik P (2019) Mathematical approach to evaluate involute gear profile and helix deviations without using special gear software. Mech Mach Theor 135:150–164

Zienkiewicz OC (1977) The finite element method, 3rd edn. McGraw-Hill Higher Education, UK

Chapter 2
Surface Durability (Pitting) of Spur and Helical Gears

Abstract In this chapter, a general survey is first done on the surface durability (pitting) of spur and helical gears, also focusing attention on pitting damage and safety factor to be used in their design. The theoretical bases of surface durability are then discussed, with particular reference to the surface and subsurface stress states that occur in sliding and rolling contacts between the mating surfaces of the conjugate flanks of the teeth in relative motion. In this framework, same brief reminder on the elastohydrodynamic lubrication theory is provided, also describing the great influence it exerts on the aforementioned contacts as well as the EDH lubrication conditions necessary to reduce or avoid surface and subsurface fatigue damages in these types of gears. Finally, the procedure for calculating the surface durability of these same gears in accordance with the ISO standards is described, highlighting when deemed necessary as the relationships used by the same ISO are anchored to the theoretical base previously discussed.

2.1 General Survey

The starting point of design calculations of a gear pair is its *performance limit*, beyond which failures occur. Basically, in most practical applications, the following four main types of damage of different nature characterize the performance limit of a gear pair, namely:

- tooth root breakage
- pitting (i.e. macropitting)
- scuffing
- wear.

It should however be noted that, for particular practical applications (for example, those concerning wind turbine gear drives), another type of surface damage, no less important than those mentioned above, can characterize the limit performance of a gear unit. This kind of damage consists of micropitting that, like macropitting (i.e., pitting in the strict sense of the word), is a fatigue damage. Although micropitting

© Springer Nature Switzerland AG 2020
V. Vullo, *Gears*, Springer Series in Solid and Structural Mechanics 11,
https://doi.org/10.1007/978-3-030-38632-0_2

and macropitting are due to the same phenomenon of fatigue, both have their own peculiarities, as we will see in Chap. 10, which deals with the important topic of micropitting. Furthermore, we should not forget tooth flank fracture, which is a frequent type of fatigue failure at the tooth flank surface, very distinct either from the classical fatigue breakage at tooth root due to bending load or classical pitting damage. This type of fatigue damage also has its own peculiarities that will be described in Chap. 11.

Here we first focus our attention on the aforementioned four classic types of damage that characterize most of the practical applications of gears (they constitute the topic discussed from Chaps. 1–9). We must however bear in mind the fact that, regardless of the type of damage, the limit performances determine the load carrying capacity of the gears; they may be more or less high, depending on the materials used for their manufacture and their heat treatments, as well as on the operating conditions (rotational speeds, lubrication conditions, etc.).

With specific reference to the above four types of damage, we must pay attention to the fact that the current state of knowledge does not allow the possibility of determining in advance which of these limit performances is the most restrictive. Therefore, in the calculation procedure, these limits must all be thoroughly assessed and compared, in order to get a complete picture that allows sizing the gear pair so that it is able to meet the design requirements.

Figure 2.1 shows qualitatively the four distribution curves of load capacity limits in terms of maximum permissible torque, T_{max} (Nm), for a given group of material and for a given heat treatment. Especially the relative position of curves of the tooth root breakage limit and pitting limit may change (one above and the other below, or vice versa), and can intersect or not. In a wider framework, all four curves of the performance limits can variously intersect each other. However, we can generally say that wear is a phenomenon related to the

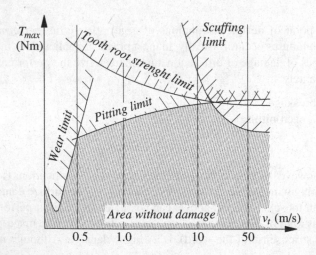

Fig. 2.1 Qualitative performance limits of a gear wheel in terms of maximum torque, T_{max}

low operating speeds, while scuffing is a phenomenon related to high operating speeds, and this determines the position of the corresponding distribution curves in Fig. 2.1 (see also Niemann and Winter 1983).

However, we must not forget here other types of fatigue damages, which determine just as many performance limits for the gears. Among these we must remember the micropitting, which is a fatigue damage phenomenon distinct for macropitting (pitting), and tooth flank fatigue breakages, which include the tooth flank fatigue fracture, tooth interior fatigue fracture and spalling. These other types of fatigue damages and the calculation methods currently available to check the load carrying capacity of the gears against them are discussed in Chaps. 10 and 11. Until then, our attention is limited to the above four main types of gear damage, which are the types of damage traditionally considered by gear designer.

The main factors affecting the performance limits related to the aforementioned four main types of damage are as follows:

- operating conditions, in terms of type of load, nominal tooth forces, additional forces due to vibrations or overloads, rotational speeds, temperature, etc.;
- materials and heat treatments;
- geometry of the gear blank and teeth;
- assembly and manufacturing accuracy;
- roughness and waviness of the tooth flank surfaces, and surface treatments;
- chemical and physical characteristics of lubricant;
- regime of lubrication, i.e. full film lubrication or elastohydrodynamic lubrication, mixed film lubrication or boundary lubrication.

In this general survey, we make only a brief mention on the four types of damage mentioned above. Information that are more detailed are instead given in the following chapters dealing with these specific topics.

As *tooth breakage*, we indicate a type of failure that can affect the entire tooth or a part of it. In this respect, a distinction is made between *overload failure* and *fatigue failure* or *fatigue fracture*. Overload failure occurs due to short (almost instantaneous) and drastic overloads acting on the gear pair. It has the characteristics of a brittle fracture, i.e. it is characterized by a fast rate of crack propagation, with no gross deformation and very little micro-deformation. This type of fracture is to be absolutely avoided, because it occurs without warning and usually causes disastrous consequences (see Shipley 1967; Rolfe and Barsom 1977; Dieter 1988).

Fatigue fracture occurs under conditions of dynamic loading, which causes repetitive or fluctuating stresses. This type of failure occurs after a more or less long service time, but it is equally insidious because it occurs without any perceptible warning. A fatigue failure can usually be recognized from the appearance of the fracture surface, which shows a smooth region (i.e., the fatigue fracture zone), due to the rubbing action as the crack propagated through the section. Furthermore, a final fracture zone is present, where the tooth has failed when the cross section was no longer able to carry the load. This zone sometimes presents evidence of plastic deformation, but usually shows the characteristic course granular surface of a brittle fracture. However, it cannot be ruled out that this final fracture zone may be either brittle, ductile or a combination of both, depending on the influences that came into play (see Shipley 1967; Frost et al. 1999; Juvinall 1967; Stephens et al. 2001; Schijve 2009; Radaj and Vormwald 2013).

Usually, a gear wheel is subjected to pulsating load, but the cases in which it is subjected to alternating loads are not uncommon. Idler gear wheels and planet gears of a planetary gear train are examples of these cases. The maximum bending stress occurs at the tooth root and varies with the pulsating or alternating load. If the stress level frequently or occasionally exceeds the permissible tooth root stress, the fatigue failure mechanism is triggered, and it can lead to fatigue fracture.

As *pitting*, we indicate a rolling and sliding contact fatigue damage, which can occur at the scale of the nominal areas subject to this type of contact or at the scale of roughness. Depending on these two scales of occurrence, we speak respectively of *macropitting* (i.e. the proper pitting, for which pitting we mean macropitting) or *micropitting*. Macropitting and micropitting are related to the same fatigue damage phenomenon, but the former sees interacting the macro-areas affected by the contact, while the latter sees only the surface asperities interact with each other. In this and the chapters that follow, we talk about pitting, while only Chap. 10 is specifically dedicate to micropitting.

As *pitting* (macropitting), we indicate a type of damage of the gear wheels characterized by the appearance of pin-holes and extended flank *spalling*, mostly below the pitch circle. Surface fatigue failures result from the repeated application of loads that cause stresses on and under the contacting surfaces, as described in Sect. 2.4. Cracks resulting from these stresses propagate until small bits of surface material become separated, producing pitting or spalling (see Juvinall 1983; Juvinall and Marshek 2012) .

Pitting originates with surface cracks, and each pit involves a relatively small area. Spalling originates with subsurface cracks, and the spalls are thin *flakes* of surface material. Depending on the assumption made, the causes of these types of failure can be surface cracks resulting from sliding, or shear stresses in the area below the tooth flank surface.

Pitting only occurs in lubricated gear drives. Resistance to pitting is influenced by material surface hardness, lubricant oil viscosity and temperature, specific sliding, flank profile deviations, surface roughness, and rotational speed.

Two different types of failure can occur when tooth flank lubrication fails, depending on the transverse tangential velocity: *scuffing*, which is distinguished in *warm* or *hot scuffing*, and *cold scuffing*, and *wear* (see also Niemann and Winter 1983) .

Warm or hot scuffing (this is the proper scuffing) arises when the lubricant film thickness breakdowns due to high temperatures or excessive loads. This leads to the direct metal-to-metal contact in correspondence of the unavoidable roughness peaks, often called asperities. Since the contact pressure and sliding frictional heat are concentrated at small local areas of contact, local temperatures and pressure are extremely high, and conditions are favorable for welding at these points, where the instantaneous temperature may locally reach the metal melting point, but with temperature gradients so steep that the part remains cool to the touch. If melting and welding of the surface asperities occur, either the weld or one of the two metals near the weld must fail in shear in such a way to permit the relative motion of the two surfaces in contact to continue. New welds, that is new adhesions and corresponding fractures continue to occur, resulting in the scuffing, which therefore is a phenomenon of *adhesive wear*. Since adhesive wear is essentially a *welding phenomenon*, metals that weld together easily are most susceptible to scuffing (see Juvinall and Marshek 2012).

Usually, scuffing causes flaking of the tooth flanks. However, also loose particles of metal and metal oxides may be formed. These loose particles and oxides resulting from adhesive wear cause further surface wear because of abrasion. The generation of scuffing damage is very complex and is due to both physical and special chemical processes. These processes occur in extremely thin layers of surface and sub-surface material, and under high pressures, while the physical phenomena are explained by elastohydrodynamic lubrication theory (Dowson and Higginson 1977).

If welding and tearing of the surface asperities cause a transfer of metal from one surface to the other, the resulting adhesive wear or surface damage is called *scoring*. If the local welding of asperities becomes so extensive that the surfaces no longer slide on each other, the resulting damage is called *seizure*. Two types of scuffing can be distinguished: scoring and scuffing. Scoring is typical of doped oils and transverse tangential velocities lower than 30 m/s: *individual scoring* or *clusters of scoring* appear in the sliding direction of the tooth flanks, varying from minor to serious. Scuffing is typical of un-doped and doped oils at transverse tangential velocities greater than 30 m/s: it is a mild adhesive wear, and occurs as individual fine lines (*scuffing lines*), as clusters (*heavy scuffing* or *galling*) or as areas across the entire face width (*scuffing zones*), and main feature of the scored areas is a matt appearance. Generally, the higher the surface hardness (more precisely, the higher the ratio of surface hardness to elastic modulus), the greater the resistance to adhesive wear.

Cold scuffing is a gear damage that occurs much more rarely than warm scuffing. It is manifested with unimportant heat development, mostly at low transverse tangential velocities (v_t less than 4 m/s), predominantly for through-hardened steel gears having teeth with coarse accuracy grade. In terms of the limiting torque as a function of the transverse tangential velocities (see Fig. 2.1), the area covered by the cold scuffing is the same as that covered by the wear. For this reason, cold scuffing and wear are often confused (see Naunheimer et al. 2011).

Finally, the term wear refers to *abrasive wear*, which is due to the rubbing of abrasive particles on a surface. This damage of gears occurs mostly at low transverse tangential velocities (below 5 m/s) on heat-treated gear wheels having teeth with coarse accuracy grade. Usually, the harder the surface the more resistant it is to abrasive wear. Hard metal surfaces are produced by heat-treatment, flame or induction hardening, carburizing, nitriding, electroplating, flame plating, or other means. The effects of abrasive wear can be exacerbated by corrosive wear, which is due to the chemical action of lubricating oils and/or additives, which are little or nothing compatible with gear materials (see Fontana 1987).

2.2   Pitting Damage and Safety Factor

Notoriously, the amount of power transmitted by a gear pair mainly depends on what we call *surface durability* or *surface endurance*, which is a measure of the ability of the gear tooth flanks to resist to the surface fatigue known as pitting, and caused by contact stresses. Since the tooth flank surfaces undergo fluctuating (or alternating), repeated and cyclic stresses of various kinds during the meshing cycle, surface fatigue failure follows, with the appearance of the characteristic cracks or pits described in the previous section.

The mechanisms that determine the phenomenon of pitting are not yet entirely clear. A plausible explanation calls into question the maximum shear stress that occurs at a point on the normal to the contact surface, localized at a certain small depth from the same contact surface. Therefore, the following two theories came into play: the Hertz theory, which allows to determine the pressure distribution on the small contact area between two bodies pressed against each other (Hertz 1882), and the Belajev theory on the stress distribution along the normal through the center of this small contact area (Belajev 1917, 1924). If the load is excessive, at this point the failure of the material begins. This failure is manifested by the generation of small cracks that, from the inside, propagate more or less rapidly to the surface (see Thomas and Hoersch 1930; Beeching and Nicholls 1948; Timoshenko and Goodier 1951; Johnson 1985).

Since the gears are generally lubricated, the lubricant under pressure penetrates within these cracks, the pressure inside the cracks increases when the same cracks are closed due to the cyclic contact between the surfaces, and then the cracks expand, resulting in the removal of metal particles and formation of pits (Giovannozzi 1965b). The pitting phenomenon depends therefore on the values reached by the surface and subsurface elastic stresses, and on the pressures that develop in the lubricant, with an interdependence between these two causes due to the fact that the elastic deflections produce a variation of the lubricant film thickness and the corresponding pressures due to lubrication affect these elastic deflections. So, the elastohydrodynamic lubrication theory comes powerfully into play (see Dowson and Higginson 1977; Jacobson 1991; Lugt and Morales-Espejel 2011; Stachowiak and Batchelor 2014) .

The experimental evidences show that pitting mostly occurs near the pitch line, on dedendum flanks of pinion and gear wheel, i.e. in the tooth flank surfaces where the specific sliding assumes a negative value. This is because the relative sliding velocity vector reverses its direction as a pair of teeth rolls through the pitch line. In fact, during approach, the chamfering sliding and related friction forces tend to compress the teeth flank surfaces, while during recess the stretching sliding and related friction forces tend to stretch them (see Fig. 3.8 in Vol.1). It follows an increase of the effect of the compressive forces, and more damaging around this region. It should also be borne in mind that this region undergoes the maximum dynamic load.

The relative sliding motion causes pitting damages of teeth on both members of the gear pair. However, in the case in which the pinion is the driving member (in this case it is equipped with a higher angular velocity and thus it performs a greater number of cycles during its service lifetime), and the material and heat treatment are the same of the wheel, the pinion is the member more vulnerable to pitting than the mating member. In any case, since the pitting phenomenon is very complex, and depends on parameters of which the usual calculation models do not take into account, the calculation of load carrying capacity of spur and helical gears in terms of surface durability must be carried out on both members, pinion and gear wheel.

It should be noted that exceeding the limits of surface durability of meshing flanks of a gear, with the consequent related generation of pitting, not always means that the gear pair must necessarily be put out of service. There are indeed cases in which pitting is tolerated, and the measure of how the generation of pits is tolerated, in terms of their size and number, varies within wide limits, particularly depending on the practical application field. In some fields, extensive pitting is accepted, while in other fields no appreciable pitting is to be accepted. In this regard, a distinction must be made between *initial pitting* and *destructive pitting*.

A linear or progressive increase of the total area of pits is generally unacceptable. The pitting is considered tolerable when the initial pitting can enlarge the effective tooth flank area, which bears the load, for which the rate of generation of pits may consequently be reduced (*degressive pitting*) or cancelled (*arrested pitting*). Instead, the *linear* or *progressive pitting* under unchanged service conditions is not tolerable. The pitting damage assessment must include the entire active area of all the tooth flanks, and the number and size of pits should be considered, since due to local stress concentration effects, they may become the origin of cracks that can lead to tooth breakage. Moreover, pitting damage is not tolerable if it is such as to put the human life in danger, or if it involves the risk of serious consequences to persons and property.

In this framework, pitting damage is unacceptable for hardened gears of aerospace transmissions, because a pit on the flank surface especially if close to the tooth root may be the starting point of a fatigue fracture. Similar considerations are true for turbine gears and gear systems for high-speed applications. In these cases, the formation of pits or a strong wear could lead to vibrations and sharp increases in dynamic secondary forces. High safety factors must be provided for in the calculations, to ensure a sufficiently low probability of failure.

On the contrary, for some slow-speed industrial gear drives having high module (e.g., 25 mm), made of low hardness steel, and intended to operate quietly for several years, an initial pitting, also extended to 100% of the working flanks and with pits of large size, can be tolerated. This provided that the apparent destructive pitting occurring in the initial stages of service tends to slow down with time. In this case, in fact, the tooth flanks become smoothed and work hardened, with an increase of the surface Brinel hardness number by 50% or more, for which pitting is regressive. Consequently, low safety factors can be used in the calculations, with correspondingly higher probabilities of tooth surface damage.

The safety factor, S_H, relative to pitting must be greater than or equal to a minimum value, S_{Hmin} (in this regard, ISO 6336-2:2006 does not provide the value), and must be carefully chosen, in reference to the field of practical application, in order to meet the required reliability at a justifiable cost. Generally, it is defined in terms of stresses, i.e. as the quotient of the pitting stress limit, σ_{HG}, divided by the calculation contact stress, σ_H. However, also a safety factor based on load (power or torque) is permitted; it is equal to the quotient of the specific calculated load divided by the specific operating load transmitted, and thus it is proportional to S_H^2.

Safety factor is affected by the following influences:

• reliability of material data, and load values used for calculations;
• variations in gear geometry due to manufacturing tolerances, and variations in alignment due to assembly errors;
• variations of material characteristics due to heat treatment or similar treatments;
• variations in lubrication and its maintenance during the service lifetime of the gears.

A corresponding safety factor is to be chosen, in dependence on the reliability of the assumptions on which the calculations are based, and according to the reliability requirements (see Giovannozzi 1965a; Burr 1982; Niemann et al. 2005).

2.3 Theoretical Basis of Surface Durability Calculations

The development of an accurate general model for calculating and evaluating the pitting load carrying capacity of the gears, which is able to follow the nucleation of the macro-pits and their progressive diffusion on the mating tooth flank surfaces, involves the solution of several complex problems. *Mutatis mutandis,* these problems are common to those we will describe in Chap. 10 about micropitting. Since macropitting, i.e. pitting, and micropitting are both a rolling and sliding contact fatigue damage, and represent two quite similar aspects of the same phenomenon, for a general theoretical treatment of pitting we can use the same models described in the aforementioned chapter, with the appropriate not relevant modifications. Here we limit ourselves to dealing with the problem concerning the surface durability (pitting) of the gears, following the conceptual lines of development over time of this important topic.

In this regard, for the assessment of the surface durability of spur and helical gears, the basic principle used by scientists and researchers, as well as the ISO standards, is the *Hertzian pressure*, σ_H, also called *contact stress*, *Hertzian stress* or *Hertz stress*. It is assumed as significant indicator of the stress generated during tooth flank engagement.

In this regard, it is first to be noted that the Hertzian pressure is not the only cause of pitting. The same thing can be said about the subsurface stress state and the corresponding subsurface shear stress, of which the Hertzian pressure does not take into account. We have already said in the previous section that the pitting is a very complex damage mechanism, which is significantly affected by other contributory influences. Between these influences, the coefficient of friction, direction and intensity of the relative sliding, and elastohydrodynamic lubrication conditions, with the consequent pressure distribution inside the lubricant film, play a very important role (see Grubin and Vinogradova 1949; Greenwood 1972; Morales-Espejel and Wemekamp 2008).

The current state of knowledge does not allow to directly take into account all these contributory influences, not included in the Hertzian pressure. Instead, we take into account these influences indirectly, through the introduction of suitable derating factors, and the careful choice of values of the mechanical characteristics of material. In this framework that highlights the intrinsic limits of the Hertzian pressure, it is taken as a basis working hypothesis, also but not only because its limiting value for a given material is obtainable with relative ease from fatigue tests on gear specimens.

According to an intuition of Buckingham (1949), two teeth of a parallel cylindrical spur gear pair that touch at a point P of the path of contact (see Fig. 2.6 in Vol. 1) can be considered as two cylinders of radii ρ_1 and ρ_2 (and therefore equal to the instantaneous radii of curvature of the two mating involute profiles). These cylinders are in contact along their common generatrix, having length equal to the face width, b, and are pressed by a force $F_t/\cos\alpha$, directed along the line of action, and therefore acting along the direction normal to the same profiles. Of course, this last assumption implies that the friction forces are neglected. In this case, the contact area between the two cylinders is theoretically a narrow rectangle, having a width $2b_H$ (b_H is the semi-width of the Hertzian contact band), and length coinciding with the face width b (Fig. 2.2).

According to the Hertz theory, the pressure distribution on the contact area is represented by a prism with semi-elliptical cross section, while the maximum value of the Hertzian pressure, σ_H, along the z-axis perpendicular to the contact area and contained in the symmetry plane (x, z) of this prismatic distribution (Fig. 2.2), which we assume as reference stress for the surface durability calculations, is given by:

$$\sigma_H = \sqrt{\frac{F_t}{b\cos\alpha}\frac{E_r}{2\pi\rho}}, \tag{2.1}$$

where ρ is the *equivalent radius of curvature* (or *effective radius of curvature*), and E_r is the *reduced modulus of elasticity* (or *equivalent modulus of elasticity*). These two quantities are given, respectively, by the following relationships:

$$\frac{1}{\rho} = \frac{1}{\rho_1} + \frac{1}{\rho_2} = \frac{\rho_1 + \rho_2}{\rho_1\rho_2} \tag{2.2}$$

$$E_r = \left\{\frac{1}{2}\left[\frac{1 - \nu_1^2}{E_1} + \frac{1 - \nu_2^2}{E_2}\right]\right\}^{-1}. \tag{2.3}$$

It is to be remembered that, according to the Hertz theory, a radius of curvature is considered positive for a convex surface, and negative for a concave surface. Therefore, for an external gear pair, both the radii of curvature ρ_1 and ρ_2 are positive, while, for an internal gear pair, ρ_1 is positive, and ρ_2 is negative.

Hertz theory is based on the following assumptions:

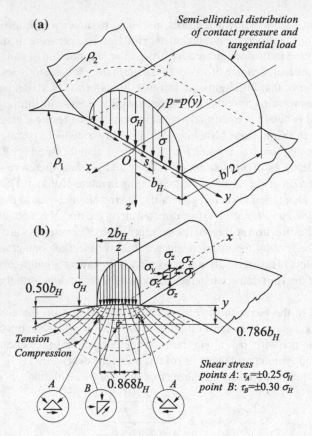

Fig. 2.2 Contact pressure and tangential load distributions between two parallel cylinders with the same length, and coordinate system $O(x, y, z)$ in the transverse mid-plane of the two cylinders

- frictionless contact between two elastic bodies, having perfectly smooth surfaces (therefore, roughness are equal to zero);
- motionless bodies;
- absence of friction forces;
- elastic and isotropic material behavior;
- load normal to the contact area;
- dry and non-lubricated contact surfaces;
- dimensions of contact area relatively small in comparison with the radii of curvature and with the sizes of the two bodies.

Since the contact area is a narrow rectangle, with one side equal to the face width, the other side of the rectangle must have a small width, $2b_H$, so that the last of the above assumptions is satisfied (see Timoshenko and Goodier 1951; Juvinall 1967; Gladwell 1980; Barber 1992; Saada 1993).

According to the first hypothesis of the Hertz theory, on the basis of which the pressure distribution on the contact area is represented by a prism with semi-elliptical cross section, the pressure distribution in any transverse section is a semi-ellipse, having σ_H and b_H as semi-axes of the semi-elliptical boundary of the prism cross section. Therefore, in the Cartesian coordinate system shown in Fig. 2.2, the intensity $\sigma = \sigma(y)$ of the local contact stress at a distance y from the origin, O, is given by:

$$\sigma(y) = \sigma_H \sqrt{1 - \frac{y^2}{b_H^2}}. \tag{2.4}$$

The total load $(F_t / \cos\alpha)$ is equal to the volume of the prism with semi-elliptical cross section, so we can write:

$$\frac{F_t}{\cos\alpha} = \frac{\pi}{2} b b_H \sigma_H. \tag{2.5}$$

From Eqs. (2.1), (2.2) and (2.5), we get:

$$b_H = \frac{2}{\sqrt{\pi}} \sqrt{\frac{2 F_t \rho}{E_r b \cos\alpha}} = \frac{2\sqrt{2}}{\sqrt{\pi}} \sqrt{\frac{F_t \rho_1 \rho_2}{E_r b \cos\alpha (\rho_1 + \rho_2)}}. \tag{2.6}$$

The equivalent radius of curvature ρ varies during the meshing cycle, since also the instantaneous radii of curvature, ρ_1 and ρ_2 (see Fig. 2.6 in Vol. 1), are continuously changing, from the beginning to the end of contact. Figure 2.3a shows the distribution curve of ρ along the line of action $T_1 T_2$ and, therefore, along the path of contact AE, for an external parallel cylindrical spur gear pair. Instead, Fig. 2.3b, which refers to an internal parallel cylindrical spur gear pair, shows the distribution curve of ρ only along the path of contact AE.

For an external parallel cylindrical spur gear pair, recalling the Eqs. (3.49)–(3.51) of Vol. 1, from Eq. (2.2) we obtain:

$$\rho = \frac{\rho_1 \rho_2}{\rho_1 + \rho_2} = \frac{(r_1 \sin\alpha + g_p)(r_2 \sin\alpha - g_P)}{(r_1 + r_2)\sin\alpha}. \tag{2.7}$$

The denominator of this equation is a constant, since it represents the length of the line of action $T_1 T_2$. Instead, the numerator is variable, and becomes equal to zero at extreme points T_1 and T_2 of the line of action, where we have respectively $g_P = -r_1 \sin\alpha$ and $g_P = r_2 \sin\alpha$. From Eq. (2.1), we get that, at these points, the Hertz stress σ_H becomes infinite. The need to exclude from contact those parts of the line of action nearest to the interference points T_1 and T_2 (see Sect. 2.4 of Vol. 1), and thus the part of profile immediately near to the base circle, remains confirmed also from the surface durability viewpoint (Giovannozzi 1965b).

The maximum value of the equivalent radius of curvature, ρ, is obtained by imposing the equality to zero of the first derivative of ρ with respect to g_P. In so doing, we infer that this maximum value of ρ, to which the minimum value of σ_H corresponds by virtue of Eq. (2.1), is obtained when the following equality is

(a)   **(b)**

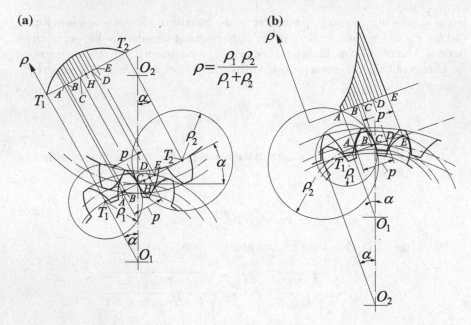

$$\rho = \frac{\rho_1 \, \rho_2}{\rho_1 + \rho_2}$$

Fig. 2.3 Distribution curves of the equivalent radius of curvature, ρ, for: **a** an external spur gear pair; **b** an internal spur gear pair

satisfied:

$$g_P = \frac{1}{2}(r_2 - r_1)\sin\alpha, \qquad (2.8)$$

that is, at the mid-point H of the length of the line of action $T_1 T_2$ (Fig. 2.3a). In any two points of path of contact, symmetrical with respect to this mid-point H, that is equidistant from point H and placed on opposite sides, ρ, and thus σ_H, assume equal values. The minimum value of ρ along the path of contact AE, and therefore the maximum value of σ_H, are those corresponding to the point of path of contact as far away the mid-point H of the line of action.

Figure 2.4 shows, with dashed line, the distribution curve of σ_H along the line of action, highlighting the two asymptotes in correspondence of points of interference T_1 and T_2, and the minimum value at the mid-point H of the line of action. This distribution curve is the theoretical one, which does not take into account the load distribution along the path of contact, shown in Fig. 1.22. However, if we consider the conventional load distribution shown in Fig. 1.22, the distribution curve of σ_H is changed. In fact, in the ideal case of absence of errors of any kind (e.g. the pitch error whose effect can be very significant, as Tanaka et al. (1991) have highlighted), it becomes the one drawn with a solid line in Fig. 2.4. This figure shows that, in the absence of errors, the maximum value of the Hertz stress σ_H takes place in correspondence of the inner point B of single pair tooth contact of the pinion. The

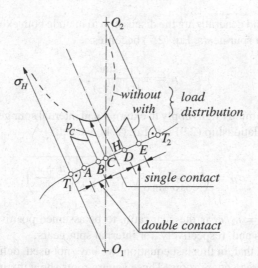

Fig. 2.4 Distribution curves of σ_H, without (dashed line) and with (solid line) load distribution

same figure highlights that the two curves, without and with load distribution, have in common the arc corresponding to the single pair tooth contact, BD, on the path of contact. Of course, the more elaborate load distribution curves can be used (see, for example, Ristivojavić et al. 2019).

However, even here we have to deal with inevitable errors, because any deviation from the assumed ideal condition of the teeth implies significant changes to this Hertzian stress distribution. Similarly to what happens for the load distribution curve (see Sect. 1.7), the deviations from the ideal condition, of any nature they are, do not affect the value of the Hertz stress in the region of the single pair tooth contact (the one between points B and D). On the contrary, their influence is very significant in the two regions of double pair tooth contact (those between points A and B, and between points D and E).

Here also, an increase in the value of the Hertzian stress in the region bounded by points A and B occurs (the curve thus moves upward), and this increase is greater the higher the deviations. Correspondingly, a decrease in the value of the Hertz stress in the region bounded by points D and E occurs (the curve thus moves downward), and this decrease is greater the higher the deviations. The reasons of this behavior are the same of those described in Sect. 1.7 for the load distribution.

In his classical work of adapting the Hertz theory to parallel cylindrical spur gears, based on numerous experimental tests, Buckingham (1949) noted that the pitting occurred predominantly near the pitch line where the sliding velocity becomes equal to zero, and consequently the elastohydrodynamic oil film thickness breaks down. *Rebus sic stantibus*, Buckingham proposed to introduce, as a reference value σ_H for calculations of surface durability, the one assumed by the Hertz stress at the pitch point, C, where $g_P = 0$.

With $g_P = 0$, and generalizing the discussion to include both external and internal parallel cylindrical spur gears, Eq. (2.7) becomes:

$$\rho = \pm \frac{r_1 r_2 \sin\alpha}{(r_1 \pm r_2)}, \tag{2.9}$$

where the plus and minus signs apply to external and internal spur gears, respectively. Substituting the relationship (2.9) into Eq. (2.1), we get:

$$\sigma_H = \sqrt{\frac{2}{\cos\alpha \sin\alpha}} \sqrt{\frac{E_r}{2\pi}} \sqrt{\frac{F_t}{bd_1}\left(\frac{u+1}{u}\right)}, \tag{2.10}$$

where $u = z_2/z_1 = r_2/r_1$ is the gear ratio, to be assumed positive or negative for external spur gears and, respectively, for internal spur gears.

It is to be noted that, in this last equation, we have not used, deliberately, a single square root, but three-square roots. These square roots show the influences exerted on the Hertzian stress respectively by the teeth geometry, in terms of pressure angle, materials of which the pinion and wheel are made, and specific load and radii of the pitch circles. The third square root is the benchmark used by the various standards concerning calculations of the surface durability.

Equation (2.10) easily allows to prove that, *ceteris paribus*, the Hertzian stress of an external cylindrical spur gear pair is greater than that which characterizes an internal cylindrical spur gear pair. The same equation also allows to show that, always with all other conditions being equal, the torque transmittable by an internal spur gear pair is greater than that transmitted by an external spur gear pair. The following two examples clearly highlight these factual circumstances.

Example 1 Let us consider two cylindrical spur gear pairs, consisting of: equal pinions having both z_1-teeth; the external pair, with gear wheel having z_2-teeth; the internal pair, with annulus also having z_2-teeth. We also assume that both cylindrical spur gear pairs share all the quantities that appear in Eq. (2.10), with the exception of the gear ratio $u = z_2/z_1$ that, despite having the same absolute value, is positive for the external pair and negative for the internal pair.

Using the Eq. (2.10), we infer that the ratio σ_{He}/σ_{Hi} between the Hertzian stresses, which characterize respectively the external and internal pairs, is given by: $\sigma_{He}/\sigma_{Hi} = \sqrt{[1 + (z_1/z_2)]/[1 - (z_1/z_2)]}$; it is therefore greater than 1, as $z_2 > z_1$, so that under the square root the denominator is less than the nominator. The greater Hertzian stress for an external spur gear compared to that which characterizes the equivalent internal spur gear is therefore demonstrated.

Example 2 With the data of the previous example, show that, *ceteris paribus*, the torque transmitted by a cylindrical internal spur gear pair is greater than that transmitted by a cylindrical external spur gear pair.

To this end, it is sufficient to first multiply the numerator and denominator under the third square root that appears in Eq. (2.10) by the transverse pitch radius (or the nominal reference radius), then to square the left-hand and right-hand sides of the same equation and, finally, to isolate the torque, T, given by the product of the nominal transverse tangential force, F_t, for the pitch (or reference) radius. By applying this procedure to both the above gear pairs, we obtain that the torque, T_e, transmitted by the external gear pair is proportional to the ratio $[z_2/(z_2 + z_1)]$, while the torque, T_i, transmitted by the internal gear pair is proportional to the ratio $[z_2/(z_2 - z_1)]$. Since the ratio $T_e/T_i = [(z_2 - z_1)/(z_2 + z_1)]$ is less than 1, it is evident that the torque transmitted by an external gear pair is less than the torque transmitted by the equivalent internal gear pair.

2.4 Surface and Subsurface Stress States

In the previous section we recalled briefly the Hertz theory in its adaptation to the involute spur gears, proposed by Buckingham (1949). As we shall see in the following sections, the results described in the previous section are the basis of the actual standard assessment methods of surface durability of parallel cylindrical spur and helical gears, but they, alone, cannot give a valid justification of the mechanisms that trigger the pitting phenomenon. In this regard, it is therefore necessary to deepen the subject, especially with reference to the surface and subsurface stress states, which characterize the contact area and the region below it.

The Hertzian stress σ_H given by Eq. (2.10) is the maximum compressive stress at the center O of the narrow rectangular contact area or contact band, having dimensions b and $2b_H$ (Fig. 2.2). This compressive stress is taken as a stress unit. In the coordinate system $O(x, y, z)$ shown in Fig. 2.2, for measurements of the distances along the z-axis, the semi-width of Hertzian contact band, b_H, is also taken as a length unit. The greatest stress is the compressive stress, σ_z, at the center O of the area of contact; it is a principal stress, and coincides with the Hertzian stress, σ_H. The other two principal stresses, σ_x and σ_y, that together with σ_z completely define the stress state at the same point O, are given by:

$$\sigma_x = 2\nu\sigma_H \qquad \sigma_y = \sigma_z = \sigma_H. \qquad (2.11)$$

All three principal stresses σ_x, σ_y, and σ_z at point O of the surface of contact are compressive stresses, and therefore they are to be taken as negative. These stresses define uniquely the surface stress state at point O, in the plane of the surface of contact. Outside the area of contact, all the stress components in the same plane of the surface of contact are equal to zero.

Now we have to define the subsurface stress state along the normal through the center O of the narrow rectangular contact band, i.e. the gradients of the stress components σ_x, σ_y, and σ_z along the z-axis. All three of these stress components are principal stresses by virtue of the load symmetry; they are compressive and therefore negative stresses, and their intensity decreases below the surface of contact when z increases. Stress components σ_x and σ_y in any plane with coordinate $z = const$

follows from Poisson's ratio related to the transverse contraction strain (see also Mott and Roland 2009). In fact, since the material is compressed in the z-axis direction due to the direct compressive stress σ_z, it tends to expand in the x- and y-directions, but this expansion is prevented by the surrounding material. Therefore, compressive stresses in the x- and y-directions are generated, with the related stress components σ_x and σ_y.

The problem of determining the subsurface stress state, i.e. the stress state below the surface of contact, has been addressed and solved for the first time by Belajev (1917). This researcher, by integrating over the loaded area of contact the Hertzian pressure distribution given by Eq. (2.4), provided the following set of three equations expressing the three principal stress components σ_x, σ_y, and σ_z (see also Johnson 1985):

$$\sigma_x = 2v\frac{\sigma_H}{b_H}\left[\left(b_H^2 + z^2\right)^{1/2} - z\right]$$
$$\sigma_y = \frac{\sigma_H}{b_H}\left[\left(b_H^2 + 2z^2\right)\left(b_H^2 + z^2\right)^{-1/2} - 2z\right]$$
$$\sigma_z = \sigma_H b_H\left(b_H^2 + z^2\right)^{-1/2}. \tag{2.12}$$

The maximum shear stresses on planes bisecting the principal planes xy, yz, and zx, which we get from the three-dimensional Mohr-circle representation of the stress state, have the following absolute values:

$$|\tau_{xy}| = \frac{1}{2}|(\sigma_x - \sigma_y)|$$
$$|\tau_{yz}| = \frac{1}{2}|(\sigma_y - \sigma_z)|$$
$$|\tau_{zx}| = \frac{1}{2}|(\sigma_z - \sigma_x)|. \tag{2.13}$$

These are principal shear stresses. Substituting Eq. (2.12) into these last equations, we obtain:

$$\tau_{xy} = \frac{1}{2}\left|\frac{\sigma_H}{b_H}\left[2v\left(b_H^2 + z^2\right)^{1/2} + 2z(1 - v) - \left(b_H^2 + 2z^2\right)\left(b_H^2 + z^2\right)^{-1/2}\right]\right|$$
$$\tau_{yz} = \left|\frac{\sigma_H}{b_H}\left[z^2\left(b_H^2 + z^2\right)^{-1/2} - z\right]\right|$$
$$\tau_{zx} = \frac{1}{2}\left|\frac{\sigma_H}{b_H}\left\{b_H^2\left(b_H^2 + z^2\right)^{-1/2} - 2v\left[\left(b_H^2 + z^2\right)^{1/2} - z\right]\right\}\right|. \tag{2.14}$$

It should be noted that the first of Eq. (2.12) follows from the condition of plane strain state in the mid-plane of the gear wheel in the direction of x-axis (Fig. 2.2), that is from the condition $\varepsilon_x = (1/E)\left[\sigma_x - v(\sigma_y + \sigma_x)\right] = 0$, so we have $\sigma_x = v(\sigma_y + \sigma_x)$. It is also to be noted that the stress components σ_y, σ_z, and τ_{yz}, and

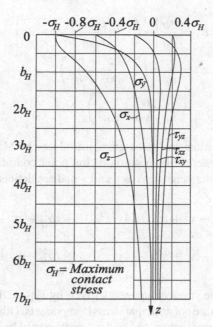

Fig. 2.5 Distribution curves of the principal normal and shear stress components along the z-axis, for $\nu = 0.3$

stress components σ_x, τ_{xy}, and τ_{zx}, are independent and, respectively, dependent on Poisson's ratio, ν.

Figure 2.5 shows the distribution curves of the three principal normal stress components σ_x, σ_y, and σ_z, and three principal shear stress components τ_{xy}, τ_{yz}, and τ_{zx}, given respectively by Eqs. (2.12) and (2.14), along the z-axis, below the surface of contact. These distribution curves refer to a Poisson's ratio $\nu = 0.3$. As we mentioned above, stress components and distance z from the origin O are referred to σ_H and b_H, respectively.

Figure 2.5 also shows that the maximum principal shearing stress is that achieved by τ_{yz} at the depth $z = 0.78 b_H$; it is inclined by 45° with respect to the surface, and its intensity is given by: $\tau_{yzmax} = 0.304\sigma_H$. This shearing stress is a fluctuating shear stress, which varies from the maximum value $0.304\sigma_H$, when the load is directly over the point under consideration, to zero, when the rolling action takes the load beyond the point under consideration. The alternating component of this fluctuating shear stress is equal only to $0.152\sigma_H$, but it can exert its influence, as notoriously the fatigue failures are mainly initiated by alternating shear stresses.

Both from Fig. 2.5 and Eqs. (2.12) and (2.14), we can deduce that the most significant stress components are σ_y, σ_z, and τ_{yz}. In the generalization of the problem in order to evaluate what happens outside of the z-axis, McEven (1949) expressed these three stress components at any point $P(y, z)$ in the yz-plane, not belonging to the z-axis ($y \neq 0$), in the following terms:

$$\sigma_y = \frac{\sigma_H}{b_H}\left[m\left(1 + \frac{z^2 + n^2}{m^2 + n^2}\right) - 2z\right]$$

$$\sigma_z = \frac{\sigma_H}{b_H}m\left(1 - \frac{z^2 + n^2}{m^2 + n^2}\right)$$

$$\tau_{yz} = \frac{\sigma_H}{b_H}n\left(\frac{m^2 - z^2}{m^2 + n^2}\right), \tag{2.15}$$

where m and n are two geometric parameters, introduced by McEven to simplify the final expression which define the position of the point considered in the xy-plane; they have respectively the same signs as z and y and are defined by:

$$m^2 = \frac{1}{2}\left\{\left[\left(b_H^2 - y^2 + z^2\right)^2 + 4y^2 z^2\right]^{1/2} + \left(b_H^2 - y^2 + z^2\right)\right\}$$

$$n^2 = \frac{1}{2}\left\{\left[\left(b_H^2 - y^2 + z^2\right)^2 + 4y^2 z^2\right]^{1/2} - \left(b_H^2 - y^2 + z^2\right)\right\}. \tag{2.16}$$

Of course, unlike the stress components given by Eqs. (2.12) and (2.14), those expressed by Eq. (2.15) are not principal stress components. Other alternative expressions of these stress components σ_y, σ_z, and τ_{yz} not reported here have been obtained by Beeching and Nicholls (1948), Poritsky (1950), and Sackfield and Hills (1983).

The subsurface stress state expressed by Belajev's equations (2.12) and (2.14), as well as the results of the additional contributions given by the aforementioned authors, have found various experimental confirmations, such as those performed with photo-elastic methods (see Dolan and Broghammer 1942; Heywood 1952; Raptis et al. 2010).

The above-mentioned analysis, based on Hertz theory and on additional contributions we have just described, although very important, is not enough by itself to explain the pitting phenomenon of gears. Therefore, a more comprehensive analysis of the stress state on and below the area of contact is necessary, because the ideal static conditions that we assumed to derive the aforementioned relationships cannot claim to represent the actual working conditions of meshing teeth.

To get a more complete picture of the surface and subsurface stress state, it is also necessary to take into account the stress state due to sliding and rolling that characterize the contact between the mating tooth flanks. Sliding and rolling are respectively due to the relative rolling velocity (or sliding velocity) of the surfaces at their point of contact, and the relative angular velocity of the two bodies about axes parallel to their common tangent plane. They determine two dangerous stress states that contribute significantly to the generation of pits and their propagation. To assess these contributions, it is here convenient to analyze separately the peculiar effects of pure sliding and pure rolling.

We do not believe that this is the place to go into these interesting topics developed starting from Fromm (1927), for which we refer the reader to specialized textbooks (see, for example, Johnson 1985), as well as the works gradually cited as references.

However, without neglecting the concepts and some basic relationships, we consider very appropriate to summarize some important results obtained by different researchers, which can make a significant contribution to the understanding of the mechanisms that trigger the pitting phenomenon.

Initially, we focus our attention only on the sliding. First it should be noted that, in condition of pure sliding, i.e. sliding without rolling, the resulting friction loads due to relative motion cause normal and shear stress components that act in the tangential direction, and that are superimposed on the stress components caused by the normal loads due to contact pressure distribution.

The second thing to do is to define the relationships between the tangential load and normal pressure in sliding contact. In this regard, we take the usual law of sliding friction, $q_0 = \mu p_0$, where $p_0 = \sigma_H$ is the maximum normal pressure, q_0 is the maximum *tangential surface force*, and μ is the coefficient of kinetic friction, which is a constant whose value depends on materials and physical conditions of the contact interface (about friction, see chapters of Vol. 1, starting from Chap. 3). As a third step, we assume that this interfacial friction law is applicable to each infinitesimal area of the contact interface, and that the distribution of the *tangential surface load* on the contact interface is also here a prism with a semi-elliptical cross section, as Fig. 2.2a conventionally shows by horizontal arrows of different length in the plane (x, y), directed along the y-axis.

Under these assumptions, the tangential friction surface load $q = q(y)$ will be given by a relationship analogous to the Eq. (2.4), for which, also taking into account that its direction is opposite to that of the sliding velocity, we can write:

$$q(y) = \mp \mu p_0 \sqrt{1 - \frac{y^2}{b_H^2}} = \mp \frac{2\mu F_t \sqrt{b_H^2 - y^2}}{\pi b b_H^2 \cos \alpha}. \tag{2.17}$$

Using a procedure similar to that developed and employed by Belajev, some researchers, such as McEven (1949), Poritsky (1950), and Sackfield and Hills (1983), integrated over the loaded area of contact the tangential surface load distribution given by Eq. (2.17), and showed that there are some analogies between the stress components due to $q(y)$ and those due to $\sigma(y)$. These analogies can be expressed in the form:

$$\frac{(\sigma_z)_q}{q_0} = \frac{(\tau_{yz})_p}{p_0}$$
$$\frac{(\tau_{yz})_q}{q_0} = \frac{(\sigma_y)_p}{p_0}, \tag{2.18}$$

where the subscripts p and q refer to the stress components respectively due to the normal load related to the contact pressure and tangential surface load acting separately, while $q_0 = \mu p_0 = \mu \sigma_H$ is the tangential friction surface load at $y = 0$. In this framework, $(\sigma_z)_q$ and $(\tau_{yz})_q$ can be obtained directly from the expressions of $(\tau_{yz})_p$ and $(\sigma_y)_p$, given by the third and the first of Eq. (2.15). Instead, the third stress component $(\sigma_y)_q$ has to be determined independently; with the notations of Eqs. (2.15) and (2.16), it is given by the following relationship:

$$(\sigma_y)_q = \frac{q_0}{b_H}\left[n\left(2 - \frac{z^2 - m^2}{m^2 + n^2}\right) - 2y\right]. \tag{2.19}$$

In the plane containing the area of contact, having coordinate $z = 0$, $(\sigma_y)_q$ is given by the following relationships:

$$(\bar{\sigma}_y)_q = -\frac{2q_0 y}{b_H} \tag{2.20}$$

$$(\bar{\sigma}_y)_q = -2q_0\left[\frac{y}{b_H} \mp \left(\frac{y^2}{b_H^2} - 1\right)^{1/2}\right], \tag{2.21}$$

which are valid for $|y| \le b_H$ and, respectively, for $|y| > b_H$. It is to be noted that the over bar is used to indicate values of quantity $(\sigma_y)_q$ referred to the surface of contact, having coordinate $z = 0$.

These stress components reach their highest values in the plane of the area of contact, and dampen rapidly with increasing z-coordinate. Figure 2.6 refers to a sliding contact where the upper surface is supposed to be motionless, while the one below, which is the driving surface, moves from left to right, so that reaction tangential friction forces are applied to it, represented by arrows directed from right to left. This assumption, fully justified as it is the relative motion that interest us here, greatly facilitates the understanding of the mechanics of sliding contact. The same Fig. 2.6 shows the distribution curves of $(\bar{\sigma}_y)_q/q_0$ and $(\bar{\tau}_{yz})_q/q_0$ along the y-axis at the surface of contact, $z = 0$. It shows that, when any given point on the tooth flank passes through the surface of contact, the shear stress component $(\bar{\tau}_{yz})_q$ varies from zero-to-maximum-to-zero, with $(\bar{\tau}_{yz})_{q\max} = q_0 = \mu p_0 = \mu\sigma_H$ at $y = 0$, while the normal stress component $(\bar{\sigma}_y)_q$ varies from zero-to-compression-to-tension-to-zero, or vice versa, depending on whether we consider the tooth flank surface of the

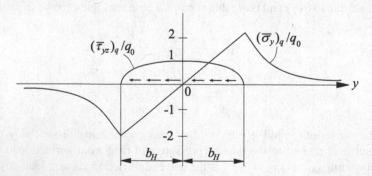

Fig. 2.6 Tooth flank surface of the driving gear wheel: distribution curves of $(\bar{\sigma}_y)_q/q_0$ and $(\bar{\tau}_{yz})_q/q_0$ along the y-axis

driving member or that of the driven member of the gear pair under consideration. The presence of a normal tensile stress on the surface of contact is undoubtedly significant in the propagation of surface fatigue cracks.

On the contact surface of the driving gear wheel (see Fig. 2.6), the shear stress component $(\bar{\tau}_{yz})_q$ is negative, in the sense that it has the opposite direction to the positive one of the y-axis, while the normal stress component $(\bar{\sigma}_y)_q$ varies linearly reaching a maximum compressive value, equal to $-2q_0 = -2\mu p_0$ at the leading edge of the same area of contact ($y = -b_H$), and a maximum tensile value, equal to $2q_0 = 2\mu p_0$, at the trailing edge of the area of contact ($y = b_H$). Instead, outside the contact area, the shear stress component vanishes, while the normal stress component is more or less rapidly damped, until it completely vanishes at a sufficient distance from the leading and trailing edges of the same area of contact.

On the contact surface of the driven member, the distribution of the shear stress component $(\bar{\tau}_{yz})_q$ does not change (however, it is positive since it has the same direction of the y-axis), while the distribution curve of the normal stress component $(\bar{\sigma}_y)_q$ changes and, to obtain it, just rotate 180° about the y-axis the distribution curve of $(\bar{\sigma}_y)_q$ inherent to the tooth flank surface of the driving member. Figure 2.7 shows the normal and shear stresses, $(\bar{\sigma}_y)_q$ and $(\bar{\tau}_{yz})_q$, due to the tangential friction load between the two instantaneous parallel cylinders pressed against one another, hypothesized by Buckingham (1949).

Contrary to what happens for the normal pressure, which gives rise to an equal compressive stress at the surface, $(\sigma_y)_p = -p(y)$, within the area of contact, but not outside of it, the tangential friction load gives rise to a variable normal stress at the surface, $(\bar{\sigma}_y)_q$, which also affects the outside of the area of contact. Moreover, whatever the value of the coefficient of friction, the maximum resulting tensile stress due to the tangential friction load occurs at the trailing edge and is equal to $2\mu p_0$ (Fig. 2.6).

As the first of Eq. (2.18) shows, the normal stress component $(\sigma_z)_q$ due to the tangential friction load is proportional to the shear stress component $(\tau_{yz})_p$ due

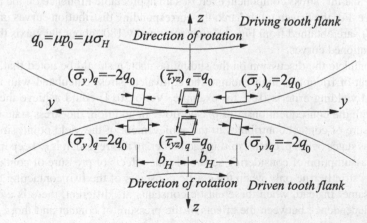

Fig. 2.7 Normal and shear stresses, $(\bar{\sigma}_y)_q$ and $(\bar{\tau}_{yz})_q$, due to the tangential friction surface load

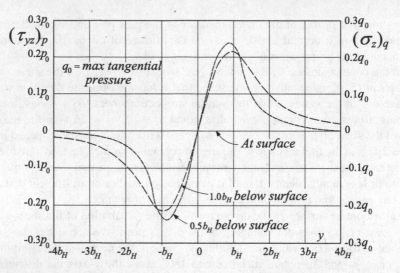

Fig. 2.8 Distribution curves of $(\tau_{yz})_p$ and $(\sigma_z)_q$ along the y-axis, in three different planes $z = const$, for the driving tooth flank

to the normal pressure, the proportionality factor being the coefficient of friction $\mu = (q_0/p_0)$. Figure 2.8 shows the distribution curves of both these stress components along the y-axis, in three different planes perpendicular to the z-axis, having respectively coordinates $z = 0$, $z = 0.5b_H$, and $z = b_H$, for the driving tooth flank. These stresses are constantly equal to zero on the interface surface ($z = 0$) as well as to any depth along the z-axis. Instead, in points at different depths $z \neq 0$, which do not lie on the z-axis, these same stresses vary according to the third of Eq. (2.15). In particular, we see that the maximum value of $(\sigma_z)_q$, approximately equal to $\mp 0.256q_0 = \mp 0.256\mu p_0$, occurs at a depth $z = 0.50b_H$, and at a distance from the z-axis of about $\mp 0.85b_H$. Therefore, we can draw the conclusion that it is unlikely that this stress component exercises an appreciable influence on the pitting resistance. For the driven tooth flank, the corresponding distribution curves of $(\sigma_z)_q$ and $(\tau_{yz})_p$ are obtained from Fig. 2.8 after rotating by 180° about the y-axis the two aforementioned curves.

To complete the discussion on the sliding contact, it should be noted that, if the coefficient of friction is high enough, the equivalent stress calculated with one of the usual yielding criteria (see, for example, Vullo 2014) could achieve the yield point, with the consequent onset of plastic flow. In addition, the stress state due to the pressure of contact contributes to the achievement of the yield point, since the two stress states overlap, often exalting each other. In this regard, it is to keep in mind that the assumption of considering separately the effects of pressure of contact and sliding is strictly true only when the elastic constants of the two contacting bodies are the same. Instead, when these elastic constants are different, there is a degree of interdependence between the effects of the pressure of contact and those of the sliding, as Bufler (1959) highlighted. On these issues, we still refer the reader to the listed specialized textbooks.

Now we turn our attention only on the rolling, without sliding. This is the so-called *pure rolling*, which is an ambiguous term, because the absence of an apparent sliding does not exclude the transmission of a tangential surface force of intensity lower than the limit friction force. The most appropriate terms in this case are those of *free rolling* or *tractive rolling*, depending on whether the tangential friction force, which characterizes the transmission of motion, is equal to or different from zero.

If the two instantaneous cylinders introduced by Buckingham, pressed one against the other, are no longer motionless, but begin to rotate about their own axes without sliding, a new alternating shear stress arises, even higher than the already highlighted alternating component of fluctuating shear stress, related to the shear stress component τ_{yz} given by the second of Eq. (2.14). We do not believe that this is the place to go into this very interesting problem, in reference to which we refer the reader to specialized textbooks. Therefore, we will not dwell on the fact for the first time highlighted by Reynolds (1876), and subsequently analyzed by other researchers, according to which the area of contact is divided into stick and micro-slip zones, depending on the correlations between the friction forces and elastic deformation.

Here, we just keep in mind that, due to rolling, a significant alternating shear stress arises, having much greater intensity than that of the already mentioned alternating component resulting from the fluctuating shear stress, related to the shear stress $(\tau_{yz})_p$ due to Hertzian pressure. This alternating shear stress occurs below the area of contact and reaches its maximum intensity in an offset position with respect to the z-axis.

Figure 2.9 shows that, due to rolling, a point A of the lower driven cylinder, located slightly below the area of contact, is brought during the meshing cycle at point B. Consequently, material is stressed first as in position A and then as in position B during each revolution. The shear stress component, τ_{yz}, increases from zero to a maximum positive value, then decreases to again down to zero for $y = 0$, to go to a maximum negative value, equal in intensity to the positive one, and then returns to decrease in absolute value up to zero.

This subsurface reversed shear stress determines an alternating shear stress cycle, which is repeated at each revolution, and is considered very significant for triggering of the subsurface fatigue cracks. This shear stress reaches, for $v = 0.3$, the maximum absolute value, equal to $0.256p_0 = 0.256\sigma_H$ at points located below the area of contact, at a depth equal to about $0.50b_H$, and at a distance from the z-axis equal to about $0.85b_H$. With the directions of rotation indicated in Fig. 2.9, the distribution curves of τ_{yz} are those shown in Fig. 2.8 for the driving tooth flank, while those of the driven tooth flank are obtained from the latter, after rotating them by 180° about the y-axis.

This alternating shear stress, which occurs at a depth of about $0.5b_H$ from the area of contact, for $v = 0.3$, has an intensity of about 68% higher than the corresponding alternating shear stress component due to the Hertzian pressure; therefore, it is considered to be much more dangerous in relation to the pitting resistance.

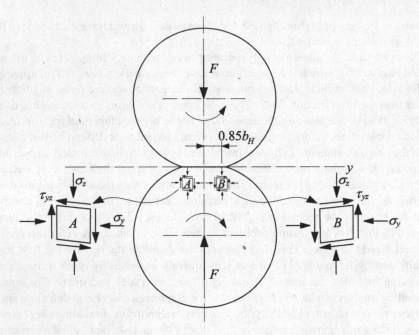

Fig. 2.9 Subsurface normal and shear stresses in rolling parallel cylinders

2.5 Other Considerations on the Pitting Generation

In the previous section, we have considered separately the various stress states that are generated in a Hertzian contact, even subjected to sliding and rolling, to better understand their individual effects. These stress states, though very significant and important, not fully interpret the actual complex stress state in the Hertzian contact of the gears, where other phenomena occur, which cannot be neglected. We begin to determine the total stress state, as if these additional phenomena were not active.

The total stress state due to normal plus tangential loads can be determined using the method of superposition, as we have assumed a linear elastic stress field (see Love 1944; Timoshenko 1940a, b; Timoshenko and Goodier 1951). To have a clear view of what is the most dangerous stress in relation to the pitting resistance, it is necessary to combine the different stress components, related to the various stress states analyzed above, in an equivalent stress, σ_e, using a *strength theory*. To this purpose, e.g. we can use the *maximum-distortion-energy theory* or *MDE theory*, *maximum-octahedral-shear-stress theory* or *MOSS theory*, *maximum-shear-stress theory* or *MSS theory*, *Mohr strength theory* and *internal-friction strength theory*, *maximum-normal-stress theory* or *MNS theory*, and so on (see, for example, Vullo 2014).

Figure 2.10 shows two distribution curves along the z-axis of the equivalent stress σ_e, made dimensionless by relating it to the Hertzian stress, σ_H, taken as a reference value, that is the quotient of the equivalent stress divided by the Hertzian stress; also

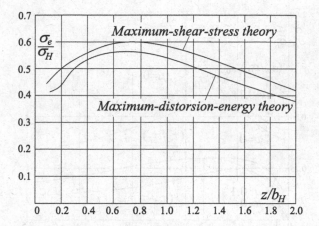

Fig. 2.10 Distribution curves of σ_e/σ_H as a function of z/b_H, for two strength theories

the z-distance of the contact area is made dimensionless by dividing it by b_H. These curves refer respectively to the MSS theory and MDE theory. The distribution curve of σ_e/σ_H calculated with the MDE theory is all below the one calculated with the MSS theory, since with the first theory the three-dimensional effect of the stress state is considered, while with the second theory this effect is neglected.

In the more general case of three-dimensional stress state, when we want to limit our analysis to only Hertzian contact, for which the stress state along the z-axis is the one described by the principal stress components given by Eq. (2.12), the equivalent stress σ_e according to the MDE theory is given by the following relationship, which also expresses the correlation with the MOSS theory:

$$\sigma_e = \frac{\sqrt{2}}{2}\left[(\sigma_x - \sigma_y)^2 + (\sigma_y - \sigma_z)^2 + (\sigma_z - \sigma_x)^2\right]^{1/2} = \frac{3}{\sqrt{2}}\tau_{oct}, \qquad (2.22)$$

where τ_{oct} is the octahedral shear stress.

When we want to extend our analysis to what happens outside the z-axis, including or not the sliding and rolling effects, the equivalent stress in terms of MDE theory and related MOSS theory is given by the following relationship:

$$\sigma_e = \frac{\sqrt{2}}{2}\left[(\sigma_x - \sigma_y)^2 + (\sigma_y - \sigma_z)^2 + (\sigma_z - \sigma_x)^2 + 6\left(\tau_{xy}^2 + \tau_{yz}^2 + \tau_{zx}^2\right)\right]^{1/2} = \frac{3}{\sqrt{2}}\tau_{oct}, \qquad (2.23)$$

in the case where the stress state is three-dimensional, or by the following relationship

$$\sigma_e = \sqrt{2}\left[\sigma_y^2 + \sigma_z^2 - \sigma_y\sigma_z + 3\tau_{yz}^2\right]^{1/2}, \qquad (2.24)$$

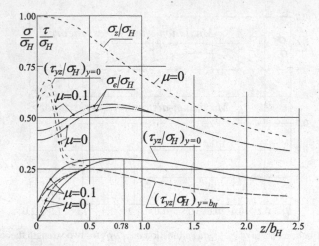

Fig. 2.11 Distribution curves of (σ_z/σ_H), (τ_{yz}/σ_H), (σ_e/σ_H) along the z-axis, and (τ_{yz}/σ_H) at a distance $y = 0.85b_H$ from the z-axis, as a function of (z/b_H)

in the case where the stress state is two-dimensional. The stress components to be substituted in these last two relationships are the ones given by Eq. (2.15).

Figure 2.11 shows the distribution curves along the z-axis of the following stresses (it is to be noted that both the stresses and distances from the surface of contact are made dimensionless by dividing them respectively by σ_H and b_H):

- normal stress component, σ_z, given by the third of Eq. (2.12);
- shear stress component, τ_{yz}, given by the second of Eq. (2.14);
- equivalent stress, σ_e, in terms of MDE theory, corresponding to the three principal stress components given by Eq. (2.12);
- shear stress component, τ_{yz}, given by the third Eq. (2.15), along an axis at a distance $y = 0.85b_H$ from the z-axis and parallel to this.

The four distribution curves show that, in absence of tangential friction loads ($\mu = 0$), the corresponding maximum values, respectively equal to $(\sigma_z/\sigma_H)_{max} = 1.00$, $(\tau_{yz}/\sigma_H)_{max} = 0.304$, $(\sigma_e/\sigma_H)_{max} = 0.545$ and $(\tau_{yz}/\sigma_H)_{max} = 0.256$, occur respectively for $z/b_H = 0$, $z/b_H = 0.78$, $z/b_H = 0.70$ (this is not shown in the figure), and $z/b_H = 0.50$.

Figure 2.11 also shows how the distribution curves of the equivalent stress and the two aforementioned shear stress components are changed, due to a tangential friction load with a coefficient of friction $\mu = 0.1$, which determines a modification of the resulting load distribution curve, as Fig. 2.12a shows. The same Fig. 2.11 shows how the distribution curves of the shear stress component along the z-axis, without and with a tangential friction load with $\mu = 0.1$, are modified due to a groove on the surface of contact, having a depth $0.1b_H$ and width $0.2b_H$, arranged symmetrically with respect to the z-axis, as Fig. 2.12b shows.

This groove is here considered to give an idea of how the surface undulations, roughnesses, various irregularities and, more generally, the geometrical deviations of the actual surface, compared to the smooth ideal surface that we have assumed

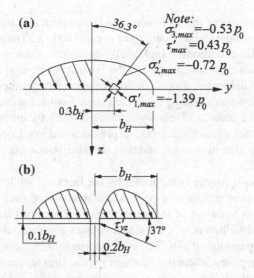

Fig. 2.12 Load distribution curves on the surface of contact: **a** continuous surface; **b** surface with a groove

as working hypothesis, can dramatically change the stress state. In fact, the groove interrupts the distribution curve of load acting on the surface, determining below the surface and in vicinity of it an intensification of the stress field, with much higher peak values than those calculated using the relationship described in the previous section. These peak values may exceed the strength limit values of materials. With reference to the specific case of Fig. 2.11, we see that, for $\mu = 0.1$ and $\mu = 0$, the peak stress exceeds, respectively, of about 126 and 107%, the one related to the Hertzian pressure (it is equal to $0.304\sigma_H$) and takes place at a depth of about $(0.08 \div 0.10)b_H$, instead of $0.78b_H$.

Of course, in addition to surface defects, which may be considered as external notches, the surface and subsurface impurities of the material structure come into play, as the inclusions of oxides and sulfides. These inclusions act as internal notches, and determine stress intensification effects, with local stress concentrations whose peak values can far exceed the stresses related to the Hertzian pressure.

For a more detailed analysis of the conditions that can cause the generation of pits, it is necessary to expand the horizons on the surface and subsurface stress states. In other words, other distribution curves of the ratio (σ_e/σ_H) must be analyzed, considering the effects of rolling and sliding, for different values of the coefficient of friction, as well as assessing what happens in offset position with respect to the z-axis. In this case, the load distribution on the surface of contact is that shown in Fig. 2.12a, to the left.

In these more general conditions of contact, which also consider the effects of rolling and sliding, for a special case ($\mu = 1/3$, and $\nu = 1/4$), Smith and Liu (1953) calculated the principal stresses σ_1', σ_2', and σ_3', and found that their maximum values occur at the same location. Figure 2.12a shows on the right both this location and the maximum values of the aforementioned three principal stresses, in the case where the maximum shear stress occurs at the contact surface. The same researchers, in the same particular conditions, found that the maximum shear stress occurs at the contact surface, when $\mu > 1/9$, and below the contact surface, when $\mu < 1/9$.

However, before the material strength limits are exceeded, other phenomena can occur. In fact, if during any loading cycle the elastic limit is exceeded, some plastic deformation will take place, and thereby residual stresses are introduced, which are additive to the contact stresses discussed in previous section. Local yielding alters the geometry of the area of contact, and thus the distribution curve of the pressure of contact is altered.

The phenomenon is further complicated by the fact that, while the load remains constant, the local plastic flow may proceed continuously during usual operation, causing a continuous variation of the actual stress distributions in the area of contact. The resulting distribution of the residual stresses can profoundly influence the generation and propagation of pits. These pits normally originate from subsurface defects in vicinity of points where the alternating shear stresses reach their maximum values, and commonly propagate along tensile stressed surfaces, just under the skin.

Two other important phenomena contribute to the generation of pits: the localized heating, and thermal expansion caused by the friction sliding, and the consequent power losses; the elastohydrodynamic pressure distribution within the lubricant oil film that is usually found between the two contacting surfaces in relative motion.

Frictional heating at sliding contacts between the two bodies results in high instantaneous local temperatures at the point of contact and in its surrounding area. These flash temperatures may be higher than $(250 \div 270)\,°C$ compared to the ambient temperature; in addition to the risk of scuffing (see Chaps. 7 and 8), they cause differential thermal expansion, which gives rise to thermal stresses that are added to the contact stresses, and may change the contact conditions through thermal distortion of their surface profiles. This is a complex phenomenon, strongly affected by the characteristics of the lubricant used.

The other important factor in explaining why pits can occur is identified in the pressures that are generated within the lubricant oil film interposed between the contacting surfaces in relative motion. These pressures can be so high as to cause elastic deformations of the surfaces, and variations in the lubricant oil viscosity, with consequent modifications of the lubricant film thickness and pressure distribution within it. The elastohydrodynamic lubrication theories come heavily into play. On this important subject, we refer the reader to the next section.

Obviously, the pitting resistance and surface durability are much higher as lower is the load and smaller the sliding. Good lubricating conditions ensure not a few advantages, such as:

- lower values of the coefficient of friction, and consequent substantial reductions of the normal and shear stresses due to the tangential friction load (see Fig. 2.7);
- a faster and more effective transfer of the heat generated by power losses due to friction, with consequent reduction of the thermal stresses;

- a more favorable pressure distribution on the area of contact, due to the presence of an adequate lubricant film thickness.

Generally, the resistance to surface fatigue increases by increasing the surface hardness. The increase of the surface hardness, however, involves a reduction in ability that the small surface imperfections have of self-adjust by wear, from which a reduction of localized contact pressures follows. To remedy this drawback, it is common practice to make one of the two members of a gear pair very hard (in general, the pinion), and the other member somewhat softer, in order to allow the running-in of the surfaces.

In addition, the surface fatigue strength takes advantage of the accuracy grade of the gear and surface smoothness. The exception is the case where sliding is significant. In this case, in fact, the surface porosity or small depressions on one of the mating surfaces may constitute a small reservoir of lubricant that helps to break down the coefficient of friction.

Finally, the surface fatigue strength also benefits from design choices, such as:

- a higher number of teeth of small module, which entails a higher transverse gear ratio, absence of zones with strong curvature of the tooth flank profiles near the base circle, as well as smaller shape deviations of the flank profile geometry, with the same accuracy grade of the teeth;
- a positive profile shift coefficient, when the number of teeth is small;
- a larger pressure angle, which, however, has the disadvantage to cause a decrease of the transverse gear ratio;
- a higher tooth depth, which, however, has the disadvantage of bringing the danger of seizure.

2.6   Some Brief Reminder on the Elastohydrodynamic Lubrication Theory

Notoriously, the lubrication of the gears is characterized not only by hydrodynamic effects, but also by deformations of the bodies, in the restricted region in which the contact takes place, and by the increases in the lubricant oil viscosity, due to the high pressures that are generated in this region.

In the case in which the pressure in the lubricant film is such as to induce, in the two bodies bounding the same lubricant film, deformations at least of the same order of magnitude of the lubricant film thickness, we speak of *elastohydrodynamic lubrication*. This type of lubrication is generally referred to by its acronym, *EHD lubrication*. For lubricated contact between two bodies made of materials with a high *Young's modulus*, the pressure in the lubricant film is such as to cause an increase of several order of magnitude of the lubricant viscosity. Therefore, the variation in the viscosity due to pressure (i.e. the *pressure-viscosity effect*, also called *piezo-viscosity effect*) plays a fundamental role in EHD lubrication. In the case in which the elastic deformation and/or the pressure-viscosity effect were neglected, we could find a strong underestimation of the lubricant film thickness, even of some orders of magnitude.

About the subject concerning lubrication and EHD lubrication, we refer the reader to specialized textbooks (see, for example: Cameron 1966; Dowson and Higginson 1977; Jacobson 1991; Stachowiak and Batchelor 2014). For our purposes, we need only mention here that the EHD lubrication is based on the combined action of the following three mechanisms:

- Elastic deformation of bodies that, in the case of metals with a high Young's modulus, is characterized by deflections of very similar magnitude of those of a Hertzian contact (this similarity of behavior is the more pronounced, the higher is the load), and in any case much more of the lubricant film thickness. In addition, the distribution curve of the pressure in the lubricant film is very similar to that typical of the Hertzian contact.
- Hydrodynamic effect, with consequent increase of pressure in the lubricant film between the two surfaces, due to the relative motion, as well as to the geometric shape of the surfaces.
- Pressure-viscosity effect, whereby the lubricant viscosity is highly depending on the pressure. For example, at a pressure of 1 GPa, corresponding to an average load, the viscosity of a common lubricant is several orders of magnitude greater.

With the assumptions that underlie the Hertz theory, and its extension to the rolling and sliding contacts (see Sects. 2.3 and 2.4), the elastic displacements of the two contacting bodies, in the z-axis direction, are given by relationships where two definite integrals appear, extended between the limits $-b_H$ and $+b_H$ (see Johnson 1985). These integrals express respectively the contributions of the normal pressure loads and tangential friction loads, $p(y)$ and $q(y)$, along the y-axis. However, the contribution due to $q(y)$ is generally neglected, because it is of a little significance in reference to the pressure distribution within the lubricant film.

The hydrodynamic lubrication theory is based on the following assumptions: perfectly rigid and smooth bodies, incompressible lubricating film, lubricant oil viscosity independent from the pressure, and thus zero pressure-viscosity effect, negligible velocity components perpendicular to the lubricant film, negligible inertia forces, etc. According to this theory, the distribution of the pressure $p = p(y)$ in the lubricant film is governed by the well-known Reynolds' equation (Reynold 1876), given by:

$$\frac{d}{dy}\left(\frac{\rho h^3}{12\eta}\frac{dp}{dy}\right) = v_\Sigma \frac{d}{dy}(\rho h), \tag{2.25}$$

where ρ and η are, respectively, the density and effective absolute (dynamic) viscosity of the oil wedge between the two cylinders at the mean wedge temperature, $h = h(y)$ is the lubricant film thickness, and v_Σ is the *cumulative velocity*, given by Eq. (3.63) of Vol. 1. It is to be noted that, in the Reynolds' equation, the meaning of symbols ρ and h is changed (unfortunately, the letters of the alphabets are not infinite).

Integrating the Reynolds' equation, and imposing the appropriate boundary conditions, we obtain the distribution curve $p = p(y)$ as a function of the minimum lubricant film thickness, h_{min}, and *cumulative semi-velocity* or *average rolling velocity* $v_w = v_\Sigma/2$, defined in Sect. 3.6 of Vol.1. A subsequent integration of function $p = p(y)$ allows to obtain the total load acting on the two cylinders. In this regard, it is to recall the fact that this result depends only on the velocity v_w, but it does not depend directly on the rolling velocities, v_{t1} and v_{t2}, given by Eqs. (3.59) and (3.60) of Vol. 1. In other words, we would get the same load carrying capacity both in the case of pure rolling and in the case of pure sliding.

The lubricant film thickness, $h = h(y)$, is easily obtained from the contact geometry, since the bodies are assumed perfectly rigid; this, of course, once that h_{min} has been determined. Near the area of contact, that is for absolute values of y much smaller than the instantaneous radii of curvature of the two cylinders, the following parabolic function of the thickness variation is usually introduced:

$$h = h_{min} + \frac{y^2}{2\rho}, \tag{2.26}$$

where ρ is the equivalent radius of curvature given by Eq. (2.2).

The boundary conditions to be imposed shall be expressed in the form:

$$[p(y)]_{y=y_i} = 0; \quad [p(y)]_{y=y_0} = 0; \quad \left[\frac{dp}{dy}\right]_{y=y_0} = 0. \tag{2.27}$$

With the first of these boundary conditions, we assume that the relative pressure, p, is equal to zero at a great distance from the center of the area of contact (for example, $y_i = -\rho$). With the second and third of conditions (2.27), we impose that, at the outgoing side of the lubricant film, i.e. at point of coordinate $y = y_0$, which is itself an unknown, the relative pressure is again equal to zero, and, to ensure the continuity in this point, also the pressure gradient is cancelled out at the same point. It is not the need to dwell on the known results obtained by integrating the Reynolds' equation, with these boundary conditions. We just have to keep in mind that, with reference to two cylinders with an interposed lubricant film in iso-viscous conditions, these results are sufficiently reliable only when deflections induced by the pressure in the two contacting bodies are much smaller than h_{min}.

As long as the pressure is sufficiently small, the lubricant oil viscosity can be regarded as a constant. However, the viscosity of most practical lubricants is very sensitive to changes in pressure and temperature, which respectively cause an increase and a decrease in viscosity. Since only the piezo-viscosity effect is of interest here, we consider the lubricant film under isothermal conditions (see Hamrock and Dowson 1977) and therefore we express the variation in viscosity with pressure by the following pressure-viscosity relationship, proposed by Barus (1893):

$$\eta = \eta_0 e^{\alpha p}, \tag{2.28}$$

where η_0 is the absolute (dynamic) viscosity at ambient pressure and temperature, and α is the *pressure-viscosity coefficient*, which significantly depends on the temperature, as the experimental data on piezo-viscosity of the same lubricants show (see Ciulli and Piccigallo 1996; Niemann et al. 2005; Lugt and Morales-Espejel 2011). For the usual lubricant oils, the variability ranges of η_0 and α are approximately the following: $(10^{-3} \leq \eta_0 \leq 10^{-1})$ Pas; $(0 \leq \alpha \leq 4 \times 10^{-8})$ Pa^{-1}. It is to be noted that Barus' relationship predicts viscosity values that are too high at high pressure. Generally, the Barus' relationship is sufficiently accurate only for pressure below $(0.1 \div 0.2)$ GPa, i.e. for pressures much lower than those typical of Hertzian contacts, which are higher at least an order of magnitude (1 GPa and beyond).

However, the use of the Barus' relationship, which is very simple, does not affect significantly the assessment of the elastohydrodynamic lubricant film thickness. The same relationship is not suitable for assessing the local temperature or the coefficient of friction. In this case, it is necessary to use more accurate relationships, such as the ones proposed by Roelands (1966), Bair (2001), and van Leeuwen (2009). It should be noted, however, that none of the aforementioned relationships is fully capable of describing the actual piezo-viscous behavior of the lubricant oils used in the practical applications that are of interest to us (see Lugt and Morales-Espejel 2011).

All the studies and researches on the subject have shown that, compared to the Hertz theory, the EHD lubrication theory simulates better what happens in a contact between deformable bodies between which a lubricant oil film is interposed, in terms of contact pressure distribution, as well as in terms of thickness and extension of the lubricant oil film. The basic equation of the EHD lubrication theory is still the Reynolds' equation (2.25), in which however the Barus' relationship or one of the other relationships proposed by the aforementioned researchers is substituted.

Usually, the modified Reynolds' equation, in which the Barus' relationship is introduced, is written in dimensionless form, by means of the introduction of the following non-dimensional numbers according to Dowson and Higginson (1977):

$$H = \frac{h}{\rho}; \quad G = \alpha E_r; \quad W = \frac{w}{\rho E_r} = 2\pi \left(\frac{\sigma_H}{E_r}\right)^2; \quad U = \frac{\eta_0 v_w}{\rho E_r}; \tag{2.29}$$

where: H is the non-dimensional number for oil film thickness; G is the non-dimensional number for materials; W is the non-dimensional number for the load; U is the non-dimensional number for speed; $w = (F_t/b \cos\alpha)$ is the specific load; $v_w = v_\Sigma/2$ is the average rolling velocity; E_r and ρ are the reduced modulus of elasticity and effective radius of curvature given respectively by Eqs. (2.3) and (2.7).

In the modified Reynolds' equation, not only the viscosity, but also the lubricant film thickness depends on the pressure. In particular, the thickness $h = h(y)$ of the lubricant oil film is given by the following equation:

$$h = h_{\min} + \frac{y^2}{2\rho} - \frac{2}{\pi E_r} \int_{y_i}^{y_0} p(s) \ln\left(\frac{y-s}{\rho}\right) ds, \tag{2.30}$$

Fig. 2.13 Lubricant film thickness in an elastohydrodynamic contact

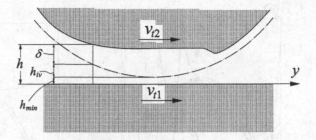

where h_{min} is the minimum lubricant film thickness, which is a unknown constant, while the other two terms represent, respectively, the thickness $h_{iv} = (y^2/2\rho)$ that we would have under iso-viscous conditions (i.e., in absence of elastic deformations), and elastic deflection of the contacting bodies, due to their compliance, i.e. the property of a material of undergoing elastic deformation (Cornell 1981).

Figure 2.13 shows the three summands that contribute to define the total lubricant film thickness, in the case in which one of the two cylinders has infinite radius. The variable, s (Fig. 2.2), is introduced for convenience of calculation, in that, under elastohydrodynamic lubrication conditions, the region of contact is significantly changed compared to that characterizing the Hertzian contact. Therefore, the integral in Eq. (2.30) is extended between the limits $s = y_i$ and $s = y_0$, related to the previously defined boundary conditions, which continue to be valid. The minimum lubricant film thickness, h_{min}, can be determined by imposing the further condition of balance between the applied load and the resulting force from the pressure distribution curve.

The Eqs. (2.25) and (2.30) provide a pair of simultaneous relationships for pressure, $p = p(y)$, and lubricant film thickness, $h = h(y)$. They may be combined together in a single integral equation for $p = p(y)$, which has been numerically solved for the first time by Herrebrugh (1968). However, to solve this integral equation, several other numerical methods may be used, which in any case allow to obtain the pressure distribution and the lubricant film thickness.

By way of example, Fig. 2.14a, b shows the distribution curves of p/p_0 and H as a function of the ratio y/b_H, and precisely: Fig. 2.14a, for $G = 5 \times 10^3$, $U = 10^{-11}$, and for four different values of the non-dimensional number for the load W (curve 1, $W = 10^{-5}$; curve 2, $W = 2 \times 10^{-5}$; curve 3, $W = 5 \times 10^{-5}$; curve 4, $W = 2 \times 10^{-4}$); Fig. 2.14b, for $G = 5 \times 10^3$, $W = 10^{-4}$, and for three different values of the non-dimensional number for speed U (curve 1, $U = 10^{-10}$; curve 2, $U = 10^{-11}$; curve 3, $U = 10^{-12}$). Both figures are taken by Ciulli and Piccigallo (1996), which obtained them using a FEM model.

For a better understanding of figures above, we must consider the correlation between the non-dimensional numbers given by the relationships (2.29), which were introduced by necessity of synthesis, and the numerous values of physical quantities used in practical applications. From this point of view, for cylinders made of steel, we have:

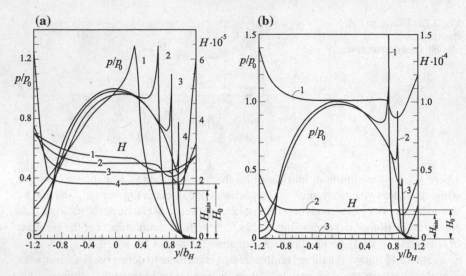

Fig. 2.14 Dimensionless distribution curves of pressure and lubricant film thickness for: **a** U and G constant, and W variable; **b** W and G constant, and U variable

- Non-dimensional number for materials: $G = 5000$ corresponds to a pressure-viscosity coefficient $\alpha = 22\ \text{GPa}^{-1}$, which is a typical value of paraffinic mineral oils at moderate operating temperature; for higher temperatures, between $(80 \div 100)\ °\text{C}$, G can assume values around 3000 or lower, while for some synthetic lubricants and some mineral oils, especially naphthenic oils, operating at low temperature, G can reach values greater than 7000.

- Non-dimensional number for the load: the variability range of W considered in Fig. 2.14 is the one corresponding to Hertzian stresses over the range $(0.29 \div 1.40)$ GPa; this is because the maximum Hertzian stress $p_0 = \sigma_H$ is related to W by the following relationship, $p_0 = E_r \sqrt{W/2\pi}$.

- Non-dimensional number for speed: U depends not only on the cumulative velocity (and thus on the rolling velocities, v_{t1} and v_{t2}), but also on the equivalent radius of curvature, ρ, and characteristics of the materials involved (and thus η_0 and E_r). Therefore, it can vary within an extremely wide range.

Figure 2.14a shows that, with the exception of a sharp pressure peak near the outgoing side, the pressure distribution curve is much closer to the Hertzian one the greater the load. The sharp pressure peak is followed by a fast pressure drop, and thinning of the lubricant film where the viscosity falls back to its ambient value, η_0. On the contrary, when the load decreases, the pressure distribution curve tends to approximate that which would obtain if the cylinders were infinitely rigid, and the operating conditions of the lubricant film were the iso-viscous conditions.

The same figure shows that, especially for medium and large loads, an appreciable fraction of the lubricant film, in the central region of contact, has an approximately constant thickness, H_0. In addition, the thickness distribution curve presents a characteristic restriction at the outgoing side, with a minimum lubricant film thickness,

H_{min}. It is also evident the low dependence on the load of this thickness minimum value, H_{min}. In fact, when W is made to increase by 20 times, H undergoes a variation of less than 30%. It is to be noted that H_{min} is about $(75 \div 80)\%$ of the thickness H_0.

Figure 2.14b shows instead that the variations of the lubricant film thickness related to changes in non-dimensional number for speed are much higher compared to those due to the change in non-dimensional number of load. It also shows that, with the increase of U, the sharp pressure peak tends to move towards the entry side, while at the same time its amplitude increases.

The variations of the non-dimensional number for material, G, which, however, covers a very narrow range, determine qualitatively similar effects to those of non-dimensional number for speed, U.

As we shall see in due course, the calculation of the minimum lubricant film thickness, h_{min}, is crucial. Several researchers have proposed various formulae, none of which, however, can be considered of general validity, since any of these formulae cannot cover all possible cases of operation. These formulae have been obtained numerically, after consideration of a wide range of values of the main variables involved, or elaborating experimental data.

Among the numerous regression formulae, as proposed by various researchers, we mention first the one developed and proposed by Ertel-Mohrenstein (1984). This formula concerns the non-dimensional number for film thickness, H_0, and has been determined for a line contact, under conditions of pure rolling (without sliding), and assuming that the film is approximately parallel. According to this formula, H_0 is given by:

$$H_0 = \frac{h_0}{\rho} = 1.95(GU)^{8/11}W^{-1/11}. \tag{2.31}$$

According to Dowson and Higginson (1977), the minimum non-dimensional number for film thickness, H_{min}, is given by:

$$H_{min} = \frac{h_{min}}{\rho} = 2.65G^{0.54}U^{0.70}W^{-0.13} \cong 0.8H_0. \tag{2.32}$$

We consider here appropriate to bring the formula developed and proposed by Jacobson (1991), since it directly highlights the influence exerted on the minimum lubricant film thickness, h_{min}, by the various main variables on which it depends:

$$h_{min} = 3.07\frac{\eta_0^{0.71}\alpha^{0.57}v_w^{0.71}\rho^{0.40}}{E_r^{0.03}w^{0.11}}. \tag{2.33}$$

This formula, which is valid for line contact, in normal operating conditions, shows that h_{min} depends mainly on the characteristics of the lubricant, η_0 and α, the average rolling velocity, v_w, and the equivalent radius of curvature, ρ, while the specific load, w, has little influence on the minimum lubricant film thickness, h_{min}.

Therefore, it is possible to reach very high pressures without that the lubrication conditions are substantially modified.

Of course, the presence of a lubricant oil film in accordance with the EHD lubrication theory cannot prevent that the surface fatigue phenomena, such as pitting (these are typical phenomena related to high pressures of contact), occur. It is easy to verify that, under iso-viscous conditions, that is neglecting the characteristic effects of the EHD lubrication, lubricant film thicknesses far lower even two orders of magnitude would be obtained, except for very low Hertzian pressures, and very high speeds.

2.7 Conditions for EHD Lubrication, and Subsurface and Surface Fatigue Damage

To prevent surface and subsurface fatigue damage and wear of the teeth, it is necessary that their working conditions be characterized by an appropriate lubrication. A good lubrication also allows to limit the friction power losses, and to facilitate the dissipation of heat generated by these power losses. The application of EHD lubrication theory allows to better assess the actual condition of sliding contact between the tooth flank surfaces, by estimating the probability of interactions between the asperities on the two surfaces due to the roughness.

So far, we have assumed that the surfaces are topographically smooth. This idealization does not correspond to reality, and the texture of the mating surfaces is to be considered, as it plays a key role in governing contact mechanics, that is to say the mechanical behavior exhibited at the interface between the same surfaces as they approach each other and transition from conditions of non-contact to full contact. To accurately describe the topography and texture of a surface, numerous areal field parameters are required (see Sect. 10.9.3). Here we limit ourselves to considering only roughness, which is notoriously just one of the profile parameters, namely the height parameter that is related to non-periodic finer irregularities in the surface texture. So, sticking to the tradition, we describe the roughness of a surface by the *average roughness*, defined as (see ISO 4287:1997; Whitehouse 2012; Morales-Espejel 2014; Sperka et al. 2016):

$$Ra = \frac{1}{L} \int_0^L |z| dx, \qquad (2.34)$$

where L is the sampling length, $|z| = |z(x)|$ the absolute value of the height of the surface above the datum line or center-line (that is the straight line, or circular arc in the case of round components, from which the mean square deviation is a minimum), and x the abscissa on the datum line from a point taken as origin of the profilometer trace.

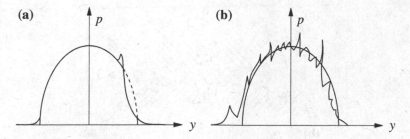

Fig. 2.15 Distribution curves of the pressure of contact with: **a** smooth surface; **b**, rough surface

We can also describe the roughness of a surface by means of the *root-mean-square roughness*, Rq, or standard deviation, σ, of the height of the surface from the datum line. It is given by:

$$Rq = \sigma = \frac{1}{L} \int\limits_0^L z^2 dx. \tag{2.35}$$

This is a statistically more meaningful quantity compared to the afore-defined average roughness. The correlation between Rq and Ra depends, to some extent, on the accuracy grade of the surface. For a Gaussian random profilometer trace and for a smooth sinusoidal profile, we have respectively: $Rq = (\pi/2)^{1/2} Ra$, and $Rq = (\pi/2\sqrt{2}) Ra$.

Generally, because of the roughness, the contact between two solid surfaces is no longer continuous, but discontinuous, and the actual area of contact is a more or less large fraction of the nominal area of contact. Its extension varies as a function of the load that, increasing, determines a flattening of the surface asperities, with a consequent extension of the bearing area. Furthermore, compared to what happens with perfectly smoothed surfaces (Fig. 2.15a), the roughness determines an irregular distribution of pressure of contact. Figure 2.15b qualitatively shows as the distribution curve of the pressure of contact is significantly changed due to the surface roughness.

For a quantitative judgment on the effectiveness of lubrication, it is necessary to estimate the probability of interaction between the asperities of the two contacting surfaces. One of the index factors of this probability that is frequently used in this regard is the *specific lubricant film thickness*, Λ, defined as the quotient of the minimum *lubricant film thickness*, h_{min}, divided by the square root of the sum of the squares of the root-mean-square roughnesses, Rq_1 and Rq_2, of the pinion and wheel, and therefore given by:

$$\Lambda = \frac{h_{min}}{\sqrt{Rq_1^2 + Rq_2^2}}. \tag{2.36}$$

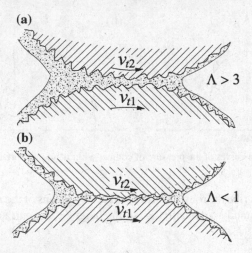

Fig. 2.16 Influence of Λ on the type of lubrication: **a** coherent thick oil film, according to EHD lubrication; **b** boundary lubrication

This specific film thickness was introduced for the first time by Bodensieck (1965), and is therefore called *Bodensieck's index* or Λ-*index*. It is an important index, but it should not be used for assessment and calculations of surface durability. About the values assumed by this specific lubricant film thickness, we must keep in mind that:

- For $\Lambda > 3$, we have *full film lubrication*, i.e. *elastohydrodynamic lubrication*. The lubrication conditions are such as to ensure the complete separation of the two bodies, the lubricant film is a thick coherent film (Fig. 2.16a), the risks of a direct contact are very low, and consequently the lifetime of the two members in relative motion is very large.
- For values of Λ less than $(1.00 \div 1.50)$, we enter the field of *mixed lubrication* and, for still lower values, we enter the field of *boundary lubrication* or *thin film lubrication*, with very high probability of metal-to-metal contact (Fig. 2.16b), and consequent wear of the materials involved.
- For intermediate values of Λ, that is included in the range $(3.00 \div 1.50)$, we have increasing contacts between the asperities of the surfaces when the specific film thickness Λ decreases.

Rebus sic stantibus, we can speak strictly of EHD lubrication, i.e. lubrication with coherent thick oil film, only when Λ is greater than 3, although in reality the roughness effect is often little relevant also for Λ-values smaller than 2.

Under boundary lubrication conditions, the lubricant oil film is very thin, and the behavior depends on the physical and chemical properties of the lubricant and materials of which the bodies are made. The thin lubricant film however provides a protective coating of the solid surfaces, reducing friction through an interfacial layer of low shear strength, and preventing the adhesion that would otherwise take place.

Several researchers, considering contact conditions both with and without sliding, have studied the effect of temperature on the viscosity. It has been shown that the viscous dissipation takes place in the entry region, with a smaller extension for pure rolling (without sliding), and with a much greater extension for simultaneous rolling and sliding. In both cases, the viscous dissipation gives rise to a rolling resistance, and an increase of temperature. Moreover, both the rolling resistance and increase in temperature grow as the viscosity and rolling speed increase. The magnitude of the dissipation forces and resulting temperature rise depend on the shear properties of the lubricant in the high-pressure region, where, among other things, a clear evidence of its non-Newtonian behavior was observed (see Johnson and Tevaarwerk 1977; Gao and Srirattayawong 2014).

Figure 2.17 shows the distribution curves of the temperature within the lubricant oil film, and temperature on the surface of bodies in relative motion, both related to the distribution curve of the local pressure of contact, also shown in the same figure. In particular, the figure highlights that the distribution curve of the temperature within the lubricant oil film has a maximum, which is slightly earlier compared with the maximum of pressure. To take into account thermal effects, several researchers (for

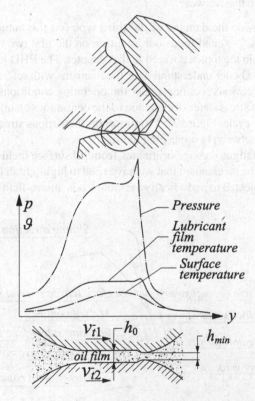

Fig. 2.17 Distribution curves of pressure of contact, and temperatures within the lubricant film and on surface of the bodies

example, Sadeghi and McClung 1991) proposed some relationships different from those described in the previous section. These relationships express the lubricant film thickness by means of more complex correlations between viscosity, pressure, and temperature.

In fact, however, the characteristics of the lubricant oil film are set in the entry region, and the shear heating in the region with a constant lubricant film thickness occurs too late to be able to appreciably influence the characteristics of the same lubricant film. Experimental measurements performed by different researchers (see Dyson et al. 1965–1966; Wymer and Cameron 1974) , with different experimental techniques, have shown that the EHD lubrication theory under isothermal conditions, both with and without sliding, provides a good model of behavior of the gears, so it is not the case here to further deepen this interesting topic.

According to Tallian (2000), two non-conforming sliding surfaces, subjected to localized contact and separated by thin lubricant films, are liable to the following type of damage, depending on the operating conditions:

- subsurface fatigue;
- surface fatigue;
- initial wear, and adhesive wear.

We here leave aside the damage of the third type (on this subject, we refer the reader to the Chaps. 7–9) and focus our attention on the first two types of damage, which are included in the topics covered in this chapter. The EHD lubrication theory can in fact help us to better understand their mechanisms with reference to gears. As we have shown in previous sections, under the operating conditions of a gear pair, a cyclic succession of stress states occurs, due to the continuous change of the contacts during the meshing cycle. Figure 2.18 summarizes the various stress components in the contact region between two meshing teeth.

The subsurface fatigue damage originates from subsurface inclusions. It is developed by following the mechanism that we have tried to highlight in Fig. 2.19a. When the surfaces are subjected to periodically variable loads, micro-fields of concentrated

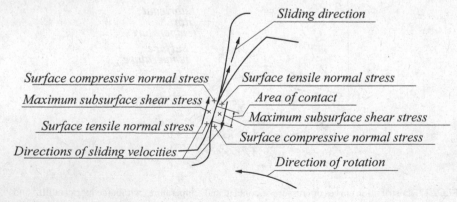

Fig. 2.18 Stress components in the contact region

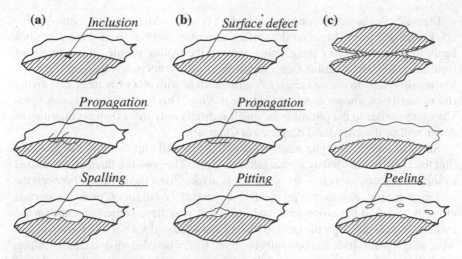

Fig. 2.19 Various types of subsurface and surface fatigue damage of gears: **a**, fatigue due to subsurface inclusions; **b** fatigue due to surface defects; **c** fatigue due to collision between surface asperities

stresses are generated in the neighborhoods of the inclusions. Resulting micro-cracks may be triggered, which tend to spread, first in a direction parallel to the surface, and subsequently toward the external surface, following the direction of the maximum shear stress, thus causing the detachment of material. Experimental tests have shown that the lubricant intrusion inside the cracks can lead to an increase of the propagation speed of the fractures.

The inclusions adversely affect the subsurface stress state, while the lubricant film thickness does not have any influence on it and, consequently, on the subsurface fatigue damage. The normal pressure distribution on the area of contact leads to a shear stress component whose maximum value occurs in correspondence of areas where the alternating normal stress components, due to the fluctuating friction load, also have their maximum value. This combination of maximum values of shear stress and alternating normal stress negatively affects surface durability, also because the load variations do not change their position, but only their intensity.

With a better choice of material, these inclusions can be avoided, but the problem of fatigue damage cannot be eliminated. It moves, so to speak, from under the skin to the surface. The surface fatigue can be originated by surface defects, such as grooves or notches, which cause dangerous stress concentrations, or by lubrication defects. Figure 2.19b shows what happens for a surface defect. This defect tends to locally interrupt the continuity of the lubricant oil film, resulting in fluctuations of the pressure therein, and then local stress intensifications that facilitate the occurrence of micro-cracks in the surrounding of the defect itself. These cracks can propagate along inclined planes corresponding to the direction of maximum shear stress, and therefore can give rise to detachment of material particles.

Figure 2.19c shows a second type of surface fatigue, which originates from micro-cracks that are formed around the surface asperities, especially if these asperities begin to repeatedly strike the existing ones on the mating profile, in conditions of insufficient EHD lubrication. Generally, these cracks do not penetrate deeply, but follow parallel paths to the surface, coming to interfere with other fractures, and giving rise to small pits, known as micropitting or *peeling*. This type of damage, conceptually quite similar to the pitting from which it differs only in its own peculiarities of detail, will be discussed and deepened in Chap. 10.

Since the surface fatigue arises from defects in EHD lubrication, the lubricant film thickness has obviously a great influence on it. The pressure fluctuations related to surface defects, as well as the shear stress arising from the collision between the surface asperities, also exert a great influence on surface fatigue. A probable thermal effect is added to the aforementioned factors. It may have some influence on the surface fatigue, although the current state of knowledge does not allow us to assess what weight it has. It should be finally observed that, when sliding overlaps to rolling, the fatigue lifetime decreases, since the sliding causes an increase in the number of collisions between opposing asperities during the meshing cycle.

2.8   Calculation of Surface Durability: ISO Basic Formulae

In previous sections, we have seen that the subsurface and surface damages in the gears are a very complex problem, and that the computed maximum value of the Hertzian contact stress, $p_0 = \sigma_H$, is not in itself a valid criterion for pitting failure. We have also seen that more rigorous procedures, able to better simulate the complexity of the phenomena that characterize the pitting, would be necessary for the assessment of surface durability. Fortunately, however, a good correlation is detected between the pitting damage and the Hertzian stress, so this is often used as an index of the severity of the contact load.

For these reasons, the ISO 6336-2:2006 assumes, as the basis for the calculation of surface durability, the contact stress, σ_H, at the pitch point or at the inner point of single pair tooth contact, with the caveat that, to define the determinant load carrying capacity, the highest of these two values is to be used. The contact stress, σ_H, is compared with the *permissible contact stress*, σ_{HP}, both for the pinion and wheel, and the comparison is expressed in terms of *safety factors*, S_{H1} and S_{H2}, which must be greater than a predetermined minimum value, $S_{H\min}$, i.e. $S_{H1} > S_{H\min}$, and $S_{H2} > S_{H\min}$. Therefore, the assessment of surface durability is carried out using the following relationships:

$$\sigma_H \leq \sigma_{HP} \tag{2.37}$$

$$\sigma_{H1} = Z_B \sigma_{H0} \sqrt{K_A K_v K_{H\beta} K_{H\alpha}} \quad \sigma_{H2} = Z_D \sigma_{H0} \sqrt{K_A K_v K_{H\beta} K_{H\alpha}} \tag{2.38}$$

$$\sigma_{H0} = Z_H Z_E Z_\varepsilon Z_\beta \sqrt{\frac{F_t}{bd_1} \frac{u+1}{u}} \tag{2.39}$$

$$\sigma_{HP} = \frac{\sigma_{Hlim} Z_{NT}}{S_{Hmin}} Z_L Z_v Z_R Z_W Z_X = \frac{\sigma_{HG}}{S_{Hmin}} \tag{2.40}$$

$$S_{H1} = \frac{\sigma_{HG1}}{\sigma_{H1}} > S_{Hmin} \quad S_{H2} = \frac{\sigma_{HG2}}{\sigma_{H2}} > S_{Hmin}. \tag{2.41}$$

The new symbols appearing in Eqs. (2.38) and (2.39), the meaning of those already introduced remaining unchanged, represent:

- σ_{H0}, the *nominal contact stress* at the pitch point C (Fig. 2.4), i.e. the contact stress in a ideal gear, without errors, due to the application of the static nominal torque;
- Z_B and Z_D, the *single pair tooth contact factors* of the pinion and wheel, which convert the contact stress at the pitch point to the contact stress at the inner point of single pair tooth contact, B, of the pinion and, respectively, to the contact stress at the inner point of single pair tooth contact, D, of the wheel (Fig. 2.4);
- Z_H, the *zone factor*, which generalizes the first square root that appears in Eq. (2.10), in order to include the helical gear wheels, and thus takes into account the flank profile curvatures at the pitch point, and transforms the tangential load at the reference cylinder to the tangential load at the pitch cylinder;
- Z_E, the *elasticity factor*, given by the second square root that appears in Eq. (2.10), and therefore depending on the reduced modulus of elasticity, E_r, expressed by Eq. (2.3);
- Z_ε, the *contact ratio factor*, which takes into account the influence of the effective length of the instantaneous lines of contact;
- Z_β, the *helix angle factor*, which takes into account the influences of the helix angle, as well as the change of the load distribution along the paths of contact.

Instead, the new symbols that appear in Eq. (2.40), which is in accordance with Method B, represent:

- σ_{Hlim}, the *allowable stress number* for contact, derived from test results on standard reference test gears with well-defined material, heat treatment and surface roughness (it is considered as the highest value of contact stress for which the material will have an endurance of at least 2×10^6 to 5×10^7 load cycles);
- Z_{NT}, the *life factor* for contact stress related to test gears, which takes into account the higher load carrying capacity for a limited number of load cycles;
- $\sigma_{HG} = \sigma_{HP} S_{Hmin}$, the *pitting stress limit*;
- Z_L, Z_R, and Z_v, are, respectively, the *lubricant factor*, *roughness factor*, and *velocity factor*, which take into account respectively the influences of the lubricant viscosity, surface roughness, and pitch point velocity (therefore, these three factors together account for the influence of the lubricant film on tooth contact stress);
- Z_W and Z_X, are, respectively, the *work hardening factor* and *size factor*, which take into account the effect of meshing of the gear wheel to be calculated with a surface

hardened or similarly hard mating gear wheel and, respectively, the influence of the tooth dimensions for the permissible contact stress.

For an appropriate use of the relationships (2.38) and (2.39), the following warnings should be remembered:

- For external and internal cylindrical spur gears with transverse contact ratio $\varepsilon_\alpha \geq 1$, σ_{H1} for the pinion is usually calculated at the inner point of single pair tooth contact. However, it is also necessary to calculate the value of σ_{H1} at the pitch point, in that, in special cases, it becomes the determinant value when it is greater than that calculated at the inner point of single pair tooth contact.
- For external spur gears with transverse contact ratio $\varepsilon_\alpha \geq 1$, σ_{H2} for the gear wheel is usually calculated at the pitch point. However, it is also necessary to calculate the value of σ_{H2} at the inner point of single pair tooth contact, in that, in special cases (particularly when the transmission ratios are small), it becomes the determinant value when it is greater than that calculated at the pitch point. Instead, for internal spur gears with transverse contact ratio $\varepsilon_\alpha \geq 1$, σ_{H2} for the gear wheel is always calculated at the pitch point.
- For helical gears with transverse contact ratio $\varepsilon_\alpha \geq 1$ and overlap ratio $\varepsilon_\beta \geq 1$, σ_{H1} and σ_{H2} for the pinion and gear wheel are always calculated at the pitch point.
- For helical gears with transverse contact ratio $\varepsilon_\alpha \geq 1$ and overlap ratio $\varepsilon_\beta < 1$, σ_H is calculated by linear interpolation between the two limit values of σ_H, corresponding respectively to spur gears and helical gears with $\varepsilon_\beta = 1$, and both calculated based on the actual number of teeth, instead of the virtual number of teeth.
- For helical gears with $\varepsilon_\alpha < 1$ and $\varepsilon_\gamma > 1$, an accurate analysis of the contact stress along the path of contact is necessary, since this case is not covered by the aforementioned ISO 6336-2:2006.

It is to be noted that, in the relationship (2.39), the term with the square root is nothing other than the third square root of Eq. (2.10). With reference to the total tangential load, F_t, that appears in this square root (this nominal value of the transverse tangential load per mesh is used even with $\varepsilon_{\alpha n} > 2$), it should be emphasized the need to introduce a mesh load factor, K_γ, following the application factor, K_A, in Eqs. (2.38), in the case of multiple-path transmission gear drives (see Sect. 1.1). Within the same square root, for double helical gears, we put $b = 2b_B$. Furthermore, the value of face width, b, is the smallest of the face widths at the root circles of pinion and wheel, ignoring any intentional transverse chamfer, end relief or tooth-tip rounding.

The determination of the permissible contact stress, σ_{HP}, can be carried out using several methods, all to be validated through accurate comparative analyses with well-documented past experiences and service histories of an adequate number of similar gears. The methods used are: Method A, Method B, and Method B_R.

With Method A, the permissible contact stress, σ_{HP} (or, in equivalent terms, the pitting stress limit, σ_{HG}), for reference stress, long and limited life and static stresses, is calculated with Eq. (2.40) from the S-N damage curve, obtained by means of tests performed on actual gear pair duplicates under appropriate service conditions. This damage curve, which is determined for the load carrying capacity at limited service life, simultaneously takes into account the materials of both members of the gear pair, heat treatments, gear dimensions such as nominal diameters and module, surface roughness of the tooth flanks, pitch line velocity, and characteristics of the lubricant used. Therefore, it is directly applicable under the conditions considered, since, already including all the above influences, factors Z_L, Z_v, Z_R, Z_W and Z_X that appear in Eq. (2.40) are all equal to 1. The permissible contact stress, σ_{HP}, can be also obtained from considerations of performance, dimensions, and service of accurately monitored reference gears. This method, very expensive, is justified only for the development of new products, whose failure could have serious consequences. We will not return to this method afterwards.

Generally, Method B is used for determining σ_{HP}, since it is considered suffi-ciently accurate whenever pitting resistance values from gear tests or special tests are available, or if the material is similar. This method, on which we will return later, is based on the availability of damage curves, which are characterized by given values of σ_{Hlim} and Z_{NT}, and are determined for a number of common gear materials and heat treatments, according to the results of gear load tests with standard reference test gears. These reference values are then converted by means of the five factors Z_L, Z_v, Z_R, Z_W and Z_X that appear in Eq. (2.40), to adapt them to the actual case study.

When stress values obtained from gear tests are not available, Method B_R can be used, with which the material characteristic values are determined by a rolling disk pair in loaded contact.

It should be noted that two special values of σ_{HP} must be considered: the *reference permissible contact stress*, σ_{HPref}, and *static permissible contact stress*, σ_{HPstat}. Both these values are determined using Eq. (2.40), the first with $Z_{NT} = 1$ and the other influence factors σ_{Hlim}, Z_L, Z_v, Z_R, Z_W, Z_X and S_{Hmin} calculated with Method B, and the second with all influence factors for static stress calculated with Method B.

2.9 Influence Factors of the Hertzian Stress

In the previous section, we have already defined the factors that appear in Eqs. (2.38) and (2.39), which affect the Hertzian stress. In this section, we describe the procedures for their determination with calculations or diagrams that allow us to read their values, as a function of determinant variables on which they depend.

2.9.1 Zone Factor, Z_H

In the more general case of cylindrical helical gears, this factor can be calculated with the following relationship:

$$Z_H = \sqrt{\frac{2 \cos \beta_b \cos \alpha_{wt}}{\cos^2 \alpha_t \sin \alpha_{wt}}}, \qquad (2.42)$$

where β_b, α_t, and $\alpha_{wt} = \alpha_t'$, are respectively the base helix angle, transverse pressure angle, and transverse pressure angle at the pitch cylinder. These quantities have been defined in Chap. 8 of Vol. 1, to which we refer the reader. It is to be noted that, in the case of cylindrical spur gears, Eq. (2.42) is reduced to the first square root that appears in Eq. (2.10) where $\alpha = \alpha_t$ is placed.

For external and internal gears with normal pressure angle $\alpha_n = 20°$, this factor can be read on the chart of Fig. 2.20, as a function of the helix angle at pitch cylinder, β, and ratio $(x_1 + x_2)/(z_1 + z_2)$. On the same chart, also the distribution curves of Z_H for $\alpha_n = 17.5°$, $\alpha_n = 22.5°$, and $\alpha_n = 25°$, and $(x_1 + x_2)/(z_1 + z_2) = 0$ are plotted. The complete charts for $\alpha_n = 22.5°$ and $\alpha_n = 25°$ can be found in ISO 6336-2:2006.

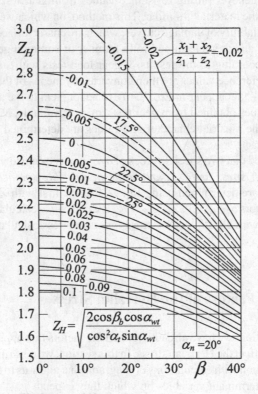

Fig. 2.20 Zone factor, Z_H, for $\alpha_n = 20°$

2.9.2 Single Pair Tooth Contact Factors, Z_B and Z_D

To determine the single pair tooth contact factors, Z_B and Z_D, in the case in which $\varepsilon_\alpha \leq 2$, we introduce the following two auxiliary quantities:

$$M_1 = \sqrt{\frac{\rho_{C1}\rho_{C2}}{\rho_{B1}\rho_{B2}}} = \frac{\tan\alpha_{wt}}{\sqrt{\left[\left(\sqrt{\left(\frac{d_{a1}^2}{d_{b1}^2}-1\right)}\right)-\frac{2\pi}{z_1}\right]\left[\left(\sqrt{\left(\frac{d_{a2}^2}{d_{b2}^2}-1\right)}\right)-(\varepsilon_\alpha-1)\frac{2\pi}{z_2}\right]}} \tag{2.43}$$

$$M_2 = \sqrt{\frac{\rho_{C1}\rho_{C2}}{\rho_{D1}\rho_{D2}}} = \frac{\tan\alpha_{wt}}{\sqrt{\left[\left(\sqrt{\left(\frac{d_{a2}^2}{d_{b2}^2}-1\right)}\right)-\frac{2\pi}{z_2}\right]\left[\left(\sqrt{\left(\frac{d_{a1}^2}{d_{b1}^2}-1\right)}\right)-(\varepsilon_\alpha-1)\frac{2\pi}{z_1}\right]}}. \tag{2.44}$$

These quantities depend on the radii of curvature at the pitch point and at inner points of single pair tooth contact, B and D, of pinion and gear wheel (Fig. 2.21). They can also be expressed by other quantities that appear in the same Eqs. (2.43) and (2.44), some of which are represented in Fig. 2.21 and, in any case, all already defined in Vol. 1. Equations (2.43) and (2.44) are not valid when undercut due to cutting interference shortens the path of contact (see Sect. 4.5 and Fig. 4.9 of Vol. 1). In addition, the Eq. (2.44) is only valid for external spur gears. It should be noted that

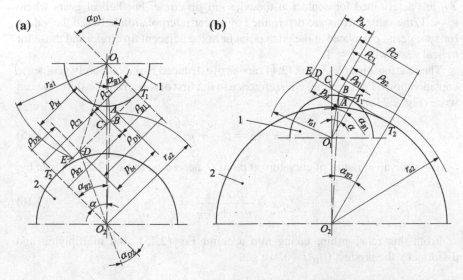

Fig. 2.21 Radii of curvature at pitch point C, and at inner points of single pair tooth contact, B and D, of the pinion and gear wheel, for: **a** external gear pair; **b** internal gear pair

Z_D should be determined only for gear pair with gear ratio $u \leq 1.5$. For $u > 1.5$, M_2 is usually less than 1 and, in this case, M_2 is made equal to 1 in Eq. (2.44). For internal gears, Z_D should be taken equal to 1.

To determine, in the various cases that may arise, the values of single pair tooth contact factors, Z_B and Z_D, it is necessary first to calculate the transverse contact ratio, ε_α, and overlap ratio, ε_β. For this purpose, we can use Eqs. (3.44)–(3.46) and Eq. (8.27) of Vol. 1. We have the following four cases:

- Spur gears with $\varepsilon_\alpha > 1$: we will take $Z_B = 1$ and $Z_D = 1$, respectively if $M_1 \leq 1$ and $M_2 \leq 1$; instead, we will take $Z_B = M_1$ and $Z_D = M_2$, respectively if $M_1 > 1$ and $M_2 > 1$.
- Helical gears with $\varepsilon_\alpha > 1$ and $\varepsilon_\beta \geq 1$: we will take $Z_B = Z_D = 1$.
- Helical gears with $\varepsilon_\alpha > 1$ and $\varepsilon_\beta < 1$: we will determine Z_B and Z_D by linear interpolation between the values for spur and helical gears with $\varepsilon_\beta \geq 1$, taking the greater of the two values $Z_B = \left[M_1 - \varepsilon_\beta(M_1 - 1)\right]$ and $Z_B \geq 1$, for Z_B, and the greater of the two values $Z_D = \left[M_2 - \varepsilon_\beta(M_2 - 1)\right]$ and $Z_D \geq 1$, for Z_D. For $Z_B = 1$ or $Z_D = 1$, Eq. (2.38) provide the values of the contact stresses at the pitch cylinders.
- Helical gears with $\varepsilon_\alpha \leq 1$ and $\varepsilon_\gamma > 1$: this last case is not covered by ISO 6336-2:2006; thus, it must be faced with a careful analysis to be carried out with refined mathematical models or with adequate experimental measurements.

It is to be noted that the considerations concerning the first three cases mentioned above, covered by ISO 6336-2:2006, refer to contact stress calculations under the usual conditions where the pitch point lies within the path of contact. In cases where the pitch point is determinant and lies outside the path of contact, factors Z_B and/or Z_D are determined for contact at the adjacent tip circle. For helical gears where $\varepsilon_\beta < 1$, the same factors are determined by linear interpolation between the values for spur gears, calculated at the pitch point or at the adjacent tip circle, and those for helical gears with $\varepsilon_\beta \geq 1$.

The relationships (2.43) and (2.44) are easily deduced based on purely geometric considerations. For example, with reference to the first of these relationships, we can write (Fig. 2.21a):

$$\rho_{1B} + \rho_{2B} = \rho_{1C} + \rho_{2C}. \tag{2.45}$$

The equivalent radius of curvature at point B, according to Eq. (2.2), is given by:

$$\rho_B = \frac{\rho_{1B}\rho_{2B}}{\rho_{1B} + \rho_{2B}}. \tag{2.46}$$

From this relationship, taking into account Eq. (2.45), and multiplying and dividing by the product $(\rho_{1C}\rho_{2C})$, we get:

$$\rho_B = \frac{\rho_{1B}\rho_{2B}}{\rho_{1C} + \rho_{2C}}\frac{\rho_{1C}\rho_{2C}}{\rho_{1C}\rho_{2C}} = \rho_C\frac{\rho_{1B}\rho_{2B}}{\rho_{1C}\rho_{2C}} = \frac{\rho_C}{M_1^2} = \frac{\rho_C}{Z_B^2}. \tag{2.47}$$

We have thus shown that $M_1 = Z_B$, and that both are expressed by the first square root that appears in Eq. (2.43). Introducing within this last square root the expressions of the four radii of curvature, as inferred by Eqs. (3.49)–(3.51) of Vol. 1, we obtain the Eq. (2.43).

In the case of high precision gear pairs with $2 < \varepsilon_\alpha \leq 2.5$, the total transverse tangential load is supported by two or three pairs of meshing teeth. For these gears, the determination of the contact stress is carried out with reference to the inner point of two pair tooth contact of the pinion.

2.9.3 Elasticity Factor, Z_E

As we have seen previously, the elasticity factor, Z_E, is given by the following relationship:

$$Z_E = \sqrt{\frac{E_r}{2\pi}} = \sqrt{\frac{1}{\pi\left(\dfrac{1 - \nu_1^2}{E_1} + \dfrac{1 - \nu_2^2}{E_2}\right)}}. \tag{2.48}$$

For $E_1 = E_2 = E$ and $\nu_1 = \nu_2 = \nu$, this relationship becomes:

$$Z_E = \sqrt{\frac{E}{2\pi(1 - \nu^2)}} \tag{2.49}$$

and, for $\nu = 0.3$ (steel and aluminum):

$$Z_E = \sqrt{0.175E}. \tag{2.50}$$

For gear pairs with pinion and wheel made of different materials ($E_1 \neq E_2$), the equivalent modulus of elasticity [see Eq. (1.93)]

$$E = \frac{2E_1 E_2}{E_1 + E_2} \tag{2.51}$$

can be used. Table 2.1 summarizes the values of Z_E for some combinations of ferrous metals of pinion and wheel, all characterized by a Poisson's ratio $\nu = 0.3$. Table 2.2 clarifies the meaning of the abbreviations used to indicate the different materials.

Table 2.1 Elasticity factor, Z_E, for some material combination, all with $\nu = 0.3$

Pinion, 1		Wheel, 2		Z_E (in $\sqrt{N/mm^2}$)
Material	Modulus of elasticity, E (in kN/mm^2)	Material	Modulus of elasticity, E (in kN/mm^2)	
St, V, Eh, IF, NT, NV	206	St, V, Eh, IF, NT, NV	206	189.8
		St (cast)	202	188.9
		GGG, GTS	173	181.4
		GG	126–118	165.4–162.0
St (cast)	202	St (cast)	202	188.0
		GGG, GTS	173	180.5
		GG	118	161.4
GGG, GTS	173	GGG, GTS	173	173.9
		GG	118	156.6
GG	126–118	GG	118	146.0–143.7

Table 2.2 Materials

Material	Type	Abbreviation
Normalized low carbon steels/cast steels	Wrought normalized low carbon steels	St
	Cast steels	St (cast)
Cast iron	Black malleable cast iron (perlitic structure)	GTS (perl.)
	Nodular cast iron (perlitic, bainitic, ferritic structure)	GGG (perl., bai., ferr.)
	Grey cast iron	GG
Through-hardened wrought steels	Carbon steels, alloy steels	V
Through-hardened cast steels	Carbon steels, alloy steels	V (cast)
Case-hardened wrought steels		Eh
Flame or induction hardened wrought or cast steels		IF
Nitrided wrought steels/nitriding steels/through-hardening steels, nitrided	Nitriding steels	NT (nitr.)
	Through hardening steels	NV (nitr.)
Wrought steels, nitrocarburized	Through hardening steels	NV (nitrocar.)

2.9.4 Contact Ratio Factor, Z_ε

In Sect. 8.7 of Vol. 1 we have seen that, in cylindrical spur gears with $1 < \varepsilon_\alpha < 2$, the total length of the line of contact, l_t, undergoes an abrupt change, passing from the value, b, when only one pair of teeth is in meshing (single contact), to the value, $2b$, when two pairs of teeth are in meshing (double contact). Of course (Fig. 1.22), the greater the transverse contact ratio, ε_α, the smaller will be the length of the single-contact portion of the path of contact, and the greater the length of the *virtual face width*, b_{vir}, which represents the instantaneous effective line of contact mediated during the meshing cycle, that is the mean value, l_m, of the length of line of contact.

In the same Sect. 8.7 of Vol. 1 we also have seen that, in cylindrical helical gears, the total length of the line of contact fluctuates between a maximum and a minimum value, and that it depends on the position of the instantaneous line of contact. With $\varepsilon_\beta > 1$, this total length of the line of contact fluctuates around the mean value that we introduce in the calculation of the load carrying capacity of the gear.

For a cylindrical spur gear, the mean value of the length of the line of contact, coinciding with the virtual face width, is given by:

$$l_m = b_{vir} = \frac{3b}{(4 - \varepsilon_\alpha)}, \tag{2.52}$$

from which we can see that, for $\varepsilon_\alpha = 1$, $b_{vir} = b$, and for $\varepsilon_\alpha = 2$, $b_{vir} = (3/2)b$.

For a cylindrical helical gear with $\varepsilon_\beta \geq 1$, the mean value of the length of the line of contact that we introduce is given by:

$$l_m = \frac{b_{vir}}{\cos \beta_b} = \frac{b\varepsilon_\alpha}{\cos \beta_b}; \tag{2.53}$$

for $0 < \varepsilon_\beta < 1$, l_m is determined by linear interpolation between the values for spur and helical gears with $\varepsilon_\beta \geq 1$, given respectively by Eqs. (2.52) and (2.53).

In this framework, the contact ratio factor, Z_ε, for calculation of the contact stress, is defined by the following relationship:

$$\frac{b_{vir}}{b} = \frac{1}{Z_\varepsilon^2}. \tag{2.54}$$

By combining these relationships into a single equation of general validity, we get:

$$Z_\varepsilon = \sqrt{\frac{4 - \varepsilon_\alpha}{3}(1 - \varepsilon_\beta) + \frac{\varepsilon_\beta}{\varepsilon_\alpha}}. \tag{2.55}$$

About the use of this general relationship, it is to be noted that it provides, for cylindrical helical gears, the value of Z_ε for $\varepsilon_\beta < 1$. When $\varepsilon_\beta \geq 1$, we always have

Fig. 2.22 Contact ratio factor, Z_ε

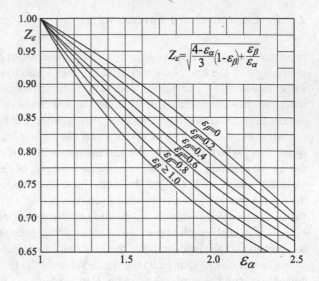

to introduce in Eq. (2.55) $\varepsilon_\beta = 1$, for which we obtain $Z_\varepsilon = \sqrt{1/\varepsilon_\alpha}$. For spur gears ($\varepsilon_\beta = 0$), since ε_α is greater than 1, we obtain values of Z_ε smaller than 1, but the conservative value $Z_\varepsilon = 1$ can be taken for a transverse contact ratio less than 2.

For known transverse contact and overlap ratios, Z_ε can be read from Fig. 2.22.

2.9.5 Helix Angle Factor, Z_β

The helix angle, β, affects both the virtual face width, and the radii of curvature. Therefore, the contact ratio factor, Z_ε, and zone factor, Z_H, which respectively depend on the virtual face width and radii of curvature, take already into account the influences exerted by the helix angle on the Hertzian stress, although not completely, but only partially. However, specific tests and practical experiences show that the Hertzian stress varies with the helix angle more strongly than the two aforementioned factors indicate.

To account for these experimental findings, the helix angle factor, Z_β, is introduced. It is defined by the following empirical relationship, which is in sufficient good agreement with the experimental evidences and service experience, and which meets the requirements of most practical applications, especially when high precision and optimum modifications are employed:

$$Z_\beta = \frac{1}{\sqrt{\cos \beta}}. \tag{2.56}$$

The helix angle factor, Z_β, can be read from Fig. 2.23.

Fig. 2.23 Helix angle factor, Z_β

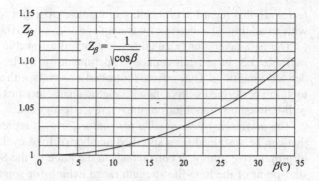

$$Z_\beta = \frac{1}{\sqrt{\cos\beta}}$$

2.10 Influence Factors of the Permissible Contact Stress

In Sect. 2.8 we defined factors and quantities that appear in Eq. (2.40), which influence the permissible contact stress. In this section, we describe the procedures and methods for their determination with calculations and diagrams that allow us to read their values, as a function of determinant variables on which they depend.

2.10.1 Allowable Stress Numbers for Contact and Bending

For reasons of uniformity and brevity, we believe it is appropriate to consider together the allowable stress numbers for contact, σ_{Hlim}, and bending, $\sigma_{FE} = \sigma_{Flim} Y_{ST}$, where σ_{Flim} is the nominal stress number for bending, and $Y_{ST} = 2$ is a stress correction factor (see next chapter). In accordance with Method B, both of these allowable stress numbers, σ_{Hlim} and σ_{FE} (and therefore σ_{Flim}), have been obtained by endurance tests of reference test gears under reference test conditions.

For the appropriate choice of the values of these quantities, three different classes of quality, ML, MQ, and ME, are provided, which depend on the type of production and quality control exercised. Synthetically, ML indicates a modest demand on the material quality and its heat treatment process during manufacturing of the gear; MQ indicates requirements that can be met at moderate cost by experienced manufacturers; ME indicates requirements that must be necessarily realized when a high degree of operational reliability is required. For more details, we refer the reader to ISO 6336-5:2016.

With Method B, the values of the *standard allowable stress numbers* are calculated by using the procedure described below. They vary depending on the material composition and surface hardness of the teeth, and therefore on the heat treatment of the material, and represent the fatigue limit in the S-N damage curves for 1% probability of damage. When other probability of damage are required by the design data sheet, the values of σ_{Hlim}, σ_{Flim}, and σ_{FE} are adjusted by an appropriate *reliability factor*. In this regard, a probability of damage different from 1% is to be indicated with an additional subscript; for example, σ_{Hlim10} for 10% probability of damage.

The parameters for calculating the standard allowable stress numbers, collected in Table 2.3, have been deduced experimentally by laboratory tests under well-defined conditions. They refer to controlled materials with appropriate heat treatment, toothing performed with certain geometric characteristics and surface finishing, and well-defined test conditions of the reference test gears.

The *allowable stress number for contact*, σ_{Hlim}, represents the contact pressure that can be sustained for a specified number of load cycles without the occurrence of progressive pitting. It corresponds to the knee of the S-N damage curve, i.e. the start-point of the long-life strength range, which for some materials is set equal to 5×10^7 stress cycles. The reference operating conditions of testing and dimensions of the reference test gear are as follows: center distance, $a = 100$ mm; helix angle, $\beta = 0$ $(Z_\beta = 1)$; module, $m = 3$ mm to 5 mm $(Z_X = 1)$; mean peak-to-valley roughness of the teeth flanks, $Rz = 3\,\mu m (Z_R = 1)$; tangential velocity, $v = 10$ m/s $(Z_v = 1)$; kinematic viscosity of the oil at 50 °C, $v_{50} = 100\,mm^2/s$ $(Z_L = 1)$; mating members of the gear made of the same material $(Z_W = 1)$; gear accuracy grade, 4 to 6, according to ISO 1328-1:2013; face width, $b = 10$ mm to 20 mm; load influence factors, $K_A = K_v = K_{H\beta} = K_{H\alpha} = 1$.

The *nominal stress number for bending*, σ_{Flim}, is determined by tests done with reference test gears, and represents the value of the bending stress limit that takes into account the influences of the material, heat treatment and surface roughness of root fillets of the test gear. Instead, the *allowable stress number for bending*, $\sigma_{FE} = \sigma_{Flim}Y_{ST}$, is the basic bending strength of the un-notched test specimen, under the assumption of a perfectly elastic behavior of the material, including the heat treatment. The *stress correction factor* of the reference test gear, Y_{ST}, is equal to 2. The value of σ_{FE} corresponds to the knee of the S-N damage curve, i.e. the start-point of the long-life strength range, which for most materials is set equal to 3×10^6 stress cycles. The reference operating conditions of testing and dimensions of the reference test gears are as follows: helix angle, $\beta = 0 (Y_\beta = 1)$; module, $m = 3-5$ mm $(Y_X = 1)$; *stress correction factor*, $Y_{ST} = 2$; *notch parameter*, $q_{ST} = 2.5 (Y_{\delta relT} = 1)$; *mean peak-to-valley roughness* of the tooth fillets, $Rz = 10\,\mu m$ $(Y_{RrelT} = 1)$; gear accuracy grade, 4–7 according to ISO 1328-1:2013; basic rack, according to ISO 53:1998; face width, $b = 10-50$ mm; load influence factors, $K_A = K_v = K_{F\beta} = K_{F\alpha} = 1$.

The allowable stress number for contact, σ_{Hlim}, and the nominal stress number for bending, σ_{Flim}, can be calculated by the following relationships:

$$\sigma_{Hlim} = Ax + B \quad \sigma_{Flim} = Ax + B, \tag{2.57}$$

where x is the surface hardness, in terms of *Brinell hardness*, HB, or *Vickers hardness*, HV (see Faupel 1964; Giovannozzi 1965a), while A and B are constants. Table 2.3 taken from ISO 6336-5:2016 allows to derive these constants, as well as the hardness ranges between the minimum and maximum hardness values related to heat treatment, for various types of materials and for the above three classes of quality.

Table 2.3 Parameters for calculation of σ_{Hlim} and σ_{Flim}

Stress	Type of material	Quality	A	B	Hardness	Min. hardness	Max. hardness
Contact	St	ML/MQ ME	1.000 1.520	190 250	HB	110 110	210 210
	St (cast)	ML/MQ ME	0.986 1.143	131 237	HB	140 140	210 210
Bending	St	ML/MQ ME	0.455 0.386	69 147	HB	110 110	210 210
	St (cast)	ML/MQ ME	0.313 0.254	62 137	HB	140 140	210 210
Contact	GTS (perl.)	ML/MQ ME	1.371 1.333	143 267	HB	135 175	250 250
	GGG	ML/MQ ME	1.434 1.500	211 250	HB	175 200	300 300
	GG	ML/MQ ME	1.033 1.465	132 122	HB	150 175	240 275
Bending	GTS (perl.)	ML/MQ ME	0.345 0.403	77 128	HB	135 175	250 250
	GGG	ML/MQ ME	0.350 0.380	119 134	HB	175 200	300 300
	GG	ML/MQ ME	0.256 0.200	8 53	HB	150 175	240 275
Contact	V (carbon steels)	ML MQ ME	0.963 0.925 0.838	283 360 432	HV	135 135 135	210 210 210

(continued)

Table 2.3 (continued)

Stress	Type of material	Quality	A	B	Hardness	Min. hardness	Max. hardness
	V (alloy steels)	ML	1.313	188	HV	200	360
		MQ	1.313	373		200	360
		ME	2.213	260		200	390
Bending	V (carbon steels)	ML	0.250	108	HV	115	215
		MQ	0.240	163		115	215
		ME	0.283	202		115	215
	V (alloy steels)	ML	0.423	104	HV	200	360
		MQ	0.425	187		200	360
		ME	0.358	231		200	390
Contact	V (cast) (carbon steels)	ML/MQ	0.831	300	HV	130	215
		ME	0.951	345		130	215
	V (cast) (alloy steels)	ML/MQ	1.276	298	HV	200	360
		ME	1.350	356		200	360
Bending	V (cast) (carbon steels)	ML/MQ	0.224	117	HV	130	215
		ME	0.286	167		130	215
	V(cast) (alloy steels)	ML/MQ	0.364	161	HV	200	360
		ME	0.356	186		200	360
Contact	Eh	ML	0.000	1300	HV	600	800
		MQ	0.000	1500		660	800
		ME	0.000	1650		660	800
Bending	Eh	ML	0.000	312	HV	600	800
		MQa	0.000	425		660	800
		MQb	0.000	461		660	800
		MQc	0.000	500		660	800
		ME	0.000	525		660	800

(continued)

Table 2.3 (continued)

Stress	Type of material	Quality	A	B	Hardness	Min. hardness	Max. hardness
Contact	IF	ML	0.740	602	HV	485	615
		MQ	0.541	882		500	615
		ME	0.505	1013		500	615
Bending	IF	ML	0.305	76	HV	485	615
		MQ	0.138	290		500	570
		MQ	0.000	369		570	615
		ME	0.271	237		500	615
Contact	NT (nitr.)*	ML	0.000	1125	HV	650	900
		MQ	0.000	1250		650	900
		ME	0.000	1450		650	900
	NV (nitr.)**	ML	0.000	788	HV	450	650
		MQ	0.000	998		450	650
		ME	0.000	1217		450	650
Bending	NT (nitr.)*	ML	0.000	270	HV	650	900
		MQ	0.000	420		650	900
		ME	0.000	468		650	900
	NV (nitr.)**	ML	0.000	258	HV	450	650
		MQ	0.000	363		450	650
		ME	0.000	432		450	650
Contact	NV (nitrocar.)***	ML	0.000	650	HV	300	650
		MQ/ME	1.167	425		300	450
			0.000	950		450	650
Bending	NV (nitrocar.)***	ML	0.000	224	HV	300	650
		MQ/ME	0.653	94		300	450
			0.000	388		450	650

It is noteworthy that, for case hardened wrought steels Eh having quality MQ, the specification letters a, b, and c indicate respectively: core hardness $\geq$ di 25 HRC with Jominy hardenability at $J = 12$ mm < 28 HRC, core hardness $\geq$ di 25 HRC with Jominy hardenability at $J = 12$ mm ≥ 28 HRC, and core hardness $\geq$ di 30 HRC. It should also be noted that asterisks for NT (nitr.), NV (nitr.) and NV (nitrocar.) steels refer to different ISO standards concerning the classification of steels, not mentioned here for brevity (see ISO 6336-5:2016).

It should also be noted that the allowable stress numbers for bending that are obtained with the calculation parameters shown in Table 2.3 are those concerning repeated, unidirectional tooth loads, i.e. repeated stress cycles characterized by *stress ratio* $R = \sigma_{\min}/\sigma_{\max} = 0$, *amplitude ratio* $A = (\sigma_a/\sigma_m) = [(1 - R)/(1 + R)] = 1$, and therefore $\sigma_a = \sigma_m$, that is alternating stress equal to the mean stress. When reversals of full load occurs, that is in the cases of completely reversed cycles of stress ($R = -1$), a reduced value of σ_{FE} is required. In most severe cases in which the full load reversal occurs for each load cycle (for example, an idler gear wheel), we can take values of σ_{Flim} and σ_{FE} equal to 70% of those related to repeated, unidirectional tooth loads. If the full load reversals do not occur for each load cycle, but are less frequent, a different reduction factor of the values related to repeated unidirectional tooth loads can be chosen, depending on the number of reversals during the expected gear lifetime.

2.10.2 Life Factor, Z_{NT}

The life factor for contact stress, Z_{NT}, takes into account the higher contact stress, including static stress, which can be tolerated in comparison with the allowable stress number for contact, σ_{Hlim}, when the number of cycles provided as lifetime is lower than that corresponding to the knee of the S-N damage curve, where $Z_{NT} = 1$. This factor is mainly influenced by: chemical composition, metallurgical structure, heat treatment, and cleanness of gear material; *number of load cycles*, or *service life*, N_L (this is defined as the number of meshing contacts under load between the teeth of the gear being analyzed); lubrication regime (full film lubrication, mixed lubrication, and boundary lubrication); pitch line velocity; failure criteria; smoothness of operation required; material ductility, fracture toughness, and residual stresses.

We here circumscribe our attention to Method *B*, by which the permissible stress or the safety factor for the limited service life are determined using the life factor Z_{NT} for the standard reference test gears. For this purpose, the normalized damage curves for pitting resistance are used. Figure 2.24 shows these curves, which express the life factor, Z_{NT}, as a function of the service life, N_L, for the four groups of materials described in Table 2.4 (note that the first two groups coincide, with the difference that for group 1 a certain degree of pitting is permissible, while for group 2 no degree of pitting is tolerated).

The values of Z_{NT} for static and reference stresses may be obtained from Fig. 2.24, as well as from Table 2.4. With reference to the values of Z_{NT} shown in this table, included in the range from 0.85 to 1.00, it is to be noted that the lower values should be

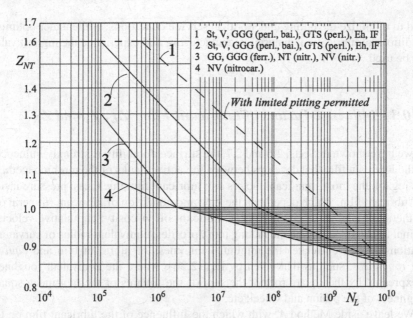

Fig. 2.24 Life factor, Z_{NT}, as a function of the service life, N_L, for standard reference test gears

Table 2.4 Life factor, Z_{NT}

Material group	Number of load cycles	Life factor, Z_{NT}
Group 1: St, V, GGG (perl., bai.), GTS (perl.), Eh, IF Only when a certain degree of pitting is permissible	$N_L \leq 6 \times 10^5$, static	1.6
	$N_L = 10^7$	1.3
	$N_L = 10^9$	1.0
	$N_L = 10^{10}$	0.85–1.00
Group 2: St, V, GGG (perl., bai.), GTS (perl.), Eh, IF No degree of pitting is tolerated	$N_L \leq 10^5$, static	1.6
	$N_L = 5 \times 10^7$	1.0
	$N_L = 10^9$	1.0
	$N_L = 10^{10}$	0.85–1.00
Group 3: GG, GGG (ferr.), NT (nitr.), NV (nitr.)	$N_L \leq 10^5$, static	1.3
	$N_L = 2 \times 10^6$	1.0
	$N_L = 10^{10}$	0.85–1.00
Group 4: NV (nitrocar.)	$N_L \leq 10^5$, static	1.1
	$N_L = 2 \times 10^6$	1.0
	$N_L = 10^{10}$	0.85–1.00

used to minimize the pitting under critical service conditions, while under optimum conditions of lubrication, material, manufacturing and experience the highest value can be used.

2.10.3 Influence Factors of Lubricant Film, Z_L, Z_v and Z_R

As we have shown in Sects. 2.6 and 2.7, the surface durability is strongly influenced by the lubricant film that develops between the active flanks of the meshing teeth. In this regard, the most significant factors are: lubricant film thickness; pressure inside the lubricant film, which depend on the type and condition of lubricant (mineral oil, synthetic oil, and its origin, age, etc.); lubricant oil viscosity; cumulative velocity; normal total load acting on the mating tooth profiles; equivalent radius of curvature; relationship that correlates the minimum thickness of lubricant film and *equivalent roughness*, such as that given by Eq. (2.36), where the equivalent roughness is expressed by the square root of the sum of the squares of the root-mean square roughness of the pinion and wheel; etc.

We leave aside Method A, with which the influence of the lubricant film on the surface durability is determined on the basis of reliable service experiences or tests on gear drives with comparable materials, dimensions, lubricants and operating conditions, and focus our attention on Method B, which is instead based on the results of tests on standard reference test gears. With this method, each of the three factors Z_L, Z_v and Z_R, which we already defined in Sect. 2.8, is made to depend on a unique influence variable. In so doing, the distribution curves and calculation equations described below are obtained.

However, these three factors are influenced not only by its specific influence variable, but also by other variables, which are not included in the calculation procedure. For this reason, the distribution curves that express the variability of each of these three factors as a function of their specific influence variable are empirical curves, i.e. curves that cannot be considered representative of physical laws. They show scatter bands related to the fact that the test results obtained by varying a single variable, with the other variables held constant, have been adjusted to take into account the field experiences with gears of different sizes and operating conditions.

It is to be noted that, generally, through-hardened gears are more sensitive than case-hardened gears to the influence of the viscosity, pitch line velocity and surface roughness. When a gear pair consists of one member of hard material and the mating member of soft material, these three factors are to be determined for the softer of the two materials. Furthermore, the influence of the lubricant film is only fully effective for long life stress levels, while it is low at the highest limited-life stress levels.

2.10.3.1 Lubricant Factor, Z_L, for Reference Stress

The lubricant factor Z_L has been obtained from tests with mineral oils, with or without anti-scuff additives (also known as EP additives). It can be determined as a

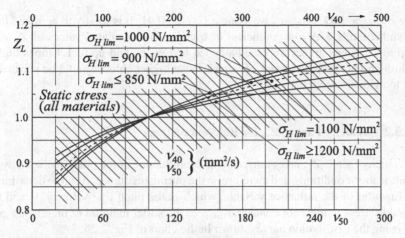

Fig. 2.25 Lubricant factor Z_L as a function of the nominal kinematic viscosity, v_{40} or v_{50}, for different values of the allowable stress number for contact, σ_{Hlim}

function of the influence variable, which is the *nominal kinematic viscosity* of the oil v_{40} at 40 °C, (or the nominal kinematic viscosity of the oil v_{50} at 50 °C), as well as of the allowable stress number for contact, σ_{Hlim}, of the softer of the two materials of the gear pair, using the distribution curves shown in the chart of Fig. 2.25. In an equivalent manner, it can be determined by calculation, using the following relationships, which are those used to plot the distribution curves of Fig. 2.25.

The influence variable of this factor is therefore the viscosity, which is expressed in terms of nominal kinematic viscosity (it is defined as the quotient of the dynamic, or absolute, viscosity divided by the density) of the oil at 40 or 50 °C. The nominal kinematic viscosity at 40 °C applies for viscosity index VI = 95 and kinematic viscosity up to 500 cSt (mm²/s) at 40 °C, while for higher kinematic viscosity we can use either the nominal kinematic viscosity obtained at 500 cSt at 40 °C or the nominal kinematic viscosity obtained at 300 cSt at 50 °C.

Lubricant factor Z_L can be calculated with the following relationships, which are consistent with the curves shown in Fig. 2.25:

$$Z_L = C_{ZL} + \frac{4(1 - C_{ZL})}{\left(1.2 + \dfrac{80}{v_{50}}\right)^2} = C_{ZL} + \frac{4(1 - C_{ZL})}{\left(1.2 + \dfrac{134}{v_{40}}\right)^2}, \qquad (2.58)$$

where C_{ZL}, in the range ($850\,\text{N/mm}^2 \le \sigma_{Hlim} \le 1200\,\text{N/mm}^2$) is given by:

$$C_{ZL} = \frac{\sigma_{Hlim}}{4375} + 0.6357. \qquad (2.59)$$

In Eq. (2.58) the nominal kinematic viscosity, v_{40} or v_{50}, is given in mm²/s (centistokes). In the range $\sigma_{Hlim} < 850\,\text{N/mm}^2$, we assume $C_{ZL} = 0.83$, while in the range $\sigma_{Hlim} > 1200\,\text{N/mm}^2$, we assume $C_{ZL} = 0.91$. It is finally to be noted that, for synthetic oils with low coefficient of friction, the calculated values of Z_L may be up to 1.1 times higher for case-hardened test gears, and up to 1.4 times higher for through-hardened test gears. However, these higher values should be verified in each individual case.

2.10.3.2 Velocity Factor, $\dot{Z}_v$, for Reference Stress

The velocity factor, Z_v, takes into account the influence of the pitch line velocity on the lubrication conditions, and therefore on the pitting resistance. It can be determined as a function of the influence variable, which is the pitch line velocity, v, and the allowable stress number for contact, σ_{Hlim}, of the softer material of the mating gear pair, using the distribution curves shown in the chart of Fig. 2.26.

In an equivalent manner, the velocity factor, Z_v, can be determined by calculation, using the following relationships, which are consistent with the curves shown in Fig. 2.26:

$$Z_v = C_{Zv} + \frac{2(1 - C_{Zv})}{\sqrt{0.8 + \dfrac{32}{v}}}, \tag{2.60}$$

where

$$C_{Zv} = C_{ZL} + 0.02, \tag{2.61}$$

and C_{ZL} is given by Eq. (2.59), with the limitations described in previous section.

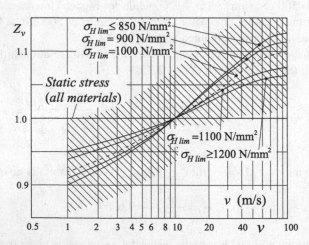

Fig. 2.26 Velocity factor, Z_v, as a function of the pitch line velocity, for different values of the allowable stress number for contact, σ_{Hlim}

2.10.3.3 Roughness Factor, Z_R, for Reference Stress

The roughness factor, Z_R, takes into account the influence of the surface roughness of the active flanks of teeth on the pitting resistance. It can be determined as a function of the influence variable, which is the *mean relative peak-to-valley roughness*, Rz_{10} (i.e., for gear pairs with relative curvature radius $\rho_{red} = 10\,\mathrm{mm}$), and the allowable stress number for contact, σ_{Hlim}, of the softer material of the mating gear pair, using the distribution curves shown in Fig. 2.27.

In an equivalent manner, the roughness factor, Z_R, can be determined by calculation, using the following relationships, which are consistent with the curves shown in Fig. 2.27:

$$Z_R = \left(\frac{3}{Rz_{10}}\right)^{C_{ZR}} \tag{2.62}$$

where C_{ZR}, in the range $(850\,\mathrm{N/mm^2} \leq \sigma_{Hlim} \leq 1200\,\mathrm{N/mm^2})$, is given by:

$$C_{ZR} = 0.32 - 2 \times 10^{-4}\sigma_{Hlim}. \tag{2.63}$$

In the range $\sigma_{Hlim} < 850\,\mathrm{N/mm^2}$, we assume $C_{ZR} = 0.15$, while in the range $\sigma_{Hlim} > 1200\,\mathrm{N/mm^2}$, we assume $C_{ZR} = 0.08$. The mean relative peak-to-valley roughness for the gear pair is given by:

$$Rz_{10} = Rz\left(\frac{10}{\rho_{red}}\right)^{1/3}, \tag{2.64}$$

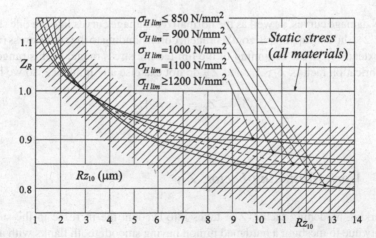

Fig. 2.27 Roughness factor, Z_R, as a function of the mean relative peak-to-valley roughness, Rz_{10}, for different values of the allowable stress number for contact, σ_{Hlim}

where Rz is the *mean peak-to-valley roughness* of the gear pair, and ρ_{red} is the *radius of relative curvature* at the pitch point, respectively given by:

$$Rz = \frac{1}{2}(Rz_1 + Rz_2) \tag{2.65}$$

and [see Eq. (2.9)]

$$\rho_{red} = \frac{r_1 r_2 \sin \alpha_{wt}}{(r_1 + r_2)} = \frac{d_{b1} d_{b2}}{2(d_{b1} + d_{b2})} \tan \alpha_{wt}, \tag{2.66}$$

with d_{b2} (or r_2) to be assumed positive or negative for external and internal gear pairs, respectively.

It is to be noted that Rz_1 for pinion and Rz_2 for gear wheel are mean values for the peak-to-valley roughness measured on several tooth flanks. When we use the average roughness Ra, given by Eq. (2.34), the following approximation may be used for conversion: $Ra = Rz/6$. Furthermore, Rz_1 and Rz_2 should be determined by considering the surface conditions of pinion and gear wheel after the manufacturing process, thus including the running-in effects, of course when we have reasonable assurance that it can take place.

Finally, it is to be noted that the curves shown in Fig. 2.27 refer to standard reference test gears having equivalent radius of curvature at the pitch point $\rho_{red} = 10\,\mathrm{mm}$. With reference to Eq. (2.64), the meaning of mean relative roughness is then referred to this radius of relative curvature at the pitch point, $\rho_{red} = 10\,\mathrm{mm}$.

2.10.3.4 Lubrication Factors Z_L, Z_v, and Z_R for Static Stress

The S-N damage curves, as well as the normalized damage curves like the one shown in Fig. 2.24, show an upper horizontal branch, corresponding to the static stress range, which extends up to the beginning of the limited life stress range. In this range, the three lubrication factors Z_L, Z_v, and Z_R do not exercise any influence, so we have

$$Z_L = Z_v = Z_R = 1. \tag{2.67}$$

2.10.4 Work Hardening Factor, Z_W

The working hardening factor, Z_W, takes into account the increase in the surface durability due to meshing a hardened pinion having smooth tooth flanks with a gear wheel made of structural steel or through-hardened steel. This increase in the surface durability of the soft gear wheel depends not only on any work hardening of the material with which it is made, but also on other influences, such as polishing effects due to lubricant, alloying elements and internal stresses in the soft material, surfaces roughness of the hardened pinion flank, contact stress, and hardening processes.

Here too we leave aside Method A, with which all influences above are determined on the basis of reliable service experiences or tests on gear systems with comparable materials, dimensions, lubricants and operating conditions. We focus our attention on Method B, which is based on the results of tests on standard reference test gears, as well as on field experience of manufactured gear units.

For the determination of the work hardening factor, Z_W, Method B distinguishes the following two cases: surface-hardened pinion with through-hardened gear wheel; through-hardened pinion and gear wheel. In both cases, the distribution curves or calculation relationships, which are proposed for the use, are to be considered as empirical curves or relationships, i.e. curves and relationships that cannot be considered representative of physical laws.

2.10.4.1 Surface-Hardened Pinion with Through-Hardened Gear Wheel

The value of the work hardening factor, Z_W, is different for reference and long-life stress, and for static stress. However, in either cases, it is made to depend on the tooth flank hardness, HB, of softer gear wheel.

For reference and long-life stress, Z_W can be determined as a function of the Brinell hardness of the tooth flanks of the softer gear wheel, HB, and the *equivalent roughness*, Rz_H, defined below, using the distribution curves shown in the chart of Fig. 2.28. In an equivalent manner, in the range ($130 \leq$ HB ≤ 470), Z_W can be

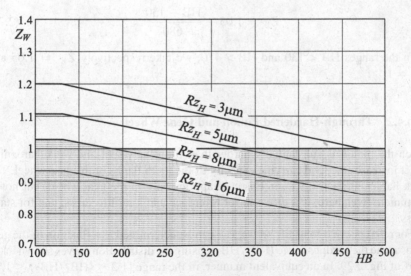

Fig. 2.28 Work hardening factor Z_W for reference and long-life stress, in the case of case-hardened pinion and through-hardened gear wheel

determined by calculation, using the following relationship:

$$Z_W = \left(1.2 - \frac{\text{HB} - 130}{1700}\right)\left(\frac{3}{Rz_H}\right)^{0.15}. \tag{2.68}$$

In the ranges HB < 130 and HB > 470, we use the same relationship with HB = 130 and, respectively, HB = 470. The calculated values of Z_W so obtained are consistent with the curves shown in Fig. 2.28.

The equivalent roughness, Rz_H, is given by:

$$Rz_H = \frac{Rz_1\left(\dfrac{10}{\rho_{red}}\right)^{0.33}\left(\dfrac{Rz_1}{Rz_2}\right)^{0.66}}{\left(\dfrac{v v_{40}}{1500}\right)^{0.33}}, \tag{2.69}$$

where the already known quantities Rz_1, Rz_2, ρ_{red}, v_{40} and v are respectively expressed in the following units: μm, μm, mm, mm^2/s, and m/s.

It is to be noted that, if $Rz_H > 16\,\mu$m or $Rz_H < 3\,\mu$m, we take respectively: $Rz_H = 16\,\mu$m and $Rz_H = 3\,\mu$m. For values of $Z_W < 1$ (this is, for example, the case of rough pinion surfaces), possible effects of wear can occur, that limit the surface durability. In these cases, it is necessary to carry out an additional analysis concerning wear, but the prudence demands that we choose $Z_W = 1$. Figure 2.28 shows these wear effects in the shaded area.

For static stress, in the range (130 $\leq$ HB $\leq$ 470), Z_W is calculated with the following relationship:

$$Z_W = 1.05 - \frac{\text{HB} - 130}{680}. \tag{2.70}$$

In the ranges HB < 130 and HB > 470, we take respectively $Z_W = 1.05$ and $Z_W = 1$.

2.10.4.2 Through-Hardened Pinion and Gear Wheel

When the pinion is substantially harder than the gear wheel, the work hardening effect increases the load carrying capacity of the tooth flanks of the gear wheel. The work hardening factor, Z_W, in this case applies only to the gear wheel, but not to the pinion. Even here, it is different for reference and long-life stress, and for static stress.

For reference and long-life stress, Z_W can be determined as a function of the gear ratio, u, and the hardness ratio (HB$_1$/HB$_2$), using the distribution curves shown in the chart of Fig. 2.29. In an equivalent manner, in the range [1.2 $\leq$ (HB$_1$/HB$_2$) $\leq$ 1.7], Z_W can be determined by calculation, using the following relationship:

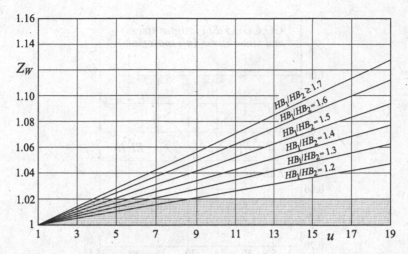

Fig. 2.29 Work hardening factor Z_W for reference and long-life stress, in the case of through-hardened pinion and gear wheel

$$Z_W = 1 + \left[898 \times 10^{-5}\left(\frac{HB_1}{HB_2}\right) - 829 \times 10^{-5}\right](u - 1), \qquad (2.71)$$

where HB_1 and HB_2 are the Brinell hardness numbers of the pinion and gear wheel, and u is the gear ratio (if $u > 20$, we take $u = 20$).

In the range $(HB_1/HB_2) < 1.2$, we take $Z_W = 1$. In the range $(HB_1/HB_2) > 1.7$, we take:

$$Z_W = 1 + 698 \times 10^{-5}(u - 1). \qquad (2.72)$$

The calculation values of Z_W are consistent with the curves shown in Fig. 2.29. For static stress, $Z_W = 1$.

2.10.5 Size Factor, Z_X

The size factor, Z_X, takes into account the statistical evidence that the stress level for which fatigue damage occurs decreases when the size of mechanical element under consideration increases. Notoriously, the statistical evidence according to which the endurance tends to decrease when size increases constitutes a factual situation in the fatigue strength field. This is probably due not to a true size effect, associated with stress gradient, but to metallurgical factors, since a uniform microstructure cannot be obtained throughout very large dimensions of heat-treated toothed members. Therefore, when the dimensions increase, the subsurface defects and weak points of the metallurgical structure also increase. Around these defects, smaller stress gradients

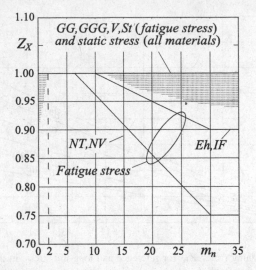

Fig. 2.30 Size factor Z_X for static and fatigue stresses, according to DIN 3990

occur, and these greatly affect the fatigue life (see Faupel 1964; Burr 1982; Budynas and Nisbett 2009).

The most significant influence parameters are: material quality, such as furnace charge, forging, and cleanliness; heat treatment, depth and distribution of hardening; radius of curvature of the transverse profile; module; depth of the hardened surface layer relative to the tooth size, and thus core supporting effect, in the case of surface hardening.

According to ISO 6336-2:2006, the size factor Z_X is taken to be 1. According to DIN 3990: 1987, Z_X is taken to be 1 for all materials and static stress, and for GG, GGG, V, St materials and fatigue stress, while for Eh and IF materials, and NT and NV materials, it is taken variable as a function of the normal module, m_n. Figure 2.30, which is taken from DIN 3990: 1987, allows us to determine the size factor value for different groups of materials, and for static and fatigue stresses, as a function of the normal module, m_n.

2.10.6   Minimum Safety Factor, $S_{H\min}$

In Sect. 2.2, we defined the safety factor, S_H, and we described the most significant influences that condition its choice. The considerations made here for pitting, however, are quite general, so that they may be extended to calculations of tooth bending strength and scuffing load capacity, that we will address in some of the following chapters.

The allowable stress numbers that we described in Sect. 2.10.1, to be used in calculations, are valid for a given probability of failure, namely for 1% probability of damage. Therefore, the safety factor to be chosen must be reported to this probability of damage. Obviously, the risk of damage reduces when the safety factor increases and vice versa.

For calculations of surface durability, the safety factor can be small. In effect, many of the influences included in the safety factor are already taken into account by the multiplying factors in Eqs. (2.38)–(2.40). Furthermore, the failure consequences are mitigated since pitting damage develops slowly, and gives warning by gradually increasing gear noise. For more, even with a surface fatigue damage of a certain extension, the gear pair can continue to operate for some period after its surface endurance life would be considered depleted. Consequently, safety factors in the range 1.1 to 1.6 are usually considered appropriate.

Table 2.5 shows the recommended minimum values of the safety factor S_{Hmin}, with some indications for the choice between the limit values, minimum and maximum. They vary depending on whether the static stress corresponding to the maximum torque (static torque; starting torque; maximum torque of the cumulative torque spectrum), the fatigue stress corresponding to the maximum torque or the nominal torque multiplied by the application factor, or the S-N damage curve or cumulative torque (or load) spectrum are considered as the design basis. They also vary according to the following three requirements:

– A, through-hardened gears and case-hardened gears calculated for fatigue stress with the maximum torque (e.g., shearing machines, mechanical presses, lifting equipment and hoisting apparatus, etc.).

Table 2.5 Safety factors

Damage limit	Static stress		Fatigue stress			S-N damage curve	
Load conditions	Maximum torque (static torque, starting torque, maximum torque of cumulative torque spectrum)		Maximum torque (work theoretical torque)	Nominal torque multiplied by K_A		Cumulative torque spectrum	
Design requirements	B	C	A	B	C	B	C
S_{Hmin} (pitting)	1.0	1.3	$(0.5 \div 0.7)$	$(1.0 \div 1.2)$	$(1.3 \div 1.6)$	1.0	$(1.2 \div 1.4)$
S_{Fmin} (bending)	1.4	1.8	$(0.7 \div 1.0)$	$(1.4 \div 1.5)$	$(1.6 \div 3.0)$	$(1.2 \div 1.4)$	$(1.4 \div 2.0)$
S_{Smin} (scuffing)	–	–	1.5	$(1.5 \div 1.8)$	$(2.0 \div 2.5)$	1.5	1.8

– B, common industrial applications, within which most of the gear units falls (the highest values refer to gearboxes that must meet the most demanding design requirements, as it is for industrial turbo-reducers).
– C, critical cases with high reliability requirements (very high duty cycles, high risk of damage with consequent high costs, lack of spare parts, no security against overloads, as it occurs in large turbo-reducers, turbine gear units, marine and aeronautical gear reducers).

2.10.7 Actual S-N Damage Curves, Obtained by Calculation

In the previous sections, from Sects. 2.10.1–2.10.6, we have described the procedure for determining the value of the permissible contact stress, σ_{HP}, to be used for the calculation of surface durability. This procedure is based on the introduction of a number of influence factors, which modify the allowable stress numbers for contact, σ_{Hlim}, obtained with standard reference test gears.

However, it is possible to use an equivalent and less dispersive procedure. This procedure consists of determining, by calculation, the S-N damage curve for pitting of the actual gear to be analyzed from the S-N damage curves for pitting obtained with standard reference test gears, or from the normalized damage curves for pitting of the standard reference test gears, such as that shown in Fig. 2.24.

When we apply this procedure, we must consider the following different conditions. In the range of long-life fatigue stress, the influence factors Z_L, Z_v, Z_R and Z_W are fully active, while the factor Z_X is active only for some materials (see Fig. 2.30) and $Z_{NT} = 1$. Instead, in the range of static stress, only the factor Z_W is fully active, but limited to the case of case-hardened pinion with through-hardened gear wheel (it is inactive for the case of through-hardened pinion and gear wheel), while factors Z_L, Z_v, Z_R and Z_W are completely inactive. Moreover, in the two above-mentioned stress ranges, also the safety factor, S_{Hmin}, are to be differentiated (see Table 2.5).

To formally simplify Eq. (2.40), we can introduce a single factor Z_N, called *life factor for contact stress*, and defined as:

$$Z_N = Z_{NT} Z_L\, Z_v\, Z_R\, Z_W\, Z_X, \qquad (2.73)$$

for which Eq. (2.40) becomes:

$$\sigma_{HP} = \frac{\sigma_{Hlim} Z_N}{S_{Hmin}} = \frac{\sigma_{HG}}{S_{Hmin}}. \qquad (2.74)$$

Figure 2.31 shows the manner in which the diagram of the life factor for contact stress, Z_N, as a function of the number of cycles, N_L, can be obtained for the curve 2 of Fig. 2.24, under the assumption that the portion of the curve corresponding to the long-life fatigue stress is horizontal. In the range of the long-life fatigue stress,

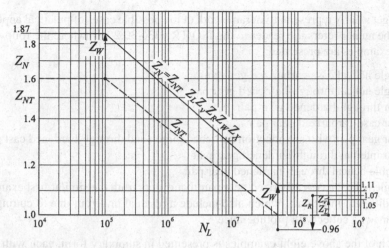

Fig. 2.31 Life factor for contact stress, Z_N

we have considered active factors Z_L, Z_v, Z_R and Z_W, with Z_R negatively working, compared to the standard reference test gear, and the other three factors operating positively. In the range of the static stress, we have considered active, and positively working, only the factor Z_W (in other words, we have considered the case of case-hardened pinion with through-hardened gear wheel).

It is evident that curves like that shown in Fig. 2.31 represent, unless a constant factor given by the allowance stress number for contact, σ_{Hlim}, the actual S-N damage curve of the gear to be examined.

2.11 Calculation Examples

For obvious reasons of space, it is not possible here to present and discuss examples regarding a complete and exhaustive calculation procedure of load carrying capacity in terms of surface durability (pitting) of spur and helical gears. This consideration is also valid for their load carrying capacity in terms of tooth root bending fatigue strength, which is the topic discussed in the next chapter.

This impossibility follows from the fact that each of the examples that could be presented and discussed would occupy a space comparable with that devoted to the theoretical aspects of the topic of interest (pitting or tooth root fatigue breakage), and such a discussion would clearly conflict with the editorial requirements of the present mono-thematic textbook. Each of these possible examples should in fact be aimed at calculating the safety factor for surface durability, SH, or the safety factor for tooth root fatigue breakage, SF, and, to define these two safety factors, it can not be overlooked any of the calculation steps that we have described in the previous sections.

To get a fairly representative framework of the possible cases of practical application, the author refers the reader to the ISO/TR 6336-30:2017, where the following eight examples are presented:

- Single helical case-carburized gear pair
- Single helical through-hardened gear pair
- Spur through-hardened gear pair
- Spur case-carburized gear pair
- Spur gear pair with an induction hardened pinion and through-hardened cast gear
- Spur internal through-hardened gear pair
- Double helical through-hardened gear pair
- Single helical case-carburized gear pair that differs from the similar first example for different input data, which also include a residual undercut due to cutting by means of a cutter with protuberance.

Each of the above eight examples is presented in summary form, each with two summary tables, the first containing the input data of the problem to be examined, and the second summarizing the output data and the related values for pinion and wheel. Each of these examples deal with the calculation of the load carrying capacity of the gear pair under consideration in parallel, that is from the point of view of both surface durability (pitting) and tooth root fatigue breakage.

Only for the first of the eight examples, the one concerning the single helical case-carburized gear pair, the ISO standard mentioned above provides, in Annex A, the complete discussion with detailed calculation, but it occupies twenty pages. Given the obvious impossibility of presenting and discussing examples of this kind here (only these examples discussed in depth would have didactic value, but not the synthetic presentation of input and output data), it only remains to refer the reader to the same ISO standard.

References

Bair S (2001) The variation of viscosity with temperature and pressure for various real lubricants. J Tribol 123(2):433–437

Barber JR (1992) Elasticity. Kluwer Academic Publishers, Dordrecht

Barus C (1893) Isothermals, isopiestics and isometrics relative to viscosity. Am J Sci 45:87–96

Beeching R, Nicholls W (1948) A theoretical discussion of pitting failures in gears. Proc Inst Mech Eng 158:317

Belajev NM (1917) Bulletin of institution engineers of ways and communications. St. Petersburg

Belajev NM (1924) Local stresses in compression of elastic bodies. In: Memoirs on theory of structures. St. Petersburg

Bodensieck EJ (1965) Specific film thickness—an index of gear tooth surface determination, In: Paper presented at 1965 aerospace gear systems committee technical division meeting. AGMA, Denver, CO, Sept 1965

Buckingham E (1949) Analytical mechanics of gears. McGraw-Hill Book Company, New York

Budynas RG, Nisbett JK (2009) Shigley's mechanical engineering design, 8th edn. McGraw-Hill Companies Inc., New York

Bufler H (1959) Zur Theorie der rollenden Reibung. Ing Arch 27(3):137–152

Burr AH (1982) Mechanical analysis and design. Elsevier Science Publishing Co., Inc., New York

Cameron A (1966) Principles of lubrication. Longmans, London

Ciulli E, Piccigallo B (1996) Fondamenti di lubrificazione elastoidrodinamica. ATA 49(3):109–116 (marzo)

Cornell RW (1981) Compliance and stress sensitivity of spur gear teeth. J Mech Des 103(2)

Dieter GE (1988) Mechanical metallurgy, SI metric edition. Adapted by David Bacon, McGraw-Hill Book Company, London

DIN 3990:1987 Tragfähigkeitsberechnung von Stirnrädern

Dolan TJ, Broghammer EL (1942) A photoelastic study of stresses in gear tooth fillets. University of Illinois Engineering Experimental Station Bulletin, p 335, Mar

Dowson D, Higginson GR (1977) Elastohydrodynamic lubrication, 2nd edn. Pergamon, London

Dyson A, Naylor H, Wilson AR (1965–1966) Measurement of oil-film thickness in elastohydrodynamic contacts. In: Proceedings of symposium on elastohydrodynamic lubrication, leeds, vol 180, Pt 3B. England Institution of Mechanical Engineers, pp 119–134

Ertel-Mohrenstein A (1984) Die Berechnung der hydrodynamischen Schmierung gekrümmter Oberflächen unter hoher Belastung und Relativbewegung. VDI - Fort - schrittsbericht Reihe 1(115)

Faupel JH (1964) Engineering design: a synthesis of stress analysis and materials engineering. Wiley, New York

Fontana MG (1987) Corrosion engineering, 3rd edn. Materials Science and Engineering Series. McGraw-Hill International Editions, New York

Fromm H (1927) Berechung des Schlupfes beim Rollen deformierbarer Scheiben. Zeitschrift für Angewandte Mathematik and Mechanik (ZAMM) 7:27–58

Frost NE, Marsh KJ, Pook LP (1999) Metal Fatigue. Dover Publications Inc., Minneola, New York

Gao S, Srirattayawong S (2014), Computational modeling of the surface roughness effects on the thermal-elastohydrodynamic lubrication problem. In: Proceedings of the international conference on heat transfer and fluid flow, Prague, Czech Republic, 11–12 Aug, Paper no. 192

Giovannozzi R (1965a) Costruzione di Macchine, vol I, Casa Editrice Prof. Riccardo Pàtron, Bologna

Giovannozzi R (1965b) Costruzione di Macchine, vol II. Casa Editrice Prof. Riccardo Pàtron, Bologna

Gladwell GML (1980) Contact problems in the classical theory elasticity. Alphen aan den Rijn, Sijthoff & Noordhoff International Publishers B.V, The Netherlands, Germantown, Maryland, USA

Greenwood JA (1972) An extension of the Grubin theory of elastohydrodynamic lubrication. J Phys D Appl Phys 5(12):2195–2211

Grubin AN, Vinogradova IE (1949) Investigation of the contact of machine component. Central Scientific Research Institute for Technology and Mechanical Engineering (TsNIIMASH), Moscow

Hamrock BJ, Dowson D (1977) Isothermal elastohydrodynamic lubrication at point contacts, part III—fully flooded results. Trans ASME Ser F J Lubric Technol 92(2):264–276

Herrebrugh K (1968) Solving the incompressible and isothermal problem in elastohydrodynamic lubrication through an integral equation. J Lubric Technol Trans ASME Ser F 90:262

Hertz HR (1882) Über die Berührung fester elastischer Körper. Journal für Reine und Angewandte Mathematik (Crelle's J) 92:156–171

Heywood RD (1952) Designing by photoelasticity. Chapman & Hall Ltd, London

ISO 1328-1:2013 Cylindrical gears—ISO system of flank tolerance classification - Part 1: Definitions and allowable values of deviations relevant to flanks of gear teeth

ISO 4287:1997 Geometrical product specifications (GPS)—surface texture: profile method—terms, definitions and surface texture parameters

ISO 53:1998 Cylindrical gears for general and heavy engineering—standard basic rack tooth profile

ISO 6336-2:2006 Calculation of load capacity of spur and helical gears—Part 2: calculation of surface durability (pitting)

ISO 6336-5:2016 Calculation of load capacity of spur and helical gears—Part 5: strength and quality of materials

ISO/TR 6336-30:2017 Calculation of load capacity of spur and helical gears—Part 30: calculation examples for the application of ISO 6336 parts 1, 2, 3, 5

Jacobson BO (1991) Rheology and elastohydrodynamic lubrication, Tribology series 19: Elsevier, Amsterdam

Johnson KL (1985) Contact mechanics. Cambridge University Press, Cambridge, United Kingdom

Johnson KL, Tevaarwerk JL (1977) Shear behavior of elastohydrodynamic oil film. Proc R Soc A 352:215

Juvinall RC (1967) Engineering considerations of stresses, strain, and strength. McGraw-Hill Book Company, New York

Juvinall RC (1983) Fundamentals of machine component design. Wiley, New York

Juvinall RC, Marshek KM (2012) Fundamentals of machine component design, 5th edn. Wiley, New York

Love AEH (1944) A treatise on the mathematical theory of elasticity, 4th edn. Dover Publications, New York

Lugt PM, Morales-Espejel GE (2011) A review of elasto-hydrodynamic lubrication theory. Tribol Trans 54(3):470–496

McEven E (1949) Stresses in elastic cylinders in contact along a generatrix. Phil Mag 40:454

Morales-Espejel GE (2014) Surface roughness effects in elastohydrodynamic lubrication: A review with contributions. Proc Inst Mechan Eng Part J: J Eng Tribol 228(11):1217–1242

Morales-Espejel GE, Wemekamp AW (2008) Ertel-Grubin methods in elastohydrodynamic lubrication—a review. Proc Inst Mechan Eng Part J: J Eng Tribol

Mott PH, Roland CM (2009) Limit to Poisson's ratio in isotropic materials. Chemistry Division, Code 6120, Naval Research Laboratory, Washington DC 20375-5342

Naunheimer H, Bertsche B, Ryborz J (2011) Automotive transmissions: fundamentals, selection, design and application, 2nd edn. Springer, Berlin, Heidelberg

Niemann G, Winter H (1983) Maschinen-Elemente Band II: Getriebe allgemein, Zahnradgetriebe-Grundlagen, Stirnradgetriebe. Springer, Berlin, Heidelberg

Niemann G, Winter H, Höhn B-R (2005) Maschinen-Elemente-Band 1: Konstruction und Berechnung von Verbindungen, Lagern, Wellen, 4th edn. Springer, Berlin, Heidelberg

Poritsky H (1950) Stresses and deflections of cylindrical bodies in contact. Trans ASME Ser E J Appl Mech 17:191

Radaj D, Vormwald M (2013) Advanced methods of fatigue assessment. Springer, Berlin, Heidelberg

Raptis KG, Costopoulos TN, Papadopoulos GA, Tsolakis AD (2010) Rating of spur gear strength using photoelasticity and the finite element method. Am J Eng Appl Sci 3(1):222–231

Reynolds O (1876) On rolling friction. Philos Trans R Soc Lond 166:155–171

Ristivojević M, Lazovic T, Vinci A (2019) Studying the load carrying capacity of spur gear tooth flanks. Mech Mach Theory 59:125–137

Roelands CJA (1966) Correlational aspects of the viscosity-temperature-pressure relationship of lubricating oils. Ph.D. thesis, Technische Hogeschool, V.R.B., Groningen

Rolfe ST, Barsom JM (1977) Fracture and fatigue control in structures: applications of fracture mechanics. Prentice-Hall Inc., Englewood Cliffs, New Jersey

Saada AS (1993) Elasticity: theory and applications, 2nd edn. Krieger Publishing Company, Malabar, Florida

Sackfield A, Hills DA (1983) Some useful results in the classical Hertz contact problem. J Strain Anal 18:101

Sadeghi F, McClung WD (1991) Formulas used in thermal elastohydrodynamic lubrication. STLE Tribol Trans 34(4):588–596

Schijve J (2009) Fatigue of structures and materials. Springer, Berlin, Heidelberg

Shipley EE (1967) Gear failures. Machine Design, 7 December

Smith JO, Liu CK (1953) Stresses due to tangential and normal loads on an elastic solid with application to some contact stress problems. Trans ASME Ser E J Appl Mech 20:157

Sperka P, Krupka I, Hartl M (2016) Surface roughness effects under high sliding EHL conditions. Jpn Soc Tribol 11(1):34–39

Stachowiak GW, Batchelor AW (2014) Engineering tribology, 4th edn. Elsevier, Amsterdam

Stephens RI, Fatemi A, Stephens RR, Fuchs HO (2001) Metal fatigue in engineering, 2nd edn. Wiley, New York

Tallian TE (2000) Failure atlas for Hertz contact machine elements, 2nd edn. ASME Press, New York

Tanaka S, Yamada T, Hattori N, Ogata K (1991) Influence of pitch errors on surface failure of spur gears. In: Proceedings of the international conference on motion and power transmissions, JSME, 23–26 Nov 23–26, pp 1084–1088

Thomas HR, Hoersch VA (1930) Stresses due to the pressure of one elastic solid upon another. University of Illinois, Engineering Experiment Station, Bulletin no. 212, July, pp 1–56

Timoshenko SP (1940a) Strength of materials, Part I: elementary theory and problems, 2nd edn. D. Van Nostrand Company Inc., Toronto

Timoshenko SP (1940b) Strength of materials, Part II: advanced theory and problems, 2nd edn. D. Van Nostrand Company Inc, Toronto

Timoshenko S, Goodier JN (1951) Theory of elasticity. McGraw-Hill Book Company Inc., New York

van Leeuwen H (2009) The determination of the pressure-viscosity coefficient of a lubricant through an accurate film thickness formula and accurate film thickness measurements. Proc Inst Mech Eng 223, Part J: J Eng Tribol 1143–1163

Vullo V (2014) Circular cylinders and pressure vessels: stress analysis and design. Springer, Cham, Heidelberg

Whitehouse D (2012) Surfaces and their measurement. Butterworth-Heinemann, Boston

Wymer DG, Cameron A (1974) Elastohydrodynamic lubrication of a line contact. Proc Inst Mech Eng 188:221–238

Chapter 3
Tooth Bending Strength of Spur and Helical Gears

Abstract In this chapter, a general survey is first done on the tooth bending strength of spur and helical gears, also focusing attention on fatigue tooth root breakage and safety factor to be used for their design. The theoretical bases of calculation of tooth bending strength are then discussed, with particular reference to the constant strength parabola introduced by Lewis and 30° tangent lines proposed and used subsequently. Stress state at the tooth root is then analyzed, considering the load application at the outer point of single pair gear tooth contact as well as the local stress concentrations, also due to combined local notch effects. Finally, the procedure for calculating the tooth bending strength of these types of gears in accordance with the ISO standards is described, highlighting when deemed necessary as the relationships used by the same ISO are founded on the theoretical bases previously discussed.

3.1 Introduction and Brief History

In Sect. 2.1 we described the four main types of damage that basically characterize the performance limit of most of the gears used in the usual practical applications, and we said that the related limits of each type of damage may be higher or lower, depending on the materials used and their heat treatment, as well as on the operating speeds. We highlighted this behavior in Fig. 2.1, which relates to a gear pair consisting of pinion and gear wheel both made of through-hardened steels.

If we considered a gear pair consisting of a pinion made of case-hardened steel, and a gear wheel made of through-hardened steel, we would get the qualitative performance limits in terms of T_{max} with somewhat substantial change compared to those shown in Fig. 2.1. In particular, we would see that:

- the four limit curves, i.e. tooth bending limit, pitting limit, wear limit, and scuffing limit, move upwards, so we have a significant general increase in the tooth load carrying capacity, *ceteris paribus*;
- the two curves of the tooth bending limit and pitting limit reverse their relative position, with the first limit curve positioned, all or in part, below the second;
- the scuffing limit curve and the wear limit curve are greatly moved upwards, for which the scuffing and wear damages would be feared.

© Springer Nature Switzerland AG 2020
V. Vullo, *Gears*, Springer Series in Solid and Structural Mechanics 11,
https://doi.org/10.1007/978-3-030-38632-0_3

However, it is to bear in mind that the state of current knowledge does not give the possibility to determine *a priori* which of the four aforementioned limits will be reached first (see Sect. 2.1). Therefore, in the calculation procedure of the gears, all these four limits must be thoroughly assessed and compared, in order to get a complete picture that allows to design and size the gear pair so that it is able to meet the design requirements.

In this chapter, we focus our attention on the calculation of the bending load capacity of spur and helical gears, that is, on their tooth bending capacity. In Sect. 2.1 we have also defined what we mean as tooth breakage, and we made a distinction between the overload failure and fatigue failure. These failures may affect the entire resistant cross section of the tooth, or a part of it. Therefore, also the failures that occur more frequently, which are those of the tooth edge, fall within this type of breakage, and can be overload failures or fatigue failures, all caused by loads irregularly distributed along the face width.

Although the importance of static failures is not to be underestimated, we mainly focus our attention on tooth bending fatigue failures, which are determined by several causes. These include: poor design of the gear drives; incorrect assembly or misalignment of the gears and shafts; overloads; inadvertent stress raisers or subsurface defects in critical areas; incorrect use of the materials; incorrect choice of heat treatments; etc. (see Shipley 1967; Fernandes 1996).

These types of failures are very dangerous, since the tooth breakage causes the total failure of the gear drive. Only in the case where the breaks are confined to a small part of one or more teeth, and a careful control with appropriate methods (for example, by PT—Penetrant Testing or similar methods) has excluded the remaining parts of the toothing are damaged, the gear unit can continue to operate, but at reduced load. Of course, before the emergency operation starts, the necessary repairs of the damaged parts must be carried out.

The bending load carrying capacity of a gear pair can be determined reliably only through a direct practical experimentation. In fact, the stress state at the tooth root, with the tooth considered as a cantilever plate, is very complex. Therefore, it can be determined with difficulty by the theoretical analysis, not only because it is of three-dimensional nature, but also because it is affected by other many influences of which it is not practically possible to take into account simultaneously. Among these, the following influences deserve a specific mention:

- shape of the teeth;
- shape of the root fillets, and related notch effects and stress concentrations in the regions of the same fillets or immediately surrounding them;
- local non-linearities of the stress state in the fillet regions, and residual stresses of various kinds, including those due to the technological processes of cutting and finishing and heat treatment;
- deviations of the tooth flank surface with respect to its theoretical geometry;
- surface finishing of the tooth flanks and root fillets;
- use or not of profile-shifted toothing;
- meshing conditions;

- manufacturing and assembly errors;
- non-uniform load distributions along the path of contact and face width;
- deflections of the shafts and bearings, and stiffnesses of all mechanical elements of the gear unit, including the housing;
- variability of the load and its peak values;
- operating conditions;
- characteristics of materials, including the effects of the heat treatment and processes might be used to improve the strength properties of the material, especially in correspondence of the root fillet (shot peening, cold rolling, and similar, etc.); etc.

It is known that gear wheels with teeth in a very rudimentary shape were built and used since the early days of human history (see Vol. 3: A Concise History). Their scientific-technical interest, however, comes only at the turn of the XVII and XVIII centuries, when the research pioneers who proposed the first scientific theories on gears began to study the possibility of making tooth flank surfaces having an appropriate kinematics, through the use of suitable conjugate profiles. The initial interest of researchers and scholars was therefore turned to geometric-kinematic aspects.

The problem of calculating the load carrying capacity of the gears was talked by researchers still later, in the late nineteenth century. In fact, the first universally recognized analysis of the stress state in the gear teeth was the one made by Lewis (1892). The *Lewis method* is in fact a turning point in the calculation of the gear load carrying capacity, and still serves as the basis of the analysis of the gear tooth bending. As we will see in more detail in Sect. 3.3, Lewis considered the gear tooth as a cantilever beam loaded at its tip by a concentrated force $(F_t/\cos\alpha)$, acting at pressure angle α, and distributed equally across the tooth face width, b. However, the real turning point proposed by Lewis, which is revolutionary compared with the past, was the introduction of the *uniform strength parabola*, the *Lewis parabola*, inscribed within the tooth outline, as Fig. 3.1 shows, and simulating the well-known *Galileo's constant strength cantilever beam* (see Galilei 1638; Buckingham 1949;

Fig. 3.1 Gear tooth as a cantilever beam, and Lewis parabola

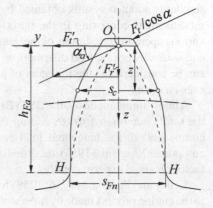

Timoshenko 1953, 1955; Giovannozzi 1965b; Juvinall 1967; Henriot 1979; Burr 1982; Juvinall and Marshek 2012; Ugural 2015).

From Fig. 3.1, it is evident the fact that the tooth is everywhere stronger than uniform strength parabola, with the exception of points in which this parabola is tangent to the tooth outline. In so doing, Lewis identified the critical section, i.e. the section where the maximum nominal bending stress occurs. *Mutatis mutandis*, this theoretical method of identifying the critical section, at least from a conceptual point of view, is still the basis of almost all currently used calculation methods of tooth bending strength.

Thirty years later, McMullen and Durkan (1922) proposed a significant change to the Lewis method, because they considered the load applied at the outer point of single pair tooth contact, instead at the tooth tip. The first studies to evaluate the dynamic effects on the bending resistance of the gear teeth followed by a few years the work made by these two researchers. In this regard, the contributions of Buckingham (1949) and his followers are above all to be mentioned, as well as those of countless other subsequent researchers who worked and still continue to work on this important topic.

Almen (1935) and Almen and Straub (1937) performed numerous fatigue tests on parallel cylindrical spur and helical gears, and straight bevel and spiral bevel gears. Applying for the first time the Wöhler curve (the S-N damage curve) to the gears, these researchers obtained a new relationship to be used for the calculation of the bending fatigue strength of the tooth. The contributions of the many subsequent researchers and scholars, who continued and continue to work on this subject, brought and continue to bring refinements and updates to the relationships that we use today to take account of the bending fatigue stresses. In this regard, we refer the reader to the two review papers of Chauhan (2016) and Aziz et al. (2017), as well as the work of Kramberger et al. (2004).

Dolan and Broghammer (1942), using experimental photoelastic techniques, focused their attention on the problem of stress concentrations in the tooth fillets and, for the first time, they determined the values of the stress concentration factors, taking into account also the radial component of the load applied at the outer point of single pair tooth contact. AGMA (American Gear Manufacturers Association) standards adopted results obtained by Dolan and Broghammer, albeit with some modifications consisting in the introduction of coefficients to take account of the various operating conditions of the gear, such as deflections of shafts, manufacturing and assembly errors, load distribution, rotational speed, etc. Even these two authors can be considered as the initiators of a line of research on the gears, still lively and ever-changing.

Almost simultaneously, the BSS (British Standards Society) standards introduced the *total compressive stress*, adding the compressive stress (due to the radial load component) to the tangential load component, this last also called bending load component. Merritt (1954) modified this method by introducing a concentration factor obtained experimentally.

Kelley and Pedersen (1956, 1958) found a significant dispersion of results related to the cutting process made by hobs with full tip radius (at the time these types of hobs were new), and therefore believed to carry out an in-depth experimental analysis, based on the Heywood photoelastic method (Heywood 1952). This analysis was aimed either to a more reliable identification of the actual weakest section (i.e. the critical section) than that theoretically identified with the uniform strength parabola, or the determination of an equally more reliable stress concentration factor than that proposed by Dolan and Broghammer. The results achieved by Kelley and Pedersen were praised for their high reliability, and were adopted by qualified US and European gear manufacturing industries.

Based on results of studies and experimental researches made by Niemann and Richter (1954, 1960) and Niemann (1965), whose works gave strong impulses to the development of calculation of gear load carrying capacity, the DIN standards introduced the method for determining the critical section of external gears by means of the chord between the points at which the 30° tangents contact the root fillets. Compared to the uniform strength parabola of the Lewis theoretical method, this mixed method, either theoretical or experimental, had the great advantage consisting in the fact that the critical section was no longer dependent on the point of load application (by Lewis applied to the tooth tip), but only on the tooth shape. Equivalently, the critical or weakest section of internal gears is determined by means of the chord between the points where the 60° tangents contact the root fillets. This method of calculating the weakest section was later acquired by the ISO standards; it is known as 30° tangent method or 60° tangent method.

Subsequently, on the basis of the results of researches carried out by Winter (1961, 1962) and Winter and Hösel (1969), the ISO standards were differentiated with respect to the DIN standards, both for the introduction of a new stress concentration factor, more reliable than that previously introduced, and for the introduction of several other coefficients, in order to parameterize all the factors influencing the gear lifetime. In this framework, even the contributions brought by Winter and Stolzle (1970), Henriot (1970), and Wellauer (1970) are to be considered and appreciated.

Unfortunately, in this brief history of the state of knowledge, it is not possible to mention all researchers and scholars who contributed to the analysis of the tooth bending strength, since it is in fact limited to the pioneers, i.e. to those who opened new lines of research on the topic. However, we cannot remain silent on the revolution brought also in this specific field, from the end of the 60 s of the 20th century, by numerical methods, most notable the FEM (Finite Element Method). This method, together with the BEM (Boundary Element Method), which was affirmed immediately afterwards, gradually became an indispensable tool to obtain results that determined new formulas to be used for a satisfactory conventional calculation method of the gears. For general textbooks on this subject, we refer the reader to Brebbia and Connor (1973), Zienkiewicz (1977), Brebbia (1978), Cook (1981), Fenner (1986) and Becker (1992), and, for special papers, to Wilcox and Coleman (1973), Garro and Vullo (1978a, b), Vijayakar and Houser (1988), and Li (2007, 2008).

3.2   Fatigue Tooth Root Breakage and Safety Factor

The load carrying capacity of spur and helical gears depends not only on surface durability, but also on the tooth bending strength. The amount of power or torque that a given gear pair can transmit, without tooth failures occur, also constitutes a measure of the ability of the gear teeth to resist static and fatigue bending stresses that arise at the teeth root. Since the teeth undergo static and fluctuating (or alternating), repeated and cyclic stresses during the meshing cycle, failures at root fillets can occur, which can lead to various types of failure (overload failures, fatigue failures, tooth edge or tooth tip failures, etc.), such as those mentioned in the previous section.

Figure 3.2 shows the fringe patterns of the stress state of four different bakelite models of gear tooth (they are designed with A, B, C and D), loaded in the photoelastic polariscope at given points of the transverse tooth profiles, with loads applied along the local normal to the same profiles, obtained by Dolan and Broghammer (1942). For details, we refer the reader to the original paper of these two authors. Here, we are only interested in highlighting the fact that, in any way the profile geometry may vary, the highest stresses are concentrated in the regions where the fringe patterns gather tightly together. This occurs in three regions: the point of contact with the mating gear wheel, simulated by the loading equipment where the load is applied, and in the entire area close to this point of contact; in the two areas corresponding to the two fillets, at the teeth root.

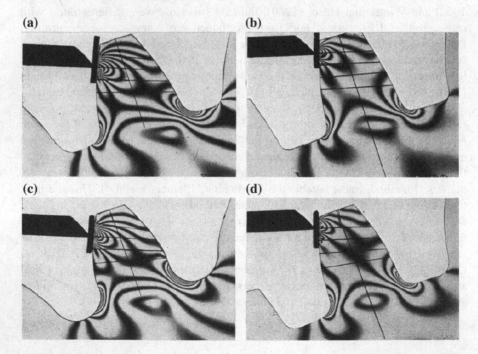

Fig. 3.2 Fringe patterns of four different bakelite models of gears tooth

We have already analyzed the problem of contact in the previous chapter. Here we focus our attention on the problem of the tooth bending strength. In this regard, a gear tooth under operating conditions can be considered as a cantilever beam under load, so the tooth ability to withstand tooth breakage at the root is also referred to as the *beam strength of gear tooth* (Buckingham 1949). With this rough approximation, the three-dimensionality due to the plate effect and to the small length of the beam compared to its transverse dimensions is neglected.

The maximum tensile stress at the tooth root, acting in the direction of the tooth depth, is the basis for assessing the bending strength of gear teeth. This maximum tensile stress occurs at the *tension fillets* of the operating tooth flanks. It must not exceed the permissible bending stress of the material. However, as we shall see in the next section, the highest value of the stress is the one that occurs at the *compression fillets* of non-operating tooth flanks. Therefore, if load-induced cracks are formed, the first of these cracks often appear at the compression fillets. When the teeth have a conventional shape, and the tooth load is unidirectional, these cracks seldom propagate up to failure. It is far more likely that the cracks that propagate up to failure are those started in the tension fillets, despite the fact that the intensity of the compressive stress at the compression fillet is larger than the tensile stress at the tension fillet. For this reason, the tooth root stress on which the tooth root strength rating calculation is based is the tensile stress at the tension fillet.

As we already mentioned in Sect. 2.10.1, the allowable stress numbers for bending of teeth subjected to a reversal of load during each revolution, such as idler gears, are less than the allowable stress numbers for bending of teeth subjected to unidirectional load. In these cases, the full range of stress is more than twice the tensile stress occurring at the root fillets of the operating flanks. In Sect. 2.10.1 we have already said how to calculate the permissible stresses.

Tooth fatigue breakage from bending cyclic load results from a crack originating at the tension fillet at the root cross section of the tooth. The entire tooth, or a part of it, more or less extended, is detached. The fatigue fracture surface is that well known, with evidence of the *fatigue fracture zone* (i.e. the zone of fatigue propagation, where often the break focal point is clearly identified), smooth and velvety to the appearance, where a pattern of beach marks are developed, and the *instantaneous fracture zone*, representing the area of final failure, with the characteristics of coarse granular fracture surface of a brittle crystalline material (see Frost et al. 1974; Fuchs and Stephens 1980; Dieter 1988).

The causes that can lead to fatigue fracture by bending load are numerous. Many of these breakages are related to excessive load on the tooth that causes, in the most stressed region, a stress state whose equivalent stress exceeds the one of endurance limit of the material. This stress state is exacerbated, and therefore it becomes even more dangerous, by the presence of stress risers, which contribute to raising the root stress levels up to values even more significantly higher than those normally predictable. These stress risers include notches in the root fillets, hob tears, inclusions, grinding burns, small heat treatment cracks, residual stresses, and similar.

To remedy this type of fatigue damage, which not only permanently endangers the operation of the entire gearbox, but can also have a negative impact on other groups of the more general mechanical system, of which the gearbox is one of the components, it is first everything necessary to limit the load in such a way that stresses generated by it remain clearly within the endurance limit of the material, or use, for a given load, a higher strength material. Moreover, to improve the fatigue strength, we can resort to finishing processes of the root fillets, essentially attributable to the cold-treatment, such as shot peening (see Straub 1953; Brooks 2016), cold rolling, etc., which generate favorable compression residual stresses (Niemann and Winter 1983). Two other effective remedies are also the following:

- a full fillet radius at tooth root, which allows a higher load carrying capacity compared with a fillet with two sharp radii;
- an adequate heat treatment of the material, suitable to minimize any harmful residual stress, especially those due to grinding operations, which are tensile residual stresses.

An overload breakage has, mainly, the characteristics of a *brittle fracture*, although often they are added to those of a *ductile fracture*. In regions of brittle fracture or *cleavage*, the fractured surface appears granular and oriented according to planes of maximum tensile stress, while in the regions of ductile fracture, the fractured surface appears fibrous and approximately aligned with planes of maximum shear stress. *Ex nomine*, it is due to overload, and originates at the tension fillet at tooth root. Overload can be due to several causes, such as: failure of the driven equipment; breakage, locking or seizure of a bearing; interposition of foreign matter between the meshing teeth; improper or incorrect use of the driven equipment; etc. These causes are unpredictable in most cases, for which it is extremely difficult to prevent these types of fractures with design choices and solutions. At best, it is possible to use overload-protection devices, such as torque-limiting couplings with suitable shear sections (Giovannozzi 1965a).

Other types of fractures are also classified as breakage fractures, even if they are not due to the bending load. The tooth tip fractures and tooth edge fractures, which have a random nature, and the rim and web failures, which generally originate between two adjacent teeth and propagate through the rim into the web, are certainly to be counted among the breakage fractures. We do not believe that this is the appropriate forum to discuss this subject, and refer the reader to Shipley (1967).

From what we have said above, it is evident that the breakage fractures (they can be fatigue fractures or overload fractures), due to the bending load, are extremely harmful. For equal operating conditions and for equal consequences of possible damages caused by these fractures, the safety factors to be used in relation to the tooth bending are higher than safety factors relating to pitting. Safety factors for bending strength represent security against tooth breakage. They must be calculated separately for pinion and gear wheel, and must be greater than or equal to a minimum value, S_{Fmin}, which is not given by ISO 6336-3:2006.

The values of safety factors for bending strength must be carefully chosen, with reference to the field of practical application, in order to meet the required reliability at a justifiable cost. Safety factor is defined as the quotient of the tooth root stress limit, σ_{FG}, divided by the calculated tooth root stress, σ_F. It is finally to be noted that safety factors for bending strength are sensitive to the same influences that we have described at the end of Sect. 2.2 for safety factors for surface durability (see Giovannozzi 1965a; Burr 1982; Niemann et al. 2005).

3.3 Lewis Analysis

We have already said that the first recognized analysis of the stress state of the gear tooth, due to the bending load, was that done by Lewis (1892). Here we consider it appropriate to make a brief summary of this analysis, as it is still the basis of the calculation of tooth bending strength of some type of gears, even if with some variation on the theme not very significant from the theoretical point of view. Even the subsequent introduction of the 30° (or 60°) tangent method, now used for most of the gears, has conceptual bases that can be traced back, at least partially, to the Lewis parabola.

Lewis considered the single gear tooth as a cantilever beam, built-into the rim at its root, and assumed as supporting the full load applied to the tooth tip. In other words, he assumed that the single tooth of the driving pinion that supports the whole load was in the position shown in Fig. 3.3, i.e. with one of its two tips (the one on the right in the figure) coinciding with the upper end point of path of contact, E, i.e. the point where the contact with the tooth of the mating gear wheel ends. In this position, the total force acting on the tooth has its maximum bending moment arm.

In dealing with this problem, in addition to the approximations highlighted in the previous section, Lewis made the following further simplifying assumptions:

(A)   The transverse load distribution between the various pairs of teeth in simultane-
      ous meshing is not considered. This is a fairly coarse approximation, since most
      of the gears are characterized by a transverse contact ratio, ε_α, greater than 1,
      usually included in the range $(1 < \varepsilon_\alpha < 2)$. In some cases, the virtual contact
      ratio of parallel cylindrical helical gear pairs, $\varepsilon_{\alpha n}$, i.e. the transverse contact
      ratio of the equivalent virtual spur gear is greater than 2. However, the most

Fig. 3.3 The most severe load condition of a single tooth, which bears the whole load

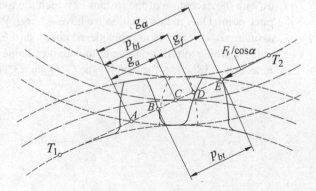

severe load condition introduced by Lewis can still be considered appropriate for gears of coarse accuracy grade. Instead, for good or high-precision gears, as those obtained with the most current gear cutting processes (these cutting processes were not available at the time of Lewis), the full load is not never applied to the tip of a single tooth. As we showed in Sect. 1.7 (see also Fig. 1.22), the total load is shared between the various tooth pairs and, for $(1 < \varepsilon_\alpha < 2)$, the most stringent condition for which the full load is supported by a single tooth is the one corresponding to point D, i.e. the outer point of single pair gear tooth contact. At this point, however, we have a shorter bending moment arm.

(B)   The full load acting on the tooth tip, equal to $F_t / \cos\alpha$, and inclined of the load application angle at tooth tip, α_a, is considered to be uniformly distributed along the whole face width. This is a non-conservative assumption, because, as we saw in Chap. 1, the load distribution along the face width is far from uniform, especially in the case of large face width and/or misaligned or flexible shafts. Figure 3.1 photographs the tooth underload in the position considered by Lewis, shown in Fig. 3.3. The angle α_a is obviously different from the pressure angle, as it is the angle between the full load vector applied to tooth tip, positioned at upper end point of path of contact, and normal to tooth axis through their point of intersection.

(C)   The radial component of the full load, F_r, acting at the tooth tip is neglected. In addition, Lewis makes a further assumption about the angle, α_a. The full load vector, $F_t / \cos\alpha$, when it is freely moved along its line of application and applied at point O (Fig. 3.1), can be resolved into the two tangential and radial components, given respectively by:

$$F'_t = F_t \frac{\cos\alpha_a}{\cos\alpha} \tag{3.1}$$

$$F'_r = F'_t \tan\alpha_a = F_t \frac{\sin\alpha_a}{\cos\alpha}. \tag{3.2}$$

During the meshing cycle with only one tooth under load, since the position of the tooth changes continuously, while the application straight line of the load remains theoretically unchanged for involute profiles, the angle α_a defined above is constantly changing, and is anywhere other than the pressure angle, α, with the exception of the instant in which the tooth axis passes through the pitch point. Only in this position we have $\alpha_a = \alpha$. Well, for his analysis, Lewis assumes $\alpha_a = \alpha$ also in the considered position (the one for which the tooth tip without profile modification up to the outside diameter is coinciding with point E, as Fig. 3.3 shows), for which Eqs. (3.1) and (3.2) become respectively:

$$F'_t = F_t \tag{3.3}$$

$$F'_r = F_r = F_t \tan\alpha. \tag{3.4}$$

It is to be noted that the first assumption, the one under which the influence of the radial component of the load is neglected, is a conservative assumption. That's why the compressive stress produced by this radial component is subtracted from the tensile stress due to bending at the tensile fillet. Instead, the fact that the same compressive stress should be added to the compressive stress due to bending at the compression fillet is unimportant, since fatigue failures always generate on the tensile fillet. The second assumption is also a conservative assumption (at point E, $\alpha_a > \alpha$), but its influence is negligible, especially when compared with that of the other assumptions.

(D) Friction forces due to sliding are neglected, and stress concentrations at the tooth fillets are not considered, but these last were not known at the time of Lewis.

(E) Shear stress due to tangential component of the load is also neglected.

Any cross section of the tooth, at a distance z from point O measured along the tooth axis, and having dimensions bs_c, where b is the face width and s_c is the local tooth chordal thickness (see Fig. 3.1), is subjected to a shear force, F_t', given by Eq. (3.1), a compression force, F_r', given by Eq. (3.2), and a bending moment, $F_t'z$. Therefore, this section is stressed by a maximum normal stress, given by the following relationship:

$$\sigma = \frac{F_t'}{bs_c} \left(\pm \frac{6z}{s_c} - \tan\alpha_a \right), \tag{3.5}$$

where the plus and minus signs of the first term within the round bracket are respectively valid in the regions subjected to tension and compression, while $(1/6)bs_c^2$ is the *section modulus*.

Therefore, the problem arises of finding the resistant cross section, and thus the value of z and that of corresponding local tooth chordal thickness, s_c, for which the normal stress given by Eq. (3.5) reaches its maximum value. Considering the assumptions for which the influence of the radial component of load is neglected, and $\alpha_a = \alpha$, the question comes down to finding the maximum value of the function

$$\sigma = 6\frac{F_t z}{bs_c^2} = \frac{M}{Z}, \tag{3.6}$$

where $M = F_t z$ is the bending moment, and $Z = (1/6)bs_c^2$ is the section modulus.

If the tooth had a constant cross-section (this extremely rough approximation was made before Lewis; in this regard, see Vol. 3, Sect. 5.3), the maximum value of the normal stress σ would certainly correspond to the maximum value of z-coordinate, that is to the built-in section of the tooth at the rim. This in fact happens quite often, although the local tooth chordal thickness is always variable.

With the intuition of Lewis, who introduced the *constant strength parabola*, the approach to the problem is radically changed. With reference to Fig. 3.1, it is evident that the maximum of the function given by Eq. (3.6) will be reached in the section in which a parabola of the type $y = k\sqrt{z}$, with $k = const$, having as its axis the tooth axis, and origin at point O, where the application straight line of the total load vector intersects this axis, is tangent to the tooth profile. In fact, a cantilever beam with a rectangular cross section of constant width, loaded by a concentrated load applied to the free end, and having characteristics of *full stressed beam* or *beam of constant strength* in relation to bending load, must have a parabolic profile of the type $s_c = k\sqrt{z}$, for which the section whose contour-line is tangent to the parabola is the most stressed (see Galilei 1638; Timoshenko 1953, 1955, 1956; Gere and Timoshenko 1997).

Figure 3.1 highlights the fact that the gear tooth is everywhere stronger than the inscribed constant strength parabola, except for the section $H - H$ where the parabola and tooth flank opposite profiles are in tangency. However, it is to be noted that the weakest cross section does not always coincide with the built-in section of the cantilever beam; it also depends on the profile shape of the tooth. Figures 3.1 and 3.4 show two different kinds of involute profiles, as well as the constant strength parabolas tangent to them. From these figures, it is evident that the most stressed cross section $H - H$ may coincide with the built-in section or be in the vicinity of it, or far from it.

According to Eq. (3.6), it is evident that, to have constant bending strength, the critical factor is constituted by the section modulus, $Z = (1/6)bs_c^2$, which is variable along the tooth axis. Since the face width, b, is a constant, the local value of the local tooth chordal thickness, $s_c = s_c(z)$, needs to vary along the tooth axis according to the following equation:

$$s_c = \left(\frac{1}{b}\right)^{1/2}\left(\frac{6M}{\sigma}\right)^{1/2} = \left(\frac{1}{b}\right)^{1/2}\left(\frac{6F_t z}{\sigma}\right)^{1/2}. \tag{3.7}$$

For the particular case shown in Fig. 3.1, for which $z = h_{Fa}$, and $s_c = s_{Fn}$, we have:

$$\sigma = \frac{M}{Z} = \frac{6F_t z}{bs_c^2} = \frac{6F_t h_{Fa}}{bs_{Fn}^2} = const; \tag{3.8}$$

Fig. 3.4 Weakest cross section away from the built-in cross section

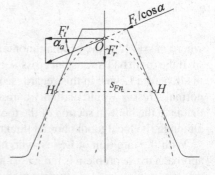

from this relationship, we get:

$$s_c = s_{Fn}\sqrt{\frac{z}{h_{Fa}}},$$ (3.9)

which is a parabolic function.

The bending deflection of point O is given by (Castigliano 1935, 1966, 1984):

$$f = \frac{\partial U}{\partial F_t} = \frac{\partial}{\partial F_t} \int_0^{h_{Fa}} \frac{M^2 dz}{2EI} = \frac{\partial}{\partial F_t} \left[\int_0^{h_{Fa}} \frac{F_t^2 z^2 dz}{2E(b/12)\left(s_{Fn}\sqrt{z/h_{Fa}}\right)^3} \right],$$ (3.10)

where U is the elastic strain energy stored within the material, due to the bending load, E the Young's modulus, and I the moment of inertia of the cross section. Thus, integrating, we get:

$$f = \frac{8F_t h_{Fa}^3}{Ebs_{Fn}^3}.$$ (3.11)

It is to be noted that the deflection at the loaded end of the idealized beam having parabolic shape is twice that of a prismatic beam having constant cross section, bs_{Fn}. Thus, the beam of constant strength having parabolic shape is considerably more flexible than the prismatic beam of constant cross section. In addition, it is to be noted that, at the loaded end ($z = 0$) of the idealized beam of parabolic shape, there is no bending moment ($M = 0$), for which the theoretical local tooth chordal thickness and cross section area are zero. This result follows from the fact that, to derive the constant strength shape, the shear stress has been neglected. Obviously, a beam of this shape would not usable, since it would be unable to withstand the shear forces near its loaded end. Fortunately, in this region, the gear tooth is developed at the outside of the idealized beam of parabolic shape, so this problem does not exist.

According to Lewis, we consider the weakest cross section, i.e. the one where the constant strength parabola is in tangency with the tooth outline. In Eq. (3.8) the term ($6h_{Fa}/s_{Fn}^2$) appears, which is a linear function of the tooth size (it can be considered as proportional to s_{Fn}), as well as of the tooth shape (it depends on the ratio s_{Fn}/h_{Fa}). Using the same geometric method proposed by Lewis, but introducing the module, m, as a measure of the tooth sizing (this is the only variation with respect to Lewis, which used the diametrical pitch, i.e. the number of teeth per inch of pitch diameter), by means of one of Euclidean theorems concerning the mean proportional in right triangles, applied to the $\widehat{OHR}$-triangle shown in Fig. 3.5, which also shows its plotting, we get:

$$\frac{x}{s_{Fn}/2} = \frac{s_{Fn}/2}{h_{Fa}}$$ (3.12)

from which

Fig. 3.5 Determination of
the Lewis form factor

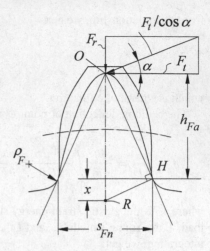

$$\frac{s_{Fn}^2}{h_{Fa}} = 4x,\tag{3.13}$$

where x is the distance, measured along the tooth axis, between the weakest cross section and point R, where the normal to the straight line OH, passing through point H, intersects the same tooth axis.

Substituting Eq. (3.13) in Eq. (3.8), and multiplying and dividing by the module, m, the result so obtained, we get:

$$\sigma = \frac{6F_t}{4bx} = \frac{3F_t m}{2bxm} = \frac{F_t}{bm\left(\dfrac{2x}{3m}\right)}.\tag{3.14}$$

From this relationship, introducing the notation

$$y = \frac{2x}{3m},\tag{3.15}$$

which is the *Lewis form factor*, we obtain:

$$\sigma = \frac{F_t}{bmy},\tag{3.16}$$

which is the basic *Lewis equation* in terms of module.

It is to be noted that, multiplying and dividing by the circular pitch $p = \pi m$, from Eq. (3.14) we would get the relationships

$$Y = \frac{2x}{3p}\qquad \sigma = \frac{F_t}{bpY}.\tag{3.17}$$

These two equations are respectively the relationships equivalent to Eqs. (3.15) and (3.16), expressed in terms of circular pitch. Of course, we have $y = \pi Y$. Similarly, we can proceed, according to Lewis, in terms of diametral pitch.

The Lewis form factor depends on the pressure angle, α, number of teeth, z, and tooth sizing. Figures (3.6) and (3.7), which relate respectively to external and internal parallel cylindrical spur gears, having standard sizing $\left(h_a = m; h_f = 1.25m\right)$, show the distribution curves of the Lewis form factor, y, as a function of the number of

Fig. 3.6 Lewis form factor for parallel cylindrical external spur gears with standard sizing

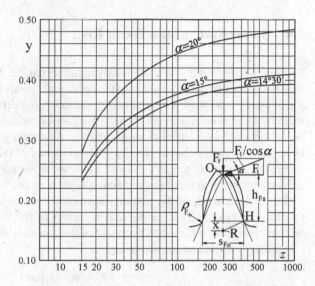

Fig. 3.7 Lewis form factor for parallel cylindrical internal spur gears with standard sizing

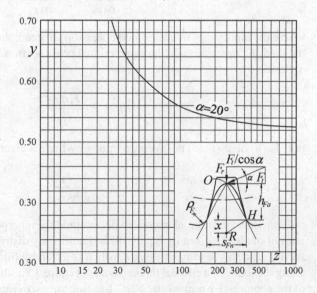

teeth, z; for the external parallel cylindrical spur gears, three different values of the pressure angle, α, are considered.

It is finally to be noted that the Lewis equation indicates that the tooth maximum bending stress varies directly with the load tangential component, and inversely with the face width, module (and thus tooth sizing), and tooth shape factor. Of course, both y and σ will vary as a function of the profile shift coefficients, because the profile shift profoundly modifies the tooth shape.

3.4 Stress State at the Tooth Root

The loads acting on a gear tooth are dynamic in nature. They are time-varying and repeated with a frequency related to the rotational speeds of the toothed members of a gear pair. The accurate assessment of the actual three-dimensional stress state at the tooth root is a problem still not solved in a rigorous manner. The many factors, which are introduced in the calculations of tooth bending strength that will be discussed in later sections of this chapter, are the clear evidence of this factual situation. In this section, according to the analysis of Lewis, we circumscribe the problem to some aspects, reserving to cover the remaining aspects in the following sections.

In the framework of the assumptions made by Lewis, and represented by the two Eqs. (3.3) and (3.4), the radial component of the load, F_r, determines, in the critical cross section identified by means of the constant strength parabola, a uniform compressive normal stress, σ_c, given by:

$$\sigma_c = -\frac{F_r}{bs_{Fn}} = -\frac{F_t \tan\alpha}{bs_{Fn}}, \tag{3.18}$$

where the minus sign indicates a compression. Instead, the tangential component of the load, F_t, determines in the same critical cross section, not only a bending stress, σ_b, given by:

$$\sigma_b = \pm\frac{6F_t h_{Fa}}{bs_{Fn}^2}, \tag{3.19}$$

but also a shear stress, the average value of which is given by:

$$\tau_m = \frac{3}{2}\frac{F_t}{bs_{Fn}}. \tag{3.20}$$

The combined normal stress can be obtained by superposition of the compressive normal stress, σ_c, given by Eq. (3.18) and uniformly distributed over the entire cross section, and the bending stress, σ_b, which is given by Eq. (3.19) and varies linearly along the tooth root chordal thickness, s_{Fn}. Figure 3.8a shows the distribution curves of the compressive normal stress, σ_c, bending stress, σ_b, and average shear stress, τ_m,

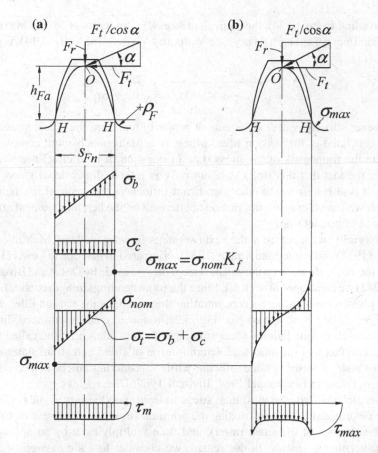

Fig. 3.8 Normal and shear stresses due to tangential and radial components of the load: **a** linear distribution curves; **b** qualitative non-linear distribution curves

in the critical cross section, as well as the final distribution curve of the total normal stress, which is equal to the algebraic sum of the normal stresses due to bending and compression, i.e. $\sigma_t = \sigma_c + \sigma_b$. The latter distribution curve highlights the fact that, due to the simultaneous action of the bending and radial loads, the neutral axis (that is, the axis in the cross section along which there are no longitudinal stresses or strains, so there is neither tension nor compression) no longer passes through the centroid of the same cross section. It is obvious that what we have described above is valid in the field of linear elasticity, as we have applied the method of superposition (Timoshenko and Goodier 1951).

In the critical cross section, in correspondence of the tension fillet, where usually the fatigue failure starts, the stress state is characterized by a total normal stress, whose absolute value is equal to $\sigma_t = \sigma_b - \sigma_c$, and by an average shear stress,

τ_m. According to Eq. (2.24), the equivalent stress, σ_e, in terms of MDE (Maximum Distortion Energy) strength theory (see Vullo and Vivio 2013; Vullo 2014), is given by:

$$\sigma_e = \sqrt{(\sigma_b - \sigma_c)^2 + 3\tau_m^2} = \sqrt{\sigma_t^2 + 3\tau_m^2}. \tag{3.21}$$

However, this equivalent stress cannot properly represent the actual stress state that is generated at the tension fillet, which is actually much more complex. To complete the framework of the stress state in question, it is in effect necessary to consider the fact that this stress state not only is a three-dimensional stress state, but that it is also sensitive to other significant influences, between which the notch effect, as well as other quantities that do not depend on the bending moment arm are playing an important role.

Temporarily, we leave aside the next two steps forward made by McMullen and Durkan (1922) and Almen and Straub (1937), compared to that due to Lewis (1892), and we focus our attention on the still next step, the one made by Dolan and Broghammer (1942) we mentioned in Sect. 3.1. Since the tooth bending strength is essentially a fatigue phenomenon, the stress concentration that occurs at the tension fillet, where the fatigue failure is generated (see Fig. 3.2), assumes great importance. In fact, notoriously, all fatigue failures occur at points of stress concentration, called *stress raisers*. The fact that the analytical determination of stress and strain states in the region of a stress raiser is often difficult, while considering an elastic behavior, is equally well known (see Faupel 1964; Juvinall 1967; Dieter 1988).

Rebus sic stantibus, the maximum stress in correspondence of a stress raiser is usually determined by first calculating the nominal stress (i.e. the stress that would be in the absence of the stress raiser), and then multiplying it by an appropriate *stress concentration factor*. In this regard, we consider first the *theoretical stress concentration factor*, K_t (also called *geometric stress concentration factor*), which refers to a hypothetical ideal material, perfectly homogeneous, isotropic and elastic. Then, to take account of the fact that a real material, subjected to fatigue load, fortunately behaves better than an ideal material (for a real material the fatigue weakening effect is reduced), especially when small notch radii in coarse-grained materials are involved, we introduce a *fatigue stress concentration factor*, K_f (also called *fatigue strength reduction factor* or *fatigue notch factor*).

The theoretical stress concentration factors have been determined with analytical, numerical and experimental methods, for various stress raiser geometries and for different types of load, from which they depend, and their values are available in the literature (e.g., Peterson 1974). In the case examined here, because we are facing a combined bending and axial load acting on the tooth, when we use these data in the literature, we must keep in mind that each nominal stress is to be multiplied by the corresponding theoretical stress concentration factor. Accordingly, σ_b must be multiplied by the bending factor, while σ_c must be multiplied by the axial factor. It should be noted that, consistent with the literature on theoretical stress concentration factors, the term axial load, with the adjective axial referred to the tooth axis, is used here as a synonym of radial load.

The tooth bending strength is further complicated in that the heat treatment, in correspondence of the root fillets, has a very significant role. In this regard, it is known that, for mechanical components that have been subjected to case-hardening (or surface hardening) processes, it is important to know not only the maximum stress, which occurs on the surface, but also the stress gradient, which tells us how the stress varies at different levels under the surface. In fact, if the material strength decays below the surface faster than stress does, the location of the critical point where the fatigue failure originates will be below the surface.

Notoriously, the fatigue stress concentration factor is defined as the quotient of the endurance limit at a given large number of cycles of an un-notched specimen divided by the endurance limit at the same number of cycles of a notched specimen. We have already recalled the fact that K_f is often less than K_t. This depends on the relationship between the stress gradient and coarseness of the material's microstructure.

The best method to determine reliable values of K_f is the experimental one, based on the above definition, i.e. on the direct comparison of the endurance limits of un-notched and notched specimens. We do not believe that it is here necessary to recall neither the Neuber's analysis of notch stresses, which led to the equation that defines the *Neuber factor* (Neuber 1946), nor the Kuhn's application of the *Neuber equation* (Kuhn 1964), which led to the modified form of the same Neuber equation. On this subject, we refer the reader to specialized textbooks (e.g., Juvinall 1967). Instead, we believe appropriate to quote here the following *Peterson equation* (Peterson 1953, 1962), based on a refinement of the Neuber approach:

$$K_f = 1 + (K_t - 1)q \tag{3.22}$$

where q is the *notch sensitivity*, which is a function of the *notch radius*, r, and *Neuber constant* or *Neuber equivalent grain half-length*, a, according to the relationship:

$$q = \frac{1}{1 + \sqrt{a/r}}. \tag{3.23}$$

Peterson developed a set of curves that allow us to calculate the notch sensitivity values for bending and axial loads that interest us here (and for torsional loads), as a function of the notch radius and the ultimate strength of the material. Even Heywood (1952, 1962), elaborating very numerous experimental data, proposed the following relationship:

$$K_f = \frac{K_t}{1 + 2[(K_t - 1)/K_t](a'/r)^{1/2}}, \tag{3.24}$$

which is similar to the combination of the two previous equations, with the variation of the empirical material constant, a', in place of the Neuber constant, a.

From the physical point of view, Eq. (3.22) reflects the influence of internal irregularities of material in reducing the severity of a notch. In this regard, also the surface (or external) irregularities attributable to the surface finishing of the tension fillet where a fatigue failure would presumably originate, reduce the additional damage that may be caused by a notch in a more or less similar way as the subsurface (or internal) irregularities do. To take account of this situation, Lipson and Juvinall (1963) proposed to modify Eq. (3.22) as follows:

$$K_f = 1 + (K_t - 1)qC_s \tag{3.25}$$

where C_s is a *surface factor*, which depends only on the surface roughness. Therefore, the endurance limit at the root fillets is estimated by multiplying the basic endurance limit by C_s, and dividing by the value of K_f obtained from Eq. (3.25).

Often, on the root fillet surfaces (remember that the root fillets are notches with large radii of curvature), we have a notch with much smaller radius of curvature, due to the fact that, for geometrical reasons, the grinding operation cannot be extended to the whole fillet profile. Paul and Faucett (1962) showed that, under the assumption that the grinding notch (i.e. the small notch) falls within the region subjected to the peak stress of the root fillet (i.e. the larger notch), the theoretical stress concentration factor at the base of the grinding notch is equal to the product of the theoretical stress concentration factors of the two notches separately considered (see Collins 1993).

The complexity of the phenomena described above, which overlap and interfere with each other, leads to the conclusion that the actual stress states at the tension and compression fillets become nonlinear, with significant deviations from the linear distribution curves shown in Fig. 3.8a. The Fig. 3.8b shows qualitatively the actual distribution curves of the total normal stress and average shear stress in the critical section of the tooth. Figure 3.8a also highlights, by means of two bold spots, the maximum values of the total normal stress at tension and compression fillets (of course, these values are equal to those shown in Fig. 3.8b), expressed as the product of the corresponding nominal stresses multiplied by the fatigue stress concentration factor, K_f.

Obviously, a more detailed analysis of the tooth stress state can be made using numerical methods, such as FEM (Finite Element Method), or BEM (Boundary Element Method). Figure 3.9 (see Garro and Vullo 1979) shows the stress distribution curves at the root critical section of a tooth that is bearing the entire load (therefore, the active profile of the tooth, in the particular instant of the meshing cycle considered, is in the region of single pair gear tooth contact). These curves were obtained using a two-dimensional FEM model, associated with a step-by-step integration program, specially developed to be able to follow the evolution of the load and related stress state during the meshing cycle; they show the principal stresses, σ_1 and σ_2, the maximum shear stress, $\tau_{max} = (1/2)(\sigma_1 - \sigma_2)$, and the normal stress, σ_z, in the direction of the tooth axis. It is evident that the two principal stresses reach their maximum values, respectively positive and negative, on the two fillet surfaces, nearly at the line bisecting the angle that delimits the tooth fillet. Figure 3.10 shows the

Fig. 3.9 Stresses at the root of a tooth carrying the entire load

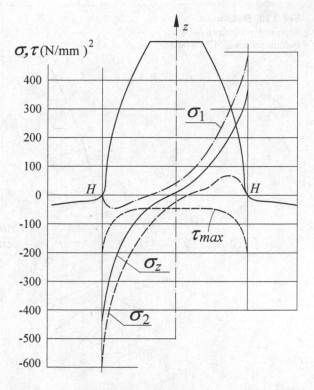

distribution curves of the total normal stress, $\sigma_t = (\sigma_b - \sigma_c)$, along the tensile and compressive tooth profiles, obtained using the same two-dimensional FEM model.

3.5 30° Tangent and Load Application at the Outer Point of Single Pair Gear Tooth Contact

Experimental evidence shows that, usually, the bending fatigue failure starts from tension fillet, roughly in the point at which the 30° tangents contact the root fillets, for external gears, or at which the 60° tangents contact the root fillets, for internal gears (see Miyachika and Oda 1991; Savage et al. 1995; Abdullah and Jweeg 2012). This experimental evidence has given rise to an empirical method for identifying the critical section of gear teeth subjected to bending loads. This empirical method of identifying the chordal thickness of the tooth to be used as a basis for bending strength calculations, has the double advantage of being simpler than that based on constant strength parabola, and to make the weakest section independent from the point of load application. For these reasons, this method is now established and acquired by most international standards, including ISO standards (see ISO 6336-3: 2006).

Fig. 3.10 Distribution curves of σ_t obtained using a two-dimensional FEM model

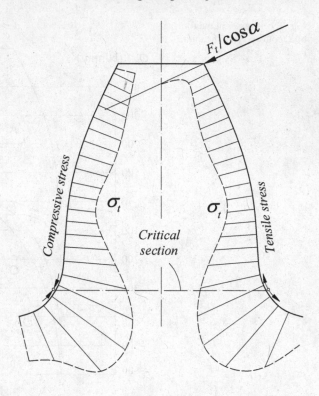

Either the experimental tests carried out with various procedures (photoelastic techniques, strain gauge techniques, etc.) or numerical models (FEM, BEM, etc.) show that the maximum values of bending and contact stresses are associated with the load applied at the outer point of single pair gear tooth contact (for driving pinion, the point D in Fig. 3.3). Therefore, in accordance with the proposal for the first time advanced by McMullen and Durkan (1922), the load condition to be considered is the one shown in Fig. 3.11 which refers to the driving pinion tooth of an external cylindrical spur gear.

For an external cylindrical spur gear wheel, without profile shift, and with a tooth whose active profile passes through the point D (see Fig. 3.3), and thus with load applied at the outer point of single pair gear tooth contact, from Eq. (3.19), with the quantities shown in Fig. 3.11, we obtain the following expression of the maximum bending stress at the tension fillet, $\sigma_{b\max}$, which we take as nominal bending stress, $\sigma_{bnom}(\sigma_{b\max} = \sigma_{bnom})$:

$$\sigma_{bnom} = \frac{F_t \cos\alpha_{Fe} h_{Fe}}{\cos\alpha\left(b s_{Fn}^2 / 6\right)} = \frac{6 F_t \cos\alpha_{Fe} m (h_{Fe}/m)}{b m^2 \left(s_{Fn}^2 / m^2\right)\cos\alpha} = \frac{F_t}{bm} Y_F, \tag{3.26}$$

where

Fig. 3.11 Driving pinion
tooth of an external virtual
cylindrical spur gear: chordal
thickness of the critical
section, obtained by 30°
tangent, and load applied at
the outer point of single pair
gear tooth contact

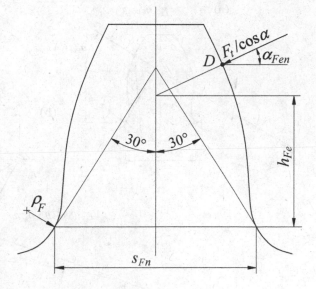

$$Y_F = \frac{6(h_{Fe}/m)\cos\alpha_{Fe}}{(s_{Fn}^2/m^2)\cos\alpha}, \tag{3.27}$$

is the *form factor*, while s_{Fn} (in mm) is the tooth root chordal thickness at the critical section, h_{Fe} (in mm) is the bending moment arm for the tooth root stress relevant to load applied at the outer point of single pair tooth contact, and α_{Fe} (in degrees) is the load direction angle, relevant to direction of application of load at the outer point of single pair tooth contact of cylindrical spur gears.

Comparing Eqs. (3.14) and (3.15) with Eqs. (3.26) and (3.27), we can deduce that a new form factor comes into play when we pass from the load applied at the tooth tip to the load applied at the outer point of single pair tooth contact. Generalizing the discussion, to extend it to the case of cylindrical external and internal helical gears and their virtual cylindrical spur gears (see Fig. 3.12a, b), we get:

$$\sigma_{bnom} = \frac{F_b\cos\alpha_{Fen}h_{Fe}}{\left(bs_{Fn}^2/6\right)} = \frac{F_b}{bm}\frac{6(h_{Fe}/m_n)\cos\alpha_{Fen}}{\left(s_{Fn}^2/m_n^2\right)\cos\alpha_n} = \frac{F_b}{bm}Y_F = \frac{F_t}{bm_n}Y_F, \tag{3.28}$$

where

$$Y_F = \frac{6(h_{Fe}/m_n)\cos\alpha_{Fen}}{\left(s_{Fn}^2/m_n^2\right)\cos\alpha_n} \tag{3.29}$$

is the tooth form factor in its more general terms, while α_{Fen} is the load direction angle, relevant to direction of application of load at the outer point of single pair tooth contact of virtual cylindrical spur gears. Clearly, the relationships (3.26) and (3.27), valid for cylindrical spur gears, are a special case of relationships (3.28) and

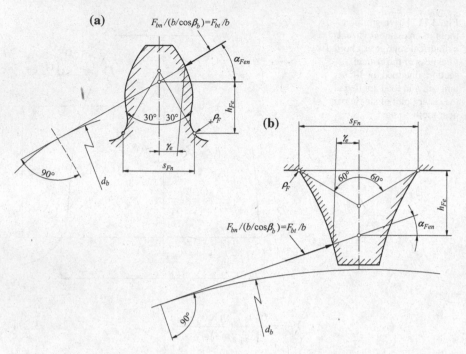

Fig. 3.12 Quantities for calculation of the form factor, Y_F, and stress concentration factor, Y_S, with Method B, for: **a** external gears; **b** internal gears

(3.29), since $\alpha_{Fen} = \alpha_{Fe}$, $m_n = m/\cos\alpha_n$, $F_{bn} = F_{bt}/\cos\beta_b$, and $F_b = F_t/\cos\alpha = F_w/\cos\alpha_w$, where F_{bn} is the nominal load, normal to the line of contact, F_{bt} is the nominal transverse load in plane of action (i.e. the base tangent plane), F_b is the nominal tangential load at the base cylinder, F_w is the nominal tangential load at the pitch cylinder, $\alpha = \alpha_t$ is the pressure angle of the basic rack profile, β_b is the base helix angle and α_w is the working pressure angle.

The form factor, Y_F, of the tooth must be determined separately for the pinion and gear wheel. For helical gears, the equivalent or virtual spur gears are to be considered; their virtual number of teeth can be determined using the exact equation (the first of Eq. 3.41), or the approximate equation [the second of Eq. 3.41]. Since this factor considers the influence of the tooth shape on the nominal bending stress, with the load applied at the outer point of single pair tooth contact, it will depend on the profile shift coefficient, x.

Determination of the form factor, Y_F, and stress correction factor, Y_S (see Sect. 3.7.1), by graphical means, is not recommended. The determination of the value of Y_F is based on the nominal tooth shape, thus neglecting the effect on the tooth bending strength of the reduction of the tooth thickness, caused by the gear finishing processes, but taking account of the profile shift. Shapes and dimensions of tooth roots of grounded or shaved gear teeth are usually generated by cutting tools such as hobs, for which they are determined by the cutting depth settings. If

Fig. 3.13 Basic rack
finished profiles with (**a**) and
without (**b**) undercut, and
their main geometrical
dimensions

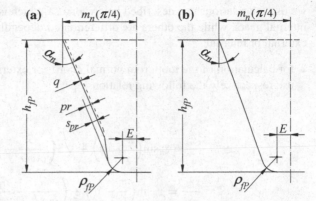

we have the warning to adjust the depth setting of the roughing tool, with respect to the gear axis, in such a way that it includes the amount of the nominal profile shift, $x m_n$, plus a tolerance selected so as to ensure that the finishing allowance is greater instead of less than the minimum requirement, the calculated values of tooth root stresses usually err on the safe side. However, if the tooth thickness deviation near the root leads to a thickness reduction greater than $0.05 m_n$, instead of the nominal profile shift, $x m_n$, we must consider the generation profile shift, $x_E m_n$, where x_E is the generation profile shift coefficient.

The calculation relationships of the form factor, Y_F, described below are in conformity with ISO 6336-3:2006, and apply to teeth that are related to all the basic rack profiles with and without protuberance, as Fig. 3.13 shows. This figure highlights the following quantities: E, auxiliary factor for tooth form factor; ρ_{fP}, cutter edge radius, equal to root fillet radius of basic rack for cylindrical gears; pr, protuberance of the cutter tool; q, machining stock; $s_{pr} = (pr - q)$, residual fillet undercut; α_n, normal pressure angle; m_n, normal module; h_{fP}, dedendum of basic rack of cylindrical gears.

Using the relationships below, the following restrictions must be taken into account:

- the root fillet radius of the basic rack must be greater than zero ($\rho_{fP} > 0$);
- generation of the teeth is made with hobs or rack-type cutters;
- for internal gears, a virtual basic rack profile is considered, which has a different root fillet radius;
- since the calculation relationships refer to the finished tooth shapes (see above), in practice we can assume that dimensions of the basic rack of the tool are those of the counterpart of the basic rack of the gear;
- the point of contact of the 30° tangent (60° tangent for internal gears) lies on the tooth root fillet generated by the root fillet of the basic rack.

For calculation of the tooth root normal chord at the critical section, s_{Fn}, tooth root radius at the critical section, ρ_F, and bending moment arm for the tooth root stress relevant to load applied at the outer point of single pair tooth contact, h_{Fe}, we use the relationships described below, one of which is common to external and internal gears, while the others are differentiated depending on whether the gear is external or internal:

- for calculation of the tooth root normal chord for external and internal gears, we use respectively the following relationships:

$$\frac{s_{Fn}}{m_n} = z_n \sin\left(\frac{\pi}{3} - \vartheta\right) + \sqrt{3}\left(\frac{G}{\cos\vartheta} - \frac{\rho_{fPv}}{m_n}\right) \tag{3.30}$$

$$\frac{s_{Fn}}{m_n} = z_n \sin\left(\frac{\pi}{6} - \vartheta\right) + \left(\frac{G}{\cos\vartheta} - \frac{\rho_{fPv}}{m_n}\right); \tag{3.31}$$

- for calculation of the radius of root fillet both for external and internal gears, we use the following relationship:

$$\frac{\rho_F}{m_n} = \frac{\rho_{fPv}}{m_n} + \frac{2G^2}{\cos\vartheta\left(z_n\cos^2\vartheta - 2G\right)}; \tag{3.32}$$

- for calculation of the bending moment arm for external and internal gears, we use respectively the following relationships:

$$\frac{h_{Fe}}{m_n} = \frac{1}{2}\left[(\cos\gamma_e - \sin\gamma_e\tan\alpha_{Fen})\frac{d_{en}}{m_n} - z_n\cos\left(\frac{\pi}{3} - \vartheta\right)\left(\frac{G}{\cos\vartheta} - \frac{\rho_{fPv}}{m_n}\right)\right] \tag{3.33}$$

$$\frac{h_{Fe}}{m_n} = \frac{1}{2}\left[(\cos\gamma_e - \sin\gamma_e\tan\alpha_{Fen})\frac{d_{en}}{m_n} - z_n\cos\left(\frac{\pi}{6} - \vartheta\right) - \sqrt{3}\left(\frac{G}{\cos\vartheta} - \frac{\rho_{fPv}}{m_n}\right)\right]. \tag{3.34}$$

In equations from (3.30) to (3.34), in addition to quantities already known, four auxiliary quantities E, G, H, and ϑ, and other quantities already defined, but still to be determined, appear. The auxiliary quantities are determined using the following relationships:

$$E = \frac{\pi}{4}m_n - h_{fP}\tan\alpha_n + \frac{s_{pr}}{\cos\alpha_n} - (1 - \sin\alpha_n)\frac{\rho_{fP}}{\cos\alpha_n} \tag{3.35}$$

$$G = \frac{\rho_{fPv}}{m_n} - \frac{h_{fP}}{m_n} + x \tag{3.36}$$

$$H = \frac{2}{z_n}\left(\frac{\pi}{2} - \frac{E}{m_n}\right) - T \tag{3.37}$$

$$\vartheta = \frac{2G}{z_n}\tan\vartheta - H, \qquad (3.38)$$

where $s_{pr} = (pr - q)$ or $s_{pr} = 0$, depending on whether the gears are with undercut or without undercut, $T = \pi/3$ or $T = \pi/6$ for external or, respectively, internal gears, and $\rho_{fPv} = \rho_{fP}$, for external gears, or

$$\rho_{fPv} \cong \rho_{fP} + m_n \frac{\left(x_0 + h_{fP}/m_n - \rho_{fP}/m_n\right)^{1.95}}{3.156 \cdot 1.036^{z_0}} \qquad (3.39)$$

for internal gears; in this last equation, x_0 and z_0 are respectively the profile shift coefficient and the number of teeth of the pinion-type cutter.

It is to be noted that Eq. (3.38) is a transcendental equation, solvable by iterative procedure. Using as initial value $\vartheta = \pi/6$ for external gears, and $\vartheta = \pi/3$ for internal gears, the solution converges to the exact value after five iterations.

Remembering that $m_n = p_{bn}/\pi\cos\alpha_n$, the other quantities of the virtual gears that appear in equations above are determined by the following relationships:

$$\beta_b = \arccos\sqrt{1 - (\sin\beta\cos\alpha_n)^2} = \arcsin(\sin\beta\cos\alpha_n) \qquad (3.40)$$

$$z_n = \frac{z}{\cos^2\beta_b\cos\beta} \quad \text{or} \quad z_n \cong \frac{z}{\cos^3\beta} \qquad (3.41)$$

$$\varepsilon_{\alpha n} = \frac{\varepsilon_\alpha}{\cos^2\beta_b} \qquad (3.42)$$

$$d_n = \frac{d}{\cos^2\beta_b} = m_n z_n, \qquad d_{bn} = d_n\cos\alpha_n, \qquad d_{an} = d_n + d_a - d \qquad (3.43)$$

$$d_{en} = 2\frac{z}{|z|}\sqrt{\left[\sqrt{\left(\frac{d_{an}}{2}\right)^2 - \left(\frac{d_{bn}}{2}\right)^2} - \frac{\pi d\cos\beta\cos\alpha_n}{|z|}(\varepsilon_{\alpha n} - 1)\right]^2 + \left(\frac{d_{bn}}{2}\right)^2} \qquad (3.44)$$

$$\alpha_{en} = \arccos\left(\frac{d_{bn}}{d_{en}}\right) \qquad (3.45)$$

$$\gamma_e = \frac{(\pi/2) + 2x\tan\alpha_n}{z_n} + \text{inv}\alpha_n - \text{inv}\alpha_{en} \qquad (3.46)$$

$$\alpha_{Fen} = \alpha_{en} - \gamma_e = \tan\alpha_{en} - \text{inv}\alpha_n - \frac{(\pi/2) + 2x\tan\alpha_n}{z_n}. \qquad (3.47)$$

When the virtual gear parameters given by the relationships above are used, the following precautions must be taken into consideration:

- In Eq. (3.44), which expresses the diameter of circle through outer point of single pair tooth contact in the normal plane, the number of teeth, z, is to be considered positive for external gears and negative for internal gears.
- If the tooth tip has been chamfered or rounded, or a tooth profile modification with tip relief has been done, instead of the tip diameter, d_a, it is necessary to use the *effective tip diameter*, d_{Na}, that is the diameter of a circle near the tip cylinder, containing the limits of the usable tooth flanks and thus delimiting the EAP (End of Active Profile).
- γ_e is the tooth thickness half angle (Fig. 3.12) at the outer point of single pair tooth contact.

All other quantities have been already defined previously.

3.6 Calculation of Tooth Bending Strength: ISO Basic Formulae

In previous sections, we saw that the stress state at the tooth root, due to the bending load, is very complex, and that the nominal bending stress at the tension fillet, given by Eq. (3.28), cannot alone explain the bending fatigue damage that generally occurs at this tension fillet. However, as the Hertzian contact stress, σ_H, for the assessment of surface durability, this nominal bending stress, calculated with load application at the outer point of single pair gear tooth contact, is the cornerstone on which the ISO standards, as well as other international and national standards, base the calculation of tooth bending strength (see Vullo and Maisano 1981).

ISO fundamental formulae, to be used for calculations of tooth bending strength of involute external and internal cylindrical spur and helical gears, consider the rim flexibility, which plays a key role in local stress distribution at the tooth root, as well as in the dynamic behavior of a gear unit, as Merritt (1954) showed for the first time, and as subsequent researchers confirmed (see Ishida 1977; Garro and Vullo 1979). These formulae, however, impose a rim thickness $s_R > 0.50 h_t$, for external cylindrical gears, and $s_R > 1.75 m_n$, for internal cylindrical gears (see Fig. 3.17a, b).

Method B of ISO 6336-3:2006 assumes, as the basis for calculations of tooth bending strength, the *tooth root stress*, σ_F, with load application at the outer point of single pair gear tooth contact, and compares this stress with the *permissible bending stress*, σ_{FP}. Tooth root stress, σ_F, is the maximum tensile stress at the tension fillet surface. This stress and the permissible bending stress, σ_{FP}, at the surface in the tooth root, must be calculated separately for pinion and gear wheel. For both toothed members of the gear pair, σ_F must be less than σ_{FP}.

The tooth load carrying capacity determined on the basis of permissible bending stress, which we call *tooth bending strength*, is assessed using the following relationships:

$$\sigma_F \leq \sigma_{FP} \tag{3.48}$$

$$\sigma_F = \sigma_{F0} K_A K_v K_{F\beta} K_{F\alpha} \tag{3.49}$$

$$\sigma_{F0} = \frac{F_t}{bm_n} Y_F Y_S Y_\beta Y_B Y_{DT} \tag{3.50}$$

$$\sigma_{FP} = \frac{\sigma_{Flim} Y_{ST} Y_{NT}}{S_{Fmin}} Y_{\delta relT} Y_{RrelT} Y_X = \frac{\sigma_{FE} Y_{NT}}{S_{Fmin}} Y_{\delta relT} Y_{RrelT} Y_X = \frac{\sigma_{FG}}{S_{Fmin}} \tag{3.51}$$

$$S_{F1} = \frac{\sigma_{FG1}}{\sigma_{F1}} \geq S_{Fmin} \qquad S_{F2} = \frac{\sigma_{FG2}}{\sigma_{F2}} \geq S_{Fmin}. \tag{3.52}$$

The meaning of symbols already introduced remaining unchanged, the new symbols appearing in Eqs. (3.49) and (3.50), which are in accordance with Method B of the aforementioned ISO standard, represent:

- σ_{F0}, the *nominal tooth root stress*, i.e. the maximum local stress at the tooth root of an ideal gear, without errors and without any pre-stress, and thus with stress ratio $R = 0$, due to the application of the static nominal torque. It is to be noted that pre-stresses such as those due to the shrink-fit of gear rims (see Vullo 2014), come into play in the calculation of permissible tooth root stress, σ_{FP}.

- Y_F, the *tooth form factor*, which modifies the tooth root stress when the outer point of single pair tooth contact is considered as the point of load application instead of the tooth tip; this factor allows to determine the nominal tooth root stress and is given by Eq. (3.29).

- Y_S, the *stress correction factor*, which corrects the nominal tooth root stress, given by Eq. (3.28), to get the local tooth root stress; it is introduced to take account of both the local stress intensification due to root fillet, and the fact that the calculation model of this nominal tooth root stress is too simple, and therefore it is not able to define, with the desired accuracy, the true three-dimensional stress state at the tooth root. In particular, the intensity of the local stress at the tooth root depends on two stress components, the first directly influenced by the bending moment, and the second increasing as the determinant position of the load application approaches more closely to the critical section.

- Y_β, the *helix angle factor*, which considers the fact that the bending moment intensity at the tooth root of a helical gear wheel is less than the corresponding value for the virtual spur gear used as the basis for calculation, as a result of the obliquity of the contact lines that characterize the helical gears.

- Y_B, the *rim thickness factor*, which corrects the calculated tooth root stress for gears with thin rim.

- Y_{DT}, the *deep tooth factor*, which corrects the calculated tooth root stress for high precision gears with a virtual contact ratio (i.e. the transverse contact ratio of the virtual spur gear) in the range ($2 \leq \varepsilon_{\alpha n} \leq 2.5$). In these cases, we assume that the determinant bending stress at the tooth root occurs with load application at the inner point of triple pair tooth contact (for helical gears, the helix angle factor, Y_β, considers deviations from these assumptions).

All the aforementioned factors are dimensionless factors. About the transverse tangential load at reference cylinder, F_t, that appears in Eq. (3.50), to be used also when $\varepsilon_{\alpha n} > 2$ (see Pedrero et al. 2007), in the case of multiple-path transmission gear drives (see Sect. 1.1), it should be emphasized the need to introduce in Eq. (3.49) a *mesh load factor*, K_γ, which follows the application factor, K_A. Inside the same equations, we will use $b = 2b_B$ for double helical gears. In addition, b is the value of the face width at the root circle, so any intentional tooth-end rounding or transverse chamfer is ignored. If then the face widths of pinion and mating gear wheel are different, for calculations relating to the gear member having the greater face width, we will assume a load bearing face width equal to that of the gear wheel having the lower face width, plus two extensions for each end of the tooth, each of which must not exceed one module (1.0 m).

The new symbols appearing in Eq. (3.51), which are also in accordance with the aforementioned Method B, represent:

- σ_{Flim}, the *nominal stress number* for bending, derived from test results on standard reference test gears, with well-defined material, heat treatment and surface roughness of the test gear root fillet. It is considered as the highest value of the tooth root stress limit for which the gear will have an endurance of at least 3×10^6 load cycles.
- Y_{ST}, the *stress correction factor*, relevant to the dimensions of the standard reference test gears, for which either $Y_{ST} = 2.0$ or for which test results are recalculated to this value.
- Y_{NT}, the *life factor* for tooth root bending stress, related to dimensions of the reference test gear, which considers the higher bending load carrying capacity for a limited number of load cycles.
- σ_{FE}, the *allowable stress number* for bending ($\sigma_{FE} = \sigma_{Flim} Y_{ST}$), which corresponds to the basic bending strength of the un-notched test piece, when we assume that the material behavior, including heat treatment, is fully elastic.
- S_{Fmin}, the minimum required value of the *safety factor* for tooth root bending stress.
- $\sigma_{FG} = \sigma_{FP} S_{Fmin}$, the *tooth root stress limit*.
- $Y_{\delta relT}$, the *relative notch sensitivity factor*, which considers the influence of the notch sensitivity of the material, and is defined as the quotient of the *notch sensitivity factor* of the gear of interest, Y_δ, divided by the *standard test gear factor*, $Y_{\delta T}$, that is $Y_{\delta relT} = Y_\delta / Y_{\delta T}$.
- Y_{RrelT}, the *relative surface factor*, which considers the influence of the relevant surface roughness of tooth root fillet, and is defined as the quotient of the *surface roughness factor* of the tooth root fillets of the gear of interest, Y_R, divided by the *tooth root fillet factor* of the reference test gear, Y_{RT}, that is $Y_{RrelT} = Y_R / Y_{RT}$.
- Y_X, the *size factor* for tooth root strength, which considers the influence of the tooth dimensions on tooth bending strength.

It should be remembered that also here all the aforementioned factors are dimensionless factors. The determination of the maximum tooth root stress, σ_F, and permissible bending stress, σ_{FP}, can be made directly by means of Method A, using appropriate techniques, such as FEM, BEM, experimental procedures by strain measurements, etc. These procedures, which are based on the direct analysis of actual gear pair duplicates, are characterized by a very high reliability, but entail a great effort, also economic, which is justifiable only in special cases of great technological importance. With the equations from (3.48–3.51), which are in conformity with Method B, we can obtain sufficiently reliable results, at lower cost, as long as the various influence factors that appear in these equations are chosen with great care.

In particular, the determination of σ_{FP} with Method B is based on the availability of S-N damage curves that are characterized by given values of σ_{Flim} and Y_{NT}, and are determined for a number of common gear materials and heat treatments, according to the results of gear load tests with standard reference test gears. These reference values are then converted by means of the three factors $Y_{\delta relT}$, Y_{RrelT}, and Y_X that appear in Eq. (3.51), to adapt them to the actual case study.

It should be noted that two special values of σ_{FP} must be considered: the *reference permissible bending stress*, σ_{FPref}, and *static permissible bending stress*, σ_{FPstat}. Both these values are determined using Eq. (3.51), the first with $Y_{NT} = 1$ and the other influence factors σ_{Flim}, Y_{ST}, $Y_{\delta relT}$, Y_{RrelT}, Y_X, and S_{Fmin} calculated with Method B, and the second with all influence factors for static stress calculated with Method B.

3.7 Influence Factors for the Nominal Tooth Stress

In the previous section, we defined the factors that appear in Eqs. (3.49) and (3.50), which influence the nominal tooth root stress, σ_{F0}. In this section, we describe the procedures for their determination, with the exception of that concerning the form factor, Y_F, which we already addressed in Sect. 3.5. Even here the determination of the remaining influence factors is made with calculations and diagrams that allow us to read their values, as a function of determinant variables on which they depend.

3.7.1 Stress Correction Factor, Y_S

In Sect. 3.4 we recalled the fundamental concepts regarding the theoretical stress concentration factor, K_t, and fatigue stress concentration factor, K_f. Here we describe the method of calculating the stress correction factor, Y_S, related to the aforementioned factors, by means of Method B, for load application at the outer point of single pair tooth contact. The relationships that follow are based on data derived from the geometry of external spur gears with pressure angle of 20°, through experimental measurements and calculations with FEM and BEM models. However, these equations can be used to obtain approximate values for gears with pressure angles different from 20°, and for internal gears.

The calculation of the stress correction factor, Y_S, is made with the following relationship, which is valid in the case in which the below defined *notch parameter*, q_s, is included in the range $(1 \leq q_s < 8)$:

$$Y_S = (1.2 + 0.13L)q_s^{\{1/[1.21+(2.3/L)]\}}, \tag{3.53}$$

where

$$L = \frac{s_{Fn}}{h_{Fe}} \qquad q_s = \frac{s_{Fn}}{2\rho_F}. \tag{3.54}$$

The quantities in Eqs. (3.53) and (3.54) are calculated in this way: s_{Fn}, according to Eq. (3.30) or Eq. (3.31), depending on whether the gear wheel is external or internal; h_{Fe}, according to Eq. (3.33) or Eq. (3.34), depending on whether the gear wheel is external or internal; ρ_F, according to Eq. (3.32).

The stress concentration at the root fillet of a gear wheel is increased in the case where a further notch is present in the same fillet near the critical section, such as a grinding notch; in this case we have a notch in the notch, this last consisting of the root fillet (Fig. 3.14). Consequently, the stress concentration factor for teeth with grinding notches in fillets, Y_{Sg}, will be greater than Y_S; it may be evaluated with sufficient accuracy, using the following relationship (see Püchner and Kamenski 1972; Niemann and Winter 1983):

$$Y_{Sg} = \frac{1.3Y_S}{1.3 - 0.6\sqrt{t_g/\rho_g}}, \tag{3.55}$$

where t_g and ρ_g are, respectively, the *maximum depth of grinding notch* and *radius of grinding notch*. The relationship (3.55), which also takes into account the reduction

Fig. 3.14 Grinding notch geometry at the root fillet and its dimensions

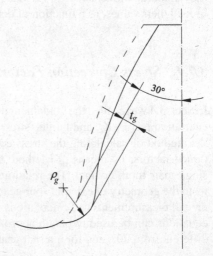

in the tooth root thickness, is valid for $\sqrt{t_g/\rho_g} < 2$. It is to be noted that, when the grinding notch is above the point of contact of the 30° tangent, for external gears, or 60° tangent, for internal gears, its effect is less than that implied in Eq. (3.55). In addition, it is to be noted that deep notches in the root fillets of surface hardened steel gears significantly reduce the tooth bending strength.

3.7.2 Helix Angle Factor, Y_β

As we already said, the helix angle factor, Y_β, converts the tooth root stress of the virtual spur gear wheel to that of the corresponding helical gear wheel. Bearing in mind that the instantaneous lines of contact are inclined of the angle β on the active flank of the tooth, three cases may arise with reference to the parameters that characterize the toothing, and precisely (see Merritt 1954; Wellauer and Seireg 1960; Niemann and Winter 1983):

(a) Total contact ratio, $\varepsilon_\gamma = (\varepsilon_\alpha + \varepsilon_\beta) < 2$. In this first case, similarly to what happens for a cylindrical spur gear, a region of single load application exists, and the maximum value of the nominal tooth root stress, σ_{F0}, occurs at the extreme position of the instantaneous line of contact that delimits superiorly this single load application region (the straight line A in Fig. 3.15a), roughly in correspondence of the point of intersection of this line with the diagonal D. To the left of this point of intersection, the bending moment arm is greater, but the stiffness of the tooth pair is smaller. Moreover, in this same part of the tooth flank surface, the strain in transverse direction is a little prevented.

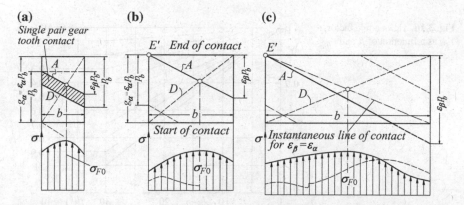

Fig. 3.15 Determinant instantaneous line of contact on the actual rectangle of contact, and σ_{F0} distribution curve along the face width, for a helical gear wheel: **a** $\varepsilon_\gamma = (\varepsilon_\alpha + \varepsilon_\beta) < 2$, with dotted region of single pair gear tooth contact; **b** $\varepsilon_\gamma = (\varepsilon_\alpha + \varepsilon_\beta) > 2$, with $\varepsilon_\beta < \varepsilon_\alpha$; (c), $\varepsilon_\gamma = (\varepsilon_\alpha + \varepsilon_\beta) > 2$, with $\varepsilon_\beta > \varepsilon_\alpha$

(b) Total contact ratio, $\varepsilon_y = (\varepsilon_\alpha + \varepsilon_\beta) > 2$, with $\varepsilon_\beta < \varepsilon_\alpha$. In this second case, a region of single load application does not exist, and the maximum value of the nominal tooth root stress, σ_{F0}, occurs when an instantaneous line of contact intersects the point E' corresponding to the tooth tip (see Vol. 1, Sects. 8.2 and 8.7). The position of the tooth transverse section where σ_{F0} is maximum is also determined coarsely by the intersection of this instantaneous line of contact with the diagonal (Fig. 3.15b).

(c) Total contact ratio, $\varepsilon_y = (\varepsilon_\alpha + \varepsilon_\beta) > 2$, with $\varepsilon_\beta > \varepsilon_\alpha$. Similar to what happens in the previous case, also in this third case a region of single load application does not exist. However, in this case, unlike the previous one, the instantaneous line of contact, which passes through the tooth tip (point E' in Fig. 3.15c), does not spread on whole face width. The portion of tooth that is not loaded by this instantaneous line of contact exerts, however, its bearing action. To account for this circumstance, we determine the bending moment arm as in the previous case, but with the limit condition for which $\varepsilon_\beta = \varepsilon_\alpha$.

The helix angle factor, Y_β, can be determined with the following relationship, which fulfills the requirements described above:

$$Y_\beta = 1 - \varepsilon_\beta \frac{\beta}{120°}, \tag{3.56}$$

where ε_β and β (in degrees) are respectively the overlap ratio and helix angle at reference cylinder. It is to be noted that, using Eq. (3.56), we assume $\varepsilon_\beta = 1$ when $\varepsilon_\beta > 1$, and $\beta = 30°$ when $\beta > 30°$. The results obtained with the use of Eq. (3.56) are consistent with the values that we can read, as a function of β and ε_β, using the curves shown in Fig. 3.16.

Fig. 3.16 Helix angle factor, Y_β, as a function of β and ε_β

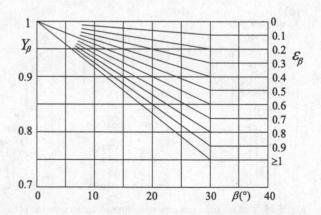

3.7.3 Rim Thickness Factor, Y_B

A more comprehensive analysis with appropriate methods (FEM, BEM, experimental measurements, etc.) is required when the *rim thickness*, s_R, is not sufficient to provide full support at the tooth root, for which the bending fatigue failure occurs in the gear rim, rather than at the tooth root fillet. When the results of such an analysis are not available, the rim thickness factor, Y_B, is a simplified derating factor of bending fatigue load capacity. This factor is a function of the *backup ratio*, s_R/h_t, for external gears, or a function of the *rim thickness ratio*, s_R/m_n, for internal gears.

For external gears, for which backup ratios less than or equal to 0.50 are to be avoided (in other words, it must always be $s_R/h_t > 0.50$), the rim thickness factor can be calculated with the following relationships:

$$Y_B = 1, \tag{3.57}$$

if $s_R/h_t \geq 1.20$, and

$$Y_B = 1.60\ln\left(2.242\frac{h_t}{s_R}\right), \tag{3.58}$$

if $0.50 < s_R/h_t < 1.20$. Equations (3.57) and (3.58) deliver results consistent with those we can read using the diagram shown in Fig. 3.17a, which also shows that the above derating effect is to be ignored for $s_R/h_t \geq 1.20$.

For internal gears, for which rim thickness ratios less than or equal to 1.75 are to be avoided (thus, it must always be $s_R/m_n > 1.75$), the rim thickness factor can be calculated with the following relationships:

$$Y_B = 1, \tag{3.59}$$

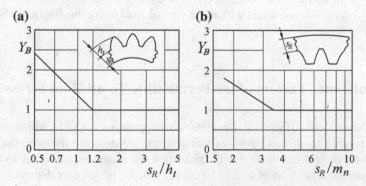

Fig. 3.17 Rim thickness factor, Y_B, as a function of the: **a** backup ratio, for external gears; **b** rim thickness ratio, for internal gears

if $s_R/m_n \geq 3.50$, and

$$Y_B = 1.15\ln\left(8.324\frac{m_n}{s_R}\right),\tag{3.60}$$

if $1.75 < s_R/m_n < 3.50$. Equations (3.59) and (3.60) deliver results consistent with those we can read using the diagram shown in Fig. 3.17b, which also shows that the above derating effect is to be ignored for $s_R/m_n \geq 3.50$.

3.7.4 Deep Tooth Factor, Y_{DT}

For high precision gears (accuracy grade ≤ 4), with virtual contact ratios in the range $2 \leq \varepsilon_{\alpha n} \leq 2.5$, and with profile modification specifically designed to get a trapezoidal load distribution along the path of contact, we need to correct the nominal tooth root stress by means of a deep tooth factor, which can be calculated with the following relationships:

$$Y_{DT} = 1,\tag{3.61}$$

if $\varepsilon_{\alpha n} \leq 2.05$ or if $\varepsilon_{\alpha n} > 2.05$ and the accuracy grade is greater than 4;

$$Y_{DT} = -0.666\varepsilon_{\alpha n} + 2.366,\tag{3.62}$$

if $2.05 < \varepsilon_{\alpha n} \leq 2.50$ and the accuracy grade is less than or equal to 4;

$$Y_{DT} = 0.7,\tag{3.63}$$

if $\varepsilon_{\alpha n} > 2.50$ and the accuracy grade is less than or equal to 4. Equations (3.61–3.63) deliver results consistent with those we can read using the diagrams shown in Fig. 3.18.

3.8 Influence Factors of the Permissible Tooth Root Stress

In Sect. 3.6 we defined factors and quantities that appear in Eq. (3.51), which influence the permissible tooth root stress. In this section, we describe the procedures and methods for their determination with calculations or diagrams that allow us to read their values, as a function of a determinant variable on which they depend.

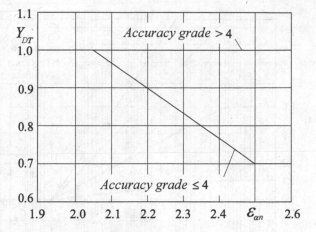

Fig. 3.18 Deep tooth factor, Y_{DT}, as a function of $\varepsilon_{\alpha n}$ and accuracy grade

3.8.1 Allowable Stress Numbers for Contact and Bending

We discussed this topic thoroughly in Sect. 2.10.1, where we clarified that, for reasons of uniformity and brevity, it would be appropriate to consider together the allowable stress numbers for contact, σ_{Hlim}, and allowable stress numbers for bending, $\sigma_{FE} = \sigma_{Flim} Y_{ST}$, where σ_{Flim} is the nominal stress number for bending, and $Y_{ST} = 2$ is a stress correction factor (see Sect. 3.6). Since we have nothing to add to what we already described in the aforementioned section, we refer the reader to that section.

3.8.2 Life Factor, Y_{NT}

The life factor for tooth root stress for reference test conditions, Y_{NT}, takes into account the higher tooth root stress that can be tolerated in comparison with the nominal stress number for bending, σ_{Flim}, when the number of cycles provided as lifetime is lower than that corresponding to 3×10^6 cycles. This factor is mainly influenced by: chemical composition, metallurgical structure, heat treatment and cleanness of gear material; numbers of load cycles or service life, N_L ($N_{L1,2} = 60hn_{1,2}$, where h is the number of hours of total lifetime of the gear, and $n_{1,2}$ is the rotational speed of pinion or gear wheel, in min^{-1}); failure criteria; smoothness of operation required; material ductility, fracture toughness, and residual stresses.

Here we circumscribe our attention to Method B, by which the evaluation of the permissible stress for limited lifetime or reliability is determined using the life factor Y_{NT} of the standard test gears. For this purpose, the normalized damage curves for tooth root bending resistance are used. Figure 3.19 shows these curves, which express the life factor, Y_{NT}, as a function of the service life, N_L, for the four groups of materials described in Table 3.1. It is to be noted that, in the more general case of gears characterized by more than one mesh contact per revolution, we must use

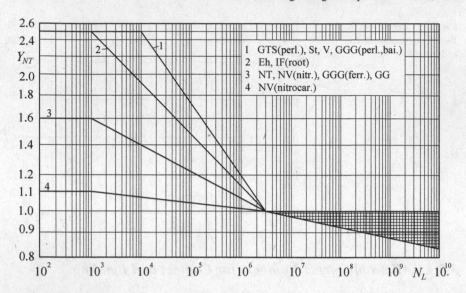

Fig. 3.19 Life factor, Y_{NT}, as a function of the service life, N_L, for standard reference test gears

Table 3.1 Life factor, Y_{NT}

Material	Number of load cycles, N_L	Life factor, Y_{NT}
St, V, GGG (perl., bai.), GTS (perl.)	$N_L \leq 10^4$, static	2.5
	$N_L = 3 \times 10^6$	1.0
	$N_L = 10^{10}$	0.85 up to 1.0
Eh, IF (root)	$N_L \leq 10^3$, static	2.5
	$N_L = 3 \times 10^6$	1.0
	$N_L = 10^{10}$	0.85 up to 1.0
GG, GGG (ferr.), NT, NV (nitr.)	$N_L \leq 10^3$, static	1.6
	$N_L = 3 \times 10^6$	1.0
	$N_L = 10^{10}$	0.85 up to 1.0
NV (nitrocar.)	$N_L \leq 10^3$, static	1.1
	$N_L = 3 \times 10^6$	1.0
	$N_L = 10^{10}$	0.85 up to 1.0

the relationship $N_{L1,2} = 60 h n_{1,2} n_c$, where n_c is the number of mesh contacts per revolution. In fact, more correctly, we can define the number of load cycles, N_L, as the number of mesh contacts, under load, of the gear tooth being analyzed.

The values of Y_{NT} for static and reference stresses may be obtained from Fig. 3.19, as well as from Table 3.1. With reference to the values of Y_{NT} shown in this table, included in the range from 0.85 to 1.00, it is to be noted that the lower value should be used to minimize the damage due to the bending load under critical service conditions, while the highest value can be used for optimum conditions of lubrication, material, manufacturing and experience.

3.8.3 Relative Notch Sensitivity Factor, $y_{\delta relT}$

The *dynamic notch sensitivity factor* and *static notch sensitivity factor*, both indicated by the symbol Y_δ, are quantities that express the extent to which the calculated tooth root stress, which caused the fatigue failure or overload breakage, has exceeded the relevant material limit stress. These quantities characterize the notch sensitivity of the material (see Sect. 3.4), and their values depend on the material, stress gradient, and type of load, dynamic or static. This concept is of general nature, and therefore can be applied not only to the *notch sensitivity factor*, Y_δ, of the actual gear, relative to a polished test piece, but also to the *sensitivity factor*, $Y_{\delta T}$, of the reference standard test gear, relative to a smooth polished test piece, as well as to the relative notch sensitivity factor, $Y_{\delta relT}$, which is the quotient of the gear notch sensitivity factor of interest divided by the standard test gear factor, i.e. $Y_{\delta relT} = Y_\delta / Y_{\delta T}$.

Even here we focus our attention on the Method B. Incidentally, it is to be noted that a careful analysis to calculate, with Method A, the relative sensitivity factor for the relevant material and relevant tooth shape has yet to be undertaken. To calculate the limit values of the reference and static stresses, the Method B uses standard reference test gears with notch parameters $q_{sT} = 2.5$. The values of the relative notch sensitivity factor $Y_{\delta relT}$ thus calculated, when they are applied to any gear, rarely deviate much from 1. The reference value $Y_{\delta relT} = 1$ for the standard reference test gear coincides with the stress correction factor $Y_S = 2$ (Fig. 3.21).

The calculation of the relative notch sensitivity factor, $Y_{\delta relT}$, is different depending on whether we consider the reference stress, or the static stress. For reference stress, this calculation is done using the following relationship:

$$Y_{\delta relT} = \frac{Y_\delta}{Y_{\delta T}} = \frac{1 + \sqrt{\rho' \chi^*}}{1 + \sqrt{\rho' \chi_T^*}}, \qquad (3.64)$$

where ρ' (in mm) is the *slip-layer thickness*, which can be taken from Table 3.2 as a function of the material (the values of ρ' for the same group of material can be interpolated for other values of the tensile strength, σ_B, yield stress, σ_s, and proof stress, $\sigma_{0.2}$), while χ^* and χ_T^* (both in mm^{-1}) are respectively the *relative stress gradient* at the notch root of the gear of interest, and the relative stress gradient at the notch root of the standard reference test gear. For the reference module $m = 5$ mm (the size effect is covered by the size factor, Y_X), the relative stress gradient can be determined by the following relationship:

$$\chi^* = \chi_P^*(1 + 2q_s), \qquad (3.65)$$

Table 3.2 Values for slip-layer thickness, ρ'

Material	ρ' (mm)
GG; $\sigma_B = 150$ N/mm^2	0.3124
GG, GGG (ferr.); $\sigma_B = 300$ N/mm^2	0.3095
NT, NV; for all hardness	0.1005
St; $\sigma_s = 300$ N/mm^2	0.0833
St; $\sigma_s = 400$ N/mm^2	0.0445
V, GTS, GGG (perl., bai.); $\sigma_s = 500$ N/mm^2	0.0281
V, GTS, GGG (perl., bai.); $\sigma_s = 600$ N/mm^2	0.0194
V, GTS, GGG (perl., bai.); $\sigma_{0.2} = 800$ N/mm^2	0.0064
V, GTS, GGG (perl., bai.); $\sigma_{0.2} = 1000$ N/mm^2	0.0014
Eh, IF (root); for all hardness	0.0030

where $\chi_P^* = 1/5$ is the relative stress gradient in a smooth polished test piece. The value of χ_T^* for the standard reference test gear is obtained with the same Eq. (3.65), by substituting in this equation $q_{sT} = 2.5$ for q_s.

Equation (3.64) delivers results consistent with those we can read using the diagrams shown in Fig. 3.20, which give $Y_{\delta relT}$ for reference stress as a function of q_s and group of material, some of which are differentiated according to their limit strength. In this figure, σ_B is the tensile strength, σ_s the yield stress, and $\sigma_{0.2}$ the proof stress (also known as 0.2% permanent set), all expressed in N/mm^2. In the same figure, the star indicates GGG with increasingly pearlitic structure.

For static stress, the calculation of $Y_{\delta relT}$ is done using the following relationships:

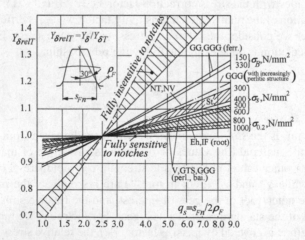

Fig. 3.20 Relative notch sensitivity factor, $Y_{\delta relT}$, for reference stress as a function of q_s and group of material

$$Y_{\delta relT} = \frac{1 + 0.82(Y_S - 1)(300/\sigma_s)^{1/4}}{1 + 0.82(300/\sigma_s)^{1/4}}, \tag{3.66}$$

for St with well-defined yield point, σ_s;

$$Y_{\delta relT} = \frac{1 + 0.82(Y_S - 1)(300/\sigma_{0.2})^{1/4}}{1 + 0.82(300/\sigma_{0.2})^{1/4}}, \tag{3.67}$$

for St with steadily increasing elongation curve and proof stress, $\sigma_{0.2}$, and for V and GGG (perl., bai.), with the caveat that the calculated values are only valid if the local stresses do not reach the yield point;

$$Y_{\delta relT} = 0.44Y_S + 0.12, \tag{3.68}$$

$$Y_{\delta relT} = 0.20Y_S + 0.60, \tag{3.69}$$

$$Y_{\delta relT} = 0.075Y_S + 0.85, \tag{3.70}$$

respectively valid for Eh and IF (root), for NT and NV, and for GTS, all with stress up to crack initiation;

$$Y_{\delta relT} = 1.0, \tag{3.71}$$

for GG and GGG (ferr.), with stress up to fracture limit.

Equations (3.66) to (3.71) deliver results consistent with those we can read using the diagrams shown in Fig. 3.21, which give $Y_{\delta relT}$ for static stress as a function of stress correction factor Y_S and group of material, some of which are differentiated according to their limit strength.

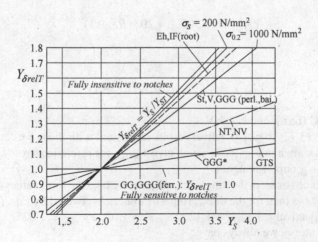

Fig. 3.21 Relative notch sensitivity factor, $Y_{\delta relT}$, for static stress as a function of Y_S and group of material

3.8.4   Relative Surface Factor, Y_{RrelT}

The *dynamic surface factor* and *static surface factor*, both indicated by the symbol Y_R, take into account the influence of the surface condition at the tooth root fillet on the tooth root stress. The *surface factor*, different for dynamic and static loads, depends on the material and surface roughness. However, this factor depends on other variables, related to the size and shape of the tooth root fillet, and especially to the presence of notches within a notch. These topics, to date, have not yet been studied with sufficient accuracy, so we just only consider the effect of surface roughness, requiring that the following relationships are valid only when stretches and similar defects deeper than $2Rz$ are not present.

The definition of surface factor is also of general nature, and therefore it can be applied not only to the tooth root surface factor, Y_R, of the actual gear, relative to a plain polished test piece, but also to the surface factor, Y_{RT}, of the reference standard test gear, relative to a smooth, plain polished test piece, as well as to the relative surface factor, Y_{RrelT}, which is the quotient of the gear tooth root surface factor of interest divided by the tooth root surface factor of the reference standard test gear, i.e. $Y_{RrelT} = Y_R / Y_{RT}$.

Even here we focus our attention on the Method B, which uses results of tests of reference standard test gears with $Rz_T = 10 \, \mu m$. The values of the relative surface factor, Y_{RrelT}, thus calculated, when they are applied to any gear of interest, differ little from 1, since $Rz_T = 10 \, \mu m$ is a common mean value of the mean peak-to-valley roughness. For static stress, we can take $Y_{RrelT} = 1$.

For reference stress, in the range $1 \, \mu m \le Rz \le 40 \, \mu m$, the calculation of Y_{RrelT} is done using the following relationships:

$$Y_{RrelT} = 1.674 - 0.529(Rz + 1)^{0.1}, \tag{3.72}$$

for V, GGG (perl., bai.), Eh and IF (root);

$$Y_{RrelT} = 5.306 - 4.203(Rz + 1)^{0.01}, \tag{3.73}$$

for St;

$$Y_{RrelT} = 4.299 - 3.259(Rz + 1)^{0.0058}, \tag{3.74}$$

for GG, GGG (ferr.), and NT, NV.

Equations (3.72) to (3.74) deliver results consistent with those we can read using the diagrams shown in Fig. 3.22, which give Y_{RrelT} for reference stress as a function of Rz and the group of material.

For reference stress, in the range $Rz < 1 \, \mu m$, we take the limit values corresponding to $Rz = 1$, so that, for the three groups of materials covered by Eqs. (3.72–3.74), we have, respectively: $Y_{RrelT} = 1.12$; $Y_{RrelT} = 1.07$; $Y_{RrelT} = 1.025$.

For static stress, we always have:

Fig. 3.22 Relative surface factor, Y_{RrelT}, for reference stress and static stress, as a function of Rz and group of material

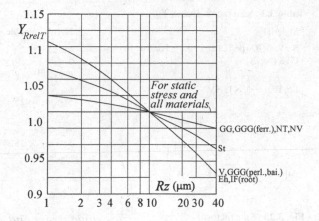

$$Y_{RrelT} = 1. \tag{3.75}$$

It is to be noted that, notoriously, other variables influence the tooth bending strength, such as: residual compressive stresses, due to shot peening, grain boundary oxidation and chemical effects. The shot peening or other similar processes, which cause a residual compressive stress state on the root fillet surface, determine an increase in the tooth bending strength, while grain boundary oxidations or other chemical effects imply a decrease in the tooth bending strength.

3.8.5 Size Factor, Y_X

The size factor, Y_X, takes into account the statistical evidence according to which the endurance tends to decrease when size increases. This factual situation in the fatigue strength field is due to the same reasons that we saw in Sect. 2.10.5, about the size factor, Z_X. The most significant influence parameters are as follows: chemical composition and cleanliness of the material, and quality of the forging process; heat treatment, depth and distribution of hardening; module in the case of surface hardening and core support effect, i.e. case depth in relation to the tooth size.

The size factor, Y_X, must be calculated separately for pinion and gear wheel. The calculation of Y_X for reference stress and for static stress is done using the relationships summarized in Table 3.3. For limited life, the size factor Y_X is obtained by means of linear interpolation between the values for reference stress and static stress.

Equations in Table 3.3 deliver results consistent with those we can read using the diagrams in Fig. 3.23, which give Y_X as a function of m_n and group of material.

Table 3.3 Size factor (root), Y_X

Material		Normal module, m_n	Size factor, Y_X
St, V, GGG (perl., bai.), GTS (perl.)	For 3×10^6 cycles	$m_n \leq 5$ $5 < m_n < 30$ $m_n \geq 30$	$Y_X = 1.0$ $Y_X = 1.03 - 0.006\,m_n$ $Y_X = 0.85$
Eh, IF (root), NT, NV		$m_n \leq 5$ $5 < m_n < 25$ $m_n \geq 25$	$Y_X = 1.0$ $Y_X = 1.05 - 0.01\,m_n$ $Y_X = 0.8$
GG, GGG (ferr.)		$m_n \leq 5$ $5 < m_n < 25$ $m_n \geq 25$	$Y_X = 1.0$ $Y_X = 1.075 - 0.015\,m_n$ $Y_X = 0.7$
All materials for static stress		–	$Y_X = 1.0$

Fig. 3.23 Size factor, Y_X, for the tooth bending strength as a function of m_n and group of material

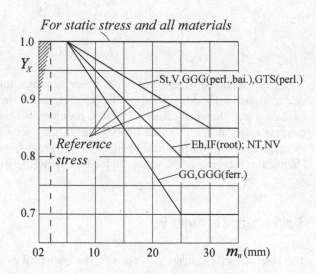

3.8.6 Minimum Safety Factor, S_{Fmin}

In Sect. 3.2, we defined the safety factor for tooth breakage, S_F, and we highlighted that its most significant influences are substantially the same as those we described in Sect. 2.2, in respect to the safety factor for pitting, S_H. We also highlighted that the tooth breakage usually puts an end to the service lifetime of a gear transmission system. The breakage of a single tooth often involves the destruction of all the gear wheels of the gear transmission, and the teeth breakage in many practical applications determines the interruption of the power flow between the input and output shafts.

For these reasons, the damages caused by the tooth breakage are considerably more significant than those due to pitting. Consequently, the chosen values of the safety factor against tooth breakage, S_F, should be larger than those of the safety factor against pitting, S_H. Table 2.5 (see Sect. 2.10.6) shows the recommended minimum values of the safety factor, S_{Fmin}, with some indications for the choice between the limit values, minimum and maximum.

3.8.7 Actual S-N Damage Curves, Obtained by Calculation

In the previous sections (from 3.8.1 to 3.8.6), we described the procedure for determining the value of the permissible tooth root stress, σ_{FP}, to be used for the calculation of tooth bending strength. This procedure is based on the introduction of a number of influence factors, which modify the nominal stress number for bending, σ_{Flim}, obtained with standard reference test gears.

However, even here it is possible to use an equivalent and less dispersive procedure, which consists of determining, by calculation, the S-N damage curve for bending of the actual gear to be analyzed from the S-N damage curve for bending obtained with standard reference test gears, or from the normalized damage curves for bending of the standard reference test gears, such as that shown in Fig. 3.19.

When we apply this procedure, we must consider that: in the range of long-life fatigue stress, the influence factors $Y_{\delta relT}$, Y_{RrelT}, and Y_X are fully active (note that Y_X is inactive for $m_n \leq 5$ mm, as Fig. 3.23 shows), and $Y_{NT} = 1$; instead, in the range of static stress, only the factor $Y_{\delta relT}$ is fully active for all the materials, with the exception of GG and GGG (ferr.) for which it is inactive (see Fig. 3.21), while factors Y_{RrelT} and Y_X are completely inactive. Moreover, in the two aforementioned stress ranges, also the safety factors S_{Fmin} are to be differentiated (see Table 2.5).

To formally simplify Eq. (3.51), we can introduce a single factor Y_N, called *life factor for tooth root stress*, and defined as:

$$Y_N = Y_{NT} Y_{\delta relT} Y_{RrelT} Y_X, \tag{3.76}$$

for which Eq. (3.51) becomes:

$$\sigma_{FP} = \frac{\sigma_{Flim} Y_{ST}}{S_{Fmin}} Y_N = \frac{\sigma_{FE}}{S_{Flim}} Y_N = \frac{\sigma_{FG}}{S_{Flim}}. \tag{3.77}$$

Figure 3.24 shows the manner in which the diagram of the life factor for tooth root stress, Y_N, as a function of the number of cycles, N_L, can be obtained for the curve 1 of Fig. 3.19, under the assumption that the portion of the curve corresponding to the long-life fatigue stress is horizontal. In the range of the long-life fatigue stress, we have considered active the factors $Y_{\delta relT}$, Y_{RrelT}, and Y_X, with $Y_{\delta relT}$ positively working, compared to the standard reference test gear, and the other two factors operating negatively. In the range of the static stress, we have considered active, and positively working, only the factor $Y_{\delta relT}$.

It is evident that curves like the modified one in Fig. 3.24 (the curve 1) represent, unless a constant factor given by the nominal stress number for bending, σ_{Flim}, the actual S-N damage curves of the gear to be examined.

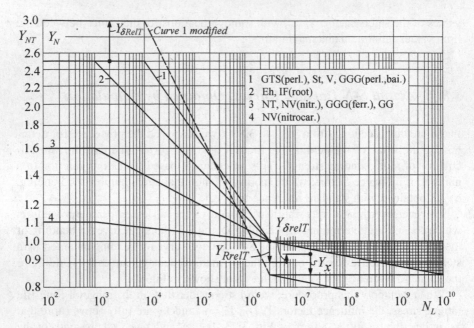

Fig. 3.24 Life factor for tooth root stress, Y_N

3.9 Design Analysis to Determine the Module

In the gear design, as well as in design of any machine element or mechanical system, we may be interested in both *response analysis* and *design analysis*. In response analysis (or *verification analysis*), the gear characteristics and material properties are given, and the response is to be determined in terms of stress, strains, deformations due to loads, load carrying capacity, allowable load, lifetime, etc. The design analysis is the inverse process with respect to response analysis, where we must choose the materials to be used and determine the gear characteristics (essentially the module) to ensure that the gear pair withstands to the known design loads during the expected operating lifetime.

Usually, in the gear design, we just make a response analysis, according to what national and international standards require (see the previous chapter, as regards the calculation of surface durability-pitting, and the previous sections of this chapter, as regards the calculation of tooth bending strength). However, to make the response analysis, the gear characteristics and material properties have been clearly defined by someone else designer, who in earlier times carried out the design analysis. Therefore, for the gears as for any other machine element or mechanical system, this last type of analysis is unavailable. We wish here to focus our attention on this aspect of the problem, i.e. the design analysis of the gears.

The design analysis is based on two different criteria, depending on whether we are faced with case-hardened gears or through-hardened gears. In fact, for case-hardened gears, the maximum bending stress at the tooth root is usually the determining factor, while, for through-hardened gears, the contact stress is usually the decisive design criterion.

In the design analysis of case-hardened gears, an appropriate determination of the normal module, m_n, can be carried out according to Eq. (3.28), which is more general than Eq. (3.26), and can therefore be used for spur and helical gears, with or without profile shift. The procedure described below assumes that the magnitude of the normal module, m_n, has no influence on the form factor Y_F, given by Eq. (3.29), as h_{Fe} and s_{Fn} increase in proportion to m_n. Furthermore, the effect of the addendum lowering, km_n (i.e. the truncation or topping) on the value of Y_F is assumed to be negligible.

The procedure used allow us to tentatively arrive to the required value of the normal module, m_n. It is in fact a so-called trial procedure, in that, once the number of teeth, z, and the nominal torque, T, applied to the gear wheel of interest are fixed, the nominal transverse tangential load at reference cylinder, F_t, depends on the unknown normal module, m_n, according to the following relationship:

$$F_t = \frac{2000T\cos\beta}{zm_n},$$ (3.78)

where T and m_n are respectively expressed in Nm and mm.

Equation (3.28) for tooth bending strength is to be verified against the permissible tooth root stress or working bending stress, σ_{FP}, of the material, the choice of which depends largely on the discretion and experience of the gear designer. Generally, the designer takes, as reference values, those of the static properties of the material. Therefore, for a material that exhibits a linear elastic-perfectly plastic behavior (in this case, the proportional limit, σ_p, the elastic limit, σ_e, and the yield point, σ_s, coincide, i.e. $\sigma_p = \sigma_e = \sigma_s$), we can assume that the permissible tooth root stress σ_{FP} is equal to the quotient of the yield point divided by a safety factor taken in the range $(2 \div 3)$, i.e. $\sigma_{FP} = \sigma_s/(2 \div 3)$. Instead, in the case in which the reference values of strength of the material are those related to the bending fatigue strength, for which the maximum tooth bending stress is chosen in accordance with the number of stress cycles that the gear is expected to undergo during its operational lifetime, the safety factor can be reduced considerably.

Considering the pinion, and substituting Eq. (3.78) into Eq. (3.28), in which we put $\sigma_{bnom} = \sigma_{FP}$, we derive the following expression of the normal module, m_n:

$$m_n = \sqrt[3]{\frac{2000T_1 Y_{F1}\cos^2\beta}{z_1^2(b/d_1)\sigma_{FP}}}.$$ (3.79)

In this equation, a suitable value of the ratio (b/d_1) is to be introduced. Among other considerations, this ratio depends on the type, quality and arrangement of the bearings. As guideline values, for straddle mounted bearings, i.e. for bearings arranged on both sides of the gear wheel, we take $(b/d_1) \leq 1.20$, while for overhung bearings, i.e. for bearings arranged on only one side of the gear wheel, we take $(b/d_1) \leq 0.75$.

Using the same procedure leading to Eq. (3.79), and introducing the proportionality factor, λ, between the face width and module, for which $b = \lambda m = \lambda m_n/\cos\beta$, we get the following expression of the normal module, m_n:

$$m_n = \sqrt[3]{\frac{2000 T_1 Y_{F1}\cos^2\beta}{\lambda z_1 \sigma_{FP}}}, \tag{3.80}$$

which is fully equivalent to Eq. (3.79). In accordance with the first of Eq. (2.72) of Vol. 1, the guideline values of λ are normally included in the range ($9 \leq \lambda \leq 14$), but for accurately machined and grounded gears, and anti-friction bearings or journal bearings fitted on rigid base and shafts of sufficient stiffness, we can reach values up to $\lambda = 30$. Usually, the face width of the pinion is made ($3 \div 4$) mm greater than that of the gear wheel, to ensure complete meshing during the service, as well as to reduce the effects of misalignment.

Apparently, Eqs. (3.79) and (3.80) allow the direct calculation of the normal module. In reality, since the permissible tooth root stress, σ_{FP}, depends on the tangential velocity, $v = \omega d/2 = \omega m z/2$, and thus on the module, generally some iteration is needed. However, since σ_{FP} appears in the two aforementioned equations under a cubic root, the variation of m_n with σ_{FP} is very slow. Therefore, with a careful choice of the initial value of σ_{FP}, dictated by experience, we can achieve an almost exact value of the normal module after one pair of iterations.

Finally, it is to be noted that, as it is apparent from Eqs. (3.79) and (3.80), for a given value of σ_{FP}, similar gear wheels undergo torques that increase with the cube of their size, i.e. the sizes of the gears are not very sensitive to a variation of the load.

In the design analysis of through-hardened gears, an appropriate determination of the normal module, m_n, can be carried out on the basis of Eq. (2.39), that expresses the nominal contact stress, σ_{H0}, in which we put $Z_\varepsilon = Z_\beta = 1$. This equation, where Z_H and Z_E are respectively given by the relationships (2.42) and (2.48), is also of a general nature, and can be used for spur and helical gears, with or without profile shift. It is to be verified against the permissible contact stress or working contact stress, σ_{HP}, of the material, the choice of which depends again largely on the discretion and experience of the gear designer.

In the case where the reference values of the surface load carrying capacity of the material are those of the surface fatigue strength, for which the maximum contact stress is chosen in accordance with the number of stress cycles that the gear is expected to undergo during its operational lifetime, for the practical purpose of interest here, we can assume the permissible contact stress, σ_{HP}, equal to the quotient of the pitting stress limit, σ_{HG}, divided by a safety factor taken equal to 1.5, i.e. $\sigma_{HP} = \sigma_{HG}/1.5$.

Considering the pinion, and elaborating Eq. (2.39), in which we put $Z_\varepsilon = Z_\beta = 1$ and replace σ_{HP} instead of σ_{H0}, we arrive at the following expression of the normal module, m_n:

$$m_n = \frac{1}{z_1} \frac{Z_H Z_E}{\sigma_{HP}} \sqrt{\frac{2000 T_1 \cos^2 \beta}{(b/d_1)} \frac{(u+1)}{u}}. \tag{3.81}$$

In this equation, the appropriate values of the ratio (b/d_1) indicated previously are to be introduced. As Eq. (3.79), or the equivalent Eq. (3.80), this equation involves some iteration for the determination of the normal module, m_n. In fact, not only the permissible contact stress, σ_{HP}, is a function of the module, through the tangential velocity, but also d_1 is a function of the module. Even here, however, with a careful choice of the initial values of σ_{HP} and (b/d_1), dictated by experience, we can achieve an almost exact value of the normal module after one pair of iterations.

Finally, it is to be emphasized that the specialization of Eqs. (3.79–3.81) to spur gears is simple and straightforward; just put in these equations $\cos\beta = 1$, for which the normal module becomes equal to the transverse module, i.e. $m_n = m_t = m$.

It is worth noting that the design criteria outlined above are among the traditional gear design criteria. They allow us to obtain gears whose features are well balanced, from the point of view of mechanical strength, efficiency, and reliability. Of course, to further improve or optimize their performance in custom applications, where higher transmission load capacity, efficiency, and reliability must be associated with lower size, weight, and cost, more sophisticated criteria must be used. In this regard we refer the reader to publications and specialized textbooks (see, for example: Kapelevich and Shekhtman 2003; Kapelevich 2013).

Example: calculate the module of a cylindrical spur pinion with the following input data, under the hypothesis that the most restrictive condition is the tooth root fatigue strength:

- transmitted power, $P = 50$ kW
- rotational speed, $n_1 = 1500 \, \text{min}^{-1}$
- number of teeth, $z_1 = 20$
- transverse pressure angle, $\alpha_t = 20°$
- face width, $b = \lambda m = 10 \, m$ (i.e., $\lambda = 10$)
- permissible tooth root stress, σ_{FP}, equal to the nominal stress number for bending, σ_{Flim}
- material: V (alloy steel), with MQ quality and hardness HV $= 280$
- tooth root factor, Y_{F1}, equal to the reciprocal of the Lewis form factor, y, for which $Y_{F1} = (1/y)$.

With the above input data, we first determine the following preliminary data:

- nominal torque at the pinion, $T_1 = 60P/2\pi n_1 = 318.31 \, \text{Nm}$;
- Lewis form factor $y = 0.332$, from Fig. 3.6 for $z_1 = 20$ and $\alpha = \alpha_t = 20°$.
- from the second of Eq. (2.57), with $x = \text{HV} = 280$, $B = 187$ and $A = 0.425$ taken from Table 2.3 for the assigned material, we obtain $\sigma_{FP} = \sigma_{Flim} = 306 \, \text{N/mm}^2$.

Then using Eq. (3.80), with a first iteration, we get the following value of the module: $m = 3.15$ mm. This is not a standard module (see ISO 54:1996). We therefore choose the module $m^* = 3.5$ mm, which is the immediately greater standard module. With this new value of the module, we calculate the transverse tangential load, $f_t = 2000 \, T_1/m^* z_1 = 9094.57$ N. Then using Eq. (3.16), we calculate the updated value of the permissible tooth root stress, $\sigma_{FP} = 223.62$ N/mm². Finally, still using Eq. (3.80) with updated data, we get $m = 3.4999$ mm. This second iteration is therefore sufficient to close the calculation loop, so the module $m = 3.5$ mm is the correct one.

3.10   Mean Stress Influence Factor, Y_M

The nominal stress number for bending, σ_{Flim}, and the allowable stress number for bending, $\sigma_{FE} = \sigma_{Flim} Y_{ST}$, are determined using standard reference test gears subjected to repeated, unidirectional tooth loading, i.e. repeated stress cycles characterized by *stress ratio* $R = (\sigma_{min}/\sigma_{max}) = 0$, *amplitude ratio* $A = (\sigma_a/\sigma_m) = [(1 - R)/(1 + R)] = 1$, and therefore an *alternating stress*, σ_a, equal to the *mean stress*, $\sigma_m = (\sigma_{max} + \sigma_{min})/2$. The values of σ_{Flim} and σ_{FE} thus determined are then usable for gears whose teeth are subjected to the same type of stress cycle, i.e. stress cycles characterized by $R = 0$ and $\sigma_a = \sigma_m$.

However, in some cases (see, for example, an idler gear wheel), we must design gears that are subjected to a completely revered stress cycle, which thus is characterized by $R = -1$, and $\sigma_m = 0$. In these cases, reduced values of σ_{Flim} and σ_{FE} are required. In most severe cases in which the full load reversal occurs for each load cycle (see again an idler gear wheel), we can take values of σ_{Flim} and σ_{FE} equal to 70% of those related to repeated, unidirectional tooth loading. If the full load reversals do not occur for each load cycle, but they are less frequent, a different reduction factor of the values related to repeated, unidirectional tooth loading can be chosen, depending on the number of reversals during the expected gear lifetime.

This reduction factor is given by the *mean stress influence factor*, Y_M, which considers the influence of working stress conditions other than pure pulsations, such as load reversing, idler gear wheels, or idler and planetary gears. The mean stress influence factor, Y_M, is to be included in Eq. (3.51), as a reduction factor of σ_{Flim} or σ_{FE}, and is defined as the quotient of the endurance (or static) strength with a stress ratio $R \neq 0$ divided by the endurance (or static) strength with $R = 0$.

For idler and planetary gears, the mean stress influence factor, Y_M, can be determined with the following relationship:

$$Y_M = \frac{1}{1 - R\dfrac{(1 - M)}{(1 + M)}}, \qquad (3.82)$$

where R is the stress ratio, in the range $(-1, 2 < R < 0)$, while M is the *mean stress ratio factor* which considers the influence of the mean stress, σ_m, on the endurance (or static) strength amplitudes. This last factor, M, which is defined as the reduction

Table 3.4 Mean stress ratio factor, M

Material	M, for endurance limit	M, for static strength
Case-hardened	$(0.80 - 0.15Y_S)$	0.7
Case-hardened and shot peened	0.4	0.6
Nitrided	0.3	0.3
Induction or flame hardened	0.4	0.6
Not surface hardened steels	0.3	0.5
Cast steels	0.4	0.6

of the endurance (or static) strength amplitude for a certain increase of the mean stress divided by this increase of the mean stress, can be read on Table 3.4.

It is to be noted that R may be assumed equal to (-1.2) for designs where the same load is applied both on forward-flank and back-flank. Instead, for designs where the applied loads are considerably different on forward-flank and back-flank, R may be assumed equal to (-1.2) multiplied by a factor given by the quotient of the load per unit face width of the lower loaded flank divided by the load per unit face width of the higher loaded flank.

In addition, it is to be noted that the values of M for endurance limit shown in Table 3.4 are independent of the root fillet shape, except for case hardening.

For case-hardened gears with full load applied periodically in both directions, i.e. gears with periodical change of rotational direction, the same Eq. (3.82) for idler gears may be used, with $R = -1$ and values of M for the endurance limit. This simplified approach is valid when the total number of load cycles exceed 3×10^6, and the number of changes of rotational direction exceeds 100.

3.11 Calculation Examples

With regard to the calculation examples, there is nothing here to add or remove from what we have already said in Sect. 2.11 of the previous chapter, to which we refer the reader. Here also, again for reasons of space, since we cannot present and discuss complete and exhaustive calculation examples concerning the load carrying capacity of spur and helical gears in term of tooth root fatigue breakage, we refer the reader to the examples concerning this subject, discussed in the ISO/TR 6336-30:2017.

References

Abdullah MQ, Jweeg MJ (2012) Analytical solution of bending stress equation for symmetric and asymmetric involute gear teeth shapes with and without profile correction. Innov Syst Des Eng 3(6):19–33

Almen JO (1935) Factors influencing the durability of spiral bevel gears for automobiles. Automot Ind 73:662–668 and pp 696–701

Almen JO, Straub JC (1937) Factors influencing the durability of automobile transmission gears, Automot Ind 77, 25th Sept, pp 426–432, and 9th Oct, 1937, pp 488–493

Aziz IAA, Idris DMN, Ghazali WM (2017) Investigation bending strength of spur gear: a review. In: MATEC web of conferences, 90:01037, AiGEV 2016

Becker AA (1992) The boundary element method in engineering: a complete course. McGraw-Hill Book Company Inc, New York

Brebbia CA (1978) The boundary element method for engineers. Pentech Press, New York

Brebbia CA, Connor JJ (1973) Fundamentals of finite element techniques. Butterworths, London

Brooks RE (2016) An introduction to shot peening for increasing gear fatigue life. Gear Solut Mag

Buckingham E (1949) Analytical mechanics of gears. McGraw-Hill Book Company Inc, New York

Burr AH (1982) Mechanical analysis and design. Elsevier Science Publishing Co., Inc, New York

Castigliano CAP (1935) SELECTA, a cura di Gustavo Colonnetti. R. Luigi Avalle Editore, Torino

Castigliano CAP (1966) The theory of equilibrium of elastic systems and its applications, translated from Théorie de l'Équilibre des Systèmes Élastiques et ses Applications by Ewart S. Dover Publications Inc, Andrews, New York

Castigliano CAP (1984) SELECTA 1984, a cura di Edoardo Benvenuto e Vittorio Nascé. Editrice Levrotto&Bella, Torino

Chauhan V (2016) A review on effect of some important parameters on the bending strength and surface durability of gears. Int J Sci Res Publ 6(3):289–298

Collins JA (1993) Failure of materials in mechanical design: analysis, prediction, prevention, 2nd edn. John Wiley & Sons Inc, New York

Cook RD (1981) Concepts and applications of finite element analysis, 2nd edn. John Wiley & Sons Inc, New York

Dieter GE (1988) Mechanical metallurgy, adapted by David Bacon, London: McGraw-Hill Book Company (UK) Limited

Dolan TJ, Broghammer EL (1942) A photoelastic study of stresses in gear tooth fillets, University of Illinois Bulletin, XXXIX (31), Engineering Experiment Station Bulletin Series No. 335

Faupel JH (1964) Engineering design: a synthesis of stress analysis and material engineering. John Wiley & Sons Inc, New York

Fenner RT (1986) Engineering elasticity: application of numerical and analytical techniques. Ellis Horwood Limited Publishers, Chichester

Fernandes PJL (1996) Tooth bending fatigue failures in gears. Eng Fail Anal 3(3):219–225

Frost NE, Marsh KJ, Pook LP (1974) Metal fatigue. Oxford University Press, London

Fuchs HO, Stephens RI (1980) Metal fatigue in engineering. John Wiley & Sons Inc, New York

Galilei G (1638) Discorsi e Dimostrazioni Matematiche intorno a due nuove scienze attinenti alla Meccanica & i Movimenti Locali. In: Leida, Appresso gli Elsevirii, M.D.C. XXXVIII

Garro A, Vullo V (1978a) Alcune considerazioni sul proporzionamento degli ingranaggi, Atti del VI Convegno Nazionale AIAS, Brescia, 22–24 giugno

Garro A, Vullo V (1978b) Note integrative sulla memoria: Alcune considerazioni sul proporzionamento degli ingranaggi, Atti del VI Convegno Nazionale AIAS, Brescia, 22–24 giugno

Garro A, Vullo V (1979) Acoustic problems in vehicle transmissions, Nauka I Motorna Vozila '79, Science and Motor Vehicles '79, Jugoslavija, 4–7 jun

Gere JM, Timoshenko SP (1997) Mechanics of materials, 4th edn. PWS Publishing Company, Boston

Giovannozzi R (1965a) Costruzione di macchine, vol I, 2nd ed. Casa Editrice Prof. Riccardo Pàtron, Bologna

Giovannozzi R (1965b) Costruzione di macchine, vol. II, 4th ed. Casa Editrice Prof. Riccardo Pàtron, Bologna

Henriot G (1970) French gear rating practices. Paper presented in commemoration of the presentation of the Edward P. Connell Award to Emeritus Professor Dr. Ing. Gustav Niemann at the AGMA Meeting, Oct., St. Louis, Mo

Henriot G (1979) Traité théorique et practique des engrenages 1, 6th edn. Bordas, Paris

Heywood RB (1952) Designing by photoelasticity. Chapman & Hall Ltd, London

Heywood RB (1962) Designing against fatigue. Chapman & Hall Ltd, London

Ishida K (1977) Computer simulation of stresses and deformations of gear case, and of gear teeth taking the influence of gear body into consideration. World Congress on Gearing, Paris 22–24 June, pp 309–323

ISO 54:1996 Cylindrical gears for general engineering and foe heavy engineering—Modules

ISO 6336-3:2006 Calculation of load capacity of spur and helical gears—part 3: calculation of tooth bending strength

ISO/TR 6336-30:2017 Calculation of load capacity of spur and helical gears—part 30: calculation examples for the application of ISO 6336 parts 1, 2, 3, 5

Juvinall RC (1967) Engineering considerations of stress, strain, and strength. McGraw-Hill Book Company, New York

Juvinall RC, Marshek KM (2012) Fundamentals of machine component design, 5th edn. John Wiley & Sons Inc, New York

Kapelevich AL (2013) Direct gear design. CRC Press, Taylor & Francis Group, Boca Raton, Florida

Kapelevich AL, Shekhtman YV (2003) Direct gear design: bending stress minimization. Gear Technol pp 44–47

Kelley BW, Pedersen R (1956) The beam strength of modern gear-tooth design. Vortragauf der Fachtagung Antriebselemente, Essen

Kelley BW, Pedersen R (1958) The beam strength of modern gear-tooth design. SAE Tech Paper 580017

Kramberger J, Šraml M, Glodez S, Flašker J, Potrč I (2004) Computational model for the analysis of bending fatigue in gears. Comput Struct 82(23):2261–2269

Kuhn P (1964) The prediction of notch and crack strength under static or fatigue loading. SAE-ASME Paper 843C

Lewis W (1892) Investigation of Strength of Gear Teeth. In: Proceedings of Engineers Club, Philadelphia, USA, October pp 16–23, and vol. 10 January 1893

Li S (2007) Finite element analyses for contact strength and bending strength of a pair of spur gears with machining errors, assembly errors and tooth modifications. Mech Mach Theory 42(1):88–114

Li S (2008) Effect of addendum on contact strength, bending strength and basic performance parameters of a pair of spur gears. Mech Mach Theory 43(12):1557–1584

Lipson C, Juvinall RC (1963) Handbook of stress and strength. The Macmillan Company, New York

McMullen FE, Durkan TM (1922) The gleason works system of bevel gears. Machinery

Merritt HE (1954) Gears, 3 ed., Sir Isaac Pitman & Sons, Ltd, London

Miyachika K, Oda S (1991) Bending strength of internal spur gears. In: Proceedings of the international conference on motion and power transmissions, Hiroshima, Japan, November 23–26, pp 781–786

Neuber H (1946) Theory of Notch Stresses. In: Edwards JW (eds) Publisher, Inc., Ann Arbor, Mich., (translation of the original German version in 1937)

Niemann G (1965) Maschinenelemente Entwerfen, Berechnen und Gestalten im Maschinenbau, vol 2. Springer-Verlag, Getriebe, Berlin Heidelberg

Niemann G, Richter W (1954) Tragfähigste evolventen-schragverzahnung. Vieweg, Zahnräder, Zahnradgetriebe, Braunschweig

Niemann G, Richter W (1960) Versuchsergebnisse zur Zahnflankentragfähigkeit, 9 Publications in Z. Konstruktion, S. 185, 236, 269, 319, 360

Niemann G, Winter H (1983) Maschinen-Elemente, Band II: Getriebe allgemein, Zahnradgetriebe-Grundlagen, Stirnradgetriebe. Springer Verlag, Berlin Heidelberg

Niemann G, Winter H, Höhn BR (2005) Maschinenelemente - Band 1: Konstruktion und Berechnung von Verbindungen, Lagern, Wellen, 4. Auflage. Springer-Verlag, Berlin Heidelberg

Paul FW, Faucett TR (1962) The superposition of stress concentration factors. J Manuf Sci Eng 84(1):129–134

Pedrero JI, Vallejo II, Pieguezuelos M (2007) Calculation of tooth bending strength and surface durability of high transverse contact ratio spur and helical gear drives. ASME J Mech Des 129(1):69–74

Peterson RE (1953) Stress concentration design factors. John Wiley & Sons Inc, New York

Peterson RE (1962) Fatigue of metals in engineering and design. Edgard Marburg Lecture, ASTM, Philadelphia

Peterson RE (1974) Stress concentration factors. John Wiley & Sons Inc, New York

Püchner O, Kamenski A (1972) Spannungskonzentration und Kerbwirkung von Kerben im Kerbrand. Konstruktion 24(4):127–134

Savage M, Rubadeux KL, Coe HH (1995) Bending strength model for internal spur gear teeth, NASA technical memorandum 107012. In: 31st joint propulsion conference and exhibit, San Diego, California, July 10–12

Shipley EE (1967) Gear failures: how to recognize them, what causes them, how to avoid them. Mach Des

Straub JC (1953) Shot peening in the design of gears. In: Proceedings of the annual meeting of the american gear manufacturers association. AGMA, Hot Spring, VA (Virginia), USA, May 31-June 3

Timoshenko SP (1953) History of strength of materials. McGraw-Hill Book Company Inc, New York

Timoshenko SP (1955) Strength of materials. D. Van Nostrand Co, Part I, Princeton, New Jersey

Timoshenko SP (1956) Strength of materials. D. Van Nostrand Co, Part II, Princeton, New Jersey

Timoshenko SP, Goodier JN (1951) Theory of elasticity, 2nd edn. McGraw-Hill Book Company Inc, New York

Ugural AC (2015) Mechanical design of machine components, 2nd edn. CRC Press, Taylor & Frencis Group, Boca Raton, Florida

Vijayakar SM, Houser DR (1988) The use of boundary elements for the determination of the AGMA geometry factor. Gear Technology, January/February

Vullo V (2014) Circular cylinder and pressure vessels, stress analysis and design. Springer International Publishing Switzerland, Cham Heidelberg

Vullo V, Maisano G (1981) Metodi normalizzati per il calcolo delle ruote dentate. Progettare 8–9:57–64

Vullo V, Vivio F (2013) Rotors: stress analysis and design. Springer-Verlag Italia, Milan Dordrecht

Wellauer EJ (1970) Comments on German, French and AGMA gear rating practices. Paper presented in commemoration of the presentation of the Edward P. Connell Award to Emeritus Professor Dr. Ing. Gustav Niemann at the AGMA Meeting, Oct., St. Louis, Mo

Wellauer EJ, Seireg A (1960) Bending strength of gear teeth by cantilever-plate theory. ASME J Eng Ind 82(3):213–220

Wilcox L, Coleman W (1973) Application of finite elements to the analysis of gear tooth stresses. ASME, J Eng Ind 95(4):1139–1148

Winter H (1961) Gear tooth strength of spur gears. Power Transm 404, 460, 516

Winter H (1962) Gear tooth strength of spur gears, Power Transm 66, 124

Winter H, Hösel Th (1969) Tragfähigkeitsberechnung von Stirnund Kegelrädern nach DIN 3990. VDI-Z 111:209

Winter H, Stolzle K (1970) German gear rating practices. Paper presented in commemoration of the presentation of the Edward P. Connell Award to Emeritus Professor Dr. Ing. Gustav Niemann at the 1970 AGMA Meeting, Oct., St. Louis, Mo

Zienkiewicz OC (1977) The finite element method, 3rd edn. McGraw-Hill Higher Education, UK

Chapter 4
Load Carrying Capacity of Bevel Gears: Factors Influencing Load Conditions

Abstract In this chapter, the main factors influencing the load carrying capacity of bevel gears in a broadest meaning, including straight, helical (or skew), spiral bevel, Zerol and hypoid gears, are first defined. Virtual cylindrical gears equivalent to the various bevel gears of interest are then determined. The various methods of calculating the influence factors (application factor, dynamic factor, face load factors and transverse load factors) are then described, focusing attention on the main quantities they depend on, and how to take them into account. In this framework, similar problems already seen for cylindrical spur and helical gears are tackled, passing over the aspect already discussed, and focusing attention on those specific of these types of gears, including those concerning the calculation of these factors based on the ISO standards.

4.1 Introduction

The basic theoretical concepts on which calculations of the load carrying capacity of bevel gears are founded, both in terms of surface durability (pitting) and tooth root strength, are similar to those relating to the strength design and resistance calculations of spur and helical gears. These basic concepts are described in all the main treatises and textbooks specifically dedicated to the gears (see Buckingham 1949; Merritt 1954; Dudley 1962; Shtipelman 1978; Henriot 1979; Stadtfeld 1993; Maitra 1994; Radzevich 2016), as well as in the widest horizon treatises concerning machine element design (see: Giovannozzi 1965; Pollone 1970; Niemann and Winter 1983; Budynas and Nisbett 2009). The basic design methods that apply these concepts are the same, but some variations are introduced, the main of which is the consideration of virtual spur and helical gears, respectively equivalent to straight bevel and spiral bevel gears under consideration.

This and the following two chapters cover the calculation of the load carrying capacity of bevel gears, in terms of pitting resistance and bending strength. In these three chapters, the term bevel gear is to be understood in its most comprehensive meaning, i.e. including straight, helical (or skew), spiral bevel, Zerol and hypoid

© Springer Nature Switzerland AG 2020
V. Vullo, *Gears*, Springer Series in Solid and Structural Mechanics 11,
https://doi.org/10.1007/978-3-030-38632-0_4

gears. Therefore, the symbols and calculation relationships that gradually we introduce, unless noted otherwise, are applicable to all these types of gears, regardless of the fact that they have tapered depth teeth or uniform depth teeth (Coleman et al. 1969).

In a similar way to what we have already seen for cylindrical spur and helical gears, the concept design of a bevel gear transmission system is generally based on prior experience, for which the main dimensions are known or can be extrapolated from these experiences. Therefore, the concept design of a bevel gear system is also set on this basis. Subsequently, the dimensions so obtained are verified, so as to ensure the required load capacity in terms of surface durability, tooth bending strength, scuffing and wear strengths, etc. The starting point is constituted by the technical specifications, which summarize the main input data of the bevel gear drive, including those of design, manufacturing and operation.

In this framework, by a calculation of first approximation, the initial dimensions, teeth characteristics and schematic structural drawing of kinematic operation, including the configuration of the housing, bearings that support the shafts, seals, type of lubrication, etc., are first determined. The load carrying capacity is therefore verified and, when this last does not give sufficient guarantees in terms of safety, the design is modified until reaching the solutions that satisfy the predetermined input data and design requirements.

As we have already said, in this and next two chapters we will focus our attention on the calculation of load carrying capacity of bevel and hypoid gears, in terms of pitting and bending strength. Therefore, this chapter and the following Chaps. 5 and 6, all three concerning bevel gears understood in the aforementioned general meaning, are respectively equivalent to Chaps. 1, 2, and 3, relating to cylindrical spur and helical gears. As of now, it is noteworthy that the relationships on pitting resistance are not applicable to other types of surface deterioration, such as micropitting, tooth interior fatigue fracture, tooth flank fracture, plastic yielding, adhesive wear and welding, abrasive wear, and case crushing. Furthermore, relationships on bending strength shall apply only to breakage at the root fillet; therefore, they cannot be applied to other types of breakage, such those on the active flank surfaces, and failures of the gear rim and gear blank through the web and hub. For these other types of failures, other theoretical, numerical, and experimental analyses are necessary.

Similar to what happens with cylindrical spur and helical gears, the tangential load that bevel gears can transmit depends on the size and shape of the teeth, tangential velocity of the gear members, lubrication and cooling conditions of the teeth, errors of manufacturing and assembly, stiffness of the teeth, shafts and housing, etc. For a bevel gear pair, it is even more difficult to carefully consider all the factors that influence the transmittable tangential load. We here follow the directives of ISO 10300-1:2014, concerning the basic principles to determine the appropriate values of the general influence factors that come into play in the calculation of the load carrying capacity of bevel gears.

It should be noted that the ISO directives can be applied within well-defined restrictions, which are related to the virtual cylindrical gear pair, equivalent to the actual bevel gear pair under consideration. These restrictions are summarized as follows: transverse contact ratio of virtual cylindrical gear, $\varepsilon_{v\alpha} < 2$; average mean spiral angle, $(\beta_{m1} + \beta_{m2})/2 \leq 45°$; effective pressure angle, $\alpha_e \leq 30°$; face width, $b \leq 13m_{mn}$, where m_{mn} is the mean normal module. Moreover, the ISO relationships are valid for bevel gears having the sum of profile shift coefficients of pinion and wheel equal to zero. For $\varepsilon_{v\alpha} \geq 2$, $(\beta_{m1} + \beta_{m2})/2 > 45°$, $\alpha_e > 30°$, and $b > 13m_{mn}$, the results obtained using the ISO relationships are not guaranteed, so they must be validated experimentally.

The general factors affecting load conditions of the bevel gears are the same as we described in Chap. 1 for cylindrical spur and helical gears. Therefore, they are as follows: application factor, K_A; dynamic factor, K_v; face load factor for contact stress, $K_{H\beta}$, and face load factor for bending stress, $K_{F\beta}$; transverse load factor for contact stress, $K_{H\alpha}$, and transverse load factor for bending stress, $K_{F\alpha}$. Even here the determination of these factors is carried out using Method A, Method B, and Method C, which have the same application peculiarities as those described in the aforementioned chapter. Therefore, as regards their accuracy and reliability, Method A is more than Method B, and Method B is more than Method C. Mixed factor rating methods can be used: for example, Method B for dynamic factor, K_{v-B}, can be used with Method C for face load factor for contact stress, $K_{H\beta-C}$.

An accurate assessment of the influence factors can be made only when the gear transmission system under consideration has been manufactured and tested, and the data obtained by direct measurements are available. However, in the preliminary design stage, the available data are limited. It is thus necessary to use approximate values of these factors, possibly based on past experiences relating to the design of similar gear drives. Furthermore, in these cases, conservative safety factors must be selected.

With Method A, the influence factors are derived by extrapolation of the test results and field data made available by previous experiences on similar transmission systems. However, these extrapolated results must be assessed by precise measurements and comprehensive mathematical analysis of the examined gear systems, or by field experience. The accuracy and reliability of this method shall be also demonstrated, through comparison with other acknowledged gear measurements.

Method B is used when the essential data of the gear transmission drive under consideration are known. For the evaluation of certain factors, even this method involves the fact that reliable results of previous operation experiences of other similar gear transmission units are available. In any case, the validity of the chosen values, for the given operating conditions, must be verified. Method B is then divided into two sub-methods, called $B1$ and $B2$, and characterized by two different calculation procedures of virtual cylindrical gears, equivalent to the to-be-designed bevel gear pair. The first sub-method, $B1$, provides only one set of formulae for both bevel and hypoid gears. In the case where the hypoid offset, a, is zero, the relationships provided by this sub-method become identical to those of the former version of the same ISO standard. Instead, the second sub-method, $B2$, provides separate sets of formulae for bevel and hypoid gears. It is to be noted that the other two parts of ISO 10300 (ISO 10300-2:2014 and ISO 10300-3:2014) follow the same division of the

Method B in sub-methods $B1$ and $B2$. On this subject we refer the reader to these parts of the aforementioned ISO standard and to the next two chapters.

Finally, Method C is a further simplified calculation method of the influence factors, to be used where suitable test results or field experience from similar designs are not available.

The influence factors and other quantities that have a key role in the calculations of surface durability and tooth root strength (i.e., the force on the tooth surface, bending moment on the tooth, etc.) are determined as a function of the *nominal tangential force*, F_{mt}, at mid-face width of the reference cone of the bevel gear under consideration. For bevel pinion and bevel wheel, this nominal tangential force is given by the following relationship:

$$F_{mt1,2} = \frac{2000 T_{1,2}}{d_{m1,2}} = \frac{1000 P}{v_{mt1,2}}, \qquad (4.1)$$

where $T_{1,2}$ (in Nm) is the *nominal torque* of pinion and wheel, P (in kW) the *nominal power*, $d_{m1,2}$ (in mm) the *mean pitch diameter* of pinion and wheel, and $v_{mt1,2}$ (in m/s) the *nominal tangential speed* at mid-face width of the reference cone of pinion and wheel. We can also write the following equations:

$$T_{1,2} = \frac{F_{mt1,2} d_{m1,2}}{2000} = \frac{1000 P}{\omega_{1,2}} \qquad (4.2)$$

$$P = \frac{F_{mt1,2} v_{mt1,2}}{1000} = \frac{T_{1,2} \omega_{1,2}}{1000} \qquad (4.3)$$

$$v_{mt1,2} = \frac{d_{m1,2} \omega_{1,2}}{2000} = \frac{\pi d_{m1,2} n_{1,2}}{60 \times 10^3} \simeq \frac{d_{m1,2} n_{1,2}}{19.099 \times 10^3} \qquad (4.4)$$

$$\omega_{1,2} = \frac{2000 v_{mt1,2}}{d_{m1,2}} = \frac{2\pi n_{1,2}}{60} \simeq \frac{n_{1,2}}{9.549}, \qquad (4.5)$$

where $\omega_{1,2}$ (in rad/s) and $n_{1,2}$ (in min^{-1}) are respectively the *angular velocity* and the *rotational speed* of pinion and wheel. The numerical coefficients that appear in Eqs. (4.1–4.5) homogenize the units of measurement in the SI-System.

The nominal tangential force of the virtual cylindrical gear, F_{vmt}, is given by the following relationship:

$$F_{vmt} = F_{mt1} \frac{\cos\beta_v}{\cos\beta_{m1}}, \qquad (4.6)$$

where β_{m1} (in degrees) is the mean spiral angle of bevel pinion, and β_v (in degrees) is the helix angle of virtual gear, when sub-method $B1$ is used, or the virtual spiral angle, when sub-method $B2$ is used.

Equation (4.6) highlights the fact that the nominal tangential force at mid-face width of the pinion reference cone, F_{mt1}, and therefore the pinion torque, T_1, is used in the basic stress calculation relationships. However, it is noteworthy that this nominal pinion torque is determined on the basis of the nominal torque of the driven machine, which is decisive for design of the gear transmission drive. As nominal torque of the driven machine we assume the operating torque to be transmitted under the most severe, regular operating conditions, over a long period of time. The nominal torque of the driving machine can be used only if it corresponds to the required torque of the driven machine.

In the case of non-uniform load, it is necessary to make an accurate analysis of the loads, in which the internal and external application factors are considered. It is also necessary to determine all the different loads that occur during the period of time corresponding to the expected lifetime of the gear unit examined, and the duration of each load. For the determination of the equivalent lifetime of the same gear unit, the Palmgren-Miner rule based on the torque spectrum can be used (see Sect. 1.8 and Palmgren 1924; Miner 1945; Fatemi and Yang 1998).

About safety factors to be used in calculations of surface durability and tooth root strength, we have nothing to add to what we said with reference to cylindrical spur and helical gears (see Sect. 2.10.6). The appropriate values to use in the selection of these safety factors depend on the required reliability of the gear transmission unit under consideration, in respect to the possible consequences of any damage that might occur in the case of failure, as well as on the reliability of the assumptions, such as those related to loads. In any case, an assembled gear drive should have a minimum safety factor for contact stress, $S_{H,min}$, not less than 1, and a minimum safety factor for bending stress, $S_{F,min}$, not less than 1.3 for spiral bevel gears, including hypoid gears, and not less than 1.5 for straight bevel gears or spiral bevel gears with $\beta_m \leq 5°$.

As we have already said before, it should remember that the ISO rating procedures for pitting resistance and tooth bending strength of bevel and hypoid gears are based on virtual cylindrical gears, equivalent to them. The reason of this choice is related to the fact that the necessary allowable stress values to be introduced in the design calculations are more readily derived from the results of several tests available for cylindrical gears. In addition, these results are more reliable than those obtained with fewer tests of bevel and hypoid gears.

As we have already said before, it should be remembered that two different computational procedures of virtual cylindrical gears are available. The first procedure is related to sub-method $B1$, while the second procedure is related to sub-method $B2$. The first calculation procedure of virtual cylindrical gears with sub-method $B1$ provides only one set of formulae for both bevel and hypoid gears. These formulae with decreasing offset values continuously approximate to those for spiral bevel gears without offset. The second calculation procedure of virtual cylindrical gears with sub-method $B2$ provides separate sets of formulae for bevel and hypoid gears.

4.2 Calculation of Virtual Cylindrical Gears with Sub-method $B1$

As we said in Vol. 1, Chap. 12, from the geometric point of view, hypoid gears are the most general type of gearing. Even for these gears, we can use the Tredgold approximation (see Buchanan 1823), similarly to what we already did for straight bevel gears (see Vol. 1, Sect. 9.5), and for spiral bevel gears that can be considered as hypoid gears without hypoid offset (see Vol. 1, Sect. 12.5). For strength calculations, the dimensions of the teeth at mean point are considered. Therefore, if a transverse section of a bevel or spiral bevel gear tooth at mid-face width is revolved in the relevant view corresponding to the plane of section, a virtual cylindrical spur or helical gear is obtained, with nearly involute tooth profiles (Fig. 4.1). For simplicity of graphic representation, in Fig. 4.1 a spiral bevel gear without hypoid offset is considered.

The equivalence between spiral bevel gear and corresponding virtual cylindrical helical gear shown in Fig. 4.1 must be understood only in the geometrical meaning. In other words, this equivalence does not mean that the thus-defined virtual cylindrical helical gear has the same meshing conditions of the actual spiral bevel gear. To obtain equivalent meshing conditions, it is necessary to introduce several appropriate correction factors, such as the hypoid factor, Z_{Hyp}, which considers the influence of the lengthwise sliding of hypoid gear teeth. However, the virtual cylindrical gears undoubtedly provide the required geometry on which to base a reliable assessment method of the load carrying capacity for all types of bevel gears.

For the determination of geometrical data of the virtual cylindrical helical gears, which are equivalent to the actual hypoid gears under consideration, it is necessary

Fig. 4.1 Spiral bevel gear without hypoid offset, and its equivalent virtual cylindrical helical gear

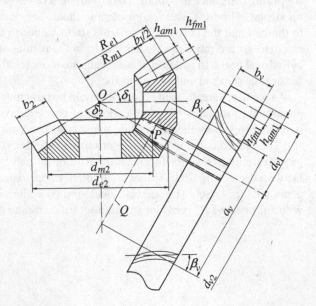

to bear in mind the quantities shown in the schematic drawing of a hypoid gear of Fig. 12.41 in Vol. 1, as well as those shown in Fig. 4.1. In some cases, the geometrical data of interest are those in the transverse section, while in other cases also those in the normal section can be of interest. To distinguish the various virtual quantities without ambiguity, quantities in transverse section and in normal section are respectively marked with subscripts v and vn.

From Fig. 12.41 in Vol. 1 and Fig. 4.1, we derive the following expressions of the *reference diameter*, d_v, *center distance*, a_v, *tip diameter*, d_{va}, and *root diameter*, d_{vf}, of virtual cylindrical gear:

$$d_{v1,2} = \frac{d_{m1,2}}{\cos\delta_{1,2}} \tag{4.7}$$

$$a_v = \frac{1}{2}(d_{v1} + d_{v2}) \tag{4.8}$$

$$d_{va1,2} = d_{v1,2} + 2h_{am1,2} \tag{4.9}$$

$$d_{vf1,2} = d_{v1,2} - 2h_{fm1,2} \tag{4.10}$$

For hypoid gears, we will have $d_{m2} \neq u d_{m1}$, while for $\Sigma = 90°$ and $a = 0$, we will have:

$$d_{v1} = d_{m1}\frac{\sqrt{1+u^2}}{u}; \qquad d_{v2} = u^2 d_{v1}. \tag{4.11}$$

In addition, from Fig. 12.41 in Vol. 1, we infer that, if the hypoid offset, a, decreases to zero $(a = 0)$, also the offset angle in pitch plane of pinion and wheel, ζ_{mp}, will be equal to zero. Therefore, the hypoid gear is reduced to the special case of a spiral bevel gear without offset, and mean cone distances, R_{m1} and R_{m2}, come to coincide. So, in this case, we will get the same geometrical parameters of virtual cylindrical helical gears that we described in Vol. 1, Chap. 12, and shown in Fig. 4.1.

For spiral bevel gears without hypoid offset, the *helix angles*, β_v, of virtual cylindrical helical gear members according to sub-method $B1$ are equal to the *mean spiral angles* of pinion and wheel, β_{m1} and β_{m2}. Since, in this case $\beta_{m1} = \beta_{m2}$, we will have $\beta_v = \beta_{m1} = \beta_{m2}$. However, as we saw in Vol. 1 Sect. 12.9, for hypoid gears with a positive hypoid offset $(a > 0)$, we have: $\beta_{m1} = \beta_{m2} + \zeta_{mp}$. Therefore, with reference to Fig. 12.41 in Vol. 1 where the bisecting line of the pinion offset angle in pitch plane, ζ_{mp}, defines the direction of the axes of virtual pinion and wheel, we obtain $\beta_{v1} = \beta_{m1} - (\zeta_{mp}/2)$ for the pinion, and $\beta_{v2} = \beta_{m2} + (\zeta_{mp}/2)$ for the wheel. Therefore, the two mean spiral angles of pinion and wheel are equal to each other, and both equal to the helix angle of the virtual cylindrical helical gear pair, β_v.

This choice is based on the results obtained by comparing the meshing conditions of the actual bevel and hypoid gears and those of the corresponding virtual cylindrical gears. The comparison was made by rigorous and advanced calculations on a wide variety of bevel and hypoid gears, using numerical methods of Tooth Contact Analysis (TCA methods), which highlighted the fact that the meshing conditions between the teeth are strongly influenced by the *inclination angle* of contact lines, β_B. This is the angle between the instantaneous line of contact (it coincides with the major axis of the Hertzian ellipse of contact under load) and the pitch line at the mean point. This inclination angle of contact line, β_B, has been assumed as a representative parameter of meshing conditions. Indeed, results of the numerical simulations based on this assumption have showed unequivocally that the values of β_B calculated by TCA models for any spiral bevel and hypoid gear were almost coinciding with those calculated for the corresponding virtual cylindrical gears having helix angle, β_v, equal to the arithmetic mean value of mean spiral angles, β_{m1} and β_{m2}. Therefore, we can write:

$$\beta_v = \frac{1}{2}(\beta_{m1} + \beta_{m2}). \tag{4.12}$$

The base diameter of virtual cylindrical gear, d_{vb}, is given by:

$$d_{vb1,2} = d_{v1,2}\cos\alpha_{vet}, \tag{4.13}$$

where α_{vet} is the transverse pressure angle of virtual cylindrical gear, given by:

$$\alpha_{vet} = \arctan(\tan\alpha_e / \cos\beta_v), \tag{4.14}$$

with $\alpha_e = \alpha_{eD}$ and $\alpha_e = \alpha_{eC}$ for drive side and, respectively, for coast side. So $\alpha_{eD,C}$ is the effective pressure angle for drive side/coast side.

The transverse module, m_{vt}, number of teeth, $z_{v1,2}$, and gear ratio, u_v, of virtual cylindrical gear are given respectively by the following relationships:

$$m_{vt} = m_{mn} / \cos\beta_v \tag{4.15}$$

$$z_{v1,2} = d_{v1,2} / m_{vt} \tag{4.16}$$

$$u_v = z_{v2} / z_{v1}. \tag{4.17}$$

For spiral bevel gears ($a = 0$), with $\Sigma = 90°$, we have:

$$z_{v1} = z_1 \frac{\sqrt{1 + u^2}}{u}; \qquad z_{v2} = z_2\sqrt{1 + u^2}. \tag{4.18}$$

The helix angle at base circle, β_{vb}, transverse base pitch, p_{vet}, and length of path of contact in transverse section, $g_{v\alpha}$, of virtual cylindrical gear are given, respectively, by the following relationships:

$$\beta_{vb} = \arcsin(\sin\beta_v \cos\alpha_e), \tag{4.19}$$

$$p_{vet} = \pi m_{mn} \cos\alpha_{vet} / \cos\beta_v \tag{4.20}$$

$$g_{va} = \frac{1}{2}\left[\left(\sqrt{d_{va1}^2 - d_{vb1}^2} - d_{v1}\sin\alpha_{vet}\right) + \left(\sqrt{d_{va2}^2 - d_{vb2}^2} - d_{v2}\sin\alpha_{vet}\right)\right]. \tag{4.21}$$

The determination of the face width of virtual cylindrical gear, b_v, involves some preliminary consideration. In fact, contrary to what happens for spiral bevel gears without hypoid offset, for which the face widths of virtual cylindrical gear and actual spiral bevel gear coincide ($b_v = b$), for hypoid gears this does not happen, and the two face widths are different ($b_v \neq b$). To determine b_v, it is necessary to first determine the effective face width of the virtual cylindrical gear pair, $b_{v,eff}$. To this end, the length of the contact pattern, $b_{2,eff}$, measured along the direction of the wheel face width is used.

In simplified way, but without compromising the discussion, the theoretical zone of action of the hypoid wheel, which actually has an arched boundary, is replaced with a parallelogram, and then projected onto the common pitch plane, T (see Vol. 1, Fig. 12.41). The parallelogram thus obtained is shown in Fig. 4.2 by dotted bold lines. The side lines of this parallelogram around the mean point, P, are perpendicular to the wheel axis, which in this view coincides with the wheel cone distance, R_{m2}. The other two boundary lines are parallel to the instantaneous axis, 1, of the helical relative motion of the hypoid gear pair, whose direction with respect to the wheel axis is detected by the auxiliary angle for virtual face width, ϑ_{mp}.

The same Fig. 4.2 shows by continuous bold lines the greatest possible parallelogram inscribed in the aforementioned theoretical zone of action, taken by us for the

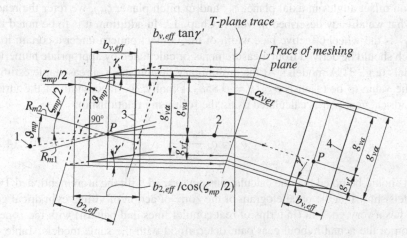

Fig. 4.2 Assumed zone of action for virtual cylindrical gears

actual hypoid gear pair. This second parallelogram, whose side lines are perpendicular to the rolling axis, 2, of the virtual cylindrical gear pair (the other two boundary lines overlap with those of the first parallelogram) is the zone of action of the corresponding virtual cylindrical gear pair. The direction of this rolling axis, 2, with respect to the wheel axis is detected by the angle $\zeta_{mp}/2$. The width of this second smallest parallelogram, which is designed by 3 in Fig. 4.2, in the projection over the tangential plane, T, represents the true length of the zone of action in the same plane, and it is therefore the effective face width, $b_{v,eff}$, of the virtual cylindrical gear pair. To get a full picture of the zone of action in true size, the given top view is also projected over the meshing plane of the virtual cylindrical gear pair, which is inclined by the effective pressure angle, α_{vet}, with respect to the plane, T (see 4 in Fig. 4.2). On this plane, which coincides with that of the tooth active flank, even the path of contact is in true size.

From Fig. 4.2, we get the following relationship of the effective face width, $b_{v,eff}$:

$$b_{v,eff} = \frac{\left[b_{2,eff}/\cos\left(\zeta_{mp}/2\right) - g_{v\alpha}\cos\alpha_{vet}\tan\left(\zeta_{mp}/2\right)\right]}{1 - \tan\gamma'\tan\left(\zeta_{mp}/2\right)}, \tag{4.22}$$

where γ' is the projected auxiliary angle for length of contact line, given by:

$$\gamma' = \vartheta_{mp} - \left(\zeta_{mp}/2\right), \tag{4.23}$$

while the auxiliary angle for virtual face width, ϑ_{mp}, is given by:

$$\vartheta_{mp} = \arctan(\sin\delta_2 \tan\zeta_m). \tag{4.24}$$

It is to be noted that the effective pressure angle of the virtual cylindrical gear pair, α_{vet} (see Eq. (4.14)), must be calculated for the active flank, i.e. for $\alpha_e = \alpha_{eD}$. As for pinion offset angles in axial plane, ζ_m, and in pitch plane, ζ_{mp}, we refer the reader to what we already described in Vol. 1, Chap. 12. In addition, it is to be noted that $b_{2,eff}$ is the wheel effective face width of the contact pattern under a certain load, which should be derived from measurements or calculated by appropriate numerical models (e.g., TCA models). At the preliminary design stage, a reasonable estimate of the value to be taken is $b_{2,eff} = 0.85b_2$. Finally, the face width of the virtual cylindrical gear pair is calculated using the following relationship:

$$b_v = b_2 \frac{b_{v,eff}}{b_{2,eff}}. \tag{4.25}$$

It should be noted that the calculations performed with the aforementioned TCA models show that the parallelogram of the zone of action of virtual cylindrical gear pair has a very good fit (in terms of real contact lines and pattern) with the zone of action of the actual hypoid gear pair, determined with the same models. Table 4.1 shows the zones of action, both projected on a plane perpendicular to the wheel

Table 4.1 Examples of action zones of virtual cylindrical gear pairs and calculated contact patterns of the bevel gears units over a projection plane perpendicular to the wheel axis

Actual flank	Hypoid offset		
	($a = 0$ mm)	($a = 15$ mm)	($a = 30$ mm)
Drive side			
Coast side			

axis, and overlapped on each other. Six sample plots are shown, related to three gear drives with different hypoid offset values ($a = 0\,\text{mm}$; $a = 15\,\text{mm}$; $a = 30\,\text{mm}$), and what happens on the drive and coast sides is highlighted. In each plot, the lines of mean cone distances of pinion, 1, and wheel, 2, as well as the rolling axis of virtual cylindrical gear pair, 3, which intersect at the respective mean point, P, are shown.

The parallelograms of the virtual zone of action contain three representative straight contact lines, drawn with bold lines. These lines fit angularly very well the calculated curved lines of contact, especially in actual contact pattern area. Furthermore, the parallelogram correlated to the corresponding virtual cylindrical gear pair covers well in size and position each of the calculated contact patterns. This circumstance gives a strong motivation to the equivalence of the meshing conditions between actual bevel gears and their corresponding cylindrical gears. Simulations with these numerical models finally demonstrate that the results obtained with the afore defined parallelograms are not worse than those obtained with the ellipse inscribed in the zone of action, previously used. However, the parallelograms are easy to use, also because it seems that the major axis of the ellipse does not always have to be parallel to the axis of the virtual cylindrical gear pair, but should at least be rotated by a certain angle, difficult to determine.

The transverse contact ratio, $\varepsilon_{v\alpha}$, and face contact ratio, $\varepsilon_{v\beta}$, for virtual cylindrical gears are given respectively by the following relationships:

$$\varepsilon_{v\alpha} = g_{v\alpha}/p_{vet} \tag{4.26}$$

$$\varepsilon_{v\beta} = \frac{b_{v,eff}\sin\beta_v}{\pi m_{mn}}. \tag{4.27}$$

The values calculated with Eqs. (4.26) and (4.27) are determinant for calculations of the load carrying capacity. However, it is possible that these values are different from those calculated on the basis of TCA models or determined on the basis of the actual dimensions of the bevel gears under consideration.

The virtual contact ratio related to Method $B1$ is given by:

$$\varepsilon_{v\gamma} = \varepsilon_{v\alpha} + \varepsilon_{v\beta}. \tag{4.28}$$

For the determination of the length of contact lines, l_b (in mm), it must be remembered that, when the tooth contact has been properly developed, the full load contact should not extend beyond the boundary of the parallelogram of the assumed zone of contact, as Fig. 4.3 shows. The meaning of the symbols already introduced remaining unchanged, the new symbols shown in this figure are as follows: f (in mm), the distance from the center point of the zone of action to a contact line (f is assumed positive for contact lines to the right of center of the zone of action, M, and negative for contact lines to the left of this point); f_{max}, the maximum distance to middle contact line; f_{maxB}, the maximum distance to middle contact line at right side of contact pattern; f_{max0}, the maximum distance to middle contact line at the left side

Fig. 4.3 Contact lines and their lengths

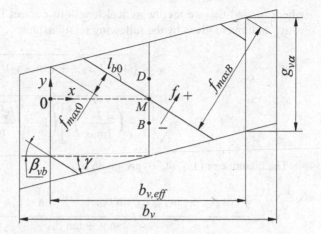

of contact pattern; l_{b0}, the theoretical length of contact line; γ (in degrees), the auxiliary angle for length of contact line calculation with Method $B1$. In the same figure, D is the outer point of single pair tooth contact, while B is the inner point of single pair tooth contact.

Usually, the contact lines are shorter than they should theoretically be, because of the crowning of the tooth flanks in profile and lengthwise directions. The introduction of a dimensionless correction factor, C_{lb}, which reduces the length of the contact lines according to an elliptical function, takes account of this fact (see Eq. 4.31 and Fig. 4.4).

The length of contact line, l_b, is given by:

$$l_b = l_{b0}(1 - C_{lb}),$$
(4.29)

Fig. 4.4 Correction factor, C_{lb}

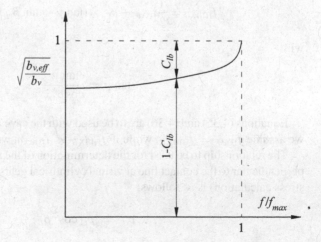

where l_{b0} and C_{lb} are the theoretical length of contact line and correction factor, which in turn are given by the following relationships:

$$l_{b0} = \sqrt{(x_1 - x_2)^2 + (y_1 - y_2)^2} \tag{4.30}$$

$$C_{lb} = \sqrt{\left[1 - \left(\frac{f}{f_{max}}\right)^2\right]\left(1 - \sqrt{\frac{b_{v,eff}}{b_v}}\right)^2}. \tag{4.31}$$

The quantities in Eq. (4.30) are given by:

$$x_1 = \frac{f \cos\beta_{vb} + \tan\beta_{vb}\left(f \sin\beta_{vb} + \frac{b_{v,eff}}{2}\right) + \frac{1}{2}\left(g_{v\alpha} + b_{v,eff}\tan\gamma\right)}{\tan\gamma + \tan\beta_{vb}} \tag{4.32}$$

$$x_2 = \frac{f \cos\beta_{vb} + \tan\beta_{vb}\left(f \sin\beta_{vb} + \frac{b_{v,eff}}{2}\right) - \frac{1}{2}\left(g_{v\alpha} + b_{v,eff}\tan\gamma\right)}{\tan\gamma + \tan\beta_{vb}} \tag{4.33}$$

$$y_{1,2} = -x_{1,2}\tan\beta_{vb} + f \cos\beta_{vb} + \tan\beta_{vb}\left(f \sin\beta_{vb} + \frac{b_{v,eff}}{2}\right). \tag{4.34}$$

Equations (4.32) and (4.33) are to be used with the caveat that, if $x_{1,2} < 0$ we assume $x_{1,2} = 0$, while if $x_{1,2} > b_{v,eff}$, we assume $x_{1,2} = b_{v,eff}$.

The maximum distances, f_{maxB} and f_{max0}, from the middle contact line are given by (see Fig. 4.3):

$$f_{maxB} = \frac{1}{2}\left[g_{v\alpha} + b_{v,eff}(\tan\gamma + \tan\beta_{vb})\right]\cos\beta_{vb} \tag{4.35}$$

$$f_{max0} = \frac{1}{2}\left[g_{v\alpha} - b_{v,eff}(\tan\gamma + \tan\beta_{vb})\right]\cos\beta_{vb}, \tag{4.36}$$

with

$$\tan\gamma = \frac{\tan\gamma'}{\cos\alpha_{vet}}. \tag{4.37}$$

Equations (4.35) and (4.36) are to be used with the caveat that, if $f_{maxB} > f_{max0}$, we assume $f_{max} = f_{maxB}$, while if $f_{maxB} < f_{max0}$, we assume $f_{max} = f_{max0}$.

The relationship to be used for the determination of the radius of relative curvature perpendicular to the contact line at virtual cylindrical gears, ρ_{rel} (it is used for contact stress calculation) is as follows:

$$\rho_{rel} = |\rho_t| \cos^2\beta_B, \tag{4.38}$$

where $|\rho_t|$ is the absolute value of relative radius of profile curvature between pinion and wheel (see next section, Method $B2$), and β_B is the inclination angle of contact line, given by:

$$\beta_B = \arctan(\tan \beta_v \sin \alpha_e), \tag{4.39}$$

with $\alpha_e = \alpha_{eD}$ and $\alpha_e = \alpha_{eC}$ for drive side and, respectively, for coast side.

The radii of relative curvature in normal section at the mean point, ρ_t, for drive side and respectively for coast side are given by:

$$\rho_t = \left[\frac{1}{\cos \alpha_{nD}(\tan \alpha_{nD} - \tan \alpha_{\lim}) + \tan \zeta_{mp} \tan \beta_B} \right. $$
$$\left. \times \frac{\cos \beta_{m1} \cos \beta_{m2}}{\cos \zeta_{mp}} \left(\frac{1}{R_{m2} \tan \delta_2} + \frac{1}{R_{m1} \tan \delta_1} \right) \right]^{-1} \tag{4.40}$$

$$\rho_t = \left[\frac{1}{\cos \alpha_{nC}\{\tan(-\alpha_{nC}) - \tan \alpha_{\lim}\} + \tan \zeta_{mp} \tan \beta_B} \right. $$
$$\left. \times \frac{\cos \beta_{m1} \cos \beta_{m2}}{\cos \zeta_{mp}} \left(\frac{1}{R_{m2} \tan \delta_2} + \frac{1}{R_{m1} \tan \delta_1} \right) \right]^{-1}. \tag{4.41}$$

For tip contact line, middle contact line (in this case, its length is indicated with l_{bm}) and root contact line, Eqs. (4.29–4.39) shall be calculated with values of f_t, f_m, and f_r summarized in Table 4.2.

It is to be noted that, due to the symmetry of the zone of contact with respect to the point M, two contact lines with the distance $\pm f$ have the same length. Therefore,

Table 4.2 Distance f of the tip, middle and root contact line from the center point of the zone of action

		Surface durability	Tooth root strength
$\varepsilon_{v\beta} = 0$	f_t	$-(p_{vet} - 0.5 p_{vet} \varepsilon_{v\alpha}) \cos \beta_{vb}$ $+ p_{vet} \cos \beta_{vb}$	$(p_{vet} - 0.5 p_{vet} \varepsilon_{v\alpha}) \cos \beta_{vb}$ $+ p_{vet} \cos \beta_{vb}$
	f_m	$-(p_{vet} - 0.5 p_{vet} \varepsilon_{v\alpha}) \cos \beta_{vb}$	$(p_{vet} - 0.5 p_{vet} \varepsilon_{v\alpha}) \cos \beta_{vb}$
	f_r	$-(p_{vet} - 0.5 p_{vet} \varepsilon_{v\alpha}) \cos \beta_{vb}$ $- p_{vet} \cos \beta_{vb}$	$(p_{vet} - 0.5 p_{vet} \varepsilon_{v\alpha}) \cos \beta_{vb}$ $- p_{vet} \cos \beta_{vb}$
$0 < \varepsilon_{v\beta} < 1$	f_t	$-(p_{vet} - 0.5 p_{vet} \varepsilon_{v\alpha}) \cos \beta_{vb}(1 - \varepsilon_{v\beta})$ $+ p_{vet} \cos \beta_{vb}$	$(p_{vet} - 0.5 p_{vet} \varepsilon_{v\alpha}) \cos \beta_{vb}(1 - \varepsilon_{v\beta})$ $+ p_{vet} \cos \beta_{vb}$
	f_m	$-(p_{vet} - 0.5 p_{vet} \varepsilon_{v\alpha}) \cos \beta_{vb}(1 - \varepsilon_{v\beta})$	$(p_{vet} - 0.5 p_{vet} \varepsilon_{v\alpha}) \cos \beta_{vb}(1 - \varepsilon_{v\beta})$
	f_r	$-(p_{vet} - 0.5 p_{vet} \varepsilon_{v\alpha}) \cos \beta_{vb}(1 - \varepsilon_{v\beta})$ $- p_{vet} \cos \beta_{vb}$	$(p_{vet} - 0.5 p_{vet} \varepsilon_{v\alpha}) \cos \beta_{vb}(1 - \varepsilon_{v\beta})$ $- p_{vet} \cos \beta_{vb}$
$\varepsilon_{v\beta} \geq 1$	f_t	$p_{vet} \cos \beta_{vb}$	$p_{vet} \cos \beta_{vb}$
	f_m	0	0
	f_r	$-p_{vet} \cos \beta_{vb}$	$-p_{vet} \cos \beta_{vb}$

the sum of the length of the tip, middle and root contact lines is independent of the sign of f. In this case, $f_{t(pitting)} = -f_{r(toothroot)}$, $f_{m(pitting)} = -f_{m(toothroot)}$, and $f_{r(pitting)} = -f_{t(toothroot)}$. Together with the symmetry of load distribution (see next Chapter, Fig. 5.2), this leads to define a load sharing factor for pitting, Z_{LS} (see Sect. 5.3), and a load sharing factor for tooth root strength, $Y_{LS} = Z_{LS}^2$ (see Sect. 6.3.5).

Finally, the following quantities of virtual cylindrical gears in normal section are of interest here, i.e.: numbers of teeth, $z_{vn1,2}$; reference diameters, $d_{vn1,2}$; tip diameters $d_{van1,2}$; base diameters $d_{vbn1,2}$; profile contact ratio, $\varepsilon_{v\alpha n}$. They are given, respectively, by the following equations:

$$z_{vn1} = \frac{z_{v1}}{\cos^2 \beta_{vb} \cos \beta_v} \tag{4.42}$$

$$z_{vn2} = u_v z_{vn1} \tag{4.43}$$

$$d_{vn1,2} = \frac{d_{v1,2}}{\cos^2 \beta_{vb}} = z_{vn1,2} m_{mn} \tag{4.44}$$

$$d_{van1,2} = d_{vn1,2} + d_{va1,2} - d_{v1,2} = d_{vn1,2} + 2h_{am1,2} \tag{4.45}$$

$$d_{vbn1,2} = d_{vn1,2} \cos \alpha_e = z_{vn1,2D,C} m_{mn} \cos \alpha_e \tag{4.46}$$

$$\varepsilon_{v\alpha n} = \frac{\varepsilon_{v\alpha}}{\cos^2 \beta_{vb}}. \tag{4.47}$$

It is to be noted that, for hypoid gears, the quantities z_{vn}, d_{van}, and d_{vbn} should be calculated separately for drive and coast flanks. This is because hypoid gears with different effective pressure angles for drive and coast sides have different virtual cylindrical gears in normal section.

4.3 Calculation of Virtual Cylindrical Gears with Method B2

In Sect. 4.1, we already saw that the sub-method $B2$ provides separate sets of formulae for bevel and hypoid gears. Below we provide the main geometric relationships of the virtual cylindrical gears, which are required for calculations of load carrying capacity of bevel and hypoid gears. The initial data of bevel gears necessary for virtual cylindrical gear calculations are those described in Vol. 1, Sect. 12.9 and following sections. The base unit used in the calculations is the outer transverse module of wheel, m_{et2}.

The relative face width, b_v, and relative mean back cone distance, $R_{mpt1,2}$, of virtual cylindrical gears (the term *relative* is used here to define quantities divided by the outer transverse module of the gear wheel, m_{et2}, which is assumed as a base unit; it therefore indicates dimensionless quantities) are given by:

$$b_v = b_2/m_{et2} \tag{4.48}$$

$$R_{mpt1,2} = \frac{R_{m1,2} \tan \delta_{1,2}}{m_{et2}}. \tag{4.49}$$

Only for hypoid gears, the angle between the direction of contact and pitch tangent is given by:

$$\cot(\zeta_R - \lambda) = \cot\zeta_R\left(1 + \frac{z_1 \cos \delta_{f2}}{z_2 \cos \delta_{a1} \cos \zeta_R}\right), \tag{4.50}$$

where, the meaning of symbols already introduced remaining unchanged, λ is the first auxiliary angle shown in Vol. 1, Fig. 12.35, given by Eq. (12.198) of the same Vol. 1, and ζ_R is the pinion offset angle in root plane, given by:

$$\zeta_R = \arcsin\left(\frac{a\cos\varphi_R \sin \delta_{f2}}{R_{m2} \cos \vartheta_{f2} - t_{z2} \cos \delta_{f2}}\right) - \varphi_R. \tag{4.51}$$

In this equation, the auxiliary angle φ_R for calculation of ζ_R is given by:

$$\varphi_R = \arctan\left(\frac{a \tan \Delta\Sigma \cos \delta_{f2}}{R_{m2} \cos \vartheta_{f2} - t_{z2} \cos \delta_{f2}}\right), \tag{4.52}$$

where $\Delta\Sigma = \Sigma - 90°$ is the shaft angle departure from $90°$.

For bevel gears and for hypoid gears, the face contact ratio, $\varepsilon_{v\beta}$, is given respectively by:

$$\varepsilon_{v\beta} = \frac{b_2}{\pi m_{mn}} \sin \beta_{m2} \tag{4.53}$$

$$\varepsilon_{v\beta} = \frac{b_2}{\pi m_{mn}}\left[\frac{\cos \beta_{m2}}{\cot(\zeta_R - \lambda)} + \sin \beta_{m2}\right]. \tag{4.54}$$

The relative mean virtual pitch radius, $r_{vn1,2}$, relative center distance, a_{vn}, relative mean virtual dedendum, $h_{vfm1,2}$, relative virtual tooth thickness, $s_{vmn1,2}$, and relative mean virtual tip radius, $r_{va1,2}$, are respectively given by:

$$r_{vn1,2} = \frac{R_{mpt1,2}}{\cos^2 \beta_{m1,2}} \tag{4.55}$$

$$a_{vn} = r_{vn1} + r_{vn2} \tag{4.56}$$

$$h_{vfm1,2} = h_{fm1,2}/m_{et2} \tag{4.57}$$

$$s_{vmn1,2} = s_{mn1,2}/m_{et2} \tag{4.58}$$

$$r_{va1,2} = r_{vn1,2} + \left(h_{am1,2}/m_{et2}\right). \tag{4.59}$$

The angular pitch of virtual cylindrical wheel, ϑ_{v2}, and the relative edge radius of tool, $\rho_{va01,2}$, can be expressed respectively as:

$$\vartheta_{v2} = \frac{\pi m_{mn}}{m_{et2} r_{vn2}} \tag{4.60}$$

$$\rho_{va01,2} = \rho_{a01,2}/m_{et2}. \tag{4.61}$$

For bevel gears and for hypoid gears, the virtual spiral angles, β_v, are given respectively by:

$$\beta_v = \beta_{m2} \tag{4.62}$$

$$\beta_v = \beta_{m1} - \lambda_r, \tag{4.63}$$

where

$$\tan(\beta_{m1} - \lambda_r) = \frac{R_{m2} \tan \delta_{f2} \tan \beta_{m1} + R_{m1} \tan \delta_{a1} \tan \beta_{m2}}{R_{m2} \tan \delta_{f2} + R_{m1} \tan \delta_{a1}}. \tag{4.64}$$

The adjusted pressure angle, α_a, i.e. the pressure angle of a tooth slot at the pitch line, when this is not passing through the wheel apex, due to tilting (see Vol. 1, Sect. 12.9 and following sections), is given by:

$$\alpha_a = \alpha_{eD} - (90° \cos \delta_2 \cos \beta_{m2}/z_2), \tag{4.65}$$

while the *base virtual helix angle*, β_{vb}, is given by:

$$\sin \beta_{vb} = \sin \beta_v \cos \alpha_a. \tag{4.66}$$

The relative mean virtual base radius, $r_{vbn1,2}$, relative length of path of contact from pinion tip to pitch circle in normal section, g_{vana}, relative length of path of contact from wheel tip to pitch circle in normal section, g_{vanr}, relative length of path of contact in normal section, g_{van}, and relative mean normal pitch of virtual cylindrical gear, p_{mn}, are given respectively by:

$$r_{vbn1,2} = r_{vn1,2} \cos \alpha_a \tag{4.67}$$

$$g_{v\alpha na} = \sqrt{r_{va1}^2 - r_{vbn1}^2} - r_{vn1} \sin \alpha_a \tag{4.68}$$

$$g_{v\alpha nr} = \sqrt{r_{va2}^2 - r_{vbn2}^2} - r_{vn2} \sin \alpha_a \tag{4.69}$$

$$g_{v\alpha n} = g_{v\alpha na} + g_{v\alpha nr} \tag{4.70}$$

$$p_{mn} = \frac{2\pi m_{mn}}{m_{et2} \cos \alpha_a \left(\cos^2 \beta_{m1} + \cos^2 \beta_{m2} + 2 \tan^2 \alpha_a\right)}. \tag{4.71}$$

The profile contact ratio of virtual cylindrical gear in mean normal section, $\varepsilon_{v\alpha n}$, and the transverse contact ratio in mean transverse section of the same virtual cylindrical gear, $\varepsilon_{v\alpha}$, are given respectively by:

$$\varepsilon_{v\alpha n} = g_{v\alpha n} / p_{mn} \tag{4.72}$$

$$\varepsilon_{v\alpha} = \varepsilon_{v\alpha n} \cos^2 \beta_{vb}. \tag{4.73}$$

Finally, the modified contact ratio for bevel gears without hypoid offset is given by:

$$\varepsilon_{v\gamma} = \sqrt{\varepsilon_{v\alpha}^2 + \varepsilon_{v\beta}^2}. \tag{4.74}$$

The tooth loading conditions related to Method $B2$ are different from those described for Method $B1$ in the previous section. In fact, in this respect, ISO standards assume that the zone of action for a bevel or hypoid gear pair is defined as a rectangle for the corresponding virtual cylindrical gears, as Fig. 4.5a shows. This rectangle has the same face width, $b_v = b_2$, and the same length, $g_{v\alpha}$, of the mean line of action.

The theoretical analyses, numerical models, and experimental measurements show that, due to deflections under load, the actual contact surface between the meshing teeth of a bevel gear pair configures an elliptical contact pattern, with the ellipse of contact tangent to the four boundary sides of the zone of action, as Fig. 4.5b shows. When the modified contact ratio, $\varepsilon_{v\gamma}$, is less than two or greater than two, the maximum load occurs at the outer point of single pair tooth contact (the point marked in Fig. 4.5c) and, respectively, at the center of the ellipse of contact (the point marked in Fig. 4.5d).

According to the Hertz theory, the pressure distribution along the major and minor axes of the ellipse of contact is represented by a prism with semi-elliptical cross section (see Fig. 2.2). Figure 4.6a shows this type of semi-ellipsoid-shaped distribution. Considering the effect of the compressive stress resulting from adjacent teeth in contact, the lines of contact and contour lines in the zone of contact are modified, as Fig. 4.6b shows. Finally, it is to be noted that, for use in calculations,

Fig. 4.5 Tooth loading:
a zone of action; **b** elliptical
contact pattern; **c** modified
contact ratio less than two;
d modified contact ratio
greater than two

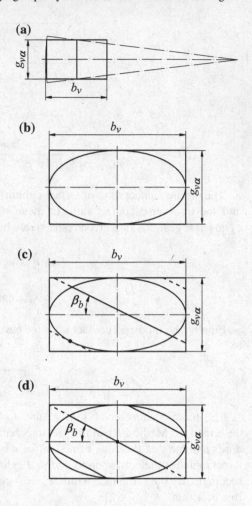

the distributed load is replaced by a concentrated load applied at a given point, called point of load application. The position of this point is shifted towards the heel, as Fig. 4.6c shows. It is calculated by means of a contact shift factor, k', given by the following empirical formula, based on etching tests:

$$k' = \frac{z_2 - z_1}{3.2 z_2 + 4 z_1}.$$ (4.75)

This factor, k', is used in both the contact stress and tooth root calculations. For the pitting resistance calculation, this factor is inserted in an iterative procedure to locate the point of highest compressive stress.

Fig. 4.6 Load distribution:
a Semi-ellipsoid, and
semi-elliptical stress
distributions; **b** Effect of
compressive stress resulting
from adjacent teeth; **c** Point
of load application, and
contact shift factor

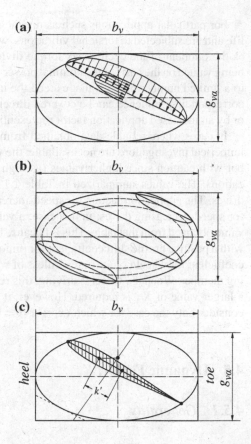

4.4 Application Factor, K_A

The application factor, K_A, takes account of the increase of the nominal tangential
force, F_{mt}, calculated with Eq. (4.1), due to external dynamic actions exerted by
the driving and driven machines on the gear set under consideration. Many possible
sources of dynamic overload are to be considered, including: vibrations of the gear
system, and critical speeds; acceleration torques, and overspeeds; braking, and sudden variations in gear system operation; negative torques, such as those produced by
retarders on vehicles, which result in loading the non-operating flanks of the gear
teeth; etc.

About the determination of values of the application factors, we have nothing
to add to what we already mentioned in Sect. 1.2, in respect to spur and helical
gears. For bevel and hypoid gears, the same considerations made in that case shall be
valid. Therefore, the values of the application factors are best determined by a thor-
ough analysis of service experience on specific applications and, when this service
experience is not available, by a through analytical and/or numerical investigations.

For particular applications such as marine gears, which are designed for infinite life and are subjected to torsional vibrations, with cyclic peak torques, K_A is defined as the quotient of the cyclic peak torque divided by the nominal torque, the latter being valued on the basis of the nominal power and speed. If then the gear is subjected to a limited number of loads that exceed the intensity of the cyclic peak torque, the corresponding influence can be covered directly by the cumulative fatigue analysis or by an increased application factor representing the influence of the load spectrum.

In cases where reliable data obtained from service experience or analytical and numerical investigations are not available, the values shown in Table 1.1 can be used, but with caution since higher values (as high as 10) have occurred in some applications. The values summarized in Table 1.1 are appropriate for speed-decreasing drives. The application factors for speed-increasing drives must be larger than those for speed-decreasing drives; in this case, a value of $0, 01u^2$ should be added to K_A value obtained from that table. This is because bevel gears are nearly always designed with a positive profile shift coefficient for pinion ($x_1 > 0$) and a negative profile shift coefficient for wheel ($x_2 < 0$), regardless of whether the pinion or wheel is the driving member. When the wheel is driving, this results in an approach action, for which a larger value of K_A is required. However, it is to be noted that the ISO standards consider only the case for which $(x_1 + x_2) = 0$.

4.5 Dynamic Factor, K_v

4.5.1 Generality

The dynamic factor, K_v, considers the effects of the gear tooth quality related to speed and load as well as other factors, such as design and manufacturing parameters, transmission error, dynamic response and resonance conditions. This factor relates the total load on the tooth, including internal dynamic effects, to the transmitted tangential load. It is expressed by the quotient of the sum of the effective internal dynamic load and the transmitted tangential tooth load divided by the transmitted tangential tooth load.

The design parameters include the tooth load, pitch line velocity, inertia and stiffness of the rotating members, tooth stiffness variation, stiffness of bearings and case structure, critical speeds and internal vibrations within the gears, and lubricant properties. Instead, the manufacturing parameters include the tooth flank and tooth spacing variations, runout of pitch surfaces with respect to the axis of rotation, balance of parts, compatibility of mating gear tooth elements, and bearing fit and preload.

As regards the transmission error, it is noteworthy that, even if the input torque and speed are constant, significant vibration of the gear masses and resulting dynamic tooth forces can occur. These dynamic forces result from the relative displacements between the mating gear wheels since they vibrate in response to an excitation known as transmission error, we have already described in Sect. 1.3.1, to which the reader is referred (see also Gosselin et al. 1995; Simon 2009). However, as regards the mesh stiffness variations during the meshing cycle of the gear, it is to borne in mind that these stiffness variations are a source of excitation especially pronounced in straight and Zerol bevel gears, while spiral bevel gears with a total contact ratio greater than two have less stiffness variations (see Litvin and Fuentes 2004; Wang et al. 2007).

As well as for the dynamic response and resonance conditions, we have nothing to add to what we said in Sect. 1.3.1, to which the reader is also referred (see von Kármán and Biot 1940; Biezeno and Grammel 1953; Ker Wilson 1956, 1963; Krall 1970a, b; Warburton 1976; Den Hartog 1985).

From a general point of view, as regards the calculation method of K_v, it is to consider that a bevel gear drive is a very complicated vibrating system. The natural frequencies of this system, which cause dynamic loads on the teeth and shafts, cannot be determined considering only the single gear pair. The alignment of the pinion shaft may change considerably depending on the assembly characteristics, backlash, and elastic deformations of gear wheels, shafts, bearings, and housing. A slight variation in alignment alters the angle of relative rotation of the gear wheels, and therefore the dynamic loads acting on them. Crowning in the lengthwise and profile directions may render ineffective the actual conjugate action, and can make difficult to determine the tooth accuracy.

In light of these issues, reliable values of the dynamic factor, K_v, may best be determined using appropriate mathematical models validated by suitable experimental measurements. Therefore, if the dynamic load calculated with these methods is added to the nominal transmitted load, the dynamic factor may be set equal to unity. Here too, as we did in Sect. 1.3.1, the several methods used to determine K_v are indicated in descending precision order, from Method A (K_{v-A}) to Method C (K_{v-C}).

It is to be noted that, when we use Method B or Method C for hypoid gears characterized by a great amount of hypoid offset, the dynamic factor may be set equal to unity. This is justified by the damping properties related to the sliding conditions in mesh of this type of gears. Instead, for smaller amounts of hypoid offset, the value of the dynamic factor is obtained by interpolation between the unit value and the value calculated as for spiral bevel gears without offset. As we saw elsewhere (see Vol. 1, Sect. 12.9 and following sections), the upper limit of the hypoid offset should not exceed $0.25\, d_{e2}$, due to lengthwise sliding (for heavy-duty applications, it should be limited to half of this value). Instead, the lower limit of the hypoid offset value is assumed to be equal to $0.05\, d_{m2}$. In relative terms, this value corresponds to $a_{rel} = 0.1$, where a_{rel} is the relative hypoid offset, given by:

$$a_{rel} = \frac{2|a|}{d_{m2}}. \tag{4.76}$$

The value of the dynamic factor, K_v, must satisfy the following inequality:

$$K_v = K_v^* - \frac{K_v^* - 1}{0,1} a_{rel} \geq 1, \tag{4.77}$$

where K_v^* is the preliminary dynamic factor for non-hypoid gears. We assume $K_v^* = K_{v-B}$ or $K_v^* = K_{v-C}$, depending on whether the Methods B (Sect. 4.5.3) or C (Sect. 4.5.4) are used.

4.5.2   Method A for Determination of K_{v-A}

With this method, the dynamic factor, K_{v-A}, is determined by a comprehensive analysis of the entire power transmission vibrational system, and using the following procedure:

- A mathematical model of the vibrating system, consisting of the gear drive, driving and driven machines, and interconnecting shafts and couplings, is first developed.
- Then, the transmission error of the bevel gear under load is measured, or calculated by a reliable mathematical model, appropriate for this type of analysis.
- Finally, the dynamic response under load of the pinion and gear shafts is analyzed with the aforementioned general system model, excited by this transmission error.

In any case, results obtained with this comprehensive analysis must be confirmed by experience of similar designs.

4.5.3   Method B for Determination of K_{v-B}

With this method, the dynamic factor, K_{v-B}, is determined under the simplifying assumption that the bevel gear pair constitutes an elementary single mass and spring vibrating system, where the mass is the combined mass, or reduced mass, of pinion and wheel, while the spring stiffness is the meshing stiffness of the contacting teeth. According to this assumption, K_{v-B} takes no account of the forces due to torsional vibrations of the shafts and coupled masses. This approximation is realistic only if the other masses (with the exception of those of the gear pair) are connected by means of shafts having relatively low torsional stiffness. For bevel gears having significant lateral flexibility of the shafts, the actual natural frequency will be less than that calculated.

Among other effects, the intensity of dynamic overloads is a function of the gear accuracy, that is of deviations of the flank shape and pitch. Furthermore, in the case of bevel gears, the flank shape deviations are not simple to measure, as it happens for the involute profiles of cylindrical gears, and the ISO tolerances do not exist. Some reference on the single-flank composite tolerances can be found in ISO 17485:2006, and the transmission error of a bevel gear set should be verified according to them, provided that a proper measuring equipment is available. On the other hand, the pitch deviations can be measured with relative simplicity. Therefore, in these cases, the simplifying assumption is made that the single pitch deviation is a representative value of the transmission error for determination of the dynamic factor.

In this framework, the following data are required for determination of K_{v-B}: gear pair accuracy, in terms of single pitch deviation; tooth stiffness; mass moment of inertia of pinion and wheel, including dimensions and material density; transmitted tangential load.

4.5.3.1 Speed Ranges and Resonance Speed

Even for bevel gears we introduce a dimensionless reference speed, N (also called *frequency ratio*, see Eq. (1.8)), defined as the quotient of the pinion rotational speed, n_1, divided by the *resonance speed* of the same pinion, n_{E1}, that is as:

$$N = \frac{n_1}{n_{E1}}, \tag{4.78}$$

where n_{E1} is given by:

$$n_{E1} = \frac{30 \times 10^3}{\pi z_1} \sqrt{\frac{c_\gamma}{m_{red}}}. \tag{4.79}$$

In this last equation, z_1 is the number of teeth of pinion, c_γ is the mean value of mesh stiffness for unit face width (in N/mm μm), and m_{red} is the mass for unit face width reduced to the line of action of dynamically equivalent cylindrical gear (in kg/mm). The comparison of Eqs. (4.78) and (4.79) with the corresponding Eqs. (1.8) and (1.7) shows that the numerical coefficient 10^3 is a factor that homogenizes the units of measurement in the SI-System. The reduced mass, m_{red}, is given by:

$$m_{red} = \frac{m_1 m_2}{m_1 + m_2}, \tag{4.80}$$

where m_1 and m_2 (both in kg/mm) are the individual gear masses per unit face width of pinion and wheel, referred to line of action.

Here also, in relation to the reference speed, N, the total range of operating speed can be divided into four ranges: subcritical range, for $N \leq 0.75$; main resonance range, for $(0.75 < N \leq 1.25)$; intermediate range, for $(1.25 < N < 1.50)$; supercritical range, for $N \geq 1.50$. The need to define a resonance range between $(0.75 < N \leq 1.25)$ is due to reasons of safety, since Eq. (4.79) does not consider the influences of the stiffnesses of shafts, bearings and gearbox as well as of the damping, so the resonance speed can be above or below the value calculated with this equation. The dynamic factor, K_v, is determined with Methods A and B for subcritical, intermediate, and supercritical ranges. For the main resonance range, operation should be avoided, but if unavoidable, a refined analysis is necessary, to be carried out with the Method A.

The mean value of mesh stiffness per unit face width, c_γ, to be introduced in Eq. (4.79) is determined by the relationship:

$$c_\gamma = c_{\gamma 0} C_F, \tag{4.81}$$

where $c_{\gamma 0}$ is the mesh stiffness for average conditions, for which a value of 20 N/mm μm is recommended, and C_F is the correction factor for non-average conditions, to be assumed equal to: $C_F = 1$, for $F_{vmt} K_A / b_{v,eff} \geq 100$ N/mm, and $C_F = \left(F_{vmt} K_A / b_{v,eff} \right) / 100$ (it should be noted that C_F is a dimensionless quantity, so that the numerical factor 100 is to be understood as a dimensional quantity, to be expressed in N/mm), for $F_{vmt} K_A / b_{v,eff} < 100$ N/mm, where F_{vmt} is the nominal tangential force of virtual cylindrical gears.

The recommended value $c_{\gamma 0} = 20 \text{N}/(\text{mm μm})$ of mesh stiffness for average conditions applies to spur gears. For helical gears, the stiffness decreases with increasing of helix angle. On the other hand, the spiral arrangement of the curved teeth around a conical blank determines a higher stiffness of the spiral bevel gears (of course, straight bevel gears are an exception). Therefore, in the absence of any better knowledge in this regard, we assume that, in average conditions given by $F_{vmt} K_A / b_{v,eff} \geq 100$ N/mm and $b_{v,eff} / b_v \geq 0.85$, the bevel gear stiffness is equal to that characterizing the spur gears.

To achieve the desired shape and position of the tooth contact pattern, it is necessary to use a tooth contact development process that, by means of subsequent modifications and refinements, is able to ensure the optimum tooth contact pattern. This process is based on the use of a suitable testing machine of bevel gears, which allow to modify the relative position of pinion and wheel according to three directions (along the pinion and wheel axes, and perpendicular to both axes) and to operate the bevel gear pair to a reasonable speed under light load.

With study and analysis of the response, which correlates the tooth contact pattern with the displacements in the three aforementioned directions, the optimum tooth contact pattern is detected and, based on this, appropriate adjustments in the cutting or grinding machine setting are defined, which are able to ensure the desired contact pattern with the bevel gear pair assembled in the gearbox according to design.

The purpose of these study and analysis is to have a tooth contact pattern under actual load as much as possible developed in the lengthwise and profile directions, and centered with respect to the centerline of face width and tooth depth. In this way, areas that are not loaded at the tooth edges and tip and root reliefs are limited. In addition, load concentrations are avoided, which would have in the following cases: toe and heel contact (Fig. 4.7a and b); cross contact (Fig. 4.7c); low and high contact (Fig. 4.7d and e); lame contact (Fig. 4.7f); pitch line narrow contact (Fig. 4.7h); profile and lengthwise bridged contact (Fig. 4.7i and l); full length long and narrow contact (Fig. 4.7j); short contact (Fig. 4.7k); bias-in and bias-out contact left hand wheel (Fig. 4.7m and n).

The type of tooth contact pattern to be obtained is the centered wide contact (Fig. 4.7g), with a zone of contact extended at the most between $(85 \div 95)\%$ in both the lengthwise and profile directions, without affecting the tooth edges, and tip and root reliefs. Typical satisfactory loaded contact patterns on bevel gear flanks are as follows: idealized 80–85% coverage of lengthwise tooth surface, relief at top and edges, no concentrations; slight cross pattern, still 80–85% coverage; slight heel

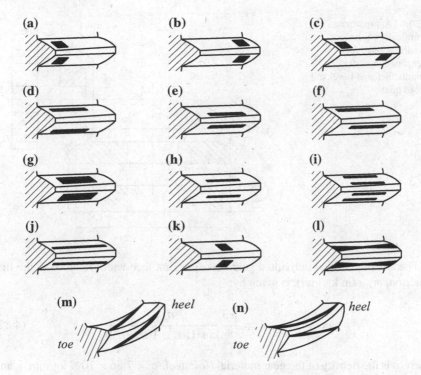

Fig. 4.7 Typical tooth contact patterns represented on the bevel wheel flanks

(or toe) pattern, still 80–85% coverage; slight lame pattern, still 80–85% coverage. Instead, typical unsatisfactory loaded contact patterns on bevel gear flanks are as follows: full length-full width, no relief at edges; lame, i.e. high contact on one side, and low contact on the other side; high contact on the heel; too much profile (or lengthwise) relief; cross contact, i.e. heel contact on one side, and toe contact on the other side; heavy toe contact on both sides.

Based on the aforesaid considerations, in the case of full load, the typical tooth contact pattern has a minimum length of 85% of the face width, b_v, of virtual cylindrical gears. In the case in which it is not possible to have information of tooth contact pattern length under load conditions, we assume $b_{v,eff} = 0.85b_v$, where $b_{v,eff}$ is the effective face width of the virtual cylindrical gears, i.e. the actual length of contact pattern.

In the design stage, where the available data are very few, or the case in which, due to cost, the determination of the mass moments of inertia for unit face width, J_1 and J_2 (see Timoshenko and Young 1951), which according to Eq. (1.9) are necessary for the calculation of masses per unit face width m_1 and m_2 is not feasible (we remember that $J_{1,2} = m_{1,2}r_{b1,2}^2$), the bevel gears of common blank design may be replaced, approximately, by dynamically equivalent cylindrical gears, as Fig. 4.8 shows. The quantities related to these cylindrical gears are designated by the subscript x.

Fig. 4.8 Approximate dynamically equivalent cylindrical gears for determination of the dynamic factor of bevel and hypoid gears

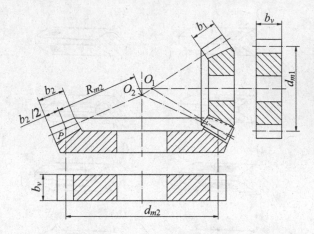

The already defined individual gear mass per unit face width reduced to the line of action, $m_{1,2}$ (in kg/mm), is given by:

$$m_{1,2} = m_{1x,2x} = \frac{\pi \rho}{8} \frac{d_{m1,2}^2}{\cos^2[(\alpha_{nD} + \alpha_{nC})/2]}, \qquad (4.82)$$

where ρ is the density of the gear material (for steel, $\rho = 7.86 \times 10^{-6}$ kg/mm^3), and α_{nD} and α_{nC} are the generated pressure angles for drive side and coast side.

4.5.3.2   Dynamic Factor K_{v-B} in Subcritical Range ($N \leq 0.75$)

For the subcritical range, which is the common operating range for industrial and vehicle gear drives, the calculation of K_{v-B} is made using the following relationship:

$$K_{v-B} = NK + 1, \qquad (4.83)$$

where K is a constant factor. This factor, by virtue of the simplifying assumptions described at the beginning of Sect. 4.5.3, can be expressed as follows:

$$K = \frac{b_v f_{p,eff} c'}{F_{vmt} K_A} c_{v1,2} + c_{v3}, \qquad (4.84)$$

where

$$f_{p,eff} = f_{pt} - y_p, \qquad (4.85)$$

with $y_p \cong y_\alpha$. In these equations, the meaning of symbols already introduced remains unchanged, while the other symbols represent: c', the single stiffness (in N/mm μm);

Table 4.3 Influence factors c_{v1} to c_{v7}

Influence factor	$1 < \varepsilon_{v\gamma} \leq 2$	$\varepsilon_{v\gamma} > 2$	
c_{v1}	0.32	0.32	$c_{v1,2} = c_{v1} + c_{v2}$
c_{v2}	0.34	$0.57/(\varepsilon_{v\gamma} - 0.30)$	
c_{v3}	0.23	$0.096/(\varepsilon_{v\gamma} - 1.56)$	
c_{v4}	0.90	$(0.57 - 0.05\varepsilon_{v\gamma})/(\varepsilon_{v\gamma} - 1.44)$	
c_{v5}	0.47	0.47	$c_{v5,6} = c_{v5} + c_{v6}$
c_{v6}	0.47	$0.12/(\varepsilon_{v\gamma} - 1.74)$	
–	$1 < \varepsilon_{v\gamma} \leq 1.5$	$1.5 < \varepsilon_{v\gamma} \leq 2.5$	$\varepsilon_{v\gamma} > 2.5$
c_{v7}	0.75	$0.125\sin[\pi(\varepsilon_{v\gamma} - 2)] + 0.875$	1

$f_{p,eff}$, the effective pitch deviation (in μm); f_{pt}, the transverse single pitch deviation (in μm); y_p, the running-in allowance for pitch deviation related to the polished test piece (in μm), which generally is assumed equal to the running-in allowance for pitch error, y_α (in μm). For determination of f_{pt} and $y_p \cong y_\alpha$, we refer the reader to the Sect. 4.7.3.

It is to be noted that Eqs. (4.83–4.85) do not take account of the possible influence of tip relief or profile crowning, for which they provide precautionary values, i.e. on the safe side, for bevel gears which normally have profile crowning. The influence factors c_{v1}, c_{v2} ($c_{v1,2} = c_{v1} + c_{v2}$ is a combined factor) and c_{v3} are empirical parameters to determine the dynamic factor, which consider, respectively, the pitch deviation effects (c_{v1} is assumed to be a constant), tooth profile deviation effects, and cyclic variation effect in mesh stiffness. The values of these empirical influence factors, and other similar factors up to c_{v7} (they are all dependent on the variability range of $\varepsilon_{v\gamma}$), can be taken from Table 4.3.

The single stiffness, c', is determined using the following relationship:

$$c' = c_0' C_F, \tag{4.86}$$

where c_0' is the single stiffness for average conditions, whose recommended value is $c_0' = 14$ N/mm μm, and C_F is a dimensionless correction factor for non-average conditions. Here also, we note that, for spur gears, a value of $c_0' = 14$ N/mm μ m applies. For helical gears, the tooth stiffness decreases with increasing of helix angle. On the other hand, the spiral arrangement of the curved teeth around a conical blank determines a higher stiffness of the spiral bevel gears (straight bevel gears are an exception). Therefore, in the absence of any better knowledge in this regard, we assume that, in average conditions given by $F_{vmt} K_A/b_{v,eff} \geq 100$ N/mm and $b_{v,eff}/b_v \geq 0.85$, the bevel gear stiffness is equal to that characterizing the spur gears.

4.5.3.3   Dynamic Factor K_{v-B} in Main Resonance Range ($0.75 < N \leq 1.25$)

For the main resonance range, which is possibly to be avoided, the calculation of K_{v-B} is made using the following relationship:

$$K_{v-B} = \frac{b_v f_{p,eff} c'}{F_{vmt} K_A} c_{v1,2} + c_{v4} + 1, \tag{4.87}$$

where c_{v4} is an empirical influence factor that considers resonant torsional oscillations of the bevel gear pair, excited by cyclic variation of the mesh stiffness (see Ker Wilson 1963). The value of this factor can be taken from Table 4.3. Equation (4.87) is also based on the simplifying assumptions described at beginning of Sect. 4.5.3.

4.5.3.4   Dynamic Factor K_{v-B} in Supercritical Range ($N \geq 1.5$)

In the supercritical range, where high-speed bevel gears and bevel gears with similar requirements operate, for calculation of K_{v-B}, we use the following relationship:

$$K_{v-B} = \frac{b_v f_{p,eff} c'}{F_{vmt} K_A} c_{v5,6} + c_{v7} + 1, \tag{4.88}$$

where the influence factors c_{v5}, c_{v6} ($c_{v5,6} = c_{v5} + c_{v6}$ is a combined factor) and c_{v7} are empirical parameters that take into account, respectively, the pitch deviation effects, tooth profile deviation effects, and component of force which, due to mesh stiffness variation, is derived from tooth bending deflections during substantially constant speed. The values of these factors (it is to be noted that c_{v5} and c_{v6} correspond respectively to c_{v1} and c_{v2} for subcritical range) can be taken from Table 4.3.

4.5.3.5   Dynamic Factor K_{v-B} in Intermediate Range ($1.25 < N < 1.50$)

Factor K_{v-B} in intermediate range is determined by linear interpolation between the values that limit, to the right, the main resonance range ($N = 1.25$), and to the left, the supercritical range ($N = 1.50$). Therefore, we use the following relationship:

$$K_{v-B} = K_{v-B(N=1.50)} + \frac{K_{v-B(N=1.25)} - K_{v-B(N=1,50)}}{0.25}(1.50 - N). \tag{4.89}$$

4.5.4 Method C for Determination of K_{v-C}

When specific knowledge of the dynamic loads acting on the bevel gear teeth are not available, for determination of dynamic factor, K_{v-C}, the curves of the diagram in Fig. 4.9 can be used or, alternatively, the relationships below, which are correlated and consistent with the same curves. Both curves in the diagram and correlated relationships are based on empirical data, and do not consider the resonance. Both from curves and correlated relationships, it is evident that the dynamic factor, K_{v-C}, is a function of the tangential velocity of the wheel at the outer pitch diameter, v_{et2}, and depends on the ISO accuracy grade, B.

Due to the approximation of the empirical curves, and the lack of measured values of tolerances at the design stage, it is good to base the choice of the curve in the diagram on experiences gained in the manufacturing processes, also considering the operating conditions that affect the design (see what we previously said in this regard). In most cases, it is then useful to compare the tooth contact pattern on the tooth flank with that obtained from previous experiences.

The choice of curves of ISO accuracy grades $B = 5$ to $B = 8$, or the hatched area, corresponding to *very accurate gearing* (these gears are characterized by an ISO accuracy grade $B < 5$), should be made on the basis of the transmission error and, when the transmission error is not available, on the basis of the tooth contact pattern on the tooth flank. In the case in which the tooth contact pattern on each tooth flank is not uniform, the pitch accuracy (i.e. the single pitch deviation) can be taken as a representative value to determine the dynamic factor.

For very accurate gearing ($B < 5$) or in cases in which design, manufacturing and application experience ensure a low transmission error, values of K_{v-C} between 1.0 and 1.1 can be used, depending on the accuracy grade actually achieved, and the experience with similar applications. These values can be used only if the gears are maintained with accurate alignment and adequate lubrication so that their overall accuracy is maintained under the operating conditions.

Fig. 4.9 Dynamic factor, K_{v-C}, as a function of the tangential velocity of the wheel at the outer pitch diameter, v_{et2}, and ISO accuracy grade, B

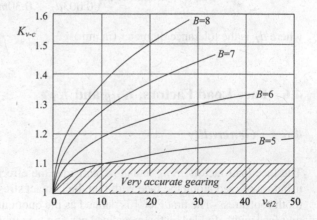

The empirical curves $B = 5$ to $B = 8$ shown in Fig. 4.9 are obtained using the following relationship:

$$K_{v-c} = \left(\frac{A}{A + \sqrt{200 v_{et2}}} \right)^{-X},$$ (4.90)

where

$$A = 50 + 56(1 - X)$$ (4.91)

$$X = 0.25(B - 4)^{2/3}$$ (4.92)

$$v_{et2} = v_{mt2}(d_{e2}/d_{m2}).$$ (4.93)

In these equations, to be used with the limitations $(5 \leq B \leq 8)$, $(1.25 \leq m_{mn} \leq 50)$, and $(6 \leq z \leq 1200$ or $3000/m_{mn})$ whichever is less, v_{mt2} is the wheel tangential speed at mid-face width of the reference cone (in m/s), and B is the ISO accuracy grade for the actual gear set. The resulting curves can be extrapolated beyond the end points shown in Fig. 4.9, on the basis of the experience and careful consideration of the factors affecting the dynamic load. The end points of these curves are defined by the maximum recommended tangential velocity, $v_{et2,max}$, for a given accuracy grade, B, determined as follows:

$$v_{et2,max} = \frac{[A + (13 - B)]^2}{200},$$ (4.94)

Finally, the accuracy grade B can be also calculated on the basis of the single pitch deviation as follows:

$$B = 4 + 2.8854 \times \ln \left(\frac{f_{pt}}{0.003 d_T + 0.30 m_{mn} + 5} \right),$$ (4.95)

where d_T is the tolerance diameter (in mm).

4.6 Face Load Factors, $K_{H\beta}$ and $K_{F\beta}$

4.6.1 Generality

The face load factors, $K_{H\beta}$ and $K_{F\beta}$, consider the effects of the non-uniform distribution of load along the face width on the contact stress and, respectively, on the tooth root stress. The first factor is defined as the quotient of the maximum load per unit face width divided by the mean load per unit face width, while the second factor is defined as the quotient of the maximum tooth root stress divided by the mean tooth root stress along the face width. The theoretical bases for calculating these two face load factors, with the appropriate variations of the case (e.g., for calculating $K_{F\beta}$, the tooth of a straight bevel gear is similar to a cantilever trapeze plate, rather than to a cantilever rectangular plate) are the same we described in Sect. 1.4 for spur and helical gears (see Timoshenko and Woinowsky-Krieger 1959; Wellauer and Seireg 1960; Kagawa 1961; Gagnon et al. 1997; Ventsel and Krauthammer 2001).

The amount of the non-uniform load distribution is influenced by: alignment of gears in their mountings; gear tooth manufacturing accuracy, tooth contact pattern, and spacing; bearing clearances; elastic deflections of the gear teeth, shafts, bearings, housing and foundation, which supports the gear system, resulting from either the internal or external gear loads; Hertzian contact deformation of the tooth flank surfaces; thermal expansion and distortion of the gear system due to operating temperatures, which are especially important for gear systems where the gear housing is made with a material different from that used for gears, shafts, and bearings; centrifugal deflections due to operating speeds.

The geometric characteristics of the tooth of a bevel gear change along its face width. Consequently, the magnitudes of the axial and radial components of the transmitted load vary with the position of the tooth contact (see Stadtfeld 1993; Gosselin et al. 1995; Litvin and Fuentes 2004; Sheveleva et al. 2007; Kolivand and Kahraman 2009). Similarly, deflections of the mountings and teeth also vary, and in turn this affects the position, size and shape of the tooth contact patterns. For applications in which the operating torque changes, the desired contact pattern should be considered ideal only for full load conditions. For intermediate loads, a satisfactory compromise should be accepted. However, it is to be noted that the ISO standards are not applicable for the typical unsatisfactory contact pattern, which we described in Sect. 4.5.3.1.

As regards the calculation of the face load factors, $K_{H\beta}$ and $K_{F\beta}$, the methods used are still Methods A, B, and C. The first method (Method A), which is used in very few cases because of its very high cost, necessarily involves an exact determination of the load distribution along the face width, by means of a general analysis of all the influence factors as well as measurements of the tooth root stress under operating conditions. Regarding the second method (Method B), it is to be noted that an approach corresponding to it has not yet been developed. Therefore, the only actually used method is the Method C.

4.6.2 Method C for Determination of the Face Load Factors, $K_{H\beta-C}$ and $K_{F\beta-C}$

For the determination of the face load factor for contact stress, $K_{H\beta-C}$, it should be noted that the load distribution along the face width of a bevel gear is essentially influenced by the crowning of the teeth and deflections occurring under service

Table 4.4 Mounting factor, $K_{H\beta-be}$

Checking of contact pattern	Mounting conditions of pinion and wheel		
	Neither member cantilever mounted	One-member cantilever mounted	Both member cantilever mounted
For each gear set in its housing under full load	1.00	1.00	1.00
For each gear set under light test load	1.05	1.10	1.25
For a sample gear set and estimated for full load	1.20	1.32	1.50

conditions. We take account of these circumstances by the length of the line of contact, which we discussed in Sect. 4.2, as well as by the load distribution, the subject of which will be detailed in the next chapter. This calculation procedure applies as long as we have to do with gear drives characterized by satisfactory tooth contact patterns.

The face load factor, $K_{H\beta-C}$, is determined using the following relationship, which is valid only for crowned gears:

$$K_{H\beta-C} = 1.5 K_{H\beta-be}.   \tag{4.96}$$

where $K_{H\beta-be}$ is the mounting factor, which takes account of the influence of the deflections including those due to the bearing arrangements. The values of this factor can be taken from Table 4.4. These values are based on optimum tooth contact as evidenced by results of a contact pattern test on gears in their mounting conditions.

It should be noted that the observed tooth contact pattern is usually an accumulated picture of each possible combination of the bevel gear pair. Equation (4.96) is valid only for small shifts of the tooth contact pattern during one revolution of the wheel, either towards the toe or heel. Otherwise, the smallest contact pattern should be chosen for determination of $b_{v,eff}$. These shifts of single contact pattern might be particularly high for gears finishing only by lapping.

For the determination of the face load factor for bending stress, $K_{F\beta-C}$, we use the following relationship:

$$K_{F\beta-C} = \frac{K_{H\beta-C}}{K_{F0}},   \tag{4.97}$$

where K_{F0} is the lengthwise curvature factor for bending stress, which considers the contact pattern shifting under different loads. This well-known pattern shifting depends on the cutter radius, r_{c0}, and mean spiral angle of the wheel, β_{m2}, and has a minimum value if the lengthwise tooth curvature at the mean point corresponds to that of an involute curve. The value of the factor K_{F0} is determined by the following two relationships:

$$K_{F0} = 1 \tag{4.98}$$

$$K_{F0} = 0.211 \left(\frac{\rho_{m\beta}}{R_{m2}} \right)^q + 0.789, \tag{4.99}$$

which are valid for straight and Zerol bevel gears as well as spiral bevel gears with large cutter radius ($r_{co} > R_{m2}$) and, respectively, for other spiral bevel and hypoid gears. In Eq. (4.99), $\rho_{m\beta}$ is the lengthwise tooth mean radius of curvature, R_{m2} the mean cone distance of the wheel, and the exponent q is a constant given by:

$$q = \frac{0.279}{\log_{10}(\sin \beta_{m2})}. \tag{4.100}$$

The lengthwise tooth mean radius of curvature, $\rho_{m\beta}$, is calculated with the following two relationships:

$$\rho_{m\beta} = r_{c0}. \tag{4.101}$$

$$\rho_{m\beta} = R_{m2} \cos \beta_{m2} \left[\tan \beta_{m2} + \frac{\tan \eta_1}{1 + \tan \nu_0 (\tan \beta_{m2} + \tan \eta_1)} \right], \tag{4.102}$$

which are valid for face milled gears and, respectively, for face hobbed gears. In Eq. (4.102), η_1 is the second auxiliary angle (see Vol. 1, Fig. 12.35), while ν_0 is the lead angle of the cutter. These two quantities are given respectively by the following relationships:

$$\eta_1 = \arccos \left[\frac{R_{m2} \cos \beta_{m2}}{\sqrt{R_{m2}^2 + r_{c0}^2 - 2R_{m2}r_{c0} \sin(\beta_{m2} - \nu_0)}} \left(1 + \frac{z_0}{z_2} \sin \delta_2 \right) \right] \tag{4.103}$$

$$\nu_0 = \arcsin \left(\frac{m_{mn} z_0}{2 r_{c0}} \right), \tag{4.104}$$

where z_0 is the number of blade groups of the cutting tool.

It is to be borne in mind that the range of validity of Eq. (4.99), which allows us to calculate the value of the lengthwise curvature factor for bending stress, K_{F0}, is limited. In this regard, if the calculated value of $K_{F0} > 1.15$, we take $K_{F0} = 1.15$, while if the calculated value of $K_{F0} < 1$, we take $K_{F0} = 1$.

4.7 Transverse Load Factors, $K_{H\alpha}$ and $K_{F\alpha}$

4.7.1 Generality

Generally, for given gear dimensions, accuracy grate and mounting condition, the distribution of the total tangential force between the various pairs of simultaneous meshing teeth depends on the same total tangential force and gear accuracy grade. The transverse load factors, $K_{H\alpha}$ and $K_{F\alpha}$, consider the effects of the load distributions on the contact stress and, respectively, on the tooth root stress. The theoretical bases to calculate these two transverse load factors, with the appropriate variations of the case, are the same we described in Sect. 1.5 for spur and helical gears. These two factors are also calculated with Methods A, B, and C.

Here also, the Method A requires a comprehensive analysis for the exact determination of the load distribution, to be carried out with theoretical and numerical models able to take account of all influence factors, or with accurate experimental measurements. In addition, the accuracy and reliability of this method must be tried and validated. However, in most cases, it is sufficient to use the approximate Methods B and C, because they are accurate enough.

Here too, for the same reasons described in Sect. 4.5.1, when we use Method B or Method C for hypoid gears characterized by a typical amount of hypoid offset, the transverse load factors may be set equal to unit, while for smaller amounts of hypoid offset the values of these factors are obtained by interpolation between the unit value and the value calculated as for bevel gears without offset. The values of these factors must satisfy the following inequality:

$$K_{H\alpha} = K_{F\alpha} = K_{H\alpha}^* - \frac{K_{H\alpha}^* - 1}{0.1} a_{rel} \geq 1, \qquad (4.105)$$

where $K_{H\alpha}^*$ is the preliminary transverse load factor for contact stress for non-hypoid gears (we assume $K_{H\alpha}^* = K_{H\alpha-B}$ or $K_{H\alpha}^* = K_{H\alpha-C}$, depending on whether the Methods B or C are used), while a_{rel} is the relative hypoid offset, given by Eq. (4.76).

4.7.2 Method B for Determination of the Transverse Load Factors, $K_{H\alpha-B}$ and $K_{F\alpha-B}$

With Method B, the determination of the transverse load factors for bevel gears having virtual cylindrical gears with contact ratio $\varepsilon_{v\gamma} \leq 2$ is carried out using the following relationship:

$$K_{H\alpha-B} = K_{F\alpha-B} = \frac{\varepsilon_{v\gamma}}{2} \left[0.9 + 0.4 \frac{c_\gamma (f_{pt} - y_\alpha)}{(F_{mtH}/b_v)} \right], \qquad (4.106)$$

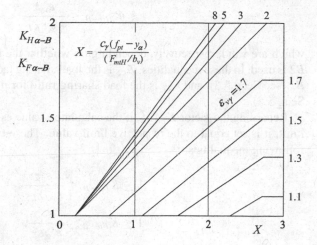

Fig. 4.10 Transverse load factors, $K_{H\alpha-B}$ and $K_{F\alpha-B}$, as a function of X and $\varepsilon_{v\gamma}$

where, the meaning of symbols already introduced remaining unchanged, y_α is the running-in allowance for pitch error (in μm), and

$$F_{mtH} = F_{vmt}K_A K_v K_{H\beta-C} \tag{4.107}$$

is the determinant tangential force at mid-face width of the pitch cone (in N). The determination of the transverse load factors, $K_{H\alpha}$ and $K_{F\alpha}$, can also be carried out using the diagram shown in Fig. 4.10 as a function of $\varepsilon_{v\gamma}$ (this is the virtual contact ratio for Method $B1$, or the modified contact ratio for Method $B2$), and the parameter of irregularity of transmission, X, given by:

$$X = \frac{c_\gamma (f_{pt} - y_\alpha)}{(F_{mtH}/b_v)}. \tag{4.108}$$

For bevel gears having virtual cylindrical gears with contact ratio $\varepsilon_{v\gamma} > 2$, the transverse load factors are instead determined using the following relationship:

$$K_{H\alpha-B} = K_{F\alpha-B} = 0.9 + 0.4 \frac{c_\gamma (f_{pt} - y_\alpha)}{(F_{mtH}/b_v)} \sqrt{\frac{2(\varepsilon_{v\gamma} - 1)}{\varepsilon_{v\gamma}}}. \tag{4.109}$$

About the values of these transverse load factors, attention should be paid to boundary conditions. Regarding the factor $K_{H\alpha-B}$, if the calculated value exceeds the lower or upper limit, it is set equal to the respective limit value. These two limits are given by the following inequalities:

$$1 \le K_{H\alpha-B} \le \frac{\varepsilon_{v\gamma}}{\varepsilon_{v\alpha} Z_{LS}^2} \tag{4.110}$$

$$1 \le K_{H\alpha-B} \le \frac{\varepsilon_{v\gamma}}{\varepsilon_{v\alpha}\varepsilon_{NI}}, \tag{4.111}$$

which are valid, respectively, depending on whether the Method $B1$ or the Method $B2$ is used. In these inequalities, Z_{LS} is the load sharing factor for pitting, for Method $B1$ (see Sect. 5.3), and ε_{NI} is the load sharing ratio for pitting, for Method $B2$ (see Sect. 5.5).

Regarding the factor $K_{F\alpha-B}$, if the calculated value exceeds the lower and upper limit, it is set equal to the respective limit value. These two limits are given by the following inequalities:

$$1 \le K_{F\alpha-B} \le \frac{\varepsilon_{v\gamma}}{\varepsilon_{v\alpha}Y_{LS}} \tag{4.112}$$

$$1 \le K_{F\alpha-B} \le \frac{\varepsilon_{v\gamma}}{\varepsilon_{v\alpha}\varepsilon_N}, \tag{4.113}$$

which are valid, respectively, depending on whether the Method $B1$ or the Method $B2$ is used. In these inequalities, Y_{LS} is the load sharing factor for bending, for Method $B1$ (see Sect. 6.3.5), and ε_N is the load sharing ratio for bending, for Method $B2$ (see Sect. 6.6.1.1).

With the aforementioned boundary conditions, the most unfavorable load distribution is considered, which corresponds to only one pair of teeth transmitting the total tangential force. Therefore, the calculation is on the safe side. The accuracy grade of bevel gears must be chosen so that neither $K_{H\alpha-B}$ nor $K_{F\alpha-B}$ exceeds the value of the transverse contact ratio of virtual cylindrical gear in mean normal section, $\varepsilon_{v\alpha n}$.

4.7.3   Method C for Determination of the Transverse Load Factors, $K_{H\alpha-C}$ and $K_{F\alpha-C}$

To determine the transverse load factors, $K_{H\alpha-C}$ and $K_{F\alpha-C}$, with this method, which generally is sufficiently accurate for industrial gears, the gear accuracy grade, specific loading, gear type and running-in behavior are required. Method C is based on the following assumption:

- The mean value of the mesh stiffness per unit face width, c_γ, given by Eq. (4.81), and the value of the single stiffness, c', given by Eq. (4.86), are set respectively equal to 20, and 14 N/mm μm.
- A transverse contact ratio in the range ($1.2 < \varepsilon_{v\alpha} < 1.9$) applies to tooth stiffness.
- A single pitch deviation is assigned to each gear accuracy grade. With this assumption, values of the load distribution factors are obtained, which are on the safe side for most applications, i.e. in cases of mean and high specific loads, as well as in cases of specific loads $F_{vmt}K_A/b_{v,eff} < 100$ N/mm.

For the various cases of interest, the determination of the transverse load factors, $K_{H\alpha-C}$ and $K_{F\alpha-C}$, is made using Table 4.5. Attention should be paid to the fact that, if the gear accuracy grades are different for pinion and wheel, the worse one shall be used.

In absence of direct experience, the running-in allowance for pitch error, y_α (it is the amount by which the mesh alignment error is reduced from the start of the operation, due to running-in), can be taken from Figs. 4.11 to 4.12, which are valid for gear pairs with a tangential speed $v_{mt2} > 10$ m/s and, respectively, $v_{mt2} \leq 10$ m/s. In these figures, the running-in allowance, y_α, is given as a function of the single pitch deviation, f_{pt}, and tangential speed, v_{mt2}, for three different groups of materials, that is: structural and through hardened steels (solid lines), case-hardened and nitrided steels (dotted line), and gray cast iron (long and short dash line). The structural and through hardened steels are differentiated according to five different values of the allowable stress number for contact stress, $\sigma_{H,\lim}$, from 400 to 1200 N/mm^2.

The curves shown in Figs. 4.11 and 4.12 have been obtained with the following relationships, which therefore can be used as an alternative to the same curves:

$$y_\alpha = \frac{160}{\sigma_{H,\lim}} f_{pt}, \tag{4.114}$$

which applies to structural and through hardened steels, without restriction for $v_{mt2} \leq 5$ m/s and with the restrictions $y_\alpha = \leq 12.8 \times 10^3 / \sigma_{H,\lim}$ and $y_\alpha = \leq 6.4 \times 10^3 / \sigma_{H,\lim}$, respectively for (5 m/s $< v_{mt2} \leq 10$ m/s) and for $v_{mt2} > 10$ m/s;

$$y_\alpha = 0.275 f_{pt}, \tag{4.115}$$

which applies to gray cast iron, without restriction for $v_{mt2} \leq 5$ m/s, and with the restrictions $y_\alpha \leq 22$ μm and $y_\alpha \leq 11$ μm, respectively for (5 m/s $< v_{mt2} \leq 10$ m/s) and for $v_{mt2} > 10$ m/s;

$$y_\alpha = 0.075 f_{pt}, \tag{4.116}$$

which applies to case-hardened and nitrided steels for all speeds, with the restriction $y_\alpha \leq 3$ μm.

If pinion and wheel are made of different materials, a mean value of y_α shall be calculated, using the following relationship:

$$y_\alpha = \frac{1}{2}(y_{\alpha 1} + y_{\alpha 2}), \tag{4.117}$$

where $y_{\alpha 1}$ and $y_{\alpha 2}$ are the values of the running-in allowance determined for materials of the pinion and wheel.

It is to be noted that the maximum value of the single pitch deviation, f_{pt}, of pinion or wheel is to be introduced in Eqs. (4.114–4.116) as well as in Eqs. (4.106), (4.108) and (4.109). For design calculations, the absolute single pitch tolerance, f_{ptT},

Table 4.5 Transverse load distribution factors, $K_{H\alpha-C}$ and $K_{F\alpha-C}$

Specific load $F_{vmt}K_A/b_{v,eff}$			≥ 100 N/mm						< 100 N/mm
Gear accuracy grade			5 and better	6	7	8	9	10 11	All accuracy grades
Surface hardened	Straight bevel gears	$K_{H\alpha}$	1.0		1.1	1.2	1.2	(B1): $1/Z_{LS}^2$ or 1.2 (✱)	(B1): $1/Z_{LS}^2$ or 1.2 (✱)
								(B2): $1/\varepsilon_{N1}$ or 1.2 (✱)	(B2): $1/\varepsilon_{N1}$ or 1.2 (✱)
		$K_{F\alpha}$						(B1): $1/Y_{LS}$ or 1.2 (✱)	(B1): $1/Y_{LS}$ or 1.2 (✱)
								(B2): $1/\varepsilon_{N}$ or 1.2 (✱)	(B2): $1/\varepsilon_{N}$ or 1.2 (✱)
	Helical and spiral bevel gears	$K_{H\alpha}$	1.0	1.1	1.2	1.4	1.4	ε_{van} or 1.4 (✱)	ε_{van} or 1.4 (✱)
		$K_{F\alpha}$							
Not surface hardened	Straight bevel gears	$K_{H\alpha}$	1.0			1.1	1.2	(B1): $1/Z_{LS}^2$ or 1.2 (✱)	(B1): $1/Z_{LS}^2$ or 1.2 (✱)
								(B2): $1/\varepsilon_{N1}$ or 1.2 (✱)	(B2): $1/\varepsilon_{N1}$ or 1.2 (✱)
		$K_{F\alpha}$						(B1): $1/Y_{LS}$ or 1.2 (✱)	(B1): $1/Y_{LS}$ or 1.2 (✱)
								(B2): $1/\varepsilon_{N}$ or 1.2 (✱)	(B2): $1/\varepsilon_{N}$ or 1.2 (✱)
	Helical and spiral bevel gears	$K_{H\alpha}$	1.0		1.1	1.2	1.4	ε_{van} or 1.4 (✱)	ε_{van} or 1.4 (✱)
		$K_{F\alpha}$							

(✱) whichever is the greater

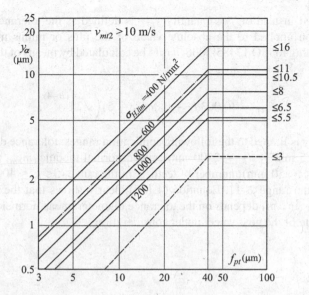

Fig. 4.11 Running-in allowance, y_α, of gear pairs with a tangential speed $v_{mt2} > 10$ m/s

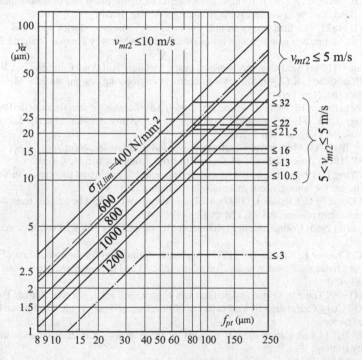

Fig. 4.12 Running-in allowance, y_α, of gear pairs with a tangential speed $v_{mt2} \leq 10$ m/s

should be used instead of f_{pt}. Quantity f_{ptT} is defined as the tolerance for single pitch deviation applied to the absolute value of the plus or minus measurement value. According to ISO 17485:2006, it is to be calculated by means of the following relationship:

$$f_{ptT} = (0.003d_T + 0.3m_{mn} + 5)\left(\sqrt{2}\right)^{(B-4)}, \qquad (4.118)$$

whose validity is limited to the following application ranges: tolerance diameter, d_T, in the range (5 mm $\leq d_T \leq$ 2500 mm); mean normal module, m_{mn}, in the range (1 mm $\leq m_{mn} \leq$ 50 mm); number of teeth, z, in the range (5$\leq z \leq$ 400); accuracy grade, B, in the range 2–11. Equation (4.118) clearly shows that the single pitch tolerance f_{ptT} (in μm) depends on the tolerance diameter, mean normal module and accuracy grade of the gear wheel under consideration.

References

Biezeno CB, Grammel R (1953) Technische Dynamik: Dampfturbinen und Brennkraftmaschinen, Zweiter Band, Zweite Erweiterte Auflage. Springer-Verlag, Berlin

Buchanan R (1823) Practical essays on mill work and other machinery, with notes and additional articles, containing new researches on various mechanical subjects by Thomas Tredgold. J. Taylor, London

Buckingham E (1949) Analytical mechanics of gears. McGraw-Hill Book Company Inc, New York

Budynas RG, Nisbett JK (2009) Shigley's mechanical engineering design, 8th edn. McGraw-Hill Companies Inc, New York

Coleman W, Lehmann EP, Mellis DW, Peel DM (1969) Advancement of straight and spiral bevel gear technology. USAAVLABS technical report 69-75, U.S. Army Aviation Material Laboratories, Fort Eustis, VA (Virginia), pp 1–267

Den Hartog JP (1985) Mechanical vibrations. Dover Publications Inc, New York

Dudley DW (1962) Gear handbook. McGraw-Hill Book Company Inc, New York

Fatemi A, Yang L (1998) Cumulative fatigue damage and life prediction theories: a survey of the state of the art for homogeneous materials. Int J Fatigue 20(1):9–34

Gagnon P, Gosselin C, Cloutier L (1997) Analysis of spur and straight bevel gear teeth deflection by the finite strip method. ASME J Mech Des 119(4):421–426

Giovannozzi R (1965) Costruzione di Macchine, vol II, 4td ed. Casa Editrice Prof. Riccardo Pàtron, Bologna

Gosselin C, Cloutier L, Nguyen QD (1995) A general formulation for the calculation of the load sharing and transmission error under load of spiral bevel and hypoid gears. Mech Mach Theory 30(3):433–450

Henriot G (1979) Traité thèorique and pratique des engrenages, vol 1, 6th edn. Bordas, Paris

ISO 10300-1:2014 Calculation of load capacity of bevel gears—part 1: introduction and general influence factors

ISO 10300-2:2014 Calculation of load capacity of bevel gears—part 2: calculation of surface durability (pitting)

ISO 10300-3:2014 Calculation of load capacity of bevel gears—part 3: calculation of tooth root strength

ISO 17485 2006 Bevel gears—ISO system of accuracy

Kagawa T (1961) Deflections and moments due to a concentrated edge-load on a cantilever plate of infinite length. In: Proceedings of 11th Japan National Congress for Applied Mechanics, pp 47–52

Ker Wilson W (1956) Practical solution of torsional vibration problems, vol I. Chapman and Hall, London

Ker Wilson W (1963) Practical solution of torsional vibration problems, vol II. Chapman and Hall, London

Kolivand M, Kahraman A (2009) A load distribution model for hypoid gears using ease-off topography and shell theory. Mech Mach Theory 44:1848–1865

Krall G (1970a) Meccanica Tecnica delle Vibrazioni: Parte Prima Sistemi Discreti. Eredi Virgilio Veschi, Roma

Krall G (1970b) Meccanica Tecnica delle Vibrazioni: Parte Seconda Sistemi Continui. Eredi Virgilio Veschi, Roma

Litvin FL, Fuentes A (2004) Gear geometry and applied theory, 2nd edn. Cambridge University Press, Cambridge

Maitra GM (1994) Handbook of gear design, 2nd edn. Tata McGraw-Hill Publishing Company Ltd., New Delhi

Merritt HE (1954) Gears, 3th edn. Sir Isaac Pitman & Sous, Ltd, London

Miner MA (1945) Cumulative damage in fatigue. J Appl Mech 12:A159–A164

Niemann G, Winter H (1983) Maschinen-Elemente, band III: Schraubrad-, Kegelrad-, Schnecken-, Ketten-, Rienem-, Reibradgetriebe, Kupplungen, Bremsen, Freiläufe. Berlin Heidelberg, Springer-Verlag

Palmgren A (1924) Die Lebensdauer von Kugellagern. Verfahrenstechinik, Berlin 68:339–341

Pollone G (1970) Il Veicolo. Libreria Editrice Universitaria Levrotto & Bella, Torino

Radzevich SP (2016) Dudley's handbook of practical gear design and manufacture, 3rd edn. CRC Press, Taylor&Francis Group, Boca Raton, Florida

Sheveleva GI, Volkov AE, Medvedev VI (2007) Algorithms for analysis of meshing and contact of spiral bevel gears. Mech Mach Theory 42(2):198–215

Shtipelman BA (1978) Design and manufacture of hypoid gears. John Wiley&Sons Canada, Limited

Simon VV (2009) Design and manufacture of spiral bevel gears with reduced transmission errors. ASME J Mech Des 131(4)

Stadtfeld HJ (1993) Handbook of bevel and hypoid gears. Rochester Institute of Engineering, Rochester, New York

Timoshenko SP, Woinowsky-Krieger S (1959) Theory of plates and shells, 2nd edn. McGraw-Hill International Editions, Singapore

Timoshenko SP, Young DH (1951) Engineering mechanics. McGraw-Hill Book Company, New York

Ventsel E, Krauthammer T (2001) Thin plates and shells: theory, analysis, and applications. Marcel Dekker. Inc, New York

von Kármán T, Biot MA (1940) Mathematical methods in engineering. McGraw-Hill Book Company, New York

Wang J, Lim TC, Li M (2007) Dynamics of a hypoid gear pair considering the effects of time-varying mesh parameters and backlash nonlinearity. J Sound Vib 308:302–329

Warburton GB (1976) The dynamical behaviour of structures, 2nd edn. Pergamon Press, Oxford

Wellauer EJ, Seireg A (1960) Bending strength of the gear teeth by cantilever-plate theory. ASME J Eng Ind 82(3):213–220

Chapter 5
Surface Durability (Pitting) of Bevel Gears

Abstract In this chapter, a general survey is first done on the surface durability (pitting) of bevel gears in the broadest meaning, including straight, helical (or skew), spiral bevel, Zerol and hypoid gears, also focusing attention on pitting damage and safety factor to be used in their design. The theoretical bases of surface durability of these types of gears are only mentioned, since they are traced back to those of their equivalent virtual cylindrical gears and therefore do not differ from those described in Chap. 2. The theoretical bases are only given for some specific factor, typical of these gears. The procedures for calculating the surface durability of these types of gears in accordance with the ISO standard are described, highlighting when deemed necessary how the relationships used by the same ISO are founded on the theoretical bases previously mentioned or discussed.

5.1 General Aspects on Pitting Damage

As we said in the initial sections of Chap. 1, the failure of gear teeth by pitting is a fatigue phenomenon (see Giovannozzi 1965; Budynas and Nisbett 2009; Juvinall and Marshek 2012; Radzevich 2016). This is not the case to go back on the general concepts relating to pitting, since they do not change, passing from the discussion of spur and helical gears to that of bevel gears. Even for these latter gears, there are two different types of pitting: the initial and destructive pitting (see Shipley 1967; Sekercioglu and Koran 2007).

In both varieties of pitting, when limits of the surface durability of the tooth flanks are exceeded, particles break out of the meshing flanks, leaving pits. Size and number of pits that can be tolerated for straight and helical (or skew), Zerol and spiral bevel gears including hypoid gears vary within wide limits, which depend largely on the materials and field of application. In some cases, extensive pitting is acceptable, while in other cases no pitting is acceptable (see Niemann and Winter 1983b; Stokes 1992; Stadtfeld 1993).

In this respect, it is necessary to distinguish those applications in which low-hardened steels or through-hardened steels are employed, from applications in which high-hardened steels and case-carburized steels are used (Dieter 1988). In the first

© Springer Nature Switzerland AG 2020 247
V. Vullo, *Gears*, Springer Series in Solid and Structural Mechanics 11,
https://doi.org/10.1007/978-3-030-38632-0_5

application case, *initial pitting* (also called *corrective pitting* or *non-progressive pitting*) frequently occurs during early use. This variety of pitting is characterized by small pits, which do not extend over the entire flank of the tooth; generally, it is not considered serious. Initial pitting occurs in localized over-stressed contact areas, and tends to redistribute the stresses, by progressively removing over-load contact spots. When the stresses have been redistributed, the pitting stops. On the contrary, in the second application case, the variety of pitting that occurs is usually a destructive pitting.

For calculations of surface durability of bevel gears, it is necessary to distinguish between the initial and destructive, acceptable and unacceptable pitting types. A linear or progressive increase of pits in the total contact zone under unchanged service conditions is generally considered to be unacceptable. However, when the effective tooth contact zone is enlarged by initial pitting, and the rate of pitting generation is subsequently decreased (*degressive pitting*) or even finished (*arrested pitting*), the pitting is considered acceptable and tolerable.

In applications where the acceptability of pitting is questionable, it is necessary to make a damage assessment including the entire active area of contact (that is extended to all the meshing tooth flanks), and taking into consideration the number and size of developed new pits on non-hardened tooth flanks. In cases where pits are formed on just one, or only a few, of the surface-hardened tooth flanks, assessment will be restricted to the flanks actually pitted. On the teeth at risk, it will be necessary to deepen the analysis in order to achieve quantitative assessment (see Coleman 1952; Höhn et al. 1992; Handschuh et al. 2001; Dempsey 2015; Petr and Vosyka 2017).

In particular cases, a first, rough assessment may be carried out evaluating the entire amount of material removed by wear (Dempsey et al. 2004). In critical cases, the condition of the flanks should be examined at least three times: the first, after at least 10^6 cycles of load, and the further ones after two periods of service established in dependence of the results of previous examinations. When surface deterioration due to pitting determines a risk of human life, or damage to property, the pitting shall not be tolerated. Due to stress concentration effects, a pit of 1 mm in diameter near the fillet of a through-hardened or case-hardened gear tooth may cause the triggering of a crack that could lead to tooth breakage (see Spievak et al. 2001; Ural et al. 2005).

For the aforementioned cases (see, for example, aerospace transmissions and turbine gears), such a pitting is unacceptable during the long life demanded for these gears, usually included between 10^{10} and 10^{11} cycles. In fact, the pitting and associated high wear of flank surfaces can lead to unacceptable vibrations and excessive dynamic loads. Since only a low probability of failure is tolerated, high safety factors must be introduced in the calculations (see Choy et al. 1994).

On the contrary, pitting over the entire surface of the working flanks may be tolerated and accepted for some low-speed industrial gear transmissions with large values of the module (e.g., module equal to or greater than 25 mm) made of low hardness steel and designed to operate at nominal power for 10 to 20 years. In these cases, individual pits can be up to 20 mm in diameter and 0,8 mm deep. The tooth flanks are progressively smoothed and work-hardened (the surface Brinell hardness number increases by 50% or more), the rate of pit generation subsequently decreases, and the apparently destructive pitting appeared during the first two or three years of service normally slows down or disappears. For such conditions, relatively low safety factors may be chosen (values around the unit or less), and a higher probability of tooth surface damage can be accepted. Anyway, ISO standards recommends to choose the minimum value of the safety factor for contact stress, S_H, equal to 1, making agree between manufacturer and customer. This choice does not interfere with that of the safety factor against tooth breakage, whose values are necessarily higher.

The theoretical bases of calculations of surface durability (pitting) of bevel gears are the same as described in Chap. 2 with regard to spur and helical gears. The specific treatises (see Buckingham 1949; Merritt 1954; Dudley 1962; Shtipelman 1978; Henriot 1979; Maitra 1994; Radzevich 2016), and the general treatises (see Giovannozzi 1965; Pollone 1970; Niemann and Winter 1983a; Budynas and Nisbett 2009; Juvinall and Marshek 2012) on this subject are also the same.

In this chapter we will describe in detail the calculation procedure for the determination of surface load carrying capacity of straight and helical (skew), Zerol and spiral bevel gears including hypoid gears. The calculation relationships shown in sections that follow comply with the ISO 10,300-2:2014, and include all the influences on surface durability for which quantitative assessments can be made. These relationships are applicable to oil lubricated bevel gears, provided that sufficient lubricant is present in meshing at all times.

The phenomenological aspects, general concepts and limits of the calculation relationships used for determination of surface load carrying capacity of bevel gears are the ones that we already described in Sect. 4.1. Given their importance, for the reader's convenience, it is useful to recall here the well-defined restrictions of these calculation relationships, which are related to the virtual cylindrical gear pairs, equivalent to the actual bevel gear pairs. These restrictions can be summarized as follows: transverse contact ratio of virtual cylindrical gears, $\varepsilon_{v\alpha} < 2$; average value of mean spiral angles, $(\beta_{m1} + \beta_{m2})/2 \leq 45°$; effective pressure angle, $\alpha_e \leq 30°$; face width, $b \leq 13 m_{mn}$; sum of profile shift coefficients of pinion and wheel, $(x_1 + x_2) = 0$, i.e. bevel gears with profile-shifted toothing without variation of shaft angle (see Vol. 1, Sect. 9.8). It is also to be noted that the relationships given in this chapter shall not apply to the assessment of other types of gear tooth surface damage, such as scratching, plastic yielding, scuffing or other similar types of damage.

Two main methods are provided for the assessment of surface durability of bevel and hypoid gears: Method $B1$ and Method $B2$. The bases of calculation are common to both methods, but the calculation procedures are specific for each of the two methods. With both methods, the ability of a gear tooth to resist pitting is determined by comparison with the following reference stresses: the *contact stress*, σ_H, which is based on the geometry of the tooth, its manufacturing accuracy, rigidity of the gear blanks, bearings and housing, and operating torque, and the permissible contact stress, σ_{HP}, which is based on the endurance limit for contact stress, $\sigma_{H,\lim}$, and effect of the operating conditions under which the gear operates. The groups of materials used for these gears are those shown in Table 2.2.

5.2   Calculation of Surface Durability: ISO Basic Formulae of Method $B1$

Here we use the term bevel gears in its broadest meaning, including straight and helical or skew gears, Zerol and spiral bevel gears, and hypoid gears. For design of these bevel gears free from destructive pitting during their working life, we use the following rating formulae, which are unique to Method $B1$. The determination of the surface load carrying capacity, that is the capability of a gear teeth to resist pitting, is performed for pinion and wheel together and, for hypoid gears, separately for drive side flank and coast side flank, using the following relationship:

$$\sigma_{H-B1} = \sigma_{H0-B1}\sqrt{K_A K_v K_{H\beta} K_{H\alpha}} \le \sigma_{HP-B1}, \qquad (5.1)$$

where σ_{H0-B1} is the *nominal contact stress* for Method $B1$, which is given by:

$$\sigma_{H0-B1} = Z_{M-B} Z_{LS} Z_E Z_K \sqrt{\frac{F_n}{l_{bm}\rho_{rel}}}. \qquad (5.2)$$

In this last equation, F_n is the *nominal normal force* of the virtual cylindrical gear at the mean point, P (Fig. 4.1), given by:

$$F_n = \frac{F_{mt1}}{\cos\alpha_n \cos\beta_{m1}}, \qquad (5.3)$$

where $\alpha_n = \alpha_{nD}$ or $\alpha_n = \alpha_{nC}$, depending on whether the generated normal pressure angle, α_n, must be calculated for the drive side or coast side (see Vol. 1, Eqs. (12.156) and (12.157)).

The meaning of symbols already introduced remaining unchanged, the other symbols appearing in equations above represent: l_{bm} (in mm), the *theoretical length of middle contact line*, i.e. the average value of the length of contact line l_b, given by Eq. (4.29), within the zone of contact (see Fig. 4.3); ρ_{rel} (in mm), the radius of relative curvature perpendicular to the contact line at virtual cylindrical gears (see Eq. (4.38)); Z_{M-B}, the dimensionless *mid-zone factor*, which takes into account the conversion of the contact stress determined at the mean point to the determinant position, as specified below; Z_{LS}, the dimensionless *load sharing factor* for Method $B1$, which accounts for the load sharing between two or more pairs of teeth; Z_E [in $(N/mm^2)^{1/2}$], the dimensional *elasticity factor*, which takes into account the influence of Young's modulus and Poisson's ratio of materials; Z_K, the dimensionless *bevel gear factor* for Method $B1$, which takes into account the influence of the bevel gear geometry.

The above described formulae express the fact that the assessment of pitting resistance is based on the Hertzian contact pressure between two curved surfaces, suitably modified to take account of the load sharing between adjacent meshing teeth, shape of the instantaneous contact area, position of the center of pressure on the teeth, and load concentration resulting from manufacturing inaccuracies. Although the conditions of contact during the tooth meshing is not well interpreted by the Hertz theory, the Hertzian relationships, adequately supported by running tests with bevel gears that include the additional influences, can be used as sufficiently reliable conventional models to convert test gear data to bevel gears of various types and sizes.

For the use of the aforementioned equations and those that follow, we consider the load distributed over the contact lines described in Sect. 4.2, and the following determinant positions of load application:

- the inner point of single pair tooth contact, if $\varepsilon_{v\beta} = 0$;
- the mid-point of the zone of contact, if $\varepsilon_{v\beta} \geq 1$;
- interpolation between the two aforementioned conditions, if $\left(0 < \varepsilon_{v\beta} < 1\right)$.

The permissible contact stress, σ_{HP-B1}, to be calculated separately for pinion and wheel, is given by:

$$\sigma_{HP-B1} = \sigma_{H,\lim} Z_{NT} Z_X Z_v Z_R Z_W Z_{Hyp}, \tag{5.4}$$

where Z_{Hyp} is the dimensionless *hypoid factor*, which takes into account the influence of lengthwise sliding on the surface durability, while the other symbols retain the same meaning as described in Sect. 2.8, that is: $\sigma_{H,\lim}$ (in N/mm^2), is the *endurance limit* or *allowable stress number for contact stress*, which takes into account the influence of material, heat treatment and surface roughness on standard reference test gears; Z_{NT}, is the dimensionless *life factor* for pitting, which considers the influence of required numbers of operating cycles; Z_X, is the dimensionless *size factor*, which takes into account the influence of the tooth dimensions in terms of module on the permissible contact stress; Z_L, Z_v, and Z_R, are respectively the dimensionless *lubricant factor*, *velocity factor*, and *roughness factor* for contact stress, which are globally known as *lubricant film factors*, since they take into account the lubrication conditions (about the individual specificity of each of these three factors, we refer the reader to the Sect. 2.8); Z_W, is the dimensionless *work hardening factor*, which considers the hardening of a softer material gear wheel running in mesh with a harder material pinion (e.g., a through hardened gear wheel running in mesh with a case-hardened pinion).

The *safety factor for contact stress*, to be checked separately for pinion and wheel if the values of permissible contact stress are different, is determined using the following relationship:

$$S_{H-B1} = \frac{\sigma_{HP-B1}}{\sigma_{H-B1}} > S_{H,\min}, \tag{5.5}$$

where $S_{H,\min}$ is the minimum safety factor against pitting, the value of which must conform to what we said in the Sect. 4.1. It is to be noted that the relationship (5.5) defines the calculated value of the safety factor against pitting, S_H, with respect to contact stress. Instead, the safety factor related to the transferable torque is equal to the square of S_H.

In place of Eqs. (5.2) and (5.4), for determination of the nominal value of contact stress, σ_{H0-B1}, and permissible contact stress, σ_{HP-B1}, the following equations may be used:

$$\sigma_{H0-B1} = Z_{M-B}Z_{LS}Z_E\sqrt{\frac{F_n}{l_{bm}\rho_{rel}}}. \tag{5.6}$$

$$\sigma_{HP-B1} = \sigma_{H,\lim}Z_{NT}Z_XZ_LZ_vZ_RZ_WZ_SZ_{Hyp}, \tag{5.7}$$

where Z_S is the dimensionless *bevel slip factor*, which considers the increase of surface durability in the flank zone of positive specific sliding versus the zone of negative specific sliding. Contrary to the bevel gear factor, Z_K, which is used for calculation of the contact stress, σ_H, the bevel slip factor, Z_S, is used for calculation of the permissible contact stress, σ_{HP}.

5.3   Contact Stress Factors for Method $B1$

In Eq. (5.2), the already defined dimensionless factors Z_{M-B}, Z_{LS} and Z_K appear, which are known to be the contact stress factors for Method $B1$. These three factors, together with the dimensional elastic factor, Z_E, adjust what we could define as a reference value, given by $(F_n/l_{bm}\rho_{rel})^{1/2}$, to provide the nominal value of the contact stress, σ_{H0-B1}.

The first of these three contact stress factors is the mid-zone factor, Z_{M-B} which takes into account the difference between the radius of relative curvature ρ_{rel} at the mean point, P, and that at the critical point of load application of the pinion. The radius of relative curvature, ρ_{rel}, at the mean point P can be directly calculated from the data of the bevel gear in meshing condition (see Eq. (4.38)). For the conversion from the mean point P to the critical point of the load application of the pinion in meshing condition, the corresponding virtual cylindrical gear is used.

Depending on the value of the face contact ratio of virtual cylindrical gear, $\varepsilon_{v\beta}$, three cases may occur: in fact, the critical point of load application of the pinion can be the inner point of single contact of the pinion, B, if $\varepsilon_{v\beta} = 0$, or point M in the middle of the path of contact, if $\varepsilon_{v\beta} \geq 1$, or a point interposed between points B and M, if $(0 < \varepsilon_{v\beta} < 1)$. Figure 5.1 shows the radii of curvature at inner point of single pair tooth contact B of the pinion, and at mid-point M of the path of contact, for determination of the mid-zone factor, Z_{M-B}. The same figure shows the radii of curvature at the pitch point, C. This figure shows the schematic view of a cylindrical gear pair in transverse section, with the line of action that is tangent to both base circles of pinion and wheel (d_{vb1} and d_{vb2} are their diameters), and the tip circles of wheel and pinion (d_{va2} and d_{va1} are their diameters) that intersect the line of action

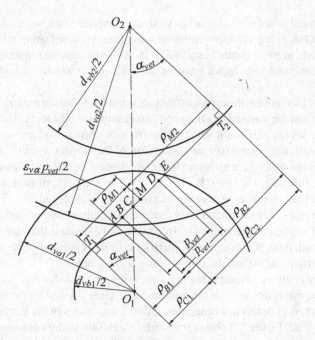

Fig. 5.1 Radii of profile curvature at pitch point, C, mid-point, M, and inner point of single tooth contact of the pinion, B

at points A and E. The different radii of profile curvature at pitch point C, mid-point M, and inner point of single contact B of the pinion are specified with the subscripts C, M, and B, for which we will have: $\rho_{C1,2}$, $\rho_{M1,2}$ and $\rho_{B1,2}$.

The mid-zone factor, Z_{M-B}, is determined using the following relationship:

$$Z_{M-B} = \frac{\tan \alpha_{vet}}{\sqrt{\left[\sqrt{\left(\frac{d_{va1}}{d_{vb1}}\right)^2 - 1} - \frac{\pi}{z_{v1}} F_1\right]\left[\sqrt{\left(\frac{d_{va2}}{d_{vb2}}\right)^2 - 1} - \frac{\pi}{z_{v2}} F_2\right]}}, \qquad (5.8)$$

where α_{vet} is the transverse pressure angle of the virtual cylindrical gear, z_{v1} and z_{v2} are the numbers of teeth of two members of the same virtual cylindrical gear, and F_1 and F_2 are the *auxiliary factors* for the mid-zone factor, whose values can be taken from Table 5.1 as a function of value of the face contact ratio $\varepsilon_{v\beta}$.

Table 5.1 Factors for calculation of mid-zone factor, Z_{M-B}

Parameters	F_1	F_2
$\varepsilon_{v\beta} = 0$	2	$2(\varepsilon_{v\alpha} - 1)$
$0 < \varepsilon_{v\beta} < 1$	$2 + (\varepsilon_{v\alpha} - 2)\varepsilon_{v\beta}$	$2(\varepsilon_{v\alpha} - 1) + (2 - \varepsilon_{v\alpha})\varepsilon_{v\beta}$
$\varepsilon_{v\beta} \geq 1$	$\varepsilon_{v\alpha}$	$\varepsilon_{v\alpha}$

It is to be noted that, for bevel gears as well as for hypoid gears, the results obtained using Eq. (5.8) show a good approximation with those obtained using reliable mathematical models of tooth contact analysis. It is also important to remember that, for hypoid gears, the mid-zone factor must be determined separately for both drive and coast flanks.

The second factor of the three contact factors is the *load sharing factor*, Z_{LS}, which takes into account the load sharing between two or more pairs of teeth. Therefore, this factor determines the maximum value of the fraction of total load acting on a single tooth. For a sufficiently approximate discussion of the subject, we first assume that the load distribution along each contact line in the zone of contact is elliptical in shape (see Johnson 1985; Kalker 1990; Barber 1992). The area, A^*, of each semi-ellipse represents the load on the respective contact line (see Fig. 5.2, and also Fig. 2.2), and the sum of all areas over all contact lines that are in simultaneous meshing represents the total load applied to the gear pair. In addition, we assume that the distribution of the peak load, p (in N/mm), over the path of contact is parabolic. To indicate this parabolic distribution, we use the dimensionless exponent, e.

Based on the above assumptions, as a measure of load sharing, we define the quotient of the maximum load on the middle contact line divided by the total load. In this framework, the contact line coincides with the major axis of the Hertzian contact ellipse under load. Figure 5.2 shows the load distribution in the contact area. In this figure, a, b, and c indicate respectively the parabolic distribution of peak loads, the elliptical load distribution, and the contact lines in simultaneous meshing.

To facilitate the calculations, dimensionless parameters are introduced for the peak load, p, and the distance, f (in mm), of the relevant contact line from the center of the zone of contact. These two quantities are therefore related to their maximum values, and the corresponding dimensionless parameters are marked with a star. So, the dimensionless *relative peak load*, p^*, for calculating the load sharing factor with Method $B1$, is given by:

$$p^* = \frac{p}{p_{\max}} = 1 - \left(\frac{|f|}{f_{\max}}\right)^e = 1 - |f^*|^e, \qquad (5.9)$$

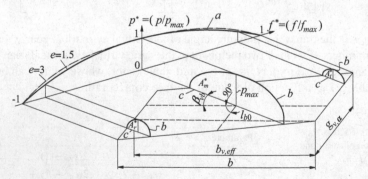

Fig. 5.2 Load distributions in the contact zone

Table 5.2 Exponent e for calculation of parabolic distribution of dimensionless peak loads, p^*

Profile crowning	Exponent e
Low (e.g. automotive gears)	3
High (e.g. industrial gears)	1.5

where $|f^*| = (|f|/f_{max})|$ is the relative distance, i.e. the dimensionless distance from the center of the zone of contact to a contact line. For calculations, the absolute value $|f|$ of f is taken from Table 4.2 for surface durability, and the exponent, e, is taken from Table 5.2.

The area A^* of each semi-elliptic load distribution over the simultaneous contact lines, which is introduced to calculate the load sharing factor, Z_{LS}, is determined by the formula of an ellipse whose minor semi-axis is given by the related peak load, p^*, and major semi-axis is half the length of the contact line, l_b. Therefore, this related area, A^*, which is expressed in mm, is given by:

$$A^* = \frac{1}{4}\pi p^* l_b, \qquad (5.10)$$

where l_b is the length of contact line, given by Eq. (4.29).

Now we introduce the dimensionless ratio, V, defined as the quotient of the maximum load over the middle contact line divided by the total load. This ratio can be expressed as:

$$V = \frac{A_m^*}{A_t^* + A_m^* + A_r^*}, \qquad (5.11)$$

where A_t^*, A_m^*, and A_r^* are respectively the areas above the tip contact line, middle contact line, and root contact line (Fig. 5.2). For each of these three contact areas, the quantities p^* and l_b that appear in Eq. (5.10) must be calculated, respectively, for f_t, f_m, and f_r, whose values are taken from Table 4.2.

Finally, the determination of the load sharing factor, Z_{LS}, is made using the following relationship:

$$Z_{LS} = \sqrt{V} = \sqrt{\frac{A_m^*}{A_t^* + A_m^* + A_r^*}}. \qquad (5.12)$$

This equation is justified by the fact that the contact stress is a function of the square root of load (see also Chap. 2), and this is necessarily also applied to the ratio of the maximum load over the middle contact line and the total load, when we determine the load sharing factor, Z_{LS}.

The third and last contact stress factor is the *bevel gear factor*, Z_K. This factor is an empirical factor which considers the differences between cylindrical and bevel gears in such a way as to agree with practical experience. It is a dimensionless constant to adjust stress, which allows the assessment of bevel gears with the use of the same allowable contact stress numbers of cylindrical gears. A reasonable approximate value of this factor is given by:

$$Z_K = 0.85. \tag{5.13}$$

A theoretical justification of this approximate value may be given based on the following considerations. The bevel gear factor is an index of the reduction of the load carrying capacity of a bevel gear wheel with respect to that of a cylindrical gear wheel having equal outer transverse module, m_{et}, and equal face width, b. For a straight bevel gear wheel, we can assume that the elastic deflections of the tooth under load are proportional to the radial distance r from the axis (Fig. 5.3). This is equivalent to assuming that the tooth axis is kept rectilinear under load, i.e. after deformation. Moreover, this assumption seems quite plausible and coherent to mechanics of deformation, given the interconnection of all resistant normal sections of the tooth between them as well as with the gear rim, which provides full support for the tooth root (see Schiebel and Lindner 1957; Giovannozzi 1965).

The total elastic deflections, φ, are equal to the sum of the elastic deflection due to tooth bending, φ_1, elastic deflection due to shearing load, φ_2, and elastic deflection due to Hertzian contact pressure, φ_3, for which due to the principle of superposition of the theory of elasticity (Timoshenko and Goodier 1953) we have $\varphi = \varphi_1 + \varphi_2 + \varphi_3$. Here we neglect the contribution of the Hertzian contact pressure, for which we can write $\varphi = \varphi_1 + \varphi_2$.

Let us consider the coordinate system shown in Fig. 5.3, where the abscissa, b (it should not be confused with the face width, for which the same symbol is used),

Fig. 5.3 Schematic axial section of a straight bevel gear wheel for approximate determination of Z_K

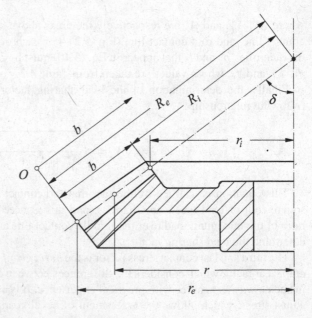

coincides with a pitch cone generatrix, and has its origin at point O, on the back cone. Consider then an infinitesimal cantilever beam, built-in in the gear rim, between two normal sections having respectively coordinates b and $(b + db)$, with length, h, corresponding to the local value of the tooth depth measured along the direction of the back cone generatrix, and with resistant cross-section at the tooth root having area $s(db)$, where s is the local tooth root thickness. If we denote with $p(db)$ the load acting on the free end of this infinitesimal cantilever beam, where $p = p(r)$ is the local load per unit length, the elastic deflections, φ_1 and φ_2, will be respectively proportional to $(ph^3/EI)db$ and $(ph/GS)db$, where E and G are respectively the Young's modulus and modulus of rigidity, while I and S are respectively the moment of inertia of the cross-section and the area of the resistant cross-section.

If we denote by $m = m(r)$ the local module corresponding to the radial distance, r, from the wheel axis, we will have: h, proportional to $m(h \equiv m)$; I, proportional to $m^3(I \equiv m^3)$; S, proportional to $m(s \equiv m)$. Therefore, $\varphi = \varphi_1 + \varphi_2$ is proportional to p, and then, due to the aforementioned assumption, p will be proportional to radial distance, r, of the intersection point between pitch and intermediate cone generatrices from the bevel gear axis; so, we can write:

$$p = p_0 \frac{r}{r_e}, \tag{5.14}$$

where p_0 is the reference local load per unit length, i.e. the local value of the load per unit length corresponding to the radius r_e, that is $p_0 = p(r_e)$.

Considering that $db = -(dr/sin\delta)$ and $b = [(r_e - r_i)/sin\delta]$ (see Fig. 5.3), the total torque transmitted by the gear wheel under consideration is given by:

$$T' = \int_{r_e}^{r_i} prdb = p_0 \frac{b}{r_e} \frac{(r_e^2 + r_e r_i + r_i^2)}{3}, \tag{5.15}$$

where r_i and r_e are the radial distances of the intersection points between pitch and inner cone generatrices and, respectively, pitch and back cone generatrices from the bevel gear wheel axis. Therefore, if we set $r_i = (r_e - \psi)$, with ψ length quite small so that its square may be neglected, and then we substitute this expression into Eq. (5.15), we get:

$$T' = p_0 b r_i. \tag{5.16}$$

A cylindrical gear wheel having the same outer transverse module, m_{et}, and the same face width, b, the elastic deflections being equal, i.e. the tooth root stresses being equal, transmits a total torque, $T = p_0 b r_e$. Therefore, the bevel gear factor is defined by the following approximate relationship:

$$Z_K \cong \frac{T'}{T} = \frac{r_i}{r_e} = \frac{R_e - b}{R_e}. \tag{5.17}$$

In order to have a not too non-uniform load distribution along the face width, we generally assume $Z_K = 0.85$. In any case, it is not convenient that the value of this factor, and therefore also the related ratio (r_i/r_e), fall below $(0.75 \div 0.65)$. The value of the bevel gear factor recommended by ISO standard (see Eq. (5.13)) constitutes a good compromise, also supported by the experience, for all types of bevel gears, that is for straight and helical (skew), Zerol and spiral bevel gears including hypoid gears. More elaborate and complex calculations substantially confirm the accuracy of the aforesaid choices (see Handschuh 1997; Litvin and Fuentes 2004; Riemann et al. 2017).

5.4 Permissible Contact Stress Factors for Method $B1$

In Eqs. (5.4) and (5.7), each to be used as an alternative to the other, the already defined dimensionless factors Z_X and Z_{Hyp}, and Z_X, Z_S, and Z_{Hyp} appear. This three factors are known to be the *permissible contact stress factors*, regardless of the fact that only two of them appear in the first of the above equations.

The size factor, Z_X, to be determined separately for pinion and wheel, considers the statistical evidence that the stress values at which fatigue damage occurs decrease with the increase of size of the components. In fact, with increasing size, the number of weak points in the structure increases. The decrease in endurance limit with increasing component size may be due to the probability that a larger piece is more likely to have a weaker grain or metallurgical defect at which a fatigue crack will start.

For surface durability problems, this decrease with increasing size is due to the influence of lower stress gradients on subsurface defects, as it is confirmed by the theoretical stress analysis, and to the gear size on material quality (variations in structure, effect on forging process, etc.). For tooth root strength problems, the same decrease with increasing size is also due to a smaller stress gradient. This means that high stresses act over a greater depth, with the greater probability of finding weak spots as well as hastening crack penetration (see Faupel 1964; Burr 1982; Budynas and Nisbett 2009).

The size factor, Z_X, is mainly influenced by the following parameters: material quality, that is furnace charge, cleanliness, and forging; heat treatment, depth and distribution of hardening; module in the case of surface hardening; depth of the hardened layer compared to the tooth size, that is core supporting effect.

A reliable procedure of calculation of this size factor has not been developed so far. Therefore, in absence of more specific knowledge, in Eqs. (5.4) and (5.7) the size factor is set equal to unity for both gear members, pinion and wheel, i.e. $Z_{X1,2} = 1$. This value is usable for most bevel gears, provided an appropriate choice of material is made.

As regards the bevel slip factor, Z_S, it is to be noted that, for bevel gears without offset, the change from positive to negative specific sliding is exact on the pitch cone that, to simplify, can be reduced to the mean point. The mid-zone factor, Z_{M-B}, (see previous section), tells us if the critical point of load application for contact stress is above or below the mean point. For hypoid gears, we assume that the change from positive to negative specific sliding also takes place at the mean point. Depending on the mid-zone factor, the bevel slip factor, Z_S, is determined for pinion and wheel. If the negative specific sliding applies to the wheel, then the positive one necessarily applies to the pinion and vice versa. The values of this factor, for $Z_{M-B} < 0.98$, are given by:

$$Z_{S1} = 1.175; \quad Z_{S2} = 1, \tag{5.18}$$

while, for $Z_{M-B} > 1$, they are given by:

$$Z_{S1} = 1; \quad Z_{S2} = 1.175. \tag{5.19}$$

In the range $(0.98 < Z_{M-B} < 1)$, the value of the bevel slip factor is obtained by linear interpolation, as Fig. 5.4 shows.

About the hypoid factor, Z_{Hyp}, it is to be noted that tests performed on a set of hypoid gears with increasing relative hypoid offset values, a_{rel}, showed that the permanent transmissible torque at first increases from zero offset to the typical offset values, to return then to decrease at very high offset values. Bevel gears without offset are characterized by the maximum value of the Hertzian pressure caused by the respective permanent transmissible torque, while this pressure immediately decreases with increasing hypoid offset. The only interpretation of this behavior seems to be that the higher sliding velocities lower the allowable contact stresses on the tooth flanks. Rising contact temperatures and decreasing lubricant film thicknesses are considered to be the main reasons of this lowering of the load carrying capacity.

Fig. 5.4 Bevel slip factor, $Z_{S1,2}$, for surface durability

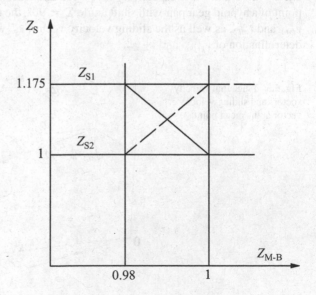

To realize the aforementioned lowering effect, different components of velocities at mean point must be considered, that is the sliding velocity component parallel to the contact line, $v_{g,par}$ (note that here, as contact line, we assume the major axis of the Hertzian contact ellipse under load), and the sum of the sliding velocity components normal to the contact line, $v_{\Sigma,vert}$ (also called *cumulative normal velocity*). The sliding velocity component parallel to the contact line is unfavorable both for the temperature effect, and for the oil film thickness, while the sum of the sliding velocity components normal to the contact line is advantageous for the oil film thickness. It should be noted that here only the absolute values of velocity components are concerned.

For bevel gears without hypoid offset, $v_{g,par}$ is negligibly small compared to $v_{\Sigma,vert}$, whereas, for hypoid gears, both $v_{g,par}$ and $v_{\Sigma,vert}$ increase with the hypoid offset value, but $v_{g,par}$ increases more than $v_{\Sigma,vert}$. Therefore, the ratio $\left(v_{g,par}/v_{\Sigma,vert}\right)$ can be taken as a good indicator of the different behavior of hypoid gears in comparison to cylindrical gears and bevel gears without hypoid offset, under the aspect of the increase in temperature and change of lubricant film thickness. In assessment relationships of surface durability featuring the Method $B1$, this indicator is made up of the hypoid factor, Z_{Hyp}; it is given by the following empirical relationship derived from experimental test results:

$$Z_{Hyp} = 1 - 0.30\left(\frac{v_{g,par}}{v_{\Sigma,vert}} - 0.15\right). \tag{5.20}$$

This equation is valid in the range $\left(0.6 \leq Z_{Hyp} \leq 1\right)$. For bevel gears without hypoid offset, we assume $Z_{Hyp} = 1$.

Figure 5.5 (in this figure, 1 and 2 respectively represent the pinion and wheel axes, while 3 and 4 are respectively the pinion and wheel flanks) shows, at the mean point of a hypoid gear pair with shaft angle $\Sigma < 90°$, the tangential velocity vectors, v_{mt1} and v_{mt2}, as well as the sliding velocity vector, v_g, which are necessary for the determination of $v_{g,par}$ and $v_{\Sigma,vert}$.

Fig. 5.5 Tangential velocity vectors and sliding velocity vector at the mean point

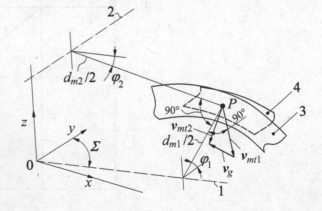

Fig. 5.6 Sliding velocity vector and its components parallel and normal to the contact line at the mean point

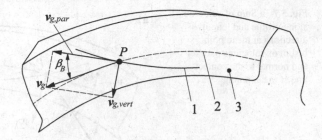

Figure 5.6 (in this figure, 1, 2, and 3 are respectively the contact line, trace of the pitch cone, and pinion flank) shows the component vectors of the sliding velocity vector, $v_{g,par}$ and $v_{\Sigma,vert}$, which are respectively tangent (parallel according to ISO nomenclature) and normal to the contact line at the mean point. From this figure, we derive the following expression of the sliding velocity component parallel to the contact line, $v_{g,par}$:

$$v_{g,par} = v_g \cos|\beta_B|, \tag{5.21}$$

where

$$v_g = v_{mt1} \cos\beta_{m1}(\tan\beta_{m1} - \tan\beta_{m2}) \tag{5.22}$$

is the magnitude of the sliding velocity vector at the mean point, P, and β_B (in degrees) is the inclination angle of contact line, which is given by Eq. (4.39).

The absolute value of the sum of velocities along the profile direction, $v_{\Sigma h}$, is given by:

$$v_{\Sigma h} = |2v_{mt1} \cos\beta_{m1} \sin\alpha_n|, \tag{5.23}$$

where $\alpha_n = \alpha_{nD}$ or $\alpha_n = \alpha_{nC}$ are the generated pressure angle for drive side and for coast side. The absolute value of the sum of velocities in the lengthwise direction, $v_{\Sigma l}$, is instead given by:

$$v_{\Sigma l} = \left| v_{mt1}\left(\sin\beta_{m1} + \frac{\sin\beta_{m2}\cos\beta_{m1}}{\cos\beta_{m2}}\right)\right|. \tag{5.24}$$

From Fig. 5.7a, we infer that the absolute value of the sum of velocities along the profile and lengthwise directions is given by:

$$v_\Sigma = \sqrt{v_{\Sigma h}^2 + v_{\Sigma l}^2}. \tag{5.25}$$

Finally, from Fig. 5.7b, we derive that the absolute value of the sum of velocities normal to the contact line is expressed by the following relationship:

Fig. 5.7 **a** Sum of velocities along profile and lengthwise directions at mean point; **b** sum of velocities parallel and normal to the contact line at mean point

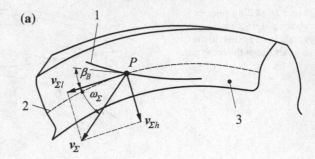

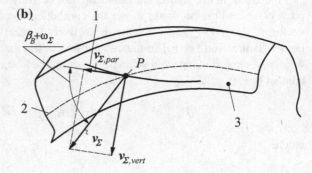

$$v_{\Sigma,vert} = v_{\Sigma}\sin(\omega_{\Sigma} + |\beta_B|), \tag{5.26}$$

where

$$\omega_{\Sigma} = |\arctan(v_{\Sigma h}/v_{\Sigma L})| \tag{5.27}$$

is the inclination angle of the sum of velocity vector with respect to the tangent to trace of pitch cone at mean point, P, expressed in degrees (see Fig. 5.7a).

5.5   Calculation of Surface Durability: ISO Basic Formulae of Method $B2$

We already said that there are no preferences in terms of when to use Method $B1$ and when to use Method $B2$. Here also we mean the term bevel gears in its broadest meaning, including straight and helical (or skew) gears, Zerol and spiral bevel gears, and hypoid gears. For design of these bevel gears with Method $B2$, the determination of the surface load carrying capacity is performed for pinion and wheel together and, for hypoid gears, only the drive side is usually considered. This determination is carried out using the following relationship:

$$\sigma_{H-B2} = \sigma_{H0-B2} Z_A \sqrt{K_A K_v K_{H\beta}} \leq \sigma_{HP-B2}, \tag{5.28}$$

where σ_{H0-B2} is the nominal value of the contact stress for Method $B2$, given by:

$$\sigma_{H0-B2} = Z_E \sqrt{\frac{F_{mt1} d_{m1} Z_{FW}}{b_2 Z_I} \left(\frac{z_2}{d_{e2} z_1}\right)^2}. \tag{5.29}$$

In these equations, the meaning of symbols already introduced remaining unchanged, the new symbols represent: Z_A, the dimensionless *contact stress adjustment factor* for Method $B2$; Z_{FW}, the dimensionless *face width factor*; Z_I, the dimensionless *pitting resistance geometry factor* for Method $B2$.

The permissible contact stress, σ_{HP-B2}, to be calculated separately for pinion and wheel, is given by the following relationship:

$$\sigma_{HP-B2} = \sigma_{H,\lim} Z_{NT} Z_L Z_v Z_R Z_W. \tag{5.30}$$

Factors Z_{NT}, Z_L, Z_v, Z_R, and Z_W that appear in this equation have the same meaning as those described for Eq. (5.4). However, in Eq. (5.30), the hypoid factor Z_{Hyp} does not appear.

The safety factor for contact stress, to be checked separately for pinion and wheel if the values of permissible contact stress are different, is determined using the following relationship:

$$S_{H-B2} = \frac{\sigma_{HP-B2}}{\sigma_{H-B2}} > S_{H,\min}, \tag{5.31}$$

where $S_{H,\min}$ is the minimum safety factor, the value of which must conform to what we said in the Sect. 5.1. Here also, it is to be noted that the relationship (5.31) defines the calculated value of the safety factor, S_H, with respect to contact stress. Instead, the safety factor related to the transferable torque is equal to the square of S_H.

5.6 Contact Stress Factors for Method $B2$

In Eqs. (5.29) and (5.28), the already defined factors Z_I, Z_{FW}, and Z_A appear, which are known to be the contact stress factors for Method $B2$. For determination of these three dimensionless factors, the base unit of the outer transverse module of the wheel, m_{et2}, is used.

The first factor of these three contact stress factors is the pitting resistance geometry factor, Z_I, which takes into account the relative radius of curvature of the mating tooth flanks, as well as the load sharing between adjacent tooth pairs at that point on the tooth surfaces where the calculated contact pressure reaches its maximum value. For the determination of Z_I, analytical or numerical calculation procedures are recommended. These calculations are, however, very complex, so the use of computerized procedures can make them more affordable. Graphs are also available for bevel gears (see ANSI/AGMA 2003-C10), but they can only be used when the data related to the problem under consideration coincide with those of these graphs.

Some initial relationships are necessary for the determination of the pitting resistance geometry factor, Z_I, and precisely:

(a) The angle between the direction of contact and tooth tangent in pitch plane, which is calculated with the equation:

$$\cot(\beta_{m1} - \lambda_1) = \frac{\cos \zeta_R}{\cos \beta_{m1} \cos \beta_{m2} \tan(\beta_{m1} - \lambda_r)} \tan \beta_{m2}, \qquad (5.32)$$

where $(\beta_{m1} - \lambda_r)$ is given by Eq. (4.64), and λ_1 (in degrees) is the angle between projection of pinion axis and direction of contact in pitch plane. Since the mean spiral angle of pinion, β_{m1}, is known, the value of λ_1 can be calculated with the following relationship:

$$\lambda_1 = \beta_{m1} - (\beta_{m1} - \lambda_1), \qquad (5.33)$$

once $(\beta_{m1} - \lambda_1)$ has been derived from Eq. (5.32). Of course, instead of using the tautological relationship (5.33), λ_1 can be directly derived from Eq. (5.32)

(b) The angle of contact line relative to the root cone, w (in degrees), which is calculated using the following relationship:

$$\tan w = \frac{\sin \alpha_a \tan(\beta_{m1} - \lambda_r)}{\cos \alpha_{\lim}}, \qquad (5.34)$$

where α_a is the adjusted pressure angle for Method $B2$, given by Eq. (4.65), and $\alpha_{\lim}$ is the limit pressure angle, given by Eq. (12.186) in Sect. 12.12.1.2 of Vol. 1.

(c) The angle between projection of wheel axis and direction of contact in pitch plane, λ_2 (in degrees), and the mean base spiral angle, β_{bm}, given respectively by the following relationships:

$$\lambda_2 = (\beta_{m1} - \lambda_r) - \beta_{m2} \qquad (5.35)$$

$$\cos \beta_{bm} = \frac{1}{\sqrt{\tan^2(\beta_{m1} - \lambda_r) \cos^2 \alpha_a + 1}}. \qquad (5.36)$$

(d) The relative mean normal base pitch, p_{nb}, and the relative base face width, b_b, given respectively by the following relationships:

$$p_{bmn} = \frac{\pi m_{mn} \cos \alpha_a \cos \beta_{bm}}{m_{et2} \cos(\beta_{m1} - \lambda_r)} \qquad (5.37)$$

$$b_b = \frac{b_2}{m_{et2} \cos \lambda_2}.$$ (5.38)

(e) The normal pressure angle at point of load application for Method B2, given by the following relationship, whose symbols have the meaning described in Sect. 4.3:

$$\cos \alpha_{L1,2} = \cos \alpha_a \left[1 - \frac{(r_{va1,2} - r_{vn1,2}) \cos^2 \beta_{m1,2}}{r_{va1,2} - r_{vn1,2} + R_{mpt1,2}} \right].$$ (5.39)

(f) The difference of radius of curvature between point of load application and mean point, $\rho_{\Delta 1,2}$, and variation of radius of curvature, $\rho_{\Delta red}$, given respectively by the following relationships:

$$\rho_{\Delta 1,2} = \frac{r_{va1,2} - r_{vn1,2} + R_{mpt1,2}}{\cos^2 \beta_{m1,2}} \cos \alpha_{L1,2} (\tan \alpha_{L1,2} - \tan \alpha_a)$$ (5.40)

$$\rho_{\Delta red} = (\rho_{\Delta 1} + \rho_{\Delta 2}) \cos \beta_{bm}.$$ (5.41)

(g) Finally, the relative length of path of contact within the contact ellipse, given by the following relationship:

$$g_\eta = \sqrt{\rho_{\Delta red}^2 \cos^2 \beta_{bm} + b_b^2 \sin^2 \beta_{bm}}.$$ (5.42)

The additional quantities required to calculate the pitting resistance geometry factor, Z_I, are the radius of relative profile curvature, ρ_0, and load sharing ratio at critical point, ε_{NI}. This critical point on the tooth surface occurs when the contact line passes through a point at a distance, y_I, from the midpoint of the length of path of contact. Then the distance y_I (in mm) defines the location of point of load application for maximum contact stress on path of contact, for Method $B2$. The value of y_I determines the minimum value of Z_I.

It is necessary to distinguish the case of straight bevel and Zerol bevel gears, and the case of spiral bevel and hypoid gears. For straight bevel and Zerol bevel gears, the contact line passes close to the lowest point of single pair tooth contact on the pinion and, in this case, the distance y_I is given by:

$$y_I = 0.5 g_{v\alpha n} - p_{bmn},$$ (5.43)

while the pitting resistance geometry factor, Z_I, is determined using relationships (5.44)–(5.53), without iteration. Instead, for spiral bevel and hypoid gears, an iteration procedure is necessary that use the same Eqs. (5.44)–(5.53).

The relative length of path of contact at critical point within the contact ellipse, $g_{\eta I}$ (this is a dimensionless quantity) is first calculated with the relationship:

$$g_{\eta I} = \sqrt{g_\eta^2 - 4y_I^2}. \tag{5.44}$$

Successively, the relative length of path of contact is determined, considering adjacent teeth, using the relationship:

$$
\begin{aligned}
g_{\eta I \Sigma} = {} & g_{\eta I}^3 + \sqrt{\left[g_{\eta I}^2 - 4p_{bmn}(p_{bmn} + 2y_I)\right]^3} + \sqrt{\left[g_{\eta I}^2 - 4p_{bmn}(p_{bmn} - 2y_I)\right]^3} \\
& + \sqrt{\left[g_{\eta I}^2 - 16p_{bmn}(p_{bmn} + y_I)\right]^3} + \sqrt{\left[g_{\eta I}^2 - 16p_{bmn}(p_{bmn} - y_I)\right]^3} \\
& + \sqrt{\left[g_{\eta I}^2 - 32p_{bmn}(2p_{bmn} + y_I)\right]^3} + \sqrt{\left[g_{\eta I}^2 - 32p_{bmn}(2p_{bmn} - y_I)\right]^3}
\end{aligned}
\tag{5.45}
$$

where however any square root term of negative value must be set equal to zero.

The load sharing ratio for pitting, ε_{NI}, is then calculated; it is given by:

$$\varepsilon_{NI} = g_{\eta I}^3 / g_{\eta I \Sigma}^3. \tag{5.46}$$

With previous quantities, the relative length of contact line, g_c is calculated by the following relationship:

$$g_c = g_{\eta I} \rho_{\Delta red} y_I / g_\eta^2. \tag{5.47}$$

Then the change of relative position along path of contact, $g_{\eta \Delta}$, is determined by the following relationship:

$$g_{\eta \Delta} = \frac{\rho_{\Delta red}^2 y_I}{g_{\eta I}^2} + k' g_c \tan \beta_{bm} + \frac{0.5 \rho_{\Delta red}}{\cos \beta_{bm}} - \rho_{\Delta 2}, \tag{5.48}$$

where k' is the contact shift factor, given by Eq. (4.75).

At this point, the following dimensionless intermediate factor should be calculated, given by:

$$X = \frac{\sin^2 w \cos \alpha_{\lim} \cos(\zeta_R - \lambda_1) \cos \lambda_1}{\sin^2(\beta_{m1} - \lambda_1) \sin \alpha_a \cos \zeta_R}. \tag{5.49}$$

Then, the relative radius of profile curvature, $\rho_{1,2}$, and the relative radius of profile curvature between pinion and wheel, ρ_t, are calculated with the following relationships:

$$\rho_{1,2} = R_{mpt1,2} X \pm g_{\eta \Delta} \tag{5.50}$$

$$\rho_t = \frac{\rho_1 \rho_2}{\rho_1 + \rho_2}. \tag{5.51}$$

As a next step, the inertia factor, Z_i, is determined as a function of the modified contact ratio, $\varepsilon_{v\gamma}$. For this purpose, the following two relationships are used:

$$Z_i = 2/\varepsilon_{v\gamma} \text{ or } Z_i = 1, \tag{5.52}$$

which are respectively valid depending on whether $\varepsilon_{v\gamma} \leq 2$ or $\varepsilon_{v\gamma} > 2$.

Finally, the pitting resistance geometry factor, Z_I, is determined using the following relationship:

$$Z_I = \frac{g_c \rho_t m_{mn} \cos \alpha_a}{b_b z_1 Z_i \varepsilon_{NI} m_{et2}}. \tag{5.53}$$

As we already said, the use of Eqs. (5.44)–(5.53) for straight bevel and Zerol bevel gears does not involve any iteration procedure, while the latter is necessary for spiral bevel and hypoid gears. For these gears, the iteration procedure is started assuming as initial value $y_I = 0$, while Z_I is calculated using Eqs. (5.44)–(5.53). These equations shall be recalculated by stepping y_I in both directions until a minimum value of Z_I is found. The iterative procedure is ended when this minimum value is reached.

The second of the three-contact stress factor is the face width factor, Z_{FW}, which reflects the non-uniformity of material properties, and depends primarily on tooth size (essentially, diameter of gear members), material characteristics, area of stress pattern, face width, and ratio of tooth size to diameter of parts.

The face width factor is evidently a function of the material strength, and therefore it should appear in the equations for permissible stress. Despite this fact, it is more practicable that this factor is included in the equations for stress calculation, in accordance with Method $B2$. In this way, it is possible to obtain S-N curves from experimental data, using a wide range of gears having different tooth sizes (otherwise, we should limit ourselves to use gears having only one tooth size).

In cases where we did not have enough experience, the face width factor for pitting resistance of bevel gears can be determined as a size width factor, which *ex nomine* depends only on the face width, using the following relationships

$$Z_{FW} = 0.50; \ Z_{FW} = 4.92 \times 10^{-3} b_2 + 0.4375; \ Z_{FW} = 0.83, \tag{5.54}$$

which are respectively valid for $b_2 < 12.7$ mm, $(12.7 \leq b_2 \leq 79.8)$, and $b_2 > 79.8$ mm. In fully equivalent terms, the diagram shown in Fig. 5.8 can be used.

The third factor of the three contact stress factors is the contact stress adjustment factor, Z_A, which adjusts the calculation results of Method $B2$ so that the contact stress numbers described in Sect. 2.10.1 can be used. The determination of Z_A is based on the comparison of ISO 6336-5: 2016 MQ grade carburized case-hardened steel, which has an allowable stress number of 1500 N/mm^2, to the equivalent ANSI/AGMA 2003-C10 grade 2 case-hardened steel, which has instead an allowable stress number of 1550 N/mm^2.

It follows that, for carburized case-hardened steel, the contact stress adjustment factor, Z_A, is given by:

Fig. 5.8 Face width factor, Z_{FW}, as a function of wheel face width, b_2

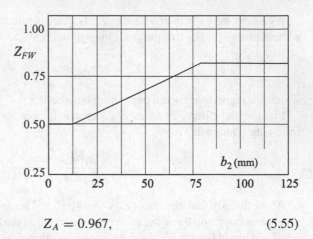

$$Z_A = 0.967, \tag{5.55}$$

while, for other specific materials and quality grades, this factor shall be calculated by taking the ratio of the allowable contact stress number in ISO 6336-5:2016 to equivalent ANSI/AGMA 2003-C10 steel.

5.7 Common Factors for Method $B1$ and Method $B2$: Factors for Contact Stress and Permissible Contact Stress

Equations (5.2) and (5.4), valid for Method $B1$, and the corresponding Eqs. (5.29) and (5.30), which instead are valid for Method $B2$, have in common the elasticity factor, Z_E, which appears in the expressions of nominal contact stress, and factors Z_L, Z_v, Z_R, Z_W and Z_{NT}, which appear in the expressions of the permissible contact stress; we already defined all these factors in Sect. 5.2. In that same section, we also said that all these factors are dimensionless factors, only with the exception of the elasticity factor, Z_E, which is a dimensional factor having the dimension of $\left(N/mm^2\right)^{1/2}$. The standardized calculation procedure of these factors is described below.

5.7.1 Elasticity Factor, Z_E

The elasticity factor, Z_E, considers the influence of the modulus of elasticity, E, and Poisson's ratio, v, of the materials of pinion and wheel on the nominal contact stress, σ_{H0}, and therefore on contact stress σ_H. It is determined using the following relationship:

$$Z_E = \sqrt{\frac{1}{\pi\left(\dfrac{1 - v_1^2}{E_1} + \dfrac{1 - v_2^2}{E_2}\right)}}. \tag{5.56}$$

For pinion and wheel made of the same material (then $v_1 = v_2 = v$, and $E_1 = E_2 = E$), Eq. (5.56) becomes:

$$Z_E = \sqrt{\frac{E}{2\pi\left(1 - v^2\right)}}, \tag{5.57}$$

and, for $v = 0, 3$, i.e. for steel and light metal alloys, we have:

$$Z_E = \sqrt{0.175E}. \tag{5.58}$$

For gear pairs made of materials with different modulus of elasticity ($E_1 \neq E_2$), the equivalent modulus of elasticity, E, may be determined using the relationship:

$$E = \frac{2E_1 E_2}{E_1 + E_2}. \tag{5.59}$$

For a steel-steel gear pair, with $E_{1,2}$ given in N/mm^2, we have $Z_E = 189.8(\text{N/mm}^2)^{1/2}$. For gear pairs made of other materials, see Table 2.1.

5.7.2   Lubricant Film Influence Factors, Z_L, Z_v, Z_R

The lubricant film influence factors, Z_L, Z_v, and Z_R consider approximately the influences on the lubricant film between the meshing tooth flanks exerted respectively by the lubricant oil viscosity (factor Z_L), tangential speed (factor Z_v), and flank surface roughness (factor Z_R). Figures 5.9, 5.10 and 5.11 show the ranges of variation of these three factors. The scattering, i.e. spread of their values, indicates the influences of other quantities besides the lubricant oil viscosity, tangential speed, and flank surface roughness, not accounted for in the approximate assumption above.

It is to be noted that, when there is no comprehensive experience or test results, as required by the Method A, factors Z_L, Z_v, and Z_R shall be determined separately according to Method B. In many cases (e.g., for most industrial gears), the shorter Method C may be used. Since these factors depend on the characteristics of the material, when the gear pair consists of one member made of hard material, and the other member made of soft material, they shall be determined for the softer of the two materials.

By Method B, the three factors above shall be calculated as described below.

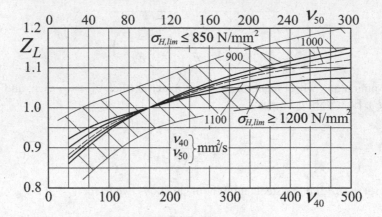

Fig. 5.9 Lubricant factor, Z_L, for mineral oils

Fig. 5.10 Speed factor, Z_v

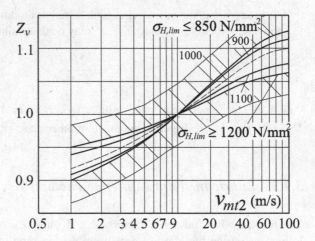

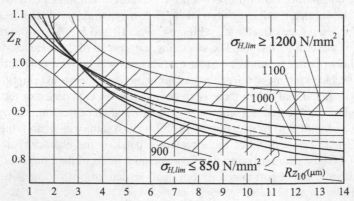

Fig. 5.11 Roughness factor, Z_R

5.7.2.1 Lubricant Factor, Z_L

With the restrictions previously mentioned, the lubricant factor, Z_L, considers the influence of the type of lubricant on the pitting resistance. The characteristic of the lubricant taken into consideration is only its viscosity. The values of this factor can be determined using the curves shown in Fig. 5.9. These curves are plotted for mineral oils, with or without EP-additives, as a function of the nominal viscosity (the nominal kinematic viscosity of the oil at 40 °C, v_{40}, or at 50 °C, v_{50}, both given in mm^2/s), and the allowable stress number for contact stress, $\sigma_{H,\lim}$ (in N/mm^2) of the softer material of the mating gear pair. For certain synthetic oils with lower coefficient of friction, larger values of Z_L than those calculated for mineral oils may be used. Oil viscosity must be chosen with reference to testing, experience or gear-lubrication publications.

Lubricant factor, Z_L, may be also calculated using the following relationship, which represents the curve courses shown in Fig. 5.9:

$$Z_L = C_{ZL} + \frac{4(1 - C_{ZL})}{\left(1.2 + \dfrac{134}{v_{40}}\right)^2}, \tag{5.60}$$

where C_{ZL}, in the range ($850\,\text{N/mm}^2 \le \sigma_{H,\lim} \le 1200\,\text{N/mm}^2$), is given by:

$$C_{ZL} = 0.08 \frac{\sigma_{H,\lim} - 850}{350} + 0.83. \tag{5.61}$$

Outside the aforementioned range of validity of Eq. (5.61), if $\sigma_{H,\lim} < 850\,\text{N/mm}^2$, we take $\sigma_{H,\lim} = 850\,\text{N/mm}^2$, while if $\sigma_{H,\lim} > 1200\,\text{N/mm}^2$, we take $\sigma_{H,\lim} = 1200\,\text{N/mm}^2$.

5.7.2.2 Speed Factor, Z_v

With the restrictions previously mentioned, the speed factor, Z_v, considers the influence of the tangential speed (in this regard, the tangential speed at the mid-face width of the reference cone of the wheel, v_{mt2}, is considered) on the pitting resistance. The values of this factor can be determined using the curves shown in Fig. 5.10. These curves are plotted as a function of the tangential speed, v_{mt2} (in m/s), and the allowable stress number for contact stress, $\sigma_{H,\lim}$ (in N/mm^2) of the softer material of the mating gear pair.

Speed factor, Z_v, may be also calculated using the following relationship, which represents the curve courses of Fig. 5.10:

$$Z_v = C_{Zv} + \frac{2(1 - C_{Zv})}{\sqrt{0,8 + \dfrac{32}{v_{mt2}}}}, \tag{5.62}$$

where C_{Zv}, in the range ($850\,\text{N/mm}^2 \le \sigma_{H,\text{lim}} \le 1200\,\text{N/mm}^2$), is given by:

$$C_{Zv} = 0.08 \frac{\sigma_{H,\text{lim}} - 850}{350} + 0.85, \tag{5.63}$$

Outside the aforementioned range of validity of Eq. (5.63), if $\sigma_{H,\text{lim}} < 850\,\text{N/mm}^2$, we take $\sigma_{H,\text{lim}} = 850\,\text{N/mm}^2$, while if $\sigma_{H,\text{lim}} > 1200\,\text{N/mm}^2$, we take $\sigma_{H,\text{lim}} = 1200\,\text{N/mm}^2$.

5.7.2.3 Roughness Factor, Z_R

Here also with restrictions previously mentioned, the roughness factor, Z_R, considers the influence of the surface roughness of the tooth flanks on the pitting resistance. In this regard, the mean roughness for gear pairs with relative curvature radius $\rho_{rel} = 10\,\text{mm}$, that is the roughness Rz_{10} is considered (see also Sect. 2.10.3.3). The values of this factor can be determined using the curves shown in Fig. 5.11, which is valid for a mating gear pair with a radius of relative curvature at the pitch point, ρ_{rel}, equal to 10 mm. These curves are plotted as a function of Rz_{10} (in μm), and the allowable stress number for contact stress, $\sigma_{H,\text{lim}}$ (in N/mm^2) of the softer material of the mating gear pair.

Roughness factor, Z_R, may be also calculated using the following relationship, which represents the curve courses shown in Fig. 5.11:

$$Z_R = \left(\frac{3}{Rz_{10}} \right)^{C_{ZR}} \tag{5.64}$$

where C_{ZR}, in the range ($850\,\text{N/mm}^2 \le \sigma_{H,\text{lim}} \le 1200\,\text{N/mm}^2$), is given by:

$$C_{ZR} = 0.12 + \frac{1000 - \sigma_{H,\text{lim}}}{5000}. \tag{5.65}$$

Here also, outside the aforementioned range of validity of Eq. (5.65), if $\sigma_{H,\text{lim}} < 850\,\text{N/mm}^2$, we take $\sigma_{H,\text{lim}} = 850\,\text{N/mm}^2$, while if $\sigma_{H,\text{lim}} > 1200\,\text{N/mm}^2$, we take $\sigma_{H,\text{lim}} = 1200\,\text{N/mm}^2$.

In Eq. (5.64), the mean roughness for gear pairs with relative curvature radius $\rho_{rel} = 10$ mm, Rz_{10}, is given by:

$$Rz_{10} = \frac{1}{2}(Rz_1 + Rz_2)\left(\sqrt[3]{\frac{10}{\rho_{rel}}} \right), \tag{5.66}$$

where Rz_1 and Rz_2 are the mean roughness of the pinion and wheel after manufacturing, while ρ_{rel} is the radius of relative curvature perpendicular to the contact line, which is expressed by Eq. (4.38).

5.7.2.4 Product of the Lubricant Film Factors, $Z_L Z_v Z_R$, for Method C

In many cases (e.g., most industrial gears), the shorter Method C may be used. It is to be considered as a simplification of Method B. This simplified method considers the product of the lubricant film factors, $Z_L Z_v Z_R$, and assumes that a lubricant viscosity has suitably been chosen for given operating conditions, in terms of tangential speed, load, and structural size.

In these cases, the recommended values of the product, $Z_L Z_v Z_R$, are as follows: 0.85, for through-hardened gear pairs without finishing process; 0.92, for gear pairs lapped after hardening; 1 or 0.92 for gear pairs ground after hardening or for hard-cut gear pairs, respectively with $Rz_{10} \leq 4\,\mu m$ or $Rz_{10} > 4\,\mu m$.

When these conditions do not apply, the three lubricant film factors must be determined separately according to Method B, as we said above.

5.7.3 Work Hardening Factor, Z_W

The work hardening factor, Z_W, takes into account the increase in pitting resistance in the case of meshing of a structural or through-hardened steel wheel with a case-hardened pinion having smooth tooth flanks ($Rz \leq 6\,\mu m$). This increase in pitting resistance of the soft wheel depends not only on work hardening, but also on other influences, such as polishing, alloying elements and residual or other internal stresses in the soft material, surface roughness of the hard pinion, hardening processes, and contact stress.

The work hardening factor, Z_W, is usually calculated by Method B, which is based on tests of different materials carried out with standard reference test gears, and on field experience with production gears. The intermediate curve between the three curves shown in Fig. 5.12 gives the factor Z_W as a function of the Brinell hardness, HB, of tooth flanks of the softer material of the mating gear pair (generally, the wheel). The extension of the sketched area between the two outer curves with respect to the

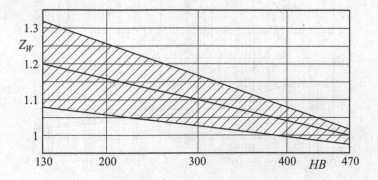

Fig. 5.12 Work hardening factor, Z_W

intermediate curve indicates the influence of other influence factors not included in the calculation procedure. The value of Z_W is an empirical value, not to be considered as an absolute value, due to approximations introduced. This value is taken as the same for endurance, limited life, and static stress.

The work hardening factor, Z_W, may be also calculated using the following relationship:

$$Z_W = 1.2 - \frac{HB - 130}{1700}, \tag{5.67}$$

which represents the course of the intermediate curve shown in Fig. 5.12. It is to be noted that $Z_W = 1.2$ for HB < 130, and $Z_W = 1$ for HB > 470. Moreover $Z_W = 1$, if pinion and wheel have the same hardness.

5.7.4   Life Factor, Z_{NT}

The life factor, Z_{NT}, considers the higher contact stress, including static stress, which can be considered acceptable for a limited life, in terms of number of cycles, as compared with the allowable stress at the knee on the curves shown in Fig. 5.13 (we define as knee the point on these curves where $Z_{NT} = 1$). For extended life, Z_{NT} can be less than 1 (see Fig. 5.13).

The main influences related to this factor are as follows: material and its heat treatment; lubrication regime; failure criteria; number of load cycles or service life, N_L; pitch line velocity; required smoothness of operation; gear material cleanliness; residual stresses; ductility and fracture toughness. Curves of Z_{NT} shown in Fig. 5.13

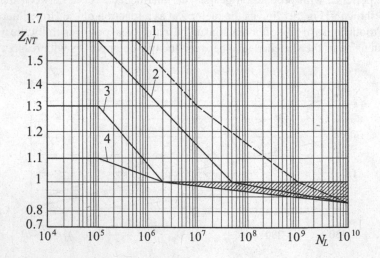

Fig. 5.13 Life factor for pitting resistance, Z_{NT} , for standard reference test gears

have been determined for standard test-gear conditions. These curves give the life factor, Z_{NT}, as a function of the number of load cycles, N_L, for the four groups of materials described in Table 5.3. The number of load cycles, N_L, is equal to the number of meshing contacts, under load, of the gear wheel under consideration. Here also it is to be noted that the first two groups of materials coincide, with the difference that for group 1 a limited pitting is permitted, while for group 2 no degree of pitting is tolerated.

It is to be noted that, with Method *B* that we are here considering, the permissible stress at limited service life, or the safety factor in the limited life stress range, shall be determined using the life factor, Z_{NT}, for standard reference test gear. The factors Z_L, Z_v, Z_R, and Z_W are not included. However, the modified effect of these factors on limited life must be considered. For static and endurance stresses, the life factor Z_{NT} can be taken from Fig. 5.13 or Table 5.3, while for limited life stresses this factor is determined by interpolation between the values for the endurance and static stresses.

Finally, it is to be noted that, if we use Method *A*, the damage curve (or S-N curve) derived from examples of the actual gear pair would be determinant for load carrying

Table 5.3 Life factor, Z_{NT}, for static and endurance stress limits

Material	Number of load cycles, N_L	Life factor, Z_{NT}
Group 1: St, V, GGG (perl., bain.), GTS (perl.), Eh, IF Only when a limited pitting is permitted	$N_L \leq 6 \times 10^5$, static	1.6
	$N_L = 10^7$, endurance	1.3
	$N_L = 10^9$, endurance	1.0
	$N_L = 10^{10}$, endurance	0.85
Group 2: St, V, GGG (perl., bain.), GTS (perl.), Eh, IF No pitting is tollerated	$N_L = 10^5$, static	1.6
	$N_L = 5 \times 10^7$, endurance	1.0
	$N_L = 10^{10}$, endurance	0.85
	$N_L = 10^{10}$, endurance with optimum lubrication, material, manufacturing and experience	1.0
Group 3: GG, GGG (ferr.), NT (nitr.), NV (nitr.)	$N_L = 10^5$, static	1.3
	$N_L = 2 \times 10^6$, endurance	1.0
	$N_L = 10^{10}$, endurance	0.85
	$N_L = 10^{10}$, endurance with optimum lubrication, material, manufacturing and experience	1.0
Group 4: NV (nitrocar.)	$N_L = 10^5$, static	1.1
	$N_L = 2 \times 10^6$, endurance	1.0
	$N_L = 10^{10}$, endurance	0.85
	$N_L = 10^{10}$, endurance with optimum lubrication, material, manufacturing and experience	1.0

capacity at limited service life. Therefore, it would be determinant for the materials of both mating gears, heat treatment, relevant diameter, module, surface roughness of tooth flanks, pitch line velocity and lubricant. Since this damage curve is directly valid for the conditions mentioned, all factors Z_L, Z_v, Z_R, Z_X, Z_W are equal to 1, because their influences are included in the curve.

5.8 Calculation Examples

Strength design, load carrying capacity assessment and determination of the safety factor against pitting of any type of bevel gear (including straight, helical or skew, Zerol, spiral bevel and hypoid gears) are reduced to those of the corresponding equivalent cylindrical gears. Therefore, the calculation examples that could be presented and discussed in this regard do not escape the space restrictions we have already highlighted in Sect. 2.11, to which we refer the reader.

References

ANSI/AGMA 2003-C10 Rating the pitting resistance and bending strength of generated straight bevel, zerol bevel and spiral bevel gear teeth

Barber JR (1992) Elasticity. Kluwer Academic Publishers, Dordrecht

Buckingham E (1949) Analytical mechanics of gears. McGraw-Hill Book Company Inc., New York

Budynas RG, Nisbett JK (2009) Shigley's mechanical engineering design, 8th edn. McGraw-Hill Companies Inc., New York

Burr AH (1982) Mechanical analysis and design. Elsevier Science Publishing Co., Inc., New York

Choy FK, Polyshchuk V, Zakrajsek JJ, Handschuh RF, Townsend DP (1994) Analysis of the effects of surface pitting and wear on the vibrations of a gear transmission system. NASA Technical Memorandum 106678, Army Research Laboratory Memorandum Report ARL-TR-520, Austrib '94, Australia, Perth, Western Australia, 5–8 Dec (1994)

Coleman W (1952) Improved method for estimating the fatigue life of bevel and hypoid gears. SAE Q Trans 2(6)

Dempsey PJ (2015) Investigation of spiral bevel gear condition indicator validation via AC-29-2C combining test rig damage progression data with fielded rotorcraft data. NASA/TM - 2015 - 218478, E-19030, GRC-E-DAA-TN20164, Technical Report, NASA Glenn Research Center, Cleveland, OH U.S.

Dempsey PJ, Lewicki DG, Decker HJ (2004) Investigation of gear and bearing fatigue damage using debris particle distributions. U.S. Army Research Laboratory, US (NASA/TM-2004-212883, ARL-TR-3133)

Dieter GE (1988) Mechanical metallurgy. McGraw-Hill Book Company Ltd., London SI Metric Edition adapted by David Bacon

Dudley DW (1962) Gear handbook. McGraw-Hill Book Company Inc, New York

Faupel JH (1964) Engineering design: a synthesis of stress analysis and materials engineering. Wiley Inc., New York

Giovannozzi R (1965) Costruzione di Macchine, vol II, 4th ed. Casa Editrice Prof. Riccardo Pàtron, Bologna

Handschuh RF (1997) Recent advances in the analysis of spiral bevel gears, NASA technical memorandum TM-107391. U.S. Army Research Laboratory, Lewis Research Center, Cleveland, Ohio, U.S.A.

Handschuh RF, Nanlawala M, Hawkins JM, Mahan D (2001) Experimental comparison of face-milled and face-hobbed spiral bevel gears. In: Proceedings of the JSME international conference on motion and power transmission, Fukuoka, Japan, 15–17 Nov (2001) (NASA/TM-2001-210940, ARL-TR-1104)

Henriot G (1979) Traité thèorique and pratique des engrenages, vol 1. 6th edn. Paris: Bordas

Höhn B-R, Winter H, Michaelis K, Vollhüter F (1992) Pitting resistance and bending strength of bevel and hypoid gear teeth. In: Proceedings of the 6th international power transmission and gearing conference, Scottsdale, Arizona, 13–16 Sept 1992

ISO 10300-2:2014 Calculation of load capacity of bevel gears—Part 2: Calculation of surface durability (pitting)

ISO 6336-5:2016 Calculation of load capacity of spur and helical gears—Part 5: Strength and quality of materials

Johnson KL (1985) Contact mechanics. Cambridge University Press, Cambridge, United Kingdom

Juvinall RC, Marshek KM (2012) Fundamentals of machine component design, 5th edn. Wiley Inc., New York

Kalker JJ (1990) Three-dimensional elastic bodies in rolling contact. Kluwer Academic Publishers, Dordrecht, The Netherlands

Litvin FL, Fuentes A (2004) Gear geometry and applied theory, 2nd edn. Cambridge University Press, Cambridge

Maitra GM (1994) Handbook of gear design, 2nd edn. Tata McGraw-Hill Publishing Company Ltd., New Delhi

Merritt HE (1954) Gears, 3rd edn. Sir Isaac Pitman & Sous Ltd, London

Niemann G, Winter H (1983a) Maschinen-Elemente, Band II: Getriebe allgemein, Zahnradgetriebe-Grundlagen, Stirnradgetriebe. Springer, Berlin Heidelberg

Niemann G, Winter H (1983b) Maschinen-Elemente, Band III: Schraubrad-, Kegelrad-, Schnecken-, Ketten-, Rienem-, Reibradgetriebe, Kupplungen, Bremsen, Freiläufe. Springer, Berlin, Heidelberg

Petr K, Vosyka J (2017) Design of test rig for bevel gears. In: 58th ICMD, Prague, Czech Republic, 6–8 Sept 2017

Pollone G (1970) Il Veicolo. Libreria Editrice Universitaria Levrotto & Bella, Torino

Radzevich SP (2016) Dudley's handbook of practical gear design and manufacture, 3rd edn. CRC Press, Taylor & Francis Group, Boca Raton, Florida

Riemann T, Karachi D, Yamanaka T, Yamamoto A, Stemplinger J-P, Stahl K (2017) Load-dependent bevel gear deflections and their impact on the pitting load carrying capacity, J-STAGE. In: The proceedings of the JSME international conference on motion and power transmissions, Session ID: 05–12

Schiebel A, Lindner W (1957) Zahnräder, Band 2: Stirn-und Kegelräder mit schrägen Zähnen Schraubgetriebe. Springer, Berlin, Heidelberg

Sekercioglu T, Koran V (2007) Pitting failure of truck spiral bevel gear. Eng Fail Anal 14(4):614–619

Shipley EE (1967) Gear failures: how to recognize them, what causes them, how to avoid them, Machine Design, 7 Dec 1967

Shtipelman BA (1978) Design and manufacture of hypoid gears. Wiley Limited, Canada

Spievak LE, Wawrzynek PA, Ingraffea AR, Lewicki DG (2001) Simulating fatigue crack growth in spiral bevel gears. Eng Fract Mech 68(1):53–76

Stadtfeld HJ (1993) Handbook of bevel and hypoid gears. Rochester Institute of Engineering, Rochester, New York

Stokes A (1992) Manual gearbox design. SAE Internal Society of Automotive Engineers, Linacre House, Jordan Hill, Oxford, Butterworth-Heinemann Ltd.

Timoshenko S, Goodier JN (1953) Theory of elasticity, 2nd edn. McGraw-Hill Book Company Inc., New York

Ural A, Heber G, Wawrzynek PA, Ingraffea AR, Lewicki DG, Neto JBC (2005) Three-dimensional, parallel, finite element simulation of fatigue crack growth in a spiral bevel pinion gear. Eng Fract Mech 72:1148–1170

Chapter 6
Tooth Root Strength of Bevel Gears

Abstract In this chapter, a general survey is first done on the tooth bending strength of bevel gears in the broadest meaning, including straight, helical (or skew), Zerol, spiral bevel and hypoid gears and the main factors affecting the tooth root stress state are defined. Attention is paid to the fatigue breakage at the tooth root of these gears and safety factor to be used for their design. The theoretical bases of calculation of tooth bending strength, already discussed in previous chapters, are then recalled, particular those concerning the Lewis constant strength parabola, 30° tangent and cantilever rectangular plate with constant thickness subject to bending load, with which the gear tooth is simulated. The methods of adaption of these theoretical bases to bevel gears are then described. A large part of the chapter is reserved for the procedures to calculate the tooth bending strength of these types of gears in accordance with the ISO standards, highlighting when deemed necessary as the relationships used by the same ISO are founded on the theoretical bases previously recalled.

6.1 Introduction

Failure of gear teeth by breakage can occur in many ways. The main types of breakage are those related to severe instantaneous overloads, tooth edging breakage, case crushing, root fracture through the rim of the gear blank, tooth bending fatigue, etc. (Shipley 1967). We described the main characteristics of these types of gear tooth failures by breakage in the initial sections of Chaps. 1 and 3, to which we refer the reader. Here we will focus our attention on the fatigue breakage at the tooth root of bevel gears due to bending load (see Coleman 1952; Höhn et al. 1992; Fernandes 1996; Handschuh et al. 2001; Vijayakar 2016). However, it is not the case to go back on the general concepts relating to tooth bending strength, because they do not change, passing from the discussion of spur and helical gears to that of bevel gears (see Giovannozzi 1965; Niemann and Winter 1983a, b; Budynas and Nisbett 2009; Juvinall and Marshek 2012; Radzevich 2016).

The theoretical bases of calculations of tooth root strength of bevel gears are the same as described in Chap. 3 with regard to spur and helical gears. The specific treatises (see Buckingham 1949; Merritt 1954; Dudley 1962; Shtipelman 1978; Henriot

1979; Stadtfeld 1993; Maitra 1994; Radzevich 2016; Klingelnberg 2016; Goldfarb and Barmina 2016), and the general treatises (see Giovannozzi 1965; Pollone 1970; Niemann and Winter 1983a, b; Budynas and Nisbett 2009; Naunheimer et al. 2011; Juvinall and Marshek 2012; Fisher et al. 2015) on this subject are also the same.

In this chapter, we will describe in detail the calculation procedure for the determination of bending load carrying capacity of straight and bevel (skew), Zerol and spiral bevel gears including hypoid gears. Therefore, here also we mean the term bevel gears in its broadest meaning, including any type of bevel gear. The calculation relationships in the following sections comply with ISO 10300-3:2014, and include all the influences on tooth root strength for which quantitative assessment can be made. The strength ratings in accordance with this ISO standard are based on cantilever beam theory suitably modified to consider:

– compressive stress at the tooth root due to radial component of the tooth load;
– non-uniform load distribution, resulting from the inclined contact lines on the tooth flank surfaces of spiral bevel gears;
– load sharing between adjacent tooth pairs in simultaneous meshing contact;
– stress concentrations at the tooth root fillets, particularly the fillets that are subject to tensile normal stress;
– lack of smoothness caused by a small total contact ratio.

However, it should be noted that, for particular issues (see Sect. 4.6.1), the cantilever rectangular or trapeze plate theory comes into play, instead of the cantilever beam theory (see Timoshenko and Woinowsky-Krieger 1959; Wellauer and Seireg 1960; Kagawa 1961; Gagnon et al. 1997; Ventsel and Krauthammer 2001).

The ISO standards do not cover fractures occurring from the tooth root through the gear rim or through the gear blank to the bore, due to insufficient rim thickness (see Lewicki and Ballarini 1997; Curà et al. 2015). In addition, ISO standards do not consider the effects of surface stresses due to pitting or wear, which occasionally may limit the bending strength, due either to stress concentration around large sharp-cornered pits, or to wear steps on the tooth flank surfaces. However, with appropriate use of the ISO formulae, the tooth root fillet fracture during the gear design life is averted. These formulae use the maximum tensile stress at the tooth root as criterion (see Vullo and Vivio 2013; Vullo 2014) for the assessment of the bending tooth root strength, as the teeth can experience breakage when the material allowable stress is exceeded.

In the case of straight bevel gears, the assumption is made that the load is applied at the tooth tip of the corresponding virtual cylindrical gear. Subsequently, the load is converted to the outer point of single tooth contact. Thus, the procedure corresponds to Method C for the bending tooth root stress of cylindrical gears. In the case of spiral bevel gears and hypoid gears with a high face contact ratio $\varepsilon_{v\beta} > 1$ (Method $B1$) or with a modified contact ratio $\varepsilon_{vy} > 2$ (Method $B2$), the mid-point in the zone of contact is considered as the critical point of load application.

It should be noted that the breakage of a tooth generally involves the end of the gear's operating life. In fact, it often happens that the breakage of a single tooth determines the consequent breakage of all the gear teeth, and therefore the power transmission between input and output shafts is interrupted. For this reason, for the bending strength assessment of the tooth, it is recommended to use a safety factor for bending stress, S_F, greater than that used in the pitting resistance assessment. As we already mentioned in Sect. 4.1, it is recommended to choose a minimum safety factor for bending stress $S_{F,min} = 1.3$ for spiral bevel gears, including hypoid gears, and $S_{F,min} = 1.5$ for straight bevel gears or spiral bevel gears with $\beta_m \leq 5°$.

The phenomenological aspects, general concepts and limits of the calculation relationships used for determination of bending load carrying capacity of bevel gears are those already described in Sect. 4.1. For the reader's convenience, it is useful to recall here the well-defined restrictions of these calculation relationships, which are related to the virtual cylindrical gear pairs, equivalent to the actual bevel gear pairs. The main restrictions are as follows: minimum rim thickness under the tooth root equal to 3.5 m_{mn}; transverse contact ratio of virtual cylindrical gear, $\varepsilon_{v\alpha} < 2$; average mean spiral angle, $(\beta_{m1} + \beta_{m2})/2 \leq 45°$; effective pressure angle, $\alpha_e \leq 30°$; face width, $b \leq 13 m_{mn}$; sum of profile shift coefficients of pinion and wheel, $(x_1 + x_2) = 0$.

It is to be remembered that the calculation relationships described in the following sections are not applicable for the assessment of tooth flank fracture (TFF) and tooth interior fatigue fracture (TIFF), also known as tooth flank breakage (TFB). In addition, these relationships do not apply for stress levels above those permitted within the lifetime rang from 0 to 10^3 cycles. This is due to the fact that stresses in this range could exceed the elastic limit of the gear tooth material. However, the same relationships consider the influences on the tooth root stress, due to the load transmitted by the gear, provided that their quantitative evaluation is feasible.

The calculation procedure takes also into account additional stresses, which overlap and add up to stresses due to tooth loads, such as those caused by the shrink fit of gear rims. About the procedures of calculating stress and strain states due to shrink fit, also including the effects of any plastic strain due to exceeding the elastic limit of the materials involved, we refer the reader to Vullo (2014). In the standardized procedure, these stresses are however considered in the calculation of the tooth root stress, σ_F, or the permissible tooth root stress, σ_{FP}.

Two main methods are provided for the assessment of tooth bending strength of bevel and hypoid gears: Method $B1$ and Method $B2$. The calculation procedures are specific for each of the two methods. With Method $B1$, the same set of relationships is used for bevel and hypoid gears; on the contrary, with Method $B2$, partly different sets of relationships are used for bevel and hypoid gears.

With both methods, the ability of a gear tooth to resist fatigue bending load is determined by comparison with the following stress values: the tooth root stress, σ_F, which is based on the geometry of the tooth, its manufacturing accuracy, rigidity of the gear blank, bearings and housing, and operating torque, and the permissible tooth root stress, σ_{FP}, which is based on the nominal stress number for bending, $\sigma_{F,lim}$, of a standard test gear, and the effect of the operating conditions under which the gear operates. The safety factor for bending, S_F, is defined as the quotient of the permissible tooth root stress divided by the calculated tooth root stress.

6.2 Calculation of Tooth Root Strength: ISO Basic Formulae of Method $B1$

For design of bevel gears (straight and helical or skew gears, Zerol and spiral gears including hypoid gears, all with a minimum rim thickness under the tooth root greater than or equal to 3.5 m_{mn}), free from tooth root breakage due to bending fatigue during their service life, we use the following rating formulae, which are unique to Method $B1$. The determination of the bending load carrying capacity, that is the capability of a gear to resist tooth bending fatigue, is performed separately for pinion and wheel and, in the case of hypoid gears, additionally for drive side flank and coast side flank, using the following relationship:

$$\sigma_{F-B1} = \sigma_{F0-B1} K_A K_v K_{F\beta} K_{F\alpha} \leq \sigma_{FP-B1}, \tag{6.1}$$

where σ_{F-B1} is the tooth root stress for Method $B1$, K_A, K_v, $K_{F\beta}$, and $K_{F\alpha}$ are the load factors already defined in Chap. 4, and σ_{F0-B1} is the nominal tooth root stress for Method $B1$, given by:

$$\sigma_{F0-B1} = \frac{F_{vmt}}{b_v m_{mn}} Y_{Fa} Y_{Sa} Y_\varepsilon Y_{BS} Y_{LS}. \tag{6.2}$$

The nominal tooth root stress, σ_{F0-B1}, is defined as the maximum bending stress at the tooth root, identified by the 30° tangent to the root fillet. In this last equation, F_{vmt} is the nominal tangential force of the virtual cylindrical gear, given by Eq. (4.6), and b_v is the face width of the virtual cylindrical gear calculated by means of Eq. (4.25) for the active flank, drive or coast side. In the same equation, the other symbols represent: Y_{Fa}, the *tooth form factor*, which takes into account the influence of the tooth shape on the nominal bending stress for load application at the tooth tip; Y_{Sa}, the *stress correction factor*, which takes into account the stress-increasing notch effect at the root fillet, as well as the radial component of the tooth load and the complex stress concentration in the critical root section (however, it does not take into account the decrease of the tooth root stress determined by the variation in the load application from the position at the tooth tip to the determinant position for pinion or wheel, from which a decrease in bending moment arm follows); Y_ε, the *contact ratio factor*, which takes into account the decrease of the root stress determined for load application at the tooth tip to the determinant position for pinion or wheel; Y_{BS}, the *bevel spiral angle factor*, which takes into account the smaller values of the theoretical length of middle contact line, l_{bm} (this can be calculated with Eq. (4.29)) compared to the total face width, b_v, and the inclined lines of contact; Y_{LS}, the *load sharing factor*, which takes into account the load distribution between two or more pairs of contacting teeth.

For the use of equations above and those that follow, we consider the load distributed over the contact lines described in Sect. 4.2, and the following determinant position of load application: (a) the outer point of single pair tooth contact, if $\varepsilon_{v\beta} = 0$;

(b) the mid-point of the zone of action, if $\varepsilon_{v\beta} \geq 1$; (c) interpolation between (a) and (b), if $\left(0 < \varepsilon_{v\beta} < 1\right)$.

The permissible tooth root stress, σ_{FP-B1}, to be calculated separately for pinion and wheel, should be preferably evaluated on the basis of the strength of standard test gears instead of prismatic specimens, which deviate too much with respect to actual gear teeth in terms of similarity in geometry, manufacture and course of movement. The permissible tooth root stress is determined with the following relationship:

$$\begin{aligned} \sigma_{FP-B1} &= \sigma_{FE} Y_{NT} Y_{\delta,relT-B1} Y_{R,relT-B1} Y_X \\ &= \sigma_{F,\lim} Y_{ST} Y_{NT} Y_{\delta,relT-B1} Y_{R,relT-B1} Y_X, \end{aligned} \qquad (6.3)$$

where $\sigma_{FE} = \sigma_{F,\lim 1,2} Y_{ST}$, is the allowable stress number for bending (it is the basic bending strength for the un-notched specimen under the assumption that the material is fully elastic, accounting also heat treatment); $\sigma_{F,\lim}$, is the nominal stress number for bending of the standard test gear, which takes into account material, heat treatment, and surface characteristics at test gear dimensions; $Y_{ST} = 2$, is the stress correction factor for the dimensions of the standard test gear; $Y_{\delta,relT-B1} = Y_\delta / Y_{\delta T}$, is the relative notch sensitivity factor for the permissible stress number, related to the conditions at the standard test gear, which takes into account the notch sensitivity of the material; $Y_{R,relT-B1} = Y_R / Y_{RT}$, is the relative surface condition factor, which takes into account the surface condition at the root fillet, related to the conditions at the test gear; Y_X, is the size factor for tooth root bending stress, which takes into account the influence of the module on the tooth root strength; Y_{NT}, is the life factor for tooth root bending stress for reference test conditions, which takes into account the influence of the required numbers of cycles of operation.

In accordance with Eq. (6.1), the evaluated tooth root stress, σ_F, must be less than or equal to the permissible tooth root stress, σ_{FP}, i.e. $\sigma_F \leq \sigma_{FP}$. The safety factor, S_F, against fatigue breakage at the tooth root due to bending load, to be determined separately for pinion and wheel with respect to the transmitted torque, is calculated using the following relationship:

$$S_{F-B1} = \frac{\sigma_{FP-B1}}{\sigma_{F-B1}} > S_{F,\min}. \qquad (6.4)$$

About the aforementioned symbols, regardless of the Method $B1$ or Method $B2$, it is to be noted that: σ_{F0}, is the nominal tooth root stress, i.e. the bending stress in the critical section of the tooth root, calculated considering the critical point of load application, for error-free gears loaded by a constant nominal torque; σ_F, is the tooth root stress, i.e. the determinant bending stress in the critical section of the tooth root, calculated considering the critical point of load application, including the load factors that take into account the static and dynamic loads and load distribution; $\sigma_{F,\lim}$, is the nominal stress number for bending, i.e. the maximum tooth root stress of standardized test gears, determined under well-defined and standardized operating conditions; σ_{FE}, is the allowable stress number for bending, i.e. the maximum bending stress of the un-notched test piece under the assumption that the material is fully elastic; σ_{FP}, is the permissible tooth root stress, i.e. the maximum tooth root stress at the gear set under consideration, including all influence factors. In addition, it is to be noted that we define as tooth root breakage a failure of gear teeth at the tooth root by static or dynamic overload.

Moreover, it must be noted that, in Eq. (6.1), K_A, K_v, $K_{F\beta}$ and $K_{F\alpha}$ are respectively the application factor, dynamic factor, face load factor for root bending stress, and transverse load factor for root bending stress already determined, using such procedures as we described respectively in Sects. 4.4–4.7. The inequality (6.1) compares the tooth root stress, σ_F, with the permissible tooth root stress, σ_{FP}, imposing that the first is less than or at most equal to the second.

6.3 Tooth Root Stress Factors for Method $B1$

In Eq. (6.2), the already defined factors Y_{Fa}, Y_{Sa}, Y_ε, Y_{BS}, and Y_{LS} appear, which are known to be the tooth root stress factors for Method $B1$. All five of these factors are dimensionless factors. These five factors adjust what we could define as a reference stress value, given by $(F_{vmt}/b_v m_{mn})$, to provide the nominal value of the tooth root stress.

6.3.1 Tooth Form Factor, Y_{Fa}

The tooth form factor, Y_{Fa}, considers the influence of the tooth shape on the nominal tooth root stress for load application at the tooth tip. This factor must be determined separately for pinion and wheel. In addition, the possibility to manufacture bevel and hypoid gears with different pressure angles at drive and coast sides must be considered. Since the load is assumed to be applied at the tooth tip, in the case of gears with tip and root relief the actual bending moment arm is slightly smaller, due to the combined effects of the two reliefs at the beginning of the meshing cycle. This issue of fact, however, is neglected, but in doing so the calculation is on the safe side.

Generally, deviations from the spherical involute profile (almost always bevel gears without offset have octoid profile, and a tip and root relief) are small, especially with reference to the calculation of the tooth root chordal thickness and bending moment arm. Thus, they may be neglected when calculating the tooth form factor, Y_{Fa}, as well as the stress correction factor, Y_{Sa}. Therefore, Y_{Fa} and Y_{Sa} are determined for nominal involute gears without profile deviations. In addition, the slight reduction in tooth thickness due to backlash between the teeth can be neglected for calculation of the bending load carrying capacity. However, the size reduction must be considered when the *outer tooth thickness allowance*, A_{sne} (in mm), is greater than $0.05 m_{mn}$ ($A_{sne} > 0.05 m_{mn}$).

The critical section to be considered for calculation is that having one side coincident with the tooth root chordal thickness, s_{Fn}, i.e. the distance between the contact

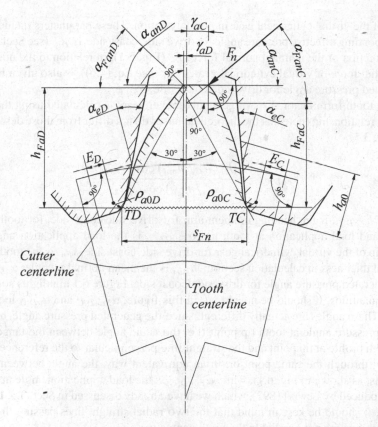

Fig. 6.1 Tooth root chordal thickness, s_{Fn}, and bending moment arm, h_{Fa} (h_{FaD} and h_{FaC}) in normal section for load application at the tooth tip of the virtual cylindrical gear

points of the 30° tangents at the root fillets of the virtual cylindrical gears (Fig. 6.1). The bending moment arm for tooth root stress and load application at tooth tip, h_{Fa}, is the distance measured along the tooth centerline between the tooth root chord, s_{Fn}, and the point where the straight line of load application at the tooth tip intersects this centerline, always with reference to the virtual cylindrical gears. Figure 6.1 shows that the bending moment arm changes depending on whether the drive side or the coast side are considered, assuming values respectively given by h_{FaD} and h_{FaC}.

The tooth form factor, Y_{Fa}, is determined by means of different procedures, depending on whether generated gears or not-generated gears are considered.

6.3.1.1   Tooth Form Factor for Generated Gears

The tooth form factor, Y_{Fa}, for generated bevel gears, and the quantities on which it depends, must be determined separately for pinion and wheel. It is to remember that the tooth form factor is to be calculated with parameters of the active tooth flank of the virtual cylindrical gear in normal section. These parameters include the corresponding effective pressure angle for drive side/coast side, $\alpha_{eD,C}$ (see Sect. 4.2). The direction of the nominal normal force, F_n (Fig. 6.1), in relation to the nominal tangential force of virtual cylindrical gear, F_{vmt} (see Eq. (4.6)), is also given by the generated pressure angle for drive side/coast side, $\alpha_{nD,C}$.

The tooth form factor, Y_{Fa}, of generated bevel gears is calculated using the following relationship (note that the theoretical bases do not differ from those described in Sect. 3.5):

$$Y_{FaD,C} = \frac{6(h_{FaD,C}/m_{mn})\cos\alpha_{FanD,C}}{(s_{Fn}^2/m_{mn}^2)\cos\alpha_{nD,C}}, \tag{6.5}$$

where: $h_{FaD,C}$, is the bending moment arm for drive side/coast side, for tooth root stress and load application at tooth tip; $\alpha_{FanD,C}$, is the load application angle at tooth tip of the virtual cylindrical gear for drive side/coast side; s_{Fn}, is the tooth root chordal thickness in calculation section; m_{mn}, is the mean normal module; $\alpha_{nD,C}$, is the generated pressure angle for drive side/coast side. Figure 6.1 highlights some of these quantities. It should be noted that, in this figure, $\alpha_{FanD,C}$ and $\alpha_{nD,C}$ look the same. These angles are actually different, since the generated pressure angle, $\alpha_{nD,C}$, is the pressure angle at tooth tip point (i.e. the acute angle between the tangent to the tooth profile at tip point and the straight line perpendicular to the reference cone passing through the same point or, in an equivalent way, the angle between their normals, as shown in Fig. 6.1), while $\alpha_{FanD,C}$ is the load application angle at tooth tip introduced by Lewis (1892), which we have already discussed in Sect. 3.3. In this regard, it should be kept in mind that the two radial straight lines passing through the tooth tips are not parallel to the tooth centerline.

For calculations of the tooth root chordal thickness, s_{Fn}, and bending moment arm, $h_{FaD,C}$, firstly the following auxiliary quantities for tooth form factor, $E_{D,C}$, $G_{D,C}$, and $H_{D,C}$, and the auxiliary quantity for tooth form and tooth correction factors, $\vartheta_{D,C}$, must be determined. It is to be noted that $E_{D,C}$ has the dimension of a length (in mm), $G_{D,C}$ and $H_{D,C}$ are dimensionless quantities, while $\vartheta_{D,C}$ has dimension of an angle (in rad). These auxiliary quantities, similar to those introduced for cylindrical spur and helical gears (see Sect. 3.5), are given respectively by:

$$E_{D,C} = \left(\frac{\pi}{4} - x_{sm}\right)m_{mn} - h_{a0}\tan\alpha_{eD,C} - \frac{\rho_{a0D,C}(1 - \sin\alpha_{eD,C}) - s_{prD,C}}{\cos\alpha_{eD,C}} \tag{6.6}$$

$$G_{D,C} = \frac{\rho_{a0D,C}}{m_{mn}} - \frac{h_{a0}}{m_{mn}} + x_{hm} \tag{6.7}$$

$$H_{D,C} = \frac{2}{z_{vnD,C}}\left(\frac{\pi}{2} - \frac{E_{D,C}}{m_{mn}}\right) - \frac{\pi}{3} \tag{6.8}$$

$$\vartheta_{D,C} = \frac{2G_{D,C}}{z_{vnD,C}}\tan\vartheta_{D,C} - H_{D,C}, \tag{6.9}$$

where: x_{hm} and x_{sm}, are respectively the profile shift coefficient and thickness modification coefficient; $\alpha_{eD,C}$, is the effective pressure angle for drive side/coast side;
$\rho_{a0D,C}$ (in mm), is the cutter edge radius for drive side/coast side; s_{pr} (in mm), is
the amount of protuberance at the cutter tool; h_{a0} (in mm), is the tool addendum;
$z_{vnD,C}$, is the number of teeth of virtual cylindrical gear in normal section for drive
side/coast side.

It is to be noted that the auxiliary quantity $E_{D,C}$ must be calculated for the magnitudes of the active tooth flank. Furthermore, for generated hypoid gears, the effective
pressure angle $\alpha_{eD,C}$ for drive side/coast side must be introduced in Eq. (6.6). In
addition, it is to be noted that the cutter edge radii, ρ_{a0D} and ρ_{a0C}, as well as the
protuberances, s_{prD} and s_{prC}, can be different, while the tool addendum, h_{a0}, does
not change. Finally, it is to be noted that Eq. (6.9) is a transcendent equation, to be
solve by iteration procedure. If, as the initial value, we put in this equation $\vartheta = \pi/6$,
in most cases the solution converges to the exact value after a few iterations. For
subsequent iterations to the initial step, it is suggested a value of the difference
$(\vartheta_{new} - \vartheta) = 1 \times 10^{-6}$ rad.

Figure 6.2 shows the shape of the teeth of the basic rack, with and without protuberance. It highlights the following quantities: E, the auxiliary quantity for tooth
form factor, which represents the distance of the center of edge radius of the basic
rack tooth profile from the tooth axis; ρ_{a0}, the cutter edge radius, which is equal to

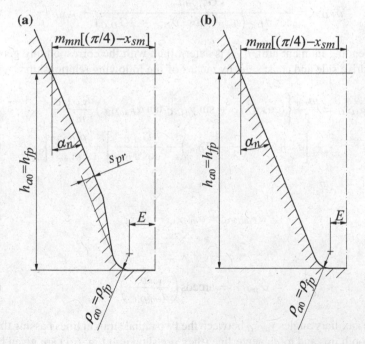

Fig. 6.2 Basic rack profile of the tooth and its dimensions: **a** with protuberance; **b** without
protuberance

root fillet radius, ρ_{fP}, of basic rack for cylindrical gears; s_{pr}, the amount of protuberance; α_n, the normal pressure angle; h_{a0}, the tool addendum, which is equal to
dedendum of the basic rack profile, h_{fP}.

Once these four auxiliary quantities have been determined, the tooth root chordal
thicknesses, s_{FnD} and s_{FnC}, are calculated for pinion and wheel, each with the
corresponding geometry data for drive side and coast side, using the following
relationship:

$$s_{FnD,C} = m_{mn}\left[z_{vnD,C} \sin\left(\frac{\pi}{3} - \vartheta_{D,C}\right) + \sqrt{3}\left(\frac{G_{D,C}}{\cos\vartheta_{D,C}} - \frac{\rho_{a0D,C}}{m_{mn}}\right)\right]. \quad (6.10)$$

Then, we get the respective tooth root chordal thickness, s_{Fn}, for pinion or wheel,
using the following relationship:

$$s_{Fn} = \frac{1}{2}(s_{FnD} + s_{FnC}). \quad (6.11)$$

To calculate the bending moment arm, h_{Fa}, we must first determine the fillet
radius, ρ_F, at the contact point of 30° tangent. This fillet radius is also determined with
the corresponding data for drive side and coast side, using the following relationship:

$$\rho_{FD,C} = \frac{2G_{D,C}^2 m_{mn}}{\cos\vartheta_{D,C}\left(z_{vnD,C}\cos^2\vartheta_{D,C} - 2G_{D,C}\right)} + \rho_{a0D,C}. \quad (6.12)$$

The bending moment arm, h_{Fa}, is determined with the corresponding geometry
data for drive side and coast side, by means of the following relationship:

$$h_{FaD,C} = \frac{m_{mn}}{2}\left\{ (\cos\gamma_{aD,C} - \sin\gamma_{aD,C}\tan\alpha_{FanD,C})\frac{d_{vanD,C}}{m_{mn}} \right.$$
$$\left. - \left[z_{vnD,C}\cos\left(\frac{\pi}{3} - \vartheta_{D,C}\right) + \frac{G_{D,C}}{\cos\vartheta_{D,C}}\right] + \frac{\rho_{a0D,C}}{m_{mn}} \right\}, \quad (6.13)$$

where

$$\alpha_{FanD,C} = \alpha_{anD,C} - \gamma_{aD,C}, \quad (6.14)$$

with

$$\alpha_{anD,C} = \arccos\left(\frac{d_{vbnD,C}}{d_{vanD,C}}\right), \quad (6.15)$$

while the auxiliary angles $\gamma_{aD,C}$ between the two radial straight lines passing through
the two tooth tips and tooth centerline (they are shown in Fig. 6.1) are given by:

$$\gamma_{aD,C} = \frac{1}{z_{vnD,C}} \left[\frac{\pi}{2} + 2\left(x_{hm} \tan \alpha_{eD,C} + x_{sm}\right) \right]$$
$$+ \mathrm{inv}\alpha_{eD,C} - \mathrm{inv}\alpha_{anD,C}. \tag{6.16}$$

In Eqs. (6.13) to (6.16), the meaning of symbols already introduced remains unchanged, while the new symbols represent: $\alpha_{anD,C}$, the normal pressure angle at tooth tip for drive side and coast side; $\gamma_{aD,C}$, the auxiliary angle for tooth form factor and tooth correction factor, for drive side and coast side; $d_{vbnD,C}$, the base diameter of virtual cylindrical gear in normal section, for drive side and coast side; $d_{vanD,C}$, the tip diameter of cylindrical gear in normal section, for drive side and coast side. Furthermore, $\alpha_e = \alpha_{eD}$ is the effective pressure angle for drive side, while $\alpha_e = \alpha_{eC}$ is the effective pressure angle for coast side. For more details, we refer the reader to Chap. 4.

Finally, it is to be noted that, at the design stage, the tooth form factor, Y_{Fa}, for bevel gears without offset can be determined for a basic rack profile of the tool characterized by the following data: $\alpha_n = 20°$; $(h_{a0}/m_{mn}) = 1.25$; $(\rho_{a0}/m_{mn}) = 0.25$ (see also Chap. 3).

6.3.1.2 Tooth Form Factor for Non-Generated Gears

In Vol. 1, Sect. 9.4, we described the cutting process for non-generated straight bevel gears, whose teeth are not cut with reference to a common generation crown wheel. In this case, the wheel of a bevel gear pair is produced without generation motion, by means of inserted-blade milling cutters, which are different depending on the cutting process used, and generally have tooth flanks with straight profile. Instead, the pinion is obtained with generation motion, by means of a profiling cutter tool able to simulate the pure rolling of its cutting surface on the operating pitch cone of the same pinion. In this way, the straight flanks of the teeth of the virtual generation wheel generate the curved flanks of the pinion teeth. So, we have a single gear pair, in the sense that only that pinion and only that wheel are able to ensure a correct kinematics, provided that the assembly and actual operating conditions strictly fulfill the cutting conditions.

These concepts are susceptible of generalization, for which they are applicable not only to straight bevel gears, but also to helical (skew), Zerol and spiral bevel gears including hypoid gears. In all these cases, the tooth form factor, Y_{Fa}, for non-generated bevel gears must be considered separately. In fact, since the wheel cutting process is done with a form cutting tool, the slot profile of the wheel is identical to the tooth profile, for which the tooth form factor can be determined directly (Fig. 6.2). Instead, the tooth form factor of the pinion, which is cut by a special generation process, can be determined approximately using the procedure for the case of generated gears, described in Sect. 6.3.1.1. In these cases, hypoid gears are also calculated with geometry data for drive side and coast side.

The tooth form factor, Y_{Fa}, of non-generated bevel wheel is calculated using the following relationship:

$$Y_{FaD,C} = \frac{6(h_{FaD,C}/m_{mn})}{(s_{Fn}^2/m_{mn}^2)},$$
(6.17)

which is obtained from Eq. (6.5) for $\alpha_{FanD,C} = \alpha_{nD,C}$.

The tooth root chordal thicknesses s_{FnD} and s_{FnC} of the wheel are calculated with the corresponding geometry data for drive side and coast side, using the following relationship:

$$s_{FnD,C} = \pi m_{mn} - 2E_{D,C} - 2\rho_{a0D,C}\cos 30°,$$
(6.18)

where

$$E_{D,C} = \left(\frac{\pi}{4} - x_{sm}\right)m_{mn} - \left[h_{a0}\tan\alpha_{nD,C} + \frac{\rho_{a0D,C}(1 - \sin\alpha_{nD,C}) - s_{prD,C}}{\cos\alpha_{nD,C}}\right].$$
(6.19)

The tooth root chordal thickness, s_{Fn}, of the wheel is given by Eq. (6.11), which continues to be valid also in this case.

For the same wheel, the fillet radius at contact point of 30° tangent, and bending moment arm are given respectively by:

$$\rho_{FD,C} = \rho_{a0D,C}$$
(6.20)

$$h_{FaD,C} = h_{a0} - \frac{\rho_{a0D,C}}{2} + m_{mn}\left[1 - \left(\frac{\pi}{4} + x_{sm} - \tan\alpha_{nD,C}\right)\tan\alpha_{nD,C}\right].$$
(6.21)

6.3.2 Stress Correction Factor, Y_{Sa}

The stress correction factor, Y_{Sa}, considers the stress increasing effect due to the root fillet notch as well as other stress components different from tooth root bending stress. This stress correction factor converts the nominal bending stress to the local tooth root stress. The calculation relationship of this factor, which is valid in the range $(1 \le q_s \le 8)$, is as follows:

$$Y_{SaD,C} = (1.2 + 0.13L_{aD,C})q_{sD,C}^{\left[\frac{1}{1.21+(2.3/L_{aD,C})}\right]},$$
(6.22)

where

$$L_{aD,C} = \frac{s_{Fn}}{h_{FaD,C}} \qquad (6.23)$$

$$q_s = \frac{s_{Fn}}{2\rho_{FD,C}} \qquad (6.24)$$

The quantities that appear in these last two equations are to be calculated in this way: s_{Fn}, according to Eq. (6.11), but considering Eq. (6.10) or Eq. (6.18), depending on whether generated gears or non-generated gears are considered; $h_{FaD,C}$, according to Eq. (6.13) or (6.21), depending on whether generated gears or non-generated gears are considered; $\rho_{FD,C}$, according to Eq. (6.12) or Eq. (6.20), depending on whether generated gears or non-generated gears are considered.

6.3.3   Contact Ratio Factor, Y_ε

The contact ratio factor, Y_ε, converts the point of load application at tooth tip, where tooth form factor, Y_{Fa}, and stress correction factor, Y_{Sa}, apply, to the determinant point of load application. For calculation of this contact ratio factor, we use the following relationships:

$$Y_\varepsilon = 0.25 + \frac{0.75}{\varepsilon_{v\alpha}} \geq 0.625 \qquad (6.25)$$

$$Y_\varepsilon = 0.25 + \frac{0.75}{\varepsilon_{v\alpha}} - \varepsilon_{v\beta}\left(\frac{0.75}{\varepsilon_{v\alpha}} - 0.375\right) \geq 0.625 \qquad (6.26)$$

$$Y_\varepsilon = 0.625, \qquad (6.27)$$

which are respectively valid for $\varepsilon_{v\beta} = 0$, $\left(0 < \varepsilon_{v\beta} \leq 1\right)$, and $\varepsilon_{v\beta} > 1$. In these relationships, $\varepsilon_{v\alpha}$ and $\varepsilon_{v\beta}$ are respectively the transverse contact ratio and face contact ratio of virtual cylindrical gears.

6.3.4   Bevel Spiral Angle Factor, Y_{BS}

The bevel spiral angle factor, Y_{BS}, considers the non-uniform distribution of the tooth root stress along the face width. This stress distribution depends on the inclination of the contact lines due to spiral angle. With an increasing spiral angle, the inclination angle also increases until the contact lines are limited by the tip and root of the teeth. Therefore, the face width is not entirely used to bear the load. This leads to a stress distribution characterized by a higher maximum value at the tooth root in the middle of the face width. Figure 6.3, where a tooth developed into a plane is replaced by a cantilever plate with constant thickness, highlights the problems that underlie tooth bending deflections, albeit in a rather rough way, both because the cantilever plate has actually a variable thickness, and because it can have not only a rectangular shape, but also a trapezoidal shape, depending on whether the teeth have constant or linearly variable depth along the face width.

The bevel spiral angle factor, Y_{BS}, is calculated in accordance with the aforementioned approximation, using the following empirical relationship:

$$Y_{BS} = 1 + \frac{a_{BS}}{c_{BS}}\left(\frac{l_{bb}}{b_a} - 1.05 b_{BS}\right)^2, \tag{6.28}$$

where a_{BS}, b_{BS}, and c_{BS} are dimensionless auxiliary factors given by the following also empirical relationships:

$$a_{BS} = -0.0182\left(\frac{b_a}{h}\right)^2 + 0.4736\left(\frac{b_a}{h}\right) - 0.320 \tag{6.29}$$

$$b_{BS} = -0.0032\left(\frac{b_a}{h}\right)^2 + 0.0526\left(\frac{b_a}{h}\right) + 0.712 \tag{6.30}$$

$$c_{BS} = -0.0050\left(\frac{b_a}{h}\right)^2 + 0.0850\left(\frac{b_a}{h}\right) + 0.540, \tag{6.31}$$

while (see Fig. 6.3) b_a, l_{bb}, and h are, respectively, the developed length of a curved tooth that is assumed as face width of the calculation model, the length part of the face width of the same model covered by the contact line, and the average tooth depth. These three quantities are given by:

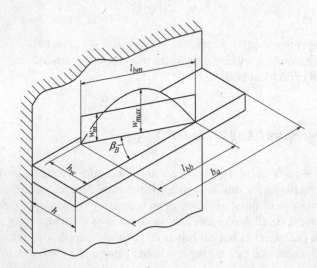

Fig. 6.3 Geometric parameters of tooth model reduced to a cantilever plate with constant thickness

$$b_a = \frac{b_v}{\cos \beta_v},$$ (6.32)

$$l_{bb} = l_{bm} \frac{\cos \beta_{vb}}{\cos \beta_v}$$ (6.33)

$$h = \frac{1}{2}(h_{m1} + h_{m2}),$$ (6.34)

where l_{bm} is the theoretical length of middle contact line, and h_m is the mean whole tooth depth (see Vol. 1, Chap. 12). The other quantities that appear in Fig. 6.3 are: h_w, the working tooth depth; β_B, the inclination angle of the contact line; w_m, the mean specific load per unit middle contact line; w_{max}, the maximum specific load per unit middle contact line.

6.3.5 Load Sharing Factor, Y_{LS}

The load sharing factor, Y_{LS}, considers the load sharing between two or more teeth pairs. For calculation of this factor we use the following relationship:

$$Y_{LS} = Z_{LS}^2,$$ (6.35)

where Z_{LS} is the load sharing factor for pitting, already defined (see Sect. 5.2).

6.4 Permissible Tooth Root Stress Factors for Method $B1$

In Eq. (6.3), various factors appear, which are called *permissible tooth root stress factors*. Among these factors, here we focus our attention on the already defined factors $Y_{RrelT-B1}$ and $Y_{\delta relT-B1}$, which are relative factors, and therefore dimensionless factors. If no surface condition factors or notch sensitivity factors determined according to Method A are available, both of these factors can be evaluated using Method B, with the procedures described below.

6.4.1 Relative Surface Condition Factor, $Y_{RrelT-B1}$

The tooth root strength depends on the surface conditions at the root fillet, which are predominantly attributable to the surface roughness. We already clarified this factual situation in Sect. 3.8.4, to which we refer the reader. In fact, from a conceptual point of view, there is nothing to add. The relative surface condition factor, $Y_{RrelT-B1}$, considers the aforementioned dependence related to standard test gear conditions with $Rz = 10\,\mu\text{m}$. It is to be determined separately for pinion and wheel. The relationships below are only valid if there are no scratches or similar defects deeper than $2Rz$, where Rz is the mean peak-to-valley roughness (in μm).

For permissible stress number relative to standard test gear dimensions, in the range of the mean peak-to-valley roughness between $(1\,\mu\text{m} \leq Rz \leq 40\,\mu\text{m})$, the determination of $Y_{RrelT-B1}$ is done using the following relationships:

$$Y_{R,relT} = \frac{Y_R}{Y_{RT}} = 1.674 - 0.529(Rz + 1)^{0.1}, \qquad (6.36)$$

for through-hardened and case-hardened steels [V, GGG(perl., bai.), Eh, IF(root)];

$$Y_{R,relT} = \frac{Y_R}{Y_{RT}} = 5.306 - 4.203(Rz + 1)^{0.01}, \qquad (6.37)$$

for non-hardened steel [St];

$$Y_{R,relT} = \frac{Y_R}{Y_{RT}} = 4.299 - 3.259(Rz + 1)^{0.005}, \qquad (6.38)$$

for gray cast iron, nitrided and nitro-carburized steels [GG, GGG(fer.), NT(nitr.), NV(nitr.), NV(nitrocar.)].

Equations (6.36) to (6.38) deliver results consistent with those we can read using the curves shown in Fig. 6.4, which give $Y_{R,relT}$ for permissible stress number relative to standard test gear dimensions as a function of Rz and material.

For permissible stress number relative to standard test gear dimensions, in the range $Rz < 1\,\mu\text{m}$, we take the limit values corresponding to $Rz = 1\,\mu\text{m}$, so that, for the three groups of materials covered by Eqs. (6.36) to (6.38), we have respectively:

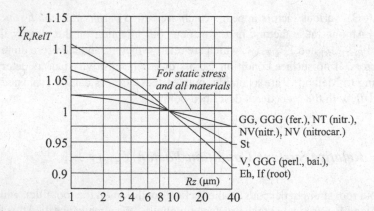

Fig. 6.4 Relative surface condition factor, $Y_{R,relT}$, for permissible stress number relative to standard test gear dimensions, as a function of Rz and material

$Y_{R,relT} = 1.12$; $Y_{R,relT} = 1.07$; $Y_{R,relT} = 1.025$. For static stress, we always have $Y_{R,relT} = 1$.

6.4.2 Relative Notch Sensitivity Factor, $Y_{\delta,relT-B1}$

The notch sensitivity factor of the actual gear relative to a polished test piece, Y_δ, expresses the amount of which the theoretical peak of the tooth root stress that determines the fatigue breakage exceeds the permissible stress number for bending. This factor (see also Sect. 3.8.3) is a function of the material and relative stress gradient. We can calculate the notch sensitivity factor on the basis of strength values determined using test gears, or un-notched or notched specimens.

The determination of the permissible tooth root stresses of bevel gears is carried out on the basis of the bending strength values determined using both bevel and cylindrical test gears. Thus, the relative notch sensitivity factor, $Y_{\delta,relT}$, is the quotient of the notch sensitivity factor of the gear under consideration divided by the notch sensitivity factor of the standard test gear, i.e. $Y_{\delta,relT} = Y_\delta / Y_{\delta T}$ (see Sect. 3.8.3).

For the calculation of the relative notch sensitivity factor, $Y_{\delta,relT-B1}$, to be done separately for pinion and wheel, we use the following relationship:

$$Y_{\delta,relT1,2} = \frac{1 + \sqrt{\rho' \chi_{1,2}^X}}{1 + \sqrt{\rho' \chi_T^X}}, \tag{6.39}$$

where ρ' (in mm) is the slip layer thickness, which can be taken from Table 3.2 as a function of the material (the values of ρ' for the same group of material can be interpolated for other values of the tensile strength, σ_B, yield stress, σ_s, and proof stress, $\sigma_{0.2}$), while χ^X (in mm^{-1}) and χ_T^X (in mm^{-1}) are respectively the relative stress gradient in notch root of the gear under consideration and the relative stress gradient in notch root of the standard test gear. For the mean normal module $m_{mn} = 5$ mm (the size influence is covered by the size factor Y_X, see Sect. 6.8.1), the relative stress gradient in notch root of the gear under consideration is given by:

$$\chi_{1,2}^X = \frac{1}{5}(1 + 2q_{s1,2}). \tag{6.40}$$

The value χ_T^X for the standard test gear is obtained with the same Eq. (6.40), by substituting in it $q_{sT} = 2,5$ for q_s; therefore, we obtain $\chi_T^X = 1,2$.

Equation (6.39) delivers results consistent with those we can read using the curves shown in Fig. 6.5, which give the relative notch sensitivity factor, $Y_{\delta,relT}$, with respect to standard test gear dimensions, as a function of the notch parameter, q_s (or as a function of the stress correction factor, Y_{Sa}) and material. In this figure, σ_B is the

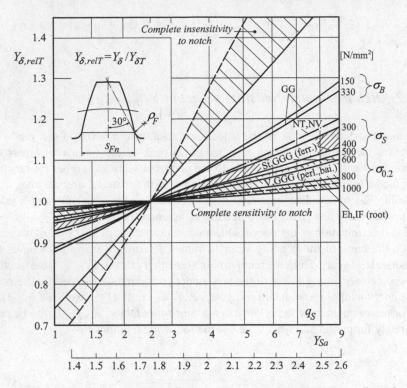

Fig. 6.5 Relative notch sensitivity factor, $Y_{\delta,relT}$, with respect to standard test gear dimensions, as a function of q_s (or Y_{Sa}) and material

tensile strength, σ_s the yield point, and $\sigma_{0.2}$ the proof stress (0.2% permanent set), all expressed in N/mm^2.

6.5 Calculation of Tooth Root Strength: ISO Basic Formulae of Method *B2*

Here too we must allow that there are no preferences in terms of when to use Method *B*1 and when to use Method *B*2. For design of bevel gears (straight and helical or skew gears, Zerol and spiral bevel gears including hypoid gears) with Method *B*2, the determination of the load carrying capacity in terms of tooth root strength is performed separately for pinion and wheel, using the following relationship:

$$\sigma_{F-B2} = \sigma_{F0-B2} K_A K_v K_{F\beta} K_{F\alpha} \leq \sigma_{FP-B2}, \tag{6.41}$$

where K_A, K_v, $K_{F\beta}$ and $K_{F\alpha}$ are the load factors already defined in Chap. 5, and σ_{F0-B2} is the nominal tooth root stress for Method *B*2 (it is defined as the maximum tensile stress at the tooth root, due to the nominal torque applied to an error-free gear), given by:

$$\sigma_{F0-B2} = \frac{F_{mt1,2}}{b_{1,2}m_{mn}}Y_{P1,2} = \frac{F_{mt1,2}}{b_{1,2}}\frac{m_{mt1,2}}{m_{et}^2}\frac{Y_{A1,2}}{Y_{J1,2}}, \tag{6.42}$$

where $Y_{P1,2}$ is a *combined geometry factor* for Method $B2$, to be determined separately for pinion and wheel (it replaces the factors Y_{Fa}, Y_{Sa}, Y_ε, Y_{BS}, and Y_{LS} of Method $B1$ in the tooth root stress equation); it is given by:

$$Y_{P1,2} = \frac{Y_{A1,2}}{Y_{J1,2}}\frac{m_{mt1,2}m_{mn}}{m_{et}^2}. \tag{6.43}$$

In Eqs. (6.42) and (6.43), $F_{mt1,2}$ is the nominal tangential force at the mid-face width of the reference cones of pinion and wheel (it is given by Eq. (4.1)), while Y_A and Y_J are respectively the *root stress adjustment factor* and the *bending strength geometry factor* for Method $B2$. The first of these two factors, Y_A, adjusts the calculation results of Method $B2$ so that it is possible to use the nominal stress numbers we can calculate by Eq. (2.57) and parameters shown in Table 2.3. The second of these factors, Y_J, takes into account the tooth shape, position at which the most damaging load is applied, stress concentration due to the root fillet geometry, load sharing between two or more pairs of mating teeth, tooth thickness balance between the wheel and mating pinion, effective face width due to lengthwise tooth crowning, buttressing effect due to an extended face width on one member of the gear pair, as well as the bending and compressive components of the load on the teeth.

The permissible tooth root stress, σ_{FP-B2}, to be calculated separately for pinion and wheel, should be determined on the basis of the strength results obtained with an actual gear (in this way, the reference value of σ_{FP-B2} in terms of geometrical similarity, manufacturing, and course of movement lies within the application field). The permissible tooth root stress is determined with the following relationship:

$$\sigma_{FP-B2} = \sigma_{FE}Y_{NT}Y_{\delta,relT-B2}Y_{R,relT-B2}Y_X$$
$$= \sigma_{F,\lim}Y_{ST}Y_{NT}Y_{\delta,relT-B2}Y_{R,relT-B2}Y_X, \tag{6.44}$$

where $\sigma_{FE} = \sigma_{F,\lim 1,2}Y_{ST}$, is the allowable stress number for bending (it is the basic bending strength for un-notched specimen under the assumption that the material is fully elastic, accounting also heat treatment); $\sigma_{F,\lim}$, is the nominal stress number for bending of the standard test gear, which takes into account the material, heat treatment, and surface characteristics at test gear dimensions; $Y_{ST} = 2$, is the stress correction factor for the dimensions of the standard test gear; $Y_{\delta,relT-B2} = Y_\delta/Y_{\delta T}$, is the relative notch sensitivity factor for the permissible stress number, related to the conditions at the standard test gear, which takes into account the notch sensitivity of the material; $Y_{R,relT-B2} = Y_R/Y_{RT}$, is the relative surface condition factor, which takes into account the surface condition at the root fillet, related to the conditions at the test gear; Y_X, is the size factor for tooth root bending stress, which takes into account the influence of the module on the tooth root strength; Y_{NT}, is the life factor for tooth root bending stress, which takes into account the influence of the required numbers of cycles of operation.

In accordance with Eq. (6.41), the evaluated tooth root stress, σ_F, must be less than or equal to the permissible tooth root stress, σ_{FP}, i.e. $\sigma_F \leq \sigma_{FP}$. The safety factor, S_F, against breakage due to bending load, to be determined separately for pinion and wheel with respect to the transmitted torque, is calculated on the basis of the bending stress number determined for the standard test gear using the following relationship:

$$S_{F-B2} = \frac{\sigma_{FP-B2}}{\sigma_{F-B2}} > S_{F,\min}.$$ (6.45)

About the choice of the minimum value of the safety factor, $S_{F,\min}$, of course to be correlated with the probability of failure, and with the risks involved, we refer the reader to Sect. 4.1.

6.6   Tooth Root Stress Factors for Method *B2*

In accordance with Eq. (6.42), when we calculate the nominal tooth root stress, σ_{F0-B2}, using the Method *B2*, we introduce the combined geometry factor, Y_P, which is expressed by Eq. (6.43). This last equation shows that, once modules m_{mt}, m_{mn}, and m_{et} are known (they are, respectively, the mean transverse module, mean normal module, and outer transverse module), in order to calculate Y_P, the bending strength geometry factor, Y_J, and the root stress adjustment factor, Y_A, should be preliminarily determined. In the more general case, the determination of factor Y_A depends on the determination of factor Y_J. Therefore, we must necessarily start from the calculation of this last factor.

6.6.1   Bending Strength Geometry Factor, Y_J

The bending strength geometry factor, Y_J, also called *bevel geometry factor*, is determined using the following relationship:

$$Y_{J1,2} = \frac{Y_{1,2}}{Y_{f1,2}\varepsilon_N Y_i} \frac{r_{my01,2}}{r_{mpt1,2}} \frac{b_{ce1,2}}{b_{1,2}} \frac{m_{mt1,2}}{m_{et}},$$ (6.46)

where $Y_{1,2}$, is the tooth form factor of pinion and wheel, to be determined in a different way for bevel gears without hypoid offset and for hypoid gears; ε_N, is the load sharing ratio for bending, also to be determined in a different way for bevel gears and hypoid gears; $r_{my01,2}$, is the mean transverse radius to point of load application for pinion and wheel, also to be determined in a different way for bevel gears and hypoid gears; $r_{mpt1,2}$, is the mean transverse pitch radius of pinion and wheel; $Y_{f1,2}$, is the stress concentration and correction factor of pinion and wheel; Y_i, is the inertia factor for bending of gears with a low contact ratio; $b_{ce1,2}$, is the calculated effective face width of pinion and wheel.

It is to be noted that the parameters for calculating the bending strength geometry factor, Y_J, are the same for bevel and hypoid gears, but the calculation procedures are different for bevel gears without offset and for hypoid gears. Furthermore, these calculation procedures are very complex, so the computerization is recommended. For straight, Zerol and spiral bevel gears characterized by face width equal to the smaller of the two conditions $b = 0, 3R_e$ and $b = 10m_{et}$, we can compute the bevel geometry factor, Y_J, using the graphs of the ANSI/AGMA 2003-C10. For hypoid gears, we can use the graphs of the AGMA 932-A05: 2005. Attention must be paid to the fact that these graphs can be used only when the design requirements of the gears under consideration correspond to those in the graphs, and this in terms of tooth proportions and thickness, face widths, tool edge radii, pressure and spiral angles, and driving with the concave side. In all other cases, we must necessarily use the calculation procedures based on Eq. (6.46).

In addition, it is to be noted that the basis for Method $B2$ is the Lewis formula (and thus the Lewis parabola, see Sect. 3.3) applied to a virtual cylindrical gear defined in transverse section as specified in Sect. 4.3, with the following modifications and additions:

- the tooth strength is considered in the normal section rather than in the transverse section;
- the position of point of load application is defined by considering the theoretical lines of contact, tooth profile modifications to better share and bear the load, and experimental evidence;
- the amount of load carried by one tooth is evaluated on the basis of the tooth profile modification and contact ratio;
- the radial component of the normal load is considered;
- a stress concentration factor based on experimental data is used;
- the effective face width is considered.

According to what Lewis proposed (Lewis 1892), the bending stress is determined by simulating the actual tooth shaped beam through a parabola tangent to the tooth profile at the critical section, i.e. the most highly stressed section, discussed by us in Sect. 3.3. Figure 6.6 shows two different constant strength parabolas according to Lewis, for the cases of no-load sharing (Fig. 6.6a) and load sharing (Fig. 6.6b). Both of these figures refer to the vast majority of cases for which the weakest cross section coincides with the built-in section of the cantilever beam represented by the gear tooth (see Sect. 3.3). However, the meaning of the geometric quantities represented in these two figures does not differ from that of the same quantities shown in Fig. 3.5.

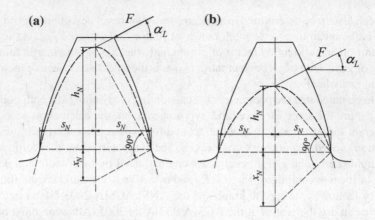

Fig. 6.6 Stress parabola according to Lewis for the cases of: **a** no-load sharing; **b** load sharing

6.6.1.1   Bending Strength Geometry Factor, Y_J, for Bevel Gears Without Hypoid Offset

To calculate the bending strength geometry factor, Y_J, for bevel gears without hypoid offset, it is first necessary to determine the location of point of load application for maximum bending stress on path of contact, given by y_J (in mm), and the location of point of load application on path of contact for maximum root stress, given by y_3 (in mm); in this regard, see Fig. 4.3. For most straight, Zerol and spiral bevel gears, the maximum tooth root stress occurs at the highest point of single pair tooth contact of the virtual cylindrical gear, when the modified contact ratio, ε_{vy}, is less than or equal to 2, i.e. $\varepsilon_{vy} \leq 2$. When $\varepsilon_{vy} > 2$, it is assumed that the contact line passes through the center of path of contact (point M in Fig. 4.3). For straight bevel and Zerol bevel gears subjected to static load, such as those used in automotive differentials, the load is applied at the tooth tip. In any case, the position of point of load application is measured along the path of contact from its center (this position is designed by the coordinate y_J); instead, the position of the same point from the beginning of the path of contact is designed by y_3.

For dynamically loaded straight, Zerol and spiral bevel gears, the location of point of load application for maximum bending stress on path of contact from its center, y_J, is given by the following relationships:

$$y_J = \frac{\pi m_{mn} \cos \alpha_a}{m_{et2}} - \frac{g_\eta}{2} \tag{6.47}$$

$$y_J = 0, \tag{6.48}$$

which are valid, respectively, for $\varepsilon_{vy} \leq 2$, and $\varepsilon_{vy} > 2$. In Eq. (6.47), α_a is the adjusted pressure angle, given by Eq. (4.65), and g_η is the relative length of contact within the contact ellipse, given by Eq. (5.42).

For statically loaded straight bevel and Zerol bevel gears, for which load is applied at the tooth tip, y_J is given by:

$$y_J = g_\eta/2, \tag{6.49}$$

where, however, g_η is given by:

$$g_\eta^2 = g_{van}^2 \cos^4 \beta_{vb} + b_v^2 \sin^2 \beta_{vb}. \tag{6.50}$$

For the determination of y_3, which depends on the type of bevel gear under consideration, we use the following three relationships:

$$y_3 = \frac{g_{van}}{2} + \frac{g_{van}^2 y_J}{g_\eta^2} \tag{6.51}$$

$$y_{31} = \frac{g_{van}}{2} + \frac{g_{van}^2 y_J \cos^2 \beta_{vb} + b_v g_{van} g_J k' \sin \beta_{vb}}{g_\eta^2} \tag{6.52}$$

$$y_{32} = \frac{g_{van}}{2} + \frac{g_{van}^2 y_J \cos^2 \beta_{vb} - b_v g_{van} g_J k' \sin \beta_{vb}}{g_\eta^2}, \tag{6.53}$$

which are valid, respectively, for straight bevel and Zerol bevel gears, for spiral bevel pinions, and for spiral bevel wheels. In Eqs. (6.52) and (6.53), g_J is the relative length of contact to point of load application, given by:

$$g_J = \sqrt{g_\eta^2 - 4y_J^2}, \tag{6.54}$$

while k' is the contact shift factor, given by Eq. (4.75).

The point of load application usually does not lie in the mean section of the tooth. Therefore, the mean transverse radius to point of load application, $r_{my01.2}$ (in mm), depends on the distance from mean section to point of load application, $x_{001.2}$ (in mm), measured in the lengthwise direction along the tooth, and is given by:

$$r_{my01,2} = r_{mpt1,2} \left(\frac{R_m + x_{001,2}}{R_m} \right) + \Delta r_{y01,2} m_{et2}. \tag{6.55}$$

For determination of $x_{001,2}$, which depends on the type of bevel gear under consideration, we use the following three relationships:

$$x_{001,2} = \frac{g_{van} g_J k'}{g_\eta^2} \tag{6.56}$$

$$x_{001} = \frac{b_v g_{van} g_J k' \cos^2 \beta_{vb} m_{et2} - b_v^2 y_J \sin \beta_{vb} m_{et2}}{g_\eta^2} \tag{6.57}$$

$$x_{002} = \frac{b_v g_{van} g_J k' \cos^2 \beta_{vb} m_{et2} + b_v^2 y_J \sin \beta_{vb} m_{et2}}{g_\eta^2},$$ (6.58)

which are valid, respectively, for straight bevel and Zerol bevel gears, for spiral bevel pinions, and for spiral bevel wheels.

The relative distance from pitch circle through the point of load application of pinion and the wheel tooth centerline, $\Delta r_{y01,2}$, which appears in Eq. (6.55), is given by:

$$\Delta r_{y01,2} = \frac{r_{vbn1,2}}{\cos \alpha_{h1,2}} - r_{vn1,2}$$ (6.59)

where $r_{vbn1,2}$ and $r_{vn1,2}$ are respectively the relative mean virtual base radius and relative mean virtual pitch radius of pinion and wheel, and

$$\alpha_{h1,2} = \alpha_{L1,2} - \xi_{h1,2}.$$ (6.60)

In this last equation, $\alpha_{L1,2}$ (Fig. 6.6) is the normal pressure angle at point of load application for pinion and wheel, and $\xi_{h1,2}$ is one half of angle subtended by normal circular tooth thickness at point of load application for pinion and wheel, also called rotation angle for pinion and wheel. These two quantities are expressed, respectively, by the following relationships:

$$\tan \alpha_{L1,2} = \frac{y_{31,2} + \alpha_{vn} \sin \alpha_n - \sqrt{r_{va1,2}^2 - r_{vbn1,2}^2}}{r_{vbn1,2}}$$ (6.61)

$$\xi_{h1,2} = \frac{s_{mn1,2}}{2r_{vn1,2}} - \text{inv} \alpha_{L1,2} + \text{inv} \alpha_n.$$ (6.62)

The load sharing ratio, ε_N, which appears in Eq. (6.46), is used to calculate the portion of the total load acting on the tooth under consideration. For statically loaded straight and Zerol bevel gears, it is given by:

$$\varepsilon_N = 1,$$ (6.63)

while for the other bevel and spiral bevel gears it is given by:

$$\varepsilon_N = \frac{g_J^3}{g_J'^3},$$ (6.64)

where the modified relative length of contact to point of load application, $g_J'^3$, is expressed by the following relationship:

$$g_J'^3 = g_J^3 + \sum_{k=1}^{k=x} \sqrt{\left[g_J^2 - 4k\frac{\pi m_{mn} \cos \alpha_a}{m_{et2}} \left(k\frac{\pi m_{mn} \cos \alpha_a}{m_{et2}} + 2y_J\right)\right]^3}$$

$$+ \sum_{k=1}^{k=y} \sqrt{\left[g_J^2 - 4k\frac{\pi m_{mn} \cos \alpha_a}{m_{et2}} \left(k\frac{\pi m_{mn} \cos \alpha_a}{m_{et2}} - 2y_J\right)\right]^3}. \tag{6.65}$$

In this last equation, k is a positive integer, with successive values from 1 to x or y, which generate all real terms in each series, i.e. positive values under the radical. Imaginary terms, i.e. negative values under the radical, shall be ignored. For most cases, x and y are not greater than 2.

The tooth form factor for pinion and wheel, $Y_{1,2}$, which appears in Eq. (6.46), considers both the radial and tangential components of the normal load. It is given by the following relationship:

$$Y_{1,2} = \frac{2}{3\left(\dfrac{1}{x_{N1,2}} - \dfrac{\tan \alpha_{h1,2}}{3s_{N1,2}}\right)} \tag{6.66}$$

where, in accordance with the Euclidean theorem already mentioned in Sect. 3.3 (Fig. 6.6), we have:

$$x_{N1,2} = \frac{s_{N1,2}^2}{h_{N1,2}}. \tag{6.67}$$

The value of the tooth form factor can be determined only by iteration, separately for pinion and wheel, because this factor defines the weakest section. This iteration procedure leads to determine the parameters $x_{N1,2}$, $s_{N1,2}$, and $h_{N1,2}$ that appear in Eqs. (6.66) and (6.67), respectively called tooth strength factor, one-half tooth thickness at critical section, and load height from critical section from ISO standards. To this purpose, the following three dimensionless auxiliary quantities are first introduced:

$$g_{01,2} = \frac{s_{vmn1,2}}{2} + h_{vfm1,2} \tan \alpha_n + \rho_{va01,2}\left(\frac{1 - \sin \alpha_n}{\cos \alpha_n}\right) \tag{6.68}$$

$$g_{yb1,2} = h_{vfm1,2} - \rho_{va01,2} \tag{6.69}$$

$$g_{f01,2(1)} = g_{01,2} + g_{yb1,2}, \tag{6.70}$$

where s_{vmn} is the relative mean normal circular thickness, while $g_{f01,2(1)}$ indicates the initial value to be taken to start the iteration procedure. The parameters $s_{N1,2}$ and $h_{N1,2}$ are calculated using the following relationships:

$$s_{N1,2} = r_{vn1,2} \sin \xi_{1,2} - \rho_{va01,2} \cos \tau_{1,2} - g_{zb1,2} \tag{6.71}$$

$$h_{N1,2} = \Delta r_{y01,2} + r_{vn1,2}\left(1 - \cos\xi_{1,2}\right) + \rho_{va01,2}\sin\tau_{1,2} + g_{za1,2}, \tag{6.72}$$

where

$$\xi_{1,2} = \frac{g_{f01,2}}{r_{vn1,2}} \tag{6.73}$$

$$\tan\tau_{1,2} = \frac{g_{za1,2}}{g_{zb1,2}} \tag{6.74}$$

with

$$g_{za1,2} = g_{yb1,2}\cos\xi_{1,2} - g_{xb1,2}\sin\xi_{1,2} \tag{6.75}$$

$$g_{zb1,2} = g_{yb1,2}\sin\xi_{1,2} + g_{xb1,2}\cos\xi_{1,2} \tag{6.76}$$

$$g_{xb1,2} = g_{f01,2} - g_{01,2}. \tag{6.77}$$

Equations (6.73) and (6.74) allow respectively to calculate the assumed angle in locating weakest section and angle between tangent of root fillet at weakest point and tooth centerline for pinion and wheel. Equations (6.75) to (6.77) complete the picture of auxiliary quantities necessary to carry out the iteration procedure.

As second step of iteration procedure, we take the value $g_{f01,2(2)} = \left(g_{f01,2(1)} + 0,005m_{et2}\right)$, while for the third and subsequent iteration steps, we take interpolated values. This iteration procedure is ended when the following result is achieved:

$$\frac{s_{N1,2}\cot\tau_{1,2}}{h_{N1,2}} = 2 \pm 0,001. \tag{6.78}$$

6.6.1.2 Bending Strength Geometry Factor, Y_J, for Hypoid Gears

The calculation procedure of the bending strength geometry factor, Y_J, for hypoid gears is much more complex than that described in the previous section for bevel gears without hypoid offset. Here too, it is necessary to introduce a great number of initial quantities, without which it is impossible to solve the problem, with the reliability required in these cases. The whole calculation procedure is based on tooth surfaces determined approximately, as sets of tooth surface points defined by the exponential function $e^{\vartheta\,\tan\alpha_f}$, where ϑ (in radiant) is the auxiliary quantity for tooth form and tooth correction factors, and α_f is the limit pressure angle in wheel root coordinates for Method $B2$.

Initial Relationships

First, we introduce the *drive flank pressure angle*, α_{Dnf}, and *coast flank pressure angle*, α_{Cnf}, in wheel root coordinates, given respectively by:

$$\alpha_{Dnf} = \alpha_{nD} - \vartheta_{f2} \sin \beta_{m2} \tag{6.79}$$

$$\alpha_{Cnf} = \alpha_{nC} + \vartheta_{f2} \sin \beta_{m2}, \tag{6.80}$$

where ϑ_{f2} is the dedendum angle of the wheel given by Eq. 12.140 in Vol. 1, and $\alpha_{nD,C}$ are the generated pressure angles on drive side/coast side (see Fig. 12.42 of Vol. 1).

Then we introduce the average pressure angle unbalance, $\Delta\alpha_1$, and the limit pressure angle in wheel root coordinates, α_f, respectively given by:

$$\Delta\alpha_1 = \frac{1}{2}\left(\alpha_{Dnf} - \alpha_{Cnf}\right) \tag{6.81}$$

$$\alpha_f = \alpha_{\lim} - \vartheta_{f2} \sin \beta_{m2}. \tag{6.82}$$

The relative distance from blade edge to centerline will be:

$$g_{rb} = \frac{1}{m_{et2}}\left[\left(h_{fm2}\tan\frac{\alpha_{nD} + \alpha_{nC}}{2} + \frac{W_{m2}}{2}\right)\cos\frac{\alpha_{nD} + \alpha_{nC}}{2}\right], \tag{6.83}$$

where W_{m2} is the wheel mean slot width (see Fig. 12.42 of Vol. 1).

Now we introduce the dimensionless intermediate factors, η_D and η_C, given by:

$$\eta_D = \tan\alpha_{Dnf}\left(\frac{g_{rb}}{\sin\alpha_{Dnf}} - h_{vfm2}\right) \tag{6.84}$$

$$\eta_C = \tan\alpha_{Cnf}\left(\frac{g_{rb}}{\sin\alpha_{Cnf}} - h_{vfm2}\right), \tag{6.85}$$

and the intermediate angles, β_a, $(\beta_D - \Delta\alpha)$ and $(\beta_C - \Delta\alpha)$, given respectively by:

$$\tan\beta_a = \frac{\dfrac{W_{m2}}{2m_{et2}} - \rho_{va02}\left(\sec\dfrac{\alpha_{nD} + \alpha_{nC}}{2} - \tan\dfrac{\alpha_{nD} + \alpha_{nC}}{2}\right)}{\left(h_{vfm2} - \rho_{va02}\right)} \tag{6.86}$$

$$(\beta_D - \Delta\alpha) = \beta_a - \Delta\alpha_1 \tag{6.87}$$

$$(\beta_C - \Delta\alpha) = -(\beta_a + \Delta\alpha_1). \tag{6.88}$$

In addition, we introduce the intermediate factor, g_1, wheel angles between centerline and fillet point on drive side/coast side, $\Delta\vartheta_D$ and $\Delta\vartheta_C$, and wheel angle between fillet points of wheel, $\Delta\vartheta_2$, which are given respectively by:

$$g_1 = \frac{h_{vfm2} - \rho_{va02}}{\cos\beta_a} \tag{6.89}$$

$$\tan\Delta\vartheta_D = \frac{g_1\sin(\beta_D - \Delta\alpha)}{r_{vn2} - g_1\cos(\beta_D - \Delta\alpha)} \tag{6.90}$$

$$\tan\Delta\vartheta_C = \frac{g_1\sin(\beta_C - \Delta\alpha)}{r_{vn2} - g_1\cos(\beta_C - \Delta\alpha)} \tag{6.91}$$

$$\Delta\vartheta_2 = \frac{\vartheta_{v2} + \Delta\vartheta_D + \Delta\vartheta_C}{2}, \tag{6.92}$$

where ϑ_{v2} is the angular pitch of virtual cylindrical wheel, given by Eq. (4.60). It should be noted that the name *wheel angle*, used here in accordance with ISO standards, could generated misunderstandings. It is to be understood as a *central angle*, that is an angle whose center is on the axis of the gear wheel under consideration, regardless of whether it is the hypoid pinion or the mating hypoid wheel.

The relative vertical and horizontal distances from pitch circle to fillet point, y_1 and x_1, are given respectively by:

$$y_1 = r_{vn2} - \frac{\left[r_{vn2} - g_1\cos(\beta_D - \alpha_f)\right]\cos(\Delta\vartheta_2 - \Delta\vartheta_D)}{\cos\Delta\vartheta_D} \tag{6.93}$$

$$x_1 = \frac{\left[r_{vn2} - g_1\cos(\beta_D - \alpha_f)\right]\sin(\Delta\vartheta_2 - \Delta\vartheta_D)}{\cos\Delta\vartheta_D}. \tag{6.94}$$

The generated pressure angle of wheel at fillet point, α_{LN2}, is given by:

$$\alpha_{LN2} = \alpha_{Dnf} - \Delta\vartheta_2. \tag{6.95}$$

At this point, we introduce the following six quantities: the relative distance from centerline to tool critical pinion drive side fillet point, μ_{1D}; the relative distance from centerline to tool critical pinion coast side fillet point, μ_{1C}; the wheel angle between centerline and critical pinion drive side fillet point, ϑ_{DLS}; the wheel angle between centerline and critical pinion coast side fillet point, ϑ_{CLS}; the relative radius from tool center to critical pinion drive side fillet point, R_{DL1}; the relative radius from tool center to critical pinion coast side fillet point, R_{CL1}. These six quantities are given respectively by:

$$\mu_{1D} = \eta_D + \tan\alpha_{Dnf}\left(h_{vfm1} + h_{vfm2}\right) + \rho_{va01}\left(\sec\alpha_{Dnf} - \tan\alpha_{Dnf}\right) \tag{6.96}$$

$$\mu_{1C} = \eta_C + \tan\alpha_{Cnf}\left(h_{vfm1} + h_{vfm2}\right) - \rho_{va01}\left(\sec\alpha_{Cnf} - \tan\alpha_{Cnf}\right) \tag{6.97}$$

$$\tan \vartheta_{DLS} = \frac{\mu_{1D}}{r_{vn2} + h_{vfm1}} \tag{6.98}$$

$$\tan \vartheta_{CLS} = \frac{\mu_{1C}}{r_{vn2} + h_{vfm1}} \tag{6.99}$$

$$R_{DL1} = \frac{r_{vn2} + h_{vfm1}}{\cos \vartheta_{DLS}} \tag{6.100}$$

$$R_{CL1} = \frac{r_{vn2} + h_{vfm1}}{\cos \vartheta_{CLS}}. \tag{6.101}$$

Now we calculate the wheel angle from centerline to pinion tip on drive side, ϑ_{D1}, by means of an iteration procedure, which starts taking $\vartheta_{D1} = \vartheta_{v2}$ as initial value, and using the following two relationships:

$$h_1 = (r_{vn2} + \Delta r_1) \sin(\alpha_{Dnf} + \vartheta_{D1}) - (r_{vn2} \sin \alpha_{Dnf} - g_{rb}) \tag{6.102}$$

$$h_{10} = \sqrt{r_{vn1}^2 - (r_{vn1} - \Delta r_1)^2 \cos^2(\alpha_{vDnf} + \vartheta_{D1})} \\ - (r_{vn1} - \Delta r_1) \sin(\alpha_{Dnf} + \vartheta_{D1}), \tag{6.103}$$

where Δr_1 is given by:

$$\Delta r_1 = r_{vn2}(e^{\vartheta_{D1} \tan \alpha_f} - 1). \tag{6.104}$$

The iteration procedure is developed changing, at each iteration step, the value of ϑ_{D1}, until we get $h_{10} = h_1$. At this point, the convergence is reached, and the procedure is ended.

Two new iteration procedures are used for the calculation of the wheel angle from centerline to tooth surface at wheel critical fillet point on drive side, ϑ_{D20}, and the wheel angle from centerline to tooth surface at wheel critical fillet point on coast side, ϑ_{C20}, using responsibility the following relationships:

$$\mu_{1D0} = r_{vn2}(e^{\vartheta_{D20} \tan \alpha_f}) \sin \vartheta_{D20} \tag{6.105}$$

$$\mu_{1C0} = r_{vn2}(e^{\vartheta_{C20} \tan \alpha_f}) \sin \vartheta_{C20}. \tag{6.106}$$

The first iteration procedure, based on Eq. (6.105), starts with the assumption $\vartheta_{D20} = (\vartheta_{v2}/2)$, and is developed chancing ϑ_{D20} until $\mu_{1D0} = \mu_{1D}$. Similarly, the second iteration procedure, based on Eq. (6.106), starts with the assumption $\vartheta_{C20} = -(\vartheta_{v2}/2)$, and is developed changing ϑ_{C20} until $\mu_{1C0} = \mu_{1C}$. Once these convergences are reached, both the iteration procedures are ended.

To obtain the wheel angle from centerline to tooth surface at pinion critical drive side fillet point, ϑ_{D10}, and the wheel angle from centerline to tooth surface at pinion critical coast side fillet point, ϑ_{C10}, we must solve the following two relationships, respectively for ϑ_{D10} and for ϑ_{C10}:

$$r_{vn2}\left(e^{\vartheta_{D20}\tan\alpha_f} - 1\right) = r_{vn1}\left(1 - e^{\vartheta_{D10}\tan\alpha_f}\right) \tag{6.107}$$

$$r_{vn2}\left(e^{\vartheta_{C20}\tan\alpha_f} - 1\right) = r_{vn1}\left(1 - e^{\vartheta_{C10}\tan\alpha_f}\right). \tag{6.108}$$

Then we calculate the wheel angle difference between tool and surface at wheel drive side fillet point, $\Delta\vartheta_{D20}$, wheel angle difference between tool and surface at wheel coast side fillet point, $\Delta\vartheta_{C20}$, wheel angle difference between tool and surface at pinion drive side fillet point, $\Delta\vartheta_{D10}$, wheel angle difference between tool and surface at pinion coast side fillet point, $\Delta\vartheta_{C10}$, and pinion angle unbalance between fillet points, $\Delta\vartheta_1$, using respectively the following expressions:

$$\Delta\vartheta_{D20} = \vartheta_{DLS} - \vartheta_{D20} \tag{6.109}$$

$$\Delta\vartheta_{C20} = \vartheta_{CLS} - \vartheta_{C20} \tag{6.110}$$

$$\tan\Delta\vartheta_{D10} = -\frac{R_{DL2}\sin\Delta\vartheta_{D20}}{r_{vn2} + r_{vn1} - R_{DL2}\cos\Delta\vartheta_{D20}} \tag{6.111}$$

$$\tan\Delta\vartheta_{C10} = -\frac{R_{CL2}\sin\Delta\vartheta_{C20}}{r_{vn2} + r_{vn1} - R_{CL2}\cos\Delta\vartheta_{C20}} \tag{6.112}$$

$$\Delta\vartheta_1 = \frac{1}{2}(\vartheta_{D10} + \vartheta_{C10} + \Delta\vartheta_{D10} + \Delta\vartheta_{C10}). \tag{6.113}$$

Now we determine the wheel angle from centerline to pinion tip on drive side, ϑ_{D0}, by solving the following equation for ϑ_{D0}:

$$\Delta r_1 = r_{vn1}\left(1 - e^{\vartheta_{D0}\tan\alpha_f}\right). \tag{6.114}$$

Subsequently, the wheel angle from centerline to tooth surface at pitch point on drive side, ϑ_D, is obtained by solving with an iteration procedure the following relationship:

$$h = (r_{vn2} + \Delta r)\sin(\alpha_{Dnf} + \vartheta_D) - (r_{vn2}\sin\alpha_{Dnf} - g_{rb}), \tag{6.115}$$

where

$$\Delta r = r_{vn2}\left(e^{\vartheta_D\tan\alpha_f} - 1\right). \tag{6.116}$$

This iteration procedure starts with the assumption $\vartheta_D = -(\vartheta_{v2}/3)$, and is developed changing ϑ_D until $h = 0$. When this result is reached, the iterations are ended.

Finally, the wheel angle from centerline to wheel fillet point on drive side, ϑ_{D2}, is determined with a last iteration procedure, using the following equations:

$$h_2 = (r_{vn2} + \Delta r_2) \sin(\alpha_{Dnf} + \vartheta_{D2}) - (r_{vn2} \sin\alpha_{Dnf} - g_{rb}) \tag{6.117}$$

$$h_{20} = \pm\sqrt{r_{va1}^2 - (r_{vn1} + \Delta r_2)^2 \cos^2(\alpha_{Dnf} + \vartheta_{D2})}$$
$$+ (r_{vn1} + \Delta r_2) \sin(\alpha_{Dnf} + \vartheta_{D2}), \tag{6.118}$$

where

$$\Delta r_2 = r_{vn2}(e^{\vartheta_{D2}\tan\alpha_f} - 1). \tag{6.119}$$

This iteration procedure starts with different initial values of ϑ_{D2}. They depend on the amount of $(r_{vn2} + \Delta r)$, and are: $\vartheta_{D2} = 0.8\vartheta_D$, for $(r_{vn2} + \Delta r) > r_{va2}$; $\vartheta_{D2} = \vartheta_D$, for $(r_{vn2} + \Delta r) = r_{va2}$; $\vartheta_{D2} = 1.2\vartheta_D$, for $(r_{vn2} + \Delta r) < r_{va2}$. In Eq. (6.118), we use the plus sign, if $(r_{vn2} \sin\alpha_{Dnf} - g_{rb}) < 0$, or the minus sign, if $(r_{vn2} \sin\alpha_{Dnf} - g_{rb}) \geq 0$. The iteration procedure is developed changing ϑ_{D2} until $h_2 = h_{20}$. At this point, the convergence is reached, and the iteration procedure is ended.

Other Peculiar Quantities of Hypoid Gears, for the Calculation of Y_J

The other quantities that are necessary to calculate the bending strength geometry factor, Y_J, which are specific for hypoid gears, are the load sharing ratio for bending, ε_N, tooth strength factor, $x_{N1,2}$, tooth form factor, $Y_{1,2}$, and mean transverse radius to point of load application, $r_{my01,2}$.

For hypoid gears, the value of the load sharing ratio, ε_N, to be introduced in Eq. (6.46), is the same used for statically loaded straight and Zerol bevel gears. Therefore, this load sharing ratio is given by Eq. (6.63).

The calculation procedure of the tooth strength factor, $x_{N1,2}$, is much more complex. In this regard, we begin to calculate the relative length of path of contact of virtual cylindrical gear in normal section, $g_{v\alpha n}$, given by:

$$g_{v\alpha n} = g_{v\alpha 1} + g_{v\alpha 2}, \tag{6.120}$$

where $g_{v\alpha 1}$ and $g_{v\alpha 2}$ are respectively the relative length of path of contact from pinion tip to pitch circle in normal section, and the relative length of path of contact from wheel tip to pitch circle in normal section. These last two quantities are given, respectively, by the following relationships:

$$g_{v\alpha 1} = \sqrt{h_1^2 + (\Delta r_1 - \Delta r)^2 - 2h_1(\Delta r_1 - \Delta r)\sin(\alpha_{Dnf} + \vartheta_{D1})} \tag{6.121}$$

$$g_{v\alpha 2} = \sqrt{h_2^2 + (\Delta r_2 - \Delta r)^2 - 2h_2(\Delta r_2 - \Delta r)\sin(\alpha_{Dnf} + \vartheta_{D2})}. \tag{6.122}$$

Then we calculate the profile contact ratio in mean normal section, $\varepsilon_{v\alpha n}$, and the modified contact ratio, $\varepsilon_{v\gamma}$, using the following relationships:

$$\varepsilon_{v\alpha n} = \frac{g_{v\alpha n}}{p_{mn}}, \tag{6.123}$$

$$\varepsilon_{v\gamma} = \sqrt{\varepsilon_{v\alpha n}^2 + \varepsilon_{v\beta}^2}. \tag{6.124}$$

Depending on whether the modified contact ratio is less than two ($\varepsilon_{v\gamma} < 2$) or greater than or equal to two $(\varepsilon_{v\gamma} \geq 2)$, the *profile load sharing factor*, ε_f, will be given, respectively, by the following relationships:

$$\varepsilon_f = 1 - \frac{\varepsilon_{v\gamma}}{2} \tag{6.125}$$

$$\varepsilon_f = 0. \tag{6.126}$$

The *lengthwise load sharing factor*, ε_b, is to be calculated with one of the following two relationships:

$$\varepsilon_b = 2\sqrt{\varepsilon_{v\gamma} - 1} \tag{6.127}$$

$$\varepsilon_b = \varepsilon_{v\gamma}, \tag{6.128}$$

which are valid, respectively, in the same two ranges of variation of the modified contact ratio, i.e. for $\varepsilon_{v\gamma} < 2$, and for $\varepsilon_{v\gamma} \geq 2$.

The relative length of path of contact from pinion tip to point of load application, $g_{v\alpha 3}$, and relative length of path of contact from wheel tip to point of load application, $g_{v\alpha 4}$, are given respectively by:

$$g_{v\alpha 3} = \left| \frac{p_{mn}\varepsilon_{v\alpha n}^2}{\varepsilon_{v\gamma}^2} \left(\frac{\varepsilon_{v\gamma}^2}{2\varepsilon_{v\alpha n}} - \frac{\varepsilon_{v\beta}\varepsilon_b k'}{\varepsilon_{v\alpha}} + \varepsilon_f \right) - g_{v\alpha 1} \right| \tag{6.129}$$

$$g_{v\alpha 4} = \left| \frac{p_{mn}\varepsilon_{v\alpha n}^2}{\varepsilon_{v\gamma}^2} \left(\frac{\varepsilon_{v\gamma}^2}{2\varepsilon_{v\alpha n}} + \frac{\varepsilon_{v\beta}\varepsilon_b k'}{\varepsilon_{v\alpha}} + \varepsilon_f \right) - g_{v\alpha 2} \right|, \tag{6.130}$$

where k' is the contact stiff factor, while the relative length of path of contact to point of load application for pinion and wheel is given by the following expression:

$$g_{J1,2} = g_{v\alpha n} - g_{v\alpha 3,4}. \tag{6.131}$$

Now we determine the wheel angle from pinion tip to point of load application, ϑ_{D3}, with an iteration procedure, using the following relationships:

$$h_3 = (r_{vn2} + \Delta r_3)\sin(\alpha_{vDnf} + \vartheta_{D3}) - (r_{vn2}\sin\alpha_{vDnf} - g_{rb}) \tag{6.132}$$

$$
\begin{aligned}
h_{30} = \sqrt{g_{va3}^2 - (\Delta r_3 - \Delta r)^2 \cos^2(\alpha_{vDnf} + \vartheta_{D3})} \\
- |\Delta r_3 - \Delta r|\sin(\alpha_{vDnf} + \vartheta_{D3}),
\end{aligned}
\tag{6.133}
$$

where

$$\Delta r_3 = r_{vn2}\left(e^{\vartheta_{D3}\tan\alpha_f} - 1\right). \tag{6.134}$$

This iteration procedure starts with the assumption $\vartheta_{D3} = -(\vartheta_{v2}/2)$, and is developed changing ϑ_{D3} until $h_3 = h_{30}$. When this convergence is reached, the iteration procedure is ended.

In the same way, we determine the wheel angle from wheel tip to point of load application, ϑ_{D4}, with a similar iteration procedure, using the following relationships:

$$h_4 = (r_{vn2} + \Delta r_4)\sin(\alpha_{vDnf} + \vartheta_{D4}) - (r_{vn2}\sin\alpha_{vDnf} - g_{rb}) \tag{6.135}$$

$$
\begin{aligned}
h_{40} = \sqrt{g_{va4}^2 - (\Delta r_4 - \Delta r)^2 \cos^2(\alpha_{vDnf} + \vartheta_{D4})} \\
- |\Delta r_4 - \Delta r|\sin(\alpha_{vDnf} + \vartheta_{D4}),
\end{aligned}
\tag{6.136}
$$

where

$$\Delta r_4 = r_{vn2}\left(e^{\vartheta_{D4}\tan\alpha_f} - 1\right). \tag{6.137}$$

This iteration procedure starts with the assumption $\vartheta_{D4} = (\vartheta_{v2}/3)$, and is developed changing ϑ_{D4} until $h_4 = h_{40}$. When this convergence is reached, the iteration procedure is ended.

The distance from pitch circle to point of load application, Δr_{LN2}, is given by:

$$\Delta r_{LN2} = \frac{(r_{vn2} + \Delta r_3)\cos(\alpha_{vDnf} + \vartheta_{D3})}{\cos\alpha_{LN2}} - r_{vn2}. \tag{6.138}$$

The angle between centerline and line from point of load application and fillet point on wheel, α_{200}, is now determined with another iteration procedure, using the following relationships:

$$s_{N2} = x_1 - \rho_{va02}\cos\alpha_{200} \tag{6.139}$$

$$y_2 = y_1 + \rho_{va02}\sin\alpha_{200} \tag{6.140}$$

$$h_{N2} = y_2 + \Delta r_{LN2} \tag{6.141}$$

$$h_{N20} = \frac{s_{N2}}{2 \tan \alpha_{200}}, \tag{6.142}$$

where: s_{N2}, is the horizontal distance from centerline to critical fillet point; y_2, is the vertical distance from pitch circle to critical fillet point; h_{N2}, is the wheel load height at weakest section; h_{N20}, is the auxiliary value of this last quantity. The iteration procedure starts with the assumption $\alpha_{200} = 2\alpha_{Dnf}$, and is developed changing α_{200} until $h_{N2} = h_{N20}$. When this convergence is reached, the iteration procedure is ended.

At this point, we are able to calculate the wheel tooth strength factor, x_{N2}, which is given by:

$$x_{N2} = \frac{s_{N2}^2}{h_{N2}}. \tag{6.143}$$

The pinion tooth strength factor, x_{N1}, is still to be calculated. To this end, we first determine the wheel angle from pitch point to pinion point of load application, ϑ_{D5}, by solving for ϑ_{D5} the following relationship:

$$\Delta r_4 = r_{vn1}\left(1 - e^{\vartheta_{D5} \tan \alpha_f}\right). \tag{6.144}$$

Then we determine the pinion pressure angle at point of load application, α_{LN1}, and the pinion radial distance to point of load application, r_{410}, using respectively the following relationships:

$$\alpha_{LN1} = \alpha_{vnf} + \vartheta_{D4} - \vartheta_{D5} + \Delta\vartheta_1 \tag{6.145}$$

$$r_{410} = \frac{(r_{vn1} - \Delta r_4) \cos(\alpha_{Dnf} + \vartheta_{D4})}{\cos \alpha_{LN1}}. \tag{6.146}$$

At this point, we activate a complex iteration procedure, which includes, in turn, an enclosed iteration. This enclosed iteration is addressed to the calculation of a preliminary value of the wheel angle between centerline and pinion fillet, ϑ_{D200}, and for this purpose we use the following relationships:

$$\mu_{D2} = \frac{r_{vn2} + h_{vfm1} - \rho_{va01} - \Delta r_5 \cos \vartheta_{D200}}{\tan \alpha_{D0}} \tag{6.147}$$

$$\mu_D = \Delta r_5 \cos \vartheta_{D200}, \tag{6.148}$$

where

$$\Delta r_5 = r_{vn2} e^{\vartheta_{D200} \tan \alpha_f}. \tag{6.149}$$

We use $\alpha_{D0} = \alpha_{nD}$ as initial value in Eq. (6.147), and for the enclosed iteration we assume $\vartheta_{D200} = (\vartheta_{v2}/2)$ to start the procedure, which is developed by changing ϑ_{D200} until $\mu_D = \mu_{D1} + \mu_{D2}$. When this convergence is reached, the enclosed iteration procedure is ended.

With the preliminary value of ϑ_{D200} thus found, which depends on the initial value $\alpha_{D0} = \alpha_{nD}$, we perform the most complex iteration procedure. This procedure is based on the use of the following eleven relationships:

$$r_{vn2}\left(e^{\vartheta_{D200}\tan\alpha_f} - 1\right) = r_{vn1}\left(1 - e^{\vartheta_{D100}\tan\alpha_f}\right) \tag{6.150}$$

$$\tan\vartheta_{L20} = \frac{\mu_{D1} - \rho_{va01}\cos\alpha_{D0}}{r_{vn2} + h_{vfm1} - \rho_{va01} + \rho_{va01}\sin\alpha_{D0}} \tag{6.151}$$

$$\Delta\vartheta_{D200} = \vartheta_{L20} - \vartheta_{D200} \tag{6.152}$$

$$r_{L20} = \frac{r_{vn2} + h_{vfm1} - \rho_{va01} + \rho_{va01}\sin\alpha_{D0}}{\cos\vartheta_{L20}} \tag{6.153}$$

$$\tan\Delta\vartheta_{D100} = -\frac{r_{L20}\sin\Delta\vartheta_{D200}}{r_{vn2} + r_{vn1} - r_{L20}\cos\Delta\vartheta_{D200}} \tag{6.154}$$

$$r_{L10} = \frac{r_{vn2} + r_{vn1} - r_{L20}\cos\Delta\vartheta_{D200}}{\cos\Delta\vartheta_{D100}} \tag{6.155}$$

$$(\Delta\vartheta_1 - \vartheta_{L10}) = \Delta\vartheta_1 - \vartheta_{D100} - \Delta\vartheta_{D100} \tag{6.156}$$

$$\alpha_1 = \alpha_{D0} - \vartheta_{D200} + \vartheta_{D100} \tag{6.157}$$

$$s_{N1} = r_{L10}\sin(\Delta\vartheta_1 - \vartheta_{L10}) \tag{6.158}$$

$$h_{N1} = r_{410} - r_{L10}\cos(\Delta\vartheta_1 - \vartheta_{L10}) \tag{6.159}$$

$$h_{N10} = \frac{s_{N1}}{2\tan\alpha_1}, \tag{6.160}$$

which allow to obtain respectively: the wheel angle between centerline and pinion fillet, ϑ_{D100}; the wheel rotation through path of contact, ϑ_{L20}; the wheel angle difference between path of contact and tooth surface at pinion fillet, $\Delta\vartheta_{D200}$; the wheel radius to pinion fillet point, r_{L20}; the wheel angle to pinion fillet point, $\Delta\vartheta_{D100}$; the pinion radius to fillet point, r_{L10}; the wheel angle from centerline to pinion fillet point, $(\Delta\vartheta_1 - \vartheta_{L10})$; the angle between centerline and line from point of load application and fillet point on pinion, α_1; the horizontal distance to pinion critical fillet point, s_{N1}; the pinion load height at weakest section, h_{N1}; the auxiliary value of this last quantity, h_{N10}.

This complex iteration procedure is developed changing the initial value $\alpha_{D0} = \alpha_{nD}$, until $h_{N1} = h_{N10}$. When this convergence is reached, the iteration procedure is ended.

Finally, at this point, we are able to calculate also the pinion tooth strength factor, x_{N1}, which is given by:

$$x_{N1} = \frac{s_{N1}^2}{h_{N1}}.$$ (6.161)

The calculation procedure of the tooth strength factor, $x_{N1,2}$, is thus completed. So only the tooth form factor, $Y_{1,2}$, and the transverse radius to point of load application, $r_{my01,2}$, are to be determined.

The calculation of the tooth form factor, $Y_{1,2}$, to be done separately for pinion and wheel, is carried out using the following relationship:

$$Y_{1,2} = \frac{2}{3\left(\dfrac{1}{x_{N1,2}} - \dfrac{\tan \alpha_{LN1,2}}{3s_{N1,2}}\right)},$$ (6.162)

the quantities of which have been already all determined.

Lastly, the calculation of the transverse radius to point of load application, $r_{my01,2}$, also to be done separately for pinion and wheel, is carried out using the following two relationships:

$$r_{my01} = \frac{r_{mpt1}(x_{001} + R_{m2})}{R_{m2}} + (r_{410} - r_{vn1})m_{et2}$$ (6.163)

$$r_{my02} = \frac{r_{mpt2}(x_{002} + R_{m2})}{R_{m2}} + \Delta r_{LN2}m_{et2},$$ (6.164)

where the *contact shift* due to load for pinion and wheel, $x_{001,2}$, i.e. the distance from mean section to point of load application for pinion and wheel, is given respectively by:

$$x_{001} = k'b_{k1} - \frac{b_1 \varepsilon_f \varepsilon_{v\beta}}{\varepsilon_{v\gamma}^2}$$ (6.165)

$$x_{002} = k'b_{k2} + \frac{b_2 \varepsilon_f \varepsilon_{v\beta}}{\varepsilon_{v\gamma}^2}.$$ (6.166)

In these last two equations, $b_{k1,2}$ is the mean face width of pinion and wheel, given by:

$$b_{k1,2} = \frac{b_{1,2}\varepsilon_{van}\varepsilon_{v\beta}}{\varepsilon_{v\gamma}^2},$$ (6.167)

while k' is the contact shift factor given by Eq. (4.75).

6.6.2 Additional Tooth Strength Parameters for Bevel and Hypoid Gears

Among the additional tooth strength parameters that apply to both bevel and hypoid gears, we include the following four parameters: the tooth fillet radius at root diameter, r_{mf}; the stress concentration and stress correction factor, Y_f; the inertia factor, Y_i; the calculated effective face width, b_{ce}.

The minimum value of the tooth fillet radius occurs at the point where the fillet is tangent to the root circles. The relative fillet radius at tooth root, which coincides with this minimum value, is given by:

$$r_{mf1,2} = \frac{\left(h_{vfm1,2} - \rho_{va01,2}\right)^2}{r_{vn1,2} + h_{vfm1,2} - \rho_{va01,2}} + \rho_{va01,2}. \tag{6.168}$$

The stress concentration and correction factor for pinion and wheel, $Y_{f1,2}$, is calculated using the following expression, derived by Dolan and Broghammer (1942):

$$Y_{f1,2} = L + \left(\frac{2s_{N1,2}}{r_{mf1,2}}\right)^M \left(\frac{2s_{N1,2}}{h_{N1,2}}\right)^O. \tag{6.169}$$

where L, M, and O are three functions of the actual pressure angle, α_n, expressed in degrees. These functions are given respectively by: $L \cong (0.3255 - 0.0073\alpha_n)$, $M \cong (0.3318 - 0.0091\alpha_n)$ and $O \cong (0.2682 + 0.0091\alpha_n)$.

It is to be noted that the stress concentration and correction factor, Y_f, depends on tooth geometry, location of the load, plasticity effects, residual stress effects, material composition effects, surface finishing resulting from gear production and subsequent service, Hertzian stress effects, size effects, and tooth edge effects. Equation (6.169), based on results obtained by Dolan and Broghammer, covers only the first two effects (tooth geometry and location of the load). The other influences listed above are included within those appearing in Eq. (6.44), and in more detail: residual stress effects, and material composition effects are included within the allowable stress number, σ_{FE}; the size effects are included within the size factor, Y_X; the tooth edge effects are included within the calculated effective face width, b_{ce}.

The inertia factor, Y_i, which considers the lack of smoothness of the tooth action in dynamically loaded gears with a relatively small contact ratio, is determined as a function of the modified contact ratio, $\varepsilon_{v\gamma}$. Depending on whether $\varepsilon_{v\gamma} < 2$ or $\varepsilon_{v\gamma} \geq 2$, we assume:

$$Y_i = \frac{2}{\varepsilon_{v\gamma}} \quad \text{or} \quad Y_i = 1. \tag{6.170}$$

Instead, for statically loaded gears (those used in vehicle drive axle differentials are considered as statistically loaded gears), we assume $Y_i = 1$ even when $\varepsilon_{v\gamma} < 2$.

Lastly, the calculated effective face width, b_{ce}, considers the effectiveness of the tooth in distributing the load over the root cross section, since the instantaneous contact line frequently does not extend over the entire face width. To determine b_{ce}, first it is necessary to introduce the *toe increment*, $\Delta b'_{i1,2}$, and the *heel increment*, $\Delta b'_{e1,2}$, which are expressed by the following relationships:

$$\Delta b'_{i1,2} = \frac{b_{1,2} - g_{K1,2}}{2\cos\beta_{m1,2}} - \frac{x_{001,2}}{\cos\beta_{m1,2}} \tag{6.171}$$

$$\Delta b'_{e1,2} = \frac{b_{1,2} - g_{K1,2}}{2\cos\beta_{m1,2}} + \frac{x_{001,2}}{\cos\beta_{m1,2}}, \tag{6.172}$$

where $x_{001,2}$ is the distance from mean section to point of load application for pinion and wheel, as given by Eqs. (6.56) to (6.58), while $g_{K1,2}$ is the projected length of the instantaneous contact line in the tooth lengthwise direction for pinion and wheel, which is given by:

$$g_{K1,2} = \frac{b_{1,2}g_{v\alpha n}g_{J1,2}\cos^2\beta_{vb}}{g_\eta^2}. \tag{6.173}$$

In this last equation, g_J is the length of path of contact from mean point to point of load application, which is given by Eq. (6.54) for bevel gears without hypoid offset, and by Eq. (6.131) for hypoid gears.

Now, we determine the calculated effective face width of pinion and wheel, $b_{ce1,2}$, using the following relationship:

$$b_{ce1,2} = 25,4h_{N1,2}\cos\beta_{m1,2}\left[\arctan\left(\frac{\Delta b_{i1,2}}{25,4h_{Na1,2}}\right)\right.$$
$$\left. + \arctan\left(\frac{\Delta b_{e1,2}}{25,4h_{Na1,2}}\right)\right] + g_{K1,2}, \tag{6.174}$$

where $\Delta b_{i1,2} = \Delta b'_{i1,2}$ and $\Delta b_{e1,2} = \Delta b'_{e1,2}$, if $\Delta b'_{i1,2}$ and $\Delta b'_{e1,2}$ are both positive; $\Delta b_{i1,2} = (b_{1,2} - g_{K1,2})/\cos\beta_{m1,2}$ and $\Delta b_{e1,2} = 0$, if $\Delta b'_{i1,2}$ is positive and $\Delta b'_{e1,2}$ is negative; $\Delta b_{i1,2} = 0$ and $\Delta b_{e1,2} = (b_{1,2} - g_{K1,2})/\cos\beta_{m1,2}$, if $\Delta b'_{i1,2}$ is negative and $\Delta b'_{e1,2}$ is positive.

6.6.3 Root Stress Adjustment Factor, Y_A

As we already mentioned in Sect. 6.5, the root stress adjustment factor, Y_A, adjusts the calculation results of Method $B2$ so that it is possible to use the nominal stress numbers we can calculate by Eq. (2.57) and parameters shown in Table 2.3.

As reference root stress adjustment factor, we assume the one referred to carburized case-hardened steel, whose average value is given by:

$$Y_A = 1.075. \tag{6.175}$$

A good quality material is the one defined by MQ grade (see Table 2.3). It meets the requirements of experienced manufacturers at moderate cost, with an allowable stress range between $(425-500)$ N/mm^2. For other specific materials and qualities, Y_A should be determined in the same manner.

6.7 Permissible Tooth Root Stress Factors for Method $B2$

In Eq. (6.44), the already defined relative surface condition factor, $Y_{R,relT-B2}$, and relative notch sensitivity factor, $Y_{\delta,relT-B2}$, appear, which are known to be the *permissible tooth root stress factors*; these are specific factors for Method $B2$.

For gears with a mean peak-to-valley roughness $Rz \leq 16\,\mu m$ at the root, the first of these two factors, i.e. the relative surface condition factor, $Y_{R,relT-B2}$, can generally be assumed as follows:

$$Y_{R,relT-B2} = 1. \tag{6.176}$$

For values of same roughness included within the range $(10\,\mu m < Rz \leq 16\,\mu m)$, the reduction of the allowable stress number is small. For $Rz < 10\,\mu m$, the value of $Y_{R,relT-B2}$ given by relationship (6.176) is on the safe side.

For gears characterized by a notch parameter $q_s \geq 1.5$, the second of the two aforementioned factors, i.e. the relative notch sensitivity factor, $Y_{\delta,relT-B2}$, is assumed as follows:

$$Y_{\delta,relT-B2} = 1. \tag{6.177}$$

For values of notch parameter $q_s < 1.5$, the expected reduction of the permissible tooth root stress may be considered assuming:

$$Y_{\delta,relT-B2} = 0.95. \tag{6.178}$$

6.8 Common Factors for Method $B1$ and Method $B2$: Factors for Permissible Tooth Root Stress

The equation (6.3), which applies to Method $B1$, and the corresponding Eq. (6.44), which applies to Method $B2$, have in common the size factor, Y_X, and the life factor, Y_{NT}. Both these factors are dimensionless factors.

6.8.1 Size Factor, Y_X

As we saw in Sect. 3.8.5, concerning the analogous subject on spur and helical gears, even here the size factor, Y_X, considers the decrease in tooth root strength of bevel gears, due to the size effect, that is the increasing size of the gear wheel under consideration. Even for bevel gears, this factor must be determined separately for pinion and wheel. In this case, however, the main factors that influence Y_X are the tooth size, diameter of the part, ratio of tooth size to diameter, material and heat treatment, area of the stress pattern, and ratio of case depth to tooth thickness.

The calculation of the size factor, Y_X, for reference stress and for static stress, can be carried out using the approximate relationships in Table 3.3, which express Y_X as a function of the mean normal module, m_n, and material. As Table 3.3 shows, for static stress and all materials, $Y_X = 1$. Instead, for reference stress, the following three groups of materials are considered:

- structural and through-hardened steels, spheroidal cast iron, and perlitic malleable cast iron [St, V, GGG (perl., bai.), GTS (perl.)];
- case-, flame-, induction-hardened steels, nitrided or nitro-carburized steels (Eh, IF, NT, NV);
- gray cast iron and spheroidal cast iron with ferritic structure [GG, GGG(ferr.)].

Equations in Table 3.3 deliver results consistent with those we can read using the curves shown in Fig. 3.23, which give Y_X as a function of m_{mn} and material. Of course, the abscissa m_n in Fig. 3.23 should be replaced with the abscissa m_{mn}, and the same substitution is necessary to do in Table 3.3. It is equally obvious that the equations in Table 3.3 and the corresponding graphs in Fig. 3.23 are to be used only when no test values or other proven experience are available regarding this size factor.

6.8.2 Life Factor, Y_{NT}

The life factor for tooth root strength of bevel gears, Y_{NT}, considers the higher tooth root stresses, which can be tolerated for a limited number of load cycles (and thus for a limited lifetime), in comparison with the allowable stress at 3×10^6 cycles.

This factor, which must be determined separately for pinion and wheel, is mainly influenced by: chemical composition, metallurgical structure, heat treatment and cleanliness of bevel gear material; number of load cycles or service life, N_L (see Sect. 3.8.2); failure criteria; smoothness of operation required; material ductility, fracture toughness, and residual stresses.

The allowable stress numbers are determined for 3×10^6 tooth load cycles, at 99% reliability. A unitary value of the life factor ($Y_{NT} = 1$) can be used for $N_L > 3 \times 10^6$ load cycles only when it is justified by a suitable experience. In these cases, however, optimum conditions for material quality and manufacturing as well as an appropriate safety factor must be guaranteed.

Here we limit our attention to Method B, by which the evaluation of the permissible tooth root stress, σ_{FP-B1} or σ_{FP-B2}, for limited life is determined using the life factor, Y_{NT}, of the standard reference test gear. With this method, the factors $Y_{\delta,relT}$, $Y_{R,relT}$, and Y_X shall be considered, and the normalized damage curves for tooth root strength are used. Figure 3.19 shows these curves, which express the life factor, Y_{NT}, as a function of the service life, N_L, for the four groups of materials described in Table 3.1. These curves apply for static, limited life, and endurance strength; they have been obtained from a large number of tests, and represent the typical damage or crack initiation curves for surface hardened and nitride hardened steels, or curves of yield stress for structural and through-hardened steels.

The values of Y_{NT} for static and endurance strengths may be obtained by calculation, using the data summarized in Table 3.1. The values of the life factor, Y_{NT}, for limited life stress are determined by means of interpolation between the values for endurance and static strength limits. With reference to the values of Y_{NT} shown in Table 3.1, included in the range from 0.85 to 1 (the corresponding range of the service life, N_L, is between 3×10^6 cycles to 1×10^{10} cycles), it is to be noted that the lower value ($Y_{NT} = 0.85$) should be used to minimize the damage due to the bending load under critical service conditions, while for optimum conditions of lubrication, material, manufacturing and experience the highest value ($Y_{NT} = 1$) can be used. It is also to be noted that stress levels above those permissible for 1×10^3 cycles should be avoided, since they might exceed the elastic limit of the gear tooth material.

Unlike for Method B, when we use the Method A, the S-N damage curves are derived from facsimiles of the actual gear. In this case, the factors $Y_{\delta,relT}$, $Y_{R,relT}$, and Y_X are already included in these damage curves. Therefore, when we calculate the permissible tooth root stress, these factors are all equal to unit.

6.9 Calculation Examples

Strength design, load carrying capacity assessment and determination of the safety factor against tooth root fatigue breakage of any type of bevel gear (including straight, helical or skew, Zerol, spiral bevel and hypoid gears) are reduced to those of the corresponding equivalent cylindrical gears. Therefore, the calculation examples that could be presented and discussed in this regard do not escape the space restrictions we have already highlighted in Sect. 3.11, to which we refer the reader.

References

AGMA 932-A05: 2005. Rating the pitting resistance and bending strength of hypoid gears

ANSI/AGMA 2003-C10. Rating the pitting resistance and bending strength of generated straight bevel, zerol bevel and spiral bevel gear teeth

Buckingham E (1949) Analytical mechanics of gears. McGraw-Hill Book Company Inc., New York

Budynas RG, Nisbett JK (2009) Shigley's mechanical engineering design, 8th edn. McGraw-Hill Companies Inc., New York

Coleman W (1952) Improved method for estimating the fatigue life of bevel and hypoid gears. SAE Q Trans 2(6)

Curà F, Mura A, Rosso C (2015) Effect of rim and web interaction on crack propagation paths in gears by means of XFEM technique. Fatigue Fract Eng Mater Struct 38(10):1237–1245

Dolan TJ, Broghammer EL (1942) A photoelastic study of stresses in gear tooth fillets. University of Illinois, Engineering Experiment Station Bulletin, p 335, Mar

Dudley DW (1962) Gear handbook. McGraw-Hill Book Company Inc., New York

Fernandes PJL (1996) Tooth bending fatigue failures in gears. Eng Fail Anal 3(3):219–225

Fisher R, Küçükay F, Jürgens G, Najork R, Pollak B (2015) The automotive transmission book. Springer International Publishing, Switzerland

Gagnon P, Gosselin C, Cloutier L (1997) Analysis of spur and straight bevel gear teeth deflection by the finite strip method. ASME J Mech Des 119(4):421–426

Giovannozzi R (1965) Costruzione di macchine, vol II. Casa Editrice Prof. Riccardo Pàtron, Bologna

Goldfarb V, Barmina N (2016) Theory and practice of gearing and transmissions, in honor of professor Faydor L. Litvin. Springer International Publishing, Switzerland

Handschuh RF, Nanlawala M, Hawkins JM, Mahan D, (2001) Experimental Comparison of fece-milled and face-hobbed spiral bevel gears, NASA/TM-2001-210940, ARL-TR-1104. In: Proceedings of the JSME international conference on motion and power transmission, Fukuoka, Japan, Nov 15–17

Henriot G (1979) Traité thèorique and pratique des engrenages, 6th edn. vol 1. Bordas, Paris

Höhn BR, Winter H, Michaelis K, Vollhüter F (1992) Pitting resistance and bending strength of bevel and hypoid gear teeth. In: Proceedings of the 6th international power transmission and gearing conference, Scottsdale, Arizona, 13–16 Sept 1992

ISO 10300-3:2014. Calculation of load capacity of bevel gears—part 3: calculation of tooth root strength

Juvinall RC, Marshek KM (2012) Fundamentals of machine component design, 5th edn. Wiley, New York

Kagawa T (1961) Deflections and moments due to a concentrated edge-load on a cantilever plate of infinite length. In: Proceedings of 11th Japan National Congress for Applied Mechanics, pp 47–52

Klingelnberg J (ed) (2016) Bevel gear: fundamentals and applications. Springer, Berlin

Lewicki DG, Ballarini R (1997) Rim thickness effects on gear crack propagation life. Int J Fract 87:59–86

Lewis W (1892) Investigation of strength of gear teeth. In: Proceedings of engineers club, Philadelphia, USA, vol 10, 16–23 Jan 1893

Maitra GM (1994) Handbook of gear design, 2nd edn. Tata McGraw-Hill Publishing Company Ltd., New Delhi

Merritt HE (1954) Gears, 3rd edn. Sir Isaac Pitman & Sous, Ltd., London

Naunheimer H, Bertsche B, Ryborz J, Novak W (2011) Automotive transmissions: fundamentals, selection, design and application, 2nd edn. Springer, Berlin

Niemann G, Winter H (1983a) Maschinen-Elemente, Band II: Getriebe allgemein, Zahnradgetriebe-Grundlagen, Stirnradgetriebe. Springer, Berlin

Niemann G, Winter H (1983b) Maschinen-Elemente, Band III: Schraubrad-, Kegelrad-, Schnecken-, Ketten-, Rienem-, Reibradgetriebe, Kupplungen, Bremsen, Freiläufe. Springer, Berlin

Pollone G (1970) Il Veicolo. Libreria Editrice Universitaria Levrotto & Bella, Torino

Radzevich SP (2016) Dudley's handbook of practical gear design and manufacture, 3rd edn. CRC Press, Taylor & Francis Group, Boca Raton

Shipley EE (1967) Gear failures: how to recognize them, what causes them, how to avoid them. Mach Des

Shtipelman BA (1978) Design and manufacture of hypoid gears. Wiley, Canada

Stadtfeld HJ (1993) Handbook of bevel and hypoid gears. Rochester Institute of Engineering, Rochester

Timoshenko SP, Woinowsky-Krieger S (1959) Theory of plates and shells, 2nd edn. McGraw-Hill International Editions, Singapore

Ventsel E, Krauthammer T (2001) Thin plates and shells: theory, analysis, and applications. Marcel Dekker Inc., New York

Vijayakar S (2016) Contact and bending durability calculation of spiral-bevel gears, NASA/CR-2016-219112. Glenn Research Center, Cleveland

Vullo V (2014) Circular cylinders and pressure vessels: stress analysis and design. Springer International Publishing, Switzerland

Vullo V, Vivio F (2013) Rotors: stress analysis and design. Springer, Dordrecht

Wellauer EJ, Seireg A (1960) Bending strength of the gear teeth by cantilever-plate theory. ASME J Eng Ind 82(3):213–220

Chapter 7
Scuffing Load Carrying Capacity of Cylindrical, Bevel and Hypoid Gears: Flash Temperature Method

Abstract In this chapter, the basic concepts of the scuffing damage of the gears are first described, framing the calculation of scuffing load carrying capacity of cylindrical, bevel and hypoid gears according to the flask temperature method. The fundamentals of the Blok's theory that ascribes the cause of scuffing to a sudden breakdown of the lubricant oil film are recalled, also lingering on the transition diagrams in lubricated contacts and on the coefficient of friction at incipient scuffing. The flash temperature is then determined for the aforementioned gears, showing how it strongly depends on the coefficient of friction and other influence factors described in detail. The contact temperature and scuffing temperature are therefore defined and indications are given on the safety factor for scuffing. A large part of the chapter is reserved for the procedures to calculate the tooth bending strength of these types of gears in accordance with the ISO standards, highlighting when deemed necessary as the relationships used by the same ISO are founded on the theoretical bases previously recalled. Finally, indications are given on cold scuffing, with insights on the proper scuffing (warm scuffing) and references to empirical formulae used in this regard in the recent past.

7.1 Introduction

In Sect. 2.1, we saw that scuffing is one of the main types of damage, characterizing the limit performance of any gear pair (see Shipley 1967; Alban 1985; Tallian 1992; Bowman and Stachowiak 1996; Bloch and Geitner 1999). In the same section, we also said that scuffing means local welding, sometimes associated with the removal of material from one another of the two mating tooth flanks in relative motion, most often occurring at roughness peaks in the flank contact surfaces, and due to insufficient or incorrect lubrication.

If gear tooth flanks are completely separated by a *thick oil film* (also called *full lubricant film*), there is no contact between the unavoidable roughness peaks, often called asperities, of the tooth flank surfaces. Usually, in this case, there is no scuffing or wear, and the coefficient of friction is very low. However, even for a thick oil film, a damage similar to scuffing can exceptionally occur, due to a sudden thermal

© Springer Nature Switzerland AG 2020

V. Vullo, *Gears*, Springer Series in Solid and Structural Mechanics 11,
https://doi.org/10.1007/978-3-030-38632-0_7

instability. This is a very particular subject, for specialists, not covered by the current international and national standards (see Barwell and Milne 1952; Blok 1974; Dyson 1976; Dowson and Toyoda 1978; Scott 1979; Miltenović et al. 2012).

For thinner elastohydrodynamic oil films, incidental asperity contacts occur, and the number of contacts between the roughness peaks increases when the mean lubricant film thickness decreases. Therefore, adhesive wear, i.e. scuffing, and abrasive wear become possible. Depending on the amount of scuffing or wear, the gears can be only partially damaged or damaged to such an extent as destroy their practical operation. The motivations of scuffing can be found in very high contact temperatures that depend on load, tangential sliding speed, characteristics of material, and lubrication conditions, including the oil sump temperature. We distinguish between *warm scuffing* or *hot scuffing*, and *cold scuffing* (see Tabor 1981; Niemann and Winter 1983; Naunheimer et al. 2011; Jelaska 2012; Goldfarb and Barmina 2016; Radzevich 2016).

Warm or *hot scuffing* is the proper scuffing. It arises when the lubricant oil film breaks down due to high contact temperature related to very high sliding velocity and surface pressures. This leads to the direct metal-to-metal contact between the unavoidable roughness peaks or asperities. Since the contact pressure and the frictional heat due to sliding velocity are concentrated at small local areas of contact, local temperatures and pressures are extremely high (Horng 1998). Therefore, conditions are favorable for welding of these points, where the instantaneous temperature may locally reach the metal melting point, but with temperature gradients so steep that the part remains cool to the touch (see Dyson 1975a, b; Juvinall and Marshek 2012; Stachowiak and Batchelor 2014).

If melting and welding of the surface asperities occur, either the weld or one of the two metals near the weld must fail in shear in such a way to permit the relative motion of the two surfaces in contact to continue. New welds, that is new adhesions and corresponding fractures continue to occur, resulting in the scuffing, which therefore is a phenomenon of *adhesive wear*. Since adhesive wear is essentially a welding phenomenon, metals that easily weld together are most susceptible to scuffing (see Rabinowicz 1980; Juvinall and Marshek 2012).

Therefore, welds that continuously formed and destroyed, with a subsequent detachment and transfer of particles from one or both meshing tooth flanks, characterize scuffing. Usually, flaking of tooth flanks occurs by scuffing, but also loose particles of metal and metal oxides may be formed. These loose particles and oxides resulting from adhesive wear, entrained by the flow of lubricant, cause further surface wear because of abrasion. In addition, abrasive wear may occur due to the rolling action of the gear teeth (see Ludema 1984; Snidle et al. 2003).

For the protection of tooth flanks from excessive adhesive wear and/or abrasive wear, the various coatings used so far, i.e. flank copper coating, flank phosphating, coating with tungsten alloys, etc., have proven in practice to be very effective and useful. However, the scuffing load capacity of gears does not only depend on the material, including heat and surface treatment, and the design of moving parts, but also on the lubricant used. If the composition of the mineral oil and additives is not right, the lubricant does not develop the required oil film thickness, and thus the required scuffing load carrying capacity is not reached (see Michalczewski et al. 2009, 2013; Habchi 2014; Gupta et al. 2017).

The generation of scuffing damage is a very complex phenomenon, due to both physical and special chemical processes. The latter occur in extremely thin layers of lubricant, adherent to the metal surfaces of the tooth flanks, and under high contact pressures, while the physical phenomena are explained by EHD-Elastohydrodynamic Lubrication Theory (Dowson and Higginson 1966), which exceeds the well-known limits of the HD-Hydrodynamic Lubrication Theory (Cameron 1952, 1954). Here we focus our attention on the traditional models of adhesive wear prediction, leaving aside the most advanced atomistic models, based on the inter-atomic potential, which show a transition in the asperity wear mechanism when contact junctions fall below a critical length scale (Aghababaei et al. 2016).

Adhesive wear as well as abrasive wear are not fatigue phenomena. Contrary to the surface durability (pitting), tooth root bending strength, micropitting, tooth flank fracture and tooth interior fatigue fracture, which are phenomena of fatigue, and then show a distinct incubation period, a short transient overload can cause a scuffing failure.

Adhesive wear and/or abrasive wear may not be harmful if they are mild, or if they are reduced in time, as happens in a common running-in process. However, usually scuffing is a severe form of adhesive wear, resulting in a more or less significant damage to the gear tooth flanks. The risk of scuffing damage is also increased by the presence in the lubricant of contaminants, such as metal particles in suspension, water or oxides due to excessive aeration. Once scuffing is triggered, particularly in gears operating at high speed, it determines high levels of dynamic loading due to vibration, which in turn leads to further damage by scuffing, pitting and tooth breakage.

It is to be noted that, in technical language, if welding and tearing of the surface asperities cause a transfer of metal from one surface to the other, the resulting adhesive wear or surface damage is called *scoring*. If the local welding of asperities becomes so extensive that the surfaces no longer slide on each other, the resulting damage is called *seizure* or *synechia* (I like to borrow this Greek word from medicine, were it is used to indicate the adherence between organs; it give a good idea of the adherence between the mating flank surfaces of gear teeth, considered almost like protruding limbs of the two members of the gear). Furthermore, two types of warm scuffing can be distinguished: scoring and scuffing.

Scoring is typical of doped oils at transverse tangential velocities lower than 30 m/s; *individual scoring* or *clusters of scoring* appear in the sliding direction of the tooth flanks, varying from minor to serious. Scuffing is typical of undoped and doped oils at transverse tangential velocities greater than 30 m/s; it is a mild adhesive wear and occurs as individual fine lines (*scuffing lines*), as clusters (*heavy scuffing* or *galling*) or as areas across the entire face width (*scuffing zones*), and the main feature of the scored areas is a matt appearance. Generally, the higher the surface hardnesses (more precisely, the higher the ratio of surface hardness to modulus of elasticity), the greater the resistance to the adhesive wear.

In most practical applications, the gear resistance to scuffing can be improved by using smaller modules, and lubricant oils with enhanced anti-scuff additives, i.e. the so-called Extreme Pressure, EP, which are lubricant oils characterized by chemically active additives. It is however be borne in mind that, in some cases, these anti-scuff lubricant oils can present some disadvantages, such as corrosion of copper, embrittlement of elastomers, lack of world-wide availability, etc.. (see Höhn et al. 1998, 1999, 2004, 2011; Martins et al. 2008; Hirani 2016).

If countermeasures are not taken, the power losses, temperature and wear increase, as well as vibration and noise, for which the danger of tooth breakage becomes looming. The warm scuffing occurs predominantly in spur and helical gears operating at high tangential velocities (over 4 m/s), highly loaded and hardened and, above all, in hypoid gears (see Lacey 1988; Conrado et al. 2007; Wink 2012; Xue et al. 2014).

In contrast to warm scuffing, cold scuffing occurs without appreciable development of heat, and is a relatively rare damage phenomenon. Generally, it is associated with relatively low tangential velocities (below 4 m/s), mainly for through-hardened and heavily loaded gears having rather poor accuracy grade.

All the sections that follow address the calculation of warm scuffing load capacity of the gears, with the exception of the last section, which instead regards a possible criterion for assessing the cold scuffing load capacity of gears. It is to be noted that here the term scuffing without adjectives means warm scuffing.

In this chapter and the next, we will focus our attention on the calculation of warm scuffing load capacity of cylindrical (spur and helical), bevel, spiral bevel and hypoid gears, made with usual gear materials, but with different heat treatments. Two different assumptions are made to approach this calculation problem, both based on the hypothesis that the high surface temperatures due to high sliding velocities and contact pressures can trigger the breakdown of lubricant oil film. The first assumption leads to the *flash temperature method*, and is based on contact temperatures, which vary along the path of contact. The second assumption leads to the *integral temperature method*, and is based on the weighted average of the contact temperatures along the path of contact. The flash temperature method is the subject of this chapter, while the next chapter will discuss the integral temperature method.

7.2   Transition Diagram and Coefficient of Friction at Incipient Scuffing

According to Blok (1937a, b, 1963a), the flash temperature method is based on the fundamental concept that scuffing is due to a sudden breakdown of the lubricant oil film, when the contact temperature has reached a certain threshold value. In other words, scuffing would not be determined by a progressive decrease of the lubricant film thickness, up to a critical value, but by the attainment of a threshold contact temperature for the given lubricant oil used. This hypothesis is still supported by the fact that several experimental evidences have confirmed the existence of a sufficient

Fig. 7.1 Example of transition diagram, where even the calculated contact temperatures are shown

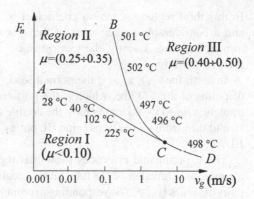

lubricant film thickness until the incipient scuffing, and then an abrupt breakdown of the same lubricant film.

The lubrication conditions of concentrated sliding contacts between surfaces of steel bodies, including a liquid lubricant film, can be described in terms of transition diagrams. These diagrams give the normal load in wear test, F_n, as a function of the sliding velocity, v_g (see Czichos 1974; Andersson and Salas-Russo 1994; Schipper and de Gee 1995; Andersson et al. 2007; Kovalchenko et al. 2011; Mang et al. 2011; Pirro et al. 2016). An example of these diagrams is that shown in Fig. 7.1, which refers to a sliding contact operating at constant oil bath temperature.

In this diagram, as well as in similar diagrams, we can identify three regions: region I, characterized either by the absence of wear or by an extremely mild wear; region II, characterized by mild wear; region III, characterized by severe wear and scuffing. Two transition lines delimit these three regions, ACD and BCD, which have in common the CD portion.

For pairs of values of normal load in wear test, F_n, and relative sliding velocity, v_g, that fall in region I, that is for pairs (v_g, F_n) below the ACD transition line, the lubrication conditions are characterized by a low coefficient of friction ($\mu \leq 0.10$). Furthermore, the lubrication conditions are characterized by a low *specific wear rate*, defined as the volume of material worn away for unit of normal force, and per unit of sliding distance (see Lipson 1967; Peterson and Winer 1980; and Chap. 9). In fact, in this region, the specific wear rate is between (10^{-6} and 10^{-2}) mm³/(Nm).

For values of the sliding velocity, v_g, lower than the abscissa $v_{g,C}$ of point C, where the above two transition lines begin to overlap, i.e. for $v_g < v_{g,C}$, if the normal load, F_n, is increased over the ordinates of points of the transition line AC, a transition takes place from region I to region II. This region II is characterized by a different condition of lubrication, which corresponds to a slight wear. In effect, in this second region, we have a coefficient of friction, μ, between (0.25 and 0.35), and a specific wear rate between (1 and 5) mm³/(Nm). Still for $v_g < v_{g,C}$, if the normal load, F_n, is further increased over the ordinates of points of the transition line BC, a new transition takes place from region II to region III, which is characterized by a still different condition of lubrication, which corresponds to a severe wear and scuffing.

In this third region we have a coefficient of friction, μ, between $(0.40$ and $0.50)$, and a considerably higher value of the specific wear rate, between $(10^2$ and $10^3)$ mm^3/(Nm); consequently, the worn surfaces show evidence of severe wear in the form of scuffing.

Instead, for $v_g \geq v_{g,C}$, if the normal load, F_n, is increased beyond the ordinates of points of the CD line, which is the portion common to the two aforementioned transition lines, we no longer have the double transition, first from region I to region II, and then from region II to region III, but a direct transition from region I to region III.

The experimental evidences show that the position of the transition line ACD changes as a function of the lubricant viscosity and Hertzian contact pressure. For pairs of values (v_g, F_n) corresponding to points located below the transition line ACD, it is assumed that the lubrication conditions are those of *partial elastohydrodynamic lubrication*. In these conditions, the surfaces are kept separated by a thin lubricant oil film, which is however penetrated by the roughness asperities (see Begelinger and de Gee 1974, 1982; Salomon 1976; Czichos 1976; Czichos and Habig 2010).

The transition line BCD is the one corresponding to the incipient scuffing. Beyond this transition line, we have the region III, in which the liquid film effects are completely absent. The experimental evidences show that, in the transition from region II to region III, for $v_g < v_{g,C}$, as well as in the transition from region I to region III, for $v_g \geq v_{g,C}$, the same transition is associated with the achievement of a critical value of the contact temperature, and this in agreement with Blok (see Bruce 2012).

Along the ACD transition line, the temperature increases when the sliding velocity increases, passing gradually from the oil bath temperature at lower speeds (this temperature coincides with the *overall bulk temperature* and with *interfacial bulk temperature*) to the *contact temperature* for the highest speeds. In the specific example shown in Fig. 7.1, the temperature varies from 28 °C at $v_g = 10^{-3}$m/s to 498 °C at $v_g = 10$ m/s; this last temperature is the interfacial contact temperature, which is the sum of the interfacial bulk temperature and *flash temperature*. This temperature behavior for $v_g \geq v_{g,C}$ highlights the fact that the collapse of partial elastohydrodynamic lubrication does not occur at a constant contact temperature; it is also attributed to the sudden destruction of chemisorbed links between lubricant oil film and asperities of the base material of which the gear is made. Instead, the pronounced decrease of the normal load, F_n, and therefore of the load carrying capacity, with increasing sliding velocity, v_g, is attributed to the decreasing viscosity.

Contrary to what happens for the transition line ACD, which is characterized by a temperature which varies as a function of v_g (only beyond the point C, that is for $v_g > v_{g,C}$, the temperature is a constant), the transition line BCD is instead characterized by a constant contact temperature, which is about equal to 500 °C for AISI 52100 steel specimens. Based on this finding, it was decided to attribute the transition from region II to region III, as well as the direct transition from region I to region III, to a downgrading of the metallurgical characteristics of the steel, and perhaps to a likely thermo-elastic instability mechanism. This last mechanism would be the main causes of the change of the wear mechanism, from the middle adhesive wear to severely adhesive wear (see Blok 1974; Dyson 1976; Miltenovic et al. 2012).

The transition diagrams described above, which summarize results of experimental tests simulating unoxidized steel contacts, that is new steel contacts just assembled, indicate that the scuffing is associated with a critical value of the contact temperature. For contact between steel bodies, lubricated with mineral oils, this critical value of the contact temperature does not depend on the load, speed and geometry, and it is equal to about 500 °C.

Figure 7.1 also shows that, during the transition from region II to region III, the coefficient of friction practically doubled, jumping from a minimum value $\mu = 0.25$ to a value $\mu = 0.50$; instead, during the direct transition from region I to region III, the coefficient of friction is much more than fivefold. The above-mentioned contact temperature, equal to about 500 °C, is the sum of the measured interfacial bulk temperature, Θ_{Mi}, equal to 28 °C, and the calculated flash temperature, Θ_{fl}, equal to 470 °C. This flash temperature is the result of the calculation obtained with $\mu = 0.35$, which is the maximum value of the coefficient of friction just before transition from region II to region III.

It is to be noted that the results described above, in terms of transition diagrams, are generally obtained with pin-and-ring tribo-testing (see Lipson 1967; Peterson and Winer 1980). Therefore, to apply these results to gear transmission systems during the design stage, it is necessary to verify that the critical value of the contact temperature is consistent with the value of the coefficient of friction to be used in the calculations.

As mentioned above, the interfacial contact temperature, often more simply called *contact temperature*, is the sum of two components, and precisely:

- The interfacial bulk temperature of the moving surface, which is a constant or varies very slowly. This interfacial bulk temperature is the equilibrium temperature of the gear tooth surface before teeth enter the contact zone. It is evaluated by making a suitable average of the two overall bulk temperatures of the mating teeth that, in turn, are determined by means of the *thermal network theory* (see Carslaw and Jaeger 1959; Blok 1969; Bathgate et al. 1970; Tanaka and Edwards 1992; Manin and Play 1999; Incropera et al. 2006; Naveros et al. 2016).
- The flash temperature of the two moving tooth flank surfaces in contact, which fluctuates rapidly due to the combined effects of load, tooth geometry, velocity, friction and material characteristics during operation. It is the calculated increase of temperature of the gear tooth surface at a given point of path of contact, and depends significantly on the coefficient of friction, which must be evaluated with special attention. In this regard, a common practice is to use a value of the coefficient of friction corresponding to regular operating conditions, despite the fact that, at the incipient scuffing, the coefficient of friction has a significantly higher value.

With reference to the coefficient of friction, the scuffing load carrying capacity of the gears can be calculated using:

- a high value of the coefficient of friction ($\mu = 0.50$), for which the result is on the safety side;

- a value of the coefficient of friction that, depending on the lubricant oil used, is included in the range ($0.25 \leq \mu \leq 0.35$), for which the result corresponds to that of a fairly accurate assessment;
- a low value of the coefficient of friction, correlated to the above-mentioned regular operating conditions, for which the result is sufficiently reliable, provided that the limit contact temperature is correspondingly low, in accordance with the aforesaid common practice.

In terms of this common practice, it is to be noted that, for non-additive and low-additive mineral lubricant oils, each combination of lubricant and sliding/rolling materials has a *critical scuffing temperature*, which is generally a constant regardless of the operating conditions, velocity, load and geometry. On the contrary, for high-additive and certain types of synthetic lubricant oils, the critical scuffing temperature can vary from one set of operating conditions to another. Therefore, in this case, the critical scuffing temperature must be determined separately, for each set of operating conditions, by means of tests that closely simulate those of the gear drive under consideration.

7.3 Contact Temperature

7.3.1 General Relationship

As we said in the previous section, the contact temperature, Θ_B (as we said above, this term simplifies the more precise term *interfacial contact temperature*), is the sum of the interfacial bulk temperature, Θ_{Mi}, and flash temperature, Θ_{fl}. Therefore, it is expressed by the following general relationship:

$$\Theta_B = \Theta_{Mi} + \Theta_{fl}. \tag{7.1}$$

Fig. 7.2 Distribution of contact temperature along the path of contact, *AE*

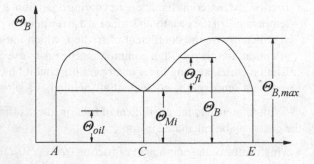

As Fig. 7.2 shows, Θ_{Mi} is a constant along the path of contact, while Θ_{fl} is a variable quantity. Therefore, the maximum contact temperature is given by:

$$\Theta_{B\max} = \Theta_{Mi} + \Theta_{fl\max},\qquad(7.2)$$

where $\Theta_{fl\max}$ is the maximum value of Θ_{fl}, which can occur either at the approach path of contact or at recess path of contact.

By comparing the calculated maximum contact temperature with a maximum achievable critical value (see Sect. 7.8), it is possible to make a reliable forecast of the probability that the scuffing occurs. This critical value of the contact temperature can be estimated by means of field investigations or by any gear-scuffing test (see, for example, Lacey 1988). However, for a reliable estimate of the risk of scuffing, it is necessary to determine accurately the value of the gear bulk temperature to be used for the analysis.

7.3.2 Interfacial Bulk Temperature and Overall Bulk Temperature

The interfacial bulk temperature, Θ_{Mi}, can be determined as an appropriate weighted average of the overall bulk temperatures, Θ_{M1} and Θ_{M2}, of the teeth of pinion and wheel in contact. For high values of the *Péclet numbers* (see next Sect. 7.3.3), a good approximation of Θ_{Mi} can be obtained using the following relationship:

$$\Theta_{Mi} = \frac{\Theta_{M1}B_{M1}\sqrt{v_{g1}} + \Theta_{M2}B_{M2}\sqrt{v_{g2}}}{B_{M1}\sqrt{v_{g1}} + B_{M2}\sqrt{v_{g2}}},\qquad(7.3)$$

where $B_{M1}\sqrt{v_{g1}}$ and $B_{M2}\sqrt{v_{g2}}$ are *weight functions*, v_{g1} and v_{g2} are the sliding velocities of pinion and wheel (in m/s), and B_{M1} and B_{M2} are the *thermal contact coefficients* of pinion and wheel [in N/(mm$^{1/2}$m$^{1/2}$s$^{1/2}$K)], while Θ_{Mi}, Θ_{M1}, and Θ_{M2} are given in °C.

Within a fairly wide range of the ratio between the two weight functions mentioned above, i.e. the ratio $(B_{M1}\sqrt{v_{g1}}/B_{M2}\sqrt{v_{g2}})$, a reasonably approximate value of the interfacial bulk temperature can be obtained as the arithmetic average of Θ_{M1} and Θ_{M2}, that is as:

$$\Theta_{Mi} = \frac{1}{2}(\Theta_{M1} + \Theta_{M2}).\qquad(7.4)$$

It is to be noted that interfacial bulk temperatures above 150 °C, persistent for long time, can determine the decay of the strength characteristics of the material of which gears are made, with consequent adverse effects on surface durability (pitting). It is also to be noted that the overall bulk temperatures, Θ_{M1} and Θ_{M2}, can be determined by any of the methods that consider the thermal interconnections of a system consisting of heat sources and sinks. These methods include the *FE methods* (Cook 1981), *bond-graph methods* (Broenink 1999), and *thermal network analogue methods* (Blok 1969).

In gear units, the main heat source, and the related friction losses, is the one due to tooth friction in the meshing area. Other sources are then to be added to this heat source, which determine related losses due to friction. Among these other sources, it is worth mentioning specifically those due to: unavoidable friction losses in the bearings, whether rolling or sliding (sliding bearings often are characterized by much higher losses than those occurring in the meshing area); equally unavoidable losses in the seals; losses associated to oil churning; windage losses, that may become very significant for pitch line velocities in excess of 80 m/s, etc. In addition, it is well to consider the losses due to mechanical pumping energy expended for sideways ejecting the superfluous lubricant, which can sometimes be far from negligible.

For the determination of the overall bulk temperatures with the above-mentioned methods, it is to be noted that, in each of these heat sources, the fluid friction depends on the lubricant viscosity under the operating conditions. Furthermore, all these heat sources are thermally interconnected by means of heat transfer elements to heat sinks, such as the ambient air or the cooling system. In this regard, the heat is transferred to the environment via the housing walls by conduction, convention and radiation, and, for forced lubrication conditions, through the oil into an external heat exchanger.

For a reliable assessment of the risk of scuffing, in most cases of practical applications, it is necessary to have an accurate value of the interfacial bulk temperature. In this regard, the methods of analysis mentioned above are fully satisfactory. In cases where a roughly approximate evaluation of the interfacial bulk temperature is considered sufficient, this temperature can be estimated as the sum of two components, according to the following relationship:

$$\Theta_M = \Theta_{oil} + 0.47 X_S X_{mp} \Theta_{flm}, \tag{7.5}$$

where: Θ_{oil} (in K or °C), is the oil temperature before reaching the meshing area; Θ_{flm} (in K or °C), is the average flash temperature along the path of contact, given by:

$$\Theta_{flm} = \frac{\int_A^E \Theta_{fl} d\Gamma_y}{\Gamma_E - \Gamma_A}; \tag{7.6}$$

X_S, is the *lubrication system factor*, and X_{mp}, is the *multiple mating pinion factor*, given by:

$$X_{mp} = \frac{1}{2}(1 + n_p). \tag{7.7}$$

In this equations, n_p is the *number of mesh contacts*. It is to be noted that X_S, X_{mp} and n_p are dimensionless quantities. The lubrication system factor, X_S, takes into account the effects of lubrication conditions on heat transfer. The values to be taken for this factor are the following: $X_S = 1.2$, for spray lubrication; $X_S = 1.0$, for dip lubrication, and for meshes with additional spray lubrication to improve the cooling conditions; $X_S = 0.2$, for gears in bath of oil, and sufficient cooling conditions. The multiple mating pinion factor, X_{mp}, takes into account the number of mesh contacts that in some cases are different for pinion and wheel (see, for example, the various members of a planetary gear train). As it regards the quantities that appear in Eq. (7.6), we refer the reader to Sect. 7.5.

7.3.3   Flash Temperature

The bases of calculation of the flash temperature are due to Blok (1937a, b, c, d, 1940). Extending the fundamentals concepts of Blok's theory to the most general case of tooth contact, represented by hypoid gears, we can assume that the successive contact areas have the shape of oval tapered bands, and that the sliding velocities, v_{g1} and v_{g2}, are directed along directions defined by unequal angles, γ_1 and γ_2, with respect to the greater of two axes of these oval areas. Figure 7.3 shows an example of the band-shaped contact area, which has a uniform width, equal to the local width, $2b_H$ (b_H, in mm, is the semi-width of the Hertzian contact band); this band-shaped contact area replaces the actual area of contact between the two surfaces of the meshing tooth flanks, which are tangent to each other at the instantaneous point of contact. The same figure also shows the tapered contact area, having the shape of an asymmetrical ovoid area, considered by this extension of Blok's theory.

In the most general case of hypoid gears, the actual Hertzian contact area can be considered approximately elliptical in shape, and the two angles of sliding velocities, γ_1 and γ_2, are neither coinciding nor perpendicular to the major axis of the elliptical contact area; they are instead oriented according to different directions that deviate slightly compared to direction of the minor axis of the ellipse. However, this contact area can be rather extended, with a rather high elliptical ratio, or it can have the shape of a somewhat tapered band. Obviously, in the simplest case, such as that of

Fig. 7.3 Band-shaped contact area in the most general case of hypoid gears, with sliding velocities differently directed

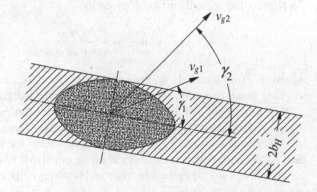

cylindrical gears, the above-mentioned angles are equal to each other, and both equal to $\pi/2$, i.e. $\gamma_1 = \gamma_2 = \pi/2$.

To determine the flash temperature as well as its maximum value according to Blok, we make the following assumptions:

- The aforementioned elliptic contact area is replaced with a band-shaped contact area, whose width, $2b_H$, equals the length of the minor axis of the ellipse.
- The contact pressure over any cross-section of the tapered contact area is distributed according to a semi-ellipse (see Sect. 2.3 and Fig. 2.2).
- The friction heat generated by the sliding contact effect is also distributed over any cross-section of the tapered contact area according to a semi-ellipse (see Sect. 2.4 and Fig. 2.2).
- The maximum contact pressure over any cross-section, including the one that has the minor axis of the ellipse as a trace, is directly proportional to the cubic root of the load, rather than the square root; this assumption entails an adaptation of the Hertz theory.
- For the actual sufficiently extended elliptic contact areas under the above-mentioned kinematic conditions, the expected actual maximum flash temperature is the one occurring at a point fairly close to the minor axis of the ellipse.

7.3.3.1 Hypoid Gears

In the most general case of hypoid gears, the flash temperature for an approximately band-shaped contact area and sliding velocities differently directed is given by the following formula of Blok:

$$\Theta_{fl} = 1.11 \frac{\mu_m X_\Gamma X_J w_{Bn}}{\sqrt{2b_H}} \frac{|v_{g1} - v_{g2}|}{\left[B_{M1}\sqrt{v_{g1}\sin\gamma_1} + B_{M2}\sqrt{v_{g2}\sin\gamma_2} \right]}. \tag{7.8}$$

In this equation, the meaning of symbols already introduced remaining unchanged, the other symbols represent: μ_m, the mean coefficient of friction (see Sect. 7.4); X_Γ, the load sharing factor (see Sect. 7.7), X_J, the *approach factor* (see Sect. 7.6); w_{Bn} (in N/mm), the *normal unit load*, given by:

$$w_{Bn} = \frac{w_{Bt}}{\cos\alpha_{wn}\cos\beta_w}, \tag{7.9}$$

where w_{Bt} (in N/mm), is the *transverse unit load*, α_{wn} (in degrees), is the *normal working pressure angle*, and β_w (in degrees), is the *working helix angle*.

It is to be noted that μ_m, X_Γ, and X_J are dimensionless quantities. Furthermore, the numerical factor 1.11 that appears in Eq. (7.8) is derived from the elliptic friction heat distribution assumed by Blok (1958), replacing the previous parabolic friction heat distribution, to which a numerical factor equal to $0.83\sqrt{2} = 1.17$ was correlated.

The normal working pressure angle and working helix angle appearing in Eq. (7.9) are given respectively by:

$$\alpha_{wn} = \arcsin(\sin \alpha_{wt} \cos \beta_b) \tag{7.10}$$

$$\beta_w = \arctan\left(\frac{\tan \beta_b}{\cos \alpha_{wt}}\right), \tag{7.11}$$

were α_{wt} and β_b are respectively the transverse working pressure angle and base helix angle.

The transverse unit load appearing in the same Eq. (7.9) for bevel gears and cylindrical gears is given respectively by:

$$w_{Bt} = K_A K_v K_{B\beta} K_{B\alpha} K_{mp} \frac{F_t}{b_{eff}} \tag{7.12}$$

$$w_{Bt} = K_A K_v K_{B\beta} K_{B\alpha} K_{mp} \frac{F_t}{b}, \tag{7.13}$$

where: F_t (in N), is the nominal tangential force on pitch circle; b_{eff} (in mm), is the effective face width, to be assumed equal to $0.85b$; b (in mm), is the face width, to be chosen equal to the smaller value for pinion and wheel; K_A, is the application factor; K_v, is the dynamic factor; $K_{B\beta} = K_{H\beta}$, is the face load factor for scuffing; $K_{B\alpha} = K_{H\alpha}$, is the transverse load factor for scuffing; K_{mp}, is the *multi-path factor*, which takes into account the misdistribution in multi-path gear transmissions, and depends on flexibility and accuracy of their various component branches. It is to be kept in mind that factors K_A, K_v, $K_{B\beta} = K_{H\beta}$, and $K_{B\alpha} = K_{H\alpha}$ are different for bevel gears and cylindrical gears. These factors are to be calculated, for the two gear families, according to the procedures described in Chaps. 4 and 1, concerning respectively bevel gears and cylindrical gears.

When data from more reliable analysis are not available, for determination of K_{mp} we can use the following relationships:

$$K_{mp} = 1 + 0.25(n_p - 3)^{1/2} \tag{7.14}$$

$$K_{mp} = 1 + (0.2/\phi) \tag{7.15}$$

$$K_{mp} = 1 + \frac{F_{ex}}{F_t \tan \beta} \tag{7.16}$$

$$K_{mp} = 1, \tag{7.17}$$

which are respectively valid for: planetary gear trains with n_p gear planets (this equation is valid for $n_p > 3$); dual tandem gears with quill shaft twist ϕ under full load (it should be remember that a quill shaft is a thin solid shaft strategically designed to carry the same torque of a larger shaft, so it works at higher stress levels; it has the particular peculiarity of reducing torsional vibrations and pulsations, as it acts as a torsional spring, twisting along its longitudinal axes); double helical gears with an external axial force F_{ex}; all other cases.

The already introduced thermal contact coefficients of pinion and wheel, $B_{Mi} = (\lambda_{Mi}\rho_{Mi}c_{Mi})^{1/2}$, with $i = (1.2)$ (see Sect. 7.3.2) are given by:

$$B_{M1} = \left(10^{-3}\lambda_{M1}\rho_{M1}c_{M1}\right)^{1/2} \tag{7.18}$$

$$B_{M2} = \left(10^{-3}\lambda_{M2}\rho_{M2}c_{M2}\right)^{1/2}, \tag{7.19}$$

where λ_{M1} and λ_{M2} [both in N/(sK)], ρ_{M1} and ρ_{M2} (both in kg/m^3), c_{M1} and c_{M2} [both in J/(kgK)] are respectively the *heat conductivity*, *density* and *specific heat per unit mass* of materials of pinion and wheel, while factor 10^{-3} aligns the units of λ_{Mi}, ρ_{Mi} and c_{Mi} (with $i = 1, 2$), which are basically those of the SI system, to the not usually unit of B_{Mi} adopted by ISO, in which a length is given in meters and another in millimeters.

It is noteworthy that all the quantities that appear in Eqs. (7.18) and (7.19) are assumed independent of temperature, and therefore constant. This approximate assumption is a source of errors, especially in the presence of considerable temperature gradients. In this regard, for general aspects, see Zudans et al. (1965), Boley and Weiner (1997), Vullo (2014), and for some specific aspects concerning thermal conductivity, Abdel-Aal (1997).

Finally, it is noteworthy that in Eq. (7.8) the absolute value $|v_{g1} - v_{g2}|$ of the difference of sliding velocities of pinion and wheel as well as the same sliding velocities v_{g1} and v_{g2} appear, and that these velocities, expressed in m/s, are given by Eq. (3.70). The semi-width, b_H, of the Hertzian contact band is expressed also in mm. The thermal contact coefficients of pinion and wheel, B_{M1} and B_{M2}, are calculated using Eqs. (7.18) and (7.19).

7.3.3.2   Cylindrical and Bevel Gears

In the case of cylindrical and bevel gears, with band-shaped contact area and parallel sliding velocities, the flash temperature is given by the following general relationship:

$$\Theta_{fl} = 1.11 \frac{\mu_m X_\Gamma X_J w_{Bn}}{\sqrt{2b_H}} \frac{|v_{g1} - v_{g2}|}{\left[B_{M1}\sqrt{v_{g1}} + B_{M2}\sqrt{v_{g2}}\right]}, \tag{7.20}$$

which is the specialization of Eq. (7.8) for $\sin\gamma_1 = \sin\gamma_2 = \sin\pi/2/ = 1$. The following equivalent equation can be used in place of the above relationship:

$$\Theta_{fl} = 2.52\mu_m \frac{X_M}{50} X_J (X_\Gamma w_{Bt})^{3/4} \left(\frac{n_1}{60}\right)^{1/2} \frac{\left|\rho_{y1}^{1/2} - (\rho_{y2}/u)^{1/2}\right|}{(\rho_{rely})^{1/4}}. \tag{7.21}$$

In this last equation, the meaning of symbols already introduced remaining unchanged, the new symbols represent, respectively: n_1 (in min^{-1}), the rotational speed of pinion; X_M (in $KN^{-3/4}$ $s^{-1/2}$ $m^{-1/2}mm$), the *thermo-elastic factor*, also called *thermo-flash factor*; ρ_{y1} and ρ_{y2} (both in mm), the local radius of curvature at arbitrary point of flanks of pinion and wheel; ρ_{rely} (in mm), the local relative radius of curvature at arbitrary point, y; $u = z_2/z_1$, the gear ratio.

It is to be noted that Eqs. (7.20) and (7.21) as well as Eq. (7.8) are valid only when the two *Péclet numbers* are sufficiently high (both greater than 5), i.e. when:

$$\text{Pé}_1 = \frac{v_{g1}b_H\rho_{M1}c_{M1}}{\lambda_{M1}\sin\gamma_1} > 5 \tag{7.22}$$

$$\text{Pé}_2 = \frac{v_{g2}b_H\rho_{M2}c_{M2}}{\lambda_{M2}\sin\gamma_2} > 5, \tag{7.23}$$

where Pé_1 and Pé_2 are the Péclet numbers of pinion and wheel, which depend mainly on material, but to a lesser extent also on sliding velocity and gear geometry. This condition is satisfied in almost all cases where scuffing may occur. For lower Péclet numbers, the heat flow from the band-shaped contact area inside the gear teeth causes a different temperature distribution, for which Eqs. (7.8), (7.20), and (7.21) are not valid.

Cylindrical Gears

To calculate the flash temperature of cylindrical gears, we can use the more general relationships (7.20) and (7.21), which are valid for cylindrical and bevel gears, or the following relationship, which is an adaptation for cylindrical gears of the above more general relationships:

$$\Theta_{fl} = \mu_m X_M X_J X_G (X_\Gamma w_{Bt})^{3/4} \frac{v_t^{1/2}}{a^{1/4}}. \tag{7.24}$$

In this equation, the new symbols represent: v_t (in m/s), the pitch line velocity; a (in mm), the center distance; X_G, the geometry factor, which is a dimensionless factor.

The already defined thermo-elastic factor, X_M, takes into account the influence of the material properties of pinion and wheel, and is given by:

$$X_M = E_r^{1/4} \frac{\left[(1+\Gamma_y)^{1/2} + (1-\Gamma_y/u)^{1/2}\right]}{\left[B_{M1}(1+\Gamma_y)^{1/2} + B_{M2}(1-\Gamma_y/u)^{1/2}\right]}, \tag{7.25}$$

where E_r (in N/mm^2) is the reduced modulus of elasticity, given by Eq. (2.3), while Γ_y is the dimensionless *parameter on the line of action at arbitrary point* (see Sect. 7.5).

In most cases, the thermal contact coefficients of pinion and wheel, B_{M1} and B_{M2}, are the same, i.e. $B_{M1} = B_{M2}$. In these cases, the thermo-elastic factor, X_M, depends only on the characteristics of material, for which it can be expressed as:

$$X_M = \frac{E_r^{1/4}}{B_M}, \tag{7.26}$$

where $B_M = B_{M1} = B_{M2}$. About the calculation of these thermal contact coefficients, it should be remembered that, for martensitic steels, the heat conductivity, λ_M, is included within the variability range (41–52) N/sK, $\rho_M = 7.8 \times 10^3$ kg/m^3 and $c_M \cong 4.9$ J/kgK $\cong 4.9$ Nm/kgK, for which the product $\rho_M c_M$ that expresses the *specific heat capacity per unit volume*, c_v, is approximately equal to 3.8×10^{-2} N/mm^{2}K. Therefore, when the thermo-elastic factor of these steels is not known, the average value of the thermal contact coefficient can be assumed to be $B_M = 43.5$ N/(mm$^{1/2}$m$^{1/2}$ s$^{1/2}$K) $= 13.8$ N/(mm s$^{1/2}$K). It follows that, for steels having $E = E_1 = E_2 = 206 \times 10^3$ N/mm^2, and $\nu = \nu_1 = \nu_2 = 0.3$, the thermo-elastic factor (or thermo-flash factor) is given by:

$$X_M = 0.50 \text{ KN}^{-3/4}\text{s}^{1/2}\text{m}^{1/2} = 15.81 \text{ KN}^{-3/4}\text{s}^{1/2}\text{mm}^{1/2}. \tag{7.27}$$

The geometry factor, X_G, of an external or internal cylindrical gear pair (in this case, the usual sign convention must be used) is given respectively by the following relationships:

$$X_G = 0.51 X_{\alpha\beta}(u + 1)^{1/2} \frac{\left|1 + \Gamma_y - 1 - \Gamma_y/u\right|}{\left(1 + \Gamma_y\right)^{1/4}\left(u - \Gamma_y\right)^{1/4}} \tag{7.28}$$

$$X_G = 0.51 X_{\alpha\beta}(u - 1)^{1/2} \frac{\left|1 + \Gamma_y - 1 + \Gamma_y/u\right|}{\left(1 + \Gamma_y\right)^{1/4}\left(u + \Gamma_y\right)^{1/4}}, \tag{7:29}$$

where $X_{\alpha\beta}$ is the *angle factor*, which takes into account the conversion of tangential velocity and load from reference circle to pitch circle. This last factor depends on the transverse working pressure angle, $\alpha_{wt} = \alpha_t'$, and base helix angle, β_b, and is given by:

$$X_{\alpha\beta} = 1.22 \frac{\sqrt[4]{\sin \alpha_{wt}} \sqrt{\cos \beta_b}}{\sqrt{\cos \alpha_{wt}}}. \tag{7.30}$$

As Table 7.1 shows, the values of the angle factor, $X_{\alpha\beta}$, for a standard rack with normal pressure angle $\alpha_n = 20°$, in the typical variability ranges of the working pressure angle, α_{wt}, and helix angle, β, that are of interest to the designer, is close to unity. Therefore, $X_{\alpha\beta}$ can be assumed approximately equal to 1, for gears with normal pressure angle, $\alpha_n = 20°$.

Table 7.1 Angle factor, $X_{\alpha\beta}$

α_{wt}	$X_{\alpha\beta}$			
	$\beta = 0°$ $\alpha_t = 20.00°$	$\beta = 10°$ $\alpha_t = 20.28°$	$\beta = 20°$ $\alpha_t = 21.17°$	$\beta = 30°$ $\alpha_t = 22.80°$
18°	0.947	–	–	–
20°	0.978	0.975	0.966	–
22°	1.007	1.004	0.995	0.981
24°	1.035	1.032	1.023	1.008
26°	1.064	1.060	1.051	1.036
28°	–	–	–	1.063

To calculate the flash temperature, Θ_{fl}, of cylindrical gears, if instead of using Eq. (7.24), we use the Eq. (7.21), which is valid both for cylindrical and bevel gears, we have to preliminary determine the local radius of curvature at arbitrary point of pinion flank, ρ_{y1}, and wheel flank, ρ_{y2}, which are given respectively by:

$$\rho_{y1} = \frac{1 + \Gamma_y}{1 + u} a \sin \alpha_{wt} \qquad (7.31)$$

$$\rho_{y2} = \frac{u - \Gamma_y}{1 + u} a \sin \alpha_{wt}, \qquad (7.32)$$

and successively determine the local relative radius of curvature, ρ_{rely}, at arbitrary point using the following relationship:

$$\rho_{rely} = \frac{\rho_{y1} \rho_{y2}}{\rho_{y1} + \rho_{y2}}. \qquad (7.33)$$

Bevel Gears

Notoriously, the successive contact areas of bevel gears have the shape of somewhat tapered bands. However, in most cases, these tapered bands can be replaced, with a good approximation, by parallel band-shaped contact areas. Furthermore, since both sliding velocities, v_{g1} and v_{g2}, have the same direction, coinciding with the straight line perpendicular to the major axis of tapered contact areas, for the calculation of the flash temperature we can use directly the Eq. (7.8), especially when correct data of the radii of curvature and Hertzian contact band are known. Obviously, in the case considered here, the Eq. (7.8) is reduced to Eq. (7.20).

For convenience of calculation, Eq. (7.20) can be rewritten with the octoid line of action approximated by a straight line, and with factors appearing in the same equation expressed by the usual quantities of bevel gears. The assumptions made for the rewritten relationship, which allow a convenient approximation of the radii of

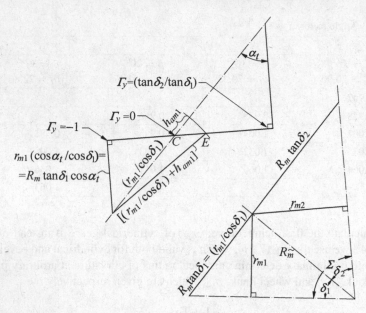

Fig. 7.4 Main quantities of a bevel gear pair, and its approximate line of action

curvature, but do not involve the replacement of the bevel gear under consideration with its virtual cylindrical gear pair, are as follows:

- Pinion and wheel of the gear have a common apex, and any arbitrary shaft angle, $\Sigma = \delta_1 + \delta_2$ (see Fig. 7.4).
- All quantities are those related to the mean cones.
- A straight line approximates the line of action, as Fig. 7.4 shows.
- The surface of action is considered to be a plane surface.

The flash temperature of a bevel gear can be calculated also considering its virtual cylindrical gear pair. To determine directly the flash temperature of a bevel gear we instead use the following relationship:

$$\Theta_{fl} = \mu_m X_M X_J X_G (X_\Gamma w_{Bt})^{3/4} \frac{v_t^{1/2}}{R_m^{1/4}}, \qquad (7.34)$$

where: R_m (in mm), is the mean cone distance; X_M, is the thermo-elastic factor, to be calculated using the relationship previously given for cylindrical gears; X_G, is the geometry factor given by:

$$X_G = 0.51 X_{\alpha\beta} (\cot \delta_1 + \cot \delta_2)^{1/4} \frac{\left| \sqrt{1 + \Gamma_y} - \sqrt{1 + \Gamma_y (\tan \delta_1 / \tan \delta_2)} \right|}{(1 + \Gamma_y)^{1/4} \left[1 - \Gamma_y (\tan \delta_1 / \tan \delta_2) \right]^{1/4}}, \qquad (7.35)$$

where $X_{\alpha\beta}$ is the angle factor, to be calculated as previously described.

To calculate the flash temperature, Θ_{fl}, of bevel gears, if instead of using the Eq. (7.34), we use the Eq. (7.21), which is valid both for cylindrical and bevel gears, we have to preliminary determine the local radius of curvature of pinion flank, ρ_{y1}, and wheel flank, ρ_{y2}, which are given respectively by:

$$\rho_{y1} = R_m \tan \delta_1 \sin \alpha_t (1 + \Gamma_y) \qquad (7.36)$$

$$\rho_{y2} = R_m \tan \delta_1 \sin \alpha_t (u - \Gamma_y), \qquad (7.37)$$

where the parameter on the line of action at arbitrary point, Γ_y, is given by:

$$\Gamma_y = \frac{\tan \alpha_{y1}}{\tan \alpha_t} - 1. \qquad (7.38)$$

In this last equation, α_t is the transverse pressure angle, while α_{y1} is the pinion pressure angle at arbitrary point.

The calculation of other quantities that appear in the equations above is described in the following section.

7.4 Coefficient of Friction

In Eqs. (7.8), (7.20), (7.21), (7.24), and (7.34), the mean coefficient of friction, μ_m, appears. This decisive quantity is difficult to determine with the desired precision, since, as it is well known, the factors affecting the friction between the gear teeth are not only numerous, but also vary throughout a meshing cycle.

As we saw in Vol. 1, Chap. 3, on one of the two mating tooth flank profiles the relative motion is uniformly accelerated, while on the other it is uniformly decelerated. In this regard, in Vol. 1, Sect. 7.3.6 and Fig. 3.5, we have highlighted that, with the exception of the pitch point, the involute arcs described by the point of contact on the two mating profiles are not equals. Therefore, we have pure rolling only at pitch point, C, while at any other current point of the path of contact combined rolling and sliding motions will occur. In addition, the load acting on the two mating tooth flanks will vary along the path of contact from one meshing position to another.

These operating conditions determine a continuous variation of the lubricant film thickness, lubrication regime, and coefficient of friction. Moreover, in the same meshing position on the path of contact, the coefficient of friction will vary for the various mating tooth pairs and different time. As coefficient of friction able to smooth various influences, the *local coefficient of friction* is considered to be an appropriate and valid quantity. However, the variation of the local coefficient of friction related to these influences, including the position of the meshing point along the path of contact, is difficult to calculate or measure, for which, in place of a local value, a mean value representative of the coefficient of friction is introduced for calculations.

The mean coefficient of friction used in calculations is defined as the mean value of the local coefficients of friction along the whole path of contact. Obviously, it differs from the actual local coefficient of friction at the pitch point. Nevertheless, the mean coefficient of friction can be expressed in terms related to the pitch point. The value of this mean coefficient of friction is also uncertain; this is because, for its determination, often influential important quantities are neglected, for example the bulk temperature, which determines the inlet oil viscosity and therefore the lubrication regime. We have been able to deepen and use these concepts, starting from Vol. 1, Chap. 3, every time we have faced the problems of efficiency of the various types of gears, which are strongly influenced by the coefficient of friction and its variability along the path of contact (see Vol. 1, Sect. 7.3.9).

According to the current state of knowledge, the mean coefficient of friction, μ_m, is primarily influenced by the tangential velocities, normal load, actual geometry of the path of contact, inlet oil viscosity, which coincides with oil viscosity at tooth bulk temperature, pressure-viscosity coefficient, reduced modulus of elasticity, surface roughness, and normal radius of curvature. Other influences on the value of the mean coefficient of friction are however not to be excluded, for which this subject is still an open problem not only for the gears that interest here, but also for many other mechanical contacts where friction phenomena come into play. In any case, the limiting contact temperature should be chosen correspondingly to the value of the mean coefficient of friction.

According to ISO/TS 6336-20:2017(E), the mean coefficient of friction can be calculated using Method A, Method B, and Method C. With Method A, the mean coefficient of friction at the onset of scuffing is determined by measurements with gear tests or pin-and-ring tribo-testing. Consequently, in this case, the limiting contact temperature is high. Instead, with Method B, which marries the common practice of using the low values of the coefficient of friction corresponding to the regular operating conditions, the mean coefficient of friction is determined by means of an appropriate relationship, in which a value of absolute viscosity, or dynamic viscosity, η_{oil}, of the lubricant corresponding to the gear bulk temperature appears. Consequently, with this method, the limiting contact temperature is low.

Finally, Method C is used when, at the beginning of a calculation, the value of the bulk temperature is not yet known. In this case, the mean coefficient of friction in the usual operating conditions is determined using the following relationship:

$$\mu_m = 60 \times 10^{-3} \left(\frac{w_{Bt}}{v_{\Sigma C} \rho_{relC}} \right)^{0.2} X_L X_R, \tag{7.39}$$

where: w_{Bt} (in N/mm), is the transverse unit load, given by Eq. (7.12) or Eq. (7.13), depending on whether cases of cylindrical gears or bevel gears should be addressed; ρ_{relC} (in mm), is the transverse relative radius of curvature at pitch point, given by Eq. (7.33), together with Eqs. (7.31) and (7.32), for $y = C$, $\Gamma_y = 0$; $v_{\Sigma C}$ (in m/s), is the absolute value of the cumulative velocity vector at pitch point that, recalling the Eq. (3.63) of Vol. 1, and in the same relationship putting $\alpha = \alpha_{wt}$ and $g_P = 0$, is equal to the sum of tangential velocities at pitch point, given by:

$$v_{\Sigma C} = 2v_t \sin \alpha_{wt}, \tag{7.40}$$

where $v_t = v_{t1} = v_{t2}$ (in m/s) is the pitch line velocity (it is to be noted that, if $v_t > 50\,\text{m/s}$, in Eq. (7.39) we must put $v_t = 50\,\text{m/s}$).

Factor, X_L, is the dimensionless lubricant factor, to be assumed equal to:

$$X_L = C(\eta_{oil})^{-0.05}, \tag{7.41}$$

where η_{oil} (in mPas) is the absolute (or dynamic) viscosity at oil temperature Θ_{oil} before reaching the mesh area, and C is a numerical constant depending on the lubricant. The values of this last numerical constant are as follows: $C = 1.0$ for mineral oils; $C = 0.6$ for water soluble polyglycols; $C = 0.7$ for non-water soluble polyglycols; $C = 0.8$ for polyalfaolefins; $C = 1.3$ for phosphate esters; $C = 1.5$ for traction fluids.

Finally, X_R is the roughness factor, given by:

$$X_R = \left(\frac{Ra_1 + Ra_2}{2}\right)^{0.25}, \tag{7.42}$$

where Ra_1 and Ra_2, (in μm), are respectively the average roughness of tooth flank surface of pinion and wheel, for newly manufactured gears. It is to be noted that, for adequately running-in gears, both Ra_1 and Ra_2 can be reduced to about 60% of their initial values.

7.5 Position of an Arbitrary Point on the Line of Action

As we saw in the previous section, the operating conditions of a lubricated contact, like the one between the mating tooth flanks of a gear pair, vary along the path of contact from one meshing position to another. Therefore, it is necessary to consider local values of all the quantities of interest. To this end, in accordance with what Wydler (1958) proposed for the first time (see also Polder 1987a, b), we introduce the parameter on the line of action at arbitrary point, Γ_y, as well as the values it assumes at particular points of the same line of action, as it is specified below.

Figure 7.5 shows, as an example, the line of action, $T_1 T_2$, and the path of contact, AE, of a cylindrical gear pair. Recalling also Fig. 1.22, we have: $AD = BE = p_{bt}$, where p_{bt} is the transverse base pitch. The parameter, Γ_y, is a dimensionless linear parameter, which defines the position of an arbitrary current point, P, on the line of action. Special points of interest on the line of action are the marked points already defined in Sect. 1.7, namely: point A, which is the lower end point of the path of contact; point E, which is the upper end point of the path of contact; point B, which

Fig. 7.5 Dimensionless linear parameter, Γ_y, defining the position of an arbitrary current point on the line of action for a cylindrical gear pair

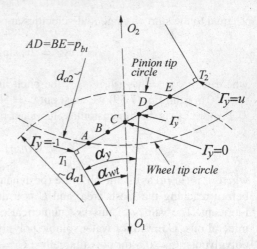

is the lower point of single pair tooth contact; point D, which is the upper point of single pair tooth contact. Of course, point, C, and points, T_1 and T_2, which delimit the line of action, are also points of interest.

About the relationships that we use to determine the position of an arbitrary point or one of the aforementioned special marked points on the line of action, we must distinguish between cylindrical gears and bevel gears. In any case, whatever the type of gear, cylindrical or bevel gear, we have respectively $\Gamma_y = -1$, $\Gamma_y = 0$, and $\Gamma_y = u$ at points T_1, C, and T_2.

7.5.1 Cylindrical Gears

For cylindrical gears, the position of an arbitrary current point on the line of action is defined by the following expression of the dimensionless parameter, Γ_y:

$$\Gamma_y = \frac{\tan \alpha_{y1}}{\tan \alpha_{wt}} - 1, \tag{7.43}$$

where α_{wt} is the transverse working pressure angle, and α_{y1} is the pinion pressure angle at arbitrary current point (Fig. 7.5).

At special points, or marked points, A, B, D, and E, the local values of parameter Γ_y are given respectively by the following relationships:

$$\Gamma_A = -\frac{z_2}{z_1}\left(\frac{\tan \alpha_{a2}}{\tan \alpha_{wt}} - 1\right), \tag{7.44}$$

$$\Gamma_B = \frac{\tan \alpha_{a1}}{\tan \alpha_{wt}} - 1 - \frac{2\pi}{z_1 \tan \alpha_{wt}}, \tag{7.45}$$

$$\Gamma_D = -\frac{z_2}{z_1}\frac{\tan \alpha_{a2}}{\tan \alpha_{wt}} - 1 + \frac{2\pi}{z_1 \tan \alpha_{wt}},$$ (7.46)

$$\Gamma_E = \frac{\tan \alpha_{a1}}{\tan \alpha_{wt}} - 1,$$ (7.47)

where α_{a1} and α_{a2} are the transverse tip pressure angles of pinion and wheel, given respectively by:

$$\tan \alpha_{a1} = \sqrt{\left(\frac{d_{a1}}{d_1 \cos \alpha_t}\right)^2 - 1}$$ (7.48)

$$\tan \alpha_{a2} = \sqrt{\left(\frac{d_{a2}}{d_2 \cos \alpha_t}\right)^2 - 1}.$$ (7.49)

7.5.2 Bevel Gears

For bevel gears, the position of an arbitrary current point on the line of action can be defined considering their virtual equivalent cylindrical gears, and then using the relationships shown in the previous section. However, for the same purpose, the following direct relationships can be used, which are valid also when the shaft angle, $\Sigma = \delta_1 + \delta_2$, is not equal to $90°$.

The position of an arbitrary current point on the line of action is defined by a relationship analogous to Eq. (7.38) where, however, the transverse pressure angle, α_t, is substituted for the transverse working pressure angle, α_{wt}. At special marked points A, B, D, and E of the path of contact, the local values of parameter Γ_Y are given respectively by the following relationships:

$$\Gamma_A = -\frac{\tan \delta_2}{\tan \delta_1}\left(\frac{\tan \alpha_{a2}}{\tan \alpha_t} - 1\right)$$ (7.50)

$$\Gamma_B = \frac{\tan \alpha_{a1}}{\tan \alpha_t} - 1 - \frac{2\pi \cos \delta_1}{z_1 \tan \alpha_t}$$ (7.51)

$$\Gamma_D = -\frac{\tan \delta_2}{\tan \delta_1}\left(\frac{\tan \alpha_{a2}}{\tan \alpha_t} - 1\right) + \frac{2\pi \delta_1 \cos \delta_1}{z_1 \tan \alpha_t}$$ (7.52)

$$\Gamma_E = \frac{\tan \alpha_{a1}}{\tan \alpha_t} - 1,$$ (7.53)

where the tip pressure angles of pinion, α_{a1}, and wheel, α_{a2}, are given by:

$$\tan \alpha_{a1} = \sqrt{\left(\frac{\cos \alpha_t}{1 + (h_{am1}/r_{m1})\cos\delta_1} \right)^2 - 1} \qquad (7.54)$$

$$\tan \alpha_{a2} = \sqrt{\left(\frac{\cos \alpha_t}{1 + (h_{am2}/r_{m2})\cos\delta_2} \right)^2 - 1}. \qquad (7.55)$$

In these equations, h_{am1} and h_{am2} (both in mm), are the tip heights in mean cones of pinion and wheel, r_{m1} and r_{m2} (both in mm), are the pitch radii in mean cones of pinion and wheel, while δ_1 and δ_2 (both in degrees), are the pitch cone angles of pinion and wheel.

7.6 Optimum Profile Modification and Approach Factor

Figure 7.6 shows two teeth 2 and 2′ of a cylindrical spur gear pair in mesh at a point, P, of the path of contact between the marked points B and D, that is in the single contact zone. We assume that the pinion is the driving member of the gear, and that the toothing of the same gear is without errors.

Let us analyze now what happens when the point of contact, P, between the two teeth under consideration coincides with the marked point, D (Fig. 7.5), which represents the beginning of the double contact. In this instant of the meshing cycle, the next pairs of teeth 1 and 1′ begins to mesh at marked point, A. If the teeth were not loaded, the contact at point A would occur in normal conditions, i.e. without any impact, as Fig. 7.6a shows. Instead, for a gear necessarily under load, the teeth

Fig. 7.6 Bending of the teeth under load, and approach impact

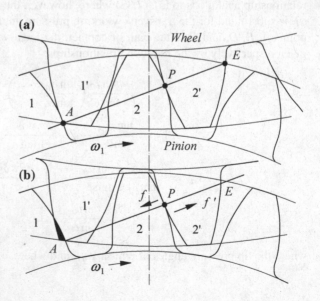

of pinion and wheel are subjected to bending, as Fig. 7.6b shows, with deflections whose components measured along the direction of the line of action are respectively f and f' (see Henriot 1979; Niemann and Winter 1983).

With the direction of rotation shown in Fig. 7.6, the driving pinion rotates right by an angle related to the sum $(f + f')$, with respect to the driven wheel, under the simplifying assumption that the body of this wheel is not deformable and therefore does not suffer a related angular displacement. In this way, the tooth 1 of the pinion, yet not loaded, would approach the tip flank of the tooth $1'$ of the wheel that is not yet loaded. Actually, however, already before the beginning of the theoretical contact, the tooth 1 of the pinion will impact against the tip of the tooth $1'$ of the wheel. This impact, called *approach impact*, occurs at the beginning of meshing of the mating tooth pair, 1 and $1'$, since the tooth tip of the tooth $1'$ of wheel tends to penetrate into the root fillet of the tooth 1 of pinion of a quantity equal to the sum $(f + f')$.

Of course, this approach impact is further enhanced by the other deformations that come into play under load, and here neglected; anyway it is a source of noise, but also it has a significant influence on the strength of the same teeth. In particular, with regard to the scuffing strength, it is found that the lubricant film that adheres to the tooth 1 of the pinion is instantaneously destroyed by the brutal interference of the tip of the tooth $1'$ of wheel inside the profile of the tooth 1 of pinion. The manufacturing errors of the teeth, such as for example the transverse single pitch deviation, f_{pt}, cause effects similar to those above-mentioned, related to the teeth deformability. It is also to be noted that, at the end of contact of a tooth (i.e. at point E in Fig. 7.6), the next tooth comes to bear the entire load, whereby a *recess impact* is determined, similar to the approach impact, but of lesser intensity.

In the case of gears subjected to heavy loads, especially when they operate at high speed, it is extremely important to make the beginning of contact smoother and more gradual, trying to mitigate the above-mentioned *impact interference* as much as possible. Quite simply, this can be achieved with an adequate tip relief, C_a (in μm), of the driven wheel teeth or with an appropriate root relief, C_f (in μm) of teeth of the driving pinion, which ensures the same result. Both the tip relief, C_a, and root relief, C_f, constitute intentional deviations of the actual profile with respect to the theoretical involute profile. These deviations are directed towards the inside of the involute profile. Therefore, in the case of tip relief, the tooth tip of the driven wheel is moved to the inside of the involute curve of an amount equal, from a theoretical point of view, to the sum of the above-mentioned elastic deflections and manufacturing errors.

Both the tip relief and root relief may cover more or less extended portions of the tooth profile. Considering, for example, the tip relief, depending on the amount of its height, we can distinguish between a short modification (*short tip relief*), and a long modification (*long tip relief*). The same thing can be said for the root relief. For cylindrical gears, the ISO standards require that the tip relieved height (that is, the portion of the tooth depth, starting from the top land, affected by the tip relief) should not reach neither the area of single pair tooth contact, nor to result in a transverse contact ratio, ε_α, less than 1, when the gear is unloaded.

To choose the appropriate tip relieved height, it is necessary to bear in mind that a long tip relief or a short tip relief, both determined as a function of the load acting on the teeth, influence the load distribution between the various pairs of mating teeth, the deformation of the teeth, as well as the regularity of the motion, in terms of deviations from the theoretical path of rotation. With a long tip relief, the modification of the teeth profile reaches up to the marked points B and D (Fig. 7.5), resulting in a transverse contact ratio $\varepsilon_\alpha < 1$. Instead, with a short tip relief of the tooth profile, the remaining portion of the involute still guarantees a transverse contact ratio $\varepsilon_\alpha = 1$. Both a long tip relief and a short tip relief ensure a continuous variation of the load acting on the teeth, while the regularity of motion, and therefore the quiet operation, are only guaranteed by a long tip relief.

In relation to the regularity of motion, it is to be noted that a long tip relief is advantageous over a short tip relief, only for full load operation. Instead, with a long tip relief, but for partial load, the regularity of motion may be less than that typical of gears without tooth profile modification. This unfavorable behavior under partial load is the reason why, in practice, the spur gears are made with an appropriate short tip relief of tooth profile, which, although it does not contribute to mitigate deviations from the theoretical path of rotation, however reduces the approach and recess impacts. Under partial load, the minimum transverse contact ratio has favorable effects. For cylindrical helical gears, a long tip relief is preferable to a short tip relief, as long as the overlap ratio, ε_β, is sufficient.

To ensure, in a wide range of operating conditions, a good regularity of motion, together with a silent operation, the amount of the tip relief is usually calculated with reference to the load corresponding to the main operating condition. With several grinding methods, this tooth profile modification is mainly applied to the pinion, which consequently is characterized by both tip relief and root relief. However, also grinding methods that allow making the tip relief of pinion and wheel exist. Very often, the recommended values of tooth profile modification are shared between pinion and wheel.

According to the ISO standards, in the case where the tooth profiles are modified, the modifications shall be designed and manufactured so as to correspond to the load sharing function request, such as those described in the next section. The optimal tip relief of pinion and wheel is given approximately by the following relationship:

$$C_{eff} = \frac{K_A K_{mp} F_t}{b c_\gamma \cos \alpha_t}, \tag{7.56}$$

where K_A and K_{mp} are respectively the dimensionless application and multi-path factors, F_t (in N) is the nominal tangential force, b (in mm) is the face width, α_t (in degrees) is the transverse pressure angle, and c_γ [in N/(mmμm)] is the mesh stiffness.

We already said that the tip relieved height for cylindrical gears should not reach neither the area of single pair tooth contact, nor to result in a transverse contact ratio $\varepsilon_\alpha < 1$ when the gear is unloaded. However, it is to point out that, in these conditions, ε_α must be calculated considering the fictitious tip diameters equal to the diameter where the relieved area starts. Furthermore, when the mating gear is manufactured with a root relief, the tip relief of the gear wheel under consideration is replaced with an *equivalent tip relief*, C_{eq}, defined as the sum of the tip relief of the same gear wheel, and a *reduced root relief* of the mating gear. For pinion and wheel, this equivalent tip relief is given respectively by:

$$C_{eq1} = C_{a1} + C_{f2}\left(\frac{H_2}{2m_n} - 1\right)^2 \tag{7.57}$$

$$C_{eq2} = C_{a2} + C_{f1}\left(\frac{H_1}{2m_n} - 1\right)^2, \tag{7.58}$$

where H_1 and H_2 (both in mm) are *auxiliary dimensions*, which have different expressions for cylindrical gears and bevel gears.

For cylindrical gears, H_1 and H_2 are given by:

$$H_1 = d_{a1} - \frac{2a}{(u+1)}\sqrt{\cos^2\alpha_{wt} + (u - \Gamma_A)^2 \sin^2\alpha_{wt}} \tag{7.59}$$

$$H_2 = d_{a2} - \frac{2a}{(u+1)}\sqrt{u^2 \cos^2\alpha_{wt} + (u - \Gamma_E)^2 \sin^2\alpha_{wt}}. \tag{7.60}$$

For bevel gears, we can calculate H_1 and H_2 by means of Eqs. (7.59) and (7.60), using data of equivalent virtual cylindrical gears. Alternatively, we can use the following relationships, which are specific for these types of gears:

$$H_1 = 2(R_m \tan\delta_1 + h_{am1}) - R_m \tan\delta_1\sqrt{\cos^2\alpha_{wt} + (1 + \Gamma_E)^2 \sin^2\alpha_{wt}} \tag{7.61}$$

$$H_2 = 2(R_m \tan\delta_2 + h_{am2}) - R_m \tan\delta_1\sqrt{u_v^2 \cos^2\alpha_{wt} + (u_v - \Gamma_E)^2 \sin^2\alpha_{wt}}, \tag{7.62}$$

where u_v is the *virtual gear ratio*, given by:

$$u_v = \frac{\tan\delta_2}{\tan\delta_1}. \tag{7.63}$$

The quantities represented by symbols appearing in Eqs. (7.59) to (7.63) are all known. In all these equations, with the exception of the dimensionless quantities u, u_v, Γ_A and Γ_E, the linear quantities are given in mm, while the angular quantities are given in degrees. Various tip reliefs appearing in Eqs. (7.56) to (7.58) are all given in micrometers.

The approach factor, X_J, which is a dimensionless factor, takes into account the increasing scuffing risk at the beginning of the path of approach, due to the mesh starting when no lubricant film is yet formed. The influence of this factor is relatively strong for large gears. For its determination, it is necessary to distinguish the case of speed reducing gears, where the pinion drives the wheel, and the case of speed increasing gears, where the wheel drives the pinion. In the first case, that of speed reducing gears, for $\Gamma_Y \geq 0$, we have:

$$X_J = 1, \tag{7.64}$$

and, for $\Gamma_Y < 0$, with the condition $X_J \geq 1$, we have:

$$X_J = 1 + \frac{C_{eff} - C_{a2}}{50}\left(\frac{-\Gamma_y}{\Gamma_E - \Gamma_A}\right)^3. \tag{7.65}$$

In the second case, that of speed increasing gears, for $\Gamma_y \leq 0$, X_j is still given by Eq. (7.64), while, for $\Gamma_y > 0$, with the condition $X_J \geq 1$, X_J is given by the following relationship:

$$X_J = 1 + \frac{C_{eff} - C_{a1}}{50}\left(\frac{\Gamma_y}{\Gamma_E - \Gamma_A}\right)^3. \tag{7.66}$$

7.7 Load Sharing Factor, X_Γ

7.7.1 General Premise

The load sharing factor, X_Γ, takes into account the load sharing between the various pairs of gear teeth in continuously variable engagement during the meshing cycle. It is given as a function of the already defined parameter on the line of action at arbitrary point, Γ_y, with the obvious limitation to the values it assumes along the path of contact. This load sharing factor depends on the type of gear (cylindrical spur gears, cylindrical helical gears, bevel gears, etc.), and on the profile modification of the teeth of the two members of the gear pair.

Because of the unavoidable inaccuracies, any pair of meshing teeth can cause a sharp increase or decrease in the theoretical value of the load-sharing factor, and this regardless of the more or less sharp increase or decrease of the same factor caused by inaccuracies of the teeth pair that immediately after enters into meshing. For cylindrical spur gears, the load sharing factor, X_Γ, does not exceed the unit value. This unit value corresponds to full transverse single tooth contact. The region of transverse single tooth contact can be more extensive than theoretically predicted, because of an irregular local variation of the dynamic load.

For cylindrical helical gears without profile modification, the *buttressing effect* comes into play, so the load-sharing factor depends not only on the tooth profile modification, but also on the buttressing effect. Therefore, for these types of gears, the actual load sharing factor is the result of the combination of the theoretical load sharing factor with the buttressing factor, $X_{but,\Gamma}$. The buttressing effect is due to the oblique contact lines, and may occur near the end points, A and E, of the path of contact.

7.7.2 Cylindrical Spur Gears

For cylindrical spur gears with an adequate profile modification on driving and driven members, the distribution of the nominal tangential force, F_t (and thus of the transverse load) along the path of contact, AE, is conventionally assumed to have a discontinuous trapezoidal shape, as the one shown in Fig. 7.7a. This same figure, with the maximum value on the ordinate axis equal to 1, represents the corresponding load sharing factor, X_Γ. The distributions of F_t and X_Γ, strictly inter-correlated between them, show that, in the two double tooth contact areas, the two quantities increase in the approach path of contact, and decrease in the recess path of contact. This is due to the unavoidable manufacturing inaccuracies.

The optimal load distribution shown in Fig. 7.7a is clearly very different from the conventional load distribution shown in Fig. 1.22, which, as we noted in Sect. 1.7, can be considered sufficiently correct for a toothing without errors. In addition, we noted that any deviation from the theoretical geometry of toothing implies significant changes to this load distribution. In fact, due to manufacturing accuracy, in the double contact region, AB, of the approach path of contact, the load and the associated load sharing factor will increase, whereas, in the double contact region, DE, of the recess path of contact, both the quantities will decrease.

For actual cylindrical spur gears with unmodified profiles and quality grade 7 or finer, a discontinuous load distribution (and an equally discontinuous corresponding load sharing factor) is conventionally considered more adherent to the practical applications. Figure 7.7b shows this type of distribution. For actual cylindrical spur

Fig. 7.7 Distribution curves of load and associated load sharing factor for cylindrical spur gears with: **a** adequate profile modification; **b** unmodified profile and quality grade 7 or finer

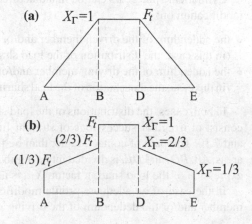

Fig. 7.8 Distribution curves of load and associated load sharing factors for cylindrical spur gears with unmodified profiles and quality grade 8 or coarser

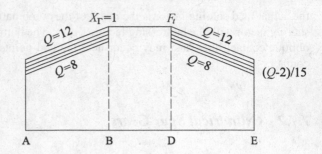

gears with unmodified profiles and quality grade 8 or coarser, the load distribution and the associated load sharing factor, which are considered representative of their actual distributions along the path of contact, are made by the envelope of possible curves, with a result like the one represented by the distribution curves shown in Fig. 7.8.

Distributions of the load sharing factor, X_Γ, along the path of contact for cylindrical spur gears with unmodified profiles, show that, regardless of the quality grade of the gears under consideration (quality grade 7 or finer, or quality grade 8 or coarser), in the single tooth contact area, i.e. for ($\Gamma_B \leq \Gamma_y \leq \Gamma_D$), the load sharing factor has unit value ($X_\Gamma = 1$). Instead, in the two double tooth contact areas, for which ($\Gamma_A \leq \Gamma_y < \Gamma_B$) and ($\Gamma_D < \Gamma_y \leq \Gamma_E$), the load sharing factor is given respectively by the following relationships:

$$X_\Gamma = \frac{Q-2}{15} + \frac{1}{3}\left(\frac{\Gamma_y - \Gamma_A}{\Gamma_B - \Gamma_A}\right) \qquad (7.67)$$

$$X_\Gamma = \frac{Q-2}{15} + \frac{1}{3}\left(\frac{\Gamma_E - \Gamma_y}{\Gamma_E - \Gamma_D}\right), \qquad (7.68)$$

where Q is the quality grade of the gear. In these equations, we put $Q = 7$, for quality grade 7 or finer, and Q equal to the actual quality grade, for quality grade 8 or coarser.

Cylindrical spur gears can be manufactured in order to have an adequate profile modification on:

- the addendum of the driven member and/or the dedendum of the driving member (in this case, the distribution of the load sharing factor is that shown in Fig. 7.9);
- the addendum of the driving member and/or the dedendum of the driven member (in this case, the distribution of the load sharing factor is the one shown in Fig. 7.10).

In both cases, the distributions of the load sharing factor along the path of contact consist of irregular successions of straight lines, like the ones shown in Figs. 7.9 and 7.10. The path of contact, rather than be fractioned according to the usual three areas, AB, BD, and DE, is divided into a number of areas, to each of which a different expression of the load sharing factor, X_Γ, is associated.

In the case of an adequate profile modification on the addendum of the driven member and/or the dedendum of the driving member, the relationships to calculate

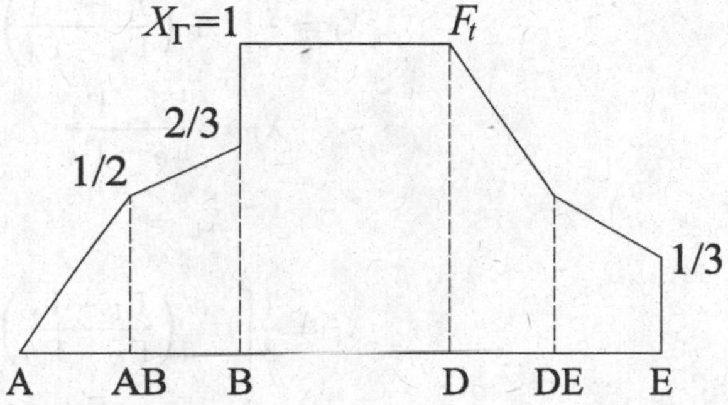

Fig. 7.9 Distribution curves of load and associated load sharing factor for a cylindrical spur gear pair with an adequate profile modification on the addendum of the driven member and/or the dedendum of the driving member

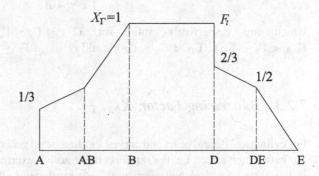

the load sharing factor corresponding to each of the five areas in which the path of contact is divided (Fig. 7.9) are as follows:

Fig. 7.10 Distribution curves of load and associated load sharing factor for a cylindrical spur gear pair with an adequate profile modification on the addendum of the driving member and/or the dedendum of the driven member

$$X_\Gamma = \frac{\Gamma_y - \Gamma_A}{\Gamma_B - \Gamma_A} \tag{7.69}$$

$$X_\Gamma = \frac{1}{3}\left[1 + \left(\frac{\Gamma_y - \Gamma_A}{\Gamma_B - \Gamma_A}\right)\right] \tag{7.70}$$

$$X_L = 1 \tag{7.71}$$

$$X_\Gamma = \frac{\Gamma_E - \Gamma_y}{\Gamma_E - \Gamma_D} \tag{7.72}$$

$$X_\Gamma = \frac{1}{3}\left[1 + \left(\frac{\Gamma_E - \Gamma_y}{\Gamma_E - \Gamma_D}\right)\right], \tag{7.73}$$

which are respectively valid for $(\Gamma_A \leq \Gamma_y \leq \Gamma_{AB})$, $(\Gamma_{AB} < \Gamma_y < \Gamma_B)$, $(\Gamma_B \leq \Gamma_y < \Gamma_D)$, $(\Gamma_D \leq \Gamma_y \leq \Gamma_{DE})$, and $(\Gamma_{DE} < \Gamma_y \leq \Gamma_E)$.

In the case of an adequate profile modification on the addendum of the driving member and/or the dedendum of the driven member, the relationships to calculate the load sharing factor corresponding to each of the five areas in which the path of contact is divided (Fig. 7.10) are as follows:

$$X_\Gamma = \frac{1}{3}\left[1 + \left(\frac{\Gamma_y - \Gamma_A}{\Gamma_B - \Gamma_A}\right)\right] \tag{7.74}$$

$$X_\Gamma = \frac{\Gamma_y - \Gamma_A}{\Gamma_B - \Gamma_A} \tag{7.75}$$

$$X_L = 1 \tag{7.76}$$

$$X_\Gamma = \frac{1}{3}\left[1 + \left(\frac{\Gamma_E - \Gamma_y}{\Gamma_E - \Gamma_D}\right)\right] \tag{7.77}$$

$$X_\Gamma = \frac{\Gamma_E - \Gamma_y}{\Gamma_E - \Gamma_D}, \tag{7.78}$$

which are respectively valid for $(\Gamma_A \le \Gamma_y \le \Gamma_{AB})$, $(\Gamma_{AB} \le \Gamma_y \le \Gamma_B)$, $(\Gamma_B < \Gamma_y < \Gamma_D)$, $(\Gamma_D < \Gamma_y \le \Gamma_{DE})$, and $(\Gamma_{DE} < \Gamma_y \le \Gamma_E)$.

7.7.3   Buttressing Factor, $X_{but,\Gamma}$

For cylindrical helical gears and curved toothed bevel gears, it is necessary to consider the buttressing effect, i.e. the reinforcement action exerted by oblique contact lines in the plane of action. Notoriously, this strengthening effect occurs independently of any tooth profile modification, and is predominantly concentrated near the end points, A and E, of the path of contact. We take into account the buttressing effect by introducing a buttressing factor, $X_{but,\Gamma}$, which, in simplified form, is represented by a sequence of straight lines along the path of contact, as Fig. 7.11 shows.

In order to consider the buttressing effect, as Fig. 7.11 highlights, the path of contact is further divided with the introduction of two new marked points, AU and EU, which are added to those defined previously. So we have two new areas, A–AU and EU–E, where the buttressing effect is evident. Therefore, the areas of subdivision of the whole path of contact become seven. The lengths of the sections A–AU and EU–E affected by the buttressing effect are supposed conventionally equal

Fig. 7.11 Simplified distribution of the buttressing factor, as a sequence of straight lines

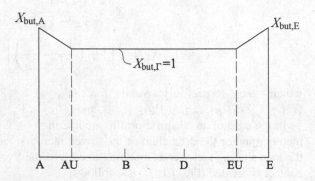

to each other, but they assume different values, depending on whether cylindrical helical gears or curved toothed bevel gears are considered. For these two families of gears, the lengths of the two aforementioned sections, expressed in mm, are given respectively by the following relationships:

$$\Gamma_{AU} - \Gamma_A = \Gamma_E - \Gamma_{EU} = 0.2 \sin \beta_b, \tag{7.79}$$

$$\Gamma_{AU} - \Gamma_A = \Gamma_E - \Gamma_{EU} = 0.2 \sin \beta_{bm}, \tag{7.80}$$

where β_b and β_{bm} are respectively the base helix angle and the base helix angle in mid-cone. Always in conventional terms, we assume that the buttressing factor, $X_{but,\Gamma}$, is equal to 1 throughout the entire AU–EU section, including the marked points AU and EU (therefore, $X_{but,\Gamma} = 1$ for $\Gamma_{AU} \leq \Gamma_Y \leq \Gamma_{EU}$, and $X_{but,AU} = X_{but,EU} = 1$), while at the end marked points, A and E, the buttressing factor is $X_{but,A} = X_{but,E} = 1.3$, when $\varepsilon_\beta \geq 1$, and $X_{but,A} = X_{but,E} = (1 + 0.3\varepsilon_\beta)$, when $\varepsilon_\beta < 1$, where ε_β is the overlap ratio. Finally, for $(\Gamma_A \leq \Gamma_y < \Gamma_{AU})$ and $(\Gamma_{EU} < \Gamma_y \leq \Gamma_E)$, the buttressing factor is given respectively by:

$$X_{but,\Gamma} = X_{but,A} - (X_{but,A} - 1) \frac{\Gamma_y - \Gamma_A}{\Gamma_{AU} - \Gamma_A} \tag{7.81}$$

$$X_{but,\Gamma} = X_{but,E} - (X_{but,E} - 1) \frac{\Gamma_E - \Gamma_y}{\Gamma_E - \Gamma_{EU}}. \tag{7.82}$$

7.7.4 Cylindrical Helical Gears

In order to calculate the load sharing factor, the cylindrical helical gear family is divided into the three following classes:

- Cylindrical helical gears with overlap ratio, $\varepsilon_\beta \leq 0.8$
- Cylindrical helical gears with overlap ratio, $\varepsilon_\beta \geq 1.2$
- Cylindrical helical gears with overlap ratio, $0.8 < \varepsilon_\beta < 1.2$.

Cylindrical helical gears with transverse contact ratio, $\varepsilon_\alpha \geq 1$, and overlap ratio, $\varepsilon_\beta \leq 0.8$, regardless of whether they have unmodified tooth profiles or modified tooth profiles, are still characterized by a small portion of single pair tooth contact. Therefore, they can be treated similarly to the cylindrical spur gears, considering the geometry and corresponding quantities in the transverse section.

The distribution of the load sharing factor for cylindrical helical gears with unmodified tooth profiles, including the buttressing effect, is the one shown in Fig. 7.12. Instead, the distributions of the load sharing factor for cylindrical helical gears with

Fig. 7.12 Distribution curves of load and associated load sharing factor for cylindrical helical gears with $\varepsilon_\alpha \geq 1$, $\varepsilon_\beta \leq 0.8$, and unmodified tooth profiles, including the buttressing effect

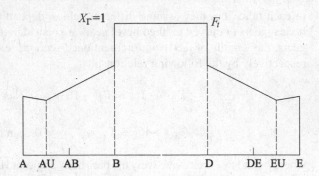

modified tooth profiles are different depending on the type of tooth profile modification. Figure 7.13 shows three different load sharing factor distributions, which refer respectively to: adequate profile modification, to be considered as the optimal profile modification; adequate profile modification on the addendum of the driven member and/or the dedendum of the driving member; adequate profile modification on the addendum of the driving member and/or the dedendum of the driven member. Regardless the cylindrical helical gears have unmodified tooth profiles or modified tooth profiles, all these load sharing factor distributions are obtained by multiplying the load sharing factor, X_Γ, of the corresponding cylindrical spur gears (see Sect. 7.7.2) with the buttressing factor, $X_{but,\Gamma}$ (see Sect. 7.7.3).

Cylindrical helical gears with transverse contact ratio, $\varepsilon_\alpha \geq 1$, and overlap ratio, $\varepsilon_\beta \geq 1.2$, i.e. wide cylindrical helical gears having a total contact ratio, $\varepsilon_\gamma \geq 2$, can have unmodified tooth profiles or modified tooth profiles. For wide cylindrical helical gears with unmodified tooth profiles, the buttressing effect, due to the local high mesh stiffness at the end of the oblique contact lines, is assumed operating near the marked points A and E of the path of contact, along the tooth helix, over a constant length given by Eq. (7.79). In this case, the distribution of the load sharing factor is quite similar to that of the buttressing factor shown in Fig. 7.11, with the variation described below.

The mean load sharing factor between the marked points AU and EU, is not equal to 1 as it occurs for cylindrical spur gears with unmodified tooth profiles, but is equal to $(1/\varepsilon_\alpha)$; this ratio represents the mean load acting in the section AU–EU of the path of contact. Instead, in sections A–AU and EU–E of the path of contact, the load sharing factor for wide cylindrical helical gears with unmodified profiles is obtained by multiplying the value $(1/\varepsilon_\alpha)$, which is a quantity proportional to the mean load, with the buttressing factor, $X_{but,\Gamma}$, which is calculated with the methods described in Sect. 7.7.3.

Distributions of the load sharing factor for wide cylindrical helical gears with modified tooth profiles are different depending on the type of tooth profile modification. Figure 7.14 shows three different load sharing factor distributions, which refer respectively to: adequate profile modification on driving and driven members of the gear pair, to be considered as an optimal profile modification; adequate profile modification on the addendum of the driven member and/or the dedendum of

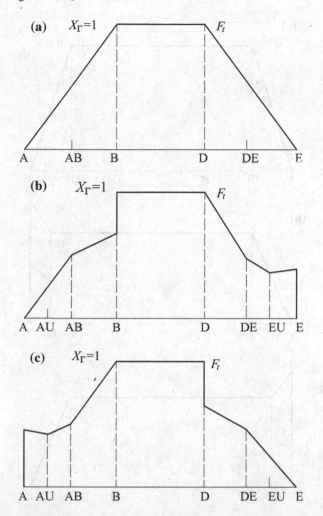

Fig. 7.13 Distribution curves of load and associated load sharing factor for cylindrical helical gears with $\varepsilon_\alpha \geq 1, \varepsilon_\beta \leq 0.8$, and: **a** adequate profile modification; **b** adequate profile modification on the addendum of the driven member and/or the dedendum of the driving member; **c** adequate profile modification on the addendum of the driving member and/or the dedendum of the driven member

the driving member; adequate profile modification on the addendum of the driving member and/or the dedendum of the driven member.

Regardless of the type and amount of the profile modification, it is to be noted that a tip relief on the pinion decreases and increases the load sharing factor, X_Γ, respectively in the range DE–E, and in the range AB–DE. Instead, a tip relief on the wheel decreases and increases the load sharing factor, X_Γ, respectively in the range A–AB, and in the range AB–DE. The extensions of tip relief at both ends of the path of contact are assumed to be equal (thus we will have lengths A–AB and DE–E equal

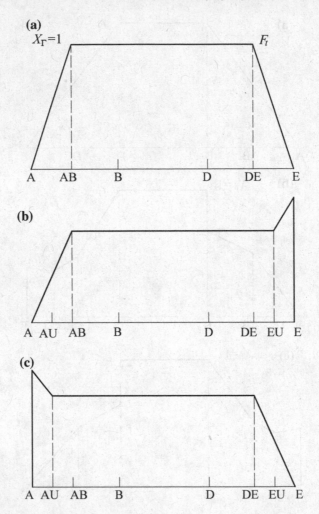

Fig. 7.14 Distribution curves of the load sharing factor for cylindrical helical gears with transverse contact ratio, $\varepsilon_\alpha \geq 1$, overlap ratio, $\varepsilon_\beta \geq 1.2$, and: **a** adequate profile modification on driving and driven members of the gear pair; **b** adequate profile modification on the addendum of the driven member and/or the dedendum of the driving member of the gear pair; **c** adequate profile modification on the addendum of the driving member and/or the dedendum of the driven member of the gear pair

to each other), and having an amount sufficient to result in a transverse contact ratio $\varepsilon_\alpha = 1$, for unloaded gears. This last condition implies that the segments A–D, B–E and AB–DE shown in Fig. 7.14a are the same.

The load sharing factor for wide cylindrical helical gears with adequate profile modification on driving and driven members of the gear pair (Fig. 7.14a) is given by the following relationships:

$$X_\Gamma = \frac{1}{\varepsilon_\alpha}\left[1 + \left(\frac{\varepsilon_\alpha - 1}{\varepsilon_\alpha + 1}\right)\right]\left(\frac{\Gamma_y - \Gamma_A}{\Gamma_{AB} - \Gamma_A}\right) \tag{7.83}$$

$$X_\Gamma = \frac{1}{\varepsilon_\alpha}\left[1 + \left(\frac{\varepsilon_\alpha - 1}{\varepsilon_\alpha + 1}\right)\right] \tag{7.84}$$

$$X_\Gamma = \frac{1}{\varepsilon_\alpha}\left[1 + \left(\frac{\varepsilon_\alpha - 1}{\varepsilon_\alpha + 1}\right)\right]\left(\frac{\Gamma_E - \Gamma_y}{\Gamma_E - \Gamma_{DE}}\right), \tag{7.85}$$

which are respectively valid in the ranges $(\Gamma_A \leq \Gamma_y \leq \Gamma_{AB})$, $(\Gamma_{AB} < \Gamma_y \leq \Gamma_{DE})$, and $(\Gamma_{DE} < \Gamma_y \leq \Gamma_E)$.

The load sharing factor for wide cylindrical helical gears with adequate profile modification on the addendum of the driven member and/or the dedendum of the driving member of the gear pair (Fig. 7.14b) is given by the following relationships:

$$X_\Gamma = \frac{1}{\varepsilon_\alpha}\left[1 + \frac{(\varepsilon_\alpha - 1)}{2(\varepsilon_\alpha + 1)}\right]\left(\frac{\Gamma_y - \Gamma_A}{\Gamma_{AB} - \Gamma_A}\right) \tag{7.86}$$

$$X_\Gamma = \frac{1}{\varepsilon_\alpha}\left[1 + \frac{(\varepsilon_\alpha - 1)}{2(\varepsilon_\alpha + 1)}\right] \tag{7.87}$$

$$X_\Gamma = \frac{1}{\varepsilon_\alpha}\left[1 + \frac{(\varepsilon_\alpha - 1)}{2(\varepsilon_\alpha + 1)}\right]X_{but,\Gamma}, \tag{7.88}$$

which are respectively valid in the ranges $(\Gamma_A \leq \Gamma_y \leq \Gamma_{AB})$, $(\Gamma_{AB} < \Gamma_y \leq \Gamma_{DE})$, and $(\Gamma_{DE} < \Gamma_y \leq \Gamma_E)$.

The load sharing factor for wide cylindrical helical gears with adequate profile modification on the addendum of the driving member and/or the dedendum of the driven member of the gear pair (Fig. 7.14c) is given by the following relationships:

$$X_\Gamma = \frac{1}{\varepsilon_\alpha}\left[1 + \frac{(\varepsilon_\alpha - 1)}{2(\varepsilon_\alpha + 1)}\right]X_{but,\Gamma} \tag{7.89}$$

$$X_\Gamma = \frac{1}{\varepsilon_\alpha}\left[1 + \frac{(\varepsilon_\alpha - 1)}{2(\varepsilon_\alpha + 1)}\right] \tag{7.90}$$

$$X_\Gamma = \frac{1}{\varepsilon_\alpha}\left[1 + \frac{(\varepsilon_\alpha - 1)}{2(\varepsilon_\alpha + 1)}\right]\left(\frac{\Gamma_E - \Gamma_y}{\Gamma_E - \Gamma_{DE}}\right), \tag{7.91}$$

which are respectively valid in the ranges $(\Gamma_A \leq \Gamma_y \leq \Gamma_{AB})$, $(\Gamma_{AB} < \Gamma_y \leq \Gamma_{DE})$, and $(\Gamma_{DE} < \Gamma_y \leq \Gamma_E)$.

Finally, to calculate the load sharing factor for cylindrical helical gears with $0.8 < \varepsilon_\beta < 1.2$, it is necessary to bear in mind that the overlap ratio changes depending on the load, as the gears have no infinite stiffness. This variation of overlap ratio affects the load sharing factor for cylindrical helical gears, which are characterized by calculated values of the overlap ratio included in the range $0.8 < \varepsilon_\beta < 1.2$.

Therefore, to take account of this behavior, the calculation of the load sharing factor for these gears is performed by interpolating the load sharing factor $X_{\Gamma(\varepsilon_\beta=0.8)}$, for $\varepsilon_\beta = 0.8$, and load sharing factor $X_{\Gamma(\varepsilon_\beta=1.2)}$, for $\varepsilon_\beta = 1.2$, both determined considering the different relationships previously described, concerning the possible cases of unmodified and modified tooth profiles. Consequently, in this framework, the equation to be used for calculation of the load sharing factor for these gears is as follows:

$$X_{\Gamma(\varepsilon_\beta)} = X_{\Gamma(\varepsilon_\beta=0.8)} \frac{1.2 - \varepsilon_\beta}{0.4} + X_{\Gamma(\varepsilon_\beta=1.2)} \frac{\varepsilon_\beta - 0.8}{0.4}. \tag{7.92}$$

7.7.5 Wide and Narrow Bevel Gears

Wide bevel gears have a total contact ratio, $\varepsilon_\gamma \geq 2$. The load sharing factor of these gears, when they are characterized by optimal profile modification, for which $C_{a1} = C_{a2} = C_{eff}$, is assumed to be parabolic, as Fig. 7.15a shows. The parameter on the line of action at midpoint, M, of the path of contact is defined by the relationship:

$$\Gamma_M = \frac{1}{2}(\Gamma_A + \Gamma_E), \tag{7.93}$$

while the distribution of the load sharing factor along the path of contact, for this optimal profile modification, is given by:

Fig. 7.15 Distribution curves of the load sharing factor for wide bevel gears with: **a** optimal profile modification; **b** undersized profile modification near point A, and oversized profile modification near point E

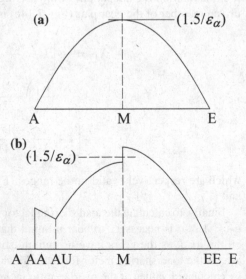

$$X_\Gamma = \frac{3}{\varepsilon_\alpha}\left[\frac{1}{2} - \frac{2(\Gamma_y - \Gamma_M)^2}{(\Gamma_E - \Gamma_A)^2}\right]. \tag{7.94}$$

In some practical application, the tip relief of the pinion, C_{a1}, differs from the tip relief of the wheel, C_{a2}. In this case, characterized by different profile modifications, the distribution curve in sections AM and ME of the path of contact has a discontinuity at midpoint M, as Fig. 7.15b shows. Therefore, the quantities related to these two sections should be calculated separately.

For undersized profile modification, the load sharing factor is obtained by linear interpolation between the value of X_Γ for optimal profile modification, given by Eq. (7.94), and the value of X_Γ for unmodified profile, the latter obtained with relationship $X_\Gamma = (1/\varepsilon_\alpha)X_{but,\Gamma}$, which takes into account the buttressing effect. This interpolation is to be made stepwise from end point A to midpoint M considering the influence of the tip relief of wheel, C_{a2}, and from midpoint M to end point E considering the influence of the tip relief of pinion, C_{a1}.

Instead, for oversized profile modification, new end points AA and EE are to be found to replace, respectively, the usual end points A and E. The parameters on the line of action that define these new end points of path of contact are as follows:

$$\Gamma_{AA} = \Gamma_A + \frac{1}{6}\varepsilon_\alpha(\Gamma_E - \Gamma_A)\left(\frac{C_{a2}}{C_{eff}} - 1\right) \tag{7.95}$$

$$\Gamma_{EE} = \Gamma_E - \frac{1}{6}\varepsilon_\alpha(\Gamma_E - \Gamma_A)\left(\frac{C_{a1}}{C_{eff}} - 1\right). \tag{7.96}$$

To calculate the load sharing factor in the various ranges that cover the path of contact, we use the following relationships:

$$X_\Gamma = \frac{3}{2\varepsilon_\alpha}\frac{3}{[4 - (C_{a2}/C_{eff})]}\left[1 - \frac{(\Gamma_y - \Gamma_M)^2}{(\Gamma_{AA} - \Gamma_M)^2}\right] \tag{7.97}$$

$$X_\Gamma = \frac{3}{2\varepsilon_\alpha}\frac{3}{[4 - (C_{a1}/C_{eff})]}\left[1 - \frac{(\Gamma_y - \Gamma_M)^2}{(\Gamma_{EE} - \Gamma_M)^2}\right] \tag{7.98}$$

$$X_\Gamma = 0. \tag{7.99}$$

Equations (7.97) and (7.98) are valid, respectively, in the ranges ($\Gamma_{AA} < \Gamma_y \leq \Gamma_M$) and ($\Gamma_M \leq \Gamma_y < \Gamma_{EE}$). Instead, Eq. (7.99) is valid in the ranges $(\Gamma_A \leq \Gamma_y \leq \Gamma_{AA})$ and $(\Gamma_{EE} \leq \Gamma_y \leq \Gamma_E)$. Of course, in the range $(\Gamma_{AA} < \Gamma_y \leq \Gamma_{AU})$, the buttressing effect must be considered; therefore, in this range, the load sharing factor is determined by multiplying the values of X_Γ determined by means of Eq. (7.97) with the buttressing factor, $X_{but,\Gamma}$, given by Eq. (7.81).

In contrast to the wide bevel gears, the narrow bevel gears have a total contact ratio, $\varepsilon_y < 2$. For narrow bevel gears with undersized profile modification ($C_a < C_{eff}$), the load sharing factor is obtained by linear interpolation between the value of X_Γ for optimal profile modification $(C_a = C_{eff})$ given by Eq. (7.94), and the value of X_Γ for unmodified profile $(C_a = 0)$. This last value is that of the narrow cylindrical helical gears (see Sect. 7.7.4), for which it is calculated by multiplying the values of X_Γ determined by means of Eqs. (7.67) and (7.68), with the buttressing factor, $X_{but,\Gamma}$.

For narrow bevel gears with optimal profile modification or with oversized profile modification $(C_a \geq C_{eff})$, the load sharing factor is calculated as for wide bevel gears.

7.8　Scuffing Temperature and Safety Factor

In Sect. 7.3 we saw that it is possible to make a reliable forecast of the probability that the scuffing occurs, comparing the calculated maximum contact temperature, Θ_{Bmax}, given by Eq. (7.2), with a critical value, which represents the scuffing temperature. Therefore, the scuffing temperature is the contact temperature at which scuffing will probably take place, for the given combination of lubricant and gear materials that characterizes the design of the gear set under consideration.

The scuffing temperature is considered to be a characteristic value for the material-lubricant-material system of the gear pair, to be determined by test gears that use the same material-lubricant-material combination. For low-additive mineral oils, the scuffing temperature is assumed to be independent of the operating conditions, in terms of gear materials and lubricant, within a fairly wide range. For mineral oils or synthetic oils, with anti-scuff or friction-reducing additives, the scuffing temperature is instead not constant, but variable depending on the operating conditions. Unfortunately, however, the knowledge regarding this variability are considered still inadequate, so further research is needed, taking into account the transition diagrams, such as that shown in Fig. 7.1, and using test gears under testing conditions that simulate correctly the actual or design conditions.

Generally, through-hardened steels are used for test gears. The scuffing temperature of low-additive mineral oils determined with these test gears may be extended to different gear steels, surface treatments or heat treatments by using the following empirical relationship:

$$\Theta_S = \Theta_{MT} + X_W \Theta_{flmaxT}, \tag{7.100}$$

where Θ_{MT} is the bulk temperature of test gears, Θ_{flmaxT} is the maximum flash temperature of the same test gears, and X_W is the *structural factor*. The values of this factor, which is a dimensionless empirical factor, can be read from Table 7.2. It is to be noted that Eq. (7.100) should be used with consistent units of temperatures that appear in it. In addition, it is to be noted that the empirical Eq. (7.100) is an approximate expression, whose validity is restricted to methods using the coefficient of friction related to common operating conditions together with an average value

Table 7.2 Structural factor, X_W

Material	X_W
Through-hardened steel	1.00
Phosphated steel	1.25
Copper-plated steel	1.50
Bath or gas nitrided steel	1.50
Hardened carburized steel, with austenite content:	
• Less than average	1.15
• Average (10–20%)	1.00
• Greater than average	0.85
Austenite steel (stainless steel)	0.45

of the thermo-elastic factor. When realistic values of the coefficient of friction and thermo-elastic factor are considered, the structural factor becomes superfluous, for which its contribution can be neglected.

Contrary to what happens for gears lubricated with low-additive mineral oils, the scuffing temperature of gears lubricated with anti-scuff oils can be affected by the *contact exposure time*, which is defined as the time during which a point on the tooth flank surface is exposed to the Hertzian contact band of meshing tooth. For a pair of meshing tooth flanks, the decisive contact exposure time is the longest contact exposure time, $t_{c}\max$, between the contact exposure time of pinion, t_{c1}, and wheel, t_{c2}, i.e. (t_{c1} and tc_2 are given in μs):

$$t_{c}\max \geq t_{c1} = \frac{2b_H}{v_{g1}} \quad \text{or} \quad t_{c}\max \geq t_{c2} = \frac{2b_H}{v_{g2}}. \tag{7.101}$$

In approximate terms, it is assumed that the dependence of scuffing temperature, Θ_S, on the contact time is represented by a curve consisting of two straight lines, as the one shown in Fig. 7.16. Therefore, we will have, for $t_{c}\max \geq t_K$:

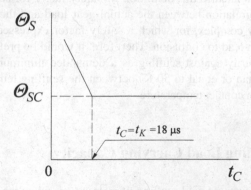

Fig. 7.16 Distribution curve of the scuffing temperature as a function of contact exposure time for anti-scuff lubricant oils

$$\Theta_S = \Theta_{Sc} = const, \tag{7.102}$$

and, for $t_{c}\text{max} < t_K$:

$$\Theta_S = \Theta_{Sc} + X_\Theta X_W (t_K - t_{c}\text{max}), \tag{7.103}$$

where: Θ_{Sc} (in °C or K), is the scuffing temperature at long contact time; X_Θ (in °C/μs or K/μs), is the gradient of the scuffing temperature; X_W, is the already defined structural factor; t_K (in μs), is the contact exposure time at the knee of the curve $\Theta_S = \Theta_S(t_c)$, as Fig. 7.16 shows; $t_{c\text{max}}$ (in μs), is the longest contact exposure time of meshing teeth.

Depending on the type of lubricant oil used, the following values of the gradient of the scuffing temperature, X_Θ, and contact exposure time at the knee of the curve, t_K, can be chosen: $X_\Theta = 0\,\text{K/μs}$, and $t_K = 0\,\mu s$, for lubricant oils without anti-scuff additives; $X_\Theta = 18\,\text{K/μs}$, and $t_K = 18\,\mu s$, for lubricant oils with anti-scuff additives.

It should be remembered that the scuffing temperature can be determined by different test methods, such as those mentioned in the ISO/TS 6336-20:2017(E), to which we refer the reader. Here we must focus our attention on the *safety factor for scuffing*, which must be carefully chosen, especially when the gears are designed to operate at high pitch line velocities. It must be borne in mind that, contrary to the longtime of development of fatigue damage, a single momentary overload can initiate a scuffing of severity such that the gears are immediately unusable.

With all the limitations of the case, we can define the safety factor for scuffing, S_B, as follows:

$$S_B = \frac{\Theta_S - \Theta_{oil}}{\Theta_{B\text{max}} - \Theta_{oil}}, \tag{7.104}$$

where all quantities have already been defined.

The limitations related to this definition of safety factor for scuffing are due to the fact that the correlation between the actual gear load and the decisive contact temperature is very complex, for which a safety factor expressed in any quotient of temperatures can lead to confusion. Therefore, it would be preferable to express the concept of security against scuffing as a demanded minimum difference (for example, greater than or equal to 50 K) between the scuffing temperature and the estimated minimum contact temperature.

7.9   Cold Scuffing Load Carrying Capacity

As warm scuffing, even the cold scuffing of gears arises when the lubricant oil film breaks down. It occurs without appreciable development of heat, and is a phenomenon of gear damage relatively rare. As we said in the introduction (see Sect. 7.1), the cold scuffing is associated with relatively low tangential velocities ($v_t < 4\,\mathrm{m/s}$), mainly for through-hardened and heavily loaded gears having rather poor accuracy grade.

The cold scuffing damage manifests itself in the form of groove-like wear on gear tooth flank surfaces and very severe material erosion, due to relatively slow sliding velocity associated with high enough tooth Hertzian pressures, related to uneven surface geometry and consequent load concentrated over discrete areas.

For the evaluation of the cold scuffing load carrying capacity of gears, there are still no validated calculation methods. For a rough estimate, we can rely on elastohydrodynamic lubrication theory. A possible solution to limit the risk of cold scuffing is to use more precise gearing, a smoother surface of the gear tooth flanks, and most appropriate lubricant oils for specific requirements. All the actions and design choices that lead to a greater lubricant film thickness (and, particularly, a greater oil viscosity), as well as minor profile deviations and lower roughness of the tooth flank surfaces, reduce the danger of cold scuffing. In addition, the profile modifications, in terms of tip relief or tip-rounded edges, that eliminate the impact of engagement, have positive effects on the cold scuffing load carrying capacity.

For a qualitative judgment on the risk of cold scuffing and, more generally, on the damage behavior of the flank surface of mating teeth, it is customary to compare the theoretical thickness of the lubricant oil film with the roughness of the tooth flank surface (see Niemann and Winter 1983). To this purpose, various indices of deterioration of the gear tooth surfaces were subsequently introduced, starting from the one proposed for the first time by Bodensieck, in 1965 (see Bodensieck 1965, 1967). All these indices are aimed at defining the elastohydrodynamic lubrication conditions between two surfaces in relative motion, with a lubricant oil film interposed. In Sect. 2.7, we have already defined one of these indices, each of which is called Bodensieck specific film thickness. Regarding the risk of cold scuffing, Niemann suggests using the following Bodensieck specific film thickness, defined as the quotient of the minimum thickness of the lubricant film at pitch point, h_c, divided by the arithmetic average roughness, Ra, of the flank surfaces of pinion and wheel, that is the ratio:

$$\lambda = \frac{h_c}{Ra}. \tag{7.105}$$

Therefore, a limiting value is imposed to this ratio, which is obviously to be determined using coherent units for h_c and Ra (both in mm or μm). The arithmetic average roughness, Ra, is defined as the mean value of the arithmetic average roughness Ra_1 and Ra_2, of pinion and wheel, which are determined according to Eq. (2.34).

By experimental evidence from numerous tests on actual gear units, the following conclusions can be drawn:

- For $\lambda < 0.7$, boundary conditions of elastohydrodynamic lubrication prevail, so superficial damage, such as cold scuffing, may occur. This is the operating range of several industrial gear sets, where the below described characteristics of lubricant to be used are important. For low transverse tangential velocities, and therefore low temperatures, it is good that polar combinations of lubricant oils, fatty acids and solid particles, i.e. graphite or molybdenum disulfide, can be physically deposited on the sliding surfaces, thus forming a stable protective layer. For high transverse tangential velocities, and therefore high temperatures, it is a positive fact that especially the anti-scuff additives may chemically react with the tooth flank surfaces, thus forming organo-metallic soaps, which act as sliding layers separating the metallic surfaces.

- For $\lambda > 2$, we have full conditions of elastohydrodynamic lubrication, actually governed by a unique determinant characteristic of the lubricant oil, that is its viscosity. Consequently, no damage of the mating surfaces is found, and the wear is almost absent. As a downside, the running-in conditions are not very favorable. It should be born in mind that, at times, for fast gear units and marine gear units, a value of this ratio, λ, greater than 1.2 is required.

According to Winter and Oster (1981), the minimum thickness of the lubricant film at pitch point, h_c, which appear in Eq. (7.105) is given by (see also Niemann and Winter 1983):

$$h_c = k\left(\frac{u}{u+1}\right)^{0.43}\left(\frac{a\sin\alpha_{wt}}{u+1}\right)^{1.13}\frac{(\omega_1\nu_M)^{0.7}}{(\cos\beta_b)^{0.3}w_{bt}^{0.13}}, \qquad (7.106)$$

where $\alpha_{wt} = \alpha_t'$ (in degrees) is the pressure angle at the pitch cylinder, $u = z_2/z_1$ is the gear ratio, a (in mm) is the center distance, ω_1 (in rad/s) is the angular velocity of the pinion, β_b (in degrees) is the base helix angle, $w_{bt} = F_{bt}/b$ (in N/mm) is the transverse tangential load for unit face width (F_{bt} is the nominal transverse load in the plane of action, i.e. in the base tangent plane), ν_M (in mm^2/s) is the kinematic viscosity of the lubricant oil at ambient pressure, and at bulk temperature, Θ_M, and k is a dimensional quantity $\left[\text{in } s^{1.4}/\left(\text{mm}^{1.66}\text{N}^{0.13}\right)\right]$ given by:

$$k = 2.65\alpha^{0.54}\rho^{0.7}\left(\frac{1}{E_r}\right)^{0.03}, \qquad (7.107)$$

where α (in mm^2/N) is the pressure-viscosity coefficient, ρ (in Ns^2/mm^4) is the lubricant density, and E_r (in N/mm^2) is the reduced modulus of elasticity, given by Eq. (2.3).

It is to be noted that, for internal gears, u, z_2, and a in Eq. (7.106) are negative quantities. From Eqs. (7.106) and (7.107), for $\alpha = 1.6 \times 10^{-2}$mm^2/N, $\rho = 0.9 \times 10^{-9}$ Ns^2/mm^4, $E_1 = E_2 = 206$ GPa, $\nu_1 = \nu_2 = 0.3$, $\alpha_{wt} = 23°$, and $\cos\beta_b \cong 1$, we obtain:

$$h_c \cong 3 \times 10^{-3}\left[\frac{au}{(u+1)^2}\right]^{0.3}(\nu_M\nu_t)^{0.7}\left(\frac{\sigma_H}{840}\right)^{-0.26}. \qquad (7.108)$$

According to FZG, the estimate value of the bulk temperature, Θ_M, before the teeth meshing, which is necessary to calculate the kinematic viscosity, ν_M, is given by the relationship:

Fig. 7.17 Tip relief factor for scuffing load carrying capacity, as a function of ε_{max} and C_a corresponding to the outer single pair contact point

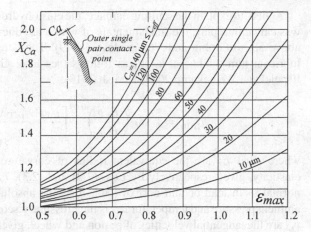

$$\Theta_M = \Theta_{oil} + 7400 \left(\frac{P_{vz}}{ab} \right)^{0.72} \frac{X_S}{1.2 X_{Ca}}. \qquad (7.109)$$

where Θ_{oil} is the oil temperature before teeth meshing (this temperature coincides with the *idling temperature*), P_{vz} (in kW) is the power loss at contact between the mating teeth (see Vol. 1, Sect. 1.1), X_S is the lubricant system factor, and X_{Ca} is the tip relief factor. The latter factor is given by the relationship:

$$X_{Ca} = 1 + 1.55 \times 10^{-2} \varepsilon_{max}^4 C_a, \qquad (7.110)$$

where ε_{max} is the maximum value of the addendum contact ratios of pinion, ε_1, and wheel, ε_2 (see Eqs. (8.22) and (8.23) in the next chapter), and C_a is the tip relief corresponding to the outer single pair contact point. The value of the tip relief factor, X_{Ca}, can also be read from Fig. 7.17, as a function of ε_{max} and C_a.

7.10 Some In-depth Information on the Warm Scuffing

In the previous sections, we saw that, in the gears, the conditions of sliding velocity and pressure of contact may become so unfavorable as to determine the development of a very high contact temperature, due to power losses. This temperature is sufficient to cause the breakdown of the lubricant oil film. In these conditions, a metal-to-metal contact occurs, with a tendency to localized welding in correspondence of the roughness peaks, which is the more pronounced, the more the tooth flank surfaces are softened by thermal effect. As we said elsewhere, the warm scuffing occurs with scratches and grooves on the tooth flank surfaces, oriented in the direction of the sliding velocity.

At first, to explain the scuffing damage, the elastohydrodynamic lubrication theory was called into play, the fundamentals of which are due to Ertel (1939, 1945) and, above all, to Grubin and Vinogradova (1949). The latter expressed the minimum lubricant film thickness, h_{min}, as follows (see also: Greenwood 1972; Morales-Espejel and Wemekamp 2008; Kudish 2013):

$$h_{min} = 1.18\rho \left(\frac{E_r\rho}{F_{n,u}}\right)^{0.09} \left[\frac{\eta_0\alpha(v_{t1} + v_{t2})}{\rho}\right]^{0.73}, \qquad (7.111)$$

where ρ and E_r are the equivalent radius of curvature and reduced modulus of elasticity, given respectively by Eqs. (2.2) and (2.3), $F_{n,u} = F_t/b$ is the specific normal tooth load or unit normal load, η_0 is the absolute or dynamic viscosity at ambient pressure and temperature, α is the pressure-viscosity coefficient, and v_{t1} and v_{t2} are the tangential velocities of pinion and wheel, given by Eqs. (3.59) and (3.60) of Vol. 1.

Evidently, the Eq. (7.111) is very complex. It is valid for any system of units, provided it is consistent. According to this equation, for a given geometry of the teeth and a given material for pinion and wheel, the minimum lubricant film thickness is greater the larger are the dynamic viscosity, tangential velocities and pressure-viscosity coefficient, and the smaller is the unit normal load. However, due to the low value of the exponent of the first parenthesis in the *Grubin-Vinogradova equation* (it is noteworthy that this exponent, equal to 0.09, has been experimentally confirmed by other researchers later), h_{min} decreases slightly as load increases. This almost insensitivity of the minimum lubricant film thickness to high loads can be explained, at least in part, with the sharp increase of the dynamic viscosity with the increase of pressure. In fact, pressure values of the magnitude of 1 GPa are not unusual in the gears.

With these levels of pressure, the lubricant oil film acts as a very stiff spring. It is also necessary to bear in mind that the Hertzian pressure changes greatly and consequently also the conditions of relative curvature change favorably. In this regard, for cylindrical gears, that is for line contact, the width of Hertzian contact band, $2b_H$, is given by Eq. (2.6). However, even taking into account these additional circumstances, which further complicate the Hertzian contact phenomenon between lubricated surfaces, from Eq. (7.111) we infer that, even for very high loads, the minimum lubricant film thickness still has a value sufficient to keep separated the two surfaces in relative motion. Therefore, the elastohydrodynamic lubrication theory may not be enough to explain in detail why scuffing damage occurs.

These difficulties of elastohydrodynamic lubrication theory in explanation of scuffing damage led to consider, at a later time, the flash temperature theory, proposed by Blok (1937a, b, c, d) more than a decade before, and further developed by the same author in the following years (1940, 1958, 1963a, b, 1969, 1974). According to this theory, the warm scuffing is due to the abrupt breakdown of the lubricant film, when the contact temperature reaches a certain critical value. In other words, the origin of the warm scuffing is not due to a progressive decrease of the lubricant film thickness, up to a critical value, but is due to the attainment of a threshold temperature for the lubricant used. On the other hand, various experiments have confirmed the existence of a lubricant film of sufficient thickness until the threshold of the incipient scuffing, and then an abrupt breakdown of the same lubricant film.

Figure 7.18 shows, in schematic form, the generation mechanism of the warm scuffing, according to the Blok's flash temperature theory. Once the critical value of the contact temperature (i.e. the scuffing temperature) is reached, the lubricant oil film adherent to the metal surfaces is locally destroyed. Therefore, metal surfaces in relative motion are directly in contact, and heat generated by friction, concentrated in extremely localized areas, is such as to determine the melting of the surface roughness asperities, and their subsequent welding. Thus, the scuffing mechanism described in Sect. 7.1 is triggered.

The distribution of the contact temperature through the area of contact between the meshing tooth flank surfaces can be represented, also this in schematic form, as Fig. 7.19 shows. Another phenomenon is associated with this distribution of the contact temperature, that of generation of thermal stresses. The metallic material of the contact area tends to expand, but the surrounding material, which is subjected to a lower temperature, prevents this thermal expansion. A compressive thermal stress is generated in the plane tangent to the contact area, oriented in the radial direction

Fig. 7.18 Generation mechanism of the warm scuffing

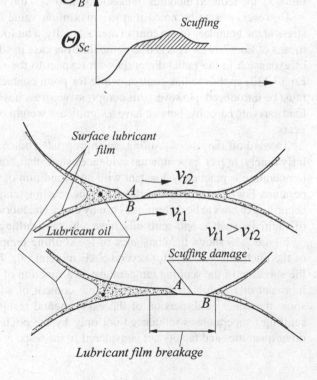

Fig. 7.19 Schematic
distribution of the contact
temperature through the area
of contact

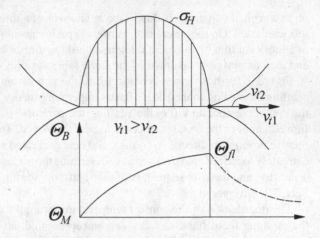

with respect to the boundary of the same contact area. The maximum value of this
compressive stress, σ_{cT}, occurs at the boundary of the contact area, and is given by:

$$\sigma_{cT} = -\alpha_T(1 + \nu)E_r\Theta_{fl}, \tag{7.112}$$

where α_T is the coefficient of linear thermal expansion of material, ν the Poisson
ratio, E_r the reduced modulus of elasticity, and Θ_{fl} the flash temperature.

However, it is to be noted that the maximum value of the actual compressive
stress at the boundary of the contact area is, really, a bit lower than that calculated by
means of Eq. (7.112). This is true in the limiting case in which there is no possibility
of expansion in the radial direction with respect to the same boundary. Moreover,
Eq. (7.112) applies to line contact, while for point contact more complex equations
must be introduced. However, this compressive stress has little influence on scuffing
load carrying capacity, but can have a significant weight on the surface durability of
gears.

To ward off the risk of scuffing, an appropriate choice of lubricant oil is particularly timely. In fact, experimental evidences show that, for mineral oils, the scuffing
temperature is practically constant with any condition of speed, load and bulk temperature. For other types of lubricant oil, the scuffing temperature can however vary
according to laws to be determined case by case. The lubricant oils with high content
of additives, such as anti-scuff oils, have a higher scuffing temperature.

Figure 7.20 shows the constancy of the scuffing temperature with the variation
of the pitch line velocity, for several SAE mineral oils. Figure 7.21 shows instead
the variation of the scuffing temperature as a function of the pitch line velocity for
lubricant oils with high, medium and low content of additives. The shaded areas
show the zones of dispersion of the experimental results, which are because the
scuffing temperature is influenced not only by the pitch line velocity, but also by
other quantities and factors not considered in the tests.

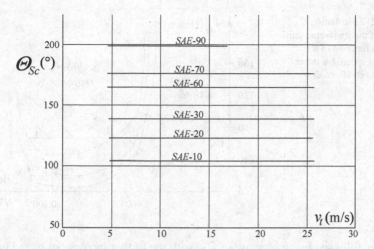

Fig. 7.20 Scuffing temperature as a function of pitch line velocity, for several SAE mineral oils

Fig. 7.21 Scuffing temperature as a function of pitch line velocity, for lubricant oils with high, medium and low content of additives

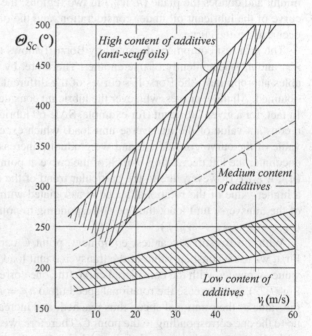

For the prediction of the probability of scuffing, the distribution curves of the transverse unit load, w_{Bt}, as a function of the rotational speed, n, for the different lubricant oils that can be used, are very important. These distribution curves, obtained for the first time by Borsoff (1959) with a special gear test equipment by himself designed, have a non-linear trend, with the concavity facing upwards. In other words, these curves are characterized by a decrease of w_{Bt} with the increase of the rotational

Fig. 7.22 Distribution curves of the transverse unit load as a function of the rotational speed, for three lubricant oils (Borsoff's curves)

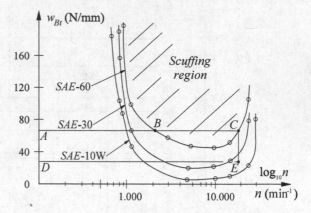

speed, n, followed by an increase of w_{Bt} with the further increase of n, as Fig. 7.22 shows. Therefore, each curve, which is specific of a given lubricant oil, has a minimum, and divides the plane (n, w_{Bt}) in two regions: the scuffing region, above the curve of the lubricant oil under consideration, and the one where scuffing does not occur, under the curve.

The gear test equipment designed by Borsoff allows to adjust the two variables, w_{Bt} and n, independently of one another. Therefore, by combining these two variables appropriately, the Borsoff's curves of the different lubricants involved can be obtained. All these curves, whatever the lubricant considered, show a singular trend. In fact, for a given lubricant (for example, SAE-60 lubricant in Fig. 7.22), if we use a constant value of the transverse unit load, which exceeds the specific minimum value of the curve considered, and we gradually increase the rotational speed, we encounter the left decreasing branch of the curve at point B, whose coordinates are $(n_B, w_{Bt,B})$. However, due to the particular trend of the curve under consideration, a higher value of the rotational speed is associated with the aforementioned value of the transverse unit load, the one corresponding to point C, whose coordinate are $(n_C > n_B, w_{Bt,C} = w_{Bt,B})$.

With the Borsoff's gear test equipment, point C can be reached in two steps. First, we use a constant value of the transverse unit load smaller than the minimum value associated with the considered curve (the one corresponding to the DE-line in Fig. 7.22), and increase the rotational speed up to $n_E = n_C$. Then we keep constant the value of the rotational speed thus reached, and increase the transverse unit load up to the one corresponding to the point C. Therefore, we infer that, for a given value of the load, we can have two different values of scuffing rotational speed.

This behavior, at least curious at that time, was explained by Borsoff and Godet (1963) and Godet (1963a, b, 1964a, b, 1965) with a theory called the *theory of the two lines*, which simultaneously gives reason of scuffing and wear. In fact, if the curve $w_{Bt} = w_{Bt}(n)$ is represented in logarithmic coordinates, as Fig. 7.23 shows, two effects are appearing:

Fig. 7.23 Curve
$w_{Bt} = w_{Bt}(n)$ in logarithmic
coordinates

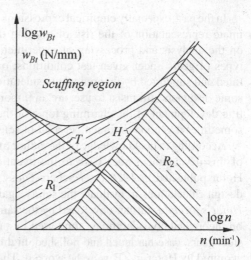

- For low values of the rotational speed, a *friction effect* is predominant, which results in a *thermal effect* with consequent heating of the tooth flank surfaces. This thermal effect requires a reduction in load with increasing speed, for which we have a *thermal line*, T, with negative slope.
- For higher values of the rotational speed, an *elastohydrodynamic effect* is generated, resulting in a load increase with increasing speed, so we have a *hydrodynamic line*, H, with a positive slope.

The curve $w_{Bt} = w_{Bt}(n)$ is obtained by combining the above two effects, that is as the sum of the two effects, or as a modification of the first effect in consequence of the second effect or vice versa. The region below the curve $w_{Bt} = w_{Bt}(n)$, the one where the scuffing does not occur, is in turn divided into two sub-regions, R_1 and R_2, by the hydrodynamic line. In the sub-region R_1, between the curve $w_{Bt} = w_{Bt}(n)$, hydrodynamic line H, and coordinate lines, the gears are subjected to abrasive wear. Instead, in the sub-region R_2, between the hydrodynamic line H and the abscissa axis, the gears do not suffer any wear or damage, since the tooth flank surfaces are separated by a lubricant oil film having a lubricant film specific thickness, Λ, sufficiently high (see Sect. 2.7).

7.11 Empirical Formulae

The scuffing of gears has been studied intensively since the early 20th century, and different criteria have gradually been proposed. The most recent criteria, that of the flash temperature that we have described in the previous sections, and that of the integral temperature to be described in the next chapter, have a sound scientific basis, as they are based on thermodynamic and elastohydrodynamic lubrication theories, moreover confirmed by experimental tests.

In the past, especially empirical expressions have been used, which give an approximate representation of the risk of scuffing damage. These expressions are based on the analysis and processing of experimental test results, obtained with certain types of gears under given test conditions; only in few cases, they were corroborated and supported by theoretical considerations. These empirical formulae, which some designers still insist to use, are an important milestone in the history of scientific development of the charming topic of the scuffing load capacity of gears, and sometimes it is good to remember a bit of scientific history, because our roots in it.

According to the knowledge of the author of this textbook, the first historical stage of this development goes back to the twenties of the XX century, when Herman Hofer published in 1926 (von Scherr-Thoss 1965) his formula to be used in the design calculations to evaluate the safety against the risk of scuffing damage. This *Hofer's formula*, further developed by the same author from 1938 to 1943, is based on theoretical considerations, supported by practical experiences with high quality steel gears, case-hardened and polished, highly loaded and operating at high speed, acquired by Hofer in ZF, were he worked. These practical experiences suggested to Hofer to consider the safety against scuffing, for high-speed operating gears, and security against the tooth root strength, for gears operating at low speed. It is to be noted that, at that time, the problem of surface durability (with the pitting treated on the basis of the Hertzian stress) had just been sketched scientifically, and was therefore ignored by gear designers.

Analyzing the results of these practical experiences based on the physical laws of heat generation by friction, Hofer developed his formula that, in its latest version, is expressed as follows (see Giovannozzi 1965):

$$S_B = 0.068 \frac{m_n z_1 b}{P}, \tag{7.113}$$

where m_n (in mm) is the normal module, z_1 the number of teeth of pinion, b (in mm) the face width (the smallest value of face width for pinion and wheel is to be used), and P (in kW) is the transmitted power.

This formula expresses the safety factor for scuffing, S_B, due to excessive heat generation by friction. Keeping in mind that $P = F_t v_t$ (F_t and v_t are respectively the nominal tangential force and pitch line velocity), and that the product $m_n z_1 b$ is proportional to the lateral surface of the smaller member of the gear pair (this is in fact the decisive value for the heat dissipation), the Hofer's formula essentially imposes a limit of product $\sigma_H v_t$.

The safety factor expressed by the Hofer's formula is not a safety factor in the proper meaning. It is a safety factor *sui generis*, in the sense that, for usual gears operating under normal lubrication conditions, values of S_B greater than or equal to 1 ($S_B \geq 1$) can be allowed, while, for high precision gears operating under optimum lubrication conditions (such as the gears of aeronautical compressors), lower values of S_B may be accepted [$S_B = (0.5 - 0.2)$]. However, it should be noted that the validity of this formula is not general, but it is limited to the types of gears, materials, and operating conditions that have constituted the experimental basis for the acquisition of data from the processing of which it was derived.

The Hofer's formula was used by gear designers, especially those of Central European School, for over twenty years, but gradually gave way to the first of the two Almen's factors, i.e. the Almen's factor $PV = \sigma_H v_g$, where $P = \sigma_H$ is the Hertzian stress, and $V = v_g$ is the tooth sliding velocity. This empirical factor was introduced in 1935 by Almen (1935), based on processing of experimental data acquired on highly-loaded spiral bevel gears of automotive rear axle, lubricated with ordinary mineral oils and running on the road at high velocity, in research laboratories of GM (General Motors Corporation), where he worked. For these gear sets (note that the gear wheels were without profile modifications), operating in the above-mentioned conditions, Almen found that the scuffing occurred when the value of the product PV exceeded 3.3 GPa(m/s). More generally, with this empirical formula, to avoid the scuffing risk, the maximum value of this factor, $(PV)_{max}$, was compared with an allowable value, characteristic of the lubricant used. Further experimental tests, carried out again on automobile rear axle gears, confirmed the validity of PV-factor for this specific application (see Almen and Boegehold 1935).

It should be noted that the Almen's factor, PV, has the dimensions of a specific power, i.e. the dimensions of a power per unit surface. Therefore, such a factor multiplied by the coefficient of friction represents the friction power lost per unit tooth contact surface, which is converted into heat. Therefore, it is reasonable to think that the higher the value of the PV-factor, the highest is the temperature increase in the area of contact between the lubricated mating teeth, so the risk of scuffing is looming.

The Almen's factor PV still represents a significant improvement compared to Hofer's formula, as the tooth sliding velocity, $V = v_g$, which appears in it, varies along the path of contact (instead v_t is a constant). Therefore, it is possible to take into account the fact that the conditions most favorable to scuffing generation will have in correspondence of the marked points A and E (Fig. 7.5). This simple Almen's method to impose a limit on the PV-product to avoid scuffing was still successful when it was applied to other rear axle gears, because the operating conditions of these gear sets in terms of load, temperature and sliding velocity cover a relatively narrow range.

The same Almen (1942) realized that this simple method was no longer valid when he tried to predict the scuffing in aircraft engine gears, for which the temperature, unit load and sliding velocities ranged over a much greater range. In fact, he found that for certain aircraft gears the scuffing did not occur even when the PV-product values were greater than 8.8 GPa(m/s), that is almost three times higher than the above-mentioned limit value.

To clarify this discrepancy, more extensive data supplied by aircraft engine manufacturers were used, and a considerable number of other empirical formulae, of the type $P^m V^n$, were proposed, none of which was successful, until another geometric factor, T, was added to the Almen's factor PV, and so the *second Almen's factor*, or *modified Almen's factor*, PVT, was introduced. Really, this modified Almen's factor should be called *factor of Almen and Straub*, as both are authors of the paper where it was introduced (1948). However, this second factor was not only able to establish a good empirical limit for scuffing of aircraft gears lubricated with normal mineral oils, but also proved to be valid, with few exceptions, for every other gear sets, for which in a short time it supplanted the first Almen's factor.

In this modified Almen's factor, PVT, $P = \sigma_H$ and $V = v_g$ are the same quantities that appear in the Almen's factor, PV, while the geometric factor, $T = g_{Pmax}$, is the maximum distance measured along the line of action of the relevant point of contact from the pitch point (see Fig. 2.4). In fact, at this point at maximum distance from the pitch point, the sliding velocity assumes its maximum value, equal to $V_{max} = v_g max = (\omega_1 \pm \omega_2)T$, where the plus and minus signs are valid for external and internal gears, respectively. The limit value of this modified Almen's factor to prevent scuffing for gears lubricated with ordinary mineral oils and with teeth without profile modification is equal to about 82.5 (kN/mm)(m/s), when P, V, and T are expressed respectively in N/mm^2, m/s, and mm.

Here it is not considered necessary to report the relationships developed by Almen and Straub to calculate the PVT-factor for pinion and wheel, and this despite the fact that some gear designer still insists to use them, even to have an estimate on the scuffing threshold of gears to be used for some specific practical applications. In this regard, we refer the reader to the aforementioned paper of Almen and Straub (1948) , as well as to Almen (1950).

However, it is noteworthy that this last empirical factor has had a lifetime a little longer than those described above as well as other similar factors that have had even more ephemeral lifetime, for which we have decided not to recall here. This slightly longer lifetime is due to the delayed assertion of new and innovative flash temperature criterion, caused by World War II. The flash temperature theory was in fact introduced by Blok in 1937 (see Blok: 1937a, b, c, d), but when the Netherlands surrendered to occupying forces in May 1940, Blok destroyed all the test results on the instruction of his employer, for which they after the war could not be published in full.

However, since the Second World War, the statement march of the Blok's theory was unstoppable and, given the strong scientific basis on which it rests (experimental confirmations become, over the years, more and more numerous), space for the empirical formulae has gradually narrowed, and today they can be considered all extinct.

In conclusion, however, it should not conceal the fact that, among the many empirical expressions that have been gradually proposed, the best empirical criterion was certainly the one related to the modified Almen's factor, PVT. This empirical factor was a bridge that led to the flash temperature criterion. In this respect, by means of the introduction of thermodynamic constants that allow to make the jump from the ratio *force/time* to *temperature*, Polder (1958) has shown that the square root of PVT-factor approximates the maximum value of the flash temperature.

References

Abdel-Aal HA (1997) A remark on the flash temperature theory. Int Commun Heat Mass Transfer 24(2):241–250

Aghababaei R, Warner DH, Molinari JF (2016) Critical length scale controls adhesive wear mechanisms. Nat Commun 7

Alban LE (1985) Systematic analysis of gear failures. American Society for Metals, Metals Park, Ohio

Almen JO, Boegehold AL (1935) Rear axle gears: factors which influence their life. Proc Am Soc Test Mater 25(2):99–146

Almen JO (1935) Durability of spiral-bevel gears for automobile, Part one. Autom Ind 73(20):662–668, and 73(21):696–701

Almen JO (1942) Facts and fallacies of stress determination. SAE Trans 50(9):52–61

Almen JO, Straub JC (1948, June 4) Aircraft Gearing, Analysis of Test and Data Service. Research Laboratories Division, General Motors Corporation, Detroit, Michigan, AGMA

Almen JO (1950) Surface determination of gear teeth. In: Burwell JT (ed) Mechanical Wear. Am Soc Metals, Cleveland, OH, pp 229–288

Andersson S, Salas-Russo E (1994) The influence of surface roughness and oil viscosity on the transition in mixed lubricated sliding steel contacts. Wear 174(1–2):71–79

Andersson S, Söderberg A, Björklund S (2007) Friction models for sliding dry, boundary and mixed lubricated contacts. Tribol Int 40(4):580–587

Barwell FT, Milne AA (1952) Criteria governing scuffing failure. J Inst Petrol 38:624–632

Bathgate J, Kendall RB, Moorhouse P (1970) Thermal aspects of gear lubrication. Wear 15(2):117–129

Begelinger A, de Gee AWJ (1974) Thin film lubrication of sliding point contacts of AISI 52100 steel. Wear 28:103–114

Begelinger A, de Gee AWJ (1982) Failure of thin film lubrication—a detailed study of the lubricant film breakdown mechanism. Wear 77:57–63

Bloch HP, Geitner FK (1999) Machinery failure analysis and troubleshooting. Gulf Professional Publishing, Gulf Publishing Company, Houston, TX

Blok H (1937a) Measurements of temperature flashes on gear teeth under extreme pressure conditions. Proc Gen Discuss Lubr Inst Mech Eng 2:18–22

Blok H (1937b) Theoretical study of temperature rise at surfaces of actual contact under oiliness lubricating conditions. Proc Inst Mech Eng Gen Discuss Lubr 2:222–235

Blok H (1937c) Les températures de surface dans des conditions de graissage sous pressions extrêmes. In: Proceedings of the 2nd World Petroleum Congress, Paris, Section IV, III, pp 151–182

Blok H (1937d) Surface temperature measurements on gear teeth under extreme pressure lubricating conditions. Power Trans 653–656

Blok H (1940) Fundamental mechanical aspects of boundary lubrication. SAE Trans 35:54–68

Blok H (1958) Lubrication as a gear design factor. In: Proceedings of international conference on gearing. Institution of Mechanical Engineers, London, pp 144–158

Blok H (1963a) The flash temperature concept. Wear 6(6):483–494

Blok H (1963b) Inverse problems in hydrodynamic lubrication and design for lubricated flexible surfaces. In: Proceedings of the international symposium on lubrication and wear, Houston, McCutchan Publishing Corporation, Berkeley, CA, pp 1–151

Blok H (1969) The thermal-network method for predicting bulk temperatures in gear transmissions. In: Proceedings of the 7th round-table discussion on marine reduction gears, stat-laval, Finspong, Sweden, pp 3–25 and 26–32

Blok H (1974) Thermal instability of flow in elasto-hydrodynamic films as a cause for cavitation, collapse and scuffing. In: Proceedings of the first leeds-lion symposium. University of Leeds, England, pp 189–197

Bodensieck EJ (1965, September) Specific film thickness—an index of gear tooth surface deterioration. Paper presented at 1965 Aerospace Gear Communication Technical Division Meeting, AGMA (Denver, CO)

Bodensieck EJ (1967) How film thickness affects gear-tooth scoring. Power Transm 23:53–56

Boley BA, Weiner JH (1997) Theory of thermal stresses. Dover Publications Inc., Mineola

Borsoff VN (1959) On the Mechanism Of Gear Lubrication. ASME Trans J Basic Eng 81(80D):79–93

Borsoff VN, Godet M (1963) A scoring factor for gears. ASLE Trans 6:147–153

Bowman WF, Stachowiak GW (1996) A review of scuffing models. Tribol Lett 2(2):113–131

Broenink JF (1999) Introduction to physical systems modelling with bond graphs. University of Twente, Department of EE, Control Laboratory, pp 1–31

Bruce RW (ed) (2012) CRC handbook of lubrication (theory and practice of tribology)—vol II theory & design, 2nd edn. CRC Press, Taylor & Frencis Group, Boca Raton, Florida

Cameron A (1952) Hydrodynamic theory in gear lubrication. J Inst Petrol 38:614

Cameron A (1954) Surface failure in gears. J Inst Petrol 40:191

Carslaw HS, Jaeger JC (1959) Conduction of heat in solids, 2nd edn. Oxford University Press, Oxford

Conrado E, Höhn B-R, Michaelis K, Klein M (2007) Influence of oil supply on the scuffing load-carrying capacity of hypoid gears. J Eng Tribol 221(8):851–858

Cook RD (1981) Concepts and applications of finite element analysis, 2nd edn. Wiley, New York

Czichos H (1974) Failure criteria in thin film lubrication-the concept of a failure surface. Tribology 7(1):14–20

Czichos H (1976) Failure criteria in thin film lubrication, investigation of the different stages of film failure. Wear 36(1):13–17

Czichos H, Habig K-H (eds) (2010) Tribologie-Handbuch: Tribometrie, Tribomaterialen, Tribotechnik. Vieweg + Teubner Verlag, Springer Fachmedien Wiesbaden GmbH

Dowson D, Higginson GR (1966) Elasto-hydrodynamic lubrication. Pergamon Press, Oxford

Dowson D, Toyoda S (1978, September) A central film thickness formula for elastohydrodynamic line contacts. In: Proceedings of leeds-lyon symposium on tribology, pp 19–22, Paper no. 11

Dyson A (1975a) Scuffing—a review: Part 1. Tribol Int 8(2):77–87

Dyson A (1975b) Scuffing—a review: Part 2. Tribol Int 8(3):117–122

Dyson A (1976) Thermal stability of models of rough elastohydrodynamic systems. J Mech Eng Sci Inst Mech Eng

Ertel AM (1939) Hydrodynamic lubrication based on new principles. Akad Nauk SSSR Prikadnaya Mathematica i Mekhanika 3:41–52

Ertel AM (1945) Hydrodynamic lubrication analysis of a contact of curvilinear surfaces, Dissertation on Proceedings of CNIITMASH, Moscow, pp 1–64

Giovannozzi R (1965) Costruzione di Macchine, vol II, 4th edn. Casa Editrice Prof. Riccardo Pàtron, Bologna

Godet M (1963a) La théorie des deux lignes, la lubrication des engrenages. C R Acad Sci 257:48–51

Godet M (1963b) Reflexion théoriques et expérimentales à propos de la recherche sir la lubrication des engrenages dans les applications de la science à l'industrie, La Machine-Outil Français: 1ère partie, n. 193

Godet M (1964a) La théorie des deux lignes. Éssais des lubricants. C R Acad Sci 258:71–74

Godet M (1964b) Reflexion théoriques et expérimentales à propos de la recherche sir la lubrication des engrenages dans les applications de la science à l'industrie, La Machine-Outil Français: 2ème partie, n. 194

Godet M (1965) Reflexion théoriques et expérimentales à propos de la recherche sir la lubrication des engrenages dans les applications de la science à l'industrie, La Machine-Outil Français: 3ème partie, n. 195

Goldfarb V, Barmina N (eds) (2016) Theory and practice of gearing and transmissions, in honor of Professor Feydor L. Springer International Publishing Switzerland, Litvin

Greenwood JA (1972) An extension of the Grubin theory of elastohydrodynamic lubrication. J Phys D Appl Phys 5(12):2195–2211

Grubin AN, Vinogradova IE (1949) Investigation of the contact of machine components. In: Ketova KF (ed) Central Scientific Research Institute for Technology and Mechanical Engineering (TsNIIMASH), Book no. 30, Moscow, (D.S.I.R. Translation no. 337)

Gupta K, Jain NK, Laubscher R (2017) Advanced gear manufacturing and finishing: classical and modern processes. Academic Press, Elsevier Inc., London

Habchi W (2014) A numerical model for the solution of thermal elastohydrodynamic lubrication in coated circular contacts. Tribol Int 73:57–68

Henriot G (1979) Traité théorique et pratique des engrenages, vol 1, 6th edn. Bordas, Paris

Hirani H (2016) Fundamentals of engineering tribology with applications. Cambridge University Press, Cambridge

Höhn B-R, Michaelis K, Eberspächer C, Schlenk L (1999) A scuffing load capacity test with the FZG gear test rig for gear lubricants with high EP performance. Tribotest J 5(4):383–390

Höhn B-R, Michaelis K, Otto HP (2011) Flank load carrying capacity and power loss reduction by minimized lubrication. Gear Technol 53–62

Höhn B-R, Oster P, Michaelis K (2004) Influence of lubricant on gear failures—test methods and application to gearboxes in practice. Tribotest J 11(1)

Höhn B-R, Oster P, Michaelis K (1998) New test methods for the evaluation of wear, scuffing and pitting capacity of gear lubricants. AGMA Technical Paper 98FTM8

Horng JH (1998) True friction power intensity and scuffing in sliding contact. J Tribol 120(4):829–834

Incropera FP, DeWitt DP, Bergmann TL, Lavine AS (2006) Fundamentals of heat and mass transfer, 6th edn. Wiley, New York

ISO/TS 6336-20: 2017 (E), Calculation of load capacity of spur and helical gears-Part 20: calculation of scuffing load capacity (also applicable to bevel and hypoid gears)—flash temperature method

Jelaska D (2012) Gears and gear drives. Wiley, U.K.

Juvinall RC, Marshek KM (2012) Fundamentals of machine component design, 5th edn. Wiley, New York

Kovalchenko A, Ajayi O, Erdemir A, Fenske G (2011) Friction and wear behavior of laser textured surface under lubricated initial point contact. Wear 271(9–10):1719–1725

Kudish II (2013) Elastohydrodynamic lubrication for line and point contacts—asymptotic and numerical approaches. CRC Press, Taylor & Sons Group, Boca Raton, Florida

Lacey PI (1988) Development of a Gear Oil Scuff Test (GOST) procedure to predict adhesive wear resistance of turbine engine lubricants. Tribol Trans 41(3):307–316

Lipson C (1967) Wear considerations in design. Prentice-Hall Inc., Englewood Cliffs, N. J.

Ludema KC (1984) A review of scuffing and running-in of lubricated surfaces, with asperities and oxides in perspective. Wear 100:315–331

Mang T, Bobzin K, Bartels T (2011) Industrial tribology: tribosystems, friction, wear and surface engineering, lubrication. Wiley-VCH Verlag GmbH&Co. KGaA, Weinheim, Germany

Manin L, Play D (1999) Thermal behavior of power gearing transmission, numerical prediction, and influence of design parameters. ASME J Tribol 121:693–702

Martins R, Cardoso N, Scabra J (2008) Influence of lubricant type in gear scuffing. Ind Lubr Tribol 60(6):299–308

Michalczewski R, Kalbarczyk M, Michalak M, Piekoszewski W, Szczerek M, Tuszynski W, Wulczynski J (2013) New scuffing test methods for the determination of the scuffing resistance of coated gears. Chapter 6 in tribology-fundamentals and advancements. Intech Open Science, pp 185–215

Michalczewski R, Piekoszewski W, Szczerek M, Tuszynski W (2009) Scuffing resistance of DLC-coated gears lubricated with ecological oil. Est J Eng 15(4):367–373

Miltenović AV, Kuzmanović SB, Miltenović VD, Tica MM, Rachov MJ (2012) Thermal stability of crossed helical gears with wheels made from sintered steel. Therm Sci 2(Suppl. 2):S607–S619

Morales-Espejel GE, Wemekamp AW (2008) Ertel-Grubin methods in elastohydrodynamic lubrication—a review. In: Proc Inst Mech Eng Part J J Eng Tribol 222:15–34

Naunheimer H, Bertsche B, Ryborz J, Novak W (2011) Automotive transmissions: fundamentals, selection, design and application, 2nd edn. Springer, Berlin, Heidelberg

Naveros I, Ghiaus C, Ordoñez J, Ruiz DP (2016) Thermal networks considering graph theory and thermodynamics. In: Proceedings of the 12th international conference on heat transfer, fluid mechanics and thermodynamics, Costa del Sol, Spain, 11–13 July, pp 1568–1573

Niemann G, Winter H (1983) Maschinen-Elemente, Band II: Getriebe allgemein, Zahradgetriebe-Grundlagen, Stirnradgetriebe. Springer, Berlin, Heidelberg

Peterson MB, Winer WO (eds) (1980) Wear control handbook. The American Society of Mechanical Engineers, New York

Pirro DM, Webster M, Daschner E (2016) Lubrication fundamentals, 3rd edn revised and expanded. CRC Press, Taylor & Frencis Group, New York

Polder JW (1987a) Influence of geometrical parameters on the gear scuffing criterion—Part I. Gear Technol 28–34

Polder JW (1987b) Influence of geometrical parameters on the gear scuffing criterion—Part II. Gear Technol 19–27

Polder JW (1958) Relation between the PVT-equation and the flash temperature equation. In: Proceedings of international conference on gears, London, p 474

Rabinowicz E (1980) Wear coefficients- metals. In: Peterson MB, Winter WO (eds) Section IV of wear control handbook. The American Society of Mechanical Engineers, New York

Radzevich SP (2016) Dudley's handbook of practical gear design and manufacture, 3rd edn. CRC Press, Taylor & Francis Group, Boca Raton, Florida

Salomon G. (1976) Failure criteria in thin film lubrication—the irg program, *Wear*, 36 (1), January, pp. 1–6

Schipper DJ, de Gee AWJ (1995) On the transitions in the lubrication of concentrated contacts. J Tribol 117(2):250–254

Scott D (ed) (1979) Wear: treatise on materials science and technology, vol 13. Academic Press, New York

Shipley EE (1967) Gear failure: how to recognize them, what causes them, how to avoid them. Mach Des

Snidle RW, Evans HP, Alanou MP, Holmes MJA (2003) Understanding scuffing and micropitting of gears. In: Proceedings of the meeting on the control and reduction of wear in military platforms, Williamsburg, USA, 7–9 June and ADM 201869, RTO-MP-AVT-109, pp 14/1–14/18

Stachowiak GW, Batchelor AW (2014) Engineering tribology, 4th edn. Elsevier, Butterworth-Heinemann, Amsterdam

Tabor BJ (1981) Failure of thin film lubrication—an expedient for the characterization of lubricants. ASME J Lubr Technol 103(4):497–501

Tallian TE (1992) The failure Atlas for Hertz contact machine elements. Mech Eng 114(3):66

Tanaka F, Edwards SF (1992) Viscoelastic properties of physically crosslinked networks. 1. Transient network theory. Macromolecules 25(5):1516–1523

von Scherr-Thoss HC (1965) Die Entwicklung der Zahrad-Technik: Zahnformen und Tragfähigkeitsberechnung. Springer, Berlin, Heidelberg

Vullo V (2014) Circular cylinders and pressure vessels: stress analysis and design. Springer International Publishing Switzerland, Cham, Heidelber

Wink CH (2012) Predicted scuffing risk to spur and helical gears in commercial vehicle transmission. Gear Technol 82–86

Winter H, Oster P (1981) Beanspruchung der Zahnflanken unter EHD-Bedingungen. Konstruktion 33:421–434

Wydler R (1958) Application of non-dimensional parameters in gear tooth design. The Institution of Mechanical Engineers, Proceedings of the International Conference on Gearing, London, 23rd–25th September, Paper 4, pp 62–71

Xue J, Li W, Qin C (2014) The scuffing load capacity of involute spur gear systems based on dynamic loads and transient thermal elastohydrodynamic lubrication. Tribol Int 79:74–83

Zudans Z, Yen TC, Steigelmann WH (1965) Thermal stress techniques in nuclear industry, The Franklin Institute Research Laboratories. American Elsevier Publishing Company, Inc, Philadelphia, New York

Chapter 8
Scuffing Load Carrying Capacity of Cylindrical, Bevel and Hypoid Gears: Integral Temperature Method

Abstract In this chapter, the basic concepts on the calculation of the scuffing load carrying capacity of cylindrical, bevel and hypoid gears are discussed, but according to the integral temperature method. For all these gears, the fundamentals of Blok's theory are then converted in terms of integral temperature, which mainly depends on the weighted average of the local values of the flash temperature distribution along the path of contact. All influence parameters are reconsidered in terms of integral temperature instead of flash temperature, including the mean coefficient of friction. The permissible integral temperature is then determined and compared with the scuffing integral temperature in order to define the corresponding scuffing safety factor. A large part of the chapter is reserved for the procedures to calculate the scuffing load carrying capacity of the aforementioned types of gears in accordance with the ISO standards, highlighting when deemed necessary as the relationships used by the same ISO are founded on the related theoretical bases. Finally, a question is asked about the use of flash temperature method or integral temperature method for assesment of the scuffing load carrying capacity of the gears.

8.1 Introduction

In the introduction of the previous chapter (see Sect. 7.1), we provided general information on the scuffing phenomenon in gears, and we described the causes that give rise to it, and the possible remedies. We have here nothing to add to what we said at the time, for which we refer the reader to the descriptions and considerations that we already did in the aforementioned Sect. 7.1. The references concerning the topic discussed in this chapter are the same mentioned in the previous chapter. Therefore, they are not explicitly mentioned here, referring to Chap. 7 in this regard. Only the new references are mentioned.

In the previous chapter, we focused our attention on the calculation of scuffing load carrying capacity of cylindrical, bevel and hypoid gears using the flash temperature method. In this chapter, as we have already said in the previous chapter, we will focus our attention on the calculation of scuffing load capacity of cylindrical, bevel and hypoid gears, by means of the integral temperature method. However, it is necessary

© Springer Nature Switzerland AG 2020

V. Vullo, *Gears*, Springer Series in Solid and Structural Mechanics 11,
https://doi.org/10.1007/978-3-030-38632-0_8

to bear in mind that all what we will say applies to the warm scuffing, as methods for the cold scuffing are not yet available.

Once the scuffing damage is triggered, it can lead to a large degradation of the tooth flank surfaces and, consequently, to the increase of power loss, dynamic load, wear, and noise. It also can lead to tooth breakage, when the severity of the operating conditions is not immediately reduced. In the occurrence of scuffing due to an instantaneous overload, immediately followed by a reduction and redistribution of load, the tooth flank surfaces can self-be healed, with smoothing of the same surfaces at least to a certain extent. However, even so, the residual damage will continue to be the cause of increased power loss, noise and dynamic load.

The continuous variation of different quantities from which the scuffing damage depends, as well as the complexity of the chemical-physical phenomena involved, and of thermo-hydro-elastic processes in the instantaneous area of contact, determine a not avoidable scatter in the calculation relationships used for the assessments of probability of scuffing risk. This fact greatly influences the choice of safety factor for scuffing, whose value must be defined accurately, especially for gears required to operate at high tangential velocities.

The approach to the evaluation of the probability of scuffing with the *integral temperature criterion*, of interest here, is similar to that described in the previous chapter, concerning the *flash temperature criterion*. In fact, the integral temperature criterion is based on the assumption that scuffing is likely to occur when the mean value of the local contact temperatures along the path of contact is equal to or exceeds a corresponding critical value. Therefore, *mutatis mutandis*, the only differences between the two criteria are in the replacement of the integral temperature to flash temperature, as well as in correspondingly different value of the critical temperature of comparison.

With this second criterion, the integral temperature is defined as the sum of the bulk temperature and the weighted average of the local values of flash temperatures integrated along the path of contact. However, as we shall see in the following sections, the bulk temperature is evaluated in a different way than we have seen about the flash temperature criterion. Furthermore, the mean value of the flash temperature is approximated considering the mean values, along the path of contact, of all the quantities that come into play for its determination, such as the coefficient of friction, dynamic load, etc. The scuffing phenomenon is greatly influenced by the actual value of the bulk temperature as well as by the mean value of the flash temperature, which is obtained by integration of the flash temperature distribution along the path of contact. To take account of these possible different influences, a weighting factor is introduced.

The critical value of temperature with which to compare the integral temperature for the assessment of the probability of scuffing is obtained by gears that have scuffed in service or by means of test gears that operate with lubricants suitable for scuffing resistance. It is however to be noted that the approach using the integral temperature criterion is based on the rig testing results, obtained with gears operating at pitch line velocities less than 80 m/s. Therefore, when the relationships that will be described in the following sections are applied to gears operating at higher velocities, the uncertainties increase with the increase of the speed exceeding the range with experimental background. These uncertainties relate to the estimation of all the quantities involved, such as coefficient of friction, bulk temperature, allowable temperatures, etc.

The approach to the evaluation of the probability of scuffing (of course, the warm scuffing) with the integral temperature criterion presents particular aspects depending on whether we consider cylindrical gears, bevel gears or hypoid gears. Some influence factors, however, are common to these three families of gears, namely the mean coefficient of friction along the path of contact reduced to pitch point, μ_{mC}, run-in factor, X_E, thermal flash factor, X_M, and pressure angle factor, $X_{\alpha\beta}$. Regarding this last influence factor, it should be noted that, contrary to what ISO/TS 6336-21: 2017(E) does, it should be called simply *angle factor*, as it depends not only on the pressure angle, but also on the helix angle. For this reason, from now on, we will call this influence angle factor, rather than pressure angle factor.

The two sections that follow have, as a subject, these common influence factors, while subsequent sections cover the particular aspects, regarding the aforementioned three categories of gears, as well as the scuffing integral temperature.

8.2 Mean Coefficient of Friction

In Sect. 7.4, we defined a mean coefficient of friction, μ_m, to be used for the approach to the evaluation of the probability of scuffing with the flash temperature criterion. In the same section, we said that the actual coefficient of friction between the lubricated tooth flank surfaces is an instantaneous and local value along the path of contact. It depends on many factors, such as the characteristics of the lubricant oils, surface roughness, material gear characteristics including heat treatment, gear dimensions, tangential velocities, loads acting on the flank surfaces, status of surface irregularities such as those left by the machining operations of the gears, etc. We also highlighted that the variation of the instantaneous and local value of the coefficient of friction related to these influences, including the position of the meshing point along the path of contact, is difficult to calculate or measure, for which, in place of a local value, a mean value representative of the coefficient of friction is introduced for calculations.

However, for the approach with the integral temperature criterion, a mean value of the coefficient of friction along the path of contact, μ_{mC}, different from that given by Eq. (7.39) is introduced. It is a *sui generis* mean coefficient of friction reduced to pitch point, as it is expressed, through an approximate relationship, as a function of parameters related to the pitch point, C (although the local coefficient of friction is almost zero at the pitch point), as well as the dynamic viscosity (or absolute viscosity) of lubricant, η_{oil}, at oil sump or spray temperature, ϑ_{oil}. This approximate relationship, which is based on processing of experimental measurements from test gears, is given by:

$$\mu_{mC} = 0.045 X_R X_L \eta_{oil}^{-0.05} \left(\frac{w_{Bt} K_{B\gamma}}{v_{\Sigma C} \rho_{redC}} \right)^{0.2}. \tag{8.1}$$

This relationship, which is derived from test gears having center distance, $a \cong$ 100 mm, takes into account the size of the gear in a different way as the mean coefficient of friction of the flash temperature criterion. It should be used only for calculation of the coefficient of friction for thermal rating, while its use should be avoided for calculations not related to this specific field. Equation (8.1) is valid for values of the specific tooth load for scuffing, w_{Bt} (in N/mm), greater than or equal to 150 N/mm ($w_{Bt} \geq 150$ N/mm), and for values of the reference line velocity, v (in m/s), in the range ($1 \leq v \leq 50$) m/s.

It is to be noted that Eq. (8.1) should not be applied outside of the above ranges of operating conditions for which it was derived. Any extrapolation of Eq. (8.1) to the outside of the aforementioned ranges can lead to deviations between the actual coefficient of friction and its calculated value, and such deviations are larger the more the operating conditions move away the above ranges. In any case, for values of the specific tooth load less than 150 N/mm, the limiting value $w_{Bt} = 150$ N/mm must be used in Eq. (8.1). In the same way, for values of the reference line velocity higher than 50 m/s, the limiting value $v = 50$ m/s must be used to calculate the corresponding limiting value of the sum of tangential speeds at pitch point, $v_{\Sigma C}$, which appears in Eq. (8.1). Instead, for reference line velocities $v < 1$ m/s, higher coefficients of friction are to be expected.

The sum of the tangential speeds $v_{\Sigma C}$ at pitch point C is given by:

$$v_{\Sigma C} = 2v \tan \alpha_t' \cos \alpha_t. \tag{8.2}$$

where α_t' and α_t are respectively the transverse working pressure angle and transverse pressure angle.

The specific tooth load for scuffing, w_{Bt}, is given by:

$$w_{Bt} = K_A K_v K_{B\beta} K_{B\alpha} \frac{F_t}{b}. \tag{8.3}$$

This equation is evidently different from the corresponding Eq. (7.13) related to the flash temperature criterion, since it lacks the multiple-path factor, K_{mp}, while all the other quantities and factors coincide. Therefore, it is not the case here to dwell on them further. We must rather dwell on the other quantities and factors that appear in Eq. (8.1).

The relative radius of curvature, ρ_{redC} (in mm), at pitch point is given by:

$$\rho_{redC} = \frac{u}{(1+u)^2} a \frac{\sin \alpha_t'}{\cos \beta_b}, \tag{8.4}$$

where $u = z_2/z_1$ is the gear ratio, a is the center distance, and β_b is the helix angle at the base circle.

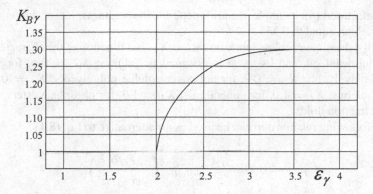

Fig. 8.1 Distribution curve of the helical load factor for scuffing as a function of the total contact ratio

The last quantity that appears in Eq. (8.1) is the absolute viscosity or dynamic viscosity, η_{oil} (in mPas), at oil temperature, ϑ_{oil}, i.e. the oil sump or spray temperature before reaching the mesh.

In Eq. (8.1), also three dimensionless factors appear, and precisely: the helical load factor for scuffing, $K_{B\gamma}$, roughness factor, X_R, and lubricant factor, X_L. The helical load factor for scuffing, $K_{B\gamma}$, takes into account the increasing friction for increasing total contact ratio, ε_γ. In the range $(2 < \varepsilon_\gamma < 3.5)$, $K_{B\gamma}$ is given by:

$$K_{B\gamma} = 1 + 0.2\sqrt{(\varepsilon_\gamma - 2)(5 - \varepsilon_\gamma)}, \tag{8.5}$$

while, for $\varepsilon_\gamma \leq 2$ and $\varepsilon_\gamma \geq 3.5$, we assume respectively $K_{B\gamma} = 1$ and $K_{B\gamma} = 1.3$. Figure 8.1 shows the distribution curve of the helical load factor for scuffing, $K_{B\gamma}$, as a function of the total contact ratio, ε_γ.

The roughness factor, X_R, is given by:

$$X_R = 2.2\left(\frac{Ra}{\rho_{redC}}\right)^{1/4}. \tag{8.6}$$

Therefore, contrary to the corresponding roughness factor that appear in Eq. (7.42), which is a function only of Ra, in this case the roughness factor is not only a function of Ra, but also of the already defined relative radius of curvature at pitch point, ρ_{redC}.

The arithmetic mean roughness, Ra (in μm), is given by:

$$Ra = \frac{1}{2}(Ra_1 + Ra_2), \tag{8.7}$$

where Ra_1 and Ra_2 are the values (in μm) of the arithmetic mean roughness of tooth flank surfaces of pinion and wheel, both measured on the new flank surfaces as manufactured (for example, the arithmetic mean roughness, Ra, of reference test gears is about equal to 0.35 μm).

Finally, the dimensionless lubricant factor, X_L, is a function of the characteristics of the lubricant oil, and its value is assumed as follows: $X_L = 0.6$, for water-soluble polyglycols; $X_L = 0.7$, for not water-soluble polyglycols; $X_L = 0.8$, for polyalfaolefins; $X_L \doteq 1.0$, for mineral oils; $X_L = 1.3$, for phosphate esters; $X_L = 1.5$, for traction fluids.

The following relationship can be used as an alternative to Eq. (8.1)

$$\mu_{mC} = 0.048 R_a^{0.25} X_L \eta_{oil}^{-0.05} \left(\frac{F_t/b}{v_{\Sigma C} \rho_{redC}} \right)^{0.2}. \tag{8.8}$$

which was obtained by processing test results with center distances included in the range ($91.5\,\text{mm} \leq a \leq 200\,\text{mm}$). It is to remember that the use of this second equation necessarily involves the consequent adjustment of the scuffing integral temperatures, ϑ_{intS}, read in the graph of Fig. 8.10 (or in other graphs mentioned in Sect. 8.9). All the quantities and factors that appear in this relationship are calculated as we said before. Only the choice of the value of the dimensionless lubricant factor, X_L, regarding the polyglycols, regardless of whether they are water-soluble or non-water-soluble polyglycols, is an exception; in fact, it is to be calculated using the relationship $X_L = 0.75(6/v_{\Sigma C})^{0.2}$. Instead, the lubricant factor values for the other lubricants do not change.

8.3 Other Common Influence Factors

The other common influence factors are the run-in factor, X_E, thermal flash factor, X_M, and angle factor, $X_{\alpha\beta}$.

As regards the first of these factors, it is first to note that the approach to the evaluation of the probability of scuffing according to the integral temperature criterion assumes that the gears should have undergone a good run-in. In practice, the scuffing failure often occurs during the first few hours of service. This happens, for example, in a full load test run, acceptance run condition of vessels and ships, or when a new gear set is inserted in a production machinery, and its gears are driven at full load, before a proper run-in.

Experimental evidence (Michaelis 1987) showed that the load carrying capacity for scuffing of newly manufactured tooth flank surfaces varies from (1/4) to (1/3) of that of the same tooth flank surfaces, after they have been subjected to a proper run-in.

The run-in factor, X_E, takes into account the aforementioned phenomena and experimental evidence. It is defined by the following relationship:

$$X_E = 1 + (1 - \varphi_E) \frac{30 Ra}{\rho_{redC}}, \tag{8.9}$$

where φ_E is the *run-in grade*. This is a dimensionless coefficient, to be chosen as follows: $\varphi_E = 0$, for newly manufactured gears, without run-in; $\varphi_E = 1$, for full rin-in gears (it is to be noted that, for carburized and ground gears, full run-in can be assumed, if the arithmetic mean roughness of the gear after the run-in is about equal to 60% of that of the same newly manufactured gear, i.e. if $Ra_{run-in} \cong 0.6 Ra_{new}$).

The second factor, i.e. the thermal flash factor or thermo-elastic factor, X_M, takes into account the influence of material characteristics of pinion and wheel on the flash temperature. Therefore, similarly to the flash temperature, it depends on the position of point considered on the path of contact, and thus it depends on the already defined parameter on the line of action at arbitrary point, Γ_y. This thermal flash factor is given by Eq. (7.25), so it is equal to that introduced for flash temperature criterion (see Sect. 7.3.3.2.1). Parameter Γ_y appearing in this last relationship is given by Eq. (7.43), for which this well is equal to that introduced for flash temperature approach. Even Eqs. (7.26) and (7.27), which characterize special cases of material combinations, continue to be valid.

The third factor is the angle factor, $X_{\alpha\beta}$, which is used to take into account the conversion of tangential velocity and load from reference circle to pitch circle. This is a common factor for both criteria, the flash temperature criterion and integral temperature criterion. However, in this case, the angle factor depends not only on the working pressure angle, α_t', and base helix angle, β_b (this is related to the helix angle, β, and normal pressure angle, α_n, by means of Eq. (8.52) of Vol. 1), but also on the transverse pressure angle, α_t. For Method A this factor can be written in the following form:

$$X_{\alpha\beta - A} = 1.22 \frac{\sqrt[4]{\sin \alpha_t'}\sqrt[4]{\cos \alpha_n}\sqrt[4]{\cos \beta}}{\sqrt{\cos \alpha_t'}\sqrt{\cos \alpha_t}}. \tag{8.10}$$

Table 8.1 shows the values of the angle factor, $X_{\alpha\beta}$, determined according to Method B, for a standard rack with normal pressure angle $\alpha_n = 20°$, in the typical range of variability of the standard transverse working pressure angle, α_t', and helix angle, β. These values are close to unit, for which the angle factor, $X_{\alpha\beta - B}$, of gears with normal pressure angle $\alpha_n = 20°$ can be assumed approximately equal to 1.

Table 8.1 Angle factor $X_{\alpha\beta - B}$

α_t'	$\beta = 0°$	$\beta = 10°$	$\beta = 20°$	$\beta = 30°$
19°	0.963	0.960	0.951	0.938
20°	0.978	0.975	0.966	0.952
21°	0.992	0.989	0.981	0.966
22°	1.007	1.004	0.995	0.981
23°	1.021	1.018	1.009	0.995
24°	1.035	1.032	1.023	1.008
25°	1.049	1.046	1.037	1.012

8.4　Scuffing Safety Factor and Permissible Integral Temperature

The experimental evidences on numerous gear drives affected by scuffing show that the assumptions underlying the calculation relationships for the evaluation of the probability of scuffing with the integral temperature criterion are not accurate. Therefore, given the uncertainties and inaccuracies related to these assumptions, it is necessary to introduce a scuffing safety factor, S_{intS}.

This scuffing safety factor must be greater than or at least equal to a recommended minimum value, S_{Smin}, which is the minimum required scuffing safety factor. This factor is defined in terms of *temperature safety*, i.e. as the ratio:

$$S_{intS} = \frac{\vartheta_{intS}}{\vartheta_{int}} \geq S_{Smin}, \tag{8.11}$$

where ϑ_{intS} is the *scuffing integral temperature* (also called *allowable integral temperature* or *scuffing integral temperature number*), and ϑ_{int} is the *integral temperature* or *integral temperature number*.

The recommended minimum value of the safety factor, S_{Smin}, must be chosen in relation to the accepted probability of scuffing risk. This probability is high for $S_{Smin} < 1$, low for $S_{Smin} > 2$, and moderate for safety factors in the range ($1 \leq S_{Smin} \leq 2$). This critical range with moderate scuffing risk is influenced by the actual gear operating conditions, such as the characteristics of gear materials and lubricant oils, gear accuracy grade including the tooth flank surface roughness, run-in effects, loads and load capacity of lubricating oil, load factors and their accurate knowledge, etc. It is to be remembered that, when the influence factors are known with reliable accuracy, a value $S_{Smin} = 1.5$ can be considered sufficient to avert the scuffing risk.

It is to be noted that the scuffing safety factor defined above in terms of temperature safety cannot be used to correlate the actual gear load, i.e. the actual gear torque, with the decisive contact temperature, in order to obtain the integral temperature number, ϑ_{int}, and the scuffing integral temperature number, ϑ_{intS}. Given the correlation between the integral temperature number and the actual gear load, the corresponding *load safety factor* against scuffing, S_{Sl}, can be expressed in approximate terms as:

$$S_{Sl} = \frac{w_{Btmax}}{w_{Bteff}} \simeq \frac{\vartheta_{intS} - \vartheta_{oil}}{\vartheta_{int} - \vartheta_{oil}}, \tag{8.12}$$

where w_{Btmax} and w_{Bteff} are respectively the maximum value and effective value of the specific tooth load for scuffing, w_{Bt}, and ϑ_{oil} is the oil temperature before reaching the mesh, which may be taken as equal to the oil sump or spray temperature. Obviously, temperatures appearing in the above equation must be measured with consistent units (Celsius degrees or Kelvin degrees).

The load safety factor, S_{Sl}, given by Eq. (8.12) is therefore the multiplication factor according to which the nominal gear load can be increased, before the scuffing occurs. More precisely, Eq. (8.12) may be derived from the correlation between the mean coefficient of friction, μ_m, and the flash temperature, Θ_{fl}.

The *permissible integral temperature*, ϑ_{intP}, is given by:

$$\vartheta_{intP} = \frac{\vartheta_{intS}}{S_{Smin}}. \tag{8.13}$$

Finally, it must be pointed out that the minimum required scuffing safety factor, S_{Smin}, is to be determined separately for each application.

8.5 Integral Temperature of Cylindrical Gears

The integral temperature of oil lubricated, involute cylindrical spur and helical gears, to be used for the assessment of the probability of warm scuffing, is given by:

$$\vartheta_{int} = \vartheta_M + C_2\vartheta_{flaint} \leq \vartheta_{intP}, \tag{8.14}$$

where ϑ_M is the bulk temperature, i.e. the temperature of the tooth flank surfaces immediately before they come into contact, C_2 is the weighting factor, obtained by experimental tests (for cylindrical spur and helical gears, $C_2 = 1.5$), and ϑ_{flaint} is the *mean flash temperature*, which is given by:

$$\vartheta_{flaint} = \vartheta_{flaE}X_\varepsilon, \tag{8.15}$$

where ϑ_{flaE} is the flash temperature at pinion tooth tip when load sharing is neglected, while X_ε is the *contact ratio factor*.

The flash temperature at pinion tooth tip, ϑ_{flaE}, is the calculated temperature increase over the tooth flank surface at a given point along the path of contact, due to combined effects of tooth geometry, friction, load, velocity and material characteristics under the operating conditions. It is given by the following relationship:

$$\vartheta_{flaE} = \mu_{mC}X_M X_{\alpha\beta}X_{BE}\frac{X_E}{X_Q X_{Ca}}\frac{\left(K_{B\gamma}w_{Bt}\right)^{3/4}v^{1/2}}{|a|^{1/4}}, \tag{8.16}$$

where $\mu_{mC}, X_M, X_{\alpha\beta}, X_E, K_{B\gamma}$ and w_{Bt} are respectively the mean coefficient of friction, thermal flash factor, angle factor, run-in factor, helical load factor for scuffing, and specific tooth load for scuffing (note that in previous sections we have already described the methods of calculation of all these quantities and factors). Quantities v and a are respectively the already introduced reference line velocity and center distance, while X_{BE}, X_Q and X_{Ca} are respectively the *geometry factor at tooth tip of pinion*, *approach factor*, and *tip relief factor*, of which we have yet to describe the methods of calculation.

It is to be pointed out that, for double helical gears, it is assumed that the total tangential force is equally distributed between the two helices, as if the double helical gear was made up of two parallel single helical gear wheels. Furthermore, the influence of any axial external force applied to the gear must be evaluated separately, with respect to that exerted by the internal axial forces.

It is then to be noted that the aforementioned equations are valid for external and internal cylindrical gear pairs, which are conjugate to the standard basic rack according to ISO 53:1998. However, for internal gear pairs, it should be remembered that the determination of the geometric factor at tooth tip of pinion, given by Eq. (8.18), must be done in compliance with the sign conventions we already defined elsewhere, and recalled whenever necessary. The same equations can still be considered valid for gear pairs defined with respect to other shapes of basic rack, whose transverse contact ratio ε_α is less or equal to 2.5 ($\varepsilon_\alpha \leq 2.5$).

The bulk temperature, ϑ_M, appearing in Eq. (8.14) can be determined by the thermal balance of the gear unit under consideration, using the various methods mentioned in Sect. 7.3.2. In the same section we have described the several heat sources that influence the thermal balance, and therefore the value of the bulk temperature.

For the determination of the bulk temperature, Method A and Method C are used, while Method B is not used for the integral temperature. With Method A, the bulk temperature, ϑ_{M-A}, is considered as a mean value of a temperature distribution over the face width, and is determined by experimental measurements or calculated by a theoretical analysis with thermal network methods, when all data are known in terms of power losses and heat transfer conditions. With Method C, an approximate value of the bulk temperature, ϑ_{M-C}, is determined as the sum of the oil temperature, ϑ_{oil} (this temperature is assumed equal to the oil sump or spray temperature) and another constant term, consisting of the mean value derived from the flash temperature over the path of contact multiplied by suitable corrective factors. Therefore, we have:

$$\vartheta_{M-C} = \vartheta_{oil} + C_1 X_{mp} \vartheta_{flaint} X_S, \qquad (8.17)$$

where C_1 is a weighting factor, which is a dimensionless constant that takes into account the heat transfer conditions (according to test results, $C_1 = 0.7$), X_{mp} is the multiple mating pinion factor, given by Eq. (7.7), ϑ_{flaint} is the mean flash temperature, and X_S is the dimensionless lubrication system factor, to be assumed equal to: $X_S = 1.2$, for spray lubrication; $X_S = 1.0$, for dip lubrication; $X_S = 0.2$, for gears submerged in oil.

The mean coefficient of friction reduced to pitch point, μ_{mC}, thermal flash factor, X_M, angle factor, $X_{\alpha\beta}$, and run-in factor, X_E, appearing in Eq. (8.16) are determined using the procedures described in Sects. 8.2 and 8.3.

The geometry factor at pinion tooth tip, X_{BE}, that appears in Eq. (8.16) takes into account the Hertzian stress and sliding velocity at the pinion tooth tip. It is a function of the gear ratio $u = z_2/z_1$, and the radius of curvature at tip of the pinion, ρ_{E1}, and radius of curvature at tip of the wheel, ρ_{E2}, according to the following relationship:

$$X_{BE} = 0.51 \frac{(\rho_{E1})^{1/2} - (\rho_{E2}/u)^{1/2}}{(\rho_{E1}|\rho_{E2}|)^{1/4}} \sqrt{(u+1)\frac{|z_2|}{z_2}}, \qquad (8.18)$$

where ρ_{E1} and ρ_{E2} are given by:

$$\rho_{E1} = 0.5\sqrt{d_{a1}^2 - d_{b1}^2} \qquad (8.19)$$

$$\rho_{E2} = a \sin\alpha_t' - \rho_{E1}. \qquad (8.20)$$

It is to remember that, for internal gears, according to the aforementioned sign convention, the gear ratio, u, center distance, a, and number of teeth of the wheel, z_2, have to be introduced as negative values.

The approach factor, X_Q, in Eq. (8.16) takes into account the impact loads at the ingoing tooth mesh in areas of high sliding. This impact occurs at tooth tip of the driven gear wheel. This factor is a function of the quotient of the approach contact ratio, ε_f, divided by the recess contact ratio, ε_a, as Fig. 8.2 shows. In the range $[1.5 < (\varepsilon_f/\varepsilon_a) < 3]$, X_Q is given by:

$$X_Q = 1.40 - \frac{4\varepsilon_f}{15\varepsilon_a}, \qquad (8.21)$$

while, for $(\varepsilon_f/\varepsilon_a) \leq 1.5$ and $(\varepsilon_f/\varepsilon_a) \geq 3$, we have respectively $X_Q = 1.00$ and $X_Q = 0.60$. It is to be noted that, when the pinion is driving, we have $\varepsilon_f = \varepsilon_2$ and $\varepsilon_a = \varepsilon_1$, while, when the pinion is driven, we have $\varepsilon_f = \varepsilon_1$ and $\varepsilon_a = \varepsilon_2$. The addendum contact ratio of the pinion, ε_1, and addendum contact ratio of the wheel, ε_2, are given respectively by:

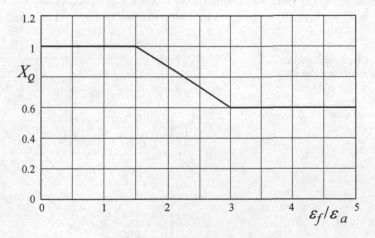

Fig. 8.2 Distribution of the approach factor as a function of the ratio $(\varepsilon_f/\varepsilon_a)$

$$\varepsilon_1 = \frac{z_1}{2\pi}\left\{\sqrt{\left[\left(\frac{d_{a1}}{d_{b1}}\right)^2 - 1\right]} - \tan\alpha_t'\right\}, \tag{8.22}$$

$$\varepsilon_2 = \frac{|z_2|}{2\pi}\left\{\sqrt{\left[\left(\frac{d_{a2}}{d_{b2}}\right)^2 - 1\right]} - \tan\alpha_t'\right\}. \tag{8.23}$$

In these relationships, the tip diameter, d_a, has to be substituted by the effective tip diameter d_{Na} at which the recess is starting, when tooth tips are rounded or chamfered.

High impact loads at tooth tips also occur in areas of relatively high sliding, due to the elastic deformations of loaded teeth. The tip relief factor, X_{Ca}, takes into account the influence of profile modifications on these impact loads. In fact, this factor is a relative factor, because it depends on the quotient of the actual amount of tip relief, C_a, divided by the effective tip relief due to elastic deformation, C_{eff}, as Fig. 8.3 shows. The distribution curves in this figure can be approximated by the following relationship obtained processing experimental data of tests carried out by Lechner (1966) and Ishikawa et al. (1972):

$$X_{Ca} = 1 + \left[0.06 + 0.18\left(\frac{C_a}{C_{eff}}\right)\right]\varepsilon_{max} + \left[0.02 + 0.69\left(\frac{C_a}{C_{eff}}\right)\right]\varepsilon_{max}^2, \tag{8.24}$$

where ε_{max} is the maximum value between the aforementioned quantities ε_1 and ε_2.

The value of the actual tip relief, C_a, to be introduced into Eq. (8.24), depends on the nominal tip relief of pinion and wheel, C_{a1} and C_{a2}, effective tip relief, C_{eff}, direction of the power flow, and ratio $(\varepsilon_1/\varepsilon_2)$ between the addendum contact ratios of pinion and wheel. In the cases of pinion driving and $\varepsilon_1 > 1.5\varepsilon_2$, and pinion driven and $\varepsilon_1 > (2/3)\varepsilon_2$, we assume $C_a = C_{a1}$ for $C_{a1} \leq C_{eff}$, and $C_a = C_{eff}$ for $C_{a1} > C_{eff}$. Instead, in the cases of pinion driving and $\varepsilon_1 \leq 1.5\varepsilon_2$, and pinion

Fig. 8.3 Distribution curves of the tip relief factor as a function of ε_{max} and ratio (C_a/C_{eff}), from experimental data

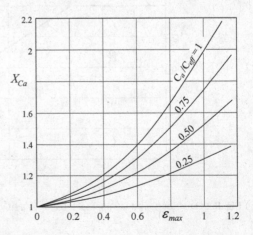

driven and $\varepsilon_1 \leq (2/3)\varepsilon_2$, we assume $C_a = C_{a2}$ for $C_{a2} \leq C_{eff}$, and $C_a = C_{eff}$ for $C_{a2} > C_{eff}$.

The effective tip relief, C_{eff}, i.e. the amount of tip relief that compensates for the elastic deformations of teeth in the single pair contact area, is given by:

$$C_{eff} = \frac{K_A F_t}{bc'} \tag{8.25}$$

$$C_{eff} = \frac{K_A F_t}{bc_\gamma}, \tag{8.26}$$

which are respectively valid for cylindrical spur gears and cylindrical helical gears. It is to be noted that, if the face width b_1 of the pinion is different from the face width b_2 of the wheel, the smaller value between b_1 and b_2 is the determinant value. It is also to be noted that the tip relief as described above applies to gears with accuracy grade 6 or better, while for gears with accuracy grade 7 or coarser, the tip relief factor, X_{Ca}, is to be set equal to 1.

The contact ratio factor, X_ε, takes into account the influence exerted by the load sharing between the meshing tooth pairs on the mean flash temperature, ϑ_{flaint}. For this purpose, X_ε converts the flash temperature value at pinion tooth tip, ϑ_{flaE}, when load sharing is neglected, to a mean value of the flash temperature along the path of contact. This contact ratio factor can be expressed in terms of addendum contact ratios, ε_1 and ε_2, as well as of their sum, $\varepsilon_\alpha = (\varepsilon_1 + \varepsilon_2)$.

To derive the relationships that express X_ε, the assumption of linear distributions of flash temperature in all the regions in which the marked points A, B, C, D and E divide the path of contact is made. However, it is to be noted that it is unlikely that the possible errors, related to this assumption, exceeds 5%, and that in any case they will always be on the safe side. The distributions of the load and temperature of contact along the path of contact are different depending on whether the transverse contact ratio, ε_α, is less than 1 ($\varepsilon_\alpha < 1$) or it is included in the range ($1 \leq \varepsilon_\alpha < 2$) or in the range ($2 \leq \varepsilon_\alpha < 3$).

For $\varepsilon_\alpha < 1$, and therefore $\varepsilon_1 < 1$ and $\varepsilon_2 < 1$, X_ε is given by:

$$X_\varepsilon = \frac{1}{2\varepsilon_1\varepsilon_\alpha}(\varepsilon_1^2 + \varepsilon_2^2). \tag{8.27}$$

In the range ($1 \leq \varepsilon_\alpha < 2$), the load distribution along the path of contact is the one shown in Fig. 7.7b, while the corresponding temperature distributions are those shown in Fig. 8.4. The relationships that express X_ε in this range are different depending on whether one of the following three cases occurs: ε_1 and ε_2 are both less than 1; ε_1 is greater than or equal to 1, and ε_2 is less than 1; ε_1 is less than 1, and ε_2 is greater than or equal to 1.

For $\varepsilon_1 < 1$ and $\varepsilon_2 < 1$ (in this case, the working pitch point is located within the single tooth contact region, as Fig. 8.4 shows), we have:

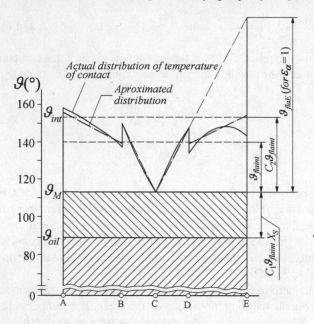

Fig. 8.4 Temperature distributions for $(1 \leq \varepsilon_\alpha < 2)$

$$X_\varepsilon = \frac{1}{2\varepsilon_1\varepsilon_\alpha}\left[0.70\left(\varepsilon_1^2 + \varepsilon_2^2\right) - 0.22\varepsilon_\alpha + 0.52 - 0.60\varepsilon_1\varepsilon_2\right]. \tag{8.28}$$

For $\varepsilon_1 \geq 1$ and $\varepsilon_2 < 1$ (in this case, the operating pitch point is located within the double tooth pair contact region), we have:

$$X_\varepsilon = \frac{1}{2\varepsilon_1\varepsilon_\alpha}\left(0.18\varepsilon_1^2 + 0.70\varepsilon_2^2 + 0.82\varepsilon_1 - 0.52\varepsilon_2 - 0.30\varepsilon_1\varepsilon_2\right). \tag{8.29}$$

For $\varepsilon_1 < 1$ and $\varepsilon_2 \geq 1$ (in this case, the operating pitch point is also located within the double tooth pair contact region), we have:

$$X_\varepsilon = \frac{1}{2\varepsilon_1\varepsilon_\alpha}\left(0.70\varepsilon_1^2 + 0.18\varepsilon_2^2 - 0.52\varepsilon_1 + 0.82\varepsilon_2 - 0.30\varepsilon_1\varepsilon_2\right). \tag{8.30}$$

In the range $(2 \leq \varepsilon_\alpha < 3)$, the actual load distribution along the path of contact is the one shown in Fig. 8.5, with continuous line. In the same figure, the dashed line indicates the approximate load distribution. The corresponding temperature distribution is that shown in Fig. 8.6. The relationships that express X_ε in this range are different depending on whether one of the two following cases occurs: ε_1 is greater than or equal to ε_2 ($\varepsilon_1 \geq \varepsilon_2$); ε_1 is less than ε_2 ($\varepsilon_1 < \varepsilon_2$). In these two cases, we have respectively:

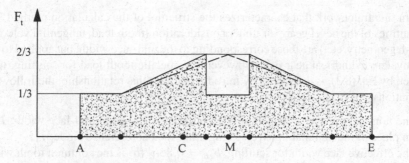

Fig. 8.5 Actual and approximate load distribution along the path of contact, for $(2 \leq \varepsilon_\alpha < 3)$

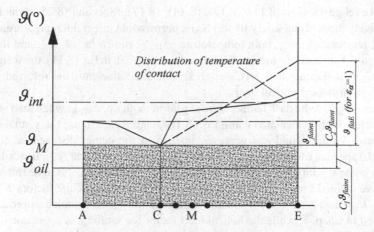

Fig. 8.6 Temperature distribution along the path of contact, for $(2 \leq \varepsilon_\alpha < 3)$

$$X_\varepsilon = \frac{1}{2\varepsilon_1\varepsilon_\alpha}\left(0.44\varepsilon_1^2 + 0.59\varepsilon_2^2 + 0.30\varepsilon_1 - 0.30\varepsilon_2 - 0.15\varepsilon_1\varepsilon_2\right) \qquad (8.31)$$

$$X_\varepsilon = \frac{1}{2\varepsilon_1\varepsilon_\alpha}\left(0.59\varepsilon_1^2 + 0.44\varepsilon_2^2 - 0.30\varepsilon_1 + 0.30\varepsilon_2 - 0.15\varepsilon_1\varepsilon_2\right). \qquad (8.32)$$

8.6 Integral Temperature of Bevel Gears

The calculation procedure of the integral temperature of bevel gears is completely analogous to that described in the previous section for cylindrical gears. In fact, it uses the same relationships quoted in that section and in the previous ones, with the only variation of approximating the bevel gear pair to be analyzed by means of its virtual equivalent cylindrical gear pair at the diameter at mid-face width, d_m.

In this framework that characterizes the structure of the calculation method, the quantities of the bevel gear pair under consideration (tooth load, tangential velocity, tooth geometry, etc.) are those corresponding to the mid-face width, but applied to the equivalent cylindrical gear pair. However, the specific tooth load for scuffing, w_{Bt}, given by Eq. (8.3), is calculated by introducing, into this relationship, the following variations:

- the nominal transverse tangential load at reference cone at mid-face width, F_{mt}, in place of the nominal transverse tangential load at reference circle, F_t;
- the effective face width for scuffing, $b_{eB} = 0.85b_2$ (b_2 is the common tooth width of pinion and wheel), in place of the face width, b (the latter is to be understood as the smaller value of face widths of pinion and wheel).

For bevel gears, Eqs. (8.11), (8.13), (8.14), (8.17), (8.9) and (8.25), which give respectively the scuffing safety factor, S_{intS}, permissible integral temperature, ϑ_{intP}, integral temperature, ϑ_{int}, bulk temperature, ϑ_{M-C}, run-in factor, X_E, and thermal flash factor, X_M, continue to be valid. It is to be noted that, in Eq. (8.14), the weighting factor C_2 is still equal to 1.5 ($C_2 = 1.5$), but this value must be referred to the equivalent virtual cylindrical gear.

For bevel gears, the flash temperature at pinion tooth tip, ϑ_{flaE}, when load sharing is neglected, is still calculated using Eq. (8.16), taking care to use the virtual center distance of virtual cylindrical gear, a_v, instead of the center distance, a, and the tangential speed at reference cone at mid-face width of bevel gear, v_{mt}, instead of the reference line velocity, v. Furthermore, in the same relationship, it is to be introduced the above defined specific tooth load for scuffing, w_{Bt}. Finally, the factors K_A, K_v, $K_{B\beta} = K_{H\beta}$, and $K_{B\alpha} = K_{H\alpha}$ should be determined by the calculation procedures described in Chap. 1, while the helical load factor for scuffing, $K_{B\gamma}$, is taken equal to 1 ($K_{B\gamma} = 1$).

The mean coefficient of friction, μ_{mC}, for bevel gears is also calculated by means of Eq. (8.1), taking care to introduce in it the above defined specific tooth load for scuffing, w_{Bt}. Even here, the helical load factor for scuffing, $K_{B\gamma}$, is taken equal to 1 ($K_{B\gamma} = 1$). Moreover, for the usual bevel gear design conditions for which the profile shift coefficients of pinion and wheel have equal value, but opposite sign ($x_1 = -x_2$), whereby the transverse working pressure angle, α_t', is equal to the transverse pressure angle of virtual cylindrical gear, α_{vt}, the sum of tangential speeds at pitch point to be introduced into Eq. (8.1) is given by:

$$v_{\Sigma C} = 2v_{mt} \sin \alpha_{vt}. \tag{8.33}$$

Even for bevel gears with normal pressure angle $\alpha_n = 20°$, the angle factor, $X_{\alpha\beta}$, can be determined using Method B, according to which we get the approximate value $X_{\alpha\beta-B} = 1$. Instead, if we use Method A, for the above-mentioned conditions of bevel gear design ($x_1 = -x_2$, and $\alpha_t' = \alpha_{vt}$), we get:

$$X_{\alpha\beta-A} = 1.22 \frac{\sqrt[4]{\sin \alpha_n}}{\sqrt[4]{\cos^3 \alpha_{vt}}}. \tag{8.34}$$

The geometry factor at tip of pinion, X_{BE}, for bevel gears is calculated using Eqs. (8.18) to (8.20), in which the following substitutions must be made: the gear ratio of virtual cylindrical gear, u_v, instead of gear ratio, u; the tip diameter of virtual cylindrical pinion, d_{va1}, instead of pinion tip diameter, d_{a1}; the base diameter of virtual cylindrical pinion, d_{vb1}, instead of pinion base diameter, d_{b1}; the transverse pressure angle of virtual cylindrical gear, α_{vt}, instead of transverse working pressure angle, α_t'.

The approach factor, X_Q, for bevel gears is calculated using Eqs. (8.21) to (8.23), while taking account of considerations made in Sect. 8.5 in relation to two possible cases of driving pinion and driven pinion. In these relationships as well as for the choice of the recess contact ratio, ε_a, and approach contact ratio, ε_f, with respect to the addendum contact ratios of the pinion and wheel, ε_1 and ε_2, the following substitutions must be made for the two cases above: the tip contact ratio of virtual cylindrical pinion, ε_{v1}, instead of addendum contact ratio of the pinion, ε_1; the tip contact ratio of virtual cylindrical wheel, ε_{v2}, instead of addendum contact ratio of the wheel, ε_2; the tip diameters of virtual cylindrical pinion and wheel, $d_{va1,2}$, instead of tip diameters of pinion and wheel, $d_{a1,2}$; the base diameters of virtual cylindrical pinion and wheel, $d_{vb1,2}$, instead of base diameters of pinion and wheel, $d_{b1,2}$; the transverse pressure angle of virtual cylindrical gear, α_{vt}, instead of transverse working pressure angle, α_t'; the number of teeth of virtual cylindrical gear, $z_{v1,2}$, instead of number of teeth, $z_{1,2}$.

The tip relief factor, X_{Ca}, for bevel gears is calculated using Eq. (8.24), with the following substitution: ε_{vmax} instead of ε_{max}, where ε_{vmax} is the maximum value between ε_{v1} and ε_{v2} defined above. In approximate terms, it is then assumed that full-load contact pattern will spread just to tip, without concentration. Optimum values of tip and root relief correspond to these operating conditions, for which $C_a = C_{eff}$. Therefore, Eq. (8.24) is simplified, since the ratio (C_a/C_{eff}) is approximately equal to 1.

Finally, the contact ratio factor, X_ε, for bevel gears is calculated using Eqs. (8.27) to (8.32), in which the following substitutions must be made: the transverse contact ratio of virtual cylindrical gear, ε_{va}, instead of contact ratio, ε_α; the tip contact ratio of virtual cylindrical pinion, ε_{v1}, instead of addendum contact ratio of the pinion, ε_1; the tip contact ratio of virtual cylindrical wheel, ε_{v2}, instead of addendum contact ratio of the wheel, ε_2.

8.7 Integral Temperature of Hypoid Gears

Even the calculation procedure of the integral temperature of hypoid gears is similar to that described in Sect. 8.5 for cylindrical gears, but with some significant variations compared to those outlined in the previous section for bevel gears. This is because the sliding and rolling motions of hypoid gears differ from those of the bevel gears, and not a little. Therefore, to detect these differences, the calculation of scuffing load carrying capacity of hypoid gears (and so the calculation of the power losses) requires the replacement of the actual hypoid gear pair with an equivalent crossed axes helical gear pair, able to simulate the same rolling and especially sliding conditions.

The use of this virtual helical gear pair with crossed axes guarantees approximate results. These results, though approximate, are however sufficiently reliable and have design value, as the experimental evidences have proved. For the calculation of these virtual crossed axes helical gears, which ensure, at point of contact, the same sliding conditions of the actual hypoid gears (thus, from the sliding point of view, the two gear pairs are equivalent), we refer the reader to the next section. Here we focus our attention on the calculation procedure of the integral temperature of the hypoid gear pairs.

For hypoid gears, Eqs. (8.11), (8.13) and (8.17), which give respectively the scuffing safety factor, S_{intS}, permissible integral temperature, ϑ_{intP}, and bulk temperature, ϑ_{M-C}, continue to be also valid.

The integral temperature of oil lubricated hypoid gears to be used for the assessment of the probability of warm scuffing is given by the following relationship:

$$\vartheta_{int} = \vartheta_M + C_{2H}\vartheta_{flainth} \leq \vartheta_{intP}. \tag{8.35}$$

This relationship differs from the corresponding Eq. (8.14), which is valid for cylindrical gears, for the following two specific substitutions: the weighting factor, C_{2H}, to be taken equal to 1.8 according to experimental results, instead of the corresponding weighting factor, C_2, for cylindrical gears; the mean flash temperature of hypoid gears, $\vartheta_{flainth}$, instead of the mean flash temperature of cylindrical gears, ϑ_{flaint}, given by (note that subscript h refers to hypoid gears):

$$\vartheta_{flainth} = 110\,\mu_{mC}\frac{X_E X_G X_\varepsilon}{X_Q X_{Ca}}\sqrt{F_n K_A K_{B\beta} v_{t1}}, \tag{8.36}$$

where

$$F_n = \frac{2 \cdot 10^3 T_1}{\cos\alpha_{mn}\cos\beta_{m1}d_{m1}}, \tag{8.37}$$

$$K_{B\beta} = 1.5 K_{B\beta be}. \tag{8.38}$$

In these last three relationships, the meaning of symbols already introduced remains unchanged, while new symbols or already known symbols, but to be calculated specifically for hypoid gears, represent: $K_{B\beta}$, the face load factor for scuffing, generally assumed equal to the face load factor for contact stress, i.e. $K_{B\beta} = K_{H\beta}$; X_G, the geometry factor of hypoid gears; $v_{t1,2}$, the tangential velocity of pinion and wheel of hypoid gear pair; α_{mn}, the normal pressure angle at mid-face width of hypoid gear; β_{m1}, the helix angle at reference cone at mid-face width of hypoid pinion; $K_{B\beta be}$, the bearing factor that, in this case, is assumed equal to the mounting factor, $K_{B\beta-be}$, as defined by ISO 10300-1:2014 (see Sect. 4.6.2).

The mean coefficient of friction, μ_{mC}, for hypoid gears is also calculated by means of Eq. (8.1), with the following substitutions: the relative radius of curvature at pitch point in normal section, ρ_{Cn}, instead of relative radius of curvature at pitch point, ρ_{redC}; the specific tooth load for scuffing, w_{Bt}, given by Eq. (8.3), must be calculated by introducing, into this relationship, the normal tooth load, F_n, instead of the nominal tangential load at reference circle, F_t; the ratio, $b_{eB} / \cos \beta_{b2}$ (here also we assume $b_{eB} = 0.85b_2$ as in Sect. 8.6, while β_{b2} is the helix angle at base circle of wheel), instead of face width, b; the helical load factor for scuffing, $K_{B\gamma}$, is assumed equal to 1, while the product $K_{B\beta} K_{B\alpha}$ is assumed equal to 2 (it is to be noted that this last approximation can be used only for calculation of μ_{mC}); the roughness factor, X_R, given by Eq. (8.6), must be calculated by introducing, in this equation, ρ_{Cn} instead of ρ_{redC}.

The run-in factor, X_E, is to be calculated by means of Eq. (8.9), also here with the aforementioned substitution of the relative radius of curvature at pitch point in normal section, ρ_{Cn}, instead of relative radius of curvature at pitch point in transverse section, ρ_{redC}.

The geometry factor of hypoid gears, X_G, takes into account the mean contact length along the path of contact, and the mean Hertzian stress. In approximate terms, it may be determined using the following relationship, where the values of the quantities ρ_{Cn} and L at pitch point appear:

$$X_G = \frac{\left(\dfrac{\sin \Sigma}{\cos \beta_{s2}}\right)\left(\dfrac{1}{\rho_{Cn}}\right)^{1/2}}{(L \sin \beta_{s1})^{1/2} + (L \cos \beta_{s1} \tan \beta_{s2})^{1/2}}. \tag{8.39}$$

In this relationship, Σ is the shaft angle of actual hypoid gear pair (see next section), β_{s1} and β_{s2} are the helix angles of pinion and wheel of the virtual crossed axes helical gear pair, ρ_{Cn} is the already defined relative radius of curvature at pitch point in normal section, and L is the dimensionless contact parameter, given by:

$$L = \frac{2}{3}\eta\xi^2, \tag{8.40}$$

where η and ξ are two dimensionless *Hertzian auxiliary coefficients* [see, for a general discussion, Timoshenko and Goodier (1951) and Juvinall (1967) and, for a specific discussion, Grekoussis and Michailidis (1981)]. In turn, these two coefficients are a function of $\cos \vartheta$ (ϑ, in degrees, is the *Hertzian auxiliary angle*), given by:

$$\cos \vartheta = \rho_{Cn}\sqrt{\frac{1}{\rho_{n1}^2} + \frac{1}{\rho_{n2}^2} + \frac{2 \cos 2\varphi}{\rho_{n1}\rho_{n2}}}, \tag{8.41}$$

where $\rho_{n1,2}$ are the radii of curvature at pitch point in normal section of pinion and wheel, and φ is the shaft angle of virtual crossed axes helical gears (see next section).

The expressions to calculate η and ξ are different depending on whether $\cos \vartheta$ is included in the range $(0 \leq \cos \vartheta < 0.949)$ or in the range $(0.949 \leq \cos \vartheta < 1)$. For $(0 \leq \cos \vartheta < 0.949)$, we have:

$$\ln \eta = \frac{\ln(1 - \cos \vartheta)}{1.525 - 0.860 \ln(1 - \cos \vartheta) - 0.0993[\ln(1 - \cos \vartheta)]^2} \tag{8.42}$$

$$\ln \xi = \frac{\ln(1 - \cos \vartheta)}{-1.530 + 0.333 \ln(1 - \cos \vartheta) + 0.0467[\ln(1 - \cos \vartheta)]^2}, \tag{8.43}$$

while, for $(0.949 \leq \cos \vartheta < 1)$, we have:

$$\ln \eta = \left\{ -0.333 + 0.2037 \ln(1 - \cos \vartheta) + 0.0012[\ln(1 - \cos \vartheta)]^2 \right\} \tag{8.44}$$

$$\ln \xi = \left\{ 0.4567 - 0.4446 \ln(1 - \cos \vartheta) + 0.1238[\ln(1 - \cos \vartheta)]^2 \right\}^{1/2}. \tag{8.45}$$

The values of the Hertzian auxiliary coefficients η and ξ can be read from Fig. 8.7 as a function of $\cos \vartheta$ or ϑ.

The approach factor, X_Q, for hypoid gears is also calculated using Eqs. (8.21) to (8.23), while taking account of considerations made in Sect. 8.5 in relation to the two possible cases of driving pinion and driven pinion. In these relationships as well as about the choice of the recess contact ratio, ε_a, and approach contact ratio, ε_f, and their correlation with the addendum contact ratios of the pinion and wheel, ε_1 and ε_2, for the two aforementioned cases, the following substitutions must be made:

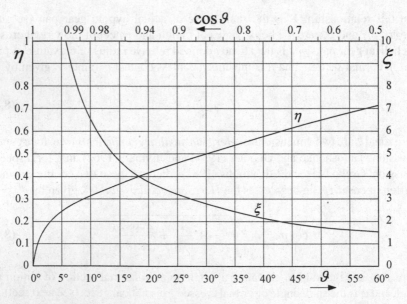

Fig. 8.7 Hertzian auxiliary coefficients, η and ξ, as a function of $\cos \vartheta$ or ϑ

the contact ratio in normal section of pinion of virtual crossed axes helical gear, ε_{n1}, instead of addendum contact ratio of the pinion, ε_1; the contact ratio in normal section of wheel of virtual crossed axes helical wheel, ε_{n2}, instead of addendum contact ratio of the wheel, ε_2.

The tip relief factor, X_{Ca}, for hypoid gears is calculated using Eq. (8.24), with the following substitution: ε_{nmax} instead of ε_{max}, where ε_{nmax} is the maximum value between ε_{n1} and ε_{n2} defined above. When adequate values of tip and root relief are used, corresponding to the operating conditions outlined in the previous section about the bevel gears, we have $C_a = C_{eff}$. Therefore, even here Eq. (8.24) is simplified, since the ratio (C_a/C_{eff}) is approximately equal to 1.

Finally, the contact ratio factor, X_ε, for hypoid gears is calculated using the following relationship:

$$X_\varepsilon = \frac{1}{\sqrt{\varepsilon_n}}\left[1 + 0.5g^*\left(\frac{v_{g\gamma 1}}{v_{gs} - 1}\right)\right], \tag{8.46}$$

where ε_n is the contact ratio in normal section of virtual crossed axes helical gear, $v_{g\gamma 1}$ (in m/s) is the maximum sliding velocity at tip of pinion, v_{gs} (in m/s) is the sliding velocity at pitch point, and g^* is the dimensionless sliding factor, given by:

$$g^* = \frac{g_{an1}^2 + g_{an2}^2}{g_{an1}^2 + g_{an1}g_{an2}}, \tag{8.47}$$

where g_{an1} (in mm) is the recess path of contact of pinion, and g_{an2} (in mm) is the recess path of contact of wheel.

For gear pairs with about the same length of recess paths, g_{an1} is approximately equal to g_{an2} ($g_{an1} \cong g_{an2}$), for which the sliding factor, g^*, is approximately equal to 1 ($g^* \cong 1$). It is to be noted that $g_{an1} = g_{fn2}$, and $g_{an2} = g_{fn1}$, where g_{fn1} (in mm) is the approach path of contact of pinion, and g_{fn2} (in mm) is the approach path of contact of wheel.

8.8 Virtual Crossed Axes Helical Gears

In the previous section, we have already provided that the calculation of the integral temperature of hypoid gears requires the preliminary determination of the virtual crossed axes helical gear pair, equivalent to the actual hypoid gear pair from the point of view of scuffing load carrying capacity. To this end, we recall here the geometrical relationships to convert a hypoid gear pair into its equivalent (or virtual) crossed axes helical gear pair. These conversion relationships are based on the geometry and operating conditions at mid-face width of actual gear pair, as Fig. 8.8 shows (see also Niemann and Winter 1983).

The normal pressure angle of virtual crossed axes helical gear pair, α_{sn}, is equal to normal pressure angle at mid-face width of actual hypoid gear pair, α_{mn}, that is

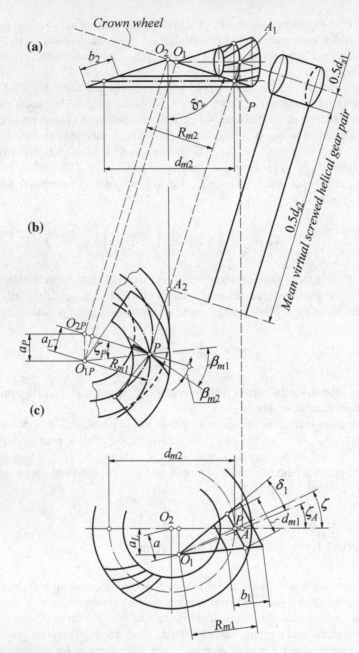

Fig. 8.8 Geometry of a hypoid gear pair and its equivalent crossed axes helical gear pair: **a** front or vertical view of the hypoid gear pair; **b** cones developed on reference plane of crown wheel; **c** view from above or plane view

$\alpha_{sn} = \alpha_{mn}$, while the shaft angle, also called crossing angle of virtual crossed axes helical gear pair, Σ, is given by:

$$\Sigma = \beta_{m1} \pm \beta_{m2}, \tag{8.48}$$

where β_{m1} and β_{m2} are the helix angles at reference cone at mid-face width of hypoid pinion and hypoid wheel respectively, and the plus and minus signs are respectively for helices that have the same directions or opposite directions. As it is well known, the most usual case is that of helix angles with opposite directions. Therefore, in Eq. (8.48), the minus sign is to be considered.

The transverse pressure angle of virtual crossed axes helical gear, α_{sti}, and the helix angle at base circle, β_{bi}, are given respectively by the following two relationships (it should be noted that from now on the subscript i, with $i = (1, 2)$, refers to pinion 1 and wheel 2):

$$\tan \alpha_{sti} = \frac{\tan \alpha_{sn}}{\cos \beta_{si}}, \tag{8.49}$$

$$\sin \beta_{bi} = \frac{\sin \beta_{si}}{\cos \alpha_{sn}}, \tag{8.50}$$

where β_{si} is the helix angle of virtual crossed axes helical gear, which is equal to the helix angle at reference cone at mid-face width of hypoid gear, β_{mi}, that is $\beta_{si} = \beta_{mi}$.

The diameter of reference circle of virtual crossed axes helical gear, d_{si}, is given by:

$$d_{si} = \frac{d_{mi}}{\cos \delta_i}, \tag{8.51}$$

where d_{mi} is the diameter at mid-face width of hypoid gear pair, and δ_i is the reference cone angle.

The tip diameter, d_{ai}, and *base diameter*, d_{bi}, of virtual crossed axes helical gear are given respectively by:

$$d_{ai} = d_{si} + 2h_{ami}, \tag{8.52}$$

$$d_{bi} = d_{si} \cos \alpha_{ti}, \tag{8.53}$$

where h_{ami} is the addendum at mid-face width of hypoid gear, and α_{ti} is the transverse pressure angle.

The above defined angles β_{bi}, β_{mi}, and α_{mn} are correlated by the following relationship:

$$\tan \beta_{bi} = \tan \beta_{mi} \sin \alpha_{mn}. \tag{8.54}$$

To calculate the Hertzian auxiliary angle, ϑ [see Eq. (8.41)], the shaft angle of virtual crossed axes helical gear, φ, is introduced, which is defined as the sum of helix angles of base circle of pinion and wheel, that is:

$$\varphi = \beta_{b1} + \beta_{b2}. \tag{8.55}$$

The normal module of virtual crossed axes helical gear, m_{sn}, is equal to normal module of hypoid gear at mid-face width, m_{mn}, that is $m_{sn} = m_{mn}$, while the normal base pitch, p_{en}, is given by:

$$p_{en} = \pi m_{sn} \cos \alpha_{sn}. \tag{8.56}$$

The radius of curvature at pitch point in normal section, ρ_{ni}, and the relative radius of curvature at pitch point in normal section, ρ_{Cn}, are given respectively by:

$$\rho_{ni} = \frac{d_{si}}{2} \frac{\sin^2 \alpha_{ti}}{\sin \alpha_{sn}} \tag{8.57}$$

$$\rho_{Cn} = \frac{\rho_{Cn1} \rho_{Cn2}}{(\rho_{Cn1} + \rho_{Cn2})}. \tag{8.58}$$

The tangential velocities, v_{ti}, of pinion and wheel of the hypoid gear pair, which coincide with tangential velocities of members of virtual crossed axes helical gear, are given by the relationships:

$$v_{t1} = \frac{\pi n_1 d_{m1}}{60 \times 10^3} \tag{8.59}$$

$$v_{t2} = v_{t1} \frac{\cos \beta_{s1}}{\cos \beta_{s2}}, \tag{8.60}$$

where n_1 (in $\min^{-1}$) is the rotational speed of pinion, while the diameter at mid-face width of hypoid pinion, d_{m1}, must be expressed in mm.

To determine the various velocities of interest, it is necessary to consider that the tooth flank surfaces of hypoid gears slide and simultaneously roll over one another at every point of contact. Contrary to what happens for bevel gears, where the relative velocity, that is the sliding velocity, is unidirectional, being directed along the tooth depth (see Fig. 8.9a), for hypoid gears the sliding velocity vector is composed of two components. The first component, v_{gs}, is directed along the longitudinal direction, and thus it is called *lengthwise velocity component*; the second component, v_{gh}, is directed along the direction of the tooth depth, and thus it is called *profile velocity component* (see Fig. 8.9b). This second component of the sliding velocity vector is also called *sliding velocity profile component*.

The lengthwise velocity component of the sliding velocity vector, v_{gs}, can be considered substantially a constant at every point of the path of contact, while the sliding velocity profile component, v_{gh}, is variable from point to point of the path

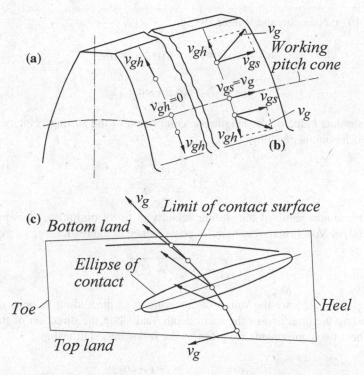

Fig. 8.9 Sliding velocity on tooth flank surfaces of bevel gears **a** and hypoid gears **b** and its variation in intensity and direction along the path of contact on tooth flank surface of the hypoid wheel **c**

of contact. The absolute value of this last component is equal to zero at pitch point, C ($v_{ghC} = 0$), while it increases as the point of contact moves away from the pitch point. Moreover, this sliding velocity profile component changes its sign: in fact, it is directed from the pitch point to the tooth tip, for addendum contact, and from the pitch point to the tooth root, for dedendum contact. Figure 8.9c shows the variation in intensity and direction of the resulting sliding velocity vector, $\boldsymbol{v}_g$ (more simply, it is called sliding velocity), obtained by the vector addition of the two above-mentioned longitudinal and profile component vectors, as a function of position of the point of contact over the path of contact.

Rebus sic stantibus, and bearing in mind the characteristics of crossed helical gears described in Chap. 10 of Vol. 1, we can express the absolute value of the resultant mean cumulative velocity, $v_{\Sigma C}$, at operating pitch point, C, as follows:

$$v_{\Sigma C} = \sqrt{v_{\Sigma s}^2 + v_{\Sigma h}^2}, \tag{8.61}$$

where $v_{\Sigma s}$ and $v_{\Sigma h}$ are the cumulative velocity along lengthwise direction and, respectively, the cumulative velocity along direction of tooth profile. These two components of the cumulative velocity are given respectively by the following relationships (see

Henriot 1979; Niemann and Winter 1983):

$$v_{\Sigma s} = v_{t1}(\sin \beta_{s1} + \cos \beta_{s1} \tan \beta_{s2}) \tag{8.62}$$

$$v_{\Sigma h} = 2v_{t1} \cos \beta_{s1} \sin \alpha_{sn}. \tag{8.63}$$

The absolute value of the lengthwise component of the sliding velocity vector, v_{gs}, at pitch point is given by:

$$v_{gs} = v_{t1} \frac{\sin \Sigma}{\cos \beta_{s2}}. \tag{8.64}$$

The maximum value of the sliding velocity at tip of pinion, $v_{g\gamma 1}$, is given by Eq. (10.62) of Vol. 1, here rewritten for convenience of the reader:

$$v_{g\gamma 1} = \sqrt{v_{g\alpha 1}^2 + v_{g\beta 1}^2}, \tag{8.65}$$

where $v_{g\alpha 1}$ and $v_{g\beta 1}$ are the values of components of the sliding velocity at tip of pinion along the direction of the tooth depth, and along the direction of the tooth trace. These two components are given by the following relationships:

$$v_{g\alpha 1} = v_{g1} \cos \gamma_1 + v_{g2} \cos \gamma_2 \tag{8.66}$$

$$v_{g\beta 1} = v_{gs} + v_{g1} \sin \gamma_1 + v_{g2} \sin \gamma_2, \tag{8.67}$$

where v_{g1} and v_{g2} are the sliding velocity of pinion and wheel, given by:

$$v_{g1} = 2v_{t1} g_{an1} \frac{\cos \beta_{b1}}{d_{s1}} \tag{8.68}$$

$$v_{g2} = 2v_{t2} g_{fn2} \frac{\cos \beta_{b2}}{d_{s2}}, \tag{8.69}$$

while γ_1 and γ_2 are auxiliary angles, given by:

$$\gamma_i = \tan \beta_{si} \sin \alpha_{sn}, \tag{8.70}$$

with ($i = (1, 2)$).

The recess path of contact of pinion, $g_{an1} = g_{fn2}$, and the recess path of contact of wheel, $g_{an2} = g_{fn1}$, which appear in Eqs. (8.68), (8.69) and (8.47), are given by Eqs. (10.39) and (10.40) of Vol. 1, also rewritten here for convenience of the reader in terms of interest, i.e. with the only replacement of the reference circle diameters of virtual crossed axes helical gear, d_{si}, in place of the diameters d_i, where as usual $i = (1, 2)$:

$$g_{an1} = g_{fn2} = \frac{g_{at1}}{\cos \beta_{b1}} = \frac{\left[\sqrt{d_{a1}^2 - d_{b1}^2} - \sqrt{d_{s1}^2 - d_{b1}^2} \right]}{2 \cos \beta_{b1}} = \overline{CE} \qquad (8.71)$$

$$g_{an2} = g_{fn1} = \frac{g_{at2}}{\cos \beta_{b2}} = \frac{\left[\sqrt{d_{a2}^2 - d_{b2}^2} - \sqrt{d_{s2}^2 - d_{b2}^2} \right]}{2 \cos \beta_{b2}} = \overline{CA}. \qquad (8.72)$$

Therefore, the length of path of contact, g, is given by:

$$g = \overline{AE} = \overline{CA} + \overline{CE} = g_{an1} + g_{an2}. \qquad (8.73)$$

The contact ratio in normal section of virtual crossed axes helical gear pair is given by:

$$\varepsilon_n = \frac{\overline{AE}}{p_{bn}} = \frac{g_{an1} + g_{an2}}{p_{bn}}, \qquad (8.74)$$

while the contact ratios in normal section of virtual pinion and wheel are given by:

$$\varepsilon_{ni} = \frac{g_{ani}}{p_{bn}}, \qquad (8.75)$$

where p_{bn} is the normal base pitch, and $i = (1, 2)$.

8.9 Scuffing Integral Temperature

The scuffing integral temperature is the limiting value of the temperature at which the scuffing occurs. As the scuffing temperature of which we have dealt in Sect. 7.8, also the scuffing integral temperature is considered a characteristic value for the material-lubricant-material system of the gear pair under consideration, to be determined by test gears that use the same material-lubricant-material combination. It may be determined also based on experimental results obtained with actual cases of scuffing damage.

The experimental methods and procedures to be recommended for determination of the scuffing integral temperature are those provided by the ISO/TS 6336-21:2017, which are valid for all types of lubricant oils (pure mineral oils, anti-scuff oils or EP-oils, and synthetic oils) for which the scuffing load capacity was determined. The scuffing integral temperature determined by these methods and procedures must be corrected, when the material-lubricant-material combinations, including heat treatment, are different for the test gear and actual gear.

According to the integral temperature criterion, the scuffing damage of gears is likely to occur when the mean flank temperature exceeds a typical value of the material-lubricant-material combination, including the heat treatment. This typical value, ϑ_{intS}, is called scuffing integral temperature number, or allowable integral temperature or, more simply, scuffing integral temperature. The scuffing integral temperature numbers can be obtained using the relationships given in Sects. 8.5, 8.6 and 8.7 for cylindrical gears, bevel gears, and hypoid gears, introducing in these relationships data and results derived from scuffing tests, carried out on any test gear characterized by a given material-lubricant-material combination.

The approximate calculation of the scuffing integral temperature numbers, ϑ_{intS}, of heat-treated or surface-treated steel gears, lubricated with mineral oils, can be derived from those of other steel gears, otherwise heat-treated or surface-treated, but lubricated with the same lubricant oils, using the following relationship:

$$\vartheta_{intS} = \vartheta_{MT} + C_2 X_{WrelT} \vartheta_{flaintT}, \tag{8.76}$$

where ϑ_{MT} is the test bulk temperature, C_2 is a dimensionless weighting factor ($C_2 = 1.5$ from experiments), X_{WrelT} is the dimensionless relative welding factor (see below), and $\vartheta_{flaintT}$ is the mean flash temperature of the test gear. Of course, Eq. (8.76) should be used with consistent units of temperatures that appear in it.

For determination of ϑ_{MT} and $\vartheta_{flaintT}$ from test results, the above-mentioned ISO standard provided three types of test, one of which is the FZG-test A/8,3/90. Here we limit ourselves to consider, as an example, only this test, and we refer the reader to the same ISO standard, with regard to the other two types of testing, respectively named Ryder and FZG-Ryder gear test R/46,5/74, and FZG L-42 test 141/19,5/110. With the FZG-test A/8,3/90, ϑ_{MT} and ϑ_{flainT} are determined using the following relationships:

$$\vartheta_{MT} = 80 + 0.23 X_L T_{1T} \tag{8.77}$$

$$\vartheta_{flaintT} = 0.20 X_L T_{1T} \left(\frac{100}{v_{40}} \right)^{0.02}, \tag{8.78}$$

where X_L is the dimensionless lubricant factor (see Sect. 8.2), v_{40} (in mm^2/s) is the kinematic viscosity of the oil at 40 °C, and T_{1T} (in Nm) is the scuffing torque at test pinion. This test pinion torque is given by: $T_{1T} \cong 3.73$ (FZG load stage)^2.

The values of temperatures ϑ_{MT} and $\vartheta_{flaintT}$, given by Eqs. (8.77) and (8.78), can also be read from Fig. 8.10, as a function of T_{1T}, shown in the lower abscissa, or as a function of the FZG load stage, shown in the upper abscissa, using the corresponding curves which mediate the experimental results scattered in the shaded areas. A single ϑ_{MT}-curve is given, while the three $\vartheta_{flaintT}$-curves refer to three different mineral oils.

The *relative welding factor*, X_{WrelT}, in Eq. (8.76), is a dimensionless empirical factor which takes into account the influence of the heat treatment or surface treatment on the scuffing integral temperature. It is defined as:

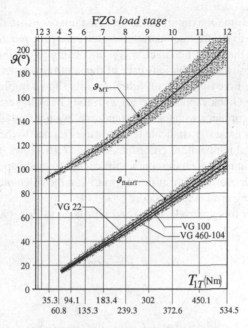

Fig. 8.10 Test bulk temperature and mean flash temperature of the test gear, as a function of the scuffing torque of test pinion (or FZG load stage) and type of mineral oil, for FZG-test A/8,3/90

$$X_{WrelT} = \frac{X_W}{X_{WT}}, \qquad (8.79)$$

where X_{WT} is the welding factor of test gear ($X_{WT} = 1$, for FZG-test A/8,3/90 as well as for Ryder gear test and FZG L-42 test), and X_W is the welding factor of the actual gear material, to be chosen in accordance with Table 8.2.

In fact, according to the CTM-(Contact-Time Criterion), the critical value of the scuffing temperature that must not be exceeded is not always a constant, but can be variable, especially at high speeds, as the contact time t_C comes into play. This

Table 8.2 Welding factor, X_W

Gear material	X_W
Through-hardened steel	1.00
Phosphated steel	1.25
Copper-plated steel	1.50
Bath and gas nitrided steel	1.50
Case-carburized steel:	
- average austenite content less than 10%	1.15
- average austenite content 10–20%	1.00
- average austenite content greater than 20–30%	0.85
Austenitic steel (stainless steel)	0.45

critical value can be the maximum local and instantaneous total contact temperature, ϑ_S, according to the flash temperature criterion, or the mean weighted surface temperature across the path of contact, ϑ_{intS}, according to the integral temperature criterion. Here we use the distribution curve of the scuffing temperature as a function of contact exposure time for anti-scuff lubricant oils shown in Fig. 7.16, were however we replace the Greek capital letter, Θ, with the Greek lowercase letter, ϑ. As a matter of fact, in accordance with the ISO standards, we use this Greek letter, respectively upper and lower case, to indicate temperatures related to the flash temperature criterion and integral temperature criterion. As Fig. 7.16 shows, the dependence of the scuffing temperature, ϑ_S, on the contact time, t_C, is usually approximated by two straight lines, which overlap with sufficient precision the calculated or experimental curves of the scuffing critical temperature.

Therefore, under this approximation, and similarly to Eqs. (7.102) and (7.103), we will have, for $t_C \geq t_K$:

$$\vartheta_S = \vartheta_{SC} = const \tag{8.80}$$

and, for $t_C < t_K$:

$$\vartheta_S = \vartheta_{SC} + C_S X_{WrelT}(t_K - t_C) \tag{8.81}$$

where: ϑ_{SC} (in °C or K), is the constant scuffing temperature at long contact times; C_S (in °C/μs or K/μs), is the gradient of the scuffing temperature; t_K (in μs), is the contact time at the knee of the curve $\vartheta_S = \vartheta_S(t_C)$, that is at the minimum of the scuffing-speed-curve; X_{WrelT}, is the relative welding factor, which is a dimensionless quantity.

The contact time, t_C, is the time that one point of the gear flank surface (pinion or wheel) needs to cross the Hertzian contact width, $2b_H$. For a pair of meshing tooth flanks, the decisive contact exposure time to be considered is the shortest contact exposure time between the contact exposure time of pinion, t_{C1}, and wheel, t_{C2}, that is

$$t_C = (t_{C1} \text{ or } t_{C2})_{min}. \tag{8.82}$$

Quantities t_{C1} and t_{C2} are given by:

$$t_{C1} = \frac{2b_H}{v_1}; \quad t_{C2} = \frac{2b_H}{v_2}, \tag{8.83}$$

where v_1 and v_2 are respectively the reference line velocities of pinion and wheel.

If oil test results at high speed are not available, the following indications can be used:

- For mineral oils, the increase of the critical temperature at high speeds can be neglected, as it is very low; therefore, for these not anti-scuff lubricant oils, we can assume $\vartheta_S = \vartheta_{SC}$.
- For anti-scuff oils, the increase of the critical temperature at high speeds can no longer be neglected and, to take this into account, in the absence of more precise experimental data, we can assume $t_K = 18\,\mu s$ and $C_S = 18\,K/\mu s$, to be introduced in Eq. (8.81), where ϑ_{SC} has to be determined with a scuffing oil test (i.e., the FZG-test A/8,3/90).

For cylindrical gears, with pinion and wheel made with the same material, and thus having the same Young's modulus, the determination of the Hertzian contact width and tangential velocities at the flank, the following relationships can be used (the meaning of symbols is already known):

$$2b_H = 3.04\sqrt{\frac{F_{bt}}{bE}\frac{\rho_1\rho_2}{(\rho_1 + \rho_2)}} \tag{8.84}$$

$$v_1 = \rho_1\omega_1 = \rho_1\frac{2\pi n_1}{60} \tag{8.85}$$

$$v_2 = \rho_2\omega_2 = \rho_2\frac{2\pi n_2}{60}. \tag{8.86}$$

8.10 Calculation Examples

Here also, for the same space reasons already highlighted in Sect. 2.11, we cannot present and discuss examples regarding a complete and exhaustive calculation procedure of the scuffing load carrying capacity in terms of flash temperature method or integral temperature method. In fact, each of the examples that could be presented and discussed should be aimed at calculating the safety factors against scuffing, S_B and S_{intS}, related to the two aforementioned methods. However, to determine these two safety factors, it is not possible to omit any of the calculation steps described in this and the previous chapter, so that the complete development of each example taken from a practical application would occupy a space comparable to that devoted to theoretical aspects concerning these two methods of calculation.

In this regard, we recall that ISO/TS 6336-20:2017, concerning the flash temperature method, does not present any practical application example. The ISO/TS 6336-21:2017 regarding the integral temperature method instead proposes, in Annex A, eight summary tables of as many practical application examples taken from the Michaelis dissertation (Michaelis 1987), having as subject respectively:

Example 1: helical gear of a turbine gear drive
Example 2: helical gear of a steel mill gear drive
Example 3: helical gear of a machine tool gear drive

Example 4: helical gear of a marine gear drive
Example 5: another helical gear of a steel mill gear drive
Example 6: another helical gear of a turbine gear drive
Example 7: yet another helical gear of a turbine gear drive
Example 8: spur gear of a vehicle gear drive.

Finally, in a ninth and last table of Annex A, the same ISO standard summarizes the results regarding a spiral bevel test gear and three hypoid test gears.

The above examples therefore concern various types of gears, which include variously configured cylindrical spur and helical gears as well as spiral bevel and hypoid test gears. However, the summary tables are limited to providing, for each example, the many input data and, as output data, only the calculated values of the integral temperature and the scuffing safety factor, with the addition of a final comment on the observed failures in terms of no scuffing, borderline scuffing or scuffing.

8.11   Flash Temperature or Integral Temperature?

The question on the use of the flash temperature criterion or the integral temperature criterion is something that must be done. The real fact that the ISO Standards propose two apparently different rating criteria of scuffing load carrying capacity of gears justifies this question. It is indeed always hoped that a given technical standard, even more so if it is of international validity, provides unequivocal directives to the designer, without any possibility of misunderstanding (Polder 1982). In fact, the ISO Standards have entrusted to gear designers the task of assessing what is the most appropriate criterion, between the two that have been standardized.

To resolve this question, it would be necessary to carry out a systematic campaign of numerical simulations, using a sufficiently large amount of data to cover the entire field of practical applications of cylindrical, bevel and hypoid gears. These systematic numerical simulations should compare the results obtained using, for the flash temperature criterion, Eq. (7.2), and, for the integral temperature criterion, Eq. (8.14). Fortunately, this comparison is a little simplified by the fact that, being the bulk temperature, ϑ_M, common to both Eqs. (7.2) and (8.14), it actually reduces to the comparison between the maximum value of the flash temperature, Θ_{flmax}, and the product $C_2\vartheta_{flaint} = C_2 X_\varepsilon \vartheta_{flaE}$. In this regard, it is in fact to be considered that, leaving aside the formalism of the ISO standards used here, the bulk temperature, Θ_{Mi}, in Eq. (7.2) is equal to the bulk temperature, ϑ_M, in Eq. (8.14).

Despite this favorable circumstance, this comparison is not trivial, but very complex, because the cases to be considered are very numerous, and the variability ranges of the factors that define the teeth geometry are very wide. To give an idea of the work to be performed, restricting our attention to the simplest case of cylindrical gears, we should compare the maximum value of the flash temperature, Θ_{flmax}, to be calculated by Eq. (7.24), with the value of the product $C_2 X_\varepsilon \vartheta_{flaE}$, to be calculated with $C_2 = 1.5$, and X_ε and ϑ_{flaE} given respectively by Eqs. (8.27) and (8.16).

The above comparison must be made for all possible geometries and operating conditions of cylindrical spur and helical gears that characterize all the possible practical applications. The same comparison should be made for bevel and hypoid gears, again taking into account all the possible practical applications of these types of gears. Of course, this is a very difficult and demanding task, which would require the help of an international task force.

Therefore, a general and comprehensive comparison between the flash temperature criterion and integral temperature criterion involves complexities and difficulties almost insurmountable. Taking into account these complexities and difficulties, it is certainly to be appreciated the comparison made by Polder (1986, 1987a, b), on the basis of the ISO Draft International Standard, DIS 6336, Part 4. The work of Polder has been done long before the ISO/TS 6336-20:2017(E) and ISO/TS 6336-21:2017(E), and the corresponding previous ISO/TR 13989-1:2000 and ISO/TR 13989-2:2000 were published. In any case, this work, even if limited to cylindrical gears, constitutes a laudable attempt to give an answer to the question of interest here, able to overcome the limitations of the short comparison in the appendix of the aforementioned Draft.

Polder began by considering that, in the Draft equations inherent to the two criteria (flash temperature and integral temperature), only the purely geometric factors differed between them. Therefore, Polder reduced the comparison to a simple mathematical comparison and, by doing numerous tests with a sufficiently large amount of data, came to the conclusion that all test results expressed in integral temperature were fully applicable to the Blok's flash temperature. Furthermore, the same results confirmed, in an unintentional way, the validity of the gear-scuffing criterion according to Blok, which he as well as other gear designers considered as the most practical.

Furthermore, Polder noted that, due to more uncertain influences of an elastohydrodynamic, thermodynamic and chemical nature, the test data of scuffing phenomena, from a statistical point of view, are characterized by a larger deviation than the corresponding test data for surface durability (pitting) and tooth bending strength. Therefore, considering the systematic dependence of both criteria to be compared on the aforementioned test data, Polder concluded, "*any assertion that the integral temperature method would be statistically superior to the flash temperature method is based on a misunderstanding*".

Despite this abrupt conclusion, the answer to the question placed at the beginning of this section remains open. This is also because, on the one hand, it would be necessary to redo the work of Polder by using the equations of the two aforementioned new ISO Standards, which are different from those of the Draft and, on the other hand, it is necessary to widen the horizons to see what happens to bevel and hypoid gears.

References

Grekoussis R, Michailidis Th (1981) Näherungsgleichungen zur Nach- und Entwurfsrechnung der Punktberührung nach Herts. Konstruktion 33

Henriot G (1979) Traité théorique et pratique des engrenages, vol 1, 6th edn. Bordas, Paris

Ishikawa J, Hayashi K, Yokoyama M (1972) Surface temperature and scoring resistance of heavy-duty gears. Institute of Technology, Tokio

ISO 10300-1:2014 Calculation of load capacity of bevel gears—part 1: introduction and general influence factors

ISO 53:1998 Cylindrical gears for general and heavy engineering—standard basic rack tooth profile

ISO/TR 13989-1:2000 Calculation of scuffing load capacity of cylindrical, bevel and hypoid gears—part 1: flash temperature method

ISO/TR 13989-2:2000 Calculation of scuffing load capacity of cylindrical, bevel and hypoid gears—part 2: integral temperature method

ISO/TS 6336-20:2017 Calculation of load capacity of spur and helical gears—part 20: calculation of scuffing load capacity (also applicable to bevel and hypoid gears)—flash temperature method

ISO/TS 6336-21:2017 Calculation of load capacity of spur and helical gears—part 21: calculation of scuffing load capacity (also applicable to bevel and hypoid gears)—integral temperature method

Juvinall RC (1967) Engineering considerations of stresses, strain, and strength. McGraw-Hill Book Company, New York

Lechner G (1966) Die Freß-Grenzlast bei Stirnrädern aus Stahl, Dissertation TH München

Michaelis K (1987) Die Integraltemperatur zur Beurteilung der Freßtragfähigkeit von Stirnrad-getrieben, Diss. TU München

Niemann G, Winter H (1983) Maschinen-Elemente, Band III: Schraubrad-, Kegelrad-, Schnecken-, Ketten-, Rienem-, Reibradgetriebe, Kupplungen, Bremsen, Freiläufe. Springer, Berlin

Polder JW (1982) Contribution to a Uniform ISO Scuffing Criterion, ISO/Technical Committee 60/Working-Group 6, Document 280, May

Polder JW (1986) Influence of geometrical parameters on the gear scuffing criterion. In: Proceedings of the 2nd World Congress on Gearing, March pp 3–5, Paris, France

Polder JW (1987a) Influence of geometrical parameters on the gear scuffing criterion—part I, Gear Technology, March/April pp 28–34

Polder JW (1987b) Influence of geometrical parameters on the gear scuffing criterion—part II, Gear Technology, May/June pp 19-27

Timoshenko S, Goodier JN (1951) Theory of elasticity. McGraw-Hill Book Company Inc., New York

Chapter 9
Wear Load Capacity Rating of Gears

Abstract In this chapter, after a brief examination of the various types of wear that can be found in the gears, the abrasive wear generation mechanism and the related influences are described. The main classic wear theories are then recalled, together with the equations that, for each of them, express the correlations between the various quantities involved. Attention is then focused on the assessment of tooth wear linear progression and, in this regard, the various criteria for checking abrasive wear in the gears are described. Finally, additional considerations on the gear wear are made.

9.1 Introduction

In the field that interests us here, the term *wear* generally describes loss of material from the contacting tooth flank surfaces of the gears and is used in relation to their load capacity rating against the *abrasive wear*. It is caused by inadequate lubrication and is due to direct rubbing of the tooth flank mating surfaces in relative motion or rubbing on these surfaces of abrasive particles entrained in the lubricant flow. The wear damage of gears occurs mostly at low transverse tangential velocities (below 5 m/s) on gear wheels made of heat-treated steel, and having coarse accuracy grade. Usually, the harder the surfaces the more resistant they are to abrasive wear.

For practical applications characterized by too low pitch line velocities, the real risk exists that the minimum value of the lubricant film thickness falls below a threshold value for which the wear becomes the dominant damage mechanism. Experimental investigations show that wear is predominant for minimum thickness of the lubricant film at the pitch point, $h_C \leq 0.1\,\mu m$.

More in detail, in a gear pair, the abrasive wear may be due to direct contact of the roughness asperities of the two mating tooth surfaces in relative motion, with a consequent volume of removed material localized predominantly on the tooth flank surface of less hard member of the kinematic pair. The abrasive wear may also be due to the interposition, between the two mating surfaces, of hard abrasive particles. These may originate from the surrounding environment outside with respect to the kinematic pair (e.g., particles that contaminate the lubricant, siliceous dusts that seep through the seals, fragments of the same seals, etc.), but more often they are originated

© Springer Nature Switzerland AG 2020

V. Vullo, *Gears*, Springer Series in Solid and Structural Mechanics 11,
https://doi.org/10.1007/978-3-030-38632-0_9

from same wear action, which sees interacting with each other the *adhesive wear*, *abrasive wear*, and *corrosion film wear*. This interaction is most often deleterious, in that the wear action in this case configures a phenomenon that exalts oneself.

Various degrees of wear may occur, ranging from light to moderate to excessive wear. Light or polishing wear consists of a slight loss of metal at a rate that has a little effect on the satisfactory performance of the gear during its lifetime. It produces beneficial consequences, as roughness asperities are progressively worn and removed away, resulting in smoother conforming surfaces that improve contact conditions. This type of wear can occur by either abrasive or adhesive wear mechanisms in gears operating under boundary lubrication conditions. Normal or moderate wear instead progresses at a slightly higher rate than that characterizing the polishing wear. It is such as to slightly influence the satisfactory performance of a gear during its expected lifetime. The metal is generally removed from the dedendum area of the tooth flank surface, although there are examples of material removed from the entire tooth flank surface. Case-hardened gears show much better behavior against wear than through-hardened gears. This type of wear depends on load, lubrication regime, surface hardness, gear accuracy grade and especially surface roughness, as well as characteristics of lubricants (in this regard, contaminants within the lubricant oil may determine abrasive or corrosive wear). Anyway, moderate wear is a normal inevitable technical event. Finally, destructive wear is an anomalous fact, absolutely to be avoid since it determines variations of the tooth surface shapes so important to irreparably compromise the meshing action, and to appreciably shorten the expected lifetime of the gear.

The removal of surface material that characterizes the abrasive wear occurs gradually, and the two members of the gear pair see progressively altered the shape of the tooth flank mating surfaces in relative motion. This progressive removal of material is due to friction, but the correlation between the wear and friction is neither simple nor unique. In fact, kinematic pairs exist that are characterized by high *wear rate* and low coefficient of friction, and vice versa. The correlation between the volume of removed material and energy lost due to friction phenomena depends on the mechanical properties of the materials in contact, surface characteristics of the tooth mating flanks in relative motion as well as on the types of wear that come into play.

The main mechanical property of materials in contact that influences the abrasive wear is the *hardness*, which is a static property (Faupel 1964). If the materials in contact have different hardness, the volume of removed material is mainly constituted by the less hard of the two materials involved. The main characteristic of surfaces in contact is the *roughness* (see ISO 4287:1997; ISO 4288:1996). The wear of the less hard of the two materials involved is much lower, the more the surface roughness of the hardest of them is low. Therefore, when all three of the aforementioned types of wear come into play, the wear phenomenon can become devastating, for the synergetic action of the adhesive wear, and corrosion film wear. In fact, fragments of hardened material can be removed from the local micro-weldings due to scuffing and, interposing itself between the surfaces in relative motion, can generate or enhance the phenomenon of the abrasive wear. Furthermore, too hard oxides, which are formed by effect of corrosion and oxidation, can be detached, causing effect similar to those of the aforementioned hard fragments.

This last phenomenon, called corrosion film wear, is due to the formation of films of chemical compounds (oxides, sulphates, etc.) which are formed by the chemical action of substances that are present in the surrounding environment. These films of chemical compounds, which are deposited on the tooth flank surfaces, normally exert a protective action on the underlying metal, but, due to the rubbing arising from local disruption of the lubricant film oil, the corroded surface film is alternatively removed by sliding, and then reformed. This chemical action, which can also be correlated with an inappropriate choice of lubricant oil, triggers a wear mechanism, which overlaps to that resulting from the mechanical action of rubbing only, for which the total wear rate is enhanced. In this regard, it is therefore to remember that only the use of appropriate lubricant oils exercises an effective protective action on the tooth flank surfaces in relative motion against corrosion film wear.

At least two other types of abrasive wear must be considered for the gears. The first of these abrasive wear types is the *run-in wear*, which is to be welcomed as long as it leads to the removal of roughness peaks (the asperities) and their leveling, without further progression. With this kind of abrasive wear, the characteristics of the working tooth flank surfaces, especially in terms of roughness, are significantly improved (see Lohner et al. 2015). Therefore, a normal run-in wear is not only accepted, but it is also desirable and, thus, it is to be favored by activating specific run-in conditions. In typical well-designed gears, the initial wear rate on rubbing tooth flank surfaces during run-in may be relatively high. As the more pronounced surface roughness peaks are worm off, causing a consequent increase of the actual contact area, the wear rate decreases up to a small constant value. However, after a certain period of operation, it returns to increase, when other factors (e.g., lubricant contamination, increased surface temperature, etc.) begin to exert their influence.

The second kind of abrasive wear to be considered for the gears is that of exogenous origin, due to impurities in the lubricant oils. In fact the lubricants, due to defects in gasket sealing as well as to inadequate deaeration, can contain abrasive particles of various kinds, such as dust, slag, grinding dust, foundry sand, etc., which interposing between the mating tooth flank surfaces, can cause wear and disfigurement of the same surfaces. The amount of damage of this kind of abrasive wear depends on grain size, hardness and concentration of the impurities in the lubricant as well as on the hardness of tooth materials. To prevent the progression of this wear damage, which is dangerous not only for the tooth mating surfaces, but also for bearings that support the shafts, timely and appropriate actions must be taken, such as:

- filter dimensioned with sufficient capacity;
- sedimentation spaces for the oil of adequate volume, where dirt can build up without going into circulation;
- continuous control of lubricant oil; etc.

The wear mechanisms of the gears are influenced mainly by the adhesive wear, which was discussed in the previous two chapters, and abrasive wear, which is the subject of this chapter. These two primary wear processes can interact among themselves, as we highlighted above. Other wear processes, such as fretting wear, cavitation, erosive wear, and corrosion and oxidation wear, have secondary importance for the gears, so we think here not having to discuss, referring the reader to more specialized textbooks (see Engel 1978; Peterson and Winer 1980; Zum Gahr 1987; Rabinowicz 1995; Sorensen et al. 1996; Stachowiak and Batchelor 2005). About a more general survey on possible wear mechanisms, we refer the reader to Burwell (1957) and Varemberg (2013).

9.2   Generation Mechanism of Abrasive Wear and Influences

In Sect. 7.10 we dutifully quoted the equation of Grubin and Vinogradova (1949), as it is unanimously considered the first historical equation of the minimum lubricant film thickness that ignited the tribology. However, we took care to mention other equations, similar to the Grubin-Vinogradova equation, whenever the topic discussed required it (see previous chapters).

The Grubin-Vinogradova equation (as well as the other similar ones, developed and proposed later) indicates that the lubricant film thickness may become critical, for gears operating at low speed and high load, especially when the pressure in the lubricant oil film is not very high; this happens for example in the case of absence of forced lubrication. Actually, with pitch line velocities below 0.5 m/s, which determine, according to the elastohydrodynamic lubrication theory, lubricant film thickness less than 0.1 μm, as usually happens in the last stage of a speed reducing gear unit, the continuous removal of material due to abrasive wear often determines the lifetime of the gear unit.

The generation mechanism of the abrasive wear is substantially different from that of the adhesive wear. In the latter mechanism, as Fig. 7.18 shows, the generation of the warm scuffing is due to the attainment of a critical value of the contact temperature, which causes the instantaneous and abrupt local breakdown of the lubricant film. In these conditions, the two mating surfaces are in direct contact, with all the consequences already described in the previous two chapters in terms of scuffing (of course, warm scuffing).

On the contrary, in the case of abrasive wear, the local breakdown of the lubricant film thickness and the resulting direct contact between the mating surfaces are not due to the temperature, the maximum value of which is maintained very below the aforementioned critical value (it should be noted that temperature also has its influence, but quite secondary). Indeed, they are to be correlated to the operating conditions (too low speed, too high load, absence of forced lubrication, etc.), which do not guarantee a sufficient elastohydrodynamic lubricant film thickness. Figure 9.1 shows, in schematic form, the aforementioned generation mechanism of the abrasive wear.

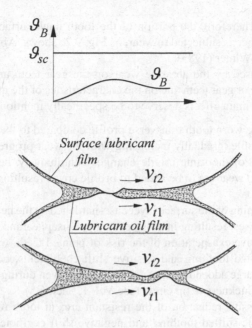

Fig. 9.1 Generation mechanism of the abrasive wear

As the actual operation time (usually it is expressed in hours) of a gear unit increases, the abrasive wear first determines a progressive variation of the operating flank shape located between the base circle and working pitch circle, that is on the dedendum flank, as Fig. 9.2a shows so deliberately exaggerated compared to the initial transverse not yet worn profile. Subsequently, also the portion of the tooth flank between the working pitch circle and the tip circle, i.e. the addendum flank, is progressively disfigured, as Fig. 9.2b shows. Still later, when the wear has reached the working pitch surface, and the tooth transverse profile is altered in such a way to present in correspondence of this surface a kind of cusp, the removal of material continues in homothetic manner, corresponding to a displacement of the worn profile

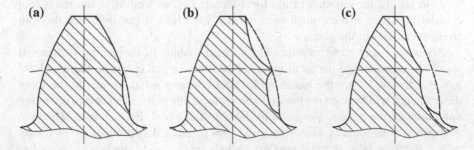

Fig. 9.2 Progression of abrasive wear: **a** only dedendum flank worn out; **b** also addendum flank worn out; **c** almost homothetic progression of wear on addendum and dedendum flanks

parallel to itself. Therefore, the portion of the tooth flank surface adjacent to the working pitch surface is subjected to wear, as Fig. 9.2c shows. About this topic, see also Niemann and Winter (1983).

The effects caused by the abrasive wear on the gear teeth are numerous, and depend on the type of gear teeth and on the characteristics of the materials involved. The following five main effects deserve to be specifically mentioned, namely:

- Deviations of the worn tooth transverse profile compared to the initial transverse not yet worn profile, gradually increasing with the wear progression. Therefore, the actual shape of the same profile changes continuously, because the profile deviations due to wear are to be added to profile errors resulting from machining operations.
- Progressive thinning of the surface layer case-hardened by the heat treatment when the wear progresses, resulting in a risk of its total disappearance.
- Further progressive exasperation of the risk of pointed teeth, when we use gears with profile-shifted toothing and positive shift coefficients, because of a tooth sizing with too large addendum that can determine, even during the cutting stage, a too low tooth thickness at tip circle (see Vol. 1, Sect. 6.5).
- Risk of an excessive reduction of the resistant area at tooth root, when we use gears with profile-shifted toothing and negative shift coefficients, because of a tooth sizing with tooth large dedendum.
- Risk of damage not only to gears, but also to bearings, due to excessive amount of metal particles removed by direct rubbing of the mating surfaces, especially when the surface hardness of one of the two materials in contact is too low compared to that of the other material.

It is to be noted that deviations of the worn profile compared to the initial profile, still not worn, determine a position change of the tooth mid-plane, i.e. the plane that divides in equal parts the tooth thickness between the operating and non-operating flanks, as Fig. 9.3 shows, again in a deliberately exaggerated manner. Furthermore, the same deviations that occur in the active profile portion adjacent to the root fillet cause consequent displacements of the 30°-tangent, as the same Fig. 9.3 shows (Henriot 1979). Therefore, taking into account the displacement of the tooth mid-plane as well as the change of direction of the 30°-tangent, it follows that the wear leads to significant variation of the form factor, Y_{Fa}, as well as of the tooth root chordal thickness at the critical section, s_{Fn}, for which it can influence the root strength capacity of the gears.

About the metal particles removed by direct rubbing of the mating surfaces, it should be noted that, if the tooth flank surfaces have equal hardness, the weight of material removed from the pinion and wheel is nearly equal. However, the more deformations will have on the pinion flank surface, due to the most severe operating conditions of the pinion, compared to those of the wheel. Furthermore, when the gear pair is constituted by a case-hardened steel pinion and a through-hardened steel wheel, the wear takes place almost exclusively on the wheel made with the softer material. Even minimal differences in hardness between pinion and wheel still lead

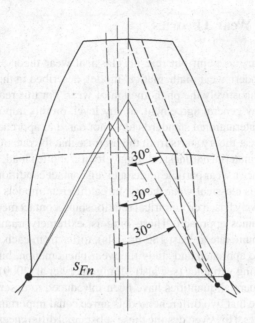

Fig. 9.3 Displacements of the tooth mid-plane and 30°-tangent due to wear

to increased wear, which is mainly concentrated on the wheel made with the softer material.

The abrasive wear is affected by any design choice, which determines a thicker lubricant oil film. In this regard, the most important role among the influence factors is that covered by the characteristics of lubricant and speed. Among the characteristics of the lubricant, the viscosity is the one that exerts the main role on the abrasive wear. At constant transverse pitch line velocity, using a lubricant oil having appropriate viscosity, the abrasive wear can be reduced by a factor of up to three. Furthermore, for the same viscosity, using some synthetic lubricant (not all), the wear can be further reduced. For low speed, a reduction of wear can also be obtained by mixing a given mineral oil with a lubricant fluid grease.

Instead, the anti-scuff additives (or EP-additives) do not always have a positive role in reducing the wear of the gears. Leaving aside other possible side effects, which we have already mentioned, the action of additives, characterizing the anti-scuff lubricant oils as well as that of synthetic oils, should be carefully evaluated by appropriate experimental laboratory tests and, when possible, monitored by tests performed on actual gear sets working under the intended design conditions.

Finally, it should be noted that the tooth profile geometry and the use of profile-shifted toothing do not exert any significant influence on the wear load capacity of gears. Only an appropriate intentional modification of profiles of the pinion teeth, involving removal of material near their tip circles, that is a certain value of tip relief of the pinion teeth, can have a favorable effect. However, this effect manifests only when the pinion is the driving member of the gear pair under consideration.

9.3 Classical Wear Theories

First, it is necessary to point out that no classical wear theory, described in this section, and no modern wear mathematical model, described in the next section, are able to explain exhaustively the phenomenon of wear. For this reason, experts have not yet reached any general agreement, at any level, on this important subject. For the same reason, international standards have not dared to approach this subject.

The classical wear theory starts from the premise that the rate of material removal is a function of the material hardness, sliding velocity, applied load, and probability of a material to produce a wear particle under a given contact condition. In the historical development of this classical wear theory, three different models can be identified, which are respectively based on empirical relationships, contact mechanics approach, and failure mechanics approach. These models, extremely numerous (Meng and Ludema (1995) enumerate at least 180 models), differ from each other not only for the method used to approach and study the wear phenomenon, but also about wear effects and governing variables (see also Montgomery et al. 2009).

About 100 influential quantities have been introduced to describe the wear processes, but it is rare that two different models agree equal importance to any of these influential quantities. However, despite these substantial differences, all models agree that friction is an important concept, which is intrinsically correlated to wear.

In the Western World, the oldest evidence on friction comes from Greece. In fact, in the 4th century B.C., the philosopher Themistius, exegete of Aristotle, observed that the rolling friction was much lower than the sliding friction, thus justifying the invention of the wheel as one of the biggest steps forward in the field of land transport. The concept of friction remained substantially a mysterious concept throughout the Middle Ages, and only in the Renaissance Leonardo da Vinci devised the basic laws of friction, providing a linear relationship between the forces related to the contact and the vertical load, regardless of the area of contact. However, the classic rules of sliding friction discovered by Leonardo da Vinci remained in his notebooks, unpublished (see Mac Curdy 1938; Panjkovic 2014).

About two centuries later, Amontons (1699) reinvented the Leonardo da Vinci laws, formulating the two laws that are still called, improperly, *Amontons' first law* and *Amontons' second law*. According to these two laws, the force of friction is directly proportional to the applied load and, respectively, independent of the apparent area of contact. Amontons formulated the well-known equation, called *Amontons' equation*:

$$F_\mu = \mu F_n. \tag{9.1}$$

which correlates the friction force, F_μ, coefficient of friction, μ, and applied normal force, F_n. The direct proportionality between friction force and applied load was unanimously accepted, but academics did not hide their skepticism on independence from the apparent area of contact, despite some experimental evidence of tests made by other researchers in this regard.

Equation 9.1 is also called *Coulomb's equation*, in honor of Coulomb (1781), which is recognized as the main architect of the friction laws. Coulomb confirmed the importance of the contact surface roughness, as emphasized by Amontons, and suggested that friction was due to the work done by dragging a surface on another surface. Coulomb also realized that in reality the contact occurred only in correspondence of the surface asperities, for which introduced the concept of actual area of contact, i.e. the common area between the mutually contacting asperities, as the areas concerned are characterized by a complex topography of asperities. Furthermore, Coulomb assumed that the kinetic friction was independent of the sliding velocity. This assumption is known as *Coulomb's law of friction*.

Coulomb, however, refused to accept the *Disaguliers' theory* (Disaguliers 1734, 1744) , according to which the frictional resistance between smooth surfaces, obviously not due to roughness, should be attributed to adhesive forces, summarized by the term *cohesion*. Subsequently, Leslie (1829) highlighted the weaknesses of theories of Amontons and Coulomb, and assumed that friction was due to surface deformations induced by the roughness. After this intuition of Leslie, nearly a century was necessary because a unified friction theory had developed, according to which the basic parameters are related to the geometry of surfaces in contact, their elastic properties, the adhesive intermolecular forces, and the energy lost due to surface deformations.

The first historical model that correlates friction and wear is due to Reye (1860); therefore, it is known as *Reye's hypothesis* (see Panetti 1930; Giovannozzi, 1965; Ferrari and Romiti 1966). This hypothesis states that the volume, V, of material removed from a body due to wear effects in a given time is proportional to the passive work, W, done in the same time by the frictional forces that have produced the wear, i.e. proportional to the energy dissipated into the body during the relative motion of the two contacting surfaces.

If we consider two contacting bodies on an area of contact, A, and denote by σ_H and v_g, respectively, the Hertian contact stress and sliding velocity at any point, P, of the contact area, and by dA the differential area around point P, the energy lost by friction at point P in the infinitesimal time dt will be equal to $\mu\sigma_H dAv_g dt$. If then we denote by s_w the thickness of material worn out away from that differential area at the same time, dt, the corresponding volume of material removed by wear will be given by $s_w dAdt$. Therefore, the Reye's hypothesis, in differential terms, can be written in the following form:

$$s_w dA = k\mu\sigma_H dAv_g, \qquad (9.2)$$

where k is a suitable coefficient of wear, which has the dimensions of a surface divided by a force and depends on materials of two contacting bodies and working conditions. This coefficient is called *Reye's wear constant*.

Integrating this equation over the whole area of contact, A, we get the following *Reye's equation* in finite terms:

$$V = kW, \qquad (9.3)$$

where V is the volume of material removed, and W is the work dissipated into the material. This Reye's model constitutes one of the first models that considered the wear phenomenon from a new point of view, the one based on energy considerations. Despite some its well-known limit (for example, the assumed shear stress distribution at boundary of the area of contact does not meet the *Betti's reciprocal theorem*), this model is a milestone. In fact, both the refined models, and new models, developed by subsequent researchers, have their roots on Reye's hypothesis.

Based on empirical investigations, Bowden and Tabor (1939) showed that the actual contact area between two bodies is really very less than the apparent contact area, since flat surfaces are held apart by small surface irregularities, which form bridges. However, to obtain quantitative results to be used for the calculation of the change of contact area with the load, they used the *Hertz theory* (1882) concerning the elastic contact deformation and considered the probability that surface asperities are to collide in a given contact situation. Moreover, by processing results of experimental measurements of conductance, performed by the method developed by Bidwell (1883), they were capable to formulate two theoretical equations. The first equation is based on the assumption of elastic behavior of the material, which provides that the conductivity between the bodies (and thus their contact area) depends on the cube root of the applied load. The second equation is based on the assumption of plastic behavior of the material, which provides that the conductivity between the bodies depends on the square root of the applied load. Further experimental measurements substantiated this second assumption, for which Bowden and Tabor stated that the total cross section of the junctions and the tangential force required to break them were directly proportional.

With reference to the relative motion between surface asperities, Holm (1946) hypothesized that the wear process was due to the collision of individual atoms on opposing asperities in the motion of one towards the other. On this basis, he established that the amount of material removed during these atomic interactions were a function of the properties of contacting materials and load applied over the contact, and so came to the following relationship, known as *Holm's wear equation*:

$$V = Z \frac{F_n}{P_m}, \qquad (9.4)$$

where V is the volume of material removed per unit sliding distance (it is to be noted that in this equation V has a different meaning with respect to Eq. (9.3)); Z is the probability of removal of an atom for atomic encounter, which depends on the properties of the contacting materials; F_n is the applied normal load; P_m is the so-called *flow pressure* of a worn surface, which is comparable to the material hardness.

Archard (1953) considered a greater number of variables that influence the wear process, i.e. wear mechanism, area of contact, contact pressure, sliding distance, surface asperities, motion and interaction of these opposing asperities, material properties, etc. Starting from the Holm's wear equation, and expanding its horizons, Archard assumed that the deformation occurring was of a plastic type, and formulated the following relationship, known as *Archard's wear equation*:

$$V = \frac{K}{H} F_n S,$$ (9.5)

where V (in mm^3) is the total volume of material worn away, K is the dimensionless *wear coefficient*, H (in MPa) is the hardness of the softest contacting materials, F_n (in N) is the compressive normal load between the contacting surfaces, and S (in mm) is the total rubbing distance. This equation implies the assumption that the volume of material worn away is independent of the area of contact.

In the same way as Reye's Eq. (9.3), also the Archard's wear Eq. (9.5) expresses that the volume of material removed by wear is proportional to the energy dissipated into the material. Since then $V = s_w A$, dividing both sides of Eq. (9.5) for time, t (in s), the Archard's wear equation can be written in terms of wear velocity, in the form:

$$v_w = \frac{s_w}{t} = \frac{K}{H} p v_g,$$ (9.6)

where $v_w = (s_w/t)$ is the wear velocity (in mm/s), P (in MPa) is the surface interface pressure, and v_g (in mm/s) is the sliding velocity. It is to be noted that also Eq. (9.6) can be obtained from Reye's hypothesis expressed by Eq. (9.2).

For two rubbing surfaces 1 and 2, Eq. (9.6) implies that the wear rate of surface 1 is proportional to the wear coefficient of material 1 when in contact with material 2, and inversely proportional to the material hardness of surface 1. Furthermore, the same wear rate, with the assumption of a constant coefficient of friction, is directly proportional to the rate of friction work. Actually, the product $p v_g = \sigma_H v_g$ that appears in this equation expresses, for a given coefficient of friction, the power lost due to friction, which is converted into heat.

The Archard's wear equation represents a simple but extremely effective model to describe the sliding wear between the asperities of two surfaces in relative motion. For this reason, it is the most used as well as the most cited in this field, also for a historical perspective error. Indeed, this equation, in both forms given by Eqs. (9.5) and (9.6), does not differ from the corresponding Eqs. (9.3) and (9.2), which express the Reye's wear equation, proposed almost a century earlier. The Reye's wear equation becomes very popular in Europe, and it is still thought in university courses of applied mechanics, but it was totally ignored in the scientific literature of the Anglo-Saxon school, where other non-new models established themselves, such as those due to Holm (1946) and Archard (1953), which have been developed much later.

However, it should be noted that, with Archard's wear equation in the form given by Eq. (9.6) or with the corresponding Reye's wear equation obtainable by Eqs. (9.3) and (9.2), we have the possibility of obtaining experimentally the values of the wear coefficient, K, for a particular design application, using for testing the same material combination and the same operating conditions of interest. The values of K are also inferable from the literature data obtained from laboratory tests for many material combinations. Typical values of K are: $\left(10^{-3} \leq K \leq 10^{-1}\right)$, for *severe wear* between two bodies without interposed abrasive particles, and $\left(10^{-6} \leq K \leq 10^{-3}\right)$, for *mild wear* between two bodies with interposed abrasive particles. In this last case, the values of K decrease from higher values to lower ones, depending on whether the concentration of interposed abrasive particles is high or low.

It is also to be noted that the same experimental equipment for measuring the wear coefficient related to abrasive wear can be used to measure the wear coefficient related to adhesive wear. The range of variability of the latter coefficient is, however, different depending on the properties of material in contact. In this case in fact K varies within the following ranges (see Rabinowicz 1995): from $K \cong 10^{-2}$ to $K \cong \left(10^{-7} - 10^{-8}\right)$, for identical metals; from $K \cong \left(10^{-2} - 10^{-3}\right)$ to $K \cong \left(10^{-7} - 10^{-8}\right)$, for compatible metals; from $K \cong \left(10^{-3} - 10^{-4}\right)$ to $K \cong \left(10^{-7} - 10^{-8}\right)$, for partly compatible metals; from $K \cong 10^{-4}$ to $K \cong \left(10^{-7} - 10^{-8}\right)$, for incompatible metals; from $K \cong 10^{-5}$ to $K \cong 10^{-7}$, for nonmetals or metal and non-metal combinations. These values of K decrease from higher values to lower ones, depending on whether the lubrication is absent, poor, good or excellent.

Finally, it should be noted that test data for wear coefficient show considerable scatter, typically ranging between a factor ± 4. This significant scatter of experimental measurements must not surprising, since adhesive wear is about proportional to the fourth or fifth power of coefficient of friction, which is itself affected by a considerable scatter. This behavior is notoriously connatural with every physical phenomenon that involves friction (Juvinall 1983).

9.4 Assessment of Tooth Wear Linear Progression

We have already said (see Sect. 9.2) that, for high load values and low pitch line velocities ($v_t < 0.5$ m/s), the minimum thickness, h_{min}, of the lubricant oil film may fall below 0.5 μm ($h_{min} < 0.5 \mu$m). In this case, the probability that the surface asperities of the mating tooth flanks come into direct contact is extremely high, whereby the abrasive wear becomes the main cause of gear damage. We have also reported that unified methods of evaluation of the wear load capacity of the gears do not yet exist. This does not mean that the gear designer should not pose the problem of how to assess the risk of this type of damage that, if not properly controlled, can have a devastating role on the service lifetime of the gears.

The assessment methods of the wear load capacity of the gears may be different, depending on whether the designer wants to check one or more effects caused by abrasive wear, which we described in Sect. 9.2. In this section, we focus our attention on the main method of the abrasive wear assessment, the one based on the imposition of a limit to the wear rate. This method ignores the run-in wear (the run-in determines an adaptation of the tooth flank surfaces and, consequently, a geometric change in the active profile), and assumes that the worn profile translates parallel to itself with respect to the initial not yet worn profile. This kind of wear model assumes a linear progression of wear with respect to time in the direction perpendicular to the profile, without any differentiation along the same profile. In other words, the thickness of material removed due to abrasive wear has no variable intensity along the active profile of the tooth, as Fig. 9.2 shows, but has a constant intensity.

The aforementioned assumptions, which form the basis of the model described below, are very approximate. In fact, as we have already said in Vol. 1, Sect. 3.8, where we discussed the profile shape variation due to wear, if we accept the validity of Reye's hypothesis or the equivalent Archard's wear equation, it is impossible to assure the constancy of the ratio (s_w/r_b) during the meshing cycle of gears having involute tooth profile. This is because the wear depth, s_w, which is proportional to the specific sliding, ζ, is like this continuously variable during the meshing cycle. Therefore, the assumptions described above cannot best interpret the actual wear phenomenon in involute profile gears and any model based on them is *a fortiori* approximate. However, the model described below attenuates the effects of these approximate assumptions, as it combines the theoretical basis highlighted above with a solid experimental basis, which processes data obtained with test gears working under operating conditions similar to the actual ones.

The basic relationship to be introduced in the assessment equations of the wear load capacity of the gears according to the method presented here is due to Plewe (1981), and is given by (see also Niemann and Winter 1983):

$$W_1 = c_{1T} \left(\frac{\sigma_H}{\sigma_{HT}} \right)^{1.4} \left(\frac{\rho_C}{\rho_{CT}} \right) \left(\frac{\zeta_w}{\zeta_{wT}} \right) 60n, \qquad (9.7)$$

where

- W_1 (in mm/h), is the *wear rate*;
- c_{1T} (in mm per revolution), is the *linear wear coefficient* of the test gear, which expresses the thickness worn away from the test gear, for each revolution and for each contact, in accordance with the aforementioned model of wear linear progression (it is therefore a factor *sui generis*, that depends on the operating conditions, summarized by the minimum thickness of the lubricant film at pitch point, h_c, calculated with Eq. (7.106), developed by Winter and Oster (1981), as well as on the material-lubricant-material combination);
- σ_{HT} (in N/mm^2), ρ_{CT} (in mm), and ζ_{wT}, which is a dimensionless quantity, are respectively the Hertzian stress, the equivalent radius of curvature at the operating pitch point, and the average specific sliding of the test gear, under the test conditions for which the wear factor, c_{1T}, was determined (as usual, the subscript, T, refers to test gear, and testing conditions);
- σ_H, ρ_C, and ζ_w are the corresponding quantities, regarding the gear pair to be calculated;
- n (in mm^{-1}), is the rotational speed of the gear wheel to be calculated.

The linear wear coefficient, c_{1T}, can be read from Fig. 9.4, as a function of the minimum thickness of the lubricant oil film at pitch point, h_c, for five different material combinations of the test gear, subjected to different lubrication conditions.

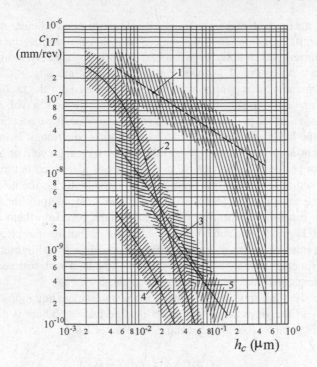

Fig. 9.4 Linear wear coefficient as a function of the minimum thickness of the lubricant film at pitch point, for five different material combinations and lubrication conditions

Each of these five material combinations is represented by a numbered curve, from 1 to 5. Each curve is obtained with a regression procedure of the experimental results, which are actually scattered within large areas, highlighted with hatching in the figure. Different lubrication conditions and different value of σ_{HT} correspond to each of these five curves, which represent respectively:

- material combination Eh/V (15CrNi6/42CrMo4), with $\sigma_{HT} = 635\,N/mm^2$, lubricated with mineral oils without anti-scuff additives (curve 1);
- material combination Eh/Eh (15CrNi6/15CrNi6), with $\sigma_{HT} = 1160\,N/mm^2$, lubricated with mineral oils without anti-scuff additives (curve 2);
- material combination V/V (42CrMo4/42CrMo4), with $\sigma_{HT} = 635\,N/mm^2$, lubricated with mineral oils without anti-scuff additives (curve 3);
- material combination NT/NT (31CrMoV9/31CrMoV9), with $\sigma_{HT} = 1160\,N/mm^2$, lubricated with mineral oils without anti-scuff additives (curve 4);
- material combination Eh/Eh (15CrNi6/15CrNi6), with $\sigma_{HT} = 1160\,N/mm^2$, lubricated with fluid grease without anti-scuff additives (curve 5).

The material symbols are as shown in Table 2.2.

For the proper use of Fig. 9.4, it should be noted that the lower and upper limits of the range of dispersion of the experimental results should be used for the gear wheel made of harder material and, respectively, for the gear wheel of softer material, when the material combination consists of materials having different hardness. The material combination corresponding to the curve 1 is an exception. The same lower and upper limits mentioned above should be used when the mating gear wheel has accuracy grade 6 or lower and, respectively, when the mating gear wheel has accuracy grade 7 or coarser. With a material combination consisting of a hard material coupled with a soft material, such as that represented by the curve 1, it is appropriate that the linear wear coefficient of the pinion made of a hard material is determined using curve 2.

The values of the linear wear coefficient, c_{1T}, shown in Fig. 9.4 are those determined by laboratory tests, using standard test gears ($m_T = 4.5$ mm, and $a = 91.5$ mm). They are important reference values, enormously useful when the actual test values concerning the to-be-designed gear unit or reliable values arising from similar past design experiences are not available. It is obvious that, when it is possible, the c_{1T} values should be determined using test gears (and therefore σ_{HT}, ρ_{CT} and ζ_{wT}) and operating conditions that are similar to those of the gear unit to be calculated. Of course, Eq. (9.7) can be directly applied when the values of the minimum lubricant film thickness at pitch point in testing and operating conditions coincide. The c_{1T} values obtained from Fig. 9.4 can be approximately used also for synthetic lubricants, for lubricant oils with anti-scuff additives, and for polluted oils.

The determination of the Hertzian stress, σ_H, which appears in Eq. (9.7), can be made using Eq. (2.38) in which we put $K_v K_{H\beta} K_{H\alpha} = 1$. This follows from the fact that, as a result of run-in wear that flattens and smoothes out the surface asperities due to roughness, the load distribution along the face width and along the tooth profile can be considered uniform, so that we have $K_{H\beta} = K_{H\alpha} = 1$. Moreover, due to the low value of the pitch line velocity, the dynamic factor K_v can be considered equal to unity ($K_v = 1$).

The equivalent radius of curvature at pitch point, ρ_C, of the gear pair to be calculated is determined using the following relationship:

$$\rho_C = \frac{1}{2} d_{b1} \tan \alpha_{wt} \frac{u}{(u+1)},\qquad(9.8)$$

where d_{b1} is the base diameter of the pinion, α_{wt} the transverse working pressure angle, i.e. the pressure angle at the pitch circle, and $u = z_2/z_1$ the gear ratio.

The *average specific sliding*, ζ_w, of the same gear pair of interest is to be determined using the following relationship:

$$\zeta_w = \zeta_{E1}\varepsilon_1 + \zeta_{A2}\varepsilon_2,\qquad(9.9)$$

where ε_1 and ε_2 are the addendum contact ratios of the pinion and wheel, given respectively by Eqs. (8.22) and (8.23), while ζ_{E1} and ζ_{A2} are the specific sliding of the pinion and wheel, respectively calculated at marked points E and A of path of contact. They are given by:

$$\zeta_{E1} = 1 - \frac{\rho_{E2}}{u\rho_{E1}} \tag{9.10}$$

$$\zeta_{A2} = 1 - \frac{u\rho_{A1}}{\rho_{A2}} \tag{9.11}$$

where ρ_{E1} and ρ_{E2} are respectively the radii of curvature of pinion and wheel at marked point E of path of contact, while ρ_{A2} and ρ_{A1} are respectively the radii of curvature of wheel and pinion at marked point A of path of contact. These radii of curvature can be calculated using the following relationship:

$$\rho_{E1} = \frac{1}{2}\sqrt{d_{a1}^2 - d_{b1}^2} \tag{9.12}$$

$$\rho_{E2} = a \sin\alpha_{wt} - \rho_{E1} \tag{9.13}$$

$$\rho_{A2} = \frac{1}{2}\frac{z_2}{|z_2|}\sqrt{d_{a2}^2 - d_{b2}^2} \tag{9.14}$$

$$\rho_{A1} = a \sin\alpha_{wt} - \rho_{A2}. \tag{9.15}$$

In these equations, the meaning of the symbols is the one already described in the previous chapters. With reference to Eq. (9.14), it is to keep in mind that z_2 is to be taken as positive, for external gear pairs, and as negative, for internal gear pairs, while $|z_2|$ is its absolute value.

It is to be noted that, once the wear rate, W_1, was calculated by Eq. (9.7), a rough estimate of the mass of worn particles referred to the unit of time, W_m (in mg/h), which has gone out to contaminate the lubricant oil, can be made using the following relationship:

$$W_m = W_1 A_{att}\rho \cong W_1 2mbz\rho, \tag{9.16}$$

where $A_{att} \cong 2mbz$ and $\rho = 7.85$ g/mm^3 are respectively the area of the active tooth flank surfaces, and density of the steel with which the gears are made. The same Plewe's Eq. (9.7) allows us to calculate the *allowable wear thickness*, W_{1P}, to be expressed in mm. To this end, it is sufficient to multiply W_1 by the number of hours of operation of the gear drive under consideration. This allowable wear thickness, W_{1P}, depends on the criterion used by the gear designer to control one or more effects of the abrasive wear. Below, we summarize some of the criteria that can be used.

1. Maximum deviation of the profile shape compared to the initial one, corresponding to the deterioration of one accuracy grade according to the ISO gear accuracy system, and therefore equal to about one third of the transverse single pitch deviation, f_{bt}, that is equal to:

$$W_{1P} \cong \frac{1}{3} f_{bt};$$

(9.17)

this criterion is to be used especially for speed reducing gear units designed to run at high speed for certain periods of time.

2. Maximum increase Δj_t of the transverse circular or circumferential backlash, j_t, and therefore equal to:

$$W_{1P} = \Delta j_t.$$

(9.18)

3. Maximum percentage of the thickness, s_h, of the hardened surface layer, corresponding to the hardening depth, when case-hardened gears are used, for which we use the following relationship:

$$W_{1P} = (\%)_{max} s_h.$$

(9.19)

4. Maximum allowable decrease of the initial thickness, s_a, at tooth top land, especially in the case of profile-shifted toothing characterized by high positive values of the profile shift coefficient; therefore, for example, if a reduction of this thickness equal to $m/10$ is admitted, we will have:

$$W_{1P} = s_a - m/10.$$

(9.20)

5. Maximum permissible mass of worn particles, W_{mP} (in mg), in the lubricant oil, for which, according to Eq. (9.16), we will have:

$$W_{1P} = \frac{W_{mP}}{(15.7\, mbz)}.$$

(9.21)

6. Minimum safety factor, S_{Flimit}, to be ensured for the worn teeth, in the case of gears permanently running at low speed, for which all of the above criteria are not usable. In this case, we use the following relationship:

$$W_{1P} \cong 2m \left(1 - \sqrt{\frac{S_{Flimit}}{S_F}} \right),$$

(9.22)

where S_F is the safety factor for tooth breakage. In the absence of specific reference values of this minimum safety factor, to be used for this type of gear assessment with respect to abrasive wear, Niemann and Winter (1983) recommend to use those concerning the tooth bending strength capacity of the gears, shown in Table 2.5.

Finally, the lifetime, L_{hw} (in h) of a gear wheel subjected to abrasive wear, can be evaluated using the following relationship:

$$L_{hw} = \frac{W_{1P}}{W_1};$$
(9.23)

this equation compares two non-homogeneous quantities, such as the wear rate, W_1, and the allowable wear thickness, W_{1P}.

9.5   Additional Considerations on the Gear Wear

The service life of the gears, as well as that of several machine elements, usually is limited by the wear of their active surfaces. Actually, the wear phenomenon causes a continuous removal of material from the tooth flanks, and thus a progressive alteration of their active surfaces due to the friction effect, which changes the dimensions and geometry of the teeth, up to the point of making them unsuitable for the design use. As the initial accuracy grade of the gear is compromised, its operation becomes increasingly irregular and disturbed, the resistant sections are reduced, and the dynamic loads increase, up to completely exhaust the gear load carrying capacity. Furthermore, the mechanical efficiency decreases, and the operation noise increases.

The friction between the mating surfaces in relative motion is the main cause of wear. As we mentioned in the previous sections, the friction that generates the wear on a given surface could be endogenous, i.e. caused from direct contact with the mating surface, itself subjected to wear, or it can be of exogenous origin, i.e. caused by solid abrasive particles that are interposed between the mating surfaces, transported by the lubricant flow. Some authors reserve the name of abrasive wear only to that related to this last exogenous phenomenon. Actually, these authors intend to highlight the fact that the nature of the endogenous wear (friction due to direct contact between the mating surfaces) differs a little from that of the exogenous wear (friction due to external abrasive particles).

The difference between the endogenous wear and exogenous wear deserves to be clarified. In the first case, two phenomena come into play: the mechanical impact between the asperities of the rough mating surfaces, and the molecular adhesion between the projecting particles of the same mating surfaces. In the second case, the molecular adhesion and the wear can be considered as the result of multiple streaks, as well as of the cutting of the metal asperities operated by the hardest abrasive particles. If then an abrasive grain penetrates into the gap between the mating surfaces, the wear of these surfaces is reinforced by the abrasive wear.

The service lifetime of a gear from the beginning of its operation until it is taken out of service for excessive wear can be divided into three stages. Figure 9.5a shows a typical wear curve, $s_w = s_w(t)$, and highlights these three stages.

The first stage is the *primary wear stage* or *run-in wear stage*, where usually the wear rate $W_1 = ds_w/dt$ is a decreasing function of the time; it corresponds to the running-in, which is especially characterized by the impact between the higher surface asperities, resulting from machining operations (Fig. 9.5b). In the run-in conditions, these asperities are cut or undergo a plastic deformation, for which their

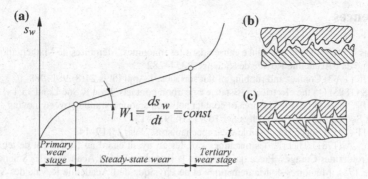

Fig. 9.5 **a** Typical $s_w = s_w(t)$ wear curve; **b** surface asperities resulting from machining operations; **c** flattened surface asperities due to running-in effect

average height decreases. The running-in is continued until the contact areas are more extensive than those at the bottom of the asperities (Fig. 9.5c). During the service lifetime of a gear unit, the running-in is a very important stage: in this stage, its operation must take place with a load as low as possible, in order not to cause an excessive heating of the contact area. Under these conditions, in fact, the scuffing phenomenon can be triggered, resulting in enhancement of the damage of the mating surfaces, as we have seen in the previous two chapters.

The second stage is the *secondary wear stage* or *steady-state wear*, and corresponds to the period of time of normal operation of the gear unit. It is characterized by a wear having a stationary trend in time, for which the wear rate $W_1 = ds_w/dt$ is a constant; it is given by the quotient of the worn material thickness removed in the time corresponding to this stage divided by this same time. Figure 9.5a shows that, during this secondary wear stage, the slope of $s_w = s_w(t)$ curve is a constant. Of course, the service lifetime of the gear unit is greater the more the wear rate is low.

The third stage is the *tertiary wear stage*, where the wear rate is growing dramatically, so this stage corresponds to the *catastrophic wear*. It is characterized by a non-permissible deviation of the tooth flank worn surfaces with respect to the initial ones, resulting in an equally unacceptable backlash between the mating surfaces, making the gear pair under consideration unsuitable to the use for which it was designed. Moreover, a too high value of the backlash greatly alters the lubrication conditions and contributes to increase the impact energy between the active tooth surfaces, for which the materials are subjected to work hardening, and their fragility increases. If adequate measures are not taken in time, the gears are put quickly out of action.

References

Amontons G (1699) De la resistance causée dans les machines. Mémoires de Mathematique et de Physique de l'Academie Royale des Sciences. 257–282

Archard JF (1953) Contact and rubbing of flat surfaces. J Appl Phys 24(8):981–988

Bidwell S (1883) On the electrical resistance of carbon contacts. Proc R Soc Lond 35:1–18

Bowden FP, Tabor D (1939) The area of contact between stationary and between moving surfaces. Proc R Soc Lond Ser A Math Phys Sci 169(938):391–413

Burwell JT (1957) Survey of possible wear mechanisms. Wear 1:119–141

Coulomb CA (1781) Théorie des machines simples en ayant égard au frottement de leur parties et a la roider des Corages. Piece qui remporté le Prix double de l'Academie des Sciences pour l'année 1781. Mémoires de Mathematique et de Physique de l'Academie Royale des Sciences, pp 145–173. A Paris, De l'Imprimerie de Montard, Imprimeur-Libraire de la Reine, de Madame, de Madame la Contesse d'Artois, & de l'Academie Royale de Sciences, 1782

Disaguliers JT (1734) A course of experimental philosophy. In: Innys W, Senex M, Longman T (eds). 1st edn, vol I. London

Disaguliers JT (1744) A course of experimental philosophy. In: Innys W, Senex M, Longman T (eds). 1st ed, vol II. London

Engel PA (1978) Impact wear of materials. Elsevier Science Publishing Co., Amsterdam

Faupel JH (1964) Engineering design: a synthesis of stress analysis and materials engineering. Wiley, New York

Ferrari C, Romiti A (1966) Meccanica applicata alle macchine. Unione Tipografica-Editrice Torinese, UTET, Torino

Giovannozzi R (1965) Costruzione di Macchine, vol I, 2nd edn. Casa Editrice Prof. Riccardo Pàtron, Bologna

Grubin AN, Vinogradova IE (1949) Investigation of the contact of machine components. In: Ketova KhF (ed) Central scientific research institute for technology and mechanical engineering (TsNIIMASH), Book no. 30, Moscow, (D.S.I.R. Translation no. 337)

Henriot G (1979) Traité théorique et pratique des engrenages 1, 6th edn. Bordas, Paris

Hertz HR (1882) Über die berehrung fester elastische korper. Journal fur die Reine und Augewandte Mathematik 156–171

Holm R (1946) Electric Contacts. Hugo Gerbers Forlag, Stockholm

ISO 4287:1997 Geometrical Product Specifications (GPS)—surface texture: profile method—terms, definitions and surface texture parameters

ISO 4288:1996 Geometrical Product Specifications (GPS)—surface texture: profile method—rules and procedures for the assessment of surface texture

Juvinall RC (1983) Fundamentals of machine component design. Wiley, New York

Leslie J (1829) Elements of natural philosophy, including mechanics and hydrostatics. In: Whittaker GB (ed), 2nd edn. Olivier and Boyd, London

Lohner T, Mayer J, Michaelis K, Höhn BR, Stahl K (2015) On the running-in behavior of lubricated line contacts. J Eng Tribol

Mac Curdy E (1938) Leonardo da Vinci notebooks. Jonathan Cape, London

Meng HC, Ludema KC (1995) Wear models and predictive equations: their form and content. Wear 181–183:443–457

Montgomery S, Kennedy D, O'Dowd N (2009) Analysis of wear models for advanced coated materials, dublin institute of technology, conference papers of school of mechanical and transport engineering, pp 1–17

Niemann G, Winter H (1983) Maschinen-Elemente, Band II: Getriebe allgemein, Zahradgetriebe-Grundlagen, Stirnradgetriebe. Springer, Berlin

Panetti M (1930) Meccanica Applicate alle Macchine, Parte Prima, Notizie Fondamentale. In: 2nd ed. La Tipografica, Torino

Panjkovic V (2014) Friction and hot rolling of steel. CRC Press, Taylor & Frencis Group, Boca Raton, FL

Peterson MB, Winer WO (1980) Wear control handbook. The American Society of Mechanical Engineers (ASME), New York

Plewe HJ (1981) Untersuchungen über den Abriebverschleiß von geschmierten, langsam laufenden Zahnradgetrieben. Diss, TU München

Rabinowicz E (1995) Friction and wear of materials, 2nd edn. Wiley, New York

Reye T (1860) Zur theorie der zapfenreibung. Der Civilingenieur 4:235–255

Sorensen MR, Jacobsen KW, Stoltze P (1996) Simulations of atomic-scale sliding friction. Phys Rev B 53:2101–2113

Stachowiak GW, Batchelor AW (2005) Engineering tribology. Elsevier Butterworth-Heinemann, Burlington

Varenberg M (2013) Towards a unified classification of wear. Friction 1(4):330–340

Winter H, Oster P (1981) Beanspruchung der Zahnflanken under EHD-Bedingungen. Konstruktion 33:421–434

Zum Gahr KH (1987) Microstructure and wear of materials. Elsevier Science Publishers B.V., Amsterdam

Chapter 10
Micropitting Load Capacity of Spur and Helical Gears

Abstract In this chapter, a general survey is first done on the micropitting damage of spur and helical gears, which manifests itself to the roughness scale. The mechanism that trigger this type of damage as well as the characteristics that distinguish it from those typical of macropitting (the classical pitting) are described. The problem to be solved for a reliable calculation procedure of micropitting load carrying capacity of gears are then analyzed and, to this end, the ideal characteristics of a general micropitting model are described. An interesting tribological-dynamic analytical model for cylindrical spur gears is then described, which also consists of a three-dimensional analytical-numerical contact sub-model and a multiaxial fatigue sub-model. The procedure for calculating the surface durability of spur and helical gears in accordance with the ISO standards is described, highlighting when deemed necessary how the formulae used by the same ISO are anchored to the theoretical bases previously discussed. Finally, for a better understanding of micropitting mechanisms, attention is drawn to the need to introduce, instead of traditional profile parameters, the areal field parameters that best describe the topography and texture of surfaces.

10.1 Generality

Micropitting is a pitting phenomenon, i.e. a rolling and sliding contact fatigue damage that, contrary to *macropitting* (the *classical pitting* or simply *pitting*), manifests itself to the roughness scale, rather than that of the nominal areas subject to rolling and sliding Hertzian contact. Therefore, micropitting and macropitting are two sides of the same coin, being related to the same fatigue damage phenomenon due to rolling and sliding contact, but the former sees only surface asperities interact with each other, while the latter sees interacting the entire macro-areas affected by contact. This phenomenological identity between micropitting and macropitting has been confirmed by SEM-(Scanning Electron Microscopy) tests, which unequivocally show that micropitting proceeds with the same fatigue damage process as the classical pitting discussed in Chaps. 2 and 5, with the variation that pits are extremely small.

© Springer Nature Switzerland AG 2020
V. Vullo, *Gears*, Springer Series in Solid and Structural Mechanics 11,
https://doi.org/10.1007/978-3-030-38632-0_10

In the gearing industry, the micropitting load carrying capacity of the gears has become a widespread problem with the growing use of hardened and heavily loaded gears (see Snidle et al. 2003; Evans et al. 2011, 2013; Clark et al. 2015). This problem, for some time well known to gear designers (Shotter 1981), has been studied and continues to be studied extensively by researches, who have tried and try to explain the mechanisms of generation and propagation of the fatigue damage induced by it, analyzing more or less general aspects and its particular characteristics (see Berthe et al. 1980; Olver et al. 1986; Höhn et al. 1996; Brandão et al. 2010; Benson et al. 2013; Moorthy and Shaw 2013; Li and Kahraman 2013a, b, 2014; Long et al. 2015; Al-Tubi et al. 2015; Clark et al. 2016; Morales-Èspejel et al. 2018; Ramdan et al. 2018; Zhou et al. 2019; Liu et al. 2019; Rycerz and Karidic 2019). The problem is particularly felt especially for specific applications, such as wind turbines gear drives (see Sheng Ed. 2010; Al-Tubi and Long 2013).

We already saw that, in gears made with relatively soft material, such as through-hardened gears, classical pitting, that is the fatigue damage due to rolling/sliding Hertzian contact, manifests itself in the form of pits having depths on the order of millimeters. With case-hardened gears, which can be carburized, nitrided, nitro-carburized, induction hardened, and flame hardened, pitting can manifest itself at a much smaller scale, with typical pits having depths on the order of microns (usually in the range 5–10 μm). With the naked eye, the areas affected by micropitting appear frosted, so micropitting is also known as *frosting*; *peeling* or *surface distress* are other names with which micropitting is called.

Generally, the micropitting damage is characterized by the formation of glazed or burnished surfaces, asperity scale microcracks, and asperity scale micropit craters. However, the numerous micropits that characterize the damaged surface can only be observed by a microscopic analysis, while a macroscopic examination with a unaided eye highlights only a gray surface. Since the light-scattering properties give the gear teeth affected by micropitting a typical gray appearance, some researchers have defined the related failure mode as *gray staining*.

Usually, with the naked eye, the microcrack openings are not distinguishable from the glazed surface. For this purpose, the use of optical microscopy techniques is necessary. When microcracks multiply and spread, the surface becomes undermined, with subsurface cavities having dimensions of the same order as the asperities, and multiple microscopic pits are formed. This micropitted surface appears visually frosted, highlighting black spots that represent micropits.

Although micropitting predominantly occurs in hardened and heavy loaded gears, it can also be found in all types of gears. Usually, the dedendum contact area of the driving gear wheel is the first to show signs of surface distress in the form of micropits, although micropitting can first manifest on the addendum contact area. In any case, micropitting process progressively alters the geometry of the tooth flank contacting surfaces, affecting the correct functionality of gear units. Unlike macropitting, micropitting occurs away from the pitch line, on the addendum and dedendum of the teeth. Most often the highest susceptibility to micropitting is the one related to both the driving and driven dedendums of the tooth flanks, where negative values of the specific sliding occur. The experimental evidence shows that fatigue failures take place about $(10 \div 20)$ μm below the theoretical tooth flank surfaces, and propagate at a shallow angle to the surface.

In some cases, micropitting is not destructive. This happens under suitable operating conditions for which the amount of generated micropits stabilizes and stops after a certain number of load cycles, also due to the effect of mild polishing wear during running-in, which determines a redistribution and consequential mitigation of the contact pressure. In these cases, micropitting is called *self-healing micropitting*. However, in other cases, unfortunately not infrequent, micropitting is often a precursor of larger surface failures. This is a *destructive micropitting*, as the continued cyclic contact between the mating tooth flanks results in fatigue failures in the form of macropitting, which usually starts from the boundaries of the micropitted areas. Thus, in these cases, micropitting has escalated into full-scale destructive pitting. It is however to remember that macropitting can also occur without the precursor of micropitting (Greco et al. 2013).

Destructive micropitting (as well as destructive macropitting) is a damage phenomenon that self-energizes, exalting itself. In fact, the variations in geometry of the contacting tooth flank surfaces due to excessive micropitting damage cause an increase in the transmission error intensity, with consequent high vibration levels and high dynamic contact forces, which further accelerate the rate of micropitting. In addition to the obvious problems of increasing noise as micropitting damage extends over the affected areas, the situation can degenerate, not only because micropitting can evolve into macropitting, but also because it can rapidly evolve into scuffing and even complete fracture of the teeth.

Micropitting damage is a type of surface fatigue damage that gradually develops when it is nucleated. Micropitting (as well as macropitting) normally initiates from some nuclei of stress raisers, e.g. indentations or asperities in the surface, or inclusions in the subsurface region. Early stages of micropitting are mainly due to surface roughness, because asperities in the tooth flank surfaces create stress risers, which are dynamically spread out over the tooth flanks during the meshing cycle. Once nucleated, micropitting damage grows and forms crack networks at a shallow angle to the tooth flank surface. This determines a sub-surface fatigue, which results in the separation of small material fragments from the tooth flank surfaces, causing flaking and leaving micropits after cracks have reached a critical depth (Sadeghi et al. 2009).

The phenomenology on which micropitting depends is very complex, because the entire tribological system between the meshing tooth flank surfaces is involved. Therefore, the topography of rolling and sliding contact surfaces, chemical and mechanical properties of materials, including those of the lubricant, as well as their mutual interactions play a key role. In this framework, the main factors influencing the micropitting occurrence are the tooth flank geometry (size, profile shift, lengthwise and profile modifications, surface finishing, roughness, waviness, etc.), gear materials (chemical composition, metallurgical structure, hardness, heat treatment, cleanliness, etc.), type of lubricant used (base oil, additives, viscosity, etc.), and operating conditions (speed, load, pressure of contact, temperature, ratio of slide to roll, etc.). In more detail, these micropitting damage factors can be grouped as follows:

- Specific film thickness, which is considered the main important influence parameter.
- Finer and larger components of the irregularities that characterize the texture of the contacting tooth flank surfaces. Between these two types of surface texture components, the finer asperities constitute the most important influence, but experimental evidence often calls into question also the larger components of the surface irregularities upon which the finer components are superimposed.
- Viscosity of lubricant, which must be as high as possible, so an amount of cool, clean, and dry lubricant must be used.
- Operating speed, which must be sufficient to guarantee an adequate oil film thickness.

According to the elastohydrodynamic theory, even under highest working loading, a thin lubricant oil film between the rolling and sliding contact surfaces is present. Plastic deformations at the contacting surface asperities are first generated, and their cumulative effect determines a particular aspect of the affected surface, which appears to be glazed when it is examined with reflected light optical instrumentation. The surface and immediately sub-surface material of these glazed surfaces is highly deformed and, due to the continued cyclic stressing, its ductility dramatically decreases and microcracks are formed. These cracks tend to swell and propagate at a shallow angle to the surface, that is at nearly parallel to the contact surface, at depth comparable to that where the asperities scale shear stresses occur with their maximum intensity.

To prevent micropitting damage, the specific film thickness should be as high as possible, using case-hardened gears characterized by an adequate surface finishing (a superfinishing, i.e. a mirrorlike finishing can effectively eliminate micropitting, or reduce it drastically), high-viscosity lubricants, and high speeds. Carburized, nitrided, and nitro-carburized gears show a better resistance to micropitting than flame-hardened and induction-hardened gears having the same hardness. Probably, this is due to the lower carbon content of the hardened surface layers related to these last two types of heat treatment.

10.2 Issues to Be Solved for a Reliable Calculation of Micropitting Load Carrying Capacity of Cylindrical Spur and Helical Gears

We have already seen that micropitting is a progressive fatigue phenomenon between two rolling and sliding contacting surfaces, which occurs on a micro-geometric scale, and is essentially due to severe localized stress concentrations at or very near the same contacting surfaces. Generally, the two contacting surfaces of a gear teeth pair are lubricated, but often the lubrication conditions are not of elastohydrodynamic type (i.e. full film lubrication), but rather of mixed and boundary type. Therefore, in these unfavorable conditions, the surface asperities collide each other and, under the combined rolling and sliding motion, cause a surface and sub-surface stress state characterized by large instantaneous peaks of both normal and shear stresses.

This stress state is a three-dimensional stress state. Furthermore, it is continuously time-varying, as it changes not only during the meshing cycle, but also during the entire lifetime of operation envisaged for the gear unit. Cyclic multi-axial stress amplitudes ensue, which result in premature crack nucleation in localized points or micro-zones within a very shallow layer of material, on the order of a few micrometers in depth. These localized cyclic stresses may be several times higher than the corresponding Hertzian stress. They sequentially generate individual micro-pits, according to their own multi-axial stress amplitudes, which may be clustered together or scattered over the mating surfaces involved, depending on the different local areal texture and topography of the same surfaces.

The number of micro-pits increases with the increasing in the number of contact cycles. In the case of gears, the micropitting phenomenon is influenced by all the factors that have an impact on the transient tribological characteristics of the lubricated contact between the mating surfaces in relative motion. The micropitting fatigue damage therefore strongly depends on the time-varying contact radii during the meshing cycle, distributions of the normal tooth load and friction force along the path of contact, rolling and sliding velocities, texture and topography of the contacting surfaces, lubrication conditions, temperature distributions, etc.

Despite the phenomenological identity between micropitting and macropitting, which we highlighted in previous section, a reliable calculation of the micropitting load carrying capacity of the gears has its own peculiarities that, at least for certain specific aspects, substantially differentiate them from those we have already described for macropitting (see Chaps. 2 and 5). To give an example, to calculate the stress state due to the cyclic contact between rough surfaces, the models used for macropitting usually use the half-space formulation based on potential theory for smooth surfaces, assuming that the difference in roughness between two points of the surface is much smaller than the distance between them. Therefore, these models ignore the effects of the local areal surface topography on the stress state to which the contacting surfaces are subjected. In the case of gears, this assumption is not satisfied, since the height of their asperities due to various finishing processes (grinding, shaving, honing, etc.) is quite significant. It follows that the aforementioned models, while show to be appropriate for macro-scale sub-surface failure nucleation (macropitting, i.e. pitting), have not proved to be equally suitable for describing surface or near surface micro-crack nucleation.

The development of a reliable general model for calculating the micropitting load carrying capacity of the gears is a very difficult problem to solve with the necessary accuracy, much more complex than the one related to the calculation of their pitting load carrying capacity. The above example is a clear evidence of this factual circumstance. A satisfactory general model that can serve this purpose should be able to simulate the following operating conditions, and correlated descriptive parameters and specific characteristic quantities, which are continuously variable (obviously, such a model, with the necessary simplifications, could also be used for surface durability calculations related to macropitting):

- The current elastohydrodynamic lubrication models, even the most sophisticated ones, have many limitations, despite the fact that they are able to handle one or more of the following aspects, such as line or point contacts, Newtonian or Non-newtonian behavior of the lubricant oils, isothermal or thermal contact conditions, relative motion between the two contacting surfaces, and surface asperities including separate or simultaneous treatment of wet and dry areas (for a summary of these models, see Li and Kahraman 2010a, b; Li et al. 2012). The numerous models developed so far, different from each other, have the common characteristic of being suitable for analyzing the contact between two rough surfaces having constant speeds, a time invariant geometry and subjected to a constant normal load. These models present certain shortcomings in handling lubricated gear contacts, which notoriously experience a number of time-varying parameters that go beyond the already highlighted transient effects of the variability of surface asperities along the path of contact and during the lifetime of the gear set to be designed. For a more reliable evaluation of the micropitting load carrying capacity of the gears, it is therefore necessary to develop more efficient elastohydrodynamic lubrication models that include all the aforementioned time-varying effects, even for the transient conditions occurring between different lubrication regimes (i.e. of elastohydrodynamic, mixed and boundary type) including the simultaneous presence of wet and dry rough contact regions and Newtonian or non-Newtonian behavior of the lubricant oils. Moreover, since the elastohydrodynamic behavior of a lubricated contact is strongly influenced by thermal effects, the EHL formulation of these models should include the use of energy equations able to evaluate the instantaneous variations of the main parameters that come into play, and thus modify, step by step, the input data of the calculations. It should be noted that the time-variability of the aforementioned elastohydrodynamic lubrication models is also linked to the time-variability of the macro-geometry and micro-geometry of the tooth flank surfaces involved.
- The operating conditions of a gear during the meshing cycle change continuously. The geometric parameters and, together with them, the kinematic parameters and normal load change instant by instant. As we have already seen in Chap. 2, the radii of curvature, ρ_1 and ρ_2 (see Fig. 2.6 of Vol. 1), of the two tooth flank profiles, which are in contact at any current point P of the path of contact, are continuously variable. In fact, at the point where the approach contact begins (for driving pinion, the point A in Fig. 2.10 of Vol. 1), they are minimum and maximum for the driving and respectively driven tooth flank profile. On the contrary, at the point where the contact ends (point E in the same Fig. 2.10 of Vol. 1), they are maximum and minimum for the driving and respectively driven tooth flank profile. Furthermore, the tangential (or rolling) velocity of a current point on the path of contact also changes continuously. During the approach contact, the contact point on the driving tooth profile is below the pitch point, and the tangential velocities of the same point, thought to belong to both mating tooth profiles, satisfy the inequality $v_{t1} < v_{t2}$ (see Fig. 3.6 of Vol. 1), while the relative or sliding velocity $v_r = v_g$ at the same point is negative, i.e. $v_r = v_{t1} - v_{t2} < 0$. On the contrary, always during the contact of approach the contact point on the driven tooth profile is above the pitch point. At

the pitch point, the two tangential velocities are the same ($v_{t1} = v_{t2}$), thus we have pure rolling, i.e. $v_r = 0$. Instead, during the recess contact, the opposite occurs in comparison to what happens for the approach contact. The contact point on the driving tooth profile is above the pitch point and tangential velocities of the same point, thought to belong to both mating surfaces, satisfy the inequality $v_{t1} > v_{t2}$, while the sliding velocity at the same point is positive, i.e. $v_r = v_{t1} - v_{t2} > 0$. On the contrary, during the contact of recess, the contact point on the driven tooth profile is below the pitch point. These time-variations in geometric parameters and related kinematic parameters (Fig. 10.1, with a different position of the current point of contact on path of contact with respect to that shown in Fig. 3.6a of Vol. 1, shows some of these time-varying quantities for a lubricated spur gear contact) influence the EHL formulation to be used for at least the following two aspects. First, the rolling velocity that appears in the Reynolds equation is time-varying. Second, the time-dependent sliding velocity, v_r, conditions the viscous shear stress that is required to determine the flow coefficient related to the non-Newtonian behavior of the lubricant.

- The areal surface texture and more generally the topography, which characterize the micro-geometry of the contacting surfaces, are also time-varying. All the quantities and parameters that are introduced to characterize the areal surface texture and topography in fact vary not only during the running-in stages, but also after the

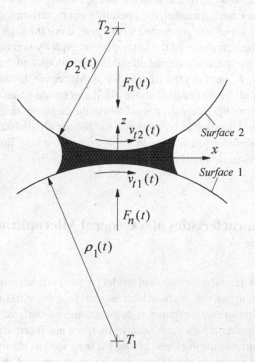

Fig. 10.1 Some time-varying quantities of a lubricated spur gear contact

running-in stages, for the entire lifetime planned or expected for the gear set to be designed. These time-variations in the areal surface texture as well as in the areal field parameters used for its description influence both the EHL formulation and the Hertzian contact stresses. In fact, for an accurate EHL formulation, it is necessary to introduce the time-variation of the areal surface texture, to be considered in accordance with the instantaneous velocities of the current point along the path of contact. Likewise, for an appropriate formulation of a contact theory, which is able to provide reliable results in terms of contact stresses between the rolling and sliding mating surfaces, with the lubricating oil film interposed, it is necessary to renounce the assumption of smooth surfaces, considering the real micro-geometry of the surfaces involved and its variability over time (see Lo 1969; Johnson 1985).

- The normal force, F_n, applied to contacting tooth surfaces in any current point along the path of contact is not constant, but variable along it. This variability depends on both static and dynamic effects. The static effects are related to the fact that the number of tooth pairs that come into play to transmit the design load fluctuates between two integers. The dynamic effects are even more complex, as they are related to both internal causes of the gear set to be designed and external to it. Among the internal causes, the elastic deformations of all the mechanical components that make up the gear drive, including its housing, the intentional tooth profile modifications, the non-intentional tooth profile modifications due to wear, and the manufacturing and assembly errors occupy a relevant place. Among the external causes, the operating characteristics of the driving and driven machines, which are connected to the gear drive through appropriate couplings, and the characteristics of the latter, play an equally significant role. All the aforementioned internal and external effects cause sudden and drastic variations in the normal force, F_n, and related friction force. Moreover, the transmission error can generate impulsive forces and, above all, it can excite resonance phenomena. Of course, an accurate calculation model should consider these impulsive and resonance effects. Unfortunately, however, the usual calculation models consider impulsive and resonance phenomena in a non-exhaustive way, limiting themselves to assuming quasi-static tooth normal forces.

10.3   Ideal Characteristics of a General Micropitting Model for Gears

The development of a satisfactory general model for the evaluation of the micropitting load carrying capacity of the gears, which is able to follow the nucleation of the micropits and their progressive expansion on the mating tooth flank surfaces, involves the solution of several complex problems. In fact, we are faced with quantities that are continuously variable during the meshing cycle as well as during the foreseeable lifetime of the gear set under consideration. Moreover, some of these quantities are difficult to quantify, as it happens notoriously for those that are related to friction phenomena, which are unavoidable in the case herein.

We must first remember that the time-varying forces applied to the teeth pairs in simultaneous meshing, due to the transmitted power and resistant power, and the time-varying friction forces, due to the combined rolling and sliding actions during the meshing cycle (these forces are the main sources of excitation of the mechanical system), determine a dynamics that is characterized by motions both in the direction of the line-of-action (LOA-motions) and in the direction perpendicular to it (off-line-of-action, OLOA-motions). These motions are coupled together, also by virtue of friction moments that stress the shafts in torsional direction. A further source of coupling between OLOA and LOA motions consists of the dissipation forces due to viscous shear stresses within the lubricating oil film, which constitute the main source of gear mesh viscous damping.

On the other hand, the fluctuations of the dynamic tooth forces and the rolling and sliding velocities involved in the meshing cycle have a strong impact on the tribological behavior of the gear pair to be designed, which is also time-dependent as well as continuously variable, given the transient conditions that are typical of mixed lubrication. This tribological behavior influences the contact pressure, lubricant film thickness, lubricant viscosity, friction forces and resulting temperature distribution. It follows that dynamic behavior and tribological behavior, both time-dependent, influence each other, greatly complicating the problem of the conception of a combined tribological-dynamic model that is able to capture with due approximation all the influences involved.

This combined tribological-dynamic model must then necessarily be coupled with a three-dimensional contact model that is able to capture the real contact conditions between the rough, lubricated or non-lubricated tooth flank surfaces, providing the three-dimensional surface and sub-surface stress state. The latter model should in turn be linked with a multi-axial fatigue model capable of providing the average values and amplitudes of stress components on the basis of which reliable micropitting severity indices can be determined, in relation to the goals of this chapter.

The scientific literature on this subject undoubtedly shows that at present there is no such a general model, consisting of a tribological sub-model combined with a dynamic sub-model and a three-dimensional contact sub-model and combined multi-axial fatigue sub-model, able to achieve the desired goals. For the conception and development of a general model having the aforementioned characteristics, the road ahead is still long and difficult, because the problems to be solved are very numerous and complex. Without neglecting these problems, some of which are also outlined below, albeit briefly, we want to give here an idea of the ideal characteristics that a general micropitting model for gears should have.

In this regard, the flowchart in Fig. 10.2 shows the block diagram of an ideal general model that could allow the achievement of the aforementioned goals. The individual blocks of which it is made describe the main aspects of the calculation procedure, aimed at determining micropitting severity indices capable of providing the gear designer with a reliable reference framework on the design choices to be

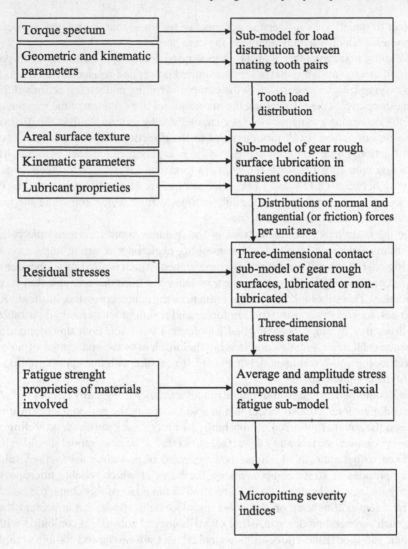

Fig. 10.2 Flowchart of an ideal general micropitting model for gears

made as well as possible ordinary and extraordinary maintenance interventions to be carried out.

Each of the blocks of this general model constitutes a sub-model, characterized by its own input and output data. The various component sub-models are then linked in a sequence designed to obtain results of sure engineering value with the minimum computational effort. It should be noted that the determination of the input data for some sub-models may involve the preparation of specific auxiliary calculation models, such as those to be used to determine the parameters describing the areal surface texture of the contacting rough surfaces starting from experimental areal measurements, the temperature distributions within the lubricating oil film and on the metallic surfaces in relative motion of the lubricated contact, etc.

As shown in Fig. 10.2, the first sub-model concerns the load distribution between the various tooth pairs in simultaneous meshing, and its output, consisting of the forces applied on the teeth, whose distributions vary along the path of contact, serves as an input data for the second sub-model. This first sub-model must be prepared considering the real operating conditions of the gear drive to be designed, and therefore within the framework of the mechanical system of which it is a part, simulating also the effects of the transmission error, which is the main source of dynamic excitation. This in order to be able to evaluate, with the greatest possible accuracy, the influence of the dynamic effects that often determine a significant increase of tooth forces. The input data of this first sub-model are the torque spectrum, the geometric characteristics of the gears involved, including the intentional profile and lengthwise modifications, the elastic characteristics of all the components of the mechanical system of which the examined gear drive is a part, and the kinematic parameters that describe the operating conditions of the same gear.

The second sub-model concerns the determination of the lubrication parameters, in transient conditions of elastohydrodynamic (or full film), mixed or boundary lubrication, and its output, consisting of the distributions of normal and tangential (or friction) forces per unit area applied to the contacting surfaces, constitutes the input of the third sub-model. In order for this sub-model to be as general as possible, it must be able to handle not only the lubricated contact between wet rough surfaces, but also the non-lubricated contact between dry rough surfaces. The input data of this second sub-model are the parameters that define the areal surface texture, the gear working kinematics, and the lubricant properties. To obtain reliable results, it is necessary to accurately determine the parameters that describe the areal surface texture of contacting rough surfaces, developing suitable auxiliary models that, starting from the experimental areal measurements on the gear tooth surfaces obtained with the usual manufacturing processes and characteristics of materials involved, are able to provide predictive indications on the time-variability of the same texture during the lifetime foreseen for the gear under consideration. It is also necessary to introduce values of the kinematic parameters that comply with the design load spectrum. Finally, it is necessary to introduce values of the parameters describing the properties of Newtonian and non-Newtonian lubricant oils in auxiliary models for determining the temperature distributions within the bodies affected by the lubricated contact, i.e. lubricant oil and metallic components in relative motion. In addition, the thermal effects due to the frictional heat flow between the rough mating surfaces in relative motion should be included.

The third sub-model concerns a three-dimensional model of contact between rough, lubricated or non-lubricated surfaces, aimed at calculating the three-dimensional stress state in the areas affected by the lubricated and non-lubricated contacts. This model, in order to accurately capture the three-dimensional stress state on the surfaces involved, and immediately below them, for a depth that includes the very shallow layers of material affected by heat treatment, should be able to consider the hardness and strength variations with depth. Moreover, since at least for a part of lifetime, the contact is extremely localized, involving the surface asperities, this sub-model should be conceived so as to be able to handle not only elastic stress fields, but also elastic-plastic and plastic stress fields. Given the known limits of the theoretical-analytical models, this sub-model can only use numerical methods (FEM-Finite Element Method, BEM-Boundary Element Method, FD-Finite Differences, etc.), possibly supported by suitable analytical methods (for example, step-by-step integration methods) and modal analysis (see Garro and Vullo 1979; Bargis et al. 1980a, b, c). The quantities that define this three-dimensional stress state and its time-variability constitute the output data of this sub-model as well as the input data of the next sub-model. Notoriously, the aforementioned three-dimensional stress state is influenced by several factors, among which the residual stresses often play a very significant role, which cannot be overlooked. To consider residual stresses, it is necessary to develop an adequate auxiliary model that is able to process the data available in this regard, usually consisting of experimental measurements on analogous gears produced with the same technologies, and to transform them into other input data of the sub-model in question.

The fourth sub-model concerns a multi-axial fatigue model, which must be able to supply, as output data, the alternating stress components, and therefore the average stresses and stress amplitudes, to be used to quantify reliable values of micropitting severity indices. In this sub-model, several multi-axial fatigue criteria can be introduced, optionally to be used alternatively, depending on the type of gear to be evaluated. This sub-model should be developed in such a way as to be able to capture the crack propagation duration from the nucleation stage, up to the formation of a micro-sized pit. The input data of this fourth sub-model consists not only of the output data of the third sub-model, but also of other data, consisting essentially of the parameters that describe the material fatigue strength properties, derived from databases or processed starting from experimental measurements carried out on gears similar to those to be designed or evaluated.

In the current state of knowledge, a general model of micropitting for gears, having the aforementioned peculiarities, is a pure illusion. Moreover, with the appropriate variations and simplifications of the case, such a model would be able to give reliable responses concerning the surface durability, i.e. the macropitting or pitting of the gears. In today's reality, it is not possible to face all the complex and numerous problems described above with the breadth of the horizon necessary to develop such a sophisticated ideal model. However, several models with a more circumscribed horizon have been developed to date. This is the topic of the next two sections.

10.4   A Tribological-Dynamic Model for Cylindrical Spur Gears

Here it is not appropriate to dwell on the numerous models described in the current scientific literature, concerning the micropitting load carrying capacity of the gears.

They are to be considered in the same way as the sub-models shown in Fig. 10.2, moreover with a limited horizon, as they almost always deal with the topic under discussion introducing simplifying hypotheses that often overlook important influences. However, we want to dwell here on an interesting tribological-dynamic model for cylindrical spur gears, developed by Li and Kahraman (2013c), which is perhaps to be considered the first model that combines a gear dynamic sub-model with a gear tribological sub-model in a unified model to capture their mutual tribological-dynamic interactions. Below we describe both the sub-models and their combined model, highlighting the basic concepts and their limitations.

10.4.1 Rigid-Body Dynamic Sub-Model

Consider the cylindrical spur gear pair shown in Figs. 2.6 and 2.10 of Vol. 1 and summarize the main geometric and kinematic parameters indicated in these two figures as shown in Fig. 10.3. This figure highlights a current point of contact, Y, on the path of contact, the corresponding radii, r_{Y1} and r_{Y2} (i.e. the distances of point Y from centers O_1 and O_2), and roll angles, ϑ_{Y1} and ϑ_{Y2}, as well as the Cartesian coordinate system $T_1(x, y)$, the instantaneous tangential tooth surface velocities,

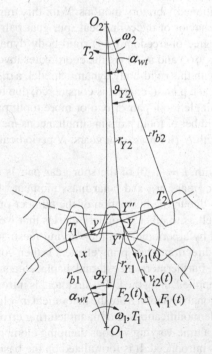

Fig. 10.3 Some basic quantities of an involute spur gear pair and roll angles related to a current contact point Y on path of contact

$v_1(t)$ and $v_2(t)$, and the instantaneous surface friction forces, $F_1(t)$ and $F_2(t)$. The meaning of the other symbols shown in the same figure remains unchanged.

The external constant torque T_1, acting in the clockwise direction, is applied to the driving pinion. It is balanced by the resistant torque T_2 applied to the driven gear wheel, which acts in the same direction as the torque T_1. At a given current point of contact, Y, which is between the end points A and E of path of contact (see Fig. 2.10 of Vol. 1), the equivalent cylinders introduced by Buckingham (1949) to describe the related instantaneous contact of the tooth pair, have time-varying radii, $\rho_1(t) = \rho_{1,Y}(t)$ and $\rho_2(t) = \rho_{2,Y}(t)$. The already defined instantaneous tangential tooth surface velocities, $v_1(t) = v_{1,Y}(t)$ and $v_2(t) = v_{2,Y}(t)$, and the instantaneous surface friction forces, $F_1(t) = F_{1,Y}(t)$ and $F_2(t) = F_{2,Y}(t)$, are also time-varying quantities and their vectors have the local OLOA-direction as application line (Fig. 10.3). It should be noted that the instantaneous tangential tooth surface velocities $v_1(t)$ and $v_2(t)$ are not to be confused with the kinematic tangential velocities $v_{t1}(t)$ and $v_{t1}(t)$; in this regard, see the following section.

The involute geometry then allows us to state that the displacements due to vibratory motions of the two gear members along the LOA-direction, denoted as $y_1(t)$ and $y_2(t)$, are coupled with the displacements due to the same vibratory motions along the OLOA-direction, $x_1(t)$ and $x_2(t)$, which are solely activated by the surface friction forces that occur at the various meshing interfaces. The rotational vibration amplitudes, $\vartheta_1(t) = \vartheta_{Y1}(t)$ and $\vartheta_2(t) = \vartheta_{Y2}(t)$, about centers O_1 and O_2, are to be added to the aforementioned vibratory motions. With this framework, for the analysis of the dynamic behavior of the cylindrical spur gear pair under consideration, we can use the six degree-of-freedom (dof) rigid-body dynamic model shown in Fig. 10.4, where $x_j(t)$, $y_j(t)$ and $\vartheta_j(t)$ are the coordinates involved (with $j = 1, 2$). It is noteworthy that, with this rigid-body dynamic model, a discrete variation of the load distribution along the path of contact is considered, due to the fact that, during the meshing cycle, a single tooth pair or two or more tooth pairs are in contact. In this regard, the total number of tooth pairs in simultaneous meshing at a given mesh position is indicated with N (for most spur gears, N periodically fluctuates between 1 and 2).

Each member j (with $j = 1, 2$) of the spur gear pair is considered as a rigid disk of radius $r_j = r_{bj}$, mass m_j, and polar mass moment of inertia J_j. The gear mesh stiffness, mainly due to the fluctuation of the number of tooth pairs in simultaneous meshing as well as to other secondary causes that we already described in Sect. 1.6, is simulated by a periodically time-varying mesh spring element, $k_m(t)$, acting along the LOA-direction. This spring element is then considered to be subject to normal backlash, j_n. Furthermore, an external displacement excitation, $\varepsilon_s(t)$, also this time-varying and applied along the LOA-direction, is introduced to simulate any intentional or unintentional deviation from the theoretical involute shape of teeth. The intentional tooth profile modifications and manufacturing errors are thus simulated. Finally, a dashpot, i.e. a time-varying viscous damping element, $c_m(t)$, acting along the OLOA-direction is introduced. It is formulated on the basis of the lubricant viscous shear power dissipation, which transforms a certain amount of kinetic energy into frictional heat, thus dissipating the power necessary to constrain the relative

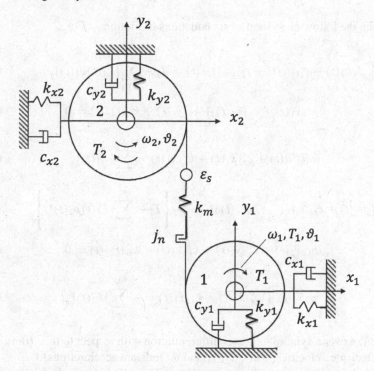

Fig. 10.4 Six degree-of-freedom rigid-body dynamic model of a spur gear

motion amplitudes at the various interfaces between the tooth pairs in simultaneous meshing.

Based on experimental results obtained by other researchers, the aforementioned authors assume quasi-static values of $k_m(t)$ and $\varepsilon_s(t)$. However, these quantities notoriously can influence the dynamic response of the system to a certain extent. The same authors, however, propose to introduce a general methodology that would allow the replacement of the rigid-body dynamic model by a deformable-body dynamic model, which obviously would have the drawbacks of significantly increased computational effort, with the related costs. As Fig. 10.4 shows, each rigid disk simulating the two members of the gear pair is restrained along the LOA-direction by a spring of stiffness, k_{yj}, and a dashpot of damping coefficient, c_{yj}, and, along the OLOA-direction, by a spring of stiffness, k_{xj}, and a dashpot of damping coefficient, c_{xj} (with $j = 1, 2$). These spring and dashpot elements simulate the flexibilities of the gear shafts and bearings. In addition, both the rigid disks are subject to torsional damping, with damping coefficient c_{tj}, to consider the viscous losses caused by the bearings supporting the shafts. As usual, the authors assume damping forces and torques respectively proportional to the linear and angular velocities and, therefore, damping coefficients of the viscous type (see Warburton 1976; Den Hartog 1985).

With reference to the positive directions of the alternating rotational displacements, $\vartheta_1(t)$ and $\vartheta_2(t)$, and assuming that the external torques, T_1 and T_2, are constant, we obtain the following system of six equations of motion:

$$J_1\ddot{\vartheta}_1(t) + c_{t1}\dot{\vartheta}_1(t) + r_{b1}k_m(t)\delta(t) = T_1 + \sum_{n=1}^{N}[F_1(t)\rho_1(t)]_n \tag{10.1}$$

$$m_1\ddot{y}_1(t) + c_{y1}\dot{y}_1(t) + k_{y1}y_1(t) + k_m(t)\delta(t) = 0 \tag{10.2}$$

$$m_1\ddot{x}_1(t) + c_{x1}\dot{x}_1(t) + k_{x1}x_1(t) = \sum_{n=1}^{N}[F_1(t)]_n \tag{10.3}$$

$$J_2\ddot{\vartheta}_2(t) + c_{t2}\dot{\vartheta}_2(t) - r_{b2}k_m(t)\delta(t) = -\left\{T_2 + \sum_{n=1}^{N}[F_2(t)\rho_2(t)]_n\right\} \tag{10.4}$$

$$m_2\ddot{y}_2(t) + c_{y2}\dot{y}_2(t) + k_{y2}y_2(t) - k_m(t)\delta(t) = 0 \tag{10.5}$$

$$m_2\ddot{x}_2(t) + c_{x2}\dot{x}_2(t) + k_{x2}x_2(t) = -\sum_{n=1}^{N}[F_2(t)]_n, \tag{10.6}$$

where a dot over a symbol indicates differentiation with respect to time (thus, $\dot{x}_j$, $\dot{y}_j$ and $\dot{\vartheta}_j$ indicate velocities, while $\ddot{x}_j$, $\ddot{y}_j$ and $\ddot{\vartheta}_j$ indicate accelerations).

Under the summation symbols that appear in Eqs. (10.1–10.6), the friction forces, $F_1(t)$ and $F_2(t)$, related to the contact of the generic tooth pair, n, where the radii of curvature are respectively $\rho_1(t)$ and $\rho_2(t)$, are to be introduced. In this regard, it should be remembered that the index of summation, n, has 1 and N as lower and respectively upper bound of summation. The displacement function, $\delta(t)$, appearing in Eqs. (10.1) and (10.4), is a non-linear function, which however can be represented by a succession of linear piecewise. It is therefore expressed with the following three relationships:

$$\delta(t) = \varepsilon_d(t) - \varepsilon_s(t) - j_n/2 \tag{10.7}$$

$$\delta(t) = 0 \tag{10.8}$$

$$\delta(t) = \varepsilon_d(t) - \varepsilon_s(t) + j_n/2, \tag{10.9}$$

which are valid respectively for $\varepsilon_d(t) - \varepsilon_s(t) > j_n/2$, $|\varepsilon_d(t) - \varepsilon_s(t)| \leq j_n/2$, and $\varepsilon_d(t) - \varepsilon_s(t) < -(j_n/2)$. In these equations, $\varepsilon_d(t)$ is the relative dynamic gear mesh displacement, i.e. the dynamic transmission error, given by:

$$\varepsilon_d(t) = r_{b1}\vartheta_1(t) + y_1(t) - r_{b2}\vartheta_2(t) - y_2(t). \tag{10.10}$$

Equations (10.7–10.9), together with the related validity conditions, represent respectively: the first, the linear motion without tooth separation; the second, the tooth separation, with single-side impact; the third, the tooth separation also associated with the back-side impact, resulting in double-sided contact. This last condition has been provided for completeness of discussion, as there is no experimental evidence that demonstrates the occurrence of double-sided impact in spur gears.

The individual dynamic normal tooth force $F_{n,dyn}(t)$, is determined as a function of the quasi-static normal tooth force, $F_{n,stat}(t)$, using the following approximate relationship:

$$F_{n,dyn}(t) = F_{n,stat}(t)\frac{F_{m,dyn}(t)}{F_{m,stat}}, \tag{10.11}$$

where $F_{m,dyn}(t) = k_m(t)\delta(t)$ is the dynamic gear mesh force and $F_{m,stat} = T_1/r_{b1}$ is the quasi-static gear mesh force, coinciding with the total static mesh force transmitted. The quasi-static normal tooth force, $F_{n,stat}(t)$, is determined using a gear load distribution program. The dynamic normal tooth force, $F_{n,dyn}(t)$, is used in the elastohydrodynamic analysis to determine the magnitudes of the gear mesh viscous damping as well as the transient tooth friction forces, $F_1(t)$ and $F_2(t)$, together with the most critical influences of the dynamic behavior of the system on the parameters that define the lubrication conditions. Finally, the dynamic bearing forces in the OLA- and OLOA-directions are defined respectively as $F_{byj}(t) = k_{yj}y_j(t) + c_{yj}\dot{y}_j(t)$ and $F_{bxj}(t) = k_{xj}x_j(t) + c_{xj}\dot{x}_j(t)$, with $j = 1, 2$.

10.4.2 EHL-Tribological Sub-model

Under dynamic conditions, the instantaneous tangential velocities, $v_1(t)$ and $v_2(t)$, at a current point Y on the path of contact (Fig. 10.3) consist of the sum of the kinematic tangential velocities, $v_{t1}(t)$ and $v_{t2}(t)$ given by Eqs. (3.59) and (3.60), and the fluctuating dynamic components due to vibratory torsional and translational motions. Therefore, these velocities can be expressed as follows:

$$v_1(t) = v_{t1}(t) + \rho_1(t)\dot{\vartheta}_1(t) - \dot{x}_1(t) \tag{10.12}$$

$$v_2(t) = v_{t2}(t) + \rho_2(t)\dot{\vartheta}_2(t) - \dot{x}_2(t), \tag{10.13}$$

where $v_{t1}(t) = \rho_1(t)\omega_1$ and $v_{t2}(t) = \rho_2(t)\omega_2 = \rho_2(t)\omega_1(z_1/z_2)$. Of course, the terms $\left[\rho_1(t)\dot{\vartheta}_1(t) - \dot{x}_1(t)\right]$ and $\left[\rho_2(t)\dot{\vartheta}_2(t) - \dot{x}_2(t)\right]$ due to vibratory motions influence the elastohydrodynamic contact behavior between the meshing teeth in addition to the dynamic tooth force. These vibratory components are included in the sub-model considered here for completeness of discussion, despite the fact that theoretical studies and experimental evidence have shown that their influence on EHL behavior is entirely secondary. Obviously, in these dynamic conditions, the average rolling velocity, v_{rt}, also known as cumulative semi-velocity, and the sliding velocity, v_{st}, are given by:

$$v_r(t) = \frac{1}{2}[v_1(t) + v_2(t)] \tag{10.14}$$

$$v_s(t) = v_1(t) - v_2(t). \tag{10.15}$$

The problem to be faced with this EHD-tribological sub-model is however a triple problem of line contact because, depending on operating conditions that mainly include load, velocity, temperature and surface asperities, the lubrication conditions can range from elastohydrodynamic lubrication (or full film lubrication) to mixed or even boundary lubrication. This often happens in most automotive and aerospace gearing applications. Therefore, this sub-model must be able to capture the transition from conditions where the contact between the mating surfaces is mediated by a lubricant film thickness up to conditions where the hydrodynamic viscous fluid film and surface asperity contacts coexist.

For line contact, the governing equation concerning the analysis of the fluid flow conditions in the contact area without asperity interactions is the one-dimensional transient Reynolds equation (see Reynolds 1876; Ferrari and Romiti 1966), which can be written in the form (see Fig. 2.13):

$$\frac{\partial}{\partial x}\left\{\frac{\rho(x,t)h^3(x,t)}{12\eta(x,t)}\cos h\left[\frac{\tau_m(x,t)}{\tau_0}\right]\frac{\partial p(x,t)}{\partial x}\right\} = v_r(t)\frac{\partial[\rho(x,t)h(x,t)]}{\partial x}$$
$$+ \frac{\partial[\rho(x,t)h(x,t)]}{\partial t}. \tag{10.16}$$

Evidently, apart from the insignificant change of the variable (x instead of y), this equation is more complex than Eq. (2.25), for a twofold reason: in fact, on the one hand, all the other variables that appear in it, already known, are also time-depending, as the same equation shows; on the other hand, to be able to capture the non-Newtonian rheological behavior of the lubricant fluid, according to what Ree and Eyring proposed (see Ree and Eyring 1955a, b; Wang et al. 1991; Ehret et al. 1998), instead of the term $(\rho h^3/12\eta)$, the approximate *Ree-Eyring fluid flow coefficient* given by $\{[\rho(x,t)h^3(x,t)]/[12\eta(x,t)]\}\cos h[\tau_m(x,t)/\tau_0]$ appears. In this approximate coefficient, τ_0 is the *lubricant reference shear stress*, $\tau_m(x,t) = \tau_0/\sin h\{\eta(x,t)v_s(t)/\tau_0 h(x,t)\}$ is the *lubricant mean viscous shear stress*, $\eta(x,t)$ is the lubricant dynamic viscosity, $h(x,t)$ is the local lubricant film thickness, and $\rho(x,t)$ is the instantaneous local lubricant density. Of course, Eq. (10.16) describes the time-varying local flow conditions of the lubricant in the contact area where $h > 0$, i.e. where a sufficient lubricant film thickness exists, which separates the two tooth flank surfaces in relative motion.

However, Eq. (10.16) is not able to describe the contact conditions between the two aforementioned surfaces when their asperities come into direct contact. It must therefore be reformulated, also because, in the case of severe interactions between the surface asperities, residual errors in the term related to the *Poiseuille flow* (i.e. the *pressure flow* given by the left-hand side in Eq. (10.16), the one for which the velocity of the lubricant fluid depends on the pressure gradient) can determine significant numerical instability. For this reason, as usual, the following reduced form of the Reynolds equation is used:

$$v_r(t)\frac{\partial[\rho(x,t)h(x,t)]}{\partial x} + \frac{\partial[\rho(x,t)h(x,t)]}{\partial t} = 0. \tag{10.17}$$

In this equation, only the terms on the right-hand side of Eq. (10.16) come into play. They represent respectively the *Couette flow* or *shear flow*, i.e. the flow induced by the average rolling velocity, and the *squeeze flow*, i.e. the flow induced by the motion of the boundary surfaces in relative motion.

Simultaneous use of both Eqs. (10.16) and (10.17) allows to analyze, albeit in an approximate way, the mixed EHL behavior of the contact, considering both the elastohydrodynamic and metal-to-metal contact pressures, under the hypothesis that the transition between these two lubrication regimes is smooth. This type of numerical approach, which obviously requires the discretization of both equations, has the advantage of a high numerical stability, but its accuracy strictly depends on the accuracy of the predictive calculation of the lubricant film thickness. In addition, an accurate model of the areal surface textures, also able to capture their evolution during the entire expected lifetime, is required, also because it is necessary to preliminary define a threshold film thickness value, $[h(x,t)]_{min}$, which constitutes the interface between the fields of use of Eqs. (10.16) and (10.17). The sub-model discussed here does not consider, at least theoretically, the evolution of the areal surface texture during the gear lifetime. Furthermore, according to considerations made by Zhu (2007), Li and Kahraman (2010a) give some indications on the aforementioned threshold film thickness value.

Limiting the analysis to the elastic deformations only (remember that in the metal-to-metal contact between the surface asperities plastic deformations also occur under not high loads), the instantaneous local film thickness can be expressed as follows:

$$h(x,t) = h_0(t) + g_0(x,t) + f(x,t) - Sa_1(x,t) - Sa_2(x,t), \tag{10.18}$$

where $h_0(t)$ is the reference film thickness, $g_0(x,t) = x^2/[2\rho_{eq}(t)]$ is the parabolic function that describes the unloaded geometric gap between the mating tooth surfaces (see Eq. 2.26), $f(x,t)$ is the instantaneous local elastic deformation at the lubricated contact, and Sa_j (with $j = 1,2$) is the arithmetical mean height of the scale-limited surface (see Sect. 10.9).

It is to be noted that $g_0(x,t)$ is time-depending through the variable equivalent radius of curvature $\rho_{eq}(t)$ given by Eq. (2.2). Since this is a line contact, and in reasonable agreement with what happens for most gear finishing processes such as grinding and shaving, constant values of Sa_j along the face widths of the two members of the gear pair under consideration can be assumed.

The instantaneous local elastic deformation, $f(x, t)$, which represents the sum of the elastic deformations of the two contacting bodies, due to the normal load related to pressure, is given by (Johnson 1985):

$$f(x, t) = \int_{x_s}^{x_e} K(x - s)p(s, t)ds, \qquad (10.19)$$

where x_s and x_e are the start and end points of the computational domain of the contact area, $K(x)$ is the coefficient of influence that tells how a displacement (or force) at a particular coordinate s influence the displacement (or force) at the coordinate x, and $p(s, t)$ is the time-varying pressure at coordinate s. This coefficient of influence is given by:

$$K(x) = -\frac{4ln|x|}{\pi E_r}, \qquad (10.20)$$

where E_r is the reduced modulus of elasticity given by Eq. (2.3).

The determination of the dependence of lubricant viscosity on pressure is performed using the two-slope viscosity-pressure model in the form developed by Goglia et al. (1984a, b). This model divides the pressure range into three regions, for each of which a special relationship for the dynamic viscosity, η, must be used. The three relationships are as follows:

$$\eta = \eta_0 e^{\alpha_1 p} \qquad (10.21)$$

$$\eta = \eta_0 e^{(c_0 + c_1 p + c_2 p^2 + c_3 p^3)} \qquad (10.22)$$

$$\eta = \eta_0 e^{[\alpha_1 p_t + \alpha_2 (p - p_t)]}, \qquad (10.23)$$

which are respectively valid for $(p < p_a)$, $(p_a \le p \le p_b)$ and $(p > p_b)$. In these relationships, η_0 is the lubricant dynamic viscosity at ambient temperature, α_1 and α_2 are respectively the pressure-viscosity coefficients for low $(p < p_a)$ and high $(p > p_b)$ pressure ranges, p_t is the transition pressure value between these two ranges, and p_a and p_b are respectively the threshold pressure values of the low and high pressure ranges. The constants c_i (with $i = 0, 1, 2, 3$) are determined in such a way that η and $\partial \eta / \partial p$ are continuous at the two threshold pressures, $p = p_a$ and $p = p_b$.

The pressure dependence of the lubricant density, ρ, is evaluated with the following relationship due to Dowson and Higginson (1977):

$$\rho = \rho_0 \frac{(1 + \lambda_1 p)}{(1 + \lambda_2 p)}, \qquad (10.24)$$

which presupposes the lubricant incompressibility. In this equation, ρ_0 is the lubricant density at ambient pressure, while $\lambda_1 = 2.266 \times 10^{-9}$ $(Pa)^{-1}$ and $\lambda_2 = 1.683 \times 10^{-9}$ $(Pa)^{-1}$ are two dimensional constants.

Finally, to determine the reference film thickness, $h_0(t)$, which appears in Eq. (10.18), a load balance equation is necessary, which states that the total contact force due to the instantaneous pressure distribution over the entire contact area (this distribution is related both to elastohydrodynamic and asperity contacts) is equal to the normal tooth load $F'_n(t)$ per unit face width applied at that instant. Such an equation can be written as follows:

$$F'_n(t) = \int\limits_{x_s}^{x_e} p(x, t)dx. \tag{10.25}$$

Here, the intensity of this normal tooth force is not constant, but varies along the path of contact according to the tooth-to-tooth load sharing characteristics of the gear meshing. This equation is not required to solve Eqs. (10.16) and (10.17). Instead, it is used as a check of their solution. In fact, if the predicted pressure distribution does not equal the applied normal force, the reference film thickness, $h_0(t)$, in Eq. (10.18) must be adjusted within a load iteration loop until the pressure distribution $p(x, t)$ satisfies Eq. (10.25).

The viscous shear stress within the lubricant also varies along the film thickness. We therefore have to do with a function also depending on z, where z is the coordinate having its origin on the pinion surface $(z = 0)$ and directed from the surface of the pinion to that of the mating gear wheel, the surface of which is defined by $z = h$. Assuming that this viscous shear stress, $\tau_v(x, z, t)$, varies linearly along the z-axis, and including both Poiseuille and Couette flows, Li and Kahraman (2013c) provide the following expression of $\tau_v(x, z, t)$:

$$\tau_v(x, z, t) = \frac{\eta^*(x, t)}{h(x, t)}[v_2(t) - v_1(t)] + \left[z - \frac{h(x, t)}{2}\right]\frac{\partial p(x, t)}{\partial x}, \tag{10.26}$$

where the *Eyring effective lubricant viscosity*, $\eta^*(x, t) = \eta(x, t)/\cos h[\tau_m(x, t)/\tau_0]$, considers the non-Newtonian effects. From this equation, we infer that the viscous shear stress acting on the pinion tooth surface, $\tau_{v1}(x, t) = \tau_v(x, 0, t)$, and the one acting on the gear wheel tooth surface, $\tau_{v2}(x, t) = \tau_v(x, h, t)$, are given respectively by:

$$\tau_{v1}(x, t) = \frac{\eta^*(x, t)}{h(x, t)}[v_2(t) - v_1(t)] - \frac{h(x, t)}{2}\frac{\partial p(x, t)}{\partial x} \tag{10.27}$$

$$\tau_{v2}(x, t) = \frac{\eta^*(x, t)}{h(x, t)}[v_2(t) - v_1(t)] + \frac{h(x, t)}{2}\frac{\partial p(x, t)}{\partial x}. \tag{10.28}$$

Within the regions where surface asperities are in direct contact, the boundary shear stress, $\tau_b(x, t)$, is determined as the product of the boundary coefficient of friction, μ_b, and the local contact pressure, i.e. as $\tau_b(x, t) = \mu_b p(x, t)$.

All the aforementioned governing differential equations for the problem herein are solved using the finite difference method with the central difference approximation (Fenner 1986). For details on applying this method, we refer the reader directly to paper of Li and Kahraman (2010a). To understand the relationships that follow, we believe it is necessary to give a brief description of the procedures that are activated. To find $\tau_{v1}(x, t)$, $\tau_{v2}(x, t)$ and $\tau_b(x, t)$ as the contact point, Y, moves along the LOA-direction, a computational domain $(-2.5b_{H,\max} \leq x \leq 1.5b_{H,\max})$ along the rolling direction (OLOA-direction or x-direction) is considered. The end points of this domain are given in terms of $b_{H,\max}$, which is the maximum half-Hertzian width of the contacts along the whole path of contact. The entire simulation starts at SAP and ends at EAP of the tooth of pinion 1 (it is to be noted that, without the profile modification at the tooth tip, the EAP terminates at the tooth tip). A refined mesh with I grid elements is used in order to capture the surface asperity geometry with sufficient accuracy. The shear stresses at each grid nodes are then determined with this model as $\tau_{v1}(x_i, t) = \tau_{v1i}(t)$, $\tau_{v2}(x_i, t) = \tau_{v2i}(t)$, and $\tau_b(x_i, t) = \tau_{bi}(t)$, with $i = 1, 2, \ldots, N$.

Denoting the area of each EHL grid element with uniform tooth surface as A, the total friction forces acting on the tooth surfaces of pinion and mating gear wheel can be expressed as follows:

$$F_1(t) = A \sum_{i=1}^{I} [\tau_{v1i}(t) + \tau_{bi}(t)] \tag{10.29}$$

$$F_2(t) = A \sum_{i=1}^{I} [\tau_{v2i}(t) + \tau_{bi}(t)]. \tag{10.30}$$

In these equations, we put $\tau_{v1i}(t) = \tau_{v2i}(t) = 0$ for grid nodes where asperity contacts occur and $\tau_{bi}(t) = 0$ for grid nodes separated by a lubricant film.

Substituting Eqs. (10.27) and (10.28) into Eqs. (10.29) and (10.30), and considering Eqs. (10.12) and (10.13), we obtain the following relationships that give the friction forces acting on the two members of the gear pair under consideration:

$$F_1(t) = c_m(t)[\rho_2(t)\dot{\vartheta}_2(t) - \dot{x}_2(t) - \rho_1(t)\dot{\vartheta}_1(t) + \dot{x}_1(t)] + [F_s(t) - F_r(t)] \tag{10.31}$$

$$F_2(t) = c_m(t)[\rho_2(t)\dot{\vartheta}_2(t) - \dot{x}_2(t) - \rho_1(t)\dot{\vartheta}_1(t) + \dot{x}_1(t)] + [F_s(t) + F_r(t)], \tag{10.32}$$

where

$$c_m(t) = A \sum_{i=1}^{I} \left[\frac{\eta^*(x_i, t)}{h(x_i, t)} \right] \tag{10.33}$$

$$F_s(t) = A \sum_{i=1}^{I} \tau_{bi}(t) + A \sum_{i=1}^{I} \left\{ \frac{\eta^*(x_i, t)}{h(x_i, t)} [v_{t2}(t) - v_{t1}(t)] \right\} \tag{10.34}$$

$$F_r(t) = A \sum_{i=1}^{I} \left[\frac{h(x_i, t)}{2} \frac{\partial p(x_i, t)}{\partial x} \right] \tag{10.35}$$

define respectively the gear mesh viscous damping, the sliding friction force and the rolling friction force.

Equation (10.33) shows that the gear mesh viscous damping, $c_m(t)$, is directly proportional to the effective lubricant viscosity and inversely proportional to the lubricant film thickness. Since the contact pressure changes with the mesh position as the gear members roll and the lubricant viscosity varies exponentially with pressure (see Eqs. 10.21–10.23), this gear mesh viscous damping becomes a time-varying parameter with period coinciding with the gear mesh period. In addition, the viscous damping intensity is also influenced by the surface asperity amplitudes through the introduction of local pressure fluctuations. Furthermore, $c_m(t)$ is also influenced by the operating speed and temperature conditions, since the surface tangential velocities and the lubricant temperature, the latter through the lubricant viscosity, condition the lubricant film thickness.

10.4.3 Tribological-Dynamic Model

The elastohydrodynamic lubrication behavior of cylindrical spur gears, especially those operating at high speed, is strongly influenced by the dynamics of the gear drive and the mechanical system of which it is a relevant part. In other words, elastohydrodynamic behavior and dynamic behavior influence each other, so the development of a model that wants to capture this interaction must necessarily take this into account (Li and Kahraman 2011b).

The tribological-dynamic model developed by Li and Kahraman (2013c) meets this requirement. It is indeed the result of the combination of the two sub-models described above in detail. The governing equations of this model are obtained by substituting Eqs. (10.31–10.35) into Eqs. (10.1–10.6). In doing so, the equations of motion become:

$$J_1 \ddot{\vartheta}_1(t) + \left\{ c_{t1} + \sum_{n=1}^{N} \left[\rho_1^2(t) c_m(t) \right]_n \right\} \dot{\vartheta}_1(t)$$

$$- \sum_{n=1}^{N} \left[\rho_1(t) \rho_2(t) c_m(t) \right]_n \dot{\vartheta}_2(t)$$

$$-\sum_{n=1}^{N} [\rho_1(t)c_m(t)]_n [\dot{x}_1(t) - \dot{x}_2(t)] + r_{b1}k_m(t)\delta(t) = T_1$$

$$+\sum_{n=1}^{N} \{[F_s(t) - F_r(t)]\rho_1(t)\}_n \tag{10.36}$$

$$m_1\ddot{y}_1(t) + c_{y1}\dot{y}_1(t) + k_{y1}y_1(t) + k_m(t)\delta(t) = 0 \tag{10.37}$$

$$m_1\ddot{x}_1(t) - \sum_{n=1}^{N'} [\rho_1(t)c_m(t)]_n \dot{\vartheta}_1(t)$$

$$+\sum_{n=1}^{N} [\rho_2(t)c_m(t)]_n \dot{\vartheta}_2(t) + \left\{ c_{x1} + \sum_{n=1}^{N} [c_m(t)]_n \right\} \dot{x}_1(t)$$

$$-\sum_{n=1}^{N} [c_m(t)]_n \dot{x}_2(t) + k_{x1}x_1(t) = \sum_{n=1}^{N} [F_s(t) - F_r(t)]_n \tag{10.38}$$

$$J_2\ddot{\vartheta}_2(t) - \sum_{n=1}^{N} [\rho_1(t)\rho_2(t)c_m(t)]_n \dot{\vartheta}_1(t)$$

$$+\left\{ c_{t2} + \sum_{n=1}^{N} [\rho_2^2(t)c_m(t)]_n \right\} \dot{\vartheta}_2(t)$$

$$+\sum_{n=1}^{N} [\rho_2(t)c_m(t)]_n [\dot{x}_1(t) - \dot{x}_2(t)]$$

$$-r_{b2}k_m(t)\delta(t) = -\left\{ T_2 + \sum_{n=1}^{N} \{[F_s(t) + F_r(t)]\rho_2(t)\}_n \right\} \tag{10.39}$$

$$m_2\ddot{y}_2(t) + c_{y2}\dot{y}_2(t) + k_{y2}y_2(t) - k_m(t)\delta(t) = 0 \tag{10.40}$$

$$m_2\ddot{x}_2(t) + \sum_{n=1}^{N} [\rho_1(t)c_m(t)]_n \dot{\vartheta}_1(t) - \sum_{n=1}^{N} [\rho_2(t)c_m(t)]_n \dot{\vartheta}_2(t)$$

$$-\sum_{n=1}^{N} [c_m(t)]_n \dot{x}_1(t) + \left\{ c_{x2} + \sum_{n=1}^{N} [c_m(t)]_n \right\} \dot{x}_2(t) + k_{x2}x_2(t)$$

$$= -\sum_{n=1}^{N} [F_s(t) + F_r(t)]_n. \tag{10.41}$$

These equations clearly show that $c_m(t)$ acts along the OLOA-direction. In fact, the gear tooth mesh damping not only contributes to motions in the OLOA-direction as Eqs. (10.38) and (10.41) show, but also couples the motions along LOA-direction with the dynamic responses along the OLOA-direction, as Eqs. (10.36), (10.38), (10.39) and (10.41) show. It is also to be noted that friction forces and moments further contribute to coupling the translations, $x_1(t)$ and $x_2(t)$, along the OLOA-direction with the rotational motions, $\vartheta_1(t)$ and $\vartheta_2(t)$.

The solution of the problem herein involves preliminary discretization in terms of finite differences of the system of differential Eqs. (10.36–10.41) that govern the tribological-dynamic model as well as an iteration loop between the gear dynamics sub-model and the gear EHL sub-model. This iteration loop is shown in the flowchart in Fig. 10.5, where the main steps of the computational method to be used are summarized. This iterative procedure starts using the input data, consisting of external torques and gear parameters, as well as the quantities $k_m(t)$ and $\varepsilon_s(t)$ determined preliminarily by means of a quasi-static gear load distribution model.

Here it is not the place to recall the numerous models for calculation of the transverse tangential load distribution between the various teeth pairs in simultaneous meshing, developed over time by many scientists and researchers who have dealt with this important topic, starting from the *Karas' model* (Karas 1941). A comprehensive

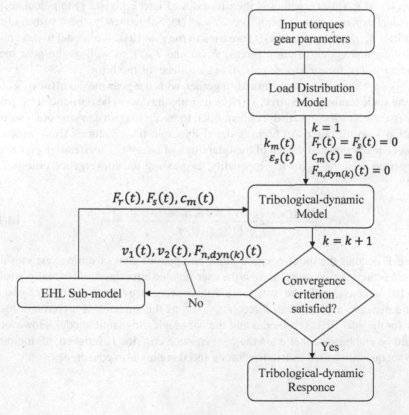

Fig. 10.5 Flowchart of the iterative computational procedure to be used with the tribological-dynamic model

overview on models of this type, developed until the beginning of third millennium, can be found in the works of Ozguven and Houser (1988) and Wang et al. (2003). An interesting model is the one proposed by Conry and Seireg (1973), but other more refined models are available today. However, it should be noted that none of the theoretical and/or numerical models developed so far has the general characteristics that would be necessary to address this topic, also for the reasons already highlighted in Sect. 1.7.

Going back to the iteration procedure, it should be remembered that, to start its loop, it is necessary to have the initial values of sliding and rolling friction forces and gear mesh damping. In this regard, for the first iteration loop, the one having iteration index $k = 0$, we can set $F_s(t) = 0$, $F_r(t) = 0$ and $c_m(t) = 0$. With these initial data, which constitute the excitation of the combined tribological-dynamic model mathematically represented by Eqs. (10.36–10.41), these equations provide the time histories of the vibratory motions of the system, from which we can obtain all the quantities that appear in Eqs. (10.12) and (10.13), which are necessary to calculate the instantaneous tangential velocities, $v_1(t)$ and $v_2(t)$. In the same way, using the aforementioned time histories, we determine the dynamic gear mesh force, $F_{m,dyn}(t) = k_m(t)\delta(t)$, which is then introduced into Eq. (10.11) to calculate the individual dynamic normal tooth force, $F_{n,dyn}(t)$. Subsequently, these values of the quantities $F_{n,dyn}(t)$, $v_1(t)$ and $v_2(t)$ are used in the gear EHL sub-model to determine the rolling and sliding friction forces, $F_r(t)$ and $F_s(t)$, as well as the gear mesh damping, $c_m(t)$, for each tooth pair, n, in simultaneous meshing.

This set of results so determined, together with the gear mesh stiffness, $k_m(t)$, and the static transmission error, $\varepsilon_s(t)$, constitute the data to be introduced by going back in the gear tribological-dynamic model, to excite it again carrying out a second iteration loop, in order to obtain updated dynamic time histories from which to extract equally updated values of the quantities of interest. This iterative procedure is repeated until the following inequality, expressing the convergence criterion, is met:

$$\frac{\sum_Y \left| F_{n,dyn(k)}(t_Y) - F_{n,dyn(k-1)}(t_Y) \right|}{\sum_Y F_{n,dyn(k)}(t_Y)} \le e, \tag{10.42}$$

where Y denotes the mesh position along the path of contact during one meshing cycle, k is the iteration index, and e is the user-defined error threshold (usually, a value $e = 10^{-5}$ is considered to be sufficient). This convergence criterion is evidently based on the dynamic normal tooth force, $F_{n,dyn}(t)$, as this quantity is a common result both for the gear EHL sub-model and the tribological-dynamic model. However, it should be emphasized that once the convergence criterion is satisfied, all the other relevant quantities such as friction forces and damping also converge.

10.5 Other Interesting Contributions for a General Micropitting Model of Spur Gears

The tribological-dynamic model described in the previous section covers only a part, although important, of the flowchart represent in Fig. 10.2. However, this model, which is to be considered one of the most advanced so far developed and perhaps the most advanced, does not cover all the needs of the desired ideal gear micropitting model outlined in Sect. 10.3, if only for the reason that it refers to cylindrical spur gears. The other important aspects that still remain to be covered, with the development of the corresponding models, are those related to the determination of the three-dimensional stress state at and near the contacting areas between teeth having rough surfaces, lubricated or non-lubricated, and to the calculation of the average and amplitude stress components, to be used in the broader framework of multi-axial fatigue, useful for defining reliable micropitting severity indices.

Leaving aside the details, we want to describe here two models, also developed by Li and Kahraman (2014), which try to cover the two aspects mentioned above. These models are only valid for cylindrical spur gears and, even if with some other limitations, they are to be considered among the most advanced in relation to the topics they face.

10.5.1 Three-Dimensional Contact Sub-model

To address the problem of the contact between gear mating tooth surfaces in three-dimensional terms, many researchers have so far used the Finite Element Method (FEM). To deepen the problem in the framework that interests here, it is however necessary a very detailed description of the 3D areal surface texture, which requires a very fine mesh grid. In this case, the FE based methods are extremely burdensome from a computational point of view, as they demand such an extremely large number of finite elements that they often configure an unsustainable effort. In this case, the BIE (Boundary Integral Equation)-method, also known as BEM (Boundary Element Method), is notoriously a sustainable alternative to the FE method. The reasons for this greater sustainability of the BEM compared to the FEM derive from the fact that it considers only the boundary instead of the entire contacting body for which, under line contact conditions, a two-dimensional problem is reduced to a one-dimensional problem. Consequently, it is possible to use such a refined mesh to capture the areal surface texture with the required accuracy.

By appropriately modifying a three-dimensional point contact model, based on BEM and developed by themselves, Li and Kahraman (2014) set up a two-dimensional line contact model, also based on BEM, to be used for three-dimensional contact analysis of cylindrical spur gears. Since the contact area is small compared to the gear members involved, the authors reduce the problem to that of a line load

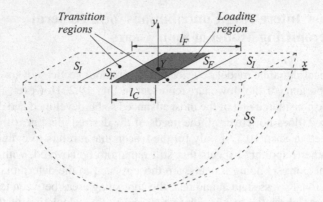

Fig. 10.6 Calculation domain corresponding to the boundary of the infinite half-space

applied to the surface of a semi-infinite body. Therefore, they address the problem by assuming the well-known half-space conditions with line load.

Thus the theoretical bases of the model are those developed by Flamant (1892) who, by modifying the three-dimensional solution of problem of the elastic body bounded by a plane (this is also known as the problem of Boussinesq and Cerruti), obtains the expressions of stresses and displacements in a linear elastic wedge loaded by forces per unit length applied at its sharp end, of which the semi-infinite body loaded with line loads normal or parallel to the free surface is a special case (see Cerruti 1882; Boussinesq 1885; see also: Love 1944; Timoshenko and Goodier 1951; Gladwell 1980; Johnson 1985; Barber 1992).

As Fig. 10.6 shows, the boundary of the computational domain corresponding to the boundary of the infinite half-space under consideration is divided into the following three regions:

- A finite region, S_F, having length l_F in the x-direction and containing the contact zone with rough surface.
- An infinite region, S_I, which is considered to have perfectly smooth surface.
- An infinite region, S_S, corresponding to the half cylindrical surface, which surrounds the half-space at infinity.

The coordinate system shown in Fig. 10.6, with origin in the current point of contact Y, has the x-axis along the direction of the common tangent to the mating tooth surfaces (see also Fig. 10.3), and the y-axis along the direction of the normal to the surface and oriented outwards. In the same figure, two solid lines delimit the finite bounds of the aforementioned finite region in the x-direction, while the dashed lines, to be considered as positioned at infinity, in mathematical terms are representing the infinite bounds of the other two infinite regions. The authors discretize the finite region with finite boundary elements and the other two regions with infinite boundary elements.

Of course, the infinite region S_I is characterized by zero traction condition, since the stress resultants (more briefly also known as tractions) applied to this field are equal to zero. Taking this condition into account and using the fundamental solution of the Kelvin problem (see Thomson 1848; Favata 2012), which expresses the displacement at a given point of an indefinitely extended, linearly elastic and isotropic body, subjected to a load applied at another point, the displacements at the load point P, given in terms of displacements and tractions at the field point Q, can be expressed by the following boundary integral equation (see Fenner 1986; Becker 1992):

$$C_{ij}(P)u_j(P) + \int_{S_F+S_I} T_{ij}(P, Q)u_j(Q)dS = \int_{S_F} U_{ij}(P, Q)t_jdS, \qquad (10.43)$$

where: $i, j = x, y$ are the directions; u_j and t_j are respectively the displacements and tractions or resultant stresses; $C_{ij}(P)$ is the free-term coefficient, which depends on the shape of the boundary and represents the jump in the integral on the left hand side, when the interior point P_i within the computational domain is taken to the boundary at P; $U_{ij}(P, Q)$ is the *integral kernel function*, i.e. the two-dimensional Kelvin fundamental solution for u_j, which defines the displacement in the i-direction at the field point Q due to a unit load applied at load point P in the j-direction; $T_{ij}(P, Q)$ is the integral kernel function, i.e. the two-dimensional Kelvin fundamental solution for t_j, which defines the traction in the i-direction at the field point Q due to a unit load applied at load point P in the j-direction. The free-term coefficient is given by:

$$C_{ij}(P) = \lim_{\varepsilon \to 0} \left[\int_{S_\varepsilon} T_{ij}(P, Q)dS \right], \qquad (10.44)$$

where S_ε is the boundary of a region that is the part of the computational domain within a circle of radius ε centered at P. If the boundary at P is smooth, $C_{ij}(P) = \delta_{ij}/2$, where δ_{ij} is the Kronecker delta ($\delta_{ij} = 1$, when $i = j$ and $\delta_{ij} = 0$, when $i \neq j$). If the boundary at P is not smooth (in this case we have rough surface condition), $C_{ij}(P)$ is more difficult to evaluate. In practice, however, $C_{ij}(P)$ contributes only to the diagonal coefficients of the relevant matrix in the numerical implementation of the method, and these coefficients can always be evaluated indirectly. However, this term can be determined either using the indirect approach of rigid body motion or as Cauchy principal value of a finite integral (see Cauchy 1828; Henrici 1988; Whittaker and Watson 1990; Bronshtein and Semendyayev 1997).

If we denote with d the distance between the load point P and the field point Q, T_{ij} and P_{ij} can be expressed as follows:

$$T_{ij} = -\frac{1}{4\pi d(1 - v')} \frac{\partial d}{\partial n} \left[(1 - 2v')\delta_{ij} + 2\frac{\partial d}{\partial i}\frac{\partial d}{\partial j} \right] + \frac{(1 - 2v')}{4\pi d(1 - v')} \left[\frac{\partial d}{\partial j}n_i - \frac{\partial d}{\partial i}n_j \right] \qquad (10.45)$$

$$U_{ij} = \frac{1}{8\pi G(1 - v')} \left[(4v' - 3)\delta_{ij} \ln d + \frac{\partial d}{\partial i}\frac{\partial d}{\partial j} \right], \qquad (10.46)$$

where G is the shear modulus (or modulus of rigidity), ν' is the effective Poisson's ratio ($\nu' = \nu$ for plane stress and $\nu' = \nu/(1 + \nu)$ for plane strain, where ν is the Poisson's ratio), and $\boldsymbol{n} = [n_x, n_y]$ is the unit outward normal vector.

The finite region S_F consists of a loaded region, represented by the dashed rectangle in Fig. 10.6, and the two adjacent transition regions, on either side of the loaded region. For the discretization, within the entire finite region S_F, isoparametric quadratic boundary elements (BE) are used, whose nodes have y-coordinates representing the surface asperity heights. In order to capture the local areal surface texture, the BE size within the loaded region is chosen to be on the same order of the surface asperities resulting from roughness measurements (thus on the order of micrometers). Moreover, since in the two transition regions the surface displacements and tractions are much smaller, a mesh size larger and gradually increasing going outward from the loaded region to the infinity is used. Then using the quadratic shape functions $N_1(\xi) = \xi(\xi - 1)/2$, $N_2(\xi) = 1 - \xi^2$ and $N_3(\xi) = \xi(\xi + 1)/2$ (Fenner 1986), where ξ is the local intrinsic coordinate (it is chosen so that the positions of the three equidistant nodes of each BE are defined by $\xi = -1$, $\xi = 0$ and $\xi = 1$, respectively), the integrals in Eq. (10.43) are discretized as follows:

$$\int\limits_{S_F} T_{ij} u_j dS = \sum_{K_F} \sum_{c=1}^{3} u_j^c I_{ij,c}^T + \sum_{K_F^*} \sum_{\substack{c=1 \\ (c \neq c^*)}}^{3} u_j^c I_{ij,c}^T + \sum_{K_F^*} u_j^{c^*} I_{ij,c^*}^T \qquad (10.47)$$

$$\int\limits_{S_F} U_{ij} t_j dS = \sum_{K_F} \sum_{c=1}^{3} t_j^c I_{ij,c}^U + \sum_{K_F^*} \sum_{\substack{c=1 \\ (c \neq c^*)}}^{3} t_j^c I_{ij,c}^U + \sum_{K_F^*} t_j^{c^*} I_{ij,c^*}^U, \qquad (10.48)$$

where

$$u_j(\xi) = \sum_{c=1}^{3} N_c(\xi) u_j^c \qquad (10.49)$$

$$t_j(\xi) = \sum_{c=1}^{3} N_c(\xi) t_j^c \qquad (10.50)$$

are respectively the displacements and tractions at cth node of the element and $I_{ij,c}^T$ and $I_{ij,c}^U$ are two integral transforms given by:

$$I_{ij,c}^T = \int\limits_{S_{BE}} T_{ij} N_c dS = \int\limits_{-1}^{+1} T_{ij}(P_c, Q(\xi)) N_c(\xi) J(\xi) d\xi \qquad (10.51)$$

$$I_{ij,c}^{U} = \int\limits_{S_{BE}} U_{ij} N_c dS = \int\limits_{-1}^{+1} U_{ij}(P_c, Q(\xi)) N_c(\xi) J(\xi) d\xi, \qquad (10.52)$$

where S_{BE} is the length of the element, P_c is the cth nodal point on the boundary element and $J(\xi) = ds/d\xi$ is the Jacobian of transformation of coordinates from the curvilinear coordinate, s, on the boundary to ξ; K_F and K_F^* are respectively the numbers of boundary elements that are away from the load point P and that contain P; c^* denote the node where P and Q coincide.

The integrals (10.51) and (10.52) that appear in Eqs. (10.47) and (10.48) are the *integral transforms* of the functions involved by the corresponding kernels, T_{ij} and U_{ij} (see Sneddon 1974). They are notoriously best evaluated numerically by the Gaussian quadrature rule (Gauss 1815; see also: Démidovitch and Maron 1973; Cook 1981; Fenner 1986). For the boundary elements K_F that are away from the load point P, there is no restriction. Instead, for the boundary elements K_F^* that contain the load point P, attention must be paid to the fact that the tensor kernel functions (10.45) and (10.46) involve singularities of the orders $(\ln d)$ and $(1/d)$ when $d \to 0$, i.e. the field point Q approaches the load point P. Therefore, in the case where $c \neq c^*$, the shape function N_c is of the order d in the vicinity of point P, for which $T_{ij} N_c$ and $U_{ij} N_c$ do not have singularities, because they have well-defined values, so that Gaussian quadrature rule can be used to evaluate the integrals. Instead, in the case where $c = c^*$, the kernel functions of I_{ij,c^*}^{T} and I_{ij,c^*}^{U} are both singular. However, the integral I_{ij,c^*}^{U}, which contains the term $(\ln d)$ can be calculated using the generalized Gaussian quadrature for integrals with logarithmic singularity (Becker 1992; see also Milovanović 2016), so that only the integral I_{ij,c^*}^{T} remains undetermined.

Isoparametric linear infinite boundary elements as those shown in Fig. 10.7a are used to discretize the indefinitely extended region S_I. Node 1 of this element, the one having coordinate $\xi = -1$, is positioned at the bound between the finite region S_F and indefinitely extended region S_I, while the node 2, the one having coordinate

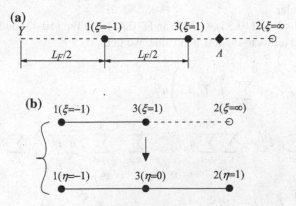

Fig. 10.7 **a** isoparametric linear infinite boundary element used to discretize the infinite region S_I; **b** transformation of the infinite boundary element to a finite boundary element

$\xi = \infty$, is located at infinity. The mid-node 3, the one having coordinate $\xi = 1$, is positioned in such a way that the distance of node 1 from the contact origin point, Y, is equal to that between node 1 and node 3, both equal to the half length l_F of the finite region S_F. The displacements u_j^A at an arbitrary point A within the infinite boundary element (Fig. 10.7a) are related to those at node 1, u_j^1, by the relationship $u_j^A = Du_j^1$ due to Moser et al., (2004), where $D = l_F/2\overline{YA}$ is the decay function starting from contact origin point Y chosen as decay origin and $\overline{YA}$ is the distance between points Y and A. On this basis, the integral over the indefinitely extended region S_I in Eq. (10.43) can be discretized as follows:

$$\int_{S_I} T_{ij} u_j \, dS = \sum_{K_I} u_j^1 \int_{S_{IBE}} T_{ij} D \, dS + \sum_{K_I^*} u_j^1 \int_{S_{IBE}} T_{ij}(D-1) \, dS + \sum_{K_I^*} u_j^1 \int_{S_{IBE}} T_{ij} \, dS,$$

(10.53)

where K_I and K_I^* are the numbers of boundary elements that are away from the load point P and respectively connected to P, and S_{IBE} is the length of the element. Considering that, in accordance with the hypothesis, the surface asperities have height equal to zero in the indefinitely extended region S_I and bearing in mind the previous definition of the mid-node 3, the geometry along the indefinitely extended region discretized by boundary elements is completely defined by the relationships $x = [x_1(1-\xi)/2 + x_3(1+\xi)/2]$, obtained by linear interpolation, and $y = 0$.

To be able to evaluate numerically the integrals that appear in Eq. (10.53) using the Gaussian quadrature technique, the infinite boundary element is transformed into a finite boundary element, through the transformation $\xi = [(1+3\eta)/(1-\eta)]$ whose Jacobian is $J_\xi^\eta = [4/(1-\eta)^2]$, as shown in Fig. 10.7b. For the infinite boundary elements K_I^* that are connected to the load point P, when Q approaches P, we have $c = c^* = 1$, $D \to 1$ and $(D-1) \to 0$, for which $(D-1)$ is on the order of d. Consequently, the singularity in the second integral on the right-hand side of Eq. (10.53) is eliminated, while the last integral appearing in the same equation remains singular.

Substituting Eqs. (10.53), (10.47) and (10.48) into Eq. (10.43) and moving all the undetermined terms to the left-hand side, we get:

$$\left(C_{ij} + \sum_{K_F^*} I_{ij,c^*}^T + \sum_{K_F^*} \int_{S_{IBE}} T_{ij} \, dS \right) u_j^{c^*}$$

$$= \sum_{K_F+K_F^*} \sum_{c=1}^{3} t_j^c I_{ij,c}^U - \sum_{K_F} \sum_{c=1}^{3} u_j^c I_{ij,c}^T - \sum_{K_F^*} \sum_{\substack{c=1 \\ (c \neq c^*)}}^{3} u_j^c I_{ij,c}^T - \sum_{K_I} u_j^1 \int_{S_{IBE}} T_{ij} D \, dS$$

$$- \sum_{K_I^*} u_j^1 \int_{S_{IBE}} T_{ij}(D-1) \, dS,$$

(10.54)

In order to evaluate the integrals with singular kernel functions and parameter C_{ij} in this last equation, the indirect approach of rigid body motion is used. In this condition, the displacements u_i (with $i = x, y$) are constant and $t_i = 0$ along the entire boundary. Applying this condition of rigid body motion and assuming $u_i = 1$, the integral boundary equation can be written as follows:

$$C_{ij}(P) + \int_{S_F+S_I+S_S} T_{ij}(P, Q)dS = 0. \tag{10.55}$$

Proceeding in the same way as for Eqs. (10.53), (10.47) and (10.48) and utilizing the azimuthal integral $\int_{S_S} T_{ij}dS = -\delta_{ij}/2$, Eq. (10.55) can be discretized as follows:

$$C_{ij} + \sum_{K_F^*} I_{ij,c^*}^T + \sum_{K_F^*} \int_{S_{IBE}} T_{ij}dS = \frac{\delta_{ij}}{2} - \sum_{K_F} \sum_{c=1}^{3} I_{ij,c}^T$$

$$- \sum_{K_F^*} \sum_{\substack{c=1 \\ (c \neq c^*)}}^{3} I_{ij,c}^T - \sum_{K_I} \int_{S_{IBE}} T_{ij}dS. \tag{10.56}$$

In this equation, no decay function appears, as the displacements in all the directions are constant under condition of rigid body motion. All the integrals that appear on the right-hand side of this equation can now be easily evaluated using the Gaussian quadrature rule. Therefore, Eq. (10.43) can be written in its final discretized form as follows, by substituting Eq. (10.56) into Eq. (10.54):

$$\left(\frac{\delta_{ij}}{2} - \sum_{K_F} \sum_{c=1}^{3} I_{ij,c}^T - \sum_{K_F^*} \sum_{\substack{c=1 \\ (c \neq c^*)}}^{3} I_{ij,c}^T - \sum_{K_I} \int_{S_{IBE}} T_{ij}dS \right) u_j^{c^*}$$

$$= \sum_{K_F+K_F^*} \sum_{c=1}^{3} t_j^c I_{ij,c}^U - \sum_{K_F} \sum_{c=1}^{3} u_j^c I_{ij,c}^T$$

$$- \sum_{K_F^*} \sum_{\substack{c=1 \\ (c \neq c^*)}}^{3} u_j^c I_{ij,c}^T - \sum_{K_I} u_j^1 \int_{S_{IBE}} T_{ij} \bar{D} dS - \sum_{K_I^*} u_j^1 \int_{S_{IBE}} T_{ij}(D-1)dS. \tag{10.57}$$

The displacement distributions along the finite region S_F are then obtained from Eq. (10.57), provided the boundary traction distributions calculated with the tribological-dynamic model described in Sect. 10.4 are available.

Therefore, having the displacement and traction distributions along the entire domain, it is now possible to determine the stress state in any boundary or interior point of the calculation domain under consideration using the following relationship:

$$\sigma_{ij}(P) = \int_{S_F} H_{kij}^t(P, Q) t_k(Q) dS - \int_{S_F+S_I} H_{kij}^u(P, Q) u_k dS, \qquad (10.58)$$

where $k = x, y$ and H_{kij}^t and H_{kij}^u are two third order kernel functions given by (Becker 1992):

$$H_{kij}^t = \frac{1}{4\pi d(1 - v')}\left[(1 - 2v')\left(\delta_{jk}\frac{\partial d}{\partial i} + \delta_{ik}\frac{\partial d}{\partial j} - \delta_{ij}\frac{\partial d}{\partial k}\right) + 2\frac{\partial d}{\partial i}\frac{\partial d}{\partial j}\frac{\partial d}{\partial k}\right]$$

$$H_{kij}^u = \frac{G}{2\pi d^2(1 - v')}\left\{n_i\left[3v'\frac{\partial d}{\partial j}\frac{\partial d}{\partial k} + (1 - 2v')\delta_{jk}\right] + n_j\left[2v'\frac{\partial d}{\partial i}\frac{\partial d}{\partial k} + (1 - 2v')\delta_{ik}\right]\right.$$

$$\left. + n_k[2(1 - 2v')\frac{\partial d}{\partial i}\frac{\partial d}{\partial j} - (1 - 4v')\delta_{ij}]\right\}$$

$$+ \frac{G}{\pi d^2(1 - v')}\frac{\partial d}{\partial n}\left[(1 - 2v')\delta_{ij}\frac{\partial d}{\partial k} + v'\left(\delta_{jk}\frac{\partial d}{\partial i} + \delta_{ik}\frac{\partial d}{\partial j}\right) - 4\frac{\partial d}{\partial i}\frac{\partial d}{\partial j}\frac{\partial d}{\partial k}\right] \qquad (10.60)$$

In this case attention must also be paid to the fact that these kernel functions involve singularities of the orders $(1/d)$ and $(1/d^2)$ when $d \to 0$, i.e. the field point Q approaches the load point P. To overcome the calculation problems related to these singularities, the stress components at the boundary points are computed using the strains and tractions according to Hooke's law. To do this, it is convenient to use a local orthogonal coordinate system (x', y'), where x' is tangent to the surface and y' has the direction of outward normal. The position of this local system (x', y') with respect to the global coordinate system (x, y) is defined by the direction cosines (α_x, α_y) and (β_x, β_y).

Therefore, by interpolating the displacements within an element using the quadratic shape function as $u_i = u_i^c N_c$, the tangential displacement in the local coordinate system (x', y') can be expressed as $u_{x'} = \alpha_i(u_i^c N_c)$. Accordingly, the local shear strain can be determined by the following relationship:

$$\varepsilon_{x'} = \frac{\partial u_{x'}}{\partial x'} = \frac{1}{J_S^\xi}\alpha_i u_i^c\left(\frac{\partial N_c}{\partial \xi}\right), \qquad (10.61)$$

where J_S^ξ is the transformation Jacobian from the boundary curve S to the intrinsic coordinate ξ. In the same way, the tractions in the local coordinate system (x', y') are defined as $t_{x'} = \alpha_i t_i$ and $t_{y'} = \beta_i t_i$. Therefore, in the same coordinate system, the stress components can be computed from the strains and tractions as follows:

$$\sigma_{x'} = \frac{E'}{\left[1 - (v')^2\right]}\varepsilon_{x'} + \frac{v'}{1 - v'}t_{y'} \qquad (10.62)$$

$$\sigma_{y'} = t_{y'} \tag{10.63}$$

$$\sigma_{x'y'} = \tau_{x'y'} = t_{x'}, \tag{10.64}$$

where E' is the effective modulus of elasticity ($E' = E$ for plane stress and $E' = [E(1 + 2\nu)/(1 + \nu)^2]$ for plane strain, where E is the Young's modulus).

Equation (10.58) can be used for the stress field computation for any interior point which is always away from the load points. However, micropitting cracks are surface or near surface cracks, whose nucleation affects a surface layer that has a very shallow depth. Any small distance between the load point P and field point Q introduces numerical errors into the Gaussian quadrature procedure due to the near singular behavior of the two third order kernel functions, H_{kij}^t and H_{kij}^u. To limit these errors, the authors use a particular progressive element subdivision technique and, to this end, discretize Eq. (10.58) similar to the discretization done for Eq. (10.34).

10.5.2 Multi-axial Fatigue Sub-model

The surface and near surface stress states determined by the calculation methods described in the previous section constitute the input data of a sub-model through which the fatigue damage due to micropitting must be evaluated and checked. For the preparation of such a sub-model, it is obviously necessary to establish a multi-axial fatigue criterion, given that the aforementioned stress states are three-dimensional stress states.

It is well known that, in the general case of mechanical components and structures, capturing the correct damage mechanism of the multi-axial fatigue is essential to define a proper damage quantification parameter for robust multi-axial fatigue life estimation. Multi-axial fatigue damage mechanisms include material microstructure, material constitutive response, non-proportional hardening, variable accumulation loading, mixed-mode crack growth, cycle counting, additional cyclic hardening of some materials under non-proportional multi-axial loading and its dependence on the load path, etc. This is not the case to recall the various multi-axial fatigue criteria so far introduced, which are the basis of the models related to them (stress-based models, strain-based models, fracture mechanics models, non-proportional models, etc.). In this regard, we refer the reader to the scientific literature on this important topic (e.g.: Fatemi and Socie 1988; Socie and Marquis 1999).

However, attention should be drawn to the fact that the critical plane damage models considering both stress and strain terms have proved to be the most appropriate models to deal with the subject in question, as they can reflect the material constitutive response under non-proportional loading. In fact, in the cases in which they have been used, these models have proven to be able to capture the crack nucleation sites and fatigue lives with the best agreement with the experimental evidences (see Fatemi and Shamsaei 2011; Yu et al. 2017).

An interesting model, which can be considered as a variation on the theme fallowing within the critical plane damage models, is the one proposed by Liu and Mahadevan (2007) for railroad wheel contacts, but claimed to be applicable to relatively wide ranges of materials and load conditions through comparison to the measured fatigue data for a variety of metals under both proportional and non-proportional loads. This model assumes that the fatigue assessment can be performed with satisfactory approximation assuming, as damage parameter, a certain stress combination on a certain plane, which does not necessarily coincide with the fatigue fracture plane, i.e. with the macro crack plane. This is the characteristic plane criterion to approach the fatigue assessment. It is defined by the following relationship:

$$\frac{1}{k_1}\left[\sigma_a^2 + \left(\frac{S_{N,b}}{S_{N,t}}\right)^2 \left(\tau_{a1}^2 + \tau_{a2}^2\right) + k_2\sigma_{a,H}^2\right]^{1/2} = S_{N,b}, \qquad (10.65)$$

where: k_1 and k_2 are two non-dimensional parameters related to the material properties; $S_{N,b}$ and $S_{N,t}$ are respectively the uniaxial fully reversed bending and torsion fatigue strengths of the materials corresponding to the finite fatigue life cycle number, N; σ_a, τ_{a1} and τ_{a2} are the normal stress amplitude and, respectively, the amplitudes of the shear stress components acting on the characteristic plane; $\sigma_{a,H}$ is the hydrostatic stress amplitude acting on the same characteristic plane.

To determine the position of this characteristic plane, the macro crack plane where the normal stress amplitude has its maximum value is taken as a reference plane. The characteristic plane is defined as the plane on which the damage introduced by the hydrostatic stress amplitude is minimal, i.e. the plane for which $k_2 = 0$.

The angle ϑ that defines the position of the characteristic plane with respect to the reference plane, the latter coinciding with the macro crack plane, and the non-dimensional material constant k_1 are then obtained by means of Eq. (10.65) applied to the two fully reversed bending and torsion fatigue problems. These two quantities are given by the following relationships:

$$\vartheta = \frac{1}{2}\cos^{-1}\left[\frac{s^2 - \sqrt{s^4 - (3s^2 - 1)(4s^4 - 5s^2 + 1)}}{(4s^4 - 5s^2 + 1)}\right] \qquad (10.66)$$

$$k_1 = \sqrt{s^2 \cos^2(2\vartheta) + \sin^2(2\vartheta)}, \qquad (10.67)$$

which are valid for non-extremely brittle materials with $s = \left(S_{N,b}/S_{N,t}\right) < 1$.

Considering that the loading characteristics of the rolling contact fatigue of the gears are rather similar to those that occur in the railroad wheel contacts studied by Liu and Mahadevan (2007), Li et al. (2012) and Li and Kahraman (2011a, 2014) took the fatigue criterion introduced by the first authors, adapting it to the gears. The same Liu and Mahadevan (2007) made Eq. (10.65) even more general, so as to consider also the mean normal stress effect; to this end, they introduced in this equation, as a multiplier of the non-dimensional parameter k_1, the term $\left[1 - \left(\sigma_{m,\max}/S_{ref}\right)\right]$. In this term, $\sigma_{m,\max}$ is the mean normal stress on the macro crack plane, which is assumed as the plane with the maximum normal stress amplitude, while S_{ref} is the reference stress whose value can be calibrated using the stress-life fatigue data with non-zero mean normal stress. In absence of such material data, S_{ref} can be approximated using the yield stress or the ultimate tensile strength. With $k_2 = 0$ and the aforementioned incorporation, Eq. (10.65) becomes:

$$\frac{1}{k_1\left(1 - \frac{\sigma_{m,\max}}{S_{ref}}\right)}\left[\sigma_a^2 + \left(\frac{S_{N,b}}{S_{N,t}}\right)^2\left(\tau_{a1}^2 + \tau_{a2}^2\right)\right]^{1/2} = S_{N,b}, \tag{10.68}$$

The methodology described above is so general that any other multi-axial fatigue criterion can be used instead of the one previously outlined. Furthermore, it overcomes computational difficulties related to the use of critical plane criteria. It is in fact to be kept in mind that these last criteria involve the calculation of the shear stress amplitude, which is defined as the radius of the minimum circle circumscribing the path described by the tip of the shear stress vector on a certain plane.

An efficient and robust algorithm to search for the minimum circumscribed circle, called randomized algorithm, is the one recommended by Bernasconi and Papadopulos (2005). However, this algorithm is still computationally unaffordable for rolling point contact problems between lubricated rough surfaces. The fatigue criterion introduced by Liu and Mahadevan (2007) has the advantage of decomposing the shear stress vector on the characteristic plane into two perpendicular components, whose amplitudes are simply defined as the half of the shear range in each of the two directions, thus eliminating the time-consuming task of the minimum circumscribed circle search. In any case, for line contacts that interest here, Eq. (10.68) is further simplified, as only τ_a^2 appears in place of $\left(\tau_{a1}^2 + \tau_{a2}^2\right)$, where τ_a is the shear stress amplitude acting on the characteristic plane.

The multi-axial fatigue criterion expressed by Eq. (10.68) thus further simplified is used to determine the fatigue lives of the material points in the shallow surface layer under consideration, once the average and amplitude values of the local stresses have been calculated using the three-dimensional contact sub-model for lubricated rough surfaces described in Sect. 10.5.1.

10.6 Calculation of Micropitting Load Carrying Capacity of Cylindrical Spur and Helical Gears: ISO Basic Formulae

The ISO standards (ISO/TS 6336-22:2018) provide only an indirect method for assessing micropitting load carrying capacity of cylindrical involute spur and helical gears with external teeth. In fact, ISO assumes as a fundamental reference quantity the *specific lubricant film thickness*, which is also referred in the literature as *film thickness ratio* or *lambda ratio*. Therefore, this quantity can serve as an evaluation criterion when it is applied as part of a suitable comparative procedure based on known gear performance.

We have already shown in previous sections that the occurrence of micropitting in cylindrical spur and helical gears can be influenced by many parameters. These include the topography of the surfaces and their texture, contact stress level, operating conditions, materials involved including the lubricant interposed between the surfaces in relative motion, lubricant chemistry, additives used in the lubricant, etc. However, in the current state of knowledge, we can only state that these parameters influence the micropitting performance of the gears. Instead, since this subject is a constantly evolving research topic, we cannot yet be sure that all aspects related to these specific parameters are comprehensively included in the calculation procedure described below. The science in this regard has not yet arrived at incontrovertible points. For example, it is certain that both tip and root relief in terms of involute modification greatly affect micropitting performance, so they must be chosen carefully, but we do not have the basis for defining their optimal values. Another example is the areal surface texture, which plays a fundamental role in micropitting resistance; it is considered introducing the effective arithmetic mean roughness, Ra, whereas today all researchers believe that it is inadequate to characterize the areal surface texture. Some basic concepts concerning the areal surface texture are summarized at the end of this chapter, in Sect. 10.9.

The procedure described below, to be used only to avoid micropitting damage, has been developed on the basis of testing and investigations on oil-lubricated cylindrical gears with external teeth, having modulus in the range between 3 and 11 mm and pitch line velocity in the range between 8 and 60 m/s. However, the procedure is applicable to any driving or driven cylindrical gear with a tooth profile in accordance with the standardized basic rack (ISO 53:1998), provided suitable reference data are available and the criteria specified below are met. It is also applicable for teeth conjugated to other basic racks where the virtual contact ratio, ε_{an}, is less than 2.5 ($\varepsilon_{an} < 2.5$). The results determined with such a procedure agree with those obtained by other methods, for normal pressure angles up to 25°, reference helix angles up to 25° and pitch line velocity higher than 2 m/s.

The ISO/TS 6336-22:2018 bases the calculation of micropitting load carrying capacity of cylindrical involute spur and helical gears with external teeth on the comparison between the minimum specific lubricant film thickness in the area of contact, $\lambda_{GF,min}$, and the permissible specific lubricant film thickness, λ_{GFP}, which constitutes a limiting value derived from specific gear testing or gears in service. To this end, the ISO standards introduce a safety factor against micropitting, S_λ, which must be equal to or higher than a minimum required safety factor against micropitting, $S_{\lambda,min}$. The safety factor against micropitting (of course, as micropitting here we mean the destructive micropitting during the working lifetime of the gear) is defined as the quotient of the minimum specific lubricant film thickness in the contact area, $\lambda_{GF,min} = (\lambda_{GF,Y})_{min}$, divided by the permissible specific lubricant film thickness, λ_{GFP}. Therefore, the following general relationship is introduced:

$$S_\lambda = \frac{\lambda_{GF,min}}{\lambda_{GFP}} \geq S_{\lambda,min}. \tag{10.69}$$

Both the minimum specific lubricant film thickness, $\lambda_{GF,min}$, and the permissible specific lubricant film thickness, λ_{GFP}, must be calculated separately for the pinion and wheel in the contact area, using the operating parameters. The minimum specific lubricant film thickness, $\lambda_{GF,min}$, is calculated by first determining the distribution curve of all calculated values of the local specific lubricant film thickness, $\lambda_{GF,Y}$, along the path of contact, and then taking its minimum value, $(\lambda_{GF,Y})_{min}$. The local specific lubricant film thickness, $\lambda_{GF,Y}$, is defined as the quotient of the local lubricant film thickness, h_Y (in μm), divided by the effective arithmetic mean roughness value, Ra (in μm); therefore, it is given by:

$$\lambda_{GF,Y} = \frac{h_Y}{Ra} = \frac{2h_Y}{(Ra_1 + Ra_2)}, \tag{10.70}$$

where Ra_1 and Ra_2 (both in μm) are respectively the arithmetic mean roughness values of pinion and wheel, while according to Dowson and Higginson (1977) the local lubricant film thickness, h_Y, is given by the following relationship:

$$h_Y = 1600(\rho_{n,Y})(G_M^{0.6})(U_Y^{0.7})(W_Y^{-0.13})(S_{GF,Y}^{0.22}), \tag{10.71}$$

where $\rho_{n,Y}$ (in mm) is the normal radius of relative curvature at point Y (Y is a current point on the path of contact), G_M is the dimensionless material parameter, U_Y is the dimensionless local velocity parameter, W_Y is the dimensionless local load parameter and $S_{GF,Y}$ is the dimensionless local sliding parameter. The theoretical bases of the quantities that appear in Eq. (10.71) are routed in the elastohydrodynamic lubrication theory, summarized by us for the aspects that are of interest here in Sect. 2.6.

The recommendation regarding the choice of the minimum safety factor against micropitting, $S_{\lambda,min}$, in Eq. (10.69) do not differ from those that we have already described each time we have faced similar problems. This choice must be made with the greatest possible care, appropriately evaluating the reliability of the assumptions on which calculations are based and the probability of failure and its consequences, in order to satisfy the required reliability at a justifiable cost.

Furthermore, it is necessary to consider the fact that micropitting can stop after a period of running-in or it can progress to macropitting (see Chap. 2) and tooth flank failure or tooth interior fatigue fracture (see next chapter). However, although there are criteria for defining micropitting failure through tests on lubricated surfaces in relative motion, there are no universally recognized criteria to define when micropitting is considered to be damaging.

For the choice of the most suitable value of $S_{\lambda,min}$, once the special requirements for specific lubricant film thickness are met, in addition to the aforementioned general recommendations, the following influences must be taken into consideration:

- If the performance of the to-be-designed gear can be appreciated by accurate testing of the actual gear drive under actual load conditions, a lower safety factor and cheaper manufacturing processes can be chosen.
- For critical applications, rigorous noise requirements and applications sensitive to wear particles in the lube, micropitting occurrence may not be tolerated, so a conservative design with higher safety factors is recommended. In these cases, the profile and lead modifications of the teeth should be designed using 3D contact analysis models, and the areal surface parameters describing the tooth flank texture should be checked to ensure that the sum of the contacting asperities is well below the lubricant film thickness.
- Larger safety factors are required in the case in which excessive tolerances due to inadequate manufacturing processes involve excessive variations in geometry and surface texture, as well as in cases of considerable alignment errors, and poor quality of materials or inadequate heat treatment, which have repercussions on chemical characteristics, on the microstructure and on the cleanliness of the same materials.
- Larger safety factors are required in cases where the applied loads or, more generally, the dynamic response to external loads and the dynamic transmission error of the mechanical system, which includes the to-be-designed gear drive, rather than being measured, are instead estimated using more or less refined models. Larger safety factors are also required in the case of variations in the lubrication and maintenance of the gear set over the expected service lifetime.
- Lower safety factors may be acceptable in cases where micropitting by itself may not be considered a failure. This happens in those industrial applications where some micropitting can be tolerated as long as it does not progress rapidly. In these cases, however, once micropitting occurrence is observed, it should be recorded and gearing service regularity checked to determine whether the pattern is growing. It is advisable to change the lubricant more frequently, filter it to remove the wear particles, or use a different lubricant with higher micropitting resistance.

From the observations above it follows that, depending on the field of application and the philosophy of the industrial sector involved, we can consider the micropitting load carrying capacity as a design-relevant aspect or not. Generally, a small amount of micropitting is often tolerated in automotive gearboxes and industrial gear drives that are subject to regular monitoring. Instead, in other fields of application, such as the one of the wind turbine industry, no micropitting on the tooth flank surfaces of the gears is usually allowed. To take account of these factual circumstances, the minimum safety factor against micropitting should be chosen according to the quality of the input variables and the tolerated probability of micropitting.

It should be noted that the aforementioned ISO standard is limited to giving only the recommendations summarized above, but does not provide the values of $S_{\lambda,min}$ to be assumed. This depends on the fact that relatively few known data are available in this regard. A guideline for the range between very low probability of micropitting (or no micropitting expected) and very high probability of micropitting is given in Fig. 10.8, taken from Hein et al. (2017), which provides the values of

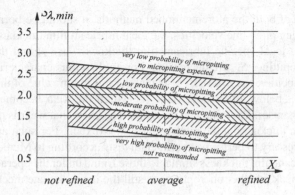

Fig. 10.8 Minimum required safety factor against micropitting, $S_{\lambda,\min}$, as a function of quality of calculation input variables and knowledge about operating conditions, X

$S_{\lambda,\min}$ differentiated into five levels of probability of micropitting as a function of the quality of calculation input variables and knowledge about operating conditions.

The calculation of the minimum specific lubricant film thickness in the contact area, $\lambda_{GF,\min}$, involves preliminary determination of the load specific lubricant film thickness, $\lambda_{GF,Y}$, which can be carried out using Method A and Method B.

With Method A, the distribution curve of $\lambda_{GF,Y}$ in the contact area is obtained using appropriate gear computing programs able to consider the main influences involved, such as the load distribution along the path of contact, local tangential and sliding velocities during the meshing cycle and actual service conditions. The minimum value, $\lambda_{GF,\min}$, read on this distribution curve is the one to be introduced into Eq. (10.69).

With Method B, which is a simplified method compared to Method A, the assumption is made that the determinant local specific lubricant film thickness occurs on the part of the tooth flank surface where sliding is negative. The simplification of this method implies that the local specific lubricant film thickness, $\lambda_{GF,Y}$, and therefore the local lubricant film thickness, h_Y, are determined in seven well-defined points on the path of contact, each of which is characterized by its own coordinate-parameter, Y. To this end, the seven points considered on the path of contact are as follows: the lower point A and upper point E on the path of contact; the lower point B and upper point D of single pair tooth contact; the pitch point C; the midway point AB between points A and B, and the midway point DE between points D and E. The minimum of the seven $\lambda_{GF,Y}$ values is the one to be introduced into Eq. (10.69). Regarding the above calculations, Method B also differentiates the case of no profile modifications from that of adequate profile modifications according to the experience of the manufacturers. Furthermore, in the case of profile modification lower than adequate profile modification, the calculation is performed as if no profile modification was done. Instead, in the case of too high-profile modification, the use of Method A is recommended.

In the use of both the aforementioned methods, it should be borne in mind that, with decreasing pitch line velocities, the local lubricant film thickness, h_Y, and consequently the local specific lubricant film thickness, $\lambda_{GF,Y}$, and the safety factor against micropitting, S_λ, are decreasing. For applications characterized by too low pitch line velocities, the real risk exists that the minimum value of the lubricant film thickness falls below a threshold value for which the wear becomes the dominant damage mechanism. Experimental investigations show that wear is predominant for lubricant film thickness at the pitch point $h_C \leq 0.1\mu m$. For such applications it is therefore necessary to perform experimental tests according to Method A or Method B, with lubricant film thickness similar to those found under the operating conditions, in order to check whether micropitting is still the main damage mechanism.

10.7   Permissible Specific Lubricant Film Thickness, λ_{GFP}

The determination of the permissible specific lubricant film thickness, λ_{GFP}, is the most difficult part of the ISO standard, as the understanding of the calculation procedures is not immediate. Several procedures can be used, which can be summarized as Method A and Method B.

Method A involves the use of experimental results or service experience, acquired on real gears, running under operating conditions where micropitting just occurs. In these just started micropitting conditions, the minimum specific lubricant film thickness, $\lambda_{GF,\min}$, can be calculated using Method A described in the previous section. This value is assumed as limiting specific lubricant film thickness, to be considered equivalent to the permissible specific lubricant film thickness, λ_{GFP}, as the correlated unit value of the safety factor, S_λ, leads to writing the following equality: $\lambda_{GF,\min} = S_{GFP}$. Obviously, experimental investigations must be carried out on gears that correctly emulate the design conditions of the real gear. In other words, gear manufacturing and accuracy, operating conditions including temperature, material and lubricant characteristics of the test gears must coincide with those of the actual gear drive. This procedure, of course expensive, is justifiable only for the development of new products or for gear drives whose failure could have serious consequences.

An alternative procedure, also included in Method A, involves the determination of the permissible specific lubricant film thickness based on considerations of dimensions, service conditions including lubrication, and performance of carefully monitored reference gears. The more accurately the dimensions and service conditions of the actual gears approximate those of the reference gears, the more effective will be the application of the calculated values of λ_{GFP} for the purpose of design analysis or response analysis calculations of micropitting load carrying capacity of gears.

Instead, Method B, which is less accurate than Method A, uses the results of well-documented histories of a number of test gears, which are carefully adapted and validated to apply them to the type, quality and manufacturing of gears under consideration. The permissible specific lubricant film thickness, λ_{GFP}, is calculated using, as a basis, the limiting (or critical) specific lubricant film thickness of the test gears, λ_{GFT}, which is a function of temperature, oil viscosity, base oil, and additive chemistry. This limiting specific lubricant film thickness, λ_{GFT}, is the result of any standardized test method that is able to evaluate the micropitting load carrying capacity of materials or lubricants by means of well-defined test gears, appropriate for the actual gears under consideration and operating under specified test conditions. The quantity λ_{GFT} can be calculated using Eq. (10.70), at the point of contact of the test gears where the minimum specific lubricant film thickness is found, under the test conditions for which the failure limit concerning micropitting in the standardized test procedure is reached.

To determine the data necessary for the application of Method B, a specific test procedure may be required and, if this is not available, any of the internationally recognized standardized test methods of micropitting assessment of gears, lubricants and materials can be used. Among these standardized methods, all of them approximate, the aforementioned ISO standard specifically mentions micropitting tests known as FVA-FZG micropitting text (FVA-Information Sheet 54/7 1993), Flender micropitting text (Theißen 1998), BGA-DU micropitting text (BGA-DU P602 2008), and Buzdygon-Cardis micropitting text (Buzdygon and Cardis 2004).

For reference purposes only, the same ISO standard, in Annex A, describes a procedure for calculation of the permissible specific lubricant film thickness, λ_{GFP}, for mineral oils, which uses a micropitting test according to the aforementioned FVA-Information Sheet 54/7. As the brief description that follows shows, it is a simple but inaccurate procedure that determines the permissible specific lubricant film thickness for mineral oils, to be considered as a generalized allowable reference, by means of the following relationship:

$$\lambda_{GFP} = 1.40 W_W \lambda_{GFT}, \tag{10.72}$$

where W_W is the *material factor* and λ_{GFT} is the already defined limiting specific lubricant film thickness ascertained by experimental tests. For mineral oils, the λ_{GFP}-values can be read from Fig. 10.9, as a function of the nominal oil viscosity and failure load stage, SKS, of the FVA-FZG micropitting test C-GF/8,3/80 with $Ra = 0.50\,\mu\text{m}$. Interpolated values among those corresponding to the four types of mineral oils in the figure can be chosen.

The SKS number, which expresses a property of the lubricant, is determined according to FVA-Information Sheet 54/7, but it can also be found on the lubricant data sheets of the leading suppliers. The variability range of the SKS number for mineral oils is as shown in Fig. 10.9. Modern lubricants like those used for wind turbine drives typically have an SKS number at the upper end of this range, i.e. $SKS = 10$. However, the SKS data need to be checked on a case-by-case basis with oil suppliers.

Figure 10.9 is valid for mineral oils. For the same viscosity and SKS number, synthetic oils have a different, typically lower, permissible specific lubricant film thickness, λ_{GFP}. It should also be noted that the same figure refers to case-carburized

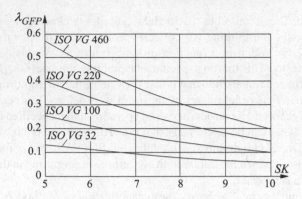

Fig. 10.9 Required minimum permissible specific lubricant film thickness, λ_{GFP}, for mineral oils as a function of the nominal oil viscosity and failure stage, SKS

steels. The values of λ_{GFP} read from Fig. 10.9 are to be taken with caution. They are in fact all the more unreliable the more the test conditions and the lubricants used differ from those to which the same figure refers.

The accurate way to determine a reliable value of λ_{GFP} is to use data of the oil performance from an FZG test rig. Many oils are tested on this type of test rig, using gear-types FZG C-GF. In this case, to determine the permissible specific lubricant film thickness, λ_{GFP}, we can use the relationship (10.72). The SKS number corresponds to the torque level at which the gear in the rig with the test oil shows micropitting (see Kissling 2012).

Finally, it is to be noted that the material factor, W_W, in Eq. (10.72) considers the influence of the gear material different from the case-carburized standardized test gear type C-GF. Table 10.1 provides values of the material factor, W_W, for materials other than the case-carburized steels.

It should be noted that micropitting load carrying capacity is often more significantly influenced by additives in the lubricant oil than by its viscosity. Since the effectiveness of the additives depends on the temperature, it is recommended to test the lubricant oil at the operating temperature of the specific application, with an

Table 10.1 Material factor, W_W

Material	Material factor, W_W
Case-carburized steel, with austenite content: – Less than 25% – Greater than 25%	 1.00 0.95
Gas nitride steel (HV >850)	1.50
Induction, flame hardened steel	0.65
Through hardened steel	0.50

approximation range included within ± 15 K. For more significant temperature differences, a specific micropitting test should be performed (usually this is carried out to the specific oil injection temperature) or an additional safety margin should be considered, however to be agreed between costumer and manufacturer.

In the absence of data determined by experimental investigations or service experience, the following guidelines can provide an informative reference to be used to calculate a value of λ_{GFP}, which however is in good correlation with the results obtained using the aforementioned test procedure according to FVA-Information Sheet 54/7. Figure 10.10a–c, taken from Annex B of the aforementioned ISO/TS 6336-22:2018 and valid for mineral oils respectively at oil temperatures $\vartheta_{oil} = 60\,°C, 90\,°C$ and $120\,°C$, allow to determine the reference guideline values of the minimum permissible specific lubricant film thickness, $\lambda_{GFP,\min}$, as a function of the nominal lubricant viscosity and lubricant micropitting performance, MP. The nominal lubricant viscosity used for this purpose is the kinematic viscosity ν_ϑ (in mm^2/s) at oil temperature, ϑ_{oil}.

Each of the three curves in the above figures corresponds to three well-defined quality grades of the lubricant micropitting performance, which are respectively designed MP-L, MP-Q and MP-E. In this regard, the following considerations must be made:

- Quality grade MP-L relates to lubricants that have not been intentionally developed to prevent micropitting. Lubricants with this quality grade usually do not exceed the failure load stage $SKS = 7$ in the FVA-FZG micropitting test at the specific oil temperature, ϑ_{oil}.
- Quality grade MP-Q relates to high quality lubricants used in standard industrial applications, which usually contain additives to prevent micropitting. Lubricants with this quality grade normally reach the failure load stage $SKS = 8$ or $SKS = 9$ in the above-mentioned micropitting test at specific oil temperature, ϑ_{oil}.
- Quality grade MP-E pertains to high quality lubricants, which are used in those applications where no micropitting at all is allowed, such as gear drive lubricants for wind turbines. Lubricants with this quality grade normally reach the failure load stage $SKS = 10$ in the above-mentioned micropitting test at specific oil temperature, ϑ_{oil}.

In case no specific information about the used lubricant is available, quality grade MP-L should be used. Interpolations between the data inferable from the above figures are allowed, but with due caution and use of an additional safety margin. Extrapolations for other nominal oil viscosities and oil temperatures than those specified in the same figures are not allowed.

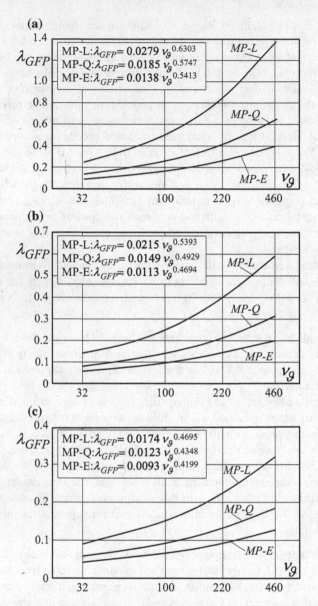

Fig. 10.10 Minimum permissible specific lubricant film thickness, $\lambda_{GFP,\min}$, for mineral oils as a function of the nominal lubricant kinematic viscosity, v_ϑ, and lubricant micropitting performance, MP, at oil temperature: **a** $\vartheta_{oil} = 60\,°\text{C}$; **b** $\vartheta_{oil} = 90\,°\text{C}$; **c** $\vartheta_{oil} = 120\,°\text{C}$

10.8 Local Specific Lubricant Film Thickness, $\lambda_{GF,Y}$

Equation (10.70) shows that, for the calculation of the local specific lubricant film thickness, $\lambda_{GF,Y}$, the effective arithmetic mean roughness value, Ra (in μm), and the local lubricant film thickness, h_Y (in μm) must first be determined.

About the determination of Ra, which is given by the half-sum of the effective mean roughness values, R_{a1} and R_{a2}, of pinion and wheel (see Eq. 10.72), we have nothing to add to what we already described in Sect. 2.7. However, we must note that the ISO standard uses, for the characterization of the texture of the lubricated surfaces involved in the contact between the teeth, a profile parameter (this is the Ra-roughness), while the most recent studies and researches on the topic have highlighted the need to use areal field parameters (see Greenwood and Williamson 1966; Greenwood and Tripp 1970–71; Chang et al. 1987; Jackson and Green 2011; Yastrebov 2013; Yastrebov et al. 2014, 2016). On the areal field parameters, which best characterize the rough surface texture and topography, we refer the reader to the last section of this chapter, which is to be considered a kind of appendix.

The determination of the local lubricant film thickness, h_Y, to be carried out along the entire meshing cycle using Eq. (10.71), constitutes a demanding and complex work because the influences on which h_Y depends are numerous and not simple to be evaluate. In Sect. 2.6 we were able to highlight the difficulties related to this determination. The procedures for calculating the various quantities that appear in Eq. (10.71) are described in detail below.

10.8.1 Normal Radius of Relative Curvature, $\rho_{n,Y}$, at Point Y

The determination of the normal radius of relative curvature, $\rho_{n,Y}$, at point Y, requires the preliminary definition of the position of the current contact point Y on the path of contact. As we already saw elsewhere, this position can vary, without interruption, between the lower point, A, and upper point, E, on the path of contact and, in a given meshing position, is defined by the coordinate, g_Y, measured along the line of action, starting from point A in the direction from point A to point E. Therefore, g_Y is the parameter on the path of contact that expresses the distance of point Y from point A (Fig. 10.11). It differs only formally (but not substantially) from that used in Sect. 7.7. Both parameters are dimensionless line parameters and define the position of an arbitrary current point of contact on the path of contact. However, they assume different points on the line of action from which the position of the arbitrary current point is measured. Different expressions of the position of this current point of contact follow, but the substance does not change. Here we conform to the ISO standard, even if for the user-designer it would be easier and more pleasant to have technical standards that always use the same coordinate systems as well as the same quantities, especially when they concern non-substantial, but not apparently different aspects of the same subject.

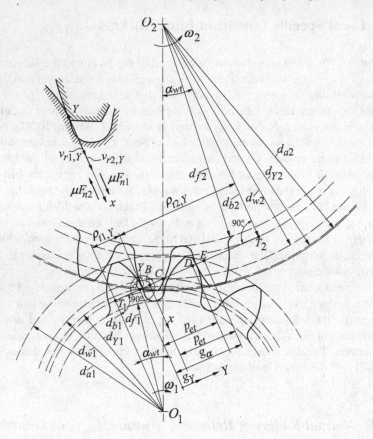

Fig. 10.11 Current contact point Y on path of contact and related quantities

We also saw that, with Method A, the continuous variability of $\rho_{n,Y}$, and therefore of Y, are of interest, while with Method B the seven well-defined points on the path of contact on which we must focus our attention are those described in Sect. 10.6. The coordinates (in mm) of these seven points A, AB, B, C, D, DE and E, measured along the path of contact, starting from point A and in the direction from point A to point E, are respectively given by the following relationships:

$$g_{Y=A} = g_A = 0 \tag{10.73}$$

$$g_{Y=AB} = g_{AB} = \frac{(g_\alpha - p_{et})}{2} \tag{10.74}$$

$$g_{Y=B} = g_B = (g_\alpha - p_{et}) \tag{10.75}$$

$$g_{Y=C} = g_C = \frac{d_{b1}}{2} \tan \alpha_{wt} - \sqrt{\frac{d_{a1}^2}{4} - \frac{d_{b1}^2}{4}} + g_\alpha \qquad (10.76)$$

$$g_{Y=D} = g_D = p_{et} \qquad (10.77)$$

$$g_{Y=DE} = g_{DE} = p_{et} + \frac{(g_\alpha - p_{et})}{2} \qquad (10.78)$$

$$g_{Y=E} = g_E = g_\alpha, \qquad (10.79)$$

where α_{wt} (in degrees) is the transverse pressure angle at the pitch cylinder, g_α (in mm) and p_{et} (in mm) are respectively the length of path of contact and transverse base pitch on the path of contact, while d_{a1} and d_{b1} (both in mm) are respectively the tip and base diameters of pinion.

The instantaneous radii in the current point of contact Y between the tooth surfaces, with point Y considered belonging to the pinion or wheel, respectively equal to $r_{Y1} = O_1 Y = d_{Y1}/2$ and $r_{Y2} = O_2 Y = d_{Y2}/2$, where d_{Y1} and d_{Y2} are the corresponding Y-circle diameters, are time-varying during the meshing cycle. In fact, they are dependent on the location coordinate, g_Y, of the contact point on path of contact. On the basis of geometric considerations (Fig. 10.11), we infer that these two Y-circle diameters can be expressed respectively by the following relationships:

$$d_{Y1} = 2\sqrt{\frac{d_{b1}^2}{4} + \left(\sqrt{\frac{d_{a1}^2}{4} - \frac{d_{b1}^2}{4}} - g_\alpha + g_Y\right)^2} \qquad (10.80)$$

$$d_{Y2} = 2\sqrt{\frac{d_{b2}^2}{4} + \left(\sqrt{\frac{d_{a2}^2}{4} - \frac{d_{b2}^2}{4}} - g_Y\right)^2}, \qquad (10.81)$$

where, the meaning of the symbols already introduced remaining unchanged, d_{a2} and d_{b2} (both in mm) are respectively the tip and base diameters of gear wheel.

As the two members of the gear rotate in mesh, the contact point Y is first initiated at the start of active profile (SAP) of the driving pinion 1 in the dedendum region, i.e. at contact point A. It moves upward toward the tip, passing through the pitch point, C, and leaving the contact at the end of active profile (EAP), at contact point E. For this driving pinion, its instantaneous transverse radius of curvature at point Y, $\rho_{t1,Y}$, is minimum when it is at the SAP and maximum when it is at the EAP. Meanwhile, the contact of the tooth of the driven wheel 2 is initiated at its EAP and moves toward its SAP as the gear members rotate in mesh. Therefore, the transverse radius of curvature of wheel at point Y, $\rho_{t2,Y}$, is maximum initially and is reduced gradually to its minimum value when the tooth leaves the contact at its own SAP. We can express the transverse radius of curvature of pinion or wheel at point Y, $\rho_{t1,2,Y}$ (both in mm) with the following relationship:

$$\rho_{t1,2,Y} = \frac{1}{2}\sqrt{d_{Y1,2}^2 - d_{b1,2}^2} = T_{1,2}Y = r_{b1,2}\vartheta_{Y1,2}, \qquad (10.82)$$

where $d_{Y1,2}$ and $d_{b1,2}$ are respectively the Y-circle and base diameters of pinion/wheel, $T_{1,2}$ are the interference points, and $\vartheta_{Y1,2}$ are the roll angles of the driving pinion and driven wheel (see also Fig. 10.3).

The transverse radius of relative curvature at point Y, $\rho_{t,Y}$, is given by:

$$\rho_{t,Y} = \frac{\rho_{t1,Y}\rho_{t2,Y}}{\rho_{t1,Y} + \rho_{t2,Y}}. \qquad (10.83)$$

The normal radius of relative curvature at the same point Y, $\rho_{n,Y}$, is given by (see Vol. 1: Sect. 8.3 and Fig. 8.3):

$$\rho_{n,Y} = \frac{\rho_{t,Y}}{\cos \beta_b}. \qquad (10.84)$$

where β_b is the base helix angle.

10.8.2   Material Parameter, G_M

The material parameter, G_M, considers the influence of the reduced modulus of elasticity, E_r, and pressure-viscosity coefficient at bulk temperature, $\alpha_{\vartheta M}$, on the local lubricant film thickness, h_Y. It is defined by the following relationship:

$$G_M = 10^6 E_r \alpha_{\vartheta M}, \qquad (10.85)$$

where E_r and $\alpha_{\vartheta M}$ are respectively given in N/mm^2 and m^2/N. Therefore, the numerical coefficient 10^6 is a factor that homogenizes the aforementioned two units of measurement in the SI-system.

In the more general case of mating gear members of different materials, with modulus of elasticity, E_1 and E_2, and Poisson's ratios, v_1 and v_2, the reduced modulus of elasticity, E_r, is determined by the following relationship:

$$E_r = \frac{2}{\left(\dfrac{1 - v_1^2}{E_1} + \dfrac{1 - v_2^2}{E_2}\right)}, \qquad (10.86)$$

which, for $E_1 = E_2 = E$ and $v_1 = v_2 = v$, becomes:

$$E_r = \frac{E}{1 - v^2}. \qquad (10.87)$$

For steel, we conventionally assume $E_1 = E_2 = E = 206 \times 10^3$ N/mm$^2 = 206$ GPa and $v_1 = v_2 = v = 0.3$.

The pressure-viscosity coefficient at bulk temperature, $\alpha_{\vartheta M}$, concerning the specific lubricant used, is determined on the basis of the data regarding this latter, when they are available. In the case this data is not available, an appropriate determination of $\alpha_{\vartheta M}$ can be carried out with the following relationship (see McGrew et al. 1970; Gupta 1984):

$$\alpha_{\vartheta M} = \alpha_{38} \left[1 + 516 \left(\frac{1}{273 + \vartheta_M} - \frac{1}{311} \right) \right], \tag{10.88}$$

where α_{38} (in m^2/N) is the pressure-viscosity coefficient of the lubricant at 38 °C and ϑ_M (in °C) is the bulk temperature. When specific values of α_{38} are not available, they can be calculated with the following approximate relationships taken from AGMA standard (AGMA 925-A03:2003):

$$\alpha_{38} = 2.657 \times 10^{-8} \eta_{38}^{0.1348} \tag{10.89}$$

$$\alpha_{38} = 1.466 \times 10^{-8} \eta_{38}^{0.0507} \tag{10.90}$$

$$\alpha_{38} = 1.392 \times 10^{-8} \eta_{38}^{0.1572}, \tag{10.91}$$

which are respectively valid for mineral oils, polyalphaolefin (PAO)-based synthetic non-VI (Viscosity Index) improved oils, and polyalkyleneglycol (PAG)-based synthetic oils. In these equations, η_{38} (in Ns/m^2) is the dynamic viscosity of the lubricant at 38 °C.

10.8.3 Local Velocity Parameter, U_Y

The local velocity parameter, U_Y, considers the proportional increase of the local lubricant film thickness, h_Y, with increasing dynamic viscosity $\eta_{\vartheta M}$ of the lubricant at bulk temperature and sum of the tangential velocities at point Y, $v_{\Sigma,Y}$, as well as with decreasing of the reduced modulus of elasticity, E_r, and normal radius of relative curvature at point Y, $\rho_{n,Y}$. It is defined by the following relationship:

$$U_Y = \frac{1}{2 \times 10^3} \frac{\eta_{\vartheta M} v_{\Sigma,Y}}{E_r \rho_{n,Y}}, \tag{10.92}$$

where $\eta_{\vartheta M}$, $v_{\Sigma,Y}$ and $\rho_{n,Y}$ are respectively given in Ns/m^2, m/s and mm, while the numerical coefficient 10^3 is a factor that homogenizes the aforementioned three units of measurement in the SI-system.

The local tangential velocities of pinion, $v_{r1,Y}$, and wheel, $v_{r2,Y}$, at a current contact point Y on the tooth flanks are respectively given by the following relationships:

$$v_{r1,Y} = \omega_1 \rho_{t1,Y} = \frac{2\pi n_1}{60} \frac{d_{w1}}{2000} \sin\alpha_{wt} \sqrt{\frac{d_{Y1}^2 - d_{b1}^2}{d_{w1}^2 - d_{b1}^2}} \qquad (10.93)$$

$$v_{r2,Y} = \frac{z_1}{z_2}\omega_1 \rho_{t2,Y} = \frac{2\pi n_1}{60u} \frac{d_{w2}}{2000} \sin\alpha_{wt} \sqrt{\frac{d_{Y2}^2 - d_{b2}^2}{d_{w2}^2 - d_{b2}^2}}, \qquad (10.94)$$

where d_{w1} and d_{w2} are the pitch diameters of pinion and wheel (both in mm), n_1 (in min^{-1}) is the rotational speed of pinion, $u = z_2/z_1$ is the gear ratio and α_{wt} (in degrees) is the pressure angle at the pitch cylinder. All other symbols retain the already known meaning. Equations (10.93) and (10.94) highlight that $v_{r1,Y}$ and $v_{r2,Y}$ (both expressed in m/s) depend respectively on the Y-circle diameters of pinion and wheel, d_{Y1} and d_{Y2}.

It is to be noted that $v_{r1,Y} < v_{r2,Y}$ when the contact point Y on the tooth flank is below the pitch point C in the dedendum region of pinion 1. As a result, the slide-to-roll ratio (briefly, SR-ratio), defined as the quotient of the local sliding velocity, $v_{g,Y} = (v_{r1,Y} - v_{r2,Y})$, divided by the local rolling velocity of contact, $v_r = (v_{r1,Y} + v_{r2,Y})/2$, is negative in this region, reaching its maximum absolute value when the contact is at the start of active profile (SAP) of the driving pinion 1. When the contact reaches the pitch point, for which $Y \equiv C, v_{g,Y} = 0$; in this instant, we have pure rolling with SR $= 0$ and $v_r = v_{r1,Y} = v_{r2,Y}$. As the contact moves up toward the tip in the addendum region, SR becomes positive (SR > 0) and reaches its maximum positive value at the end of active profile (EAP) of the driving pinion 1.

The sum of the tangential velocities at point Y, to be introduced in Eq. (10.92), is given by:

$$v_{\Sigma,Y} = v_{r1,Y} + v_{r2,Y}. \qquad (10.95)$$

The dynamic viscosity $\eta_{\vartheta M}$ of the lubricant at bulk temperature can be calculated by the following relationship:

$$\eta_{\vartheta M} = 10^{-6} v_{\vartheta M} \rho_{\vartheta M}, \qquad (10.96)$$

where $v_{\vartheta M}$ (in mm^2/s) and $\rho_{\vartheta M}$ (in kg/m^3) are respectively the kinematic viscosity and density of lubricant at bulk temperature, while 10^{-6} is a numerical factor that homogenizes the units of measurement of the quantities used in the SI-system.

The kinematic viscosity, $v_{\vartheta M}$, at bulk temperature, ϑ_M (in °C), can be determined from the kinematic viscosity v_{40} of the lubricant at 40 °C and kinematic viscosity v_{100} of the same lubricant at 100 °C, using the following relationship:

$$\log[\log(v_{\vartheta M} + 0.7)] = A \log(\vartheta_M + 273) + B, \qquad (10.97)$$

where:

$$A = \frac{\log[\log(\nu_{40} + 0.7)/\log(\nu_{100} + 0.7)]}{\log(313/373)} \qquad (10.98)$$

$$B = \log[\log(\nu_{40} + 0.7)] - A\log(313). \qquad (10.99)$$

It is to be noted that the values of kinematic viscosity obtained by extrapolation for temperatures higher than 140 °C should be verified by experimental measurements.

The density, $\rho_{\vartheta M}$, of lubricant at bulk temperature, ϑ_M, should be determined by experimental measurements. If this experimental data is not available, $\rho_{\vartheta M}$ can be determined using the following approximate relationship, which is based on the density ρ_{15} of lubricant at 15 °C:

$$\rho_{\vartheta M} = \rho_{15}\left[1 - 0.7\frac{(\vartheta_M + 273) - 288}{\rho_{15}}\right]. \qquad (10.100)$$

If no data for ρ_{15} is available, for mineral oils the following approximate relationship can be used:

$$\rho_{15} = 43.37\log\nu_{40} + 805.5. \qquad (10.101)$$

10.8.4 Local Load Parameter, W_Y

The local load parameter, W_Y, considers the influence of the reduced modulus of elasticity, E_r, and the local Hertzian contact stress, $p_{dyn,Y}$, including the load factors K, on the local lubricant film thickness, h_Y. It is defined by the following relationship:

$$W_Y = \frac{2\pi p_{dyn,Y}^2}{E_r^2}. \qquad (10.102)$$

The quantity $p_{dyn,Y}$, to be expressed in N/mm^2, can be calculated with Method A or with Method B. The calculation method used is indicated by adding the subscripts A or B, for which we will have respectively $p_{dyn,Y,A}$ or $p_{dyn,Y,B}$.

For the determination of the local Hertzian contact stress, $p_{dyn,Y,A}$, according to Method A, a suitable three-dimensional mesh contact program and a load distribution analysis procedure should be used. With this 3D mesh contact and load distribution model, the local nominal Hertzian contact stress, $p_{H,Y,A}$, is first determined. Once this last quantity has been determined, the local Hertzian contact stress, $p_{dyn,Y,A}$, is calculated using the following relationship:

$$p_{dyn,Y,A} = p_{H,Y,A}\sqrt{K_A K_v}, \qquad (10.103)$$

where, as usual, K_A and K_v are respectively the application and dynamic factors. Obviously, in case the influences related to one or both of these two factors were already included in the 3D mesh contact model (this can be an elastic model, but it can be of a more general nature, to evaluate not only the elastic effects, but also the elastic-plastic and plastic effects (see Johnson 1985)) one or both factors in Eq. (10.103) must be assumed as having value equal to 1.

For the determination of the local Hertzian contact stress, $p_{dyn,Y,B}$, according to Method B, we use the following relationship:

$$p_{dyn,Y,B} = p_{H,Y,B}\sqrt{K_A K_\gamma K_v K_{H\alpha} K_{H\beta}}, \qquad (10.104)$$

where the local nominal Hertzian contact stress, $p_{H,Y,B}$, is given by:

$$p_{H,Y,B} = Z_E \sqrt{\frac{F_t X_Y}{b\rho_{n,Y}\cos\alpha_t \cos\beta_b}}, \qquad (10.105)$$

with

$$Z_E = \sqrt{\frac{E_r}{2\pi}}, \qquad (10.106)$$

where Z_E and Z_Y are respectively the elastic factor and local sharing factor, while the meaning of the other symbols already introduced elsewhere remains unchanged. However, it should be noted that the transverse load factor, $K_{H\alpha}$, does not include the effects of profile modifications, which are instead considered with the load sharing factor, X_Y. Instead, the face load factor, $K_{H\beta}$, includes the effects of lead modifications. It should also to be noted that Eq. (10.105), where $\cos\beta_b$ appears, differs from the corresponding formula of the ISO/TS 6336-22: 2018. Actually, Hertzian contact pressure, $p_{H,Y,B}$, in the case of cylindrical helical gears depends on normal force, F_n, given by Eq. (8.81) of Vol. 1. Of course, in the particular case of cylindrical spur gears, we have $\cos\beta_b = 0$.

When we use Eq. (10.104), we must also bear in mind that, in the case of gear trains with multiple transmission paths or planetary gear trains, the total load is not quite evenly distributed over the individual tooth pairs in simultaneous meshing. To keep this fact in due consideration, we introduce into Eq. (10.104) a mesh load factor, K_γ, which adjusts the average load per mesh. Of course, this factor is assumed equal to 1 in the cases of single transmission path. Finally, the load sharing factor, X_Y, considers the influence of different profile modifications (see Sect. 10.8.6).

10.8.5 Local Sliding Parameter, $S_{GF,Y}$

The local sliding parameter, $S_{GF,Y}$, considers the influence of local sliding on the local temperature, which affects both the local pressure-viscosity coefficient, $\alpha_{\vartheta B,Y}$

(also known as pressure-viscosity coefficient at local contact temperature, $\vartheta_{B,Y}$), and the local dynamic viscosity, $\eta_{\vartheta B,Y}$ (also known as dynamic viscosity at local contact temperature, $\vartheta_{B,Y}$) and, therefore, the local lubricant film thickness, h_Y. The local contact temperature, $\vartheta_{B,Y}$, is the sum of the local flash temperature, $\vartheta_{fl,Y}$, and the bulk temperature, ϑ_M. All these temperatures are given in degrees. The sliding parameter, $S_{GF,Y}$, is defined by the following relationship:

$$S_{GF,Y} = \frac{\alpha_{\vartheta B,Y} \eta_{\vartheta B,Y}}{\alpha_{\vartheta M} \eta_{\vartheta M}}. \tag{10.107}$$

The pressure-viscosity coefficient $\alpha_{\vartheta B,Y}$ at local contact temperature $\vartheta_{B,Y}$, concerning the specific lubricant used, is also determined on the basis of the data regarding this latter, when they are available. In the case this data is not available, an approximate determination of $\alpha_{\vartheta B,Y}$ can be carried out with the following relationship:

$$\alpha_{\vartheta B,Y} = \alpha_{38} \left[1 + 516 \left(\frac{1}{273 + \vartheta_{B,Y}} - \frac{1}{311} \right) \right]. \tag{10.108}$$

This equation differs from Eq. (10.88) for the replacement of $\alpha_{\vartheta B,Y}$ and $\vartheta_{B,Y}$ respectively in place of $\alpha_{\vartheta M}$ and ϑ_M.

The dynamic viscosity at local contact temperature, $\eta_{\vartheta B,Y}$, is determined using the following relationship:

$$\eta_{\vartheta B,Y} = 10^{-6} \nu_{\vartheta B,Y} \rho_{\vartheta B,Y}, \tag{10.109}$$

where $\nu_{\vartheta B,Y}$ and $\rho_{\vartheta B,Y}$ are respectively the kinematic viscosity and density of the lubricant at local contact temperature. This equation differs from Eq. (10.96) for the substitution of quantities calculated at local contact temperature instead of the same quantities calculated at bulk temperature.

The kinematic viscosity at local contact temperature, $\nu_{\vartheta B,Y}$, is determined from the kinematic viscosity ν_{40} at 40 °C and kinematic viscosity ν_{100} at 100 °C using the following relationship:

$$\log[\log(\nu_{\vartheta B,Y} + 0.7)] = A \log(\vartheta_{B,Y} + 273) + B, \tag{10.110}$$

where $\vartheta_{B,Y}$ is the local contact temperature. This equation also differs from Eq. (10.97) for the substitution of quantities calculated at local contact temperature instead of the same quantities calculated at bulk temperature, while A and B are still given by Eqs. (10.98) and (10.99). Here also it is to be noted that the values of the kinematic viscosity at local contact temperature obtained by extrapolation for temperatures higher than 140 °C should be verified by experimental measurements.

Finally, the density of lubricant at local contact temperature, $\vartheta_{B,Y}$, should be determined by experimental investigations. If this experimental data is not available, $\rho_{\vartheta B,Y}$ can be determined using the following relationship:

$$\rho_{\vartheta B,Y} = \rho_{15}\left[1 - 0.7\frac{(\vartheta_{B,Y} + 273) - 288}{\rho_{15}}\right], \qquad (10.111)$$

which also differs from Eq. (10.100) for the usual substitution of quantities calculated at the local contact temperature instead of the same quantities calculated at bulk temperature.

10.8.6   Load Sharing Factor, X_Y

10.8.6.1   General Premise

The load sharing factor, X_Y, considers the load sharing between teeth pairs that came into simultaneous meshing in gradual succession. This factor is expressed as a function of the linear parameter g_Y on the path of contact, defined in Sect. 10.8.1. The load sharing factor depends on the type of gear (cylindrical spur or helical gears) and on the profile modification of the teeth of the two members of the gear pair under consideration.

As we already highlighted in Sect. 7.7.1, due to the unavoidable inaccuracies elsewhere described in detail, any tooth pair entering into meshing can cause a sharp, almost instantaneous, increase or decrease in the theoretical value of the load sharing factor, and this regardless of the more or less sharp, almost instantaneous, increase or decrease of the same factor caused by inaccuracies of the subsequent meshing tooth pair at a later time. For cylindrical gears, the load sharing factor, X_Y, does not exceed the unit value. The region of transverse single pair tooth contact can be more extensive than theoretically predicted, because of an irregular local variation of the dynamic load.

For cylindrical helical gears without profile modification, the buttressing effect comes into play, so the load sharing factor depends not only on the tooth profile modification, but also on the buttressing effect. Therefore, for these types of gears, the actual load sharing factor is the result of the combination of the theoretical load sharing factor with the buttressing effect, $X_{but,Y}$. This buttressing effect is due to the oblique contact lines, and may occur near the end points, A and E, of the path of contact.

It should be noted that, as a consequence of the non-substantial difference of the dimensionless linear parameter that defines the arbitrary current point of contact on the path of contact with respect to that used in Chap. 7, highlighted in Sect. 10.8.1, also the expressions that define the load sharing factor in the various cases described below, corresponding to those discussed in Chap. 7, differ only in the mathematical formalism. It follows that the figures shown below, which represent the distribution curves of the load sharing factor related to the same cases, are identical to those shown in Chap. 7, even if expressed with different parameters. In order to comply

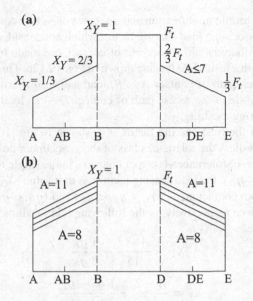

Fig. 10.12 Load distribution and associated load sharing factor for cylindrical spur gears without profile modification: **a** tolerance class 7 or finer; **b** tolerance class 8 or coarser

with the ISO standard mentioned above, we have decided here to follow the different formalism proposed by it.

10.8.6.2 Cylindrical Spur Gears

For the determination of the load sharing factor of cylindrical spur gears, two cases need to be distinguished: the case of spur gears with unmodified profiles and the one of spur gears with profile modification.

Spur Gears with Unmodified Profiles

As we already said in Sect. 7.7.2, in the two double tooth contact areas the distribution curves of the nominal transverse tangential load at reference cylinder per mesh, F_t, and the local load sharing factor, X_Y, which are closely inter-correlated between them, increase in the approach path of contact $A - B$, and decrease in the recess path of contact, $D - E$. This is due to the unavoidable inaccuracies, of whatever nature they may be.

To consider the influence of these inaccuracies in cylindrical spur gears without profile modification, the discontinuous conventional trapezoidal load distribution and the associated load sharing factor shown in Fig. 10.12a are assumed, which can be considered sufficiently correct for tolerance class 7 or finer. For actual cylindrical spur gears without profile modification and tolerance class 8 or coarser, the load distribution and the associated load sharing factor, which are considered representative of their actual distributions along the path of contact, are made by the envelope of possible curves, with a result like the one shown in Fig. 10.12b. This figure highlights that, in the approach path of contact $A - B$, load and load sharing factor increase during meshing, while in the recess path of contact $D - E$, load and load sharing factor decrease during meshing.

Distributions of the load sharing factor, X_Y, shown in Fig. 10.12a and 10.12b, highlight that, regardless the tolerance class of the gears under consideration (tolerance class 7 or finer, or tolerance class 8 or coarser), in the single tooth contact area, i.e. for $(g_B \leq g_Y \leq g_D)$, the load sharing factor has unit value ($X_Y = 1$). Instead, in the two double tooth contact areas $(g_A \leq g_Y < g_B)$ and $(g_D < g_Y \leq g_E)$, the load sharing factor is given respectively by the following relationships:

$$X_Y = \frac{A - 2}{15} + \frac{1}{3}\frac{g_Y}{g_B} \qquad (10.112)$$

$$X_Y = \frac{A - 2}{15} + \frac{1}{3}\frac{g_\alpha - g_Y}{g_\alpha - g_D}, \qquad (10.113)$$

where A is the tolerance class of the gear. In these equations, we put $A = 7$, for tolerance class 7 or finer, and A equal to the actual tolerance class, for tolerance class 8 or coarser.

Spur Gears with Profile Modification

For cylindrical spur gears with an adequate profile modification in driving and driven members, the distribution of the nominal transverse tangential load at reference cylinder per mesh, F_t, along the path of contact, $A - E$, is conventionally assumed to have a discontinuous trapezoidal shape, as Fig. 10.13a shows. This same figure, with the maximum value equal to 1, represents the corresponding load sharing factor, X_Y. The distribution curves of F_t and X_Y, strictly inter-correlated between them, show that, in the two double tooth contact areas, both these quantities increase from 0 to the corresponding maximum values (respectively, F_t or 1) in the approach path of contact, and decrease from the maximum values to 0 in the recess path of contact; both these quantities instead retain their maximum values in the single tooth contact area.

The load sharing factor of cylindrical spur gears manufactured in order to have an adequate profile modification on both their driving and driven members is given by the following relationships:

$$X_Y = \frac{g_Y}{g_B} \qquad (10.114)$$

$$X_Y = 1 \qquad (10.115)$$

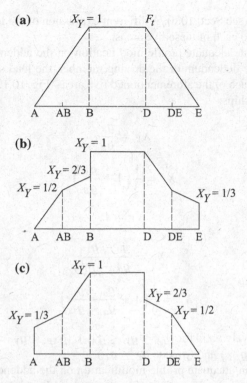

Fig. 10.13 Load sharing factor, X_Y, for cylindrical spur gears with adequate profile modification on: **a** both driving and driven members; **b** the addendum of the driven member and/or the dedendum of the driving member; **c** the addendum of the driving member and/or the dedendum of the driven member

$$X_Y = \frac{g_\alpha - g_Y}{g_\alpha - g_D},\qquad(10.116)$$

which are respectively valid for $(g_A \leq g_Y \leq g_B)$, $(g_B < g_Y < g_D)$ and $(g_D \leq g_Y \leq g_E)$.

However, cylindrical spur gears can be also manufactured in order to have an adequate profile modification on:

- the addendum of the driven member and/or the dedendum of the driving member (in this case, the distributions of F_t and X_Y are those shown in Fig. 10.13b);
- the addendum of the driving member and/or the dedendum of the driven member (in this case, the distributions of F_t and X_Y are those shown in Fig. 10.13c).

In both these cases, the distribution curve of the load sharing factor, X_Y, along the path of contact is constituted by an irregular succession of straight lines, like the ones shown in Fig. 10.13b, c. The path of contact, rather than be fractioned according to the usual three areas $A - B$, $B - D$ and $D - E$, is divided into five areas with the introduction of the midway point AB between A and B, and the midway point DE

between D and E (see Sect. 10.6). A different expression of the load sharing factor, X_Y, is associated to each of these five areas.

In the case of an adequate profile modification on the addendum of the driven member and/or the dedendum of the driving member, the load sharing factor, X_Y, corresponding to each of the aforementioned five areas (Fig. 10.13b), is given by the following relationships:

$$X_Y = \frac{g_Y}{g_B} \tag{10.117}$$

$$X_Y = \frac{1}{3}\left(1 + \frac{g_Y}{g_B}\right) \tag{10.118}$$

$$X_Y = 1 \tag{10.119}$$

$$X_Y = \frac{g_\alpha - g_Y}{g_\alpha - g_D} \tag{10.120}$$

$$X_Y = \frac{1}{3}\left(1 + \frac{g_\alpha - g_Y}{g_\alpha - g_D}\right), \tag{10.121}$$

which are respectively valid for $(g_A \leq g_Y \leq g_{AB})$, $(g_{AB} < g_Y < g_B)$, $(g_B \leq g_Y < g_D)$, $(g_D \leq g_Y \leq g_{DE})$ and $(g_{DE} < g_Y \leq g_E)$.

In the case of an adequate profile modification on the addendum of the driving member and/or the dedendum of the driven member, the load sharing factor, X_Y, corresponding to each of the aforementioned five areas (Fig. 10.13c) is given by the following relationships:

$$X_Y = \frac{1}{3}\left(1 + \frac{g_Y}{g_B}\right), \tag{10.122}$$

$$X_Y = \frac{g_Y}{g_B} \tag{10.123}$$

$$X_Y = 1 \tag{10.124}$$

$$X_Y = \frac{1}{3}\left(1 + \frac{g_\alpha - g_Y}{g_\alpha - g_D}\right) \tag{10.125}$$

$$X_Y = \frac{g_\alpha - g_Y}{g_\alpha - g_D}, \tag{10.126}$$

which are respectively valid for $(g_A \leq g_Y \leq g_{AB})$, $(g_{AB} < g_Y \leq g_B)$, $(g_B < g_Y \leq g_D)$, $(g_D < g_Y \leq g_{DE})$ and $(g_{DE} < g_Y \leq g_E)$.

10.8.7 Local Buttressing Factor, $X_{but,Y}$

For cylindrical helical gears, it is necessary to consider the buttressing effect, that is the reinforcement action exerted by oblique contact lines in the plane of action (see Sect. 7.7.3). This strengthening effect occurs independently of any tooth profile modification, and is predominantly concentrated near the end points, A and E, of the path of contact. We consider the buttressing effect for cylindrical helical gears without profile modification by introducing a local buttressing factor, $X_{but,Y}$, whose distribution along the path of contact is given by a sequence of straight lines, as Fig. 10.14 shows.

Figure 10.14 shows that, in relation to the local buttressing effect, the path of contact is further divided, with the introduction of two new marked points, AU and EU, which are added to those defined previously. Therefore, we have two new areas, $A - AU$ and $EU - E$, where the buttressing effect is evident. The lengths of the portions $A - AU$ and $EU - E$ of the path of contact that are affected by the buttressing effect are assumed conventionally equal to each other; they are expressed in mm and are given by the following relationship:

$$g_{AU} - g_A = g_E - g_{EU} = 0.2\sin\beta_b, \tag{10.127}$$

where β_b is the base helix angle, while $g_A = 0$ and $g_E = g_\alpha$ (see Sect. 10.8.1).

In conventional terms, we assume that the local buttressing factor, $X_{but,Y}$, is equal to 1 throughout the entire $AU - EU$ area, including the marked points AU and EU (therefore, $X_{but,Y} = 1$ for $g_{AU} \le g_Y \le g_{EU}$, and $X_{but,AU} = X_{but,EU} = 1$), while at the end marked points, A and E, the local buttressing factor is $X_{but,A} = X_{but,E} = 1.3$, when $\varepsilon_\beta \ge 1$, and $X_{but,A} = X_{but,E} = (1 + 0.3\varepsilon_\beta)$ when $\varepsilon_\beta < 1$, where ε_β is the overlap ratio. Finally, for $(g_A \le g_Y < g_{AU})$ and $(g_{EU} < g_Y \le g_E)$, the local buttressing factor is given respectively by the following relationships:

$$X_{but,Y} = X_{but,A} - \frac{g_Y}{0.2\sin\beta_b}(X_{but,A} - 1) \tag{10.128}$$

$$X_{but,Y} = X_{but,E} - \frac{g_\alpha - g_Y}{0.2\sin\beta_b}(X_{but,E} - 1). \tag{10.129}$$

Fig. 10.14 Distribution of the local buttressing factor, $X_{but,Y}$, along the path of contact

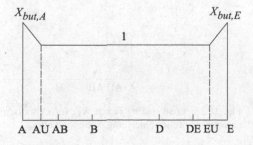

10.8.8   Cylindrical Helical Gears

For the determination of the load sharing factor of cylindrical helical gears, these gears are divided into the following three classes:

- Cylindrical helical gears with overlap ratio, $\varepsilon_\beta \leq 0.8$.
- Cylindrical helical gears with overlap ratio, $\varepsilon_\beta \geq 1.2$.
- Cylindrical helical gears with overlap ratio within the range $0.8 < \varepsilon_\beta < 1.2$.

10.8.8.1   Cylindrical Helical Gears with Overlap Ratio $\varepsilon_\beta \leq 0.8$

Cylindrical helical gears with transverse contact ratio, $\varepsilon_\alpha \geq 1$, and overlap ratio, $\varepsilon_\beta \leq 0.8$, can have unmodified profiles or profile modification. This class of cylindrical helical gears, regardless of whether they have unmodified or modified tooth profiles, is still characterized by a small portion of single pair tooth contact. Therefore, these gears can be discussed similarly to the cylindrical spur gears, considering the geometry and corresponding quantities in the transverse section.

Regardless the cylindrical helical gears under consideration have unmodified or modified tooth profiles, the load sharing factor distributions are obtained by multiplying the load sharing factor, X_Y, of the corresponding cylindrical spur gear (see Sect. 10.8.6.2) with the buttressing factor, $X_{but,Y}$ (see Sect. 10.8.7). The distribution of the load sharing factor for cylindrical helical gears with $\varepsilon_\beta \leq 0.8$ and unmodified profiles, including the buttressing effect, is the one shown in Fig. 10.15.

Instead, the distributions of the load sharing factor of the same class of cylindrical helical gears, but with profile modification, are different depending on the type of tooth profile modification. Figure 10.16 shows three different load sharing factor distributions, which refer respectively to: adequate profile modification on teeth of both gear members, to be considered as the optimal profile modification; adequate profile modification on the addendum of the driven member and/or the dedendum of the driving member; adequate profile modification on the addendum of the driving member and/or the dedendum of the driven member.

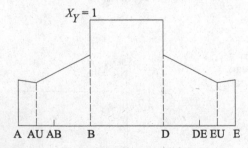

Fig. 10.15 Load sharing factor, X_Y, for cylindrical helical gears with $\varepsilon_\alpha \geq 1$, $\varepsilon_\beta \leq 0.8$ and unmodified profiles, including buttressing effect

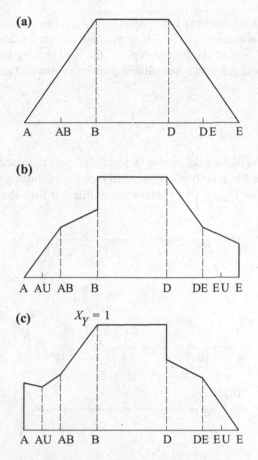

Fig. 10.16 Load sharing factor, X_Y, for cylindrical helical gears with $\varepsilon_\alpha \geq 1$, $\varepsilon_\beta \leq 0.8$ and: **a** adequate profile modification on teeth of both gear members; **b** adequate profile modification on the addendum of the driven member and/or the dedendum of the driving member; **c** adequate profile modification on the addendum of the driving member and/or the dedendum of the driven member

10.8.8.2 Cylindrical Helical Gears with $\varepsilon_\beta \geq 1.2$

Cylindrical helical gears with transverse contact ratio, $\varepsilon_\alpha \geq 1$, and overlap ratio, $\varepsilon_\beta \geq 1.2$, i.e. wide cylindrical helical gears having a total contact ratio, $\varepsilon_\gamma \geq 2.2$, can have unmodified or modified tooth profiles.

For cylindrical helical gears with unmodified profiles included in this class, the buttressing effect, due to the local high mesh stiffness at the end of the oblique contact lines, is assumed acting near the marked points A and E of the path of contact, along the tooth helix, over a constant length given by Eq. (10.127). In this case, the distribution of the load sharing factor is quite similar to that of the buttressing factor shown in Fig. 10.14, with the variation that the mean load sharing factor between the marked points AU and EU is not equal to 1 (as it happens for cylindrical spur gears with unmodified profiles), but is equal to $(1/\varepsilon_\alpha)$; this ratio also represents the mean load acting in the area $AU - EU$ of the path of contact. Instead, in the areas $A - AU$ and $EU - E$ of the path of contact, the load sharing factor for this class of cylindrical helical gears with unmodified profiles is obtained using the following relationship:

$$X_Y = \frac{1}{\varepsilon_\alpha} X_{but,Y}. \tag{10.130}$$

The distributions of the load sharing factor of the same class of cylindrical helical gears, but with profile modification, are different depending on the type of tooth profile modification. Figure 10.17 shows three different load sharing factor distri-

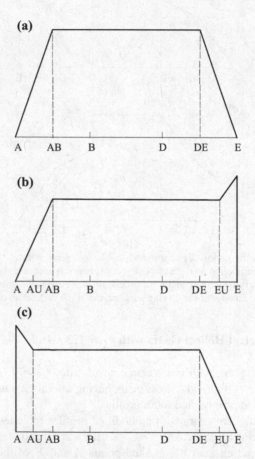

Fig. 10.17 Load sharing factor X_Y for cylindrical helical gears with $\varepsilon_\alpha \geq 1$, $\varepsilon_\beta \geq 1.2$ and: **a** adequate profile modification on teeth of both gear members; **b** adequate profile modification on the addendum of the driven member and/or the dedendum of the driving member; **c** adequate profile modification on the addendum of the driving member and/or the dedendum of the driven member

butions, which refer respectively to: adequate profile modification on teeth of both gear members, to be considered as the optimal profile modification; adequate profile modification on the addendum of the driven member and/or the dedendum of the driving member; adequate profile modification on the addendum of the driving member and/or the dedendum of the driven member.

Regardless of the type and amount of the profile modification, it is to be noted that a tip relief on the pinion decreases and increases the load sharing factor, X_Y, respectively in the area $DE - E$ and area $AB - DE$. Instead, a tip relief on the wheel decreases and increases the load sharing factor, X_Y, respectively in the area $A - AB$ and area $AB - DE$. The extensions of the tip relief at both ends of the path of contact are assumed to be equal (thus we will have lengths $A - AB$ and $DE - E$ equal to each other), and having an amount sufficient to result in a transverse contact ratio $\varepsilon_\alpha = 1$, for unloaded gears. This last condition implies that the lengths $A - D$, $B - E$, and $AB - DE$ shown in Fig. 10.17a are the same.

The load sharing factor for this class of cylindrical helical gears with adequate profile modification on teeth of both gear members (Fig. 10.17a) is given by the following relationships:

$$X_Y = \frac{1}{\varepsilon_\alpha}\left(1 + \frac{\varepsilon_\alpha - 1}{\varepsilon_\alpha + 1}\right)\frac{g_Y}{g_{AB}} \tag{10.131}$$

$$X_Y = \frac{1}{\varepsilon_\alpha}\left(1 + \frac{\varepsilon_\alpha - 1}{\varepsilon_\alpha + 1}\right) \tag{10.132}$$

$$X_Y = \frac{1}{\varepsilon_\alpha}\left(1 + \frac{\varepsilon_\alpha - 1}{\varepsilon_\alpha + 1}\right)\frac{g_\alpha - g_Y}{g_\alpha - g_{DE}}, \tag{10.133}$$

which are respectively valid in the areas $(g_A \leq g_Y \leq g_{AB})$, $(g_{AB} < g_Y \leq g_{DE})$ and $(g_{DE} < g_Y \leq g_E)$.

The load sharing factor for the same class of cylindrical helical gears with adequate profile modification on the addendum of the driven member and/or the dedendum of the driving member (Fig. 10.17b) is given by the following relationships:

$$X_Y = \frac{1}{\varepsilon_\alpha}\left[1 + \frac{\varepsilon_\alpha - 1}{2(\varepsilon_\alpha + 1)}\right]\frac{g_Y}{g_{AB}} \tag{10.134}$$

$$X_Y = \frac{1}{\varepsilon_\alpha}\left[1 + \frac{\varepsilon_\alpha - 1}{2(\varepsilon_\alpha + 1)}\right] \tag{10.135}$$

$$X_Y = \frac{1}{\varepsilon_\alpha}\left[1 + \frac{\varepsilon_\alpha - 1}{2(\varepsilon_\alpha + 1)}\right]X_{but,Y}, \tag{10.136}$$

which are respectively valid in the areas $(g_A \leq g_Y \leq g_{AB})$, $(g_{AB} < g_Y \leq g_{EU})$ and $(g_{EU} < g_Y \leq g_E)$.

The load sharing factor for the same class of cylindrical helical gears with adequate profile modification on the addendum of the driving member and/or the dedendum of the driven member (Fig. 10.17c) is given by the following relationships:

$$X_Y = \frac{1}{\varepsilon_\alpha}\left[1 + \frac{\varepsilon_\alpha - 1}{2(\varepsilon_\alpha + 1)}\right]X_{but,Y} \tag{10.137}$$

$$X_Y = \frac{1}{\varepsilon_\alpha}\left[1 + \frac{\varepsilon_\alpha - 1}{2(\varepsilon_\alpha + 1)}\right] \tag{10.138}$$

$$X_Y = \frac{1}{\varepsilon_\alpha}\left[1 + \frac{\varepsilon_\alpha - 1}{2(\varepsilon_\alpha + 1)}\right]\frac{g_\alpha - g_Y}{g_\alpha - g_{DE}}, \tag{10.139}$$

which are respectively valid in the areas $(g_A \leq g_Y \leq g_{AU})$, $(g_{AU} < g_Y \leq g_{DE})$ and $(g_{DE} < g_Y \leq g_E)$.

10.8.8.3   Cylindrical Helical Gears with $0.8 < \varepsilon_\beta < 1.2$

To calculate the load sharing factor for cylindrical helical gears with $0.8 < \varepsilon_\beta < 1.2$, it is necessary to bear in mind that the overlap ratio, ε_β, changes depending on the load, as the gears have not an infinite stiffness. This overlap ratio variation affects the load sharing factor for cylindrical helical gears characterized by calculated values of ε_β included in the range $0.8 < \varepsilon_\beta < 1.2$.

Therefore, to consider this behavior, the calculation of the load sharing factor for gears included in this class is performed by interpolating the load sharing factor $X_{Y(\varepsilon_\beta=0.8)}$, for $\varepsilon_\beta = 0.8$, and the load sharing factor $X_{Y(\varepsilon_\beta=1.2)}$, for $\varepsilon_\beta = 1.2$, both determined considering the different relationships previously described, concerning the possible cases of unmodified and modified tooth profiles. Consequently, in this framework, the equation to be used for calculation of the load sharing factor for these types of gears is as follows:

$$X_{Y(\varepsilon_\beta)} = X_{Y(\varepsilon_\beta=0.8)}\frac{1.2 - \varepsilon_\beta}{0.4} + X_{Y(\varepsilon_\beta=1.2)}\frac{\varepsilon_\beta - 0.8}{0.4}. \tag{10.140}$$

10.8.9   Local Contact Temperature, $\vartheta_{B,Y}$

In accordance with the general definition of contact temperature already given elsewhere (see Sects. 7.2 and 7.3), also the local contact temperature, $\vartheta_{B,Y}$, which is of interest here, is defined as the sum of the bulk temperature, ϑ_M, and the local flash temperature, $\vartheta_{fl,Y}$. Therefore, it is expressed by the following general relationship:

$$\vartheta_{B,Y} = \vartheta_M + \vartheta_{fl,Y}. \tag{10.141}$$

In a simplified way, the bulk temperature, ϑ_M, is assumed to be constant along the whole path of contact, while the local flash temperature, $\vartheta_{fl,Y}$, varies along it due to the localized friction contact losses during the meshing cycle. Thus, the local flash temperature is a function of position of the contact point, Y, on the path of contact and must therefore be determined for each point of the path of contact that is of interest. In any case, to complete the picture of the quantities that come into play in Eq. (10.71), we must here calculate the two quantities, $\vartheta_{fl,Y}$ and ϑ_M, in the specific terms involved in the micropitting problem we are facing.

10.8.9.1 Local Flash Temperature, $\vartheta_{fl,Y}$

The continuous variation of the local flash temperature, $\vartheta_{fl,Y}$, along the path of contact is due to the more or less continuous and/or abrupt variation of the influences on which it depends. In this regard, just remember the fact that, at each position of the contact point, Y, on the path of contact, continuously variable values of the rolling and sliding velocities and of the local contact loads and related friction forces are associated. In addition, the elastohydrodynamic, mixed or boundary contact conditions are also continuously variable. Consequently, the distribution of the local flash temperature will be variable, with fluctuation along the path of contact depending on the local influences. The local value of this quantity can be calculated with the following relationship, which is based on the results of Blok's researches (Blok 1937a, b, c, d, 1940):

$$\vartheta_{fl,Y} = \frac{10^6 \sqrt{\pi}}{2} \frac{\mu_m p_{dyn,Y} |v_{g,Y}|}{B_{M1}\sqrt{v_{r1,Y}} + B_{M2}\sqrt{v_{r2,Y}}} \sqrt{\frac{8\rho_{n,Y} p_{dyn,Y}}{10^3 E_r}}, \tag{10.142}$$

where, the meaning of the symbols already introduced remaining unchanged, the other symbols represent: μ_m, the mean coefficient of friction; $|v_{g,Y}|$, the absolute value of the local sliding velocity, $v_{g,Y} = (v_{r1,Y} - v_{r2,Y})$; B_{M1} and B_{M2} [in $N/(ms^{1/2}K)$], the thermal contact coefficients of pinion and wheel, which are given respectively by:

$$B_{M1} = \sqrt{\rho_{M1} c_{M1} \lambda_{M1}} \tag{10.143}$$

$$B_{M2} = \sqrt{\rho_{M2} c_{M2} \lambda_{M2}}. \tag{10.144}$$

In these equations, ρ_{M1} and ρ_{M2} (in kg/m^3) are the densities, c_{M1} and c_{M2} [in $J/(kgK)$] are the specific heat capacities, and λ_{M1} and λ_{M2} [in $W/(mK)$] are the specific heat conductivities of material of pinion and wheel. Table 10.2 summarizes the value of these three quantities to be assumed for steels normally used for the gear manufacturing. Of course, the numerical coefficients appearing in Eq. (10.142) are factors that homogenize the units of measurement in the SI-system.

Table 10.2 Values of material properties for steel

Material	Density, ρ_M (kg/m^3)	Specific heat capacity, c_M [J/(kgK)]	Specific heat conductivity, λ_M [W/(mK)]
Steel	7800	440	45

10.8.9.2　Bulk Temperature, ϑ_M

The bulk temperature, ϑ_M, of the two mating tooth flank surfaces in relative motion, also called interfacial bulk temperature, is very slowly variable or constant along the path of contact, for which it is usually assumed to be constant. It is defined as the equilibrium temperature of the gear tooth flank surfaces before teeth enter the contact zone. This temperature should be obtained by means of suitable experimental measurements, or it should be calculated using appropriate numerical models based on thermal network theory (see Carslaw and Jaeger 1959; Blok 1969; Bathgate et al. 1970; Tanaka and Edwards 1992; Manin and Play 1999; Incropera et al. 2006; Naveros et al. 2016). In the case where these two procedures are not possible, an appropriate value of the bulk temperature, ϑ_M, can be calculated using the following relationship due to Oster (1982):

$$\vartheta_M = \vartheta_{oil} + 7.4 \times 10^3 \left(\frac{\mu_m P H_v}{ab} \right)^{0.72} \frac{X_S}{1.2 X_{Ca}}, \qquad (10.145)$$

with

$$P = \omega_1 T_1 = \frac{2\pi n_1}{60} \frac{T_1}{10^3}, \qquad (10.146)$$

where: ϑ_{oil} (in °C) is the lubricant inlet or oil sump temperature; μ_m is the mean coefficient of friction; a and b (both in mm) are respectively the center distance and face width; H_v is the dimensionless load losses factor; X_S is the dimensionless lubrication factor; X_{Ca} is the dimensionless tip relief factor; P (in kW) is the transmitted power; T_1 (in Nm) is the nominal torque at the pinion; ω_1 (in rad/s) and n_1 (in min^{-1}) are respectively the angular velocity and rotational speed of the pinion.

Mean Coefficient of Friction, μ_m

In Eqs. (10.142) and (10.145), the mean coefficient of friction, μ_m, appears. As we said in the Sect. 7.4, this important and decisive quantity is difficult to determine with the desired precision. This is because the numerous factors that influence the friction between the gear teeth vary throughout the meshing cycle. This is not the place to dwell further on this topic, which we discussed not only in the Sect. 7.4, but also elsewhere. Here it is enough to remember that an approximate calculation of this mean coefficient of friction, μ_m, can be carried out using the following relationship:

$$\mu_m = 45 \times 10^{-3} \left(\frac{K_A K_v K_{H\alpha} K_{H\beta} F_{bt} K_{B\gamma}}{bv_{\Sigma,C}\rho_{n,C}} \right)^{0.2} \left(10^3 \eta_{\vartheta oil}\right)^{-5/100} X_R X_L, \quad (10.147)$$

where

$$X_R = 2.2 \left(\frac{Ra}{\rho_{n,C}} \right)^{1/4}. \quad (10.148)$$

In these equations, the meaning of symbols already introduced remaining unchanged, the other symbols represent: F_{bt} (in N), the nominal transverse load in plane of action (i.e. the base tangent plane); $K_{B\gamma}$, the dimensionless helical load factor; $v_{\Sigma,C}$ (in mm/s), the sum of the tangential velocities at the pitch point (it is calculated using Eq. (10.95) for $Y = C$); $\rho_{n,C}$ (in mm), the normal radius of relative curvature at pitch diameter (it is calculated using Eq. (10.84) for $Y = C$); $\eta_{\vartheta oil}$ (in Ns/m^2), the dynamic viscosity at inlet or oil sump temperature; X_L, the dimensionless lubricant factor.

The helical load factor, $K_{B\gamma}$, considers the increasing friction for increasing total contact ratio, ε_γ. It does not differ from the equivalent factor we already introduced in Sect. 8.2 for calculation of scuffing load carrying capacity of cylindrical gears using the integral temperature method. In the range $(2 < \varepsilon_\gamma < 3.5)$, $K_{B\gamma}$ is given by the following relationship:

$$K_{B\gamma} = 1 + 0.2\sqrt{(\varepsilon_\gamma - 2)(5 - \varepsilon_\gamma)}, \quad (10.149)$$

while, for $\varepsilon_\gamma \leq 2$ and $\varepsilon_\gamma \geq 3.5$, we have respectively $K_{B\gamma} = 1$ and $K_{B\gamma} = 1.3$. Figure 8.1 shows the distribution curve of the helical load factor, $K_{B\gamma}$, as a function of the total contact ratio, ε_γ.

The dimensionless lubricant factor, X_L, is a function of the characteristics of the lubricant oil, and its value is assumed as follows: $X_L = 0.6$ for water-soluble polyglycols; $X_L = 0.7$ for not water-soluble polyglycols; $X_L = 0.8$ for polyalphaolefin; $X_L = 1.0$ for mineral oil; $X_L = 1.3$ for phosphate ester; $X_L = 1.5$ for traction fluid.

Load Losses Factor, H_v

The load losses factor, H_v, is determined using the following relationships:

$$H_v = \left(\varepsilon_1^2 + \varepsilon_2^2 + 1 - \varepsilon_\alpha\right)\left(\frac{1}{z_1} + \frac{1}{z_2}\right)\frac{\pi}{\cos\beta_b} \quad (10.150)$$

$$H_v = \frac{\varepsilon_\alpha}{2}\left(\frac{1}{z_1} + \frac{1}{z_2}\right)\frac{\pi}{\cos\beta_b}, \quad (10.151)$$

which are respectively valid for $\varepsilon_\alpha < 2$ and $\varepsilon_\alpha \geq 2$. In these equations, ε_1 and ε_2 are the addendum contact ratio of the pinion and wheel, ε_α is the transverse contact

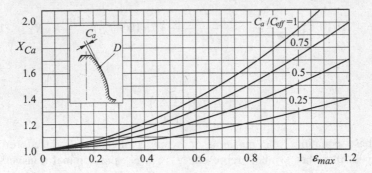

Fig. 10.18 Tip relief factor, X_{Ca}, defined according to Method A

ratio, z_1 and z_2 are the number of teeth of pinion and wheel, and β_b is the base helix angle.

Tip Relief Factor, X_{Ca}

The elastic deformations due to the loads applied to the gear teeth, which we analyzed in Chaps. 2 and 3, result in overload on the tooth tips in the high sliding area. An adequate profile modification determines a more than appreciable decrease in the intensity of this overload. The tip relief factor, X_{Ca}, considers the positive influence of this profile modification.

The tip relief factor, X_{Ca}, applies to gears with flank tolerance class 6 or finer ($A \leq 6$, according to ISO 1328-1:2013). For gears having flank tolerance class 7 or coarser ($A \geq 7$), this factor is to be taken equal to 1 ($X_{Ca} = 1$). For the determination of the values of this factor, to be assumed for gears having $A \leq 6$, two methods can be used; in order of decreasing accuracy, they are Method A and Method B.

With Method A, the curves shown in Fig. 10.18 are used. These curves allow to determine the value of the tip relief factor, X_{Ca}, as a function of ε_{max} (this is the maximum value between the addendum contact ratio of the pinion and wheel, ε_1 and ε_2), for four different values of the ratio C_a/C_{eff}, which is the quotient of the tip relief, C_a, divided by the effective tip relief, C_{eff}, both expressed in μm. In the same figure, point D is the upper point of single pair tooth contact on the tooth profile.

The curves shown in Fig. 10.18 are approximately described by the following relationship:

$$X_{Ca} = 1 + \left(0.06 + 0.18\frac{C_a}{C_{eff}}\right)\varepsilon_{max} + \left(0.02 + 0.69\frac{C_a}{C_{eff}}\right)\varepsilon_{max}^2. \quad (10.152)$$

The value of the tip relief C_a to be introduced in Eq. (10.152) depends on the actual value of the tip relief of pinion and wheel, C_{a1} and C_{a2}, the effective tip relief,

C_{eff}, the direction of the power flow and the ratio $\varepsilon_1/\varepsilon_2$ between the addendum contact ratios of pinion and wheel.

For pinion driving the wheel and $(\varepsilon_1/\varepsilon_2) > 1.5$ or for pinion driven by the wheel and $(\varepsilon_1/\varepsilon_2) > (2/3)$, we assume:

$$C_a = C_{a1} \tag{10.153}$$

$$C_a = C_{eff}, \tag{10.154}$$

depending on whether $C_{a1} \leq C_{eff}$ or $C_{a1} > C_{eff}$.

Instead, for pinion driving the wheel and $(\varepsilon_1/\varepsilon_2) \leq 1.5$ or pinion driven by the wheel and $(\varepsilon_1/\varepsilon_2) \leq (2/3)$, we assume:

$$C_a = C_{a2} \tag{10.155}$$

$$C_a = C_{eff}, \tag{10.156}$$

depending on whether $C_{a2} \leq C_{eff}$ or $C_{a2} > C_{eff}$.

The effective tip relief, C_{eff}, which is the particular value of the tip relief that compensates for the elastic deformation of the teeth in single pair contact area, is given by the following relationships:

$$C_{eff} = \frac{K_A F_t}{bc'} \tag{10.157}$$

$$C_{eff} = \frac{K_A F_t}{bc_{\gamma\alpha}}, \tag{10.158}$$

which are respectively valid for spur gears and helical gears. In these relationships, the meaning of symbols already introduced remaining unchanged, the other symbols represent: c' (in N/mm μm), the single stiffness of a tooth pair per unit face width (i.e., the maximum tooth stiffness per unit face width of a tooth pair); $c_{\gamma\alpha}$ (in N/mm μm), the mean value of mesh stiffness per unit face width.

With Method B, the curves shown in Fig. 10.19 are used. These curves allow to determine the value of the tip relief factor, X_{Ca}, as a function of $\varepsilon_{\max}$, in the following two conditions: gear drive with adequate tip relief according to the experience of manufacturers (curve 1), and without tip relief or non-adequate tip relief (curve 2). As in the previous figure, also in this figure point D is the upper point of single pair tooth contact on the tooth profile.

The curves shown in Fig. 10.19 are approximately described by the following relationship:

$$X_{Ca} = 1 + 0.24\varepsilon_{\max} + 0.71\varepsilon_{\max}^2, \tag{10.159}$$

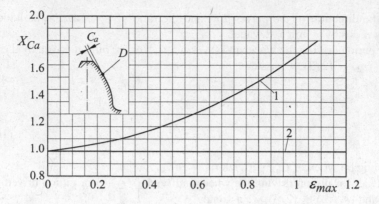

Fig. 10.19 Tip relief factor, X_{Ca}, defined according to Method B

which is valid for:

- pinion driving the wheel, $(\varepsilon_1/\varepsilon_2) > 1.5$ and driving gear with adequate tip relief;
- pinion driving the wheel, $(\varepsilon_1/\varepsilon_2) \leq 1.5$ and driven gear with adequate tip relief;
- pinion driven by the wheel, $(\varepsilon_1/\varepsilon_2) > (2/3)$ and driving gear with adequate tip relief;
- pinion driven by the wheel, $(\varepsilon_1/\varepsilon_2) \leq (2/3)$ and driven gear with adequate tip relief.

In all other cases, we assume:

$$X_{Ca} = 1. \tag{10.160}$$

Lubricant Factor, X_S

The lubricant factor, X_S, considers the better heat transfer for dip-lubricated gear sets than for injection-lubricated gear sets (see Liu et al. 2018). The following values of X_S are recommended: $X_S = 1.2$ for injection lubrication; $X_S = 1$ for dip (or dipped or splash) lubrication; $X_S = 0.2$ for gears submerged in oil.

10.9 Calculation Examples

Here also, for the same space reasons highlighted elsewhere, it is not possible to present and discuss examples regarding a complete and exhaustive calculation procedure of the micropitting load carrying capacity of spur and helical gears. As a matter of fact, each of the examples that could be presented and discussed would be occupy a space comparable to that devoted to the theoretical aspects of micropitting fatigue failure and, as we have already said in Sect. 2.11, this is not possible here. Each of these possible examples should in fact be aimed at calculating the safety factor against micropitting, S_λ, and, to evaluate this safety factor, any of the calculation steps described in the previous sections cannot be overlooked.

To get a reference picture on this topic, we refer the reader to ISO/TR 6336-31:2018(E), where the following four examples of application of the calculation procedure described in the previous sections of this chapter are carried out. The four examples concern practical applications of spur and helical gears working under both high-speed and low-speed operating conditions and, for each of them, the safety factor against micropitting is determined. They are:

Example 1: spur gear working at high speed ($n_1 = 3000$ min^{-1}), where micropitting damages were found consisting of: profile deviations on the driving pinion located between points A and C of the path of contact, with peak values of about ($8 \div 10$) μm near point A; profile deviations on the driven wheel located between points C and E of the path of contact, with the same peak values above near point E.

Example 2: spur gear working at low speed ($n_1 = 1000$ min^{-1}), where micropitting damages were found only on the driving pinion, consisting of profile deviations located between points A and B of the path of contact, with a peak of about 15 μm, about halfway between these points.

Example 3: helical gear working at high speed ($n_1 = 3000$ min^{-1}), where micropitting damages were found consisting of: profile deviations on the driving pinion located between the point A and an intermediate point between points B and C of the path of contact, with a peak value of about 10 μm about halfway between the points delimiting the damaged area; profile deviations on the driven wheel located between points E and D of the path of contact, with a peak value of about 5 μm about halfway between these two points.

Example 4: helical gear speed increaser, with driving wheel and driven pinion working at low rotational speed ($n_1 = 530.9$ min^{-1}), where micropitting damages were found consisting of: profile deviations on the driven pinion located between points A and C of the path of contact, with a peak value of about 12 μm about halfway between these points; profile deviations on the driving wheel located between points E and D of the path of contact, with a peak value of about 3 μm about halfway between these points.

Each of the aforementioned examples is calculated using Method B. Examples 1, 3 and 4 are also calculated using Method A.

10.10   Topography and Texture of Surfaces

10.10.1   Basic Concepts

In the previous sections we have repeatedly point out that micropitting is a fatigue phenomenon occurring in rolling and sliding contacts of the Hertzian type, which, as far as the gears are concerned, operate under elastohydrodynamic, mixed or boundary lubrication conditions. Therefore, micropitting is to be considered in close correlation with Hertz theory. This theory, as we already saw in Sect. 2.3, is based on the assumptions of frictionless contact between two elastic bodies having perfectly smooth and dry surfaces, motionless and continuous contacting surfaces, elastic and isotropic material behavior, load normal to the contacting area, and dimensions of this area relatively small in comparison with the radii of curvature and sizes of the two bodies.

However, due to the deviations of the actual tooth flank surfaces with respect to their theoretical geometry, the rolling and sliding contact between the two surfaces delimiting the two contacting solid bodies in relative motion is discontinuous, and the actual area of contact is a small fraction of the nominal contact area. Therefore, the actual contact between the two bodies, to which both normal and tangential loads are applied, is very localized. Thus, it should not be surprising that plastic deformations not only between the contacting peak asperities of surfaces involved, but also between the contacting peak waviness of the same surfaces can occur (Johnson 1985).

Of course, due to the aforementioned deviations of the actual tooth flank surfaces with respect to their theoretical ideal geometry, the conventional contact theory, even if it includes both normal and tangential loads, but obtained on the basis of smooth surfaces that ensure a continuous contact, must be deeply revised. We have repeatedly called attention to the fact that, to better capture the interactions between the asperities of the surfaces, regardless of the lubrication conditions (full film, mixed or boundary lubrication conditions), it would be necessary to introduce areal field parameters. This parameters in fact describe the texture and topography of the surfaces involved more accurately than the traditional profile parameters do.

Moreover, with the introduction of the areal field parameters, the dichotomy related to the distinction between roughness and waviness would be overcome, as we will be able to specify below. Furthermore, a lot of attention must be paid to the characteristics of isotropy or anisotropy of the contacting surfaces that are of interest. These surfaces do not always present an isotropic asperity distribution along any direction. Very often they have a strong privileged orientation of asperities characterizing their topography. In the most general case, surface asperities are randomly distributed along any of the endless cross sections through the normal at a current point $P(u, v)$ of each of the two contacting surfaces (see Vol. 1, Fig. 11.15).

To accurately define the topography of a surface, it is necessary to measure and specify its texture in three-dimensional space (see Leach 2009; Blateyron 2013; Aver'yanova et al. 2017). Traditional profile parameters (see ISO 4288: 1996; ISO

4287: 1997; ISO 12085: 1996; ISO 13565-1: 1996; ISO 13565-2: 1996; ISO 13565-3: 1998; ISO 1302: 2002), which are 2D-parameters, are not sufficient to characterize and define the 3D surface texture. For this purpose, areal field parameters, i.e. 3D-parameters, are required. The first measuring instruments introduced for the detection of the areal surface texture used methods mainly based on the simple extrapolation of those related to 2D methods, for roughness measurements. In addition, the surface texture parameters introduced in this regard were sometimes calculated as the simple average of profile parameters, evaluated for each line on the surface under consideration or for radial profiles extracted from a circle with its origin at the center of the detected area.

Therefore, in this reference framework that photographs the current situation, for the development of more general and reliable methods for calculating the micropitting load carrying capacity of the gears, it is necessary to characterize the texture and, more generally, the topography of the surface involved, in the three-dimensional space. The sections that follow summarize what established by the ISO standards regarding the main surfaces to be considered and the areal surface parameters that characterize the texture and topography of the same surfaces.

10.10.2 Main Surfaces

Here we want to summarize the parts of the ISO 25178-2:2012 that is of great interest to the topic under discussion. In fact, it redefines the basic concepts of surface texture, recognizing for the first time the intrinsically three-dimensional nature of it. This standard defines the 3D surface texture parameters and the associated specification operators, and describes the applicable measurement technologies, greatly enlarging the horizons of this important subject. This is no longer restricted to contact measurements, using contact instruments such as stylus profilometers to detect profile parameters, but also covers non-contact measurement techniques, using optical instrumentation, such as confocal chromatic probe, phase-shifting interferometric microscopy, coherence scanning interferometry, point autofocus probe, focus variation instruments, and confocal microscopy.

Given the aforementioned three-dimensional nature of the surface texture, at least the following surfaces must be considered:

- Skin model, i.e. non-ideal surface model of a workpiece, which defines the physical interface of the same workpiece with the surrounding environment.
- Mechanical surface or measured surface, which is obtained as locus of the centers of an ideal tactile spherical ball with a given radius rolled over the skin model above.
- Primary surface, given by a surface portion representing a specified primary mathematical model with specified nesting index (this specifies the inclusion of a block of data in another block). An S-filter is used to derive this primary surface.

- Primary extracted surface, defined as finite set of data points sampled from the primary surface.
- S-F surface, which is the surface obtained from the primary surface, through the use of an F-operation that removes the nominal form from the same primary surface; its range is defined by the S-filter and F-operation.
- S-L surface, which is the surface obtained from the S-F surface after applying an L-filter that removes the large-scale components; its range is defined by the S-filter and L-filter.
- Scale-limited surface, given by the S-F surface or S-L surface.
- Reference surface, which is the surface associated to the scale-limited surface according to a predefined criterion. It is used for definition of surface texture parameters and include plane, cylinder and sphere.
- Evaluation area, i.e. a portion of the scale-limited surface used as area under evaluation.
- Definition area, i.e. a portion of the evaluation area used to define the parameters characterizing the scale-limited area.

It is noteworthy that, in the transition from two-dimensional to three-dimensional problem, industry-specific taxonomies such as roughness and waviness are replaced by the more general concept of scale-limited surface, and cutoff is replaced by nesting index. It is also to be noted that some surfaces defined above involve filtering operations of sampling data. The following two types of surface filter are used as a filtration operator for the primary surface. The first type of surface filter is the S-filter, by which the smallest scale lateral components (or the shortest wavelengths for a linear filter) are removed from the primary surface. The second type of surface filter is the L-filter, by which the largest scale lateral components (or the longest wavelengths for a linear filter) are removed from the primary surface or S-F surface. Finally, F-operation is the operation by which the nominal form is removed from the primary surface.

It should be noted that some F-operations, as well as those associated with them, have a very different action compared to that of filtration. Although their action may limit the larger lateral scales of a surface, this action is very fuzzy, so it follows that the line for the action of the F-operation does not have a well-defined position. Therefore, this line is also a fuzzy line. Furthermore, many L-filters are sensitive to the shape of the surface, so they require an F-operation first as pre-filter before being applied.

ISO 16610-1:2015 describes the various types of filters that can be used, classified as profile filters and areal filters. The profile filters include linear profile filters (Gaussian filters, spline filters and spline wavelets), robust profile filters (Gaussian regression filters and spline filters) and morphological profile filters (disc and horizontal line-segment filters, and segmentation filters). The specific standards concerning these filters are already available, except those concerning the segmentation filters, still to be published. The areal filters include linear areal filters (Gaussian filters, spline filters and spline wavelets), robust areal filters (Gaussian regression filters) and morphological areal filters (sphere and horizontal planar segment filters and segmentation filters). Only some specific standards concerning these filters are already available, while the other are still to be published. Generally, Gaussian filters and spline filters are used for areal surface measurements. Gaussian filters are smoothing filters that eliminate noise using the Gaussian function. Spline filters are used to obtain a smooth profile by interpolating the sections between adjacent points using the spline function.

A value equal to or more than 3 times the measurement resolution in the (x, y)-plane is used as cutoff wavelength for S-filter. If the value set on the measuring instrument is not sufficiently effective, it must be increased until the scale-limited surface noise is eliminated. The cutoff wavelength for L-filter is difficult to specify uniformly based on lens magnification or stylus tip diameter. Therefore, it is to be adjusted with reference to the real surface. Generally, a value equal to or more than 5 times the directional length of the profile in the (x, y)-plane that we wish to remove as waviness can be considered satisfactory.

In Vol. 1, Sect. 11.15, we saw that a surface, σ, in the three-dimensional space is defined by the position vector, $r(u, v)$. Generally, the evaluation area is so small that it can be thought of as belonging to the plane tangent to the surface in one of its current point $P(u, v)$. If the nominal surface is a plane, or portion of a plane, we use Cartesian coordinates instead of curvilinear coordinates. In this case, in accordance with the aforementioned ISO standard, we use a rectangular coordinate system $O(x, y, z)$, the axes of which form a right-handed Cartesian set. The x-axis and y-axis of this coordinate system lie on the nominal surface, while the z-axis is directed towards the outside of the surface, i.e. from the material towards the surrounding medium. To define the autocorrelation length (see next section), a cylindrical polar coordinate system $O(r, \vartheta, z)$ is also used. This system has the same origin and the same z-axis of the Cartesian coordinate system $O(x, y, z)$, while a current point in the (x, y)-plane is defined by the angular coordinates (r, ϑ). The angle, ϑ, is considered positive in a counterclockwise direction from the positive x-axis, which is defined by the angle $\vartheta = 0$.

Another general concept must be kept in mind. In fact, about the definition of the areal field parameters and measurement techniques on which they are based, it is necessary to distinguish between isotropic surfaces and anisotropic surfaces. A surface is called isotropic, when it has the same characteristics regardless of the direction of measurement. This requirement is satisfied by surfaces that have a random texture, i.e. surfaces that do not have any texture that stands out. This type of surface is unhappily fairly rare. Most surfaces resulting from industrial processes are anisotropic surfaces. In fact, these surfaces have an oriented texture, as in turned, ground, and brushed surfaces, or have a periodic texture, as in Electron Beam Technology (EBT) impacts, and grained plastics.

10.10.3 Surface Texture Parameters

The areal surface texture is described by a number of 3D areal parameters, which are designed by the capital letter S (or the capital letter V), followed by a suffix of one or two small letters. This abbreviated way of indicating the areal parameters must not be confused with their corresponding symbols, used in the equations by which they are defined. In fact, the symbol consists of the same capital letter (S or V), with a subscript of one or two small letters, equal to those of the aforementioned suffix in the abbreviated term. All 3D areal surface texture parameters are determined on the entire area under consideration. Therefore, we are faced with a substantial difference compared to the 2D parameters, which are instead determined by averaging estimations carried out on a number of base lengths.

In contrast to the naming conventions used with 2D parameters, 3D areal parameters do not reflect the nature of the surface, which is related to the filtering context. Therefore, they apply regardless of the nature of the surface and corresponding filtering context, not distinguishing between roughness and waviness, as is the case with 2D parameters, where we use the abbreviated terms Pa, Ra or Wa depending on whether we consider the primary, roughness or waviness profiles. For surfaces, we use only one abbreviated term, Sa, which can therefore represent a parameter of roughness or waviness, or even a parameter calculated on the primary surface, depending on the pre-filtering carried out before the parameter is calculated. This way of proceeding follows from the multiplicity of processing and filtering methods that are available to extract information from a surface. These methods do not necessarily separate the two components (roughness and waviness) of the surface texture, but in some cases alter the surface in a subtler manner.

All the 3D areal surface texture parameters refer to scale-limited surface, but are defined over the definition area, A. They are divided into three classes: *height parameters*, *spatial parameters* and *hybrid parameters*. The height parameters involve only the statistical distribution of height values along the z-axis; spatial parameters involve the spatial periodicity of the data, specifically its direction; hybrid parameters relate to the spatial shape of the data.

The height parameters are as follows:

- Arithmetic mean height at the scale-limited surface, Sa, defined as the arithmetic mean of the absolute value of the ordinates within the definition area, A, and given by:

$$S_a = \frac{1}{A} \iint\limits_A |z(x, y)| dx dy, \qquad (10.161)$$

where $z(x, y)$ is the ordinate value, i.e. the height of the scale-limited surface at position (x, y). This parameter is used as a global evaluation of the asperity amplitude on a surface. It does not say anything on the spatial frequency of the irregularities or the shape of the surface. It is meaningful for random surface irregularities (stochastic) machined with tools that do not leave marks on the surface, such as sand blasting, milling and polishing.

- Root mean square height of the scale-limited surface, Sq, defined as the root mean square value of the values of ordinates within the definition area, A, and given by:

$$S_q = \sqrt{\frac{1}{A} \iint\limits_{A} z^2(x, y)dxdy}. \tag{10.162}$$

This parameter, which corresponds to the standard deviation of the height distribution, defined on the definition area, provides the same information as Sa

- Skewness of the scale-limited surface, Ssk, defined as the quotient of the mean cube value of the values of ordinates divided by the cube of Sq within the definition area, A, and given by:

$$S_{sk} = \frac{1}{S_q^3}\left[\frac{1}{A} \iint\limits_{A} z^3(x, y)dxdy\right]. \tag{10.163}$$

This parameter is important because it gives information on the morphology of the surface texture, especially on the asymmetry of the height distribution. It is to be noted that positive values of this parameter correspond to high peaks spread on a regular surface (distribution skewed towards bottom), while negative values are found on surfaces with scratches and pores (distribution skewed towards top). Contrary to Sa, this parameter does not give any information on the absolute height of the surface, but it is interesting when contact or lubrication function are required.

- Kurtosis of the scale-limited surface, Sku, defined as the quotient of the mean quartic value of the values of ordinates divided by the fourth power of Sq within the definition area, A, and given by:

$$S_{ku} = \frac{1}{S_q^4}\left[\frac{1}{A} \iint\limits_{A} z^4(x, y)dxdy\right]. \tag{10.164}$$

This parameter gives information on the sharpness of the height distribution.

- Maximum peak height of the scale-limited surface, Sp, defined as the largest peak height value within the definition area, A. It is designed with the symbol, S_p, in equations.
- Maximum pit height of the scale-limited surface, Sv, defined as minus the smallest pit height value within the definition area, A. It is designed with the symbol, S_v, in equations.

- Maximum height on the scale-limited surface, Sz, defined as the sum of the maximum peak height value and the maximum pit height value within the definition area, A. It is designed with the symbol, S_z, in equations.

The spatial parameters are as follows:

- Autocorrelation length, Sal, defined as the horizontal distance of the autocorrelation function, $f_{ACF}(t_x, t_y)$, which has the fastest decay to a specific value, s, with $0 \leq s \leq 1$. The autocorrelation function, $f_{ACF}(t_x, t_y)$, describes the correlation between a surface and the same surface translated by (t_x, t_y), and is given by the following relationship:

$$f_{ACF}(t_x, t_y) = \frac{\iint\limits_A z(x, y)z(x - t_x, y - t_y)dxdy}{\iint\limits_A z(x, y)z(x, y)dxdy}, \tag{10.165}$$

where A is the definition area. This autocorrelation function gives a measure of how similar the texture is at a given distance from the original location. The surface texture is similar along a given direction, if $f_{ACF}(t_x, t_y)$ is near to 1 for a given amount of shift along the same direction. Instead, if $f_{ACF}(t_x, t_y)$ falls rapidly to zero along a given direction, then the surface is different and thus uncorrelated with the original measurement location. For example, for a surface with a dominant lay, like a turned surface, the autocorrelation function $f_{ACF}(t_x, t_y)$ in direction normal to the turning grooves falls to zero quickly as the peaks of the shifted surface align with the mean plane, and becomes negative as the peaks of the surface align with the pits of the shifted surface. Instead, shifting along the direction of the turning grooves, the surface is near identical to the original one, and $f_{ACF}(t_x, t_y)$ remains near to 1.

To determine the autocorrelation length (and simultaneously the texture aspect ratio, defined below), we must have the autocorrelation function image, which always includes a central peak, whose height is normalized to 1. In some cases, such as surfaces that show periodic or pseudo-periodic lays, this image includes secondary peaks, which indicate a certain correlation between portions of the surface under consideration with the surface itself. The shape of the central peak is instead an indicator of the isotropy of the surface.

To characterize the shape of the central peak, we perform a threshold autocorrelation with a value $s = 0.2$ and then determine the threshold boundary of the central zone of the image, corresponding to the portion of the central peak that remains after thresholding. To define the geometry of this threshold boundary, which has an oval shape (Fig. 10.20), we introduce a Cartesian coordinate system $O(x, y)$ associated with a polar coordinate system $O(\vartheta, R)$. Both these systems have the same origin O, coinciding with the center of the oval, and are such that the x-axis of the first coincides with the axis $\vartheta = 0$ of the second. The distance $R = \sqrt{t_x^2 + t_y^2}$

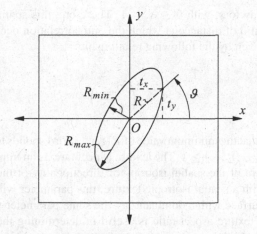

Fig. 10.20 Threshold boundary of the central threshold oval, coordinate systems, and autocorrelation lengths in different directions

of a current point $P(t_x, t_y)$ on the threshold boundary is a point function, and will be characterized by a minimum value, R_{min}, and a maximum value, R_{max}.

As an autocorrelation length, Sal (in μm), we defined the minimum value of $R(t_x, t_y)$ corresponding to the autocorrelation function $f_{ACF}(t_x, t_y) \leq s$, that is:

$$S_{al} = (R_{min})_s. \tag{10.166}$$

It is to be noted that, if the surface under consideration has the same characteristics in every direction, the oval shape of the aforementioned threshold boundary will become approximately circular, and R_{min} and R_{max} will be approximately equal ($R_{min} \cong R_{max}$). Instead, if the surface has a strong privileged orientation, the threshold boundary will have a very stretched out shape, and R_{max} will be much larger than R_{min} ($R_{max} \gg R_{min}$). In addition, it is to be noted that the threshold value $s = 0.2$ is a default value. However, in certain cases, to ensure that the boundary of the central oval is well defined and does not touch the edges of the image, it may be advisable to choose a higher or lower threshold value.

Thus, the autocorrelation length gives a measure of the surface for which the new location has the minimal correlation with the original location, and therefore a very different texture from the statistical point of view. Since the autocorrelation function has the fastest decrease along the direction where the minimum value R_{min} occurs, this spatial parameter can be used to determine whether there is a point where the surface height changes abruptly. In addition, it is able to define the tribological characteristics of a surface such as friction and wear.

- Texture aspect ratio, Str, defined as the quotient of the horizontal distance of the autocorrelation function, $f_{ACF}(t_x, t_y)$, which has the fastest decay to a specified value, s, divided by the horizontal distance of the same $f_{ACF}(t_x, t_y)$, which has the slowest decay to s, with $0 \leq s \leq 1$. Therefore, this spatial parameter that expresses the ratio of distance at which the autocorrelation decreases the fastest and lowest, is given by the following relationship:

$$S_{tr} = \left(\frac{R_{\min}}{R_{\max}} \right)_s .$$ (10.167)

It is to be noted that the minimum value of $R(t_x, t_y)$ corresponds to the autocorrelation function $f_{ACF}(t_x, t_y) \geq s$. The texture aspect ratio is an important parameter, as it is a measure of the spatial isotropy or directionality of the surface texture. For a surface with a spatial isotropic texture, this parameter will tend towards 1, whereas for a surface with a dominant lay the same parameter will tend towards zero. Thus, the texture aspect ratio is useful in determining the presence of lay in any direction. In the cases where a surface is produced by multiple process, this spatial parameter may be used to detect the presence of underlying surface modifications and subtle directionality on an otherwise isotropic texture.

The hybrid parameters are as follows:

- Root mean square gradient of the scale-limited surface, Sdq, defined as the root mean square of the surface gradient within the definition area, A, and given by:

$$S_{dq} = \sqrt{ \frac{1}{A} \iint_A \left[\left(\frac{\partial z(x, y)}{\partial x} \right)^2 + \left(\frac{\partial z(x, y)}{\partial y} \right)^2 \right] dx \, dy },$$ (10.168)

where $\partial z(x, y)/\partial x$ and $\partial z(x, y)/\partial y$ are the components of the local gradient vector of the scale-limited surface at position (x, y). It is often associated with the angle of the steepest gradient, $\alpha(x, y)$, and direction of the steepest gradient, $\beta(x, y)$, this last measured counterclockwise from the x-axis, both expressed in degrees and given by the following relationships:

$$\alpha(x, y) = \arctan \sqrt{ \left(\frac{\partial z(x, y)}{\partial x} \right)^2 + \left(\frac{\partial z(x, y)}{\partial y} \right)^2 }$$ (10.169)

$$\beta(x, y) = \arctan \left[\left(\frac{\partial z(x, y)}{\partial y} \right)^2 \middle/ \left(\frac{\partial z(x, y)}{\partial x} \right)^2 \right]$$ (10.170)

The angle α characterizes the steepest gradient in the vertical plane passing through the z-axis, and is included in the range $0° \leq \alpha \leq 90°$, whose extreme values, $0°$ and $90°$, correspond respectively to horizontal and vertical facets. Instead, the angle β, when calculated on the entire scale-limited surface, characterizes the mean orientation of the surface facets, thus constituting a parameter of evaluation of the texture direction. It is included in the range $0° \leq \beta \leq 360°$, (the value

$\beta = 0°$ corresponds to an orientation in the direction of the x-axis) and is measured counterclockwise. From the two angles, $\alpha(x, y)$ and $\beta(x, y)$, it is possible to obtain the *gradient density function* of the scale-limited surface, i.e. the density function showing the relative frequency of occurrences against the angles $\alpha(x, y)$ and $\beta(x, y)$. This gradient density function is a vector function, being defined as the cross product of the two angles given by Eqs. (10.169) and (10.170), also considered as vector functions.

- Developed interfacial area ratio of the scale-limited surface, Sdr, defined as the ratio of the increment of the interfacial area of the scale-limited surface within the definition area, A, over the definition area, and given by:

$$S_{dr} = \frac{1}{A} \left\{ \iint\limits_A \left(\sqrt{\left[1 + \left(\frac{\partial z(x, y)}{\partial x} \right)^2 + \left(\frac{\partial z(x, y)}{\partial y} \right)^2 \right]} - 1 \right) dx dy \right\}.$$
(10.171)

This parameter expresses, in continuous terms, the sum of the local area when following the surface curvature. Since most scale-limited surfaces are globally flat, the developed area is usually only slightly larger than the projected area. Then this parameter, which expresses in percentage terms (or in an equivalent way, e.g. with a unit-less positive number) the excess value of the developed area compared to the projected area, is usually included in the range $0\% \leq Sdr \leq 10\%$. For a perfectly flat and smooth surface, we have $Sdr = 0\%$. This parameter is used as a measure of surface complexity, especially to compare the characteristics related to the several stages of its manufacturing process.

Another useful tool, whose parameters do not fall within the three classes defined above, is the *polar spectrum*, which considers the power spectrum in each direction. By integrating the Fourier spectrum in polar coordinates, it is possible to determine the privileged direction of surface textures. To this end, we introduce the Fourier transform (see Sneddon 1974; Smirnov 1975; Bernardini et al. 1993), i.e. the operator that transforms the scale-limited surface into the Fourier space and is defined by the following relationship:

$$F(p, q) = \iint\limits_A z(x, y) e^{-(ipx + iqy)} dx dy,$$
(10.172)

where A is the definition area. The angle with the largest power spectrum corresponds in fact to the privileged texture direction.

As a characterizing parameter we use the angular spectrum, $f_{APS}(s)$, defined as the power spectrum for a given direction with respect to a specified direction ϑ in the plane of the definition area, and given by:

$$f_{APS}(s) = \int_{R_2}^{R_1} r |F[r\sin(s - \vartheta), r\cos(s - \vartheta)]|^2 dr, \qquad (10.173)$$

where R_1 and R_2 are the limits of the range of integration in the radial direction and s is the specified direction. It is to be noted that the positive x-axis is defined as the zero angle ($\vartheta = 0$) and that the angle ϑ is positive in the counterclockwise direction from the x-axis. The representation of the angular spectrum (or polar spectrum) clearly shows the privileged directions. The angle corresponding to the maximum value of the angular spectrum makes it possible to define the texture direction, *Std*, which is a parameter expressed in degrees.

10.10.4 Function Related Parameters

For the characterization of the texture and topography of a surface, other parameters are necessary, especially those related to the functions that a surface must be able to perform. These are the function related parameters, for whose definition the height distribution and material ratio curve, also called *Abbott-Firestone curve* (Abbott and Firestone 1933) or bearing area curve (BAC), must be available. The height distribution, usually represented by a histogram that expresses the number of points on the surface that lie at a given height, serves to determine the material ratio curve. This curve is the cumulative curve of height distribution, which is known as the cumulative probability density function and expresses the sample cumulative probability of the ordinates $z(x, y)$ within the evaluation area, A. Therefore, the highest ordinate corresponds to a probability of 0%, while the lowest ordinate corresponds to the probability of 100% (Fig. 10.21).

In the areal field analysis, the material ratio parameters are calculated using a specified height or cutting depth, c, which is counted from the reference plane, rather than from the highest peak, as in the case of profile analysis. This reference provides a more robust definition of these parameters, as it avoids the well-known outliers, which are typical of the reference assumed for the profile analysis. Among the various function related parameters introduced by the ISO 25178-2:2012, we consider it opportune to dwell on the following parameters:

- Areal material ratio of the scale-limited surface, $Smr(c)$, defined as the quotient of the area of the material at a specified height, c, divided by the evaluation area, and expressed as a percentage (Fig. 10.21). The heights are taken from the reference plane, while the $Smr(c)$ function expresses the material ratio, mr, corresponding to a cutting depth, c, given as a parameter.
- Inverse areal material ratio of the scale-limited surface, $Smc(mr)$, defined as the height, c, at which a given areal material ratio, mr, is satisfied. Therefore, the $Smc(mr)$ function gives the height value, c, corresponding to a material ratio, mr, given as a parameter (Fig. 10.21).

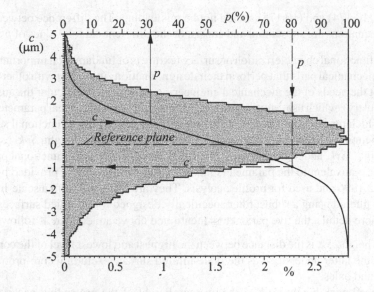

Fig. 10.21 Material ratio curve, reference plane, specified height, areal material ratio, and inverse areal material ratio

- Peak extreme height, Sxp, defined as the difference in height between two material ratios, p and q, and therefore given by:

$$S_{xp} = S_{mc}(p) - S_{mc}(q).$$ (10.174)

This parameter is aimed at characterizing the upper part of the surface, from the middle plane to the highest peak, excluding outliers given by a small percentage of the highest peaks whose influence is not considered significant. Therefore, as default values of p and q the ISO 25178-3:2011 suggests to take $p = 2.5\%$ and $q = 50\%$. For special applications, other values of p and q can be taken, but these values should however be close to the aforementioned default values, as this parameter is specifically defined for peak characterization.

- Surface section difference, Sdc, defined by the following relationship:

$$S_{dc} = S_{mc}(p) - S_{mc}(q),$$ (10.175)

where p and q are two material ratios to be chosen according to the application under consideration. This parameter, not foreseen by the ISO 25178-2:2012, corresponds to the Rdc parameter of the profile analysis. It can be used to define the maximum height of the surface, excluding outliers given by small percentages of the largest peaks and smallest pits whose influences are not considered significant. To this end, as an example, the following values of p and q can be chosen : $p = 2\%$ and $q = 98\%$. However, the formal equality of the right-hand sides of

Eqs. (10.174) and (10.175) should not be misleading. The difference between the two equations are substantial and consist of the different default values of p and q.

The functional characterization of surface texture is of fundamental importance for all the mechanical parts that perform their design functions in contact with other parts. To meet the needs of the mechanical engineering industry (in particular the automotive industry), which uses stratified functional surfaces, five other areal parameters are used, which represent the areal material ratio of the scale-limited functional surface as a function of height. They are: core height, Sk, reduced peak height, Spk, reduced dale height, Svk, and material ratios, $Smr1$ and $Smr2$. These areal functional parameters are equivalent to the parameters Rk, Rpk, Rvk, $Mr1$ and $Mr2$, provided by ISO 13565-2:1996 and used for profile analysis. They are extracted from a scale-limited surface filtered using a robust filter specifically designed for stratified surfaces.

In more details, the five parameters mentioned above are defined as follows:

- Core height, Sk, is the distance between the highest and lowest level of the core surface, the latter understood as the scale-limited surface excluding core-protruding hills and dales.
- Reduced peak height, Spk, is the average height of the protruding peaks above the core surface. It is used to characterize peaks that could be eliminated during operation.
- Reduced dale height, Svk, is the average height of the protruding dales below the core surface. It is used to characterize pits and valleys that could retain lubricant or worn-out materials.
- Material ratios, $Smr1$ and $Smr2$, are respectively the quotients of the peaks or dales of the area of the material at the intersection line that separates the protruding hills or dales from the core surface, divided by the evaluation area. They are expressed in percentage.

Figure 10.22 shows, through an intuitive graphic representation, the methods of calculation of the aforesaid five parameters, starting from the material ratio curve. Of course, this last curve is to be considered together with the associated scale-limited surface, given by the S-F surface or S-L surface.

Then expressing the areal material area ratio as a Gaussian probability in terms of standard deviation values plotted linearly on horizontal axis, we obtain the areal material probability curve. Notoriously, in this type of curve, the rectilinear parts represent a Gaussian distribution. Figure 10.23 represents an example of areal material probability curve. It exhibits two straight line regions, so it refers to a stratified surface composed of two Gaussian distributions. More in detail, we have: a plateau region, 1; a dale region, 2; a region showing debris or outlying peaks in the data, 3; a region with scratches or outlying dales in the data, 4; a unstable region, 5, highlighted by the curvature, introduced at the plateau-to-dale transition point, due to the combination of two distributions.

Three further areal parameters can be extracted from the above described areal material probability curve, which are used for the evaluation of plateau honed surfaces. They are:

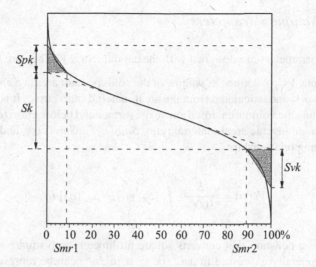

Fig. 10.22 Graphical representation of the calculation method of the areal parameters *Sk, Spk, Svk, Smr*1 and *Smr*2

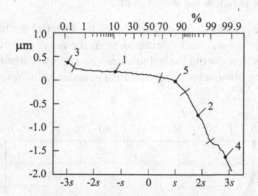

Fig. 10.23 Example of areal material probability curve

- Dale root mean square deviation, *Svq*, which is the slope of a linear regression performed through the dale region. This parameter can be interpreted as the *Sq*-value (in μm) of the random process that generated the dale component of the surface.
- Plateau root mean square deviation, *Spq*, which is the slope of a linear regression performed through the plateau region. This parameter can also be interpreted as the *Sp*-value (in μm) of the random process that generated the plateau component of the surface.
- Material ratio, *Smq*, which is the plateau-to-dale areal material ratio at the plateau-to-dale intersection. It is expressed in percentage.

10.10.5   Volume Parameters

All volume parameters are designed with the capital letter, V. They are:

- Void volume, $Vv(p)$, defined as volume of the voids per unit area at a given material ratio, $p = mr$, and calculated from the areal material ratio curve. It is determined by integrating the volume enclosed above the surface and below a horizontal cutting plane set at an inverse areal material ratio, $Smc(p) = Smc(mr)$, and is given by the following relationship:

$$V_v(p) = \frac{K}{100\%} \int\limits_{p}^{100\%} [S_{mc}(p) - S_{mc}(q)]dq, \qquad (10.176)$$

where K is a constant that converts square millimeters into square meters. Void volume, generally expressed in $\mu m^3/mm^2$ or m^3/m^2, can be represented on the areal material ratio curve as shown in Fig. 10.24. For $p = mr = 100\%$, the void volume is zero, while for $p = mr = 0\%$, the void volume is a maximum, since the cutting plane is below the lowest point.

- Material volume, $Vm(p)$, defined as volume of the material per unit area at a given material ratio, $p = mr$, and calculated from the areal material ratio curve. It is determined by integrating the volume enclosed below the surface and above a horizontal cutting plane set at an inverse areal material ratio, $Smc(p) = Smc(mr)$,

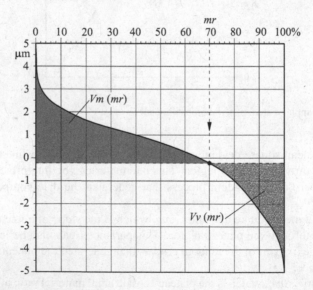

Fig. 10.24 Void volume, Vv, and material volume, Vm, below and respectively above a section height defined by a material ratio $p = mr$

and is given by the following relationship:

$$V_m(p) = \frac{K}{100\%} \int_0^p [S_{mc}(p) - S_{mc}(q)]dq, \qquad (10.177)$$

where K is the same constant defined for the previous parameter. The material volume can also be represented on the areal material ratio curve as shown in Fig. 10.24. For $p = mr = 100\%$, the material volume is a maximum, while for $p = mr = 0\%$, the material volume is zero, since the cutting plane is above the highest point. Void and material volumes are useful to evaluate the surface texture of machine elements that are in contact with other surfaces of mechanical components.

- Peak material volume of the scale-limited surface, Vmp, defined as the material volume at a given material ratio, $p = mr$, i.e. $V_{mp} = V_m(p)$. The default value of p is taken equal to 10%, i.e. $p = 10\%$. This default value of p can be changed for specific applications. In any case, it must be specifically indicated together with the value of Vmp. This parameter, which expresses the volume of material that is likely to be removed during running-in of a mechanical part, is used for the same purpose as the Spk parameter.
- Core material volume of the scale-limited surface, Vmc, defined by the following relationship:

$$V_{mc} = V_m(q) - V_m(p), \qquad (10.178)$$

i.e. as the difference in material volume between two material ratios, q and p. The default values of q and p are taken respectively as follows: $q = 80\%$ and $p = 10\%$. This parameter represents the portion of the surface material which does not interact with other surface in contact, and therefore does not play any role on the effectiveness of the lubrication between two contacting surfaces.

- Dale void volume of the scale-limited surface, Vvv, defined as the dale volume at a given material ratio, $p = mr$, i.e. $V_{vv} = V_v(p)$. The default value of p is taken equal to 80%, i.e. $p = 80\%$.
- Core void volume of the scale-limited surface, Vvc, defined by the following relationship:

$$V_{vc} = V_v(p) - V_v(q), \qquad (10.179)$$

i.e. as the difference in void volume between two material ratios, p and q. The default values of p and q are taken respectively as follows: $p = 10\%$ and $q = 80\%$.

Figure 10.25 shows the four volume parameters, Vmp, Vmc, Vvv and Vvc, defined on the bearing areal ratio curve and calculated for the default values of the material ratio, p and q, previously described. These four parameters represent an evaluation

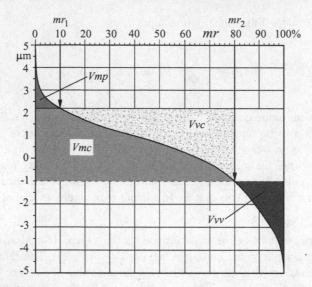

Fig. 10.25 Volume parameters, *Vmp, Vmc, Vvv* and *Vvc*, defined on the bearing area ratio curve, and calculated for default values of the material ratios, *p* and *q*

of the three *functional indices*, *Sbi* (Surface bearing index), *Sci* (Surface core fluid retention index) and *Svi* (Surface valley fluid retention index), previously used to characterize surface zones involved in lubrication, wear and contact phenomena. These three functional indices were introduced to characterize respectively the upper zone of the surface involved in wear phenomena, the main volume acting as a lubricant reserve, and the void volume of the deepest valleys.

The four volume parameters perform the same task as the three functional indices, but with greater reliability as they further improve the correlation with the functional phenomena involved. More particularly, the *Vvv* parameter is not affected by wear processes of surfaces and better characterizes the volume of material located at the highest peaks of the surface that is removed during wear process. After several hours of operation, the highest peaks of the surfaces of mechanical components are cut away or plastically deformed, and the corresponding particles of material detached from the surfaces are captured by the deepest valleys, so the surface behavior is probably better described by the *Vmc* and *Vvc* parameters.

10.10.6 Roughness

Traditionally, to define as fully as possible the topography of a surface, i.e. its texture in both two or three dimensions, a certain number of profile parameters have been used. These parameters can be grouped in the following four basic categories:

roughness, waviness, spacing and hybrid parameters (hybrid parameters are combination of spacing and roughness parameters). Thus, surface roughness (or shortly roughness, which is the term usually used) is only a component of the surface texture.

Roughness is quantified by the deviations in direction of the normal vector of the real surface from its ideal shape. We define a rough or smooth surface, depending on whether these deviations that characterize roughness are large or small. In the general terms of surface metrology, roughness represents the high-frequency, short-wavelength component of a measured surface. However, in practical applications as well as in manufacturing processes, it is often necessary to know not only the amplitude of these deviations, but also their frequency.

To measure the surface profile parameters that define the topography of a surface, only contact type profilometers, which are mechanical stylus profilometers, are used. With these types of profilometers, the tip of a diamond point stylus directly touches the surface of the sample. As the stylus traces across the sample, it rises and falls together with the roughness on the sample surface. The movement in the stylus is pitched up and used to measure surface roughness. The stylus moves closely with the sample surface, so data is highly reliable. Instead, with non-contact profilometers generally used for areal surface measurements, the stylus is an optical stylus or a laser stylus, that is a ray of light. Light emitted from the instrument is reflected and read, to measure without touching the sample.

Surface texture measurements using contact type profilometers are two-dimensional measurements, affecting a slice through the surface area involved. In other words, they are profile measurements, referring to a given cross section of the area of interest. Obviously, for a three-dimensional evaluation of the surface texture, it is necessary to scan the areal according to a sufficient number of profiles through the surface area. To this end, at least the following types of profiles must be considered:

- Actual profile, which is the profile obtained by intersecting the workpiece surface with a plane normal to the same surface, and directed along a direction that maximizes roughness (usually, this direction is at right angles to the lay of machining marks). Actual profile can also be considered as the cross section of the real surface, which separates a body from the surrounding medium.
- Traced profile, which is the enveloping profile of the real surface acquired by a stylus profilometer. This profile consists of form deviations, waviness and roughness components.
- Measured profile, which is the profile obtained by scanning the actual profile with a probe that filters the same profile, acquired by a stylus profilometer having a given tip radius, r_{tip}, also filtering errors due to the detection system. Surface imperfections, such as cracks, scratches and dents, should not be included in the recording, since they are not part of the profile.
- Primary profile or P-profile, which is the profile obtained by electronic low-pass filtering of the measured profile, with a cutoff wavelength, λ_s. This filtering process eliminates the shortest wavelength components that are considered irrelevant for roughness measurements. The corresponding parameters are evaluated within the sampling length, Lr, and are designed with capital letter, P. Measurements are

instead made along the evaluation length, Ln, i.e. the total length of the surface profile recorded.

- Roughness profile or R-profile, which is the profile obtained by electronic high-pass filtering of the primary profile, with a cutoff wavelength, λ_c. This filtering process eliminates the longer wavelength components that influence waviness, but are irrelevant to roughness. The corresponding parameters are evaluated within the evaluation length, Ln, and are designed with capital letter, R. The evaluation length usually consists of five sampling lengths, Lr, each of which corresponds to the cutoff wavelength, λ_c, of the filtering process.

- Waviness profile or W-profile, which is the profile obtained by electronic low-pass filtering of the primary profile with a cutoff wavelength, λ_c, followed by a high-pass filtering with a cutoff wavelength, λ_f. The corresponding parameters are evaluated within the evaluation length, Ln, and are designed with capital letter W. The evaluation length consists of several sampling lengths, Lw, each of which corresponds to the cutoff wavelength, λ_f, of the high-pass filtering process. Usually the evaluation length is chosen between five and ten times λ_f.

About the transmission parameters of the filtering process used to separate roughness and waviness characteristics, it should be noted that the wavelengths included between the cutoff wavelengths in the range ($\lambda_s < \lambda < \lambda_c$) represent the region of roughness, while the wavelengths included between the cutoff wavelengths in the range ($\lambda_c < \lambda < \lambda_f$) represent the region of waviness. Filtered-out short-wavelength components, in the region where $\lambda < \lambda_s$, and filtered-out long-wavelength components, in the region where $\lambda > \lambda_f$, are not required for the definition of roughness and waviness characteristics. Regardless of the type of filtering used (low-pass or high-pass filtering), the filter response is considered as a Gaussian response.

From the foregoing, we infer that some profile parameters are calculated on sampling length, also known as cutoff length, Lr (i.e. the profile segment selected for assessment and evaluation of the roughness parameters having the cutoff wavelength), and then averaged. Other parameters are defined and calculated with reference to the evaluation length, which is the profile from which data is obtained, usually coinciding with the profile length after filtering. Therefore, since the surface texture problem is dealt with in two-dimensional space, the evaluation length, Ln, is the two-dimensional profile corresponding to the length of the slice considered. Usually, the sampling length is defined as the cutoff length, i.e. the wavelength, λ_c, of the filter used to separate roughness and waviness. Any surface irregularities spaced father apart than the sampling length are to be considered waviness. Therefore, waviness parameters are larger components of surface texture upon which roughness is superimposed.

It is not the case here to dwell further on this subject and, in this regard, we refer the reader to the already mentioned ISO Standards as well as to specialized textbooks (e.g., Blateyron 2013). However, we consider it useful to provide the reader with general synoptic framework, given by Table 10.3, which comparatively summarizes the height, spatial, hybrid and function parameters that describe the surface texture

Table 10.3 Surface texture: areal field and profile parameters

ISO Standard		25178-2:2012	13565-1:1996
Measuring instruments		Contact-type and non-contact type profilometers	Contact-type profilometers
Evaluation target		S-F surface	Cross-sectional profile
Filter		S-filter	λs-filter
Evaluation target		S-L surface	Roughness profile
Filter		S-filter, L-filter	λs-filter, λc-filter
Height parameters	Maximum peak height	Sp	Rp
	Maximum pit height	Sv	Rv
	Maximum height	Sz	Rz
	Arithmetical mean height	Sa	Ra
	Root mean square height	Sq	Rq
	Skewness	Ssk	Rsk
	Kurtosis	Sku	Rku
Spatial parameters		Sal, Str, Std	–
Hybrid parameters		Sdq, Sdr	$R\Delta q$
Function parameters	Level difference on core surface	Sk	Rk
	Reduced peak height	Spk	Rpk
	Reduced valley depth	Svk	Rvk
	Peak material portion	$Smr1$	$Mr1$
	Valley material portion	$Smr2$	$Mr2$

based on the available data that, depending on the measuring instruments used, may be related to areal field measurements or profile measurements.

Finally, it must be remembered that some of the profile parameters have been provided in Sect. 2.7. The expressions of the other profile parameters, when existing, are formally similar to dose given before for the areal field parameters, of course with the necessary obvious variations of the case.

References

AGMA 925-A03:2003 Effect of lubrication on gear surface distress

Abbott EJ, Firestone FA (1933) Specifying surface quality: a method based on accurate measurement and comparison. Mech Eng 55:569–572

Al-Tubi IS, Long H (2013) Prediction of wind turbine gear micropitting under variable load and speed conditions using ISO/TR 15144-1:2010. Proc Inst Mech Eng, Part C: J Mech Eng Sci 227(9):1898–1914

Al-Tubi IS, Long H, Zhang J, Shaw B (2015) Experimental and analytical study of gear micropitting initiation and propagation under varying load conditions. Wear 328–329:8–16

Aver'yanova IO, Bogomolov DY, Porishin VV (2017) ISO 25178 standard for three-dimensional parametric assessment of surface texture. Russ Eng Res 37(6):513–516

Barber JR (1992) Elasticity. Kluwer Academic Publishers, Dordrecht

Bargis E, Garro A, Vullo V (1980a) Crankshaft design and evaluation. Part 1—Critical analysis and experimental evaluation of current methods. In: ASME, reliability, stress analysis and failure prevention methods in mechanical design. New York, pp 181–201

Bargis E, Garro A, Vullo V (1980b) Crankshaft design and evaluation. Part 2—a modern design method: modal analysis. In: ASME, reliability, stress analysis and failure prevention methods in mechanical design. New York, pp 203–211

Bargis E, Garro A, Vullo V (1980c) Crankshaft design and evaluation. Part 3—a modern design method: direct integration. In: ASME, reliability, stress analysis and failure prevention methods in mechanical design. New York, pp 213–218

Bathgate J, Kendall RB, Moorhouse P (1970) Thermal aspects of gear lubrication. Wear 15(2):117–129

Becker AA (1992) The boundary element method in engineering: a complete course. McGraw-Hill Book Company, New York

Benson R, Sroka GJ, Bell M (2013) The effect of the roughness profile on micropitting. GearSolutions, March pp 47–53

Bernardini C, Ragnisco O, Santini PM (1993) Metodi matematici della fisica. Carocci Editore SpA, Roma

Bernasconi A, Papadopulos IV (2005) Efficiency of algorithms for shear stress amplitude calculation in critical plane class fatigue criteria. Comput Mater Sci 34:355–368

Berthe D, Flamand L, Foucher D, Godet M (1980) Micro-pitting in Hertzian contacts, Transactions of the ASME. J Lubr Technol 102:478–489

BGA-DU P602 (2008) Gear micropitting procedure. Test procedure for the evaluation of micropitting performance of spur and helical gears

Blateyron F (2013) The areal field parameters. In: Leach Richard (ed) Characterization of areal surface texture. Springer-Verlag, Berlin Heidelberg

Blok H (1937a) Theoretical study of temperature rise at surfaces of actual contact under oiliness lubricating conditions. Proc Inst Mech Eng (General Discussion on Lubrication) 2:222–235

Blok H (1937b) Les températures de surface dans des conditions de graissage sous pressions extrêmes. In: Proceedings of the 2nd world petroleum congress, Paris, Section IV, vol. III, pp 151–182

Blok H (1937c) Measurements of temperature flashes on gear teeth under extreme pressure conditions. Inst Mech Eng (Proceedings of the general discussion on lubrication) 2:18–22

Blok H (1937d) Surface temperature measurements on gear teeth under extreme pressure lubricating conditions. Power Transm, 653–656

Blok H (1940) Fundamental mechanical aspects of boundary lubrication. SAE Trans 35:54–68

Blok H (1969) The thermal-network method for predicting bulk temperatures in gear transmissions. In: Proceedings of the 7th round-table discussion on marine reduction gears, stat-laval, Finspong, Sweden, pp 3–25 and 26–32

Boussinesq J (1885) Application des Potentiels à l'étude de l'équilibre et du mouvement des solides élastiques avec des notes étendues sur divers points de physique mathématique et d'analyse. Gauthier-Villars, Paris

Brandão JA, Scabra JHO, Castro MJD (2010) Gear micropitting: model and validation. WIT Trans Eng Sci 66:25–36

Bronshtein IN, Semendyayev KA (1997) Handbook of mathematics, 3rd edn. Springer-Verlag, New York

Buckingham E (1949) Analytical mechanics of gears. McGraw-Hill Book Company, New York

Buzdygon KJ, Cardis AB (2004) A short procedure to evaluate micropitting using the new AGMA designed gears. AGMA Fall Tech Meet

Carslaw HS, Jaeger JC (1959) Conduction of heat in solids, 2nd edn. Oxford University Press, Oxford

Cauchy A-L (1828) Sur les équations qui expriment les conditions d'équilibre ou les lois du mouvement intérieur d'un corps solide élastique ou non élastique. Exerc. de Mathématiques 3:160–187

Cerruti V (1881–82) Ricerche intorno all'equilibrio dei corpi elastici isotropi, Atti della Reale Accademia dei Lincei, Memorie della Classe di Scienze Fisiche, Matematiche e Naturali, Serie 3, Annata 279, vol. 13, pp 81–122; reprint with the same title, Roma, Salviucci, 1882

Chang WR, Etsion I, Bogy DB (1987) An elastic-plastic model for the contact of rough surfaces. J Tribol 109(2):257–263

Clark A, Evans HP, Snidle RW (2015) Understanding micropitting in gears. In: Part C: J Mech Eng Sci (Proceedings of the institution of mechanical engineers)

Clark A, Weets IJJ, Snidle RW, Evans HP (2016) Running-in and micropitting behavior of steel surfaces under mixed lubrication conditions. Tribol. Int, 101:59–68

Conry TF, Seireg A (1973) A mathematical programming technique for the evaluation of load distribution and optimal modification for gear systems. J Eng Ind 95:1115–1122

Cook RD (1981) Concepts and applications of finite element analysis, 2nd edn. Wiley, New York

Démidovitch B, Maron I (1973) Éléments de Calcule Numérique. Éditions MIR-Moscou, Moscou

Den Hartog JP (1985) Mechanical vibrations, 4th edn. Dover Publications Inc, New York

Dowson D, Higginson GR (1977) Elastohydrodynamic lubrication, 2nd edn. Pergamon, London

Ehret P, Dowson D, Taylor CM (1998) On the lubricant transport conditions in elastohydrodynamic conjunctions. Proc R Soc Lond, Series A, 454

Evans HP, Snidle RW, Sharif KJ (2011) Analysis of micro-elastohydrodynamic lubrication and surface fatigue in gear micropitting tests. In: ASME proceedings of 11th international power transmission and gearing conference, and 13th international conference on advanced vehicle and tire technologies, Washington, DC, USA, August 28–31, Vol. 8, Paper No. DETC2011-47714, pp 585–591

Evans HP, Snidle RW, Sharif KJ, Shaw BA, Zhang J (2013) Analysis of micro-elastohydrodynamic lubrication and prediction of surface fatigue damage in micropitting tests on helical gears. J Tribol 135(1)

Fatemi A, Shamsaei N (2011) Multiaxial fatigue: an overview and some approximation models for life estimation. Int J Fatigue 33(8):948–958

Fatemi A, Socie DF (1988) A critical plane to multiaxial fatigue damage including out-of-phase loading. Fatigue Fracture Eng Mater Struct 11(3):149–165

Favata A (2012) On the Kelvin problem. J Elast 109(2):189–204

Fenner RT (1986) Engineering elasticity: application of numerical and analytical techniques. Ellis Horwood Limited Publishers, Chichester

Ferrari C, Romiti A (1966) Meccanica applicata alle macchine. Torino: Unione Tipografica–Editrice Torinese (UTET)

Flamant A-A (1892) Sur la répartition des pressions dans un solide rectangulaire chargé trnsversalement. Comptes Rendus des Séances de l'Academie des Sciences, Paris 114:1465–1468

FVA-Information Sheet 54/7 (1993) Test procedure for the investigation of the micropitting capacity gear lubricants

Garro A, Vullo V (1979) Acoustic problems of vehicle transmission. Nauka I Motorna Vozila '79, Bled, Slovenija, Jugoslavija, June 4–7

Gauss CF (1815) Methodus nova integralium valores per approximationem inveniendi: auctore Carolo Friderico Gauss. H. Dicterich, Gottingae

Gladwell GML (1980) Contact problems in the classical theory elasticity, Sijthoff & Noordhoff International Publishers B.V., Alphen aan den Rijn, the Netherlands Germantown, Maryland, USA

Goglia PR, Cusano C, Conry TF (1984a) The effects of surface irregularities on the elasto-hydrodynamic lubrication of sliding line contacts. Part I-Single IrregulIties, ASME J Tribol 106(1):104–112

Goglia PR, Cusano C, Conry TF (1984b) The effects of surface irregularities on the elasto-hydrodynamic lubrication of sliding line contacts. Part II-Wavy SurfS, ASME J Tribol 106(1):113–119

Greco A, Sheng S, Keller J, Erdemir A (2013) Material wear and fatigue in wind turbine systems. Wear 302:1583–1591

Greenwood JA, Tripp JH (1970–71) The contact of two nominally flat rough surfaces. Proc Inst Mech Eng 185(48/71): 625–634

Greenwood JA, Williamson, J-BP (1966) Contact of nominally flat surfaces. In: Proc R Soc Lond, Ser A, Math Phys Sci 295(1442):300–319

Gupta PK (1984) Advanced dynamics of rolling elements. Springer-Verlag, Berlin Heidelberg

Hein M, Stahl K, Tobie T (2017) Practical use of micropitting test results according to FVA 54/7 for calculation of micropitting load capacity according to ISO/TR 15144-1. In: International conference on gears Sept. 13-15, 2017, Technische Universität München (TUM), Garching/Munich, Germany

Henrici P (1988) Applied and computational complex analysis. Power series, integration, conformed mapping, vol 1. John Wiley & Sons, Inc, Location of Zeros, New York

Höhn B-R, Oster P, Emmert S (1996) Micropitting in case-carburized gears - FZG micropitting test. In: International Conference on Gears, Dresden, Germany, VDI Berichte Nr. 1230, pp 331–334

Incropera FP, DeWitt DP, Bergmann TL, Lavine AS (2006) fundamentals of heat and mass transfer, 6th edn. John Wiley & Sons Inc, New York

ISO 12085:1996 Geometrical product specifications (GPS)—surface texture: Profile method—motif parameters

ISO 1302:2002 Geometrical product specifications (GPS)—indication of surface texture in technical product documentation

ISO 1328-1:2013, Cylindrical gears—ISO system of flank tolerance classification—part 1: definitions and allowable values of deviations relevant to flanks of gear teeth

ISO 13565-1:1996, Geometrical product specifications (GPS), surface texture: profile method; surfaces having stratified functional properties—part. 1: filtering and general measurement conditions

ISO 13565-2:1996 Geometrical product specifications (GPS)—surface texture: profile method; surfaces having stratified functional properties—part 2: height characterization using the linear material ratio curve

ISO 13565-3:1998 Geometrical product specifications (GPS)—surface texture: profile method; surfaces having stratified functional properties—part 3: height characterization using the material probability curve

ISO 16610-1:2015, Geometrical product specifications (GPS)—filtration—part 1: overview and basic concepts

ISO 4288:1996 Geometrical product specifications (GPS)—surface texture: profile method—rules and procedures for the assessment of surface texture

ISO 4287:1997 Surface roughness testing: surface texture: profile method—terms, definitions and surface texture parameters

ISO 53:1998 Cylindrical gears for general and heavy engineering—standard basic rack tooth profile

ISO 25178-2:2012 Geometrical product specifications (GPS)—surface texture: areal-part 2: terms, definitions and surface texture parameters

ISO 25178-3:2011 Geometrical product specifications (GPS)—surface texture: areal—part 3: specifications operators

ISO/TS 6336-22:2018 Calculation of load capacity of spur and helical gears—part 22: calculation of micropitting load capacity

ISO/TS 6336-31:2018(E) Calculation of load capacity of spur and helical gears—part 31: calculation examples of micropitting load capacity

Jackson RL, Green I (2011) On the modeling of elastic contact between rough surfaces. Tribol Trans 54:300–314

Johnson KL (1985) Contact mechanics. Cambridge University Press, Cambridge, United Kingdom

Karas F. (1941) Elastische formänderung und lastverteilung beim doppeleingriff gerader stirnradzähne, VDI—forschungheft 406, B, Bd. 12

Kissling U. (2012) Application of the first international calculation method for micropitting. Gear Technol 54–60

Leach RK (2009) Fundamental principles of engineering nanometrology. Elsevier, Amsterdam

Li S, Kahraman A (2010a) A transient mixed elastohydrodynamic lubrication model for spur gear pair. ASME J Tribol 132(1), 011501-1-9

Li S, Kahraman A (2010b) Prediction of spur gear mechanical power losses using a transient elastohydrodynamic lubrication model. Tribol Trans 53(4):554–563

Li S, Kahraman A (2011a) A fatigue model for contacts under mixed elastohydrodynamic lubrication condition. Int J Fatigue 33(3):427–436

Li S, Kahraman A (2011b) Influence of dynamic behavior on elastohydrodynamic lubrication of spur gears. Proc Inst Mech Eng, Part J, J Eng Tribol 225:740–753

Li S, Kahraman A (2013a) A physics-based model to predict micro-pitting lives of lubricated point contacts. Int J Fatigue 47:205–215

Li S, Kahraman A (2013b) Micro-pitting fatigue lives of lubricated point contacts: Experiments and model validation. Int J Fatigue 48:9–18

Li S, Kahraman A (2013c) A tribo-dynamic model for a spur gear pair. J Sound Vib 332:4963–4978

Li S, Kahraman A (2014) A micro-pitting model for spur gear contacts. Int J Fatigue 59:224–233

Li S, Kahraman A, Klein M (2012) A Fatigue model for spur gear contacts operating under mixed elastohydrodynamic lubrication conditions. ASME J Mech Des 134(4):041007-1-11

Liu Y, Mahadevan S (2007) A unified multiaxial fatigue damage model for isotropic and anisotropic materials. Int J Fatigue 29:347–359

Liu H, Lohner T, Jurkschat T, Stahl K (2018) Detailed investigation on the oil flow on dip-lubricant gearboxes by the finite volume GFD method. Lubricants 6:47

Liu H, Liu H, Zhu C, Zhou Y (2019) A review on micropitting studies of steel gears. Coatings 9(1):42

Lo CC (1969) Elastic contact of rough cylinders. Int J Mech Sci 11(1), 105–106, IN7-IN8, 107–115

Long H, Al-Tubi IS, Martinze MTM (2015) Analytical and experimental study of gear surface micropitting due to variable loading. Appl Mech Mater 750:96–103

Love AEH (1944) A treatise on the mathematical theory of elasticity, 4th edn. Dover Publications, New York

Manin L, Play D (1999) Thermal behavior of power gearing transmission, numerical prediction, and influence of design parameters, ASME. J Tribol 121:693–702

McGrew JM, Gu A, Cheng HS, Murray SF (1970) Elastohydrodynamic lubrication—preliminary design manual, U.S. Air force technical report # AFAPL-TR-70-27, air force aero-propulsion laboratory, air force systems command, Wright-Patterson Air Force Base, Ohio

Milovanović GV (2016) Generalized gaussian quadratures for integrals with logarithmic singularity. Filomat 30(4):1111–1126

Moorthy B, Shaw BA (2013) An observation on the initiation of micro-pitting damage in as-ground and coated gears during contact fatigue. Wear 297(1):878–884

Morales-Èspejel GE, Rycerz P, Kadiric A (2018) Prediction of micropitting damage in gear teeth contacts considering the concurrent effects of surface fatigue and mild wear. Wear 398–399:99–115

Moser W, Duenser C, Beer G (2004) Mapped infinite elements for three-dimensional multi-region boundary element analysis. Int J Numer Meth Eng 61(3):317–328

Naveros I, Ghiaus C, Ordoñez J, Ruiz DP (2016) Thermal networks considering graph theory and thermodynamics. In: Proceedings of the 12th international conference on heat transfer, fluid mechanics and thermodynamics, Costa del Sol, Spain, 11–13 July, pp 1568–1573

Olver AV, Spikes HA, MacPherson PB (1986) Wear in rolling contacts. Wear 112(2):121–144

Oster P (1982) Beanspruchung der Zahnflanken unter Bedingungen der Elastohydrodynamic, Dissertation Technische Universität München, Forschungsheft/Forschungsvereinigung Antriebstechnik, vol. 131, Frankfurt, M, Frankfurt A.M

Ozguven HN, Houser DR (1988) Mathematical models used in gear dynamics—a review. J Sound Vib 121:383–411

Ramdan RD, Setiawan R, Sasmita F, Suratman R, Taufiqulloh T (2018) Determination on damage mechanism of the planet gear of heavy vehicle final drive. IOP Conf Ser, Mater Sci Eng 307:1–7

Ree T, Eyring H (1955a) Theory of non-newtonian flow. I. solid plastic system. J Appl Phys 26(7):793–800

Ree T, Eyring H (1955b) Theory of non-newtonian flow. II? solution system of high polymers. J Appl Phys 26(7):800–809

Reynolds O (1876) On rolling friction. Philos Trans R Soc Lond 166:155–171

Rycerz P, Kadiric A (2019) The influence of slide-roll ratio on the extend of micropitting damage in rolling-sliding contacts pertinent to gear applications. Tribol Lett 67(2):1

Sadeghi F, Jalalahmadi B, Slak TS, Raje N, Arakere NK (2009) A review of rolling contact fatigue. J Tribol 141:041403-1–04140304140315

Sheng S Ed (2010) Wind turbine micropitting workshop: a recap. National Renewable Energy Laboratory, Technical Report NREL/TP-500-46572, Febr

Shotter BA (1981) Micropitting: its characteristics and implications on the test requirements of gear oils. In: Performance testing of gear oils and transmission fluids, Institute of Petroleum, pp 53–59 and 320–323

Smirnov V (1975) Cours de Mathématiques Supérieurs. Tome IV, Première partie, MIR Éditions de Moscou, Moscou

Sneddon IN (1974) The use of integral transforms, TMH edn. Tata McGraw-Hill Publishing Company Ltd, New Delhi

Snidle RW, Evans HP, Alanou MP, Holmes MJA (2003) The control and reduction of wear in military platforms. Williamsburg, USA, 7–9 June also published in RTO-AVT-190

Socie DF, Marquis G (1999) Multiaxial fatigue, R-234. SAE International Publisher, New York

Tanaka F, Edwards SF (1992) Viscoelastic properties of physically crosslinked networks. 1. Transient network theory, Macromolecules, 25(5):1516–1523

Theißen J (1998) Eignungsnachweise von Schmierölen für Industriegetriebe, 11th International Colloquium, 13–15.1.1998, Technische Akademie Esslingen

Thomson W, Lord Kelvin - (1848) A note on the integration of the equation of the equilibrium of an elastic solid. Camb Dublin Math J 3:87–89

Timoshenko S, Goodier JN (1951) Theory of elasticity. McGraw-Hill Book Company Inc, New York

Wang S, Cusano C, Conry TF (1991) Thermal analysis of elastohydrodynamic lubrication of line contacts using the ree-eyring fluid model. Trans ASME, J Tribol 113:232–244

Wang J, Li R, Peng X (2003) Survey of nonlinear vibration of gear transmission systems. ASME, Applied Mechanics Reviews 56(3):309–329

Warburton GB (1976) The dynamical behaviour of structures, 2nd edn. Pergamon Press, Oxford

Whittaker ET, Watson GN (1990) A course in modern analysis, 4th edn. Cambridge University Press, Cambridge, England

Yastrebov VA (2013) Numerical methods in contact mechanics. John Wiley&Sons, New York

Yastrebov VA, Anciaux G, Molinari J-F (2014) From infinitesimal to full contact between rough surfaces: evolution of the contact area, arXiv: 1401.3800v1, [physics.class-ph] 16 Jan

Yastrebov VA, Anciaux G, Molinari J-F (2016) On the accurate computation of the true contact-area in mechanical contact of random rough surfaces. Tribology International, vol. 114

Yu Z-Y, Zhou S-P, Liu Q, Liu Y (2017) Multiaxial fatigue damage parameter and life prediction without any additional material constants, Materials, MDPI Basel Switzerland, 10(8)

Zhou Y, Zhu C, Liu H (2019) A micropitting study considering rough sliding and mild wear. Coatings 9:639–653

Zhu D (2007) On some aspects of numerical solutions of thin-film and mixed elastohydrodynamic lubrication. Proc Inst Mech Eng, Part J, J Eng Tribol 221(5):561–579

Chapter 11
Tooth Flank Breakage Load Carrying Capacity of Spur and Helical Gears

Abstract In this chapter, a general survey is first done on the tooth fatigue breakage of spur and helical gears, which can manifest itself in the form of tooth flank fatigue fracture (TFF) or tooth interior fatigue fracture (TIFF). The mechanics that trigger these two types of tooth fatigue breakage as well as the characteristics that distinguish one type of damage from the other are described. The fundamentals on the TIFF and TFF calculation methods developed so far are then briefly recalled. Attention is then focused mainly on the fatigue crack initiation criterion, on a refined TFF-risk assessment model as well as on a practical-oriented TFF calculation approach. Some insights on the multiaxial stress state that may originate TIFF and TFF crack initiation as well as the weakest link theory and classical multiaxial criteria are described, focusing attention on a general fatigue criterion for multiaxial stress to be included in the framework of the shear stress intensity hypothesis (SIH). Finally, the procedure for calculating the tooth flank fracture load capacity of cylindrical spur and helical case-carburized gears with external teeth in accordance with the ISO standards is described, highlighting when deemed necessary how the formulae used by the same ISO are rooted in the theoretical and experimental bases previously discussed.

11.1 Introduction

We have already discussed extensively about flank surface damages, such as pitting and micropitting as well as tooth root bending fracture, which are all fatigue failure phenomena. However, the types of flank surface and subsurface fatigue damages in gears do not end with those referable to pitting and micropitting. Actually, at least the so-called surface fatigue damages with crack initiation below the surface of loaded gear flanks as *tooth flank fatigue fracture* (TFF) and *tooth interior fatigue fracture* (TIFF), also known as *tooth flank fatigue breakage* (TFB), are to be added to the above types of surface fatigue damages of teeth flanks. These types of flank surface fatigue damages are so severe that the affected gear drives become completely unusable in almost all cases in which they occur. These surface fatigue damages, which would be more correct to call subsurface fatigue damages, are the topic of this chapter.

© Springer Nature Switzerland AG 2020

V. Vullo, *Gears*, Springer Series in Solid and Structural Mechanics 11,
https://doi.org/10.1007/978-3-030-38632-0_11

Both TFF and TIFF are flank surface fatigue damages that do not fall into the description of either surface contact fatigue or tooth root bending fatigue. These types of flank surface fatigue failures have been especially but not exclusively observed in highly loaded case-carburized gears for power transmission used in many practical applications (spur and helical gears for car, truck and bus transmissions, wind turbines and turbo transmissions as well as bevel and hypoid gears for water turbines). Contrary to surface contact fatigue failures, such as pitting and micropitting, which occur as a result of cracks initiated at or just below the flank surface, these types of failures are due to deeper primary crack initiation, typically located in the area of the case-core interface at approximately mid-height of the tooth. However, they have been found not only in case-carburized cylindrical gears, but also in bevel and hypoid gears used for different applications as well as in nitride and induction-hardened gears.

The crack propagation through the material that is typical of these flank surface damages can result in a tooth flank failure, which is sometimes referred to as a subsurface fatigue failure. It can lead to the total tooth breakage, with consequent irreparable destruction of the gear drive where this type of damage occurs. These subsurface fatigue failures were also observed in gears stressed by loads below the rated allowable ones calculated with the load carrying capacity evaluation procedures against pitting, micropitting and tooth root bending strength. This factual circumstance unequivocally demonstrates that their failure mechanism is substantially different from those described for the fatigue failures due to pitting, micropitting and tooth root bending fatigue loads.

The risk of tooth flank fatigue failures has increased enormously with the use of case-hardened gears. The surface hardening heat treatment, especially the one known as case-hardening, is nowadays widely used to produce compressive residual stresses at the tooth flank surfaces, from which noticeable improvements in wear resistance, contact fatigue strength and tooth root bending fatigue strength ensue. These beneficial compressive residual stresses at the tooth flank surfaces are however balanced by tensile residual stresses within the core, and particularly at the case-core interface. This leads to an increased risk of crack initiation and propagation below the hardened surface, especially in correspondence with a preexisting material fault, such as local inclusions and/or other defects of the metallurgical structure, which act as stress risers. The development of a tooth flank fatigue breakage, either a tooth flank fatigue fracture or a tooth interior fatigue fracture, presents the well-known classic three stages, namely crack initiation, crack propagation and final fracture.

So far, the risk of tooth flank fatigue breakage has not been the subject of standardized international calculation procedures. This is due to a still inadequate understanding of its failure mechanism and of the main factors that influence it. The need to deepen this topic, through adequate and extensive theoretical and experimental investigations, has become a categorial imperative with the use of the aforementioned case-hardened and especially case-carburized gears, operating at high pitch line velocities under equally high loads. These investigations began in the 1960s (see Pedersen and Rice 1961) and continued until the end of the last century (see Sharma et al. 1977; Sandberg 1981; Oster 1982; Elkholy 1983; Lang 1988; Thomas

1997), but gradually intensified in the present beginning of the third millennium (see MackAldener and Olsson 2000a, b, 2001, 2002; MackAldener 2001; Klein et al. 2011; Ghribi and Octrue 2014; Boiadjiev et al. 2014; Beerman and Kissling 2015; Hein et al. 2017; Al et al. 2017; Octrue et al. 2018; Stahl et al. 2018; Ristivojević et al. 2019), leading to the Technical Specification of ISO/TS 6336-4:2019, which however is limited only to the calculation of the tooth flank fatigue fracture load capacity of cylindrical involute spur and helical core-carburized gears with external teeth.

Since TIFF and TFF failures of gears can appear at loads lower than the allowable ones correlated with fatigue failure modes for pitting, micropitting and tooth root bending strength, it is clear the need for a deeper understanding of the phenomena that determine these types of fatigue failures at the design stage. This understanding is required to avoid gear durability losses compared to those related to pitting, micropitting and tooth root bending strength. This important topic must necessarily be addressed within the framework of the incomplete tooth knowledge, which make the TIFF and TFF risk dependent on the microgeometry, material, loading and hardening properties.

Finally, it should be noted that complex interaction of stress fluctuations and material inhomogeneities can greatly influence some aspects of these two types of fatigue damages. Actually, the presence of retained austenite in the carburized case and its transformation during service, resulting in the associated volumetric variation, can cause certain distortion of the teeth and loss of the original contact quality; this can cause variation in the localized stress distribution. Furthermore, the presence of localized *white etching areas* that determine local work hardening can determine crack initiation and propagation, which further aggravate the phenomenon analyzed here, whose complete understanding requires the performance of demanding investigations, both theoretical and experimental.

11.2 TFF and TIFF Failure Mechanisms

Either TFF and TIFF are flank fatigue damages that, as we have already seen above, have crack initiation below the surface of the loaded gear flanks. The crack initiation of both these two types of fatigue damage is substantially similar, while their crack propagation mechanism is very different, as the two fractures follow different paths. Therefore, despite the fact that both types of fractures present the three classic stages of fatigue crack development recalled in the previous section, the microgeometry of these two types of fracture is substantially different, like Fig. 11.1a, b show.

Usually, TFF occurs on the driven member of a gear with one-sided flank loading. It is characterized by a crack initiation under the active flank surface due to shear stresses caused by the Hertzian contact pressure as well as by various contributing causes, among which local stress risers consisting of non-metallic inclusions and metallurgical defects play a prominent role. After a crack is initiated under the active surface of the tooth flank, it slowly propagates during operation in the direction of

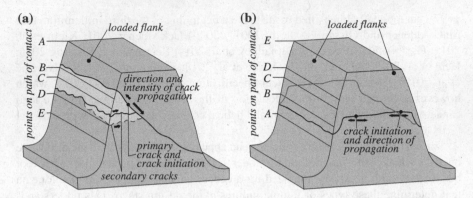

Fig. 11.1 Crack propagation of: **a** TFF (tooth flank fatigue fracture); **b** TIFF (tooth interior fatigue fracture)

both the active flank surface and the core area, with a primary crack that has no direct connection to the surface and is oriented at an angle of (40–50)° with respect to the tooth flank profile. During crack propagation, occasionally the so-called *fish eye* may form, due to relative microscopic movements between cracked surfaces. Starting from the crack initiation, the primary crack propagates on one side towards the surface of the loaded flank and on the other towards the tooth core in the direction of the opposite tooth root section.

Due to the higher hardness that characterizes the case-hardened layer, the crack propagation rate towards the loaded flank surface is however smaller compared to that towards the tooth core. After the primary crack has propagated enough, swelling to such an extend as to cause an appreciable reduction in tooth stiffness, secondary cracks can occur under load. A distinctive feature of these secondary cracks is that they normally start on the flank surface and propagate parallel to the top land surface of the tooth into the material. After one of these secondary cracks has reached the primary one, particles from the active flank can break off. Then, when the primary crack has reached sufficiently large dimensions, the load carrying capacity of the tooth rapidly decreases so that, once a resistant critical cross section is reached, the upper piece of the tooth is separated from its portion adjacent to the gear rim. The final breakage of the tooth is due to forced rupture; therefore, it is an overload breakage.

Figure 11.1a shows the primary and secondary cracks as well as the propagation direction of the primary crack, which are typical of the tooth flank fatigue fracture. Examining the fracture area, the usual three stages of fatigue crack development already mentioned above can be observed, with the typical shiny crack lens around the crack starter and the area of rough overload breakage. It is evident that a fracture of the tooth of this type, which generally occurs after 10^7 load cycles, constitutes the end of the lifetime of the gear drive that is affected by it. The damage described above cannot be repaired, so the gear drive can no longer be used and must therefore be used for scrap.

Contrary to TFF, which is typical of driven gears with one-sided flank loading, the TIFF is instead most common in gears with alternating loads, where both tooth flanks are used to transmit design torque, such as occurs in idler gears with external teeth and planet gears of planetary gear trains. Tooth interior fatigue fracture is also characterized by a crack initiation below the active flank surface, but has a different fracture shape due to two potential crack initiation points related to the reverse loading. A characteristic crack propagation path of TIFF is that shown in Fig. 11.1b. It shows that the fatigue fracture propagates nearly parallel to the top land surface, with small branches and drops sub-parallel to the two tooth profiles that veer towards them.

The same Fig. 11.1b shows the final shape of the TIFF-type fracture, which is characterized by a well distinguishable plateau in the central part along the tooth thickness, approximately located at half tooth height. Along this fracture, the separation between the upper and lower tooth segments occurs. In a close-up of the cross section of a TIFF-type fracture, small wing cracks can be observed. Their presence indicates that the main crack has propagated from the tooth center towards the tooth flanks. Several researchers (see Suresh 1998; Sponzilli et al. 2000) have correlate these types of fracture to the presence of large non-metallic inclusions in the interior of the tooth, which act as initiation sites for fatigue failure. Also, in this type of fatigue failure, which also occurs after about 10^7 load cycles, the fracture area shows the usual three stages of fatigue crack development already mentioned for TFF-type fracture. Even the severity in the damage related to this type of fatigue failure is no less than that described for the tooth flank fracture.

The causes that determine the TIFF-type fractures are not only the alternating loads on both sides of the tooth, but also the tensile residual stresses within the tooth core resulting from the heat treatment. All the stresses that come into play then undergo significant increases due to the inevitable presence of non-metallic inclusions and metallurgical defects, which act as stress risers with consequent dangerous stress concentrations.

Tooth flank breakages, regardless of whether they are TFF-type fractures or TIFF-type fractures, can sometimes be accompanied by other types of fatigue failures, such as pitting, micropiotting and tooth root bending failures, as well as non-fatigue fractures, such as scuffing, but most often they appear without the presence of these other surface damages. Although tooth flank breakage damage is sometimes found in only one gear tooth, it usually occurs in several teeth. This type of damage occurs more frequently in gears having large wheel diameter and module than in smaller gear wheels. The run time up to the fracture is strongly variable but, in any case, the fatigue lifetime is never less than $(10^7–10^8)$ load cycles (see Klein et al. 2011).

The diagram in Fig. 11.2, taken from Townsend (1992), shows the likely different failure types of which a gear can be affected, depending on the transmitted torque and pitch line velocity. As we have already pointed out elsewhere, it is obvious that the higher the pitch line velocity, the thicker the lubricant oil film between the mating teeth surfaces. This type of diagram is specific to each gear design, so the *spalling* and breakage lines can switch depending on the gear design under consideration.

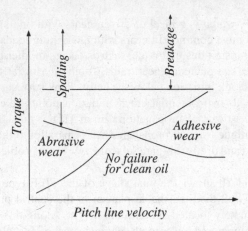

Fig. 11.2 Expected gear failure type depending on transmitted torque and pitch line velocity

Here it is necessary to clarify the difference between pitting and spalling, which are two types of surface contact fatigue. In the scientific literature there is no common recognized definition to distinguish pitting from spalling. Often pitting and spalling terms have been used and are still used indiscriminately, while at other times they have been used and are used to distinguish a different severity of surface contact fatigue. Tallian (1992) defined spalling as a *macro-scale contact fatigue failure* caused by fatigue crack propagation, while reserving the term pitting to indicate surface contact damage caused by sources other than crack propagation (see also Ding and Rieger 2003). The reasons for this lack of clarity and uniformity of views is probably due to the fact that the physical causes of pitting and spalling are not yet established in a convincing and exhaustive way.

In general terms, we can say that we are faced with three types of surface contact fatigue, namely pitting, micropitting and spalling. Let's leave the micropitting aside, and focus our attention on pitting and spalling. Pitting appears as shallow craters at contact surfaces, with pits that reach a maximum depth equal to about the thickness of the work-hardened layer (about 10–20 μm). Spalling instead appears as deeper cavities at contact surfaces, with spalls having a depth typically equal to (50–150) μm, roughly corresponding to (0.25–0.35) b_H, where b_H is half of the Hertzian contact width.

Figure 11.3 shows a schematic representation of pitting and spalling fatigue failures. It highlights the typical contact fatigue fracture due to spalling, which starts at the tooth flank surface, in correspondence with a subsurface defect that acts as a stress riser, propagates in the work-hardened layer in a direction inclined by about 60° with respect to the local normal, until it reaches a depth of about (0.25–0.35) b_H, and then rises towards the tooth flank surface in a direction inclined of about 30° with respect to the local normal. The sizes of spalls that can be separated from the tooth flank surfaces are considerably larger than those of pits due to pitting. The always destructive nature of spalling is evident.

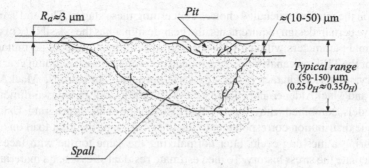

Fig. 11.3 Schematic representation of pitting and spalling fatigue failures

Although both pitting and spalling constitute a common form of surface contact fatigue failure, spalling results in a faster deterioration of surface durability compared to pitting. Furthermore, spalling often induces early failure by severe secondary damage. Consequently, spalling has been considered as the most destructive surface failure mode of a gear.

We do not want to go into the details of the existing theories that ascribe the origin of contact fatigue failures to surface defects or subsurface defects. It is enough for us to mention that, according to the failure theories most accredited today, pitting fatigue failures are mainly (but not always) developed from surface defects, while spalling fatigue failure are mainly (but not always) developed from subsurface defects. For a brief description of these theories, we refer the reader to the scientific literature (see Fernandes and McDuling 1997; Ding and Rieger 2003). The most important thing to keep in mind here, however, is the fact that, to date, we do not have reliable calculation criteria of surface durability against spalling.

11.3 Fundamentals on the TIFF and TFF Calculation Methods

In this section, a brief summary of the TIFF and TFF calculation methods available in scientific literature is provided, in accordance with what is proposed by Al et al. (2017). These methods have a common denominator, as they share the approach procedures referable to the following four stages:

- calculation of stress history;
- specification or calculation of residual stresses;
- calculation of equivalent stresses by means of some strength criterion;
- comparison of the maximum equivalent stress with some threshold value of stress that triggers the fatigue crack, based on experimental results or field experiences.

Each of the available calculation methods differs in some particular details concerning these four stages. Thus, their applicability depends on the assumptions made as well as on the particular detailed choices concerning these stages. Therefore, in order to achieve certain design requirements, the gear designer has the possibility of choose the optimal parameters within a wide spectrum of different possible permutations.

As for the TIFF risk analysis and the related determination of optimum gear micro-geometry, material characteristics and case-hardening properties, MackAldener (2001) and MackAldener and Olsson (2000a, b, 2001, 2002) used two-dimensional FEA models, combined with a specific load distribution analysis program. Using then the torque distribution corresponding to the distribution of the total load on a single tooth during a meshing cycle, after normalizing the same torque with face width, they calculate the stress history. To then estimate residual stresses and material properties, they first used a procedure where transformation strain and material fatigue properties were assumed to be constant throughout the case, while subsequently they introduced a procedure where transformation strain and material fatigue properties were considered to have non-homogeneous profiles along the case thickness.

Given the complexity of setting up the numerical-analytical model used and the computationally expensive calculation running, the same authors proposed the use of a simpler semi-analytical method. This method allowed faster calculations, design parameter studies as well as a satisfactory optimization of the parameters involved, but at the expense of accuracy of the numerical results, as the crack initiation risk factor was over-predicted by about 20% when compared with that of the previous numerical-analytical model. To assess the influence of gear design parameters in the TIFF risk, the authors used a factorial planning, examining a wide range of variability of the numerous parameters considered, and concluded that the TIFF could be avoided if the slenderness ratio and tensile residual stresses were reduced, the gear non used as a idler gear and optimum case and core properties were used.

To reduce the drawbacks, Al and Langlois (2015) introduced a modification to the method described above, consisting in the fact that a specialized three-dimensional elastic contact analysis was carried out, which constituted a separate gear loaded tooth contact analysis (LTCA). Results of this separate specialized analysis were then used to determine the load boundary conditions for the TIFF analysis. In this way, the calculation procedure became computationally less expensive, with a reliability of results comparable with that of the method described above.

The first model concerning the TFF risk analysis was the one developed by FZG scholars and researchers, such as: Witzig (2012), Tobie et al. (2013), Boiadjiev et al. (2014). This method is based on the *hypothesis of shear stress intensity*, introduced by Hertter (2003), whose calculation procedure relies on the determination of local stress history. Its field of applicability is however limited, since the equation for calculating the local exposure of the material to the FTT risk is eminently empirical, as it also uses significant empirical contributions. Another limitation is the fact that the method was developed only for single-flank loading, although, at least from a theoretical point of view, it is possible to extend it to double-flank loading (for example, idler gears and planet gears); this extension of the method is however not trivial. Finally, the assumptions underlying the residual stress calculation entail a further restriction of this method, which can only be used for case-carburized gears.

This model uses as input the Hertzian contact stress, which is calculated using a gear load distribution program or a simplified analytical procedure like the standardized one, described in Chaps. 2 and 5. The method neglects the tensile residual stresses within the core, for which it can appreciably underestimate the critical fatigue stress when these residual stresses are not negligible. The assumption of negligible tensile residual stresses within the core is sufficiently valid when the tooth core section is much larger compared to the thickness of the case. Instead, it entails great limitations when the method is applied to slender teeth and extensive case hardening depths.

A more general method than the one described above, which can be applied to both TFF and TIFF, is the one proposed by Ghribi and Octrue (2014). It uses a multi-axial fatigue criterion and takes into account the tensile residual stresses within the core, which are very important for the accuracy of the results. Stress history is determined by using Hertzian contact stress together with Belajev's theoretical analysis (see Belajev 1917, 1924; Johnson 1985), described in Sect. 2.4, which allows us to derive the distributions of the subsurface stress components. The Hertzian contact stress is calculated using the procedures of ISO/TR 15144-1:2014, which was subsequently replaced with the ISO 6336-22:2018. The influence of bending stresses, which certainly influence the subsurface stress state at different material depths inside the tooth, are not considered.

Both of these methods do not use FEA models. They also do not consider the critical effects of the material quality and presence of inclusions, which certainly constitute a key factor of the problem discussed here. This omission represents a non-negligible limit, which could affect the results obtained. To take into account the material quality and presence of inclusions, it would be necessary to introduce a factor that could be applied to characteristic threshold quantities of the material. However, to do this, a systematic campaign of wide-ranging studies and considerable experimental work is necessary as well as an adequate accumulation of experience in the field.

11.4 An Interesting Method of TIFF and TFF Risk Evaluation

11.4.1 General Concepts and Main Assumptions

Here we want to summarize the method developed by Al and Langlois (2015) and Al et al. (2016, 2017). In comparison with the two methods mentioned in the previous section, this method uses the whole stress tensor, including bending stresses, and considers the tensile residual stresses within the core, whose effects increase with torque. It also introduces a different procedure for calculating TIFF and TFF failure thresholds and, in the wake of MackAldener and Olsson (2002), is based on the FE analysis. However, unlike MackAldener and Olsson, loading conditions calculated with a specialized load tooth contact analysis (LTCA) are used in place of a full FE tooth contact analysis, too complex to perform and computationally expensive. The procedure is used not only for calculating crack initiation risk factor, but also for investigating the crack propagation mechanism.

The specialized LTCA model is a hybrid model, developed by Langlois et al. (2016), which combines a Hertzian contact sub-model for the local contact stiffness and a FE sub-model of bending and base rotation stiffness of gear teeth and blank. The model also considers the extension of the path of contact and the consequent increase in the transverse contact ratio, due to the bending of the non-involute tooth tip part beyond the effective outside diameter, whose effects are particularly important for slender gear teeth that are characterized by a higher risk of TIFF.

About the details of the aforementioned model, which also combines 2D and 3D FE sub-models, we refer to the original work of Al et al. (2017). We would like to point out here that it is used to determine the load boundary conditions at a selected number of time steps through the meshing cycle. At each of these time steps, the load distribution between and across the teeth is calculated and, at each contact line corresponding to the specific time step, load positions, load intensities and Hertzian half widths are determined.

The model then assumes that the material properties are variable within the case and core of the gear teeth. These properties play an important role in TIFF, especially for case-hardened gears, which are notoriously characterized by their variability along any local normal to tooth profile. As required by the critical plane criterion, critical shear stress and fatigue sensitivity to normal stress are also assumed to be variable along any local normal to tooth profile. In accordance with what has been done by MackAldener and Olsson (2001), the variability law of all these quantities is assumed to coincide with the one corresponding to the hardness profile.

11.4.2 Hardness Profile and Material Properties

The aforementioned researchers use, as a hardness profile, the one proposed by Mack-Aldener and Olsson (2001), which is described by the following two relationships that however are written using the local coordinate system introduced in Chap. 2 (see Fig. 2.2, where z-axis and y-axis respectively represent the local normal and local tangent at any load point of the tooth profile, and x-axis has the same direction of the tooth axis), while symbols of ISO/TS 6336-4:2019 are used to represent the other quantities, but with some other exceptions:

$$HV(z) = HV_{surface}g\left(\frac{z}{\bar{z}}\right) + HV_{core}\left[1 - g\left(\frac{z}{\bar{z}}\right)\right] \tag{11.1}$$

$$g\left(\frac{z}{\bar{z}}\right) = 1 - 3\left(\frac{z}{\bar{z}}\right)^2 + 2\left(\frac{z}{\bar{z}}\right)^3, \tag{11.2}$$

where: $HV_{surface}$ and HV_{core} (both in HV) are respectively the surface hardness and core hardness; $HV(z)$ is the hardness at the material depth, z (in mm), at the contact point under consideration; $\bar{z}$ (in mm) is the total case depth; $g(z/\bar{z})$ is a depth function that determines the variation between the case and core defined by MackAldener and Olsson (2001).

It should be noted that all parameters depending on the z-coordinate are defined as local values. It should also be noted that the aforementioned method relies on measurement of the total case depth, $\bar{z}$, which is often neither measured nor known. In these cases, the hardness measured at a defined effective case depth is used as an alternative to the total case depth. For these same cases, a different hardness profile can be used, namely the one proposed by Lang (1979) and subsequently adopted by Witzig (2012), which is described by the following two relationships also written using the aforementioned local coordinate system and symbols of ISO/TS 6336-4:2019, when they exist:

$$HV(z) = HV_{surface}g\left(\frac{z}{z_{CHD}}\right) + HV_{core}\left[1 - g\left(\frac{z}{z_{CHD}}\right)\right] \tag{11.3}$$

$$g\left(\frac{z}{z_{CHD}}\right) = 10^{-\left[0.0381\left(\frac{z}{z_{CHD}}\right) + 0.2662\left(\frac{z}{z_{CHD}}\right)^2\right]}, \tag{11.4}$$

where z_{CHD} (in mm) is the *case hardening depth* at 550 HV, while the meaning of the symbols already introduced remains unchanged. Indeed, z_{CHD} represents the *effective case depth*, that is the reference material depth where hardness is equal to 550 HV; beyond this material depth, hardness drops below 550 HV. It should be noted that, to indicate the case hardening depth at 550 HV, we have introduced here the symbol z_{CHD} instead of *CHD* used by ISO and other authors, since this quantity concerns a particular material depth so we believe that this symbol is more in keeping with the local coordinate system used in this textbook.

In addition to the two models mentioned above, other empirical models of hardness profiles are available in the literature, such as those proposed by Tobe et al. (1986) and Thomas (1997). Figure 11.4 shows a comparison of the four empirical models mentioned here with the experimental results obtained by MackAldener and Olsson

Fig. 11.4 Hardness profile models and comparison with experimental results

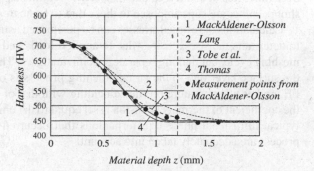

Material depth z (mm)

(2001). Three of these models (those of MackAldener and Olsson, Thomas, and Tobe et al.) show a substantial agreement between them and with the experimental results, while the *Lang's model* highlights appreciable deviations, especially for material depths between 0.5 and 1.6 mm. The same figure shows, marked by a vertical dashed line, a total case depth equal to 1.2 mm, while the effective case depth where hardness drops below 550 HV is equal to about 0.68 mm. The *MackAldener-Olsson hardness profile* was obtained by fitting experimental results.

Due to the differences highlighted above, the Lang's model, adopted by Witzig, could determine differences in the risk prediction of TIFF and TFF. This is because it is foreseeable that fatigue properties of the material and residual stresses may vary near the case-core interface, as experimental evidence and specific investigations in the literature have already shown.

The multi-axial fatigue analysis then requires the preliminary determination of the properties of the material. For the cases dealt with by them, where these material properties represented by *critical shear stress*, $\tau_{crit}(z)$, and *fatigue sensitivity to normal stress*, $a_{cp}(z)$, were not known, the authors assume that they are variable with continuity between case and core and that their variability conforms to that of the hardness profile. The relationship used by the same authors to define these two quantities are as follows:

$$\tau_{crit}(z) = \tau_{crit,surface} g\left(\frac{z}{\bar{z}}\right) + \tau_{crit,core}\left[1 - g\left(\frac{z}{\bar{z}}\right)\right] \tag{11.5}$$

$$a_{cp}(z) = a_{cp,surface} g\left(\frac{z}{\bar{z}}\right) + a_{cp,core}\left[1 - g\left(\frac{z}{\bar{z}}\right)\right]. \tag{11.6}$$

11.4.3 Residual Stresses

Residual stresses play a decisive role in the problem examined here, as they influence stress states on tooth flank surfaces and below these surfaces, within teeth bodies. These residual stresses depend only on the heat and surface treatments on the gear, while they are independent of the load, so they are generally assumed to be constant over time. Residual stresses due to specific treatments, such as case-hardening and shot peening, are superimposed to those due to basic treatments.

Calculations of residual stresses can be performed using the same FE models used to determine the stress history due to applied loads and their variability during the meshing cycle, but making separate specific analyses. The volume expansion within the surface layer due to the case-hardening process is obtained by applying to the FE models a temperature profile coinciding with that of transformation strain when the coefficient of thermal expansion is set equal to the unity. With this procedure, the typical diffusional phase transformations that occurred during the case-hardening process are adequately taken into account.

For the analyzes carried out by the authors, all side nodes of the FE models are allowed to move only in the radial direction. Furthermore, to carry out the calculations of their interest, the authors assume, as transformation strain profile, that used by MackAldener and Olsson (2001), which is an isotropic profile, measured relative to the core. It is expressed by the following piecewise polynomial with smooth connections:

$$\varepsilon_t(z) = \varepsilon_1 + 4(\varepsilon_2 - \varepsilon_1)\left[\left(\frac{z}{\bar{z}}\right) - \left(\frac{z}{\bar{z}}\right)^2\right] \tag{11.7}$$

$$\varepsilon_t(z) = -4\varepsilon_2\left[1 - 6\left(\frac{z}{\bar{z}}\right) + 9\left(\frac{z}{\bar{z}}\right)^2 - 4\left(\frac{z}{\bar{z}}\right)^3\right] \tag{11.8}$$

$$\varepsilon_t(z) = 0, \tag{11.9}$$

which apply respectively for material depth values, z, included within the following ranges: $(0 \leq z \leq \bar{z}/2)$, $(\bar{z}/2 \leq z \leq \bar{z})$ and $(z \geq \bar{z})$. In these relationships, $\varepsilon_t(z)$ is the instantaneous local transformation strain, while ε_2 and ε_1 are respectively the maximum transformation strain and transformation strain at the surface.

It is here also to be remembered the calculation method of residual stresses proposed by Lang (1979) that, to determine the local residual shear stress, $\tau_{residual}(z)$, simply requires that the type of heat treatment and depth from surface are known. This method of analysis uses the following two relationships:

$$\tau_{residual}(z) = -1.25[HV(z) - HV_{core}] \tag{11.10}$$

$$\tau_{residual}(z) = 0.2857[HV(z) - HV_{core}] - 460; \tag{11.11}$$

these two relationships apply respectively according to whether $[HV(z) - HV_{core}] \leq 300$ or $[HV(z) - HV_{core}] > 300$. Of course, in these equations the Lang's hardness profile, $HV(z)$, must be introduced. It should be noted that both these equations clearly show that this method only considers the compressive residual normal stresses.

This Lang's analysis model was used by both the TFF calculation methods proposed by Witzig (2012) and Ghribi and Octrue (2014). However, the method proposed by these two last researchers also allows us to calculate tensile residual normal stresses using a force balance across the teeth. Nevertheless, this improvement to the Lang's model has not been considered by the authors of the contributions we are describing here, who have instead used the above described model proposed by MackAldener and Olsson (2001).

Figure 11.5, taken from Al et al. (2017), shows the residual stress profiles with increasing depth, obtained with the MackAldener and Olsson model and with the Lang's model with the gear drive used by them for experimental tests to support their investigations. The residual stress profile obtained with the MackAldener-Olsson

Fig. 11.5 Residual stress
profiles and comparison with
experimental results

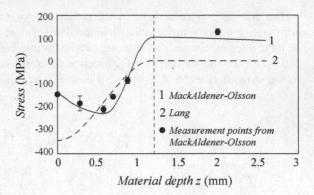

Material depth z (mm)

model results from the strain profile obtained using Eqs. (11.7) to (11.9), with $\varepsilon_2 = 114 \times 10^{-5}$ and $\varepsilon_1 = 833 \times 10^{-6}$.

The same figure shows, marked by a vertical dashed line, a total case depth equal to 1.2 mm. From the same figure, it is evident that the MackAldener-Olsson residual stress profile is in a good agreement with the experimental results, while the Lang residual stress profile, used by Witzig, differs appreciably from both the previous profile and experimental results. The exact reason for this behavior cannot be identified at the current state of knowledge, which deserves to be investigated. However, some significant material dependence not considered in the model cannot be excluded. Finally, it should be noted that the resulting calculated residual stresses may vary from one position to another along the path of contact due to the variation in tooth thickness.

11.4.4 Fatigue Crack Initiation Criterion

Whatever the fatigue crack initiation criterion to be used for the TIFF and TFF risk assessment, it is first necessary to determine the effective stress state within the gear teeth during the meshing cycle, considering the load distribution along the path of contact. This effective stress state is determined by superimposing the calculated stress history states and the initial estimated residual stresses that, as we have already said, are assumed to be constant over time and independent of the load.

To analyze the stress history and to verify that a fracture may occur, the authors then use the multiaxial fatigue criterion proposed by Findley (1959), which is a criterion based on the critical plane method. With this fatigue criterion, which Findley developed for a torsion cyclic loading combined with in-phase bending, the sum $(\tau_{\varphi\psi,a} + k\sigma_{\varphi\psi,\max})$, is computed in each plane; in this sum, $\tau_{\varphi\psi,a}$ is the shear stress amplitude, $\sigma_{\varphi\psi,\max}$ is the maximum normal stress, taken over a whole cycle, and k is the Findley's parameter (for the meaning of the symbols, see Fig. 11.8 and the following section). The critical plane is the one where the maximum of this sum is reached. Using FE models, which albeit with some variations substantially follow those of MackAldener and Olsson (2000b), the same authors determine the equivalent stresses, which are expressed in terms of Findley stress, σ_F, as follows:

$$\sigma_F = \tau_a + a_{cp}\sigma_{n,\max},$$

(11.12)

where τ_a is the shear stress amplitude, $\sigma_{n,\max}$ is the maximum normal stress and a_{cp} is the already defined fatigue sensitivity to normal stress.

Material properties, including the critical plane stress and fatigue sensitivity to normal stress, are assumed to vary continuously in accordance with the hardness profile. The risk threshold of crack initiation is calculated using, as criterion, the quotient of the maximum Findley critical plane stress divided by the critical shear stress that defines the permissible stress at a given point. This ratio is called crack initiation risk factor (CIRF).

11.5 A Refined TFF-Risk Assessment Model

The TFF-risk assessment model developed by FZG (Furschungsstelle für Zahnräder und Getriebebau) deserves to be briefly described here. It is a material-physicality based model, whose fundamental contributions are to be ascribed to Oster (1982) and Hertter (2003). This model allows the assessment of the risk of tooth flank fatigue fracture and other surface and subsurface initiated fatigue failures on gear flank surfaces. Essentially, the model is based on the comparison of a local shear stress occurring in a volume element positioned at or below the flank surface and the local material strength at the considered material depth (see also Boiadjiev et al. 2014).

As we have already pointed out in Sect. 11.2, the primary crack often begins at points corresponding to non-metallic inclusions or metallurgical defects, where the Young's modulus differs more or less significantly compared to that of the normal material structure. These imperfections below the surface act as stress risers during the roll off of the flank, due to the notch effect. If the material strength is locally exceeded, a crack with growth potential could be initiated into the material that can lead to tooth flank fatigue fracture later. Attention must therefore be focused on the stress state at the critical points represented by these imperfections.

Considering an actual tooth flank surface, which is a rough and lubricated surface, the general stress state at any point inside the tooth is mainly given by the superposition of the following individual stress states:

- stress state related to normal contact force due to the applied torque, which determines a pressure distribution over the contact area in accordance with the Hertz theory;
- stress state related to the tangential load caused by the friction forces that develop over the contact area due to the relative sliding between the two gear flank surfaces;

- stress state related to modified pressure distribution due to the lubricated contact in accordance with the elastohydrodynamic theory;
- stress state related to thermal load due to localized heating over the contact area caused by friction power losses;
- stress state related to residual stresses due to heat treatment, which is usually characterized by compressive residual stresses in the case-hardened area compensated by tensile residual stresses in the tooth core;
- stress state related to the dynamic loads that develop between the asperity peaks of the two tooth flank surfaces in relative motion.

All the aforementioned single stress states are dependent on the load, with the exception of that related to residual stresses, which can be considered load-independent and therefore constant over time if the load is not too high. However, the current determination of the total stress state by overlapping these individual stress states involves certain precautions. Residual stresses and load induced stresses can only be superimposed if the different time dependence of the stress components is considered. To this end, it is necessary to consider how the stress distribution changes during the relative motion between the rolling and sliding tooth flank surfaces.

On a loaded tooth flank surface, the rolling and sliding direction y (see Figs. 2.2 and 2.12a), can also be considered as time axis, t. Figure 11.6 shows schematically a rolling and sliding lubricated contact like that of Fig. 2.12a; it highlights how any stress component of the three-dimensional stress state that is generated below the surface changes during the aforementioned relative motion. Under the assumption of constant normal force and equivalent radius of curvature, all volume elements at the same material depth, z/b_H, are exposed to equal stresses, but to different times during the roll-over process related to the meshing cycle. Furthermore, a single volume element is subjected to different stresses at different times, so that, when the material exposure is calculated in a considered volume element below the surface, the stresses have to be considered over the entire t-axis. The problem is then further complicated by the fact that the principal coordinate system rotates during the meshing cycle, so the absolute stress values and direction of the principal stresses change for each volume element.

Fig. 11.6 Time-dependent stress in a rolling and sliding contact

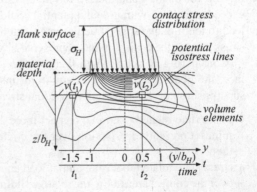

Fig. 11.7 Shear stress below
the surface: influence of the
equivalent radius of
curvature

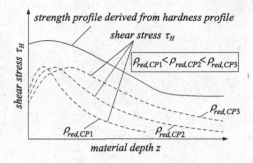

The stress distribution along the z-axis (Fig. 2.2) is mainly influenced by the maximum Hertzian contact stress, $\sigma_H = p_H$, resulting from the Hertzian contact load (this is variable along the path of contact), as well as by the local equivalent normal radius of curvature or local normal radius of relative curvature, $\rho_{red,CP}$, at arbitrary point of contact, CP (Fig. 11.13). Figure 11.7 clearly shows how shear stress, τ_H, below the surface affects growing material depths with an increasing local equivalent normal radius of curvature, $\rho_{red,CP}$, although the maximum contact pressure, σ_H, and the maximum shear stress, τ_{Hmax}, remain the same. Since the relative radius of curvature increases with increasing center distance, pressure angle and addendum modification factor, it is obvious that in these cases increased values of shear stress, τ_H, at increasing material depths below the surface must be expected. Considering then that the local strength profile obtained from the hardness profile remains constant, the remaining strength reserve is eroded, as the curves in Fig. 11.7 clearly show; therefore, the crack initiation risk below the surface increases.

To determine the equivalent stress of a multiaxial stress state, the model uses the so-called shear stress intensity hypothesis (SIH), better described in Sect. 11.7. Theoretical investigations made by Lang (1988) had indeed shown that this hypothesis, for lubricated rolling and sliding contacts, lent itself better than the common equivalent stress hypotheses (such as the distortion energy hypothesis, normal stress hypothesis, shear stress hypothesis, octahedral shear stress hypothesis, etc.) to capture a three-dimensional stress state also characterized by alternating stress and stress conditions with a rotating principal coordinate system. In accordance with this hypothesis, the model introduces the *shear stress intensity*, τ_{eff}, given by the following relationship:

$$\tau_{eff} = \left[\frac{1}{4\pi} \int\limits_{\varphi=0}^{\pi} \int\limits_{\psi=0}^{2\pi} \tau_{\varphi\psi} \sin\varphi d\psi d\varphi \right]^{1/2}, \tag{11.13}$$

which considers all maximum shear stresses, $\tau_{\varphi\psi}$, in each section plane, $\varphi\psi$, of the considered volume element. Figure 11.8 shows a spherical surface centered on the considered volume element representing the defect where crack has initiation and the related spherical coordinate system $O(r, \varphi, \psi)$ to define the position of the intersection plane of the same defect. In this coordinate system, a current point on

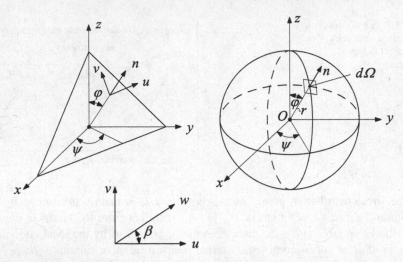

Fig. 11.8 Spherical surface centered on the defect and spherical coordinate system to define the position of the defect intersection plane

the spherical surface is defined by the radial distance, r, from the origin, the zenith angle or colatitude, φ, and the azimuthal angle, ψ.

To assess the risk of tooth flank fracture, the FZG-model considers different volume elements below the surface and, for each of them, determines the local material exposure by comparing the local occurring equivalent shear stress and the local material strength. The local occurring equivalent shear stress is based on the aforementioned SIH and is calculated using the following relationship:

$$\tau_{eff}(z) = \tau_{eff,EL,RS}(z) - \tau_{eff,RS}(z), \tag{11.14}$$

where $\tau_{eff,EL,RS}(z)$ is the local equivalent shear stress due to external loads, which also considers the influence of residual stresses, and $\tau_{eff,RS}(z)$ is the quasi-stationary residual shear stress. Both of these two shear stresses are calculated using Eq. (11.13), considering however different stress components. For the calculation of $\tau_{eff,EL,RS}(z)$, all stress components due to external loads as well as stress components due to residual stresses are considered. Instead, for the calculation of $\tau_{eff,RS}(z)$ only the quasi-stationary residual stresses are considered.

The comparison term of the local occurring equivalent shear stress, $\tau_{eff}(z)$, is the local material strength, $\tau_{per}(z)$, which limits the local carrying capacity of the gears in terms of TFF. This type of fatigue fracture can actually occur if the material strength below the surface is exceeded. The determination of this material strength can be based on in-depth physical and metallurgical considerations of the material involved, able to consider the influences due to grain size and segregation, as well as to suitable experimental tests. A simpler alternative method, which does not consider the aforementioned influence, but which has proved to be sufficiently reliable, assumes that the material strength is directly proportional to the local hardness. In this framework, the FZG-model assumes that the local material strength below the surface changes in accordance with the hardness profile, $HV(z)$, for which it is given by the following relationship:

$$\tau_{per}(z) = K_{\tau,per} HV(z). \tag{11.15}$$

It should be noted that from the theoretical point of view there is no valid relationship between permissible shear strength and local hardness. Gear running tests and field results obtained with industry gear drives, however, have shown that, especially for case-hardened gears, material strength can be assumed directly proportional to local hardness. These same experimental tests have provided a constant value of the linear proportional factor, $K_{\tau,per}$, between the material permissible shear strength, $\tau_{per}(z)$, and Vickers hardness, $HV(z)$, approximately equal to 0.4. However, this value of $K_{\tau,per}$ is to be considered valid only for case-hardened gears made with case-hardening steels and appropriate heat treatments.

Finally, the local material exposure, A_{FF}, to be calculated for each volume element below the surface, is given by the following relationship that expresses the quotient of the local occurring equivalent shear stress divided by the local material strength:

$$A_{FF} = \frac{\tau_{eff}(z)}{\tau_{per}(z)}. \tag{11.16}$$

Using this equation for all points of the effective tooth flank surface, included in the face width between tooth root and tooth tip, and considering different meshing positions to which different values of the local equivalent radius of curvature, local Hertzian contact stress and local hardness correspond, a complete mapping of the local material exposure is obtained. On the basis of results derived from gear running tests on test gears and calculation checks on gear drives from industry applications, the FZG-model considers $A_{FF} = 0.8$ as the critical limit value above which tooth flank fatigue fracture should be expected.

For further details on the models described above, we refer the reader to the specific literature (see Oster 1982; Hertter 2003; Höhn et al. 2010; Witzig 2012; Tobie et al. 2013; Boiadjiev et al. 2014) . It should be noted, however, that this model allows to check whether the maximum material exposure occurs near the surface of the loaded tooth flank or at greater material depth as well as whether or not it has exceeded its critical limit. The model is also able to assess the risk of tooth flank fracture and to predict the depth below surface of a crack initiation and, consequently, what type of failure can be expected. However, the calculation procedure requires comprehensive and detailed input quantities and some integrations and iteration steps are necessary. Therefore, it is very complex, not compatible with the needs of a practical-oriented calculation approach required by designers.

11.6   A Practical-Oriented Calculation Approach for TFF-Risk Assessment

To overcome the difficulties and complexity of calculation procedure of the FZG-model, Witzig (2012), has derived from it a simple approximate model, here called (FZG/Witzig)-model, which allows a practical-oriented calculation approach. It adds to the advantage of a simpler calculation procedure, which does not require integrations and iteration steps, the additional advantage of allowing to assess the risk of tooth flank fatigue fracture already in the design phase of a new gear drive, as it requires a few specific input quantities that are usually available at the design stage.

However, in comparison with the FZG-model, which allows assessing fatigue damages both on the tooth flank surface and in the material depth below surface, the (FZG/Witzig)-model is only able to assess the risk of tooth flank fatigue fracture in the material depth. It does not consider tensile residual stresses or shear stresses induced by friction, elastohydrodynamic contact, surface asperities or thermal load, but has the additional advantage of expressing the calculation equations in closed form. However, it has also the disadvantage of being valid only for case-carburized gears. It should be noted that both the above models, which belong to the same chain, use very similar basic relationships.

The decisive parameter used by the (FZG/Witzig)-model to assess, for case-carburized gears, the risk of tooth flank fatigue fracture is still the local material exposure, A_{FF}, which however is defined by the following relationship:

$$A_{FF} = \frac{\tau_{eff}(z)}{\tau_{per}(z)} + 0.04, \qquad (11.17)$$

where also the local equivalent shear stress, $\tau_{eff}(z)$, and the local material strength, $\tau_{per}(z)$, have a different formulation. Indeed, these quantities are respectively defined by the following expressions:

$$\tau_{eff}(z) = \tau_{eff,L}(z) - \Delta\tau_{eff,L,RS}(z) - \tau_{eff,RS}(z) \qquad (11.18)$$

$$\tau_{per}(z) = K_{\tau,per} K_{material} HV(z). \qquad (11.19)$$

The calculation of the local equivalent shear stress, $\tau_{eff}(z)$, is then carried out considering separately the local equivalent shear stress without consideration of residual stresses, $\tau_{eff,L}(z)$, the influence of the residual stresses on the local equivalent shear stress, $\Delta\tau_{eff,L,RS}(z)$, and the quasi-stationary residual shear stress, $\tau_{eff,RS}(z)$. Equation (11.19) also differs from Eq. (11.15), as a material factor, $K_{material}$, is further introduced to calculate the local material strength.

The local equivalent shear stress without consideration of residual stresses, $\tau_{eff,L}(z)$, is calculated using the following approximate relationship:

$$\tau_{eff,L}(z) = \frac{0.1488\, p_H + \dfrac{zE_r}{4\rho_{red}} - \dfrac{z^2 E_r}{16\rho_{red}^2 \sqrt{\left(\dfrac{p_H}{E_r}\right)^2 + \left(\dfrac{z}{4\rho_{red}}\right)^2}}}{0.4\dfrac{zE_r}{4\rho_{red}\, p_H} + 1.54}. \tag{11.20}$$

This equation is based on the calculation of stress state in a Hertzian line contact between semi-cylinder and half-plane, according to Föppl procedure (Föppl and Föppl 1947). Witzig (2012) has shown that the computed principal shear stress is converging to the shear stress intensity in accordance with the FZG-model, particularly at greater material depths were the crack initiation of tooth flank fracture damages is usually found. This convergence has justified the assumption on the basis of which, to deduce the aforementioned approximate equation, the residual stresses and shear stresses induced by friction, EHL contact, surface asperities and thermal load have not been considered. Equation (11.20) shows however that $\tau_{eff,L}(z)$ is dependent on the Hertzian stress, $\sigma_H = p_H$, reduced modulus of elasticity, E_r, local radius of relative curvature, ρ_{red}, and material depth, z.

The influence of residual stresses on the local equivalent shear stress, $\Delta\tau_{eff,L,RS}(z)$, is calculated using the following relationship:

$$\Delta\tau_{eff,L,RS}(z) = 32 K_1 \frac{|\sigma_{RS}(z)|}{100} \tanh\left(9z^{1.1}\right) - K_2, \tag{11.21}$$

where $\sigma_{RS}(z)$ is the residual stress depth profile and K_1 and K_2 are two adjustment factors that allow to write the expression of $\Delta\tau_{eff,L,RS}(z)$ in a closed form. These factors are given by the following expressions:

$$K_1 = \left(1 - K_{p_H,\sigma_{RS,max}}\right)\tanh\left(K_{z_{CHD}} z^{4.58}\right) + K_{p_H,\sigma_{RS,max}} \tag{11.22}$$

$$K_2 = \left[1 - \tanh\left(\frac{\rho_{red}}{10} - 1\right)\right]\left\{\frac{z_{CHD}^2}{16}\left[z\left(\frac{|\sigma_{RS,max}|}{10}\tanh\left(\frac{-2\left(\dfrac{p_H}{100} - 200\right)}{100}\right)\right.\right.\right.$$
$$\left.\left.\left. + \frac{|\sigma_{RS,max}|}{10}\right)\right]\right\} \tag{11.23}$$

where $\sigma_{RS,max}$ is the maximum value of the residual stresses, given by $\sigma_{RS,max} = \max|\sigma_{RS}(z)|$, z_{CHD} is the case hardening depth at 550 HV, while $K_{p_H,\sigma_{RS,max}}$ and $K_{z_{CHD}}$ are given respectively by:

$$K_{p_H,\sigma_{RS,max}} = -1.98\tanh\left[-0.07\left(\frac{p_H}{\sigma_{RS,max}}\right)^{2.385}\right] - 0.98 \tag{11.24}$$

$$K_{z_{CHD}} = 11.25 e^{-5.151 z_{CHD}} + 0.115 e^{-1.834 z_{CHD}}. \tag{11.25}$$

All the aforementioned relationships, even if obtained with sophisticated calculation methods, are empirically determined formulae. It should be remembered that this calculation procedure of $\Delta\tau_{eff,L,RS}(z)$ considers only compressive residual stresses, while it neglects the tensile residual stresses in the core region, since they are generally assumed to be small for typical tooth profiles. Higher tensile residual stresses in the core region may, however, appreciably increase the risk of tooth flank fatigue fracture, so that their neglecting configures a limit of the calculation approach. A further limitation of the model consists in the fact that the compressive residual stress components oriented in the normal direction to the tooth flank are neglected, while the compressive residual stress component oriented according to the tangent to the tooth flank and according to the tooth axis are assumed of equal value.

The quasi-stationary residual shear stress, $\tau_{eff,RS}(z)$, is determined using the following relationship, which is based on the same assumptions described above for the calculation of $\Delta\tau_{eff,L,RS}(z)$.:

$$\tau_{eff,RS}(z) = |\sigma_{RS}(z)|\sqrt{\frac{2}{15}}. \tag{11.26}$$

Figure 11.9 shows, as an example, the distribution curve of the local equivalent shear stress, $\tau_{eff}(z)$, as well as the distribution curves of the three shear stress components, $\tau_{eff,L}(z)$, $\Delta\tau_{eff,L,RS}(z)$ and $\tau_{eff,RS}(z)$, whose algebraic sum gives $\tau_{eff}(z)$. These distribution curves have been taken from Hein et al. (2017). They have been obtained using the (FZG/Witzig)-model, with a given test gear photographed in the position of the lower point of single tooth contact, B, and loaded in such a way as to determine a Hertzian stress $\sigma_H = p_H = 1475$ N/mm^2 and a Hertzian contact half-width, $b_H = 0.37$ mm.

Fig. 11.9 Example of distribution curves of $\tau_{eff}(z)$, $\tau_{eff,L}(z)$, $\Delta\tau_{eff,L,RS}(z)$ and $\tau_{eff,RS}(z)$

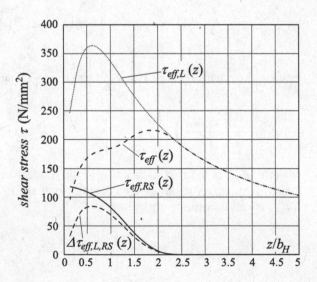

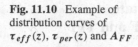

Fig. 11.10 Example of distribution curves of $\tau_{eff}(z)$, $\tau_{per}(z)$ and A_{FF}

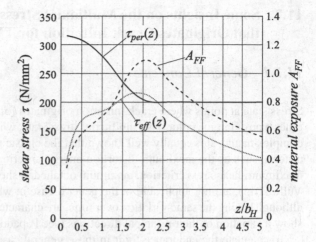

Figure 11.10, also taken from Hein et al. (2017), for the same example as in the previous figure, shows the distribution curves of quantities $\tau_{eff}(z)$, $\tau_{per}(z)$ and A_{FF}. The distribution curve of the material strength, $\tau_{per}(z)$, which is a function of hardness depth profile, tracing its course, shows that, for this particular example, the case-core interface is localized at a material depth $z/b_H \cong 2$. The distribution curve of $\tau_{eff}(z)$ then shows that, approximately at this same depth, it reaches its maximum value, $\tau_{eff,max}$. Consequently, at the same depth also the local material exposure, A_{FF}, reaches its maximum value, in this example equal to about 1,1 ($A_{FF} \cong 1, 1$). On the basis of what we have previously noted, we can conclude the gear in question has a high calculated risk of failure by tooth flank fatigue fracture for the given load condition, something confirmed by experimental evidence.

A detailed parametric study carried out by Witzig (2012), where the Hertzian stress, $\sigma_H = p_H$, radius of relative curvature, ρ_{red}, case hardening depth, z_{CHD}, and residual shear stress depth profile were varied, showed that the results obtained with the (FZG/Witzig)-model differ slightly from those obtained with the more sophisticated FZG-mode. In particular, evaluating the local material exposure for all parameter combinations in the wide material depth range ($0 \le z \le 9b_H$), Witzig has shown that the results obtained with the two models in terms of material exposure depth profile, A_{FF}, maximum material exposure, $A_{FF,max}$, and depth coordinate of the maximum material exposure z_{max}, are in good agreement for material depths $z \ge b_H$. Indeed, the difference between the results provided by the two methods do not exceed ±10%.

In conclusion it should be remembered that the (FZG/Witzig)-model is only applicable for case-carburized gears, in the material depth range, $z \ge b_H$. Furthermore, this model is validated only for parameters included within the following ranges: (500 N/mm² $\le p_H \le$ 3000 N/mm²), (5 mm $\le \rho_{red} \le$ 150 mm) and (0.3 mm $\le z_{CHD} \le$ 4.5 mm). It should also be remembered that $A_{FF,max} = 0.8$ represents the critical threshold value of the maximum material exposure, beyond which tooth flank fatigue failures may occur for gear materials with quality MQ according to ISO 6336-5:2016, which are typical quality and cleanness materials used for gears.

11.7 Some Insights on the Multiaxial Stress State that Originates Crack Initiation for TIFF and TFF

11.7.1 General Concepts

Stress state at points where crack initiation originates for TIFF and TFF (or for any other type of fatigue failure) is a multiaxial stress state, which is known to have a very complex nature. It is equally well known that the classic three-dimensional strength criteria, such as the maximum distortion energy criterion (or von Mises criterion), maximum shear stress criterion, maximum octahedral shear stress criterion, etc. (see Vullo 2014), are not applicable in the general case in which the principal stresses, although having the same variation over time, are characterized by time histories that show a variation of their principal directions (see Papadopoulos 1994).

To calculate the endurance limit in these general cases of multiaxial stresses, a number of multiaxial criteria has been developed over the past few years (see Liu and Zenner 1993; Martens and Hahn 1993; Häfele and Dietmann 1994; Yu 2002). These criteria differ considerably in formulation, applicability range, predictive reliability and physical interpretation of the phenomena involved in relation to the hypotheses formulated. In principle, the known multiaxial fatigue criteria can be divided according to the hypotheses formulated, so they can be grouped into the following three families: criteria based on the integral approach hypothesis; criteria based on the critical plane approach hypothesis; empirical criteria.

Leaving aside the empirical criteria, it should be remembered that, in the case of the integral approach, the equivalent stress is calculated as an integral of stresses over all the intersection planes of a volume element, while in the case of the critical intersection plane approach, only the intersection plane with critical stress combination is considered. Whatever the family of the classical multiaxial criteria, they can be considered as special cases of the more general Weakest Link Theory (WLT). More specifically, the integral approach and critical intersection plane approach constitute two limiting cases of the WLT.

11.7.2 Weakest Link Theory and Classical Multiaxial Criteria

The WLT was originally conceived and developed by Weibull (1939) to describe the static strength of brittle materials and was later extended by other authors (Batdorf and Crose 1974; Batdorf 1977; Batdorf and Heinisch 1978; Evans 1978; Lamon 1988; Munz and Fett 1989) to evaluate the probability of failure of ceramic materials under multiaxial static load. Here we are interested in ductile materials, but to better understand the calculation equations described below it is necessary to conform to the historical development, which started with brittle materials. To do this in the best possible way, we here follow the more than appreciable synthesis made by Liu (1999).

According to the WLT, the probability of survival, P_S, of a mechanical component can be expressed by the following relationship:

$$P_S = \exp\left[-\frac{1}{4\pi}\int_V\int_\Omega\left(\frac{\sigma_{\varphi\psi e}}{\sigma_0}\right)^k d\Omega dV\right],\tag{11.27}$$

where $\sigma_{\varphi\psi e}$ is the local equivalent stress in the intersection plane of the defect, σ_0 is the local static strength of the material, k is the Weibull's exponent, Ω is the spherical surface area (Fig. 11.8) and V is the volume of the machine component.

The equivalent stress, σ_{eff}, is then determined as follows:

$$\sigma_{eff} = \left[\frac{1}{4\pi}\int_\Omega\left(\sigma_{\varphi\psi e}\right)^k d\Omega\right]^{1/k}.\tag{11.28}$$

In the more general case of an inhomogeneous stress distribution within the volume, the following stress integral, I, is introduced, which considers the statistical size effect:

$$I = \int_V\left(\frac{\sigma_{eff}}{\sigma_{eff,max}}\right)^k dV.\tag{11.29}$$

Therefore, the probability of survival for the whole mechanical component is expressed as follows:

$$P_S = \exp\left[-I\left(\frac{\sigma_{eff,max}}{\sigma_0}\right)^k\right].\tag{11.30}$$

Equations (11.29) and (11.30), which describe the statistical size effect, are obtained assuming that the same probability exists that failure may originate at the interior of the volume or at the surface. In cases where failure occurs at the surface, in the above equations the surface area, A, must be replaced instead of the volume, V.

The choice of the local failure criterion, and therefore that of the related equivalent stress, must necessarily conform to the characteristics of the material, which can have brittle or ductile behavior. In the case of brittle materials, such as ceramics, the defect can be considered in a first approximation as a crack. For these materials, the decisive stress is the normal stress, $\sigma_{\varphi\psi}$, which is perpendicular to the crack plane. This normal stress, if the crack is not sensitive to shear stress, is chosen as equivalent stress. It is given by the following relationship:

$$\sigma_{eff} = \left[\frac{1}{4\pi} \int\limits_{\varphi=0}^{\pi} \int\limits_{\psi=0}^{2\pi} \left(\sigma_{\varphi\psi} \right)^k \sin\varphi\, d\psi\, d\varphi \right]^{1/k} ; \qquad (11.31)$$

Considering a planar stress state defined by the principal stresses, σ_1 and σ_2, and making Weibull's exponent vary in the range $(2 \le k \le \infty)$, it is easy to show that the above equation specializes in some well-known strength criteria (see Liu 1999). Actually, for $k = \infty$, we get the same failure limit that we would have obtained using the principal normal stress criterion, with an associated equivalent stress equal to the major principal stress. Instead, for $k = 2$, the well-known elliptic curve is obtained, which is representative of the distortion energy criterion. Of course, other well-known stress states can be examined, such as those characterized by biaxial tension $(\sigma_1 = \sigma_2)$ and pure shear $(\sigma_2 = -\sigma_1 = \tau)$, calculating for each of them the corresponding failure limits for various Weibull exponents. It is possible to find other well-known results. For $(10 \le k \le 30)$, results are obtained that differ little from those obtained with the principal stress criterion. For $k = 2$, we obtain that the quotient of the tolerable stress divided by the tolerable uniaxial stress, σ_A, is respectively equal to 0.61 and 0.866 for biaxial tension and pure shear.

However, here we are interested in ductile materials, such as steels. For these materials, the beginning of the plastic deformation, that is the beginning of the slip motion of the so-called slip system under shear stress, is usually assumed as failure limit. The slip system comprises the slip direction, i.e. the direction most densely occupied by atoms, and the slip plane, i.e. the most densely occupied intersection plane. The local failure criterion depends on the orientation of the slip plane, $\varphi\psi$, as well as on the orientation of the slip direction. The shear stress, $\tau_{\varphi\psi\beta}$, along direction w (see Fig. 11.8) is chosen as local equivalent stress. Therefore, Eq. (11.28) must be extended by integration over the angular range $(0 \le \beta \le \pi)$, so we get:

$$\tau_{eff} = \left[\frac{1}{4\pi} \int\limits_{\varphi=0}^{\pi} \int\limits_{\psi=0}^{2\pi} \int\limits_{\beta=0}^{\pi} \left(\tau_{\varphi\psi\beta} \right)^k \sin\varphi\, d\psi\, d\varphi\, d\beta \right]^{1/k} . \qquad (11.32)$$

Here also, considering a planar stress state defined by the principal stresses, σ_1 and σ_2, and making Weibull's exponent vary in the range $(2 \le k \le \infty)$, it is possible to show that Eq. (11.32) specializes in some well-known strength criteria (see Liu 1999). For $k = \infty$, we get the same failure limit that we would have obtained using the maximum shear stress criterion. For $k = 2$, the elliptic failure limit is obtained, which is representative of the von Mises criterion. All other failure limits corresponding to Weibull's exponents between 2 and infinity are located between the aforementioned two limiting curves. Here also, other stress states can be examined, such as those characterized by biaxial tension $(\sigma_1 = \sigma_2)$ and pure shear $(\sigma_2 = -\sigma_1 = \tau)$, calculating for each of them the corresponding failure limits for various Weibull's exponents. In particular we find that, for biaxial tension $(\sigma_1 = \sigma_2)$, the failure limit is independent of the Weibull's exponent. Instead, for pure shear ($\sigma_2 = -\sigma_1 = \tau$), we obtain that the quotient of the tolerable stress divided by the uniaxial tolerable stress, σ_A, varies from 0.5 at $k = \infty$ (maximum shear stress criterion) to 0.577 at $k = 2$ (von Mises criterion). Other well-known numerical ratios between these quantities are thus obtained (Timoshenko and Goodier 1951).

Traditionally, the von Mises criterion, which would be better to call (Maxwell/Huber/von Mises/Henchy)-criterion, since starting from Maxwell's initial contribution dating back to 1856 (Maxwell 1856), it was progressively enriched by the contributions of Huber (1904), von Mises (1913) and Henchy (1924), was predominantly interpreted as distortion energy criterion, according to what Henchy did. This denomination, introduced by the same Henchy, remedies an error of historical perspective, which removes the due merit to Maxwell, who was the first promotor of the criterion, as well as that of the other two authors mentioned above who, together with von Mises, enriched it with others more than appreciable contributions (see Hill 1950; Timoshenko 1953; Shames and Cozzarelli 1997; Vullo 2014). In 1933 Nadai (1933) conceived the octahedral shear stress criterion and demonstrated that it can be traced back to the distortion energy criterion (see also Nadai 1937, 1950, 1963). Subsequently, in 1952, Novozhilov (1958) demonstrated that the distortion energy criterion was interpretable as the root mean square of the shear stresses for all intersection planes. Still later, in 1968, Paul (1968), in the wake of Nadai, interpreted the distortion energy criterion as the root mean square of the principal shear stresses.

Here we are particularly interest in the interpretation given by Novozhilov, according to which the root mean square of the shear stresses for all planes of intersection is identified with the distortion energy criterion, so we can write the following equation that correlates the shear stress integral, τ_{int}, to the second deviator invariant for principal axes, J_2:

$$\tau_{int} = \left[\frac{1}{4\pi} \int_{\varphi=0}^{\pi} \int_{\psi=0}^{2\pi} \left(\tau_{\varphi\psi} \right)^2 \sin\varphi \, d\psi \, d\varphi \right]^{1/2} = (3J_2)^{1/2}. \tag{11.33}$$

This deviator invariant is given by the following relationship:

$$J_2 = I_2 + (1/3)I_1^2, \tag{11.34}$$

which correlates it to the first and second tensor invariants for principal axes, given respectively by $I_1 = (\sigma_1 + \sigma_2 + \sigma_3)$ and $I_2 = -(\sigma_1\sigma_2 + \sigma_1\sigma_3 + \sigma_2\sigma_3)$, where σ_1, σ_2 and σ_3 are the principal stresses of the three-dimensional stress state. This interpretation of Novozhilov expressed by Eq. (11.33) is important as it led to the development of the effective shear stress hypothesis and shear stress intensity hypothesis. However, it can be considered as a special case of the more general Eq. (11.32). It is in any case possible to show that the integration over the angle β (Fig. 11.8) allows to obtain the resultant shear stress, $\tau_{\varphi\psi}$, so we have:

$$\tau_{\varphi\psi} = \left[\int\limits_{\beta=0}^{\pi} \left(\tau_{\varphi\psi\beta} \right)^{k} d\beta \right]^{1/k}. \tag{11.35}$$

11.7.3 A General Fatigue Criterion for Multiaxial Stress

In the previous subsection, we have shown that classical multiaxial criteria, such as the distortion energy criterion, principal normal stress criterion and maximum shear stress criterion, can be considered as special cases of the more general weakest link theory. This fact allows us to formulate the below described general strength hypothesis to which an equally general fatigue criterion for multiaxial stress is correlated.

It is known that a multiaxial fatigue criterion must first satisfy the condition of invariance, so that the calculated equivalent stress must be independent of the chosen coordinate system rigidly connected with the examined structural component. Furthermore, it must be able to take into account time histories of principal stresses that show a variation of their principal directions. A multiaxial fatigue criterion that satisfies these conditions can be formulated in two ways, that is as integral approach hypothesis or as critical plane approach hypothesis. Here we focus our attention on the general fatigue criterion for multiaxial stress developed by Liu (1999).

For multiaxial fatigue stresses characterized by a periodically variable stress tensor, $\sigma_{ij}(t)$, the normal and shear stress components can be calculated in an intersection plane arbitrarily oriented with respect to the coordinate axes at any time (Fig. 11.8). These normal and shear stress components, which are time-dependent, are as usual described by their mean values and their amplitudes. The amplitude and mean values of the normal stress, $\sigma_{\varphi\psi}$, in the intersection plane, and of the shear stress, $\tau_{\varphi\psi\beta}$, in the direction w, can be easily calculated from their maximum and minimum values during a period. If the local failure criterion is chosen independently of the w-direction in the intersection plane, the maximum values of $\tau_{\varphi\psi\beta a}$ and $\tau_{\varphi\psi\beta m}$ can be used. Four stress components act in the intersection plane: they are the mean normal and shear stress components, $\sigma_{\varphi\psi m}$ and $\tau_{\varphi\psi m}$, and the normal and shear stress amplitudes, $\sigma_{\varphi\psi a}$ and $\tau_{\varphi\psi a}$, as Fig. 11.11 shows.

Fig. 11.11 Stress components acting in an intersection plane

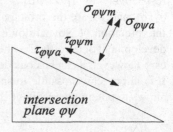

If we indicate with $\Sigma_{\varphi\psi}$ and $T_{\varphi\psi}$ two stress components or two arbitrary combinations of the above four stress components in the intersection plane, in terms of weakest link theory we can write the following relationship of the equivalent stress, $\Sigma_{\Sigma eff}$:

$$\Sigma_{\Sigma eff} = \left[\int_{\Omega} \left(\Sigma_{\varphi\psi} \right)^{\mu} d\Omega \right]^{1/\mu}, \tag{11.36}$$

with $\Sigma_{\varphi\psi} > 0$. This general form of equation allows us to capture the effect of stress components that are decisive for damage triggering, such as the normal stress amplitude for brittle and *defective* materials and the shear stress amplitude for ductile and *flawless* materials.

For $\mu \to \infty$, the resulting equivalent stress is the maximum stress, $\Sigma_{\max}$. In this special case, the formulation of the multiaxial criterion according to Eq. (11.36) is used for the critical intersection plane approach, so the stresses in the maximum stress intersection plane are the decisive ones for fatigue failure. Instead, for μ equal to a defined real number, Eq. (11.36) is used for the fatigue hypothesis of the integral approach. For simplicity, Liu (1999) suggests choosing $\mu = 2$ for the shear stress intensity hypothesis.

Notoriously, the only mean stresses cannot cause fatigue failures. However, it is also known that, in the presence of fatigue stresses, they contribute to increasing or decreasing the tolerable stress amplitude. The assessment of the effect of mean normal and mean shear stress components on the risk of fatigue failure is carried out using the following relationship:

$$T_{\Sigma eff} = \left[\frac{\int_{\Omega} \left(\Sigma_{\varphi\psi} \right)^{\nu} \left(T_{\varphi\psi} \right)^{\mu} d\Omega}{\int_{\Omega} \left(\Sigma_{\varphi\psi} \right)^{\nu} d\Omega} \right]^{1/\mu}, \tag{11.37}$$

with $\Sigma_{\varphi\psi} > 0$ and $T_{\varphi\psi} > 0$. If both the exponents μ and ν approach infinity or ν is greater than μ, the stress component T in the intersection plane corresponds to the maximum stress component, Σ.

The two Eqs. (11.36) and (11.37), which express the equivalent stresses of stress components or combinations of the latter, define the fatigue failure criterion for multiaxial stress proposed by Liu (1999). It is a fatigue failure criterion of general validity, since all previous known multiaxial criteria can be defined from it. In this regard, Liu provides a significant example concerning the critical shear stress criterion, which was studied with interesting contributions by Nøkleby (1981). Anyway, from this general fatigue criterion, arbitrary fatigue criteria can be formulated. To this end, it is sufficient to define differently the exponents that appear in the aforementioned equations (in the examples, for simplicity, they are set equal to 1, 2 and ∞), or to choose stress components and their combinations in different ways.

11.7.4 Modification of the Shear Stress Intensity Hypothesis

The Shear Stress Intensity Hypothesis (SIH) was introduced by Simbürger (1975) and Zenner and Richter (1977). It is an integral hypothesis, able to take into account the mean stresses. This hypothesis was further developed by Heindenreich et al. (1983) and Liu and Zenner (1993). In the framework of the aforementioned general fatigue criterion, the below described modification of the shear stress intensity hypothesis has been proposed by Liu (1999). For this modification, which broadens the SIH-horizons, the shear stress amplitude and normal stress amplitude are determined as the integral of stresses over all intersection planes. Furthermore, the mean shear stress and mean normal stress are weighted respectively over the shear stress amplitude and normal stress amplitude. In this way, an equivalent stress is defined for each of the four stress components in the intersection plane (Fig. 11.11). They are given by:

$$\tau_{effa} = \left[\frac{15}{8\pi} \int_{\varphi=0}^{\pi} \int_{\psi=0}^{2\pi} \left(\tau_{\varphi\psi a} \right)^{\mu_1} \sin\varphi d\psi d\varphi \right]^{1/\mu_1} \tag{11.38}$$

$$\sigma_{effa} = \left[\frac{15}{8\pi} \int_{\varphi=0}^{\pi} \int_{\psi=0}^{2\pi} \left(\sigma_{\varphi\psi a} \right)^{\mu_2} \sin\varphi d\psi d\varphi \right]^{1/\mu_2} \tag{11.39}$$

$$\tau_{effm} = \left[\frac{\int_{\varphi=0}^{\pi} \int_{\psi=0}^{2\pi} \left(\tau_{\varphi\psi a} \right)^{\mu_1} \left(\tau_{\varphi\psi m} \right)^{\nu_1} \sin\varphi d\psi d\varphi}{\int_{\varphi=0}^{\pi} \int_{\psi=0}^{2\pi} \left(\tau_{\varphi\psi a} \right)^{\mu_1} \sin\varphi d\psi d\varphi} \right]^{1/\nu_1} \tag{11.40}$$

$$\sigma_{effm} = \left[\frac{\int_{\varphi=0}^{\pi} \int_{\psi=0}^{2\pi} \left(\sigma_{\varphi\psi a} \right)^{\mu_2} \left(\sigma_{\varphi\psi m} \right)^{\nu_2} \sin\varphi d\psi d\varphi}{\int_{\varphi=0}^{\pi} \int_{\psi=0}^{2\pi} \left(\sigma_{\varphi\psi a} \right)^{\mu_2} \sin\varphi d\psi d\varphi} \right]^{1/\nu_2}. \tag{11.41}$$

In the shear stress intensity hypothesis, exponents μ_1 and μ_2 are again set equal to 2. For simplicity, a value $\nu_1 = 2$ is then chosen, while to evaluate the mean normal stress, σ_{effm}, the exponent ν_2 is set equal to unity. In this way, the difference between a positive and a negative mean stress is considered. As failure condition, the following combination of equivalent stresses is then used:

$$k_1 \tau_{effa}^2 + k_2 \sigma_{effa}^2 + k_3 \tau_{effm}^2 + k_4 \sigma_{effm}^2 = \sigma_{lim}^2. \tag{11.42}$$

The left-hand side of this equation is made up of four terms, whose sum can be interpreted as the square of a resultant equivalent stress that, in the condition of incipient failure, must equal the square of the fatigue alternating strength, σ_{falim}^2. As it is known, in reality often the fatigue limit, σ_{falim}, under fully reversed uniaxial loading does not exists (Sonsino 2007). The first two terms that appear in the left-hand side of Eq. (11.42) consist of shear and normal stress amplitudes, while the other two terms consist of shear and normal mean stresses.

Coefficients k_1, k_2, k_3 and k_4 are determined by imposing the condition that failure criterion is satisfied for a uniaxial stress state. By doing so, the following relationships are obtained:

$$k_1 = \frac{1}{5}\left[3\left(\frac{\sigma_{falim}}{\tau_{falim}}\right)^2 - 4\right] \tag{11.43}$$

$$k_2 = \frac{1}{5}\left[6 - 2\left(\frac{\sigma_{falim}}{\tau_{falim}}\right)^2\right] \tag{11.44}$$

$$k_3 = \frac{\sigma_{falim}^2 - \left(\frac{\sigma_{falim}}{\tau_{falim}}\right)^2\left(\frac{\tau_{fplim}}{2}\right)^2}{\frac{4}{7}\left(\frac{\tau_{fplim}}{2}\right)^2} \tag{11.45}$$

$$k_4 = \frac{\sigma_{falim}^2 - \left(\frac{\sigma_{fplim}}{2}\right)^2 - \frac{4}{21}k_3\left(\frac{\sigma_{fplim}}{2}\right)^2}{\frac{5}{7}\left(\frac{\sigma_{fplim}}{2}\right)^2} \tag{11.46}$$

where $\sigma_{falim}, \tau_{falim}, \sigma_{fplim}$ and τ_{fplim} are respectively the alternating tensile strength, alternating torsional strength, pulsating tensile strength and pulsating torsional strength.

The fatigue strength ratio $(\sigma_{falim}/\tau_{falim})$, which is typical of a given material, imposes well defined limits of applicability of the shear stress intensity hypothesis. In reality, the coefficients k_1 and k_2, given respectively by Eqs. (11.43) and (11.44), cannot have negative values, so the application of SIH is limited to the following variability range of the fatigue strength ratio $(\sigma_{falim}/\tau_{falim})$:

$$\sqrt{\frac{4}{3}} < \frac{\sigma_{falim}}{\tau_{falim}} < \sqrt{3}. \tag{11.47}$$

To be able to extend this range of applicability of SIH, the exponents μ_1 and μ_2 must be taken larger than 2.

To calculate the resultant equivalent stress in accordance with SIH, an integration is necessary in the general case of multiaxial stress state having the characteristics described at the beginning of the previous subsection. For a synchronous biaxial stress state like the one shown in Fig. 11.12, the following relationships can be written:

$$\sigma_y = \sigma_{ym} + \sigma_{ya} \sin \omega t$$
$$\sigma_z = \sigma_{zm} + \sigma_{za} \sin \omega t$$
$$\tau_{yz} = \tau_{yzm} + \tau_{yza} \sin \omega t \tag{11.48}$$

In this case, the sum of the first two terms in the left-hand side of Eq. (11.42) and the following two terms can be calculated analytically using the following equations:

Fig. 11.12 Biaxial stress state under the tooth flank surface

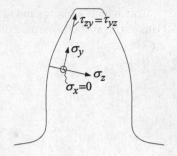

$$k_1 \tau_{effa}^2 + k_2 \sigma_{effa}^2 = \sigma_{ya}^2 + \sigma_{za}^2 + \left[2 - \left(\frac{\sigma_{falim}}{\tau_{falim}} \right)^2 \right] \sigma_{ya} \sigma_{za} + \left(\frac{\sigma_{falim}}{\tau_{falim}} \right)^2 \tau_{yza}^2$$

$$\tag{11.49}$$

$$\tau_{effm}^2 = \frac{1}{21} \big[A_{11} \sigma_{ym}^2 + A_{12} \sigma_{zm}^2 + A_{13} \sigma_{ym} \sigma_{zm} + A_{21} \tau_{yzm}^2$$
$$+ A_{22} \sigma_{ym} \tau_{yzm} + A_{23} \sigma_{zm} \tau_{yzm} \big] \tag{11.50}$$

$$\sigma_{effm} = \frac{3}{7} \big[A_{31} \sigma_{ym} + A_{32} \sigma_{zm} + A_{33} \tau_{yzm} \big], \tag{11.51}$$

where coefficients A_{ij} depend only on the mutual ratios between the stress amplitudes, σ_{ya}, σ_{za} and τ_{yza}. These coefficients are summarized in Table 11.1, where $\sigma_{ya} = x$, $\sigma_{za} = y$ and $\tau_{yza} = z$.

Table 11.1 Coefficients appearing in Eqs. (11.49) to (11.51)

i	j		
	1	2	3
1	$\dfrac{4x^2+3y^2-4xy+7z^2}{x^2+y^2-xy+3z^2}$	$\dfrac{3x^2+4y^2-4xy+7z^2}{x^2+y^2-xy+3z^2}$	$\dfrac{-4x^2-4y^2+6xy-6z^2}{x^2+y^2-xy+3z^2}$
2	$\dfrac{7x^2+7y^2-6xy+36z^2}{x^2+y^2-xy+3z^2}$	$\dfrac{10xz-6yz}{x^2+y^2-xy+3z^2}$	$\dfrac{-6xz+10yz}{x^2+y^2-xy+3z^2}$
3	$\dfrac{5x^2+y^2+2xy+4z^2}{3x^2+3y^2+2xy+4z^2}$	$\dfrac{x^2+5y^2+2xy+4z^2}{3x^2+3y^2+2xy+4z^2}$	$\dfrac{8z(x+y)}{3x^2+3y^2+2xy+4z^2}$

11.8 Calculation of Tooth Flank Fracture Load Carrying Capacity of Cylindrical Spur and Helical Gears: ISO Basic Formulae

11.8.1 General

The ISO standards (ISO/TS 6336-4:2019) provides a method for assessing the risk of tooth flank fracture of cylindrical involute spur and helical gears, to be considered provisional since the subject is susceptible to further development. The method is based on the shear stress intensity hypothesis (SIH), whose fundamentals have been summarized in the previous section. It allows to calculate, with equations written in a closed form, the local material exposure using only a small set of parameters concerning gear geometry, gear material and gear load condition.

The calculation model of the tooth flank fracture load carrying capacity proposed by the ISO standard mentioned above does not differ substantially from the practical-oriented calculation approach for TFF-risk assessment, which we have called (FZG/Witzig)-model and which we have described in Sect. 11.6. Therefore, the ISO calculation model has the same limitations as the (FZG/Witzig)-model, so among other things it only applies to case-carburized gears.

The calculation is performed for defined contact points, CP, to be identified in the active tooth contact area. Each contact point is defined by the tooth width coordinate, b^*, measured from a lateral end plane of the gear wheel under consideration, and the tooth height coordinate, r_{CP}, which is the local contact radius (Fig. 11.13); this second coordinate is correlated with the current contact point on the path of contact, Y (see Fig. 10.11). For each specific point of contact, the material depth, z, is oriented along the local normal to the tooth flank surface, in the direction towards the inside of the material, like Fig. 11.13 shows. It should be noted that the aforementioned ISO standards uses the y-coordinate instead of z-coordinate. The author apologizes to the reader if, for once, he does not follow the ISO, but it would have been a big problem to standardize everything written so far with the choice of the ISO. Of

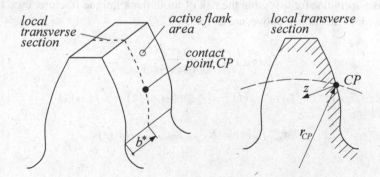

Fig. 11.13 Material depth, z, and parameters to locate the local contact point, CP

course, the following formulae differ from ISO ones for the substitution of z to y. All the quantities depending on z are indicated by the notation (z).

For each specific calculation point, defined by the contact point CP and material depth z (Fig. 11.13), the local material exposure $A_{FF,CP}(z)$ is calculated, considering a reasonably chosen number of contact points. For each calculation point, a local comparison is made between the total occurring stress, due to the load induced stresses and residual stresses and expressed by the local equivalent shear stress, $\tau_{eff,CP}(z)$, and the local material strength described by the local material shear strength, $\tau_{per,CP}(z)$. The local material exposure for each calculation point is determined as the quotient of the local equivalent shear stress, $\tau_{eff,CP}(z)$, divided by the local material shear strength, $\tau_{per,CP}(z)$.

The risk of tooth flank fatigue fracture and the resulting related safety factor S_{FF} are determined by considering the maximum calculated material exposure, $A_{FF,max}$ for all the analyzed contact points CP over the material depth z, where z is equal to or greater than b_H, this being the half of the Hertzian contact width. This determination of the maximum material exposure and resulting safety factor can be performed by Method A or Method B where, as usual, Method A is the most accurate method, which however requires more input quantities and more complex calculations or measurements with respect to Method B.

Method A is used when detailed information about the local Hertzian contact stress calculated with a 3D load distribution program is available; it requires that the calculations be made for discrete contact points in the contact area along the face width and tooth height, and for each considered material depth, z. Method B is instead used when this detailed information is not available, so the analysis is limited to some specified points of contact reasonably distributed on the contact area. Anyway, the influence of tooth flank modifications on the contact pressure distribution should be adequately considered.

11.8.2   ISO Formulae for TFF Assessment

The basic formulae for assessing the risk of tooth flank fatigue fracture, used by the ISO standard mentioned above, are as follows:

$$A_{FF,CP}(z) = \frac{\tau_{eff,CP}(z)}{\tau_{per,CP}(z)} + c_1 \tag{11.52}$$

$$\tau_{eff,CP}(z) = \tau_{eff,L,CP}(z) - \Delta\tau_{eff,L,RS,CP}(z) - \tau_{eff,RS}(z) \tag{11.53}$$

$$\tau_{per,CP}(z) = K_{\tau,per} K_{material} HV(z) \tag{11.54}$$

$$\tau_{eff,L,CP}(z) = \cfrac{0.149 p_{dyn,CP} + \cfrac{zE_r}{4\rho_{red,CP}} - \cfrac{z^2 E_r}{16\rho_{red,CP}^2 \sqrt{\left(\cfrac{p_{dyn,CP}}{E_r}\right)^2 + \left(\cfrac{z}{4\rho_{red,CP}}\right)^2}}}{0.4\cfrac{zE_r}{4\rho_{red,CP}\, p_{dyn,CP}} + 1.54} \tag{11.55}$$

$$\Delta\tau_{eff,L,RS,CP}(z) = 32 K_1 \frac{|\sigma_{RS}(z)|}{100}\tanh\left(9z^{1.1}\right) - K_2 \tag{11.56}$$

$$K_1 = \left(1 - K_{p_H,\sigma_{RS,\max}}\right)\tanh\left(K_{z_{CHD}} z^{4.58}\right) + K_{p_H,\sigma_{RS,\max}} \tag{11.57}$$

$$K_2 = \left[1 - \tanh\left(\frac{\rho_{red,CP}}{10} - 1\right)\right]\left\{\frac{z_{CHD}^2}{16}\left[z\left(\frac{|\sigma_{RS,\max}|}{10}\tanh\left(\frac{-2\left(\dfrac{p_{dyn,CP}}{100} - 200\right)}{100}\right)\right.\right.\right.$$
$$\left.\left.\left. + \frac{|\sigma_{RS,\max}|}{10}\right)\right]\right\} \tag{11.58}$$

$$K_{p_H,\sigma_{RS,\max}} = -1.98\tanh\left[-0.07\left(\frac{p_{dyn,CP}}{\sigma_{RS,\max}}\right)^{2.385}\right] - 0.98 \tag{11.59}$$

$$K_{z_{CHD}} = 11.25 e^{-5.151 z_{CHD}} + 0.115 e^{-1.834 z_{CHD}} \tag{11.60}$$

$$\tau_{eff,RS}(z) = |\sigma_{RS}(z)|\sqrt{\frac{2}{15}}. \tag{11.61}$$

By comparison with the corresponding Eqs. (11.17) to (11.26) of the (FZG/Witzig)-model, described in Sect. (11.6), it is possible to easily see that they at most differ only in the symbolism used.

By means of Eq. (11.52), where c_1 is the material exposure calibration factor, to be assumed equal to 0.04 for case-carburized steels, it is necessary to capture the maximum material exposure, given by:

$$A_{FF,\max} = \left[A_{FF,CP}(z)\right]_{\max}. \tag{11.62}$$

To this end, it is necessary to consider material depths $z \geq b_{H,CP}$, where the half of the Hertzian contact width at the contact point CP is related to the reduced modulus of elasticity, E_r, and to the local Hertzian contact stress, $p_{dyn,CP} = \sigma_{H,CP}$, and local normal radius of relative curvature, $\rho_{red,CP}$, at the contact point CP by the following relationship:

$$b_{H,CP} = 4\rho_{red,CP}\frac{p_{dyn,CP}}{E_r}. \tag{11.63}$$

For case-carburized gears, on the basis of experimental evidence, which still has to be confirmed or acquired for certain practical applications, the maximum calculated material exposure, $A_{FF,max}$, must have values less than 0.8. Actually, for $A_{FF,max} \geq 0.8$, tooth flank fatigue fractures can already occur for a constant input torque. For calculated maximum local material exposure that has to occur near the surface, i.e. for $z \leq b_{H,CP}$, a potential damage with near-surface initiation, such as pitting, can be feared (see Chap. 2). In this case, additional influences can come into play, such as surface roughness, lubricant oil and lubrication conditions. These influences do not come into play when a failure mode with a crack initiation at a considerable material depth, like the one considered here, is examined.

The reduced modulus of elasticity, E_r, which appears in Eq. (11.63) as well as in Eq. (11.55) is calculated using Eqs. (10.86) and (10.87). The local normal radius of relative curvature, $\rho_{red,CP}$, is determined by the following relationship:

$$\rho_{red,CP} = \frac{\rho_{red,t,CP}}{\cos \beta_b}, \tag{11.64}$$

where β_b is the base helix angle and $\rho_{red,t,CP}$ is the local transverse radius of relative curvature at the contact point CP, which is given by:

$$\rho_{red,t,CP} = \frac{\rho_{t1,CP} \rho_{t2,CP}}{\rho_{t1,CP} + \rho_{t2,CP}}, \tag{11.65}$$

where $\rho_{ti,CP}$, with $i = (1, 2)$, is given by:

$$\rho_{ti,CP} = \frac{1}{2} \sqrt{d_{CPi}^2 - d_{bi}^2}; \tag{11.66}$$

in this last equation, d_{CPi} are the diameters of pinion and wheel at the contact point CP, while d_{bi} are the base diameters of pinion and wheel.

The local Hertzian contact stress, $p_{dyn,CP}$, can be calculated as we have already said with Method A and Method B. With Method A, which involves a detailed contact analysis carried out for example by suitable 3D elastic contact models, $p_{dyn,CP,A}$ is determined using the following relationship:

$$p_{dyn,CP,A} = p_{H,CP,A} \sqrt{K_A K_v}, \tag{11.67}$$

where K_A and K_v are respectively the application factor and dynamic factor (see Chap. 1), while $p_{H,CP,A}$ is the local nominal Hertzian contact stress determined by means of a 3D load distribution program, for which it takes into account the elastic deflections under load, static displacements and stiffnesses of the entire elastic system. If then the three-dimensional elastic contact model used for calculation of $p_{H,CP,A}$ is able to consider also the influence of both factors or of one of the two factors, K_A and K_v, either K_A or K_v or both should be set equal to 1 in Eq. (11.67).

With method B, with which a detailed contact analysis is not carried out, $p_{dyn,CP,B}$ is determined for several defined contact points, using the following relationship, which is valid for $\varepsilon_\alpha \leq 2$ (for $\varepsilon_\alpha > 2$, the local Hertzian contact stress can only be calculated with Method A):

$$p_{dyn,CP,B} = p_{H,CP,B}\sqrt{K_A K_\gamma K_v K_{H\alpha} K_{H\beta}}, \tag{11.68}$$

where factors that appear under the square root are those that we have already described in Chap. 1 (with the limits there highlighted), while $p_{H,CP,B}$ is the local nominal Hertzian contact stress for Method B, given by:

$$p_{H,CP,B} = Z_E \sqrt{\frac{F_t X_{CP}}{b\rho_{red,CP}\cos\alpha_t \cos\beta_b}}. \tag{11.69}$$

It should first to be noted that Eq. (11.69), where $\cos\beta_b$ appears, differs from the corresponding formula of the ISO/TS 6336-4:2019. Actually, Hertzian contact pressure, $p_{H,CP,B}$, in the case of cylindrical helical gears depends on normal force, F_n, given by Eq. (8.81). Of course, in the particular case of cylindrical spur gears, we have $\cos\beta_b = 0$. It should also to be noted that all factors and quantities that appear in this equation are already known. The elasticity factor, Z_E, is given by Eq. (2.48). The load sharing factor, X_{CP}, is introduced to take into account the influence of different profile modifications. The contact points, CP, to be considered are all located on the path of contact. They are seven in all and coincide with those defined by Eqs. (10.73) to (10.79), in which it is sufficient to replace Y with CP to obtain the corresponding formulae of the ISO standard mentioned above. As we have already seen for other gear problems, these seven points, to which as many meshing positions correspond, are considered sufficiently representative of the meshing evolution during the meshing cycle. As shown in Fig. 10.11, these points are located between that start point of active profile, SAP (contact point A or E, for driving pinion or driving wheel, respectively), and the end point of active profile, EAP (contact point E or A, for driving pinion or driving wheel, respectively). Diameters d_{CPi} in Eq. (11.66), which depend on the location of the contact point CP, can be determined using Eqs. (10.80) and (10.81), where however d_{Yi} and g_Y are respectively replaced with d_{CPi} and g_{CP}, with $i = (1, 2)$.

About the calculation of the local load sharing factor, X_{CP}, which appears in Eq. (11.69), we have nothing to add to or subtract from what we have described in Sects. 10.8.6, 10.8.7 and 10.8.8. The same figures and the same equations described there are still fully valid. These last ones allow to obtain the corresponding formulae of the ISO standard mentioned above with the simple formal substitution of X_Y, g_Y and $X_{but,Y}$ with X_{CP}, g_{CP} and $X_{but,CP}$ respectively.

As regard the calculation of the local occurring equivalent stress, $\tau_{eff,CP}(z)$, by means of Eq. (11.53), it must be remembered that, when its calculated value is negative, it is necessary to set $\tau_{eff,CP}(z) = 0$. This quantity can also be determined using alternative methods based on SIH, such as those develop by Oster (1982) and Hertter (2003), upon agreement between supplier and customer; this provided that results are in line with those deriving from the calculation described above.

It should also be noted that Eq. (11.55), which expresses the local equivalent stress without consideration of residual stresses, can be written in the following alternative form, which is referenced on half of the Hertzian contact width, $b_{H,CP}$:

$$\tau_{eff,L,CP}(z) = \frac{0.149 p_{dyn,CP} + z \dfrac{p_{dyn,CP}}{b_{H,CP}} - \left(1 - \dfrac{z}{\sqrt{b_{H,CP}^2 + z^2}}\right)}{0.4\left(\dfrac{z}{b_{H,CP}}\right) + 1.54}. \tag{11.70}$$

This relationship can be obtained from Eq. (11.55), considering Eq. (11.63), which allows to express the second and third terms of the numerator and the first term of the denominator of Eq. (11.55), causing the half Hertzian contact width $b_{H,CP}$ to appear in them.

It is then to remember that the calculation of the quasi-stationary residual stress, $\tau_{eff,RS}(z)$, which is performed with Eq. (11.61), assumes that the tensile residual stresses on the core for typical tooth profiles are small, so they are neglected in the calculation approach. Of course, this assumption is strongly limiting, as higher tensile residual stresses in the core region contribute to increasing the risk of tooth flank fracture; on the other hand, they are hardly determinable by means of the existing measuring methods. Another hypothesis underlying the calculation approach assumes that the compressive residual stress components oriented according to the tangent to the tooth flank profile and according to the tooth axis have similar values and that the compressive residual stress component normal to the same tooth flank profile can be neglected. Evidently, Eq. (11.61) is used with the condition that the tangential component $\sigma_{RS}(z)$ of the residual stress (so it is named by ISO) at the material depth, z, is less than or equal to zero, i.e. $\sigma_{RS}(z) \leq 0$. Therefore, only the compressive residual stresses are considered.

The determination of the tangential component of the residual stress, $\sigma_{RS}(z)$, can be done with Method A or Method B. With Method A, $\sigma_{RS}(z)$ is measured with adequate instrumentation, such as the X-ray diffractometer, in order to obtain the residual stress depth profile, that is the distribution curve of $\sigma_{RS}(z)$ along the material depth, z. Method B instead determines the residual stress depth profile correlating it with the hardness depth profile, $HV(z)$, this following the procedure proposed by Lang (1988). The correlated calculation relationships, which derive from measurements on test gears and are only applicable for case-carburized gear are as follows:

$$\sigma_{RS}(z) = -1.250[HV(z) - HV_{core}] \tag{11.71}$$

$$\sigma_{RS}(z) = -0.286[HV(z) - HV_{core}] - 460; \tag{11.72}$$

they are applied respectively for $[HV(z) - HV_{core}] \leq 300$ and $[HV(z) - HV_{core}] > 300$, where HV_{core} and $HV(z)$ are the core hardness and local hardness at the material depth, z. The distribution curve of $HV(z)$ along the material depth z identifies the hardness depth profile. Both Eqs. (11.71) and (11.72) show that the calculated value of $\sigma_{RS}(z)$ depends of the hardness depth profile, whose accuracy in turn depends on the calculation method, which must therefore be as precise as possible (see Sect. 11.4).

Finally, some indications must be given on the determination of $\Delta\tau_{eff,L,RS,CP}(z)$, which can influence the total local equivalent stress, $\tau_{eff,CP}(z)$, in a significant way. Calculation of $\Delta\tau_{eff,L,RS,CP}(z)$ should be done with Eq. (11.56) that, together with the adjustment factors, K_1 and K_2 that appear in it, shows that this quantity depends on the residual stress depth profile as well as on the local Hertzian stress, $p_{dyn,CP}$, at the considered contact point, CP. Adjustment factors, K_1 and K_2, which were obtained with the methods proposed by Hertter (2003) and Witzing (2012), allow expressing $\Delta\tau_{eff,L,RS,CP}(z)$ in a closed form. The other two quantities not yet defined that appear in Eqs. (11.56) to (11.60) are the maximum residual stress, $\sigma_{RS,max} = \max[|\sigma_{RS}(z)|]$, and the case hardening depth at 550 HV, z_{CHD}, which is assumed as reference depth. When reliable measured data concerning the residual stress depth are not available, $\sigma_{RS}(z)$ is calculated with Eqs. (11.71) and (11.72). In order to determine the maximum residual stress, $\sigma_{RS,max}$, the effect of surface hardening treatments, such as shot peening, which are effective only near the surface, so they are not significant for the calculation of tooth flank fracture load capacity, is neglected.

11.9 Local Material Strength

For case-carburized gears made of steels that fully meet the requirements of quality grade MQ according to ISO 6336-5:2003 (see also Sect. 2.10), the determination of the local material strength, $\tau_{per,CP}(z)$, is carried out using Eq. (11.54) where, the meaning of symbols already introduced remaining unchanged, $K_{\tau,per}$ and $K_{material}$ are respectively the hardness conversion factor and material factor.

For these gears, in accordance with the proposal of Witzig (2012) made on the basis of experimental evidence and available practical experiences, ISO assumes $K_{\tau,per} = 0.4$. The material factor $K_{material}$, which for the same gears considers the influences of ductility, hardenability and other microstructural characteristics, is also the result of experimental tests. This factor is made to depend on the relevant tensile strength of the material, R_m, as well as on the chordal tooth thickness in transverse section at the diameter corresponding to the middle between points B and D on path of contact, $s_{t,B-D}$. This second dependence is justified by the fact that material hardenability depends on the tooth size, which is considered by the aforementioned averaged value of the chordal tooth thicknesses corresponding to points B and D on path of contact.

The determination of $K_{material}$ is carried out with the following relationships:

$$K_{material} = 0.70 + 10^{-2}\left(C_{K_{material}} - s_{t,B-D}\right) \tag{11.73}$$

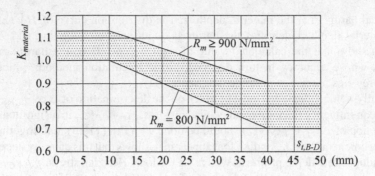

Fig. 11.14 Distribution curves of $K_{material}$ as a function of R_m and $s_{t,B-D}$

$$K_{material} = 0.90 + 7.7 \times 10^{-3}\left(C_{K_{material}} - s_{t,B-D}\right), \qquad (11.74)$$

which are respectively valid for materials having $R_m = 800$ N/mm^2 and $R_m \geq 900$ N/mm^2. $C_{K_{material}}$ is a dimensional constant, to be assumed equal to 40 mm. About the use of Eqs. (11.73) and (11.74), it must be remembered that for $s_{t,B-D} > 40$ mm and $s_{t,B-D} < 10$ mm, the following values are to be taken respectively: $s_{t,B-D} = 40$ mm and $s_{t,B-D} = 10$ mm. Finally, for values of the tensile strength R_m between 800 N/mm^2 and 900 N/mm^2, a linear interpolation between Formulae (11.73) and (11.74) can be made. Figure 11.14 shows, as a function of R_m and $s_{t,B-D}$, the distribution curves of $K_{material}$, corresponding to Eqs. (11.73) and (11.74).

11.9.1 Hardness Depth Profile

In accordance with the ISO standard mentioned above, the determination of the hardness depth profile, $HV(z)$, is carried out with Method B, Method $C1$ and Method $C2$, since at the present time the intended Method A is not yet specified.

Method B uses a hardness depth profile obtained by discrete measurements, performed in accordance with the standardized Vickers hardness test, considering adequately chosen depth increments. Experimental planning and related measurements must be such as to allow a reliable evaluation of the following parameters: surface hardness, $HV_{surface}$; core hardness, HV_{core}; case hardening depth at 550 HV, z_{CHD}; z-coordinate of the maximum hardness, $z_{HV,max}$. The determination of the hardness depth profile, $HV(z)$, is carried out by inserting these four parameters in the equations of the model developed by Thomas (1997), which are described below. This calculation procedure allows to obtain a smooth and continuous hardness depth profile, which is characterized by hardness gradually decreasing from the surface to the core. Discontinuities are thus avoided that could cause errors in the entire calculation procedure, which is the subject of this section.

Method $C1$ was developed by Lang (1988), based on the previous model he introduced in 1979 (Lang 1979), which has already been described in Sect. 11.4.2 This Method uses only three of the aforementioned parameters, namely $HV_{surface}$, HV_{core} and z_{CHD}. It is valid only for case-carburized gears and provides approximate results for typical hardness depth profiles as measured on gears. The calculation relationship is as follows:

$$HV(z^*) = HV_{core} + (HV_{surface} - HV_{core}) f(z^*), \qquad (11.75)$$

where $z^* = (z/z_{CHD})$ and

$$f(z^*) = 10^{(a+bz^*)z^*}, \qquad (11.76)$$

with $a = -3.81 \times 10^{-2}$ and $b = -26.62 \times 10^{-2}$.

Method $C2$, which was developed by Thomas (1997), is more accurate then Method $C1$. It uses all the four parameters specified above and defines the hardness depth profile by means of a distribution curve composed of three contiguous sections, each of which is derived using the following relationships:

$$HV(z) = a_a z^2 + b_a z + c_a \qquad (11.77)$$

$$HV(z) = a_b z^2 + b_b z + c_b \qquad (11.78)$$

$$HV(z) = HV_{core}, \qquad (11.79)$$

which are applied respectively for $(0 \leq z < z_{CHD})$, $(z_{CHD} \leq z < z_{core})$ and $(z \geq z_{core})$. The six constants that appear in Eqs. (11.77) and (11.78) are defined as follows:

$$a_a = \frac{550 - HV_{surface}}{z_{CHD}^2 - 2z_{HV,max} z_{CHD}} \qquad (11.80)$$

$$b_a = -2a_a z_{HV,max} \qquad (11.81)$$

$$c_a = HV_{surface} \qquad (11.82)$$

$$a_b = \frac{H'(z_{CHD})}{2(z_{CHD} - z_{core})} \qquad (11.83)$$

$$b_b = -2a_b z_{core} \qquad (11.84)$$

$$c_b = 550 - a_b z_{CHD}^2 - b_b z_{CHD}, \qquad (11.85)$$

where

$$H'(z_{CHD}) = 2a_a z_{CHD} + b_a. \tag{11.86}$$

In these equations, the meaning of symbols already introduced remaining unchanged, $z_{HV,\max}$ is the z-coordinate where hardness reaches its maximum value (when this value is not known, we take $z_{HV,\max} = 0$), while z_{core} is the z-coordinate where $HV(z) = HV_{core}$. It is given by:

$$z_{core} = \frac{-B + \sqrt{B^2 - 4AC}}{2A}, \tag{11.87}$$

where

$$A = -H'(z_{CHD}) \tag{11.88}$$

$$B = 2z_{CHD}H'(z_{CHD}) + 2(HV_{core} - 550) \tag{11.89}$$

$$C = -H'(z_{CHD})z_{CHD}^2 - 2(HV_{core} - 550)z_{CHD}. \tag{11.90}$$

References

Al BC, Langlois P (2015) Analysis of tooth interior fatigue fracture using boundary conditions from an efficient and accurate loaded tooth contact analysis. In: British Gears Association (BGA) Gears 2015 Technical Awareness Seminar, 12th of November 2015, Nottingham, U.K. (also *Gear Solutions*, Feb. 2016)

Al BC, Patel R, Langlois P (2016) Finite element analysis of tooth flank fracture using boundary conditions from LTCA. In: CTI Symposium USA, Novi, MI, 11–12 May 2016

Al BC, Patel R, Langlois P (2017) Comparison of tooth interior fatigue fracture load capacity to standardized gear failure models. Gear Solutions, pp 47–57

Batdorf SB (1977) Some approximate treatments of fracture statistics for polyaxial tension. Int J Fract 13:5–11

Batdorf SB, Crose JG (1974) Statistical theory for the fracture of brittle structures subjected to nonuniform polyaxial stresses. J Appl Mech 41:459–464

Batdorf SB, Heinisch HL (1978) Weakest link theory reformulated for arbitrary fracture criterion. J Am Ceram Soc 61:355–358

Beermann S, Kissling U (2015) Tooth flank fracture—a critical failure mode. Influence of Macro and Micro Geometry. In: KISSsoft User Conference India

Belajev NM (1917) Bulletin of Institution Engineers of Ways and Communications, St. Petersburg

Belajev NM (1924) Local stresses in compression of elastic bodies, in memoirs on theory of structures, St. Petersburg

Boiadjiev I, Witzig J, Tobie T, Stahl K (2014) Tooth flank fracture-basic principles and calculation model for a sub-surface-initiated fatigue mode of case-hardened gears. In: International Gear Conference, Lyon, France. Also Gear Technology, 2015 26–28 August 2014

Ding Y, Rieger N (2003) Spalling formation mechanism for gears. Wear 254(12):1307–1317

Elkholy A (1983) Case depth requirements in carburized gears. Wear 88:S233–S244

Evans AG (1978) A general approach for the statistical analysis of multiaxial fracture. J Am Ceram Soc 61:302–308

Fernandes PJL, McDuling C (1997) Surface contact fatigue failures in gears. Eng Fail Anal 4(2):99–107

Findley WN (1959) A theory for the effect of mean stress on fatigue of metals under combined torsion and axial load or bending. J Eng Indus 301–306

Föppl L, Föppl A (1947) Drang und Zwang, Band III—Eine höhere Festigkeitslehre für Ingenieure, Leibnitz Verlag, München

Ghribi D, Octrue M (2014) Some theoretical and simulation results on the study of the tooth flank breakage in cylindrical gears. In: International Gear Conference 2014, Lyon, France, 26–28 August 2014

Häfele P, Dietmann H (1994) Weiterentwicklung der Modifizierten Oktaederschubspannungshypothese (MOSH). Konstruktion 46:52–58

Hein M, Tobie T, Stahl K (2017) Parameter study on the calculated risk of tooth flank fracture of case-hardened gears. In: Bulletin of the JSME, Journal of Advanced Mechanical of Design, Systems, and Manufacturing, II, (6)

Heindenreich R, Zenner H, Richter I (1983) Dauerschwingfestigkeit bei mehrachsiger Beanspruchung, Forschungshefte FKM, Heft 105

Henchy H (1924) Zur Theorie plastischen Deformationen und hierdurch in Material hervorgerufenen Nebenspannungen. In: Proceedings 1st International Congress for Applied Mechanics, Deft, pp 312–317

Hertter T (2003) Rechnirischer Festigkeitsnachweis der Ermüdungstragfähigkeit vergüteter und einsatzgehärteter Zahnräder. Ph. D. thesis, Technical University of Munich

Hill R (1950) The mathematical theory of plasticity. Oxford University Press, Oxford

Höhn BR, Oster P, Hertter T (2010) A calculation Model for rating the gear load capacity based on local stresses and local properties of the gear material. In: International Conference on Gears, Garching, Germany, VDI

Huber MT (1904) A contribution to fundamentals of the strength of materials. Czasopismo Tow. Technicze Krakow 22:81 (in Polish)

ISO 6336-5:2016 Calculation of load capacity of spur and helical gears—Part 5: Strength and quality of materials

ISO/TR 15144-1:2014 Calculation of micropitting load capacity of cylindrical spur and helical gears—Part 1: Introduction and basic principles

ISO/TS 6336-22:2018 Calculation of load capacity of spur and helical gears—Part 22: Calculation of micropitting load capacity

ISO/TS 6336-4:2019 Calculation of load capacity of spur and helical gears—Part 4: Calculation of tooth flank fracture load capacity

Johnson KL (1985) Contact Mechanics, Cambridge University Press, Cambridge, United Kingdom

Klein M, Höhn BR, Michaelis K, Annast R (2011) Theoretical and experimental investigations about flank breakage in bevel gears. Indus Lubrication Tribol 63(1):5–10

Lamon J (1988) Statistical approaches to failure for ceramic reliability assessment. J Am Ceram Soc 71:106–112

Lang OR (1979) The dimensioning of complex steel members in the range of endurange strength and fatigue life. Zeitschrift fuer Werkstofftechnik 10:24–29

Lang OR (1988) Berechnung und Auslegung induktiv gehärteter Bauteile, Berichtsband zur AWT-Tagung, Induktives Randschichthärten, Darmstadt

Langlois P, Al BC, Harris O (2016) Hybrid hertzian and FE-based helical gear-loaded tooth contact analysis and comparison with FE. Gear Technology, pp 54–63

Liu J (1999) Weakest Link Theory and Multiaxial Criteria. In: Macha E, Bedkowki W, Lagoda T (eds) Multiaxial fatigue and fracture. Elsevier, New York, pp 55–68

Liu J, Zenner H (1993) Berechnung der Dauerschwingfestigkeit bei mehrachsiger Beanspruchung, Mat.-Wiss. und Werkstofftech., 24, part 1: pp 240–249; part 2: pp 296–303 and part3: pp 339–347

MackAldener M (2001) Tooth interior fatigue fracture and robustness of gears. Doctoral thesis, KTH Stockholm

MackAldener M, Olsson M (2000a) Interior fatigue fracture of gear teeth. Fatigue Fract Eng Mater Struct 23(4):283–292

MackAldener M, Olsson M (2000b) Design against tooth interior fatigue fracture. Gear Technolog, pp 18–24

MackAldener M, Olsson M (2001) Tooth interior fatigue fracture. Int J Fatigue 23:329–340

MackAldener M, Olsson M (2002) Analysis of crack propagation during tooth interior fatigue fracture. Eng Fracture Mech 69(18):2147–2162

Martens H, Hahn M (1993) Vergleichspannungshypothese zur Schwingfestigkeit bei Zweiachsiger Beanspruchung ohne und mit Phasenverschiebungen, Konstruktion, 45, pp 196–202

Maxwell JC (1856) Letter to Lord Kelvin, Dec. 18 (pertinent portion of letter quoted by Nadai, Theory of flow and Fracture of solids, Vol II, p 43)

Mises RV (1913) Mechanik der festen Körper im Plastisch-detormablem Zustand. Göttinger Nachrichten, Akad. Wiss, Math-Physik., Kl., pp 582–592

Munz D, Fett T (1989) Mechanisches Verhalten Keramischer Werkstoffe. Springer, Berlin

Nadai A (1933) Theories of strength. ASME J Appl Mech 1(3):111–129

Nadai A (1937) Plastic behavior of metals in the strain hardening range. J Appl Phys 8:205–213

Nadai A (1950) Theory of flow and fracture of solids, vol I. McGraw-Hill Book Company Inc., New York

Nadai A (1963) Theory of flow and fracture of solids, vol II. McGraw-Hill Book Company Inc., New York

Nøkleby JO (1981) Fatigue under multiaxial stress conditions. Report MD-81001, Div. Masch. Elem., The Norwey Institute of Thecnology, Trondheim/Norwegen

Novozhilov VV (1958) Theory of elasticity. Sudpromgiz (in Russian, Teoriya uprugosti), Moscow

Octrue M, Ghribi D, Sainsot P (2018) A contribution to study the Theory Flank Fracture (TFF) in cylindrical gears. Procedia Eng 213:215–226

Oster P (1982) Beanspruchung der Zahnflanken unter Bedingungen der Elastohydrodynamik. Doctoral thesis, Technical Univesity of Munich

Papadopoulos IV (1994) A new criterion of fatigue strength for out-of-phase bending and torsion of hard metal. Int J Fatigue 16:377–384

Paul P (1968) Generalized pyramidal fracture and yield criteria. Int J Solids Struct 4:175–196

Pedersen R, Rice RL (1961) Case crushing of carburized and hardened gears. SAE Technical Paper 610031 and SAE Transactions, SAE Transactions, Warrendale, PA, pp S360–370

Ristivojević M, Lazovic T, Venci A (2019) Studying the load carrying capacity of spur gear tooth flanks. Mech Mach Theory 59:125–137

Sandberg E (1981) A calculation method for subsurface fatigue. In: International symposium in gearing and power transmissions, Tokyo, I, pp S429–434

Shames IH, Cozzarelli FA (1997) Elastic and inelastic stress analysis. Revised Printing, Taylor & Francis Ltd., Philadelphia

Sharma VK, Breen DH, Walter GH (1977) An analytical approach for establishing case depth requirements in carburized and hardened gears. In: Transaction of ASME for presentation at the design engineering technical conference, Chicago, IL, pp 26–30

Simbürger A (1975) Festigkeitsverhalten zäher Werkstoffe bei einer mehrachsigen phasenver-schobenen Schwingbeanspruchung mit körperfesten und veränderlichen Hauptspannungsrich-tungen, Diss., TH Darmstadt

Sonsino CM (2007) Course of SN-curves especially in the high-cycle fatigue regime with regard to component design and safety. Int J Fatigue 29(12):2246–2258

Sponzilli JT, Remus GE, Clarke TM, Sawdo EJ (2000) Steel quality requirements for heavy duty off-highway gearing, SAE Technical Paper 2000-01-2566

Stahl K, Hein M, Tobie T (2018) Calculation of tooth flank fracture load capacity. Gear Solution, Sri Lanka

Suresh S (1998) Fatigue of materials, 2nd edn. Cambridge University Press, Cambridge

Tallian TE (1992) Failure Atlas for Hertz contact machine elements. ASME Press, New York

Thomas J (1997) Flankentragfähigkeit and Laufverhalten von hart-feinbearbeiteten Kegelrädern. Doctoral thesis, Technical University of Munich

Timoshenko SP (1953) History of strength of materials. McGraw-Hill Book Company Inc, New York

Timoshenko SP, Goodier JN (1951) Theory of elasticity, 2nd edn. McGraw-Hill Book Company Inc, New York

Tobe T, Kato M, Inoe K, Takatsu N, Morita I (1986) Bending strength of carburized C42OH spur gear teeth. ISME, pp 273–280

Tobie T, Höhn B-R, Stahl K (2013) Tooth flank breakage-influences on subsurface initiated fatigue failures of case-hardened gears. In: Proceedings of the ASME 2013 power transmission and gearing conference, Portland, OR, 4–7 August 2014

Townsend DP (1992) Gear handbook. McGraw-Hill Book Company inc., New York

Vullo V (2014) Circular cylinders and pressure vessels: stress analysis and design. Springer International Publishing Switzerland, Cham-Heidelberg

Weibull W (1939) A statistical theory of strength of materials, Ingeniörs Vatenskaps Akademiens Handlinger, Nr. 151, Generalstabens Litografiska Anstalts Förlag, Stockholm

Witzig J (2012) Flankenbruch Eine Grenze der Zahnradtragfähigkeit in der Werkstofftiefe. Ph.D. thesis, Technical University of Munich

Yu M (2002) Advances in strength theories for materials under complex stress state in the 20th Century. ASME Appl Mech Rev 55(3):169–218

Zenner H, Richter I (1977) Eine Festigkeitshypothese für die Dauerfestigkeit bei beliebigen Beanspruchungskombinationen. Konstruktion 29:11–18

Appendix
Load Carrying Capacity of Worm Gears

A.1 General Information

As we have already said in *Preface*, in this textbook we did not consider it necessary to reserve, to the subject regarding the load carrying capacity and strength design of the worm gears, the same space dedicated to that of the other types of gears discussed in the previous chapters. The present Annex in place of a chapter complies with this choice, which is not to be correlated with the more or less great interest of these types of gears. In fact, this interest is to be considered at least comparable to that of the other gears in terms of practical applications, but perhaps even greater and more stimulating from the point of view of the new knowledge to be acquired for the optimal solution of their strength problems.

In the same *Preface* mentioned above, we gave the reasons of our choice. Synthetically, these reasons are rooted in the fact that the various aspects concerning the load carrying capacity and strength design of these types of gears, which are more numerous and more complex than those of the other types of gears, have so far addressed and solved with methods to be considered essentially empirical. The only theoretical contributions used in this specific field, moreover almost always in support of empirical methods, do not differ from those discussed in the previous chapters concerning all other types of gears. Now, notoriously, theoretical methods have general validity once the fundamentals of the theory on which they are based are satisfied, while empirical methods, even the most elaborate ones, have a range of validity only circumscribed to that of the experimental reference framework, for which the errors deriving from their application are the greater the larger the deviations from the reference test conditions.

In Chap. 11 of Vol. 1, we have already extensively discussed the most qualifying theoretical fundamentals that characterize the geometric and kinematic design of these gears. The few theoretical contributions characterizing the load carrying capacity and strength design aspects of the same types of gears, which could be extrapolated from what is summarized below, would not compare with those discussed in the aforementioned chapter. It should also be considered that the most

© Springer Nature Switzerland AG 2020

V. Vullo, *Gears*, Springer Series in Solid and Structural Mechanics 11,
https://doi.org/10.1007/978-3-030-38632-0

recent scientific literature on this subject, as well as the previous one, is very lacking in this regard, and does not even deserve to be mentioned.

Rebus sic stantibus, we decided to very briefly discuss the various load carrying capacity and strength design problems of these gears, limiting ourselves here to summarize the main empirical and/or semi-empirical formulae collected in ISO/TR 14521:2010 (R2018) and referring the reader to the same Technical Report for the details of the case. This technical report, also revised in 2018, contains the most up-to-date information we have today on the subject in question. The same problems described below in accordance with this ISO standards have been addressed by other appreciable national and international standards, such as AGMA, BS and DIN. For a historical overview in this regard, we refer the reader to the often-mentioned textbooks by Pollone (1970), Henriot (1979), Niemann and Winter (1983), Maitra (1994), Budynas and Nisbett (2009) and Radzevich (2016).

Generally, the load carrying capacity of worm gears, in terms of torque (or power) that can be transmitted without the occurrence of tooth breakage or the appearance of excessive damages on the active flanks of worm threads and worm wheel teeth during their planned design lifetime, is subject to various limiting conditions, the main of which can be summarized as follows:

- Wear (of course, abrasive wear), which usually appears on the active tooth flanks of bronze worm wheels and is affected by the influence factors described below, but also by the number of starts per hour.
- Shear tooth breakage, which is a shear failure of worm threads or worm wheel teeth that can occur when threads or teeth become thin due to wear or when overload reaches threshold values.
- Pitting, which can appear on the active flanks of worm wheel teeth; its development is strongly influenced by the load sharing conditions and load transmitted, and is also affected by the influence factors described below.
- Worm thread and worm shaft breakage, with this last occurring due to bending fatigue or overload.
- Worm shaft deflection, due to excessive deformation under load that modifies the contact pattern between worm threads and worm wheel teeth as well as the surface of contact.
- Scuffing, which often appears suddenly and is strongly influenced by transmitted load, sliding velocity and conditions of lubrication as well as by the influence factors described below.
- Working temperature that, when becomes excessively high, can lead to accelerated degradation of the lubricant oil used.

It is important to bear in mind that wear, pitting, scuffing and efficiency (see Chap. 11 of Vol. 1) of a worm gear are strongly affected by the Hertzian pressure, worm rotational speed, type and characteristics of lubricant oil, lubrication regime and therefore oil film thickness, contact pattern and worm surface roughness. Tooth breakage is also influenced by Hertzian pressure, contact pattern and shear strength value. Also, the worm shaft deflection is affected by the Hertzian pressure. It should also be noted that different distribution curves expressing the output torque from the worm wheel, T_2, as a function of the rotational speed of the worm shaft, n_1, correspond to each of the damages due to wear, pitting, shear tooth breakage, worm shaft deflection and working temperature. The curves corresponding to these five types of damage intersect variously with each other and change their relative position with the variation of the center distance, a.

Therefore, the designer of these types of gears cannot ignore any of the aforementioned aspects, for which he must perform the assessment of their load carrying capacity in terms of safety factor against any possible type of damage, choosing as permissible torque the minimum value among those calculated for the various types of damage considered.

A.2   Security Against the Main Types of Damage

We summarize below the formulae that the technical standard ISO/TR 14521:2010 (R2018) proposes for the assessment of the load carrying capacity of worm gears. They have a well-defined range of validity (sliding velocity ranging between 0.1 and 25 m/s; contact ratio equal to or greater than 2.1; differentiated calculation procedures depending on whether the center distance is equal to or greater than 50 mm or less than 50 mm), which is closely correlated with that of the experimental planning used for the elaboration of the same formulae. It is in fact to be kept in mind that the methods proposed by the aforementioned technical standard are based in part on investigations carried out with test gears and in part on application experience. It should be also noted that the formulae of the calculation procedure can be expressed using absolute parameters or relative parameters. Absolute parameters are used when results of specific tests are not available, while relative parameters are used when the designer has the availability of direct investigation results.

A.2.1   Safety Against Wear

Under the hypothesis of full contact pattern between worm threads and worm wheel teeth (this hypothesis can be satisfied only with a high accuracy grade of both members of the worm gear under consideration as well as respecting stringent conditions for their assembly), and assuming that the wear load capacity is independent of pitting load capacity, the safety against the abrasive wear of worm gears is guaranteed by requiring that the wear safety factor, S_W, satisfies the following inequality:

$$S_W \geq S_{W\min}, \tag{A.1}$$

where $S_{W\min}$ is the minimum wear safety factor, to be assumed equal to 1.1.

The safety factor against wear is defined as the quotient of the limiting value of flank loss in normal section, $\delta_{W\lim n}$ (in mm), divided by the flank loss from worm wheel through abrasive wear in normal section, δ_{Wn} (in mm), i.e. as:

$$S_W = \frac{\delta_{W\lim n}}{\delta_{Wn}}. \tag{A.2}$$

The aforementioned ISO standard then describes the procedure for calculating the quantities that appear in Eq. (A.2) for worm gear pairs differently lubricated (mineral oils, polyalphaolefines, polyglicols), with worm made of case-hardened steel, 16MnCr5, and worm wheels made of different materials (GZ-CuSn12Ni2, GC-CuSn12Ni2, GZ-CuSn12, GZ-CuAl10Ni, GGG40, GG25).

A.2.2 Safety Against Surface Durability (Pitting)

The pitting load carrying capacity assessment of worm gears is done under the hypothesis that the Hertzian stress is an essential variable of influence for the pitting damage development mechanism, while other important variables of influence are neglected, such as tangential friction forces and effects of sliding and rolling motions (see Chap. 2). This last assumption is motivated by the fact that the formulae for determining the limit value of the pitting load carrying capacity of these gears were obtained by tests on worm gears or by evaluation of relevant operational results. Under these assumptions, the safety against the pitting of worm gears is guaranteed by requiring that the pitting safety factor, S_H, satisfies the following inequality:

$$S_H \geq S_{H\min}, \tag{A.3}$$

where $S_{H\min}$ is the minimum pitting safety factor, to be assumed equal to 1 (only for type C worm it may be appropriate to choose a higher safety factor).

The safety factor against pitting is defined as the quotient of the limiting value for the mean contact stress, σ_{HG} (in N/mm^2), divided by the mean contact stress, σ_{Hm} (in N/mm^2), i.e. as:

$$S_H = \frac{\sigma_{HG}}{\sigma_{Hm}}. \tag{A.4}$$

The limiting value for the mean contact stress is determined by the following relationship:

$$\sigma_{HG} = \sigma_{H\lim T} Z_h Z_v Z_S Z_u Z_{oil}, \tag{A.5}$$

where $\sigma_{H\lim T}$ (in N/mm^2) is the pitting strength obtained with test gears, and Z_h, Z_v, Z_S, Z_u and Z_{oil} are respectively the life factor, velocity factor, size factor, gear ratio factor and lubricant factor.

The mean contact stress is determined using the following relationship:

$$\sigma_{Hm} = \frac{4}{\pi} \left(\frac{10^3 T_2 p_m^* E_{red}}{a^3} \right)^{0.5},$$ (A.6)

where T_2 (in Nm) is the output torque from the worm wheel, E_{red} (in N/mm^2) is the equivalent modulus of elasticity, a (in mm) is the center distance, and p_m^* is a dimensionless parameter for the mean Hertzian stress.

ISO/TR 14521:2010 (R2018) describes in detail the procedures to calculate the quantities that appear in the above equations, not included in the input data.

A.2.3 Security Against Shear Tooth Breakage

Tooth root strength of the worm wheel teeth (but also the one of worm threads, albeit to a lesser extent) can be irreparably compromised by plastic deformations or breaks due to too high tooth root shear stresses, related to the noticeable reduction in the thickness of teeth (or threads) caused by wear, overload or other similar events. The safety against shear tooth root breakage of worm gears is guaranteed by requiring that the shear tooth breakage safety factor, S_F, satisfies the following inequality:

$$S_F \geq S_{Fmin},$$ (A.7)

where S_{Fmin} is the minimum shear tooth breakage safety factor, to be assumed equal to 1.1. S_F is the shear tooth breakage safety factor related to the transferable torque.

In the most general case, the safety factor against shear tooth breakage is defined as the quotient of the limiting value for shear stress at tooth root, τ_{FG} (in N/mm^2), divided by the nominal shear stress at tooth root, τ_F (in N/mm^2), i.e. as:

$$S_F = \frac{\tau_{FG}}{\tau_F}.$$ (A.8)

The nominal shear stress at tooth root is determined using the following relationship:

$$\tau_F = \frac{F_{tm2}}{b_{2H} m_{x1}} Y_\varepsilon Y_F Y_\gamma Y_K,$$ (A.9)

where: F_{tm2} (in N) is the tangential force applied to the worm wheel; b_{2H} (in mm) is the effective worm wheel face width; m_{x1} (in mm) is the worm axial module; Y_ε, Y_F, Y_γ and Y_K are respectively the contact factor, form factor, lead factor and rim thickness factor. All four of these dimensionless factors are specific to this type of tooth breakage.

The limiting value for shear stress at tooth root is determined by the following relationship:

$$\tau_{FG} = \tau_{FlimT} Y_{NL},$$ (A.10)

where τ_{FlimT} (in N/mm^2) is the shear endurance strength obtained with test gears, and Y_{NL} is the life factor for shear tooth breakage.

Also, for this type of damage, the ISO standard mentioned above provides all the information necessary to calculate the quantities that appear in the above equations.

A.2.4 Security Against Worm Shaft Deflection

In Sect. 11.1 of Vol. 1, but also in the sections following it, we have highlighted that a worm gear pair is characterized by only one correct relative position between the axes of the two members, which is theoretically the only one that can guarantee optimal operating conditions. The greater the deviations from this theoretical relative position, the greater the malfunctioning. Under equal geometry and assembly conditions of the kinematic pair under consideration, the contribution of deviations related the worm shaft, which is the most deformable mechanical component of the same worm gear, is certainly the most relevant. It follows the need to limit the bending and torsional deflections of the worm shaft as much as possible.

To limit the risk that the meshing interferences due to these deviations further increase the effects of wear, but considering only the bending deflections associated with symmetric (see Figs. 11.38 and 11.40 of Vol. 1) and asymmetric geometry, loading and assembly conditions, the aforementioned technical standard imposes a safety factor against worm shaft deflection, S_δ, that satisfies the following inequality:

$$S_\delta \geq S_{\delta min},$$ (A.11)

where $S_{\delta min}$ is the minimum value of worm shaft deflection safety factor, to be assumed equal to 1. S_δ is the deflection safety factor related to the transmittable torque.

The deflection safety factor is defined as the quotient of the limiting value of worm shaft deflection, δ_{lim} (in mm), divided by the experienced worm shaft deflection, δ_m (in mm), i.e. as:

$$S_\delta = \frac{\delta_{lim}}{\delta_m}.$$ (A.12)

The limiting value of worm shaft deflection, determined in accordance with the operating experience, is given by:

$$\delta_{lim} = 0.04\sqrt{m_{x1}}.$$ (A.13)

Depending on whether the assembly of the worm is symmetric or asymmetric with respect to the plane perpendicular to the worm shaft axis passing through the worm wheel axis, the resultant deflection is determined using respectively the following two expressions:

$$\delta_m = \frac{2 \times 10^{-6} l_1^3 F_{tm2}}{d_{m1}^4} \sqrt{\tan^2(\gamma_{m1} + \arctan \mu_{zm}) + \left(\frac{\tan^2 \alpha_0}{\cos^2 \gamma_{m1}}\right)} \qquad (A.14)$$

$$\delta_m = \frac{3.2 \times 10^{-5} l_{11}^2 l_{12}^2 F_{tm2}}{l_1 d_{m1}^4} \sqrt{\tan^2(\gamma_{m1} + \arctan \mu_{zm}) + \left(\frac{\tan^2 \alpha_0}{\cos^2 \gamma_{m1}}\right)}, \qquad (A.15)$$

where, the meaning of the symbols already introduced remaining unchanged, the other symbols represent:

- l_1 (in mm), the spacing between the worm shaft bearing;
- l_{11} and l_{12} (both in mm), the two portions in which the aforementioned plane perpendicular to the worm shaft axis passing through the worm wheel axis divides the bearing spacing;
- d_{m1} (in mm), the worm reference diameter;
- γ_{m1} (in degrees), the reference lead angle of the worm;
- μ_{zm}, the dimensionless mean tooth coefficient of friction;
- α_0 (in degrees), the pressure angle of the cutter.

Also, for this type of assessment, the same ISO standard in question provides all the information necessary to calculate the quantities appearing in the above equations.

A.2.5 Security Against Working Temperature

The well-known non-high efficiency of worm gear drives causes high power losses, which results in heat with a consequent increase in temperature. When a given threshold value is exceeded, this temperature increase causes degradation of the lubricant oil, which sees its life expectancy shortened, as well as accelerated decomposition of the additives and damage to the sealing rings. It is therefore necessary that the working temperature does not reach this threshold value.

The calculation procedures to ensure a suitable safety factor against this temperature increase are different depending on whether the splash lubrication or the oil spray lubrication is used to lubricate the worm gear drive under consideration.

A.2.5.1 Security Against Working Temperature for Splash Lubrication

With this type of lubrication, the safety against uncontrolled temperature increases is guaranteed by requiring that the working temperature safety factor, S_T, satisfies the following inequality:

$$S_T \geq S_{T\lim}, \tag{A.16}$$

where $S_{T\lim}$ is the minimum working temperature safety factor, to be assumed equal to 1.1.

The temperature safety factor for splash lubrication is defined as the quotient of the limiting value of the oil sump temperature, $\vartheta_{S\lim}$ (in °C), divided by the oil sump temperature, ϑ_S (in °C), i.e. as:

$$S_T = \frac{\vartheta_{S\lim}}{\vartheta_S}. \tag{A.17}$$

In the absence of specific information, the following limiting values of the oil sump temperature, ϑ_{Slim}, can be assumed: ~90 °C, for mineral oils; ~100 °C, for polyalphaolefines; ~100–120 °C (preferably, ~100^C), for polyglicols.

The oil sump temperature, ϑ_S, is determined using the following relationship:

$$\vartheta_S = \vartheta_0 + \left[a_0 + \frac{a_1 T_2}{(a/63)^3} \right] a_2, \tag{A.18}$$

where, according to the aforementioned ISO standard, ϑ_0 (in °C) is the ambient temperature, a_0 and a_1, should be non-dimensional oil sump temperature coefficients, to be calculated with different formulae, depending on whether the gear set housings are equipped with fans or are without fans, and a_2 should be a specific dimensionless coefficient of the three types of lubricant oils mentioned above, also to be calculated with appropriate formulae.

However, if the three aforementioned coefficients are dimensionless factors according to ISO, then the relationship (A.18) is not dimensionally consistent. On the other hand, the relationships provided by ISO to calculate the same coefficients for the different cases considered show that they are dimensional coefficients, but their dimensions are unfortunately not such as to make the same relationship (A.18) consistent, even in the case where the plus sign that appears in it be eliminated. Perhaps this inconsistency is to be ascribed to the bad habit, unfortunately frequent but irrational in the deduction of empirical formulae, of using quantities having certain dimensions as if they were pure numbers (see Giovannozzi 1965). This leads to misunderstandings, and is often a source of errors.

In any case, provided the relationship (A.18) is correct, its validity range has the following application limits: $a = (63 - 400)$ mm; $n_1 = (60 - 3000)$ min^{-1}; $u = (10 - 40)$. The use of the same equation, which provides approximate values of the oil temperature within $\pm 10°$C, presupposes a well ribbed housing made out of cast iron.

A.2.5.2 Security Against Working Temperature for Oil Spray Lubrication

With this other type of lubrication, the safety against uncontrolled temperature increases is guaranteed by requiring that the temperature safety factor, S_T, satisfies the same inequality (A.16). The safety factor for oil spray lubrication is however defined as the quotient of the cooling capacity of the oil with spray lubrication, P_K (in W), divided by the total power loss of the worm gear unit. P_V (in W), i.e. as:

$$S_T = \frac{P_K}{P_V}. \tag{A.19}$$

The cooling capacity of the oil with spray lubrication, P_K, is determined using the following relationship:

$$P_K = c_{oil} \rho_{oil} Q_{oil} \Delta \vartheta_{oil}, \tag{A.20}$$

where: c_{oil} (in Ws/kgK), is the specific heat capacity of the oil (usually, we assume $c_{oil} = 1.9 \times 10^3$ Ws/kgK); ρ_{oil} (in kg/dm^3), is the lubricant density; Q_{oil} (in m^3/s), is the oil spray quantity; $\Delta \vartheta_{oil} = (\vartheta_{out} - \vartheta_{in})$ is the difference between the oil exit temperature and oil entrance temperature, expressed in °C.

The total power loss of the worm gear drive under consideration, P_V, is determined using the following relationship:

$$P_V = P_{Vz} + P_{V0} + P_{VLP} + P_{VD}, \tag{A.21}$$

where: P_{Vz} (in W), is the mesh power loss (for reducers, P_{Vz1-2}; for increasers, P_{Vz2-1}); P_{V0} (in W), is the idle running power loss; P_{VLP} (in W), is the bearing power loss through loading; P_{VD} (in W), is the sealing power loss.

The ISO standard mentioned above provides all the information necessary to calculate the quantities appearing in these equations.

A.3 Security Against Other Types of Damage

In Sect. A.1 we have seen that other types of damage can occur in worm gears, no less important than those discussed in the previous section. Among these damages, the scuffing has a relevant aspect, as its almost sudden appearance makes the gear drive where it occurs no longer suitable for use. It is a very complex phenomenon, which becomes even more complex for the worm gears that interest here, as it is strongly affected not only by the factors that influence the other types of damage described above, but also and/or to a greater extent by the transmitted load, sliding velocity and lubrication conditions.

The progress made in this specific field is not significant and the reference technical standards are practically absent. This factual evidence is then to be extended to other types of damage that, moreover, can interact with the damages described above, with synergistic effects that generally lead to their exaltation.

References

Budynas RG, Nisbett JK (2009) Shigley's mechanical engineering design, 8th edn. McGraw-Hill Companies Inc, New York

Giovannozzi R (1965) Costruzione di Macchine, vol II, 4td edn. Casa Editrice Prof. Riccardo Pàtron, Bologna

Henriot G (1979) Traité thèorique and pratique des engrenages, vol 1, 6th ed. Bordas, Paris

ISO/TR 14521:2010 (R2018), Geras—calculation of load capacity of wormgears

Maitra GM (1994) Handbook of gear design, 2nd edn. Tata McGraw-Hill Publishing Company Ltd, New Delhi

Niemann G, Winter H (1983) Maschinen-Elemente, Band III: Schraubrad-, Kegelrad-, Schnecken-, Ketten-, Rienem-, Reibradgetriebe, Kupplungen, Bremsen, Freiläufe. Springer-Verlag, Berlin, Heidelberg

Pollone G (1970) Il Veicolo. Torino, Libreria Editrice Universitaria Levrotto & Bella

Radzevich SP (2016) Dudley's handbook of practical gear design and manufacture, 3rd edn. CRC Press, Taylor&Francis Group, Boca Raton, Florida

Index of Standards

AGMA 925-A03:2003, *Effect of Lubrication on Gear Surface Distress.*

AGMA 932-A05:2005, *Rating the Pitting Resistance and Bending Strength of Hypoid Gears.*

ANSI/AGMA 2003-C10, *Rating the Pitting Resistance and Bending Strength of Generated Straight Bevel, Zerol Bevel and Spiral Bevel Gear Teeth.*

BGA-DU P602, (2008), *Gear Micropitting Procedure. Test Procedure for the Evaluation of Micropitting Performance of Spur and Helical Gears.*

DIN 3990:1987, *Tragfähigkeitsberechnung von Stirnrädern.*

FVA-Information Sheet 54/7, (1993), *Test procedure for the investigation of the micropitting capacity gear lubricants.*

ISO 10300-1:2014, *Calculation of load capacity of bevel gears—Part 1: Introduction and general influence factors.*

ISO 10300-2:2014, *Calculation of load capacity of bevel gears—Part 2: Calculation of surface durability (pitting).*

ISO 10300-3:2014, *Calculation of load capacity of bevel gears—Part 3: Calculation of tooth root strength.*

ISO 10825:1995, *Gears—Wear and damage to gear teeth—Terminology.*

ISO 12085:1996, *Geometrical Product Specifications (GPS)—Surface texture: Profile method—Motif parameters.*

ISO 1302:2002, *Geometrical Product Specifications (GPS)—Indication of surface texture in technical product documentation.*

ISO 1328-1:2013, *Cylindrical gears—ISO system of flank tolerance classification—Part 1: Definitions and allowable values of deviations relevant to flanks of gear teeth.*

ISO 13565-1:1996, *Geometrical Product Specifications (GPS), Surface texture: Profile method; Surfaces having stratified functional properties—Part. 1: Filtering and general measurement conditions.*

ISO 13565-2:1996, *Geometrical Product Specifications (GPS)—Surface texture: Profile method; Surfaces having stratified functional properties—Part 2: Height characterization using the linear material ratio curve.*

ISO 13565-3:1998, *Geometrical Product Specifications (GPS)—Surface texture: Profile method; Surfaces having stratified functional properties—Part 3: Height characterization using the material probability curve.*

© Springer Nature Switzerland AG 2020
V. Vullo, *Gears*, Springer Series in Solid and Structural Mechanics 11,
https://doi.org/10.1007/978-3-030-38632-0

Author Index

© Springer Nature Switzerland AG 2020
V. Vullo, *Gears*, Springer Series in Solid and Structural Mechanics 11,
https://doi.org/10.1007/978-3-030-38632-0

Subject Index

© Springer Nature Switzerland AG 2020

V. Vullo, *Gears*, Springer Series in Solid and Structural Mechanics 11,
https://doi.org/10.1007/978-3-030-38632-0

Printed in the United States
By Bookmasters